MATH 1050

The Ohio State University

Richard N. Aufmann | Joanne S. Lockwood

CENGAGE Learning

Australia • Brazil • Japan • Korea • Mexico • Singapore • Spain • United Kingdom • United States

CENGAGE Learning

MATH 1050: The Ohio State University

Executive Editors:
Maureen Staudt
Michael Stranz

Senior Project Development Manager:
Linda deStefano

Marketing Specialist:
Courtney Sheldon

Senior Production/Manufacturing Manager:
Donna M. Brown

Production Editorial Manager:
Kim Fry

Sr. Rights Acquisition Account Manager:
Todd Osborne

Printed at CLDPC, USA, 05-19

Beginning Algebra, Eighth Edition
Richard N. Aufmann | Joanne S. Lockwood

Beginning Algebra with Applications, Seventh Edition
Richard N. Aufmann | Vernon C. Barker | Joanne S. Lockwood

ISBN-13: 978-1-285-10860-5

ISBN-10: 1-285-10860-4

Cengage Learning
5191 Natorp Boulevard
Mason, Ohio 45040
USA
Cengage Learning is a leading provider of customized learning solutions with office locations around the globe, including Singapore, the United Kingdom, Australia, Mexico, Brazil, and Japan. Locate your local office at: **international.cengage.com/region.**

Cengage Learning products are represented in Canada by Nelson Education, Ltd.
For your lifelong learning solutions, visit **www.cengage.com/custom.**
Visit our corporate website at **www.cengage.com.**

Contents

Digital Vision

Chapter 8 Factoring 349

Chapter 9 Rational Expressions 395

Digital Vision

Preface

Among the many questions we ask when we begin the process of revising a textbook, the most important is, "How can we improve the learning experience for the student?" We find answers to this question in a variety of ways, but most commonly by talking to students and instructors and evaluating the written feedback we receive from our customers. As we set out to create the eighth edition of *Beginning Algebra,* bearing in mind the feedback we received, our ultimate goal was to increase our *focus on the student.*

In the eighth edition, as in previous editions, popular features such as "Take Note" and "Point of Interest" have been retained. We have also retained the worked Examples and accompanying Problems, with complete worked-out solutions to the Problems given at the back of the textbook. New to this edition is the "Focus on Success" feature that appears at the beginning of each chapter. "Focus on Success" offers practical tips for improving study habits and performance on tests and exams.

Also new to the eighth edition are "How It's Used" boxes. These boxes present real-world scenarios that demonstrate the utility of selected concepts from the text. New "Focus On" examples offer detailed instruction on solving a variety of problems. "In the News" exercises are new application exercises appearing in many of the exercise sets. These exercises are based on newsworthy data and facts and are drawn from current events. The definition/key concept boxes have been enhanced in this edition; they now include examples to show how the general case translates to specific cases.

We trust that the new and enhanced features of the eighth edition will help students to engage more successfully with the content. By narrowing the gap between the concrete and the abstract, between the real world and the theoretical, students should more plainly see that mastering the skills and topics presented is well within their reach and well worth the effort.

Updates to This Edition

- NEW! Chapter Openers have been revised and now include "Prep Tests" and "Focus on Success" vignettes.
- NEW! "Try Exercise" prompts are included at the end of each Example/Problem pair.
- NEW! "How It's Used" boxes are featured in each chapter.
- NEW! "Focus On" examples provide detailed instructions for solving problems.
- NEW! "Concept Check" exercises have been added to the beginning of each exercise set.
- NEW! "In the News" applications appear in many of the end-of-section exercise sets.
- NEW! "Projects or Group Activities" exercises are included at the end of each exercise set.
- Definition/key concept boxes have been enhanced with examples.
- Revised exercise sets include new applications.
- Improved Chapter Summaries now include a separate column containing an objective and page number for quick reference.

Organizational Changes

We have made the following organizational changes, based on the feedback we received, in order to improve the effectiveness of the textbook and enhance the student's learning experience.

- Chapter 3 has been reorganized. Section 3.1 of the previous edition, *Introduction to Equations,* has been separated into two sections as suggested by reviewers. Now students will have the opportunity to master the skill of solving equations of the form $ax = b$ in Section 3.1 before they solve the percent problems and uniform motion problems that require this skill in Section 3.2.

  3.1 *Introduction to Equations*
  3.2 *Applications of Equations of the Form* $ax = b$
  3.3 *General Equations*
  3.4 *Inequalities*

  The eighth edition contains expanded coverage of solving an equation by clearing denominators, including more exercises on this concept in the Section 3.3 exercise set. In Section 3.3, students are now warned that clearing denominators is a method of solving equations and that the process is never applied to expressions.

- Chapter 4 has been reorganized. Objective 4.2.2 of the previous edition, *Coin and stamp problems,* has been deleted. Objective 4.2.1 of the previous edition, *Consecutive integer problems,* has been incorporated into the new Objective 4.1.1, *Translate a sentence into an equation and solve.*

  4.1 *Translating Sentences into Equations*
  4.2 *Geometry Problems*
  4.3 *Markup and Discount Problems*
  4.4 *Investment Problems*
  4.5 *Mixture Problems*
  4.6 *Uniform Motion Problems*
  4.7 *Inequalities*

- Section 5.3, *Slopes of Straight Lines,* now includes the topic of perpendicular lines in the rectangular coordinate system. Also in Section 5.3, the approach to graphing equations using the slope and y-intercept has changed so that students are instructed first to move up or down from the y-intercept, and then to move right or left to plot a second point.

- Section 8.4 of the previous edition, *Special Factoring,* has been separated into two sections. Section 8.4 now covers factoring the difference of two squares and perfect square trinomials, and Section 8.5 is devoted solely to factoring polynomials completely using the techniques introduced in the first four sections.

  8.1 *Common Factors*
  8.2 *Factoring Polynomials of the Form* $x^2 + bx + c$
  8.3 *Factoring Polynomials of the Form* $ax^2 + bx + c$
  8.4 *Special Factoring*
  8.5 *Factoring Polynomials Completely*
  8.6 *Solving Equations*

- A new section on variation has been added to Chapter 9. Section 9.6, *Variation,* covers direct variation and inverse variation.

  9.1 *Multiplication and Division of Rational Expressions*
  9.2 *Expressing Fractions in Terms of the LCD*
  9.3 *Addition and Subtraction of Rational Expressions*
  9.4 *Complex Fractions*
  9.5 *Equations Containing Fractions*
  9.6 *Variation*
  9.7 *Literal Equations*
  9.8 *Application Problems*

Beginning Algebra is organized around a carefully constructed hierarchy of OBJECTIVES. This "objective-based" approach provides an integrated learning environment that allows both the student and the instructor to easily find resources such as assessment tools (both within the text and online), videos, tutorials, and additional exercises.

NEW! FOCUS ON SUCCESS appears at the start of each Chapter Opener. These tips are designed to help you make the most of the text and your time as you progress through the course and prepare for tests and exams.

Each Chapter Opener outlines the learning OBJECTIVES that appear in each section. The list of objectives serves as a resource to guide you in your study and review of the topics.

Complete each PREP TEST to determine which topics you may need to study more carefully in order to be ready to learn the new material.

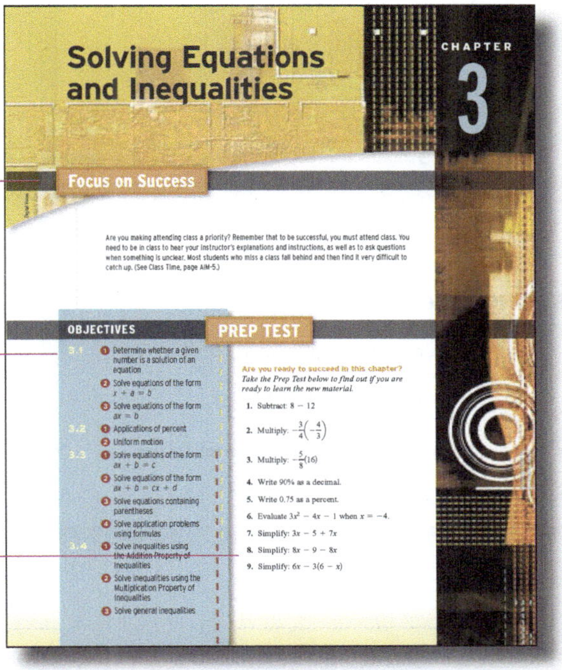
Solving Equations and Inequalities

CHAPTER 3

Focus on Success

Are you making attending class a priority? Remember that to be successful, you must attend class. You need to be in class to hear your instructor's explanations and instructions, as well as to ask questions when something is unclear. Most students who miss a class fall behind and then find it very difficult to catch up. (See Class Time, page AIM-5.)

OBJECTIVES

3.1
1 Determine whether a given number is a solution of an equation
2 Solve equations of the form $x + a = b$
3 Solve equations of the form $ax = b$

3.2
1 Applications of percent
2 Uniform motion

3.3
1 Solve equations of the form $ax + b = c$
2 Solve equations of the form $ax + b = cx + d$
3 Solve equations containing parentheses
4 Solve application problems using formulas

3.4
1 Solve inequalities using the Addition Property of Inequalities
2 Solve inequalities using the Multiplication Property of Inequalities
3 Solve general inequalities

PREP TEST

Are you ready to succeed in this chapter? Take the Prep Test below to find out if you are ready to learn the new material.

1. Subtract: $8 - 12$
2. Multiply: $-\frac{3}{4}\left(-\frac{4}{3}\right)$
3. Multiply: $-\frac{5}{8}(16)$
4. Write 90% as a decimal.
5. Write 0.75 as a percent.
6. Evaluate $3x^2 - 4x - 1$ when $x = -4$.
7. Simplify: $3x - 5 + 7x$
8. Simplify: $8x - 9 - 8x$
9. Simplify: $6x - 3(6 - x)$

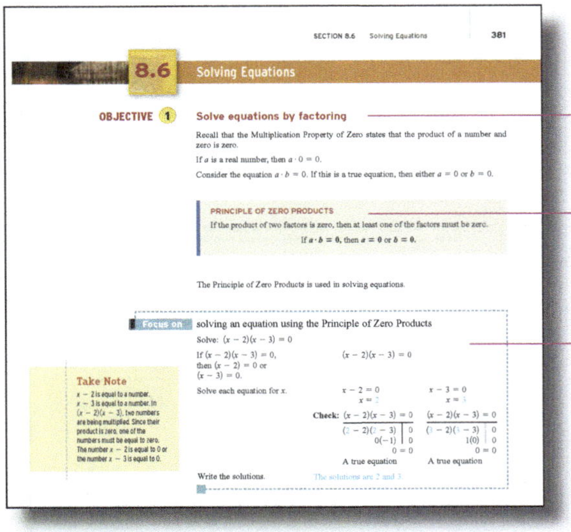
SECTION 8.6 Solving Equations 381

8.6 Solving Equations

OBJECTIVE 1 Solve equations by factoring

Recall that the Multiplication Property of Zero states that the product of a number and zero is zero.

If a is a real number, then $a \cdot 0 = 0$.

Consider the equation $a \cdot b = 0$. If this is a true equation, then either $a = 0$ or $b = 0$.

PRINCIPLE OF ZERO PRODUCTS

If the product of two factors is zero, then at least one of the factors must be zero.

If $a \cdot b = 0$, then $a = 0$ or $b = 0$.

The Principle of Zero Products is used in solving equations.

Focus on solving an equation using the Principle of Zero Products

Solve: $(x - 2)(x - 3) = 0$

If $(x - 2)(x - 3) = 0$, then $(x - 2) = 0$ or $(x - 3) = 0$. $\quad (x - 2)(x - 3) = 0$

Solve each equation for x. $\quad x - 2 = 0 \quad x = 2 \qquad x - 3 = 0 \quad x = 3$

Check: $(x - 2)(x - 3) = 0$; $(2 - 2)(2 - 3) = 0$; $0(-1) = 0$; $0 = 0$ A true equation. $(x - 2)(x - 3) = 0$; $(3 - 2)(3 - 3) = 0$; $1(0) = 0$; $0 = 0$ A true equation.

Write the solutions. The solutions are 2 and 3.

Take Note

$x - 2$ is equal to a number. $x - 3$ is equal to a number. In $(x - 2)(x - 3)$, two numbers are being multiplied. Since their product is zero, one of the numbers must be equal to zero. The number $x - 2$ is equal to 0 or the number $x - 3$ is equal to 0.

In each section, OBJECTIVE STATEMENTS introduce each new topic of discussion.

NEW! Many of the DEFINITION/KEY CONCEPTS boxes now contain examples to illustrate how each definition or key concept applies in practice.

NEW! FOCUS ON boxes alert you to the specific type of problem you must master in order to succeed with the homework exercises or on a test. Each FOCUS ON problem is accompanied by detailed explanations for each step of the solution.

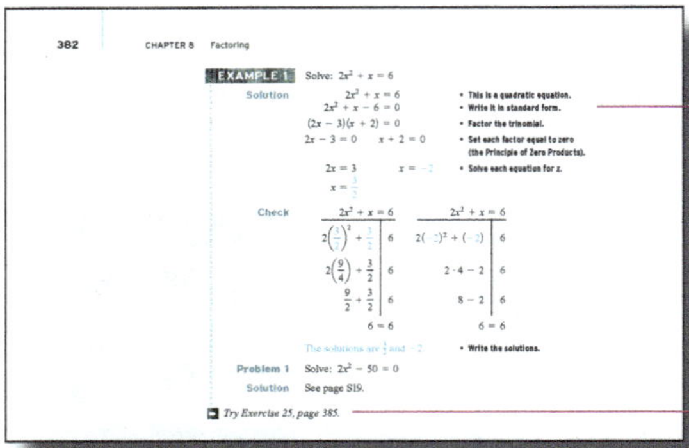
382 CHAPTER 8 Factoring

EXAMPLE 1 Solve: $2x^2 + x = 6$

Solution

$2x^2 + x = 6$ • This is a quadratic equation.

$2x^2 + x - 6 = 0$ • Write it in standard form.

$(2x - 3)(x + 2) = 0$ • Factor the trinomial.

$2x - 3 = 0 \quad x + 2 = 0$ • Set each factor equal to zero (the Principle of Zero Products).

$2x = 3 \quad x = -2$ • Solve each equation for x.

$x = \frac{3}{2}$

Check

$2x^2 + x = 6$		$2x^2 + x = 6$	
$2\left(\frac{3}{2}\right)^2 + \frac{3}{2}$	6	$2(-2)^2 + (-2)$	6
$2\left(\frac{9}{4}\right) + \frac{3}{2}$	6	$2 \cdot 4 - 2$	6
$\frac{9}{2} + \frac{3}{2}$	6	$8 - 2$	6
$6 = 6$		$6 = 6$	

The solutions are $\frac{3}{2}$ and -2. • Write the solutions.

Problem 1 Solve: $2x^2 - 50 = 0$

Solution See page S19.

Try Exercise 25, page 385.

The EXAMPLE/PROBLEM matched pairs are designed to actively involve you in the learning process. The Problems are based on the Examples. They are paired so that you can easily refer to the steps in the Example as you work through the accompanying Problem.

NEW! TRY EXERCISE prompts are given at the end of each Example/Problem pair. They point you to a similar exercise at the end of the section. By following the prompts, you can immediately apply the techniques presented in the worked Examples to homework exercises.

SECTION 8.6

Problem 1

$2x^2 - 50 = 0$ • **A quadratic equation**

$2(x^2 - 25) = 0$ • **Factor out 2.**

$x^2 - 25 = 0$ • **Divide each side by 2.**

$(x + 5)(x - 5) = 0$ • **Factor.**

$x + 5 = 0 \quad x - 5 = 0$ • **Set each factor equal to zero.**

$x = -5 \quad x = 5$

The solutions are -5 and 5.

Complete WORKED-OUT SOLUTIONS to the Problems are found in an appendix at the back of the text. Compare your solution with the one given in the appendix to obtain immediate feedback and reinforcement of the concept(s) you are studying.

Beginning Algebra contains a WIDE VARIETY OF EXERCISES that promote skill building, skill maintenance, concept development, critical thinking, and problem solving.

NEW! CONCEPT CHECK exercises promote conceptual understanding. Completing these exercises will deepen your understanding of the topics in the section.

GETTING READY exercises appear in most end-of-section exercise sets. These exercises provide guided practice and test your understanding of the underlying concepts in a lesson. They act as stepping stones to the remaining exercises for the objective.

NEW! IN THE NEWS application exercises help you see the usefulness of mathematics in our everyday world. They are based on information culled from popular media sources, including newspapers, magazines, and the Internet.

THINK ABOUT IT exercises promote conceptual understanding. Completing these exercises will deepen your understanding of the concept being addressed.

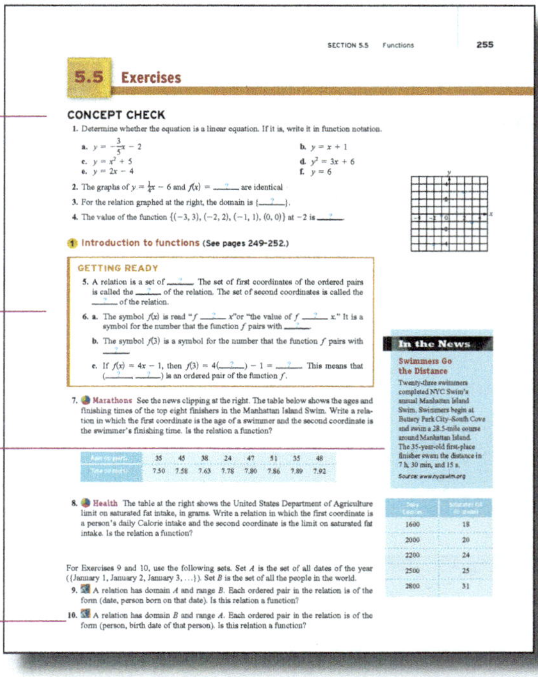
SECTION 5.5 Functions 255

5.5 Exercises

CONCEPT CHECK

1. Determine whether the equation is a linear equation. If it is, write it in function notation.
   a. $y = -\frac{3}{5}x - 2$ b. $y = x + 1$
   c. $y = x^2 + 5$ d. $y^2 = 3x + 6$
   e. $y = 2x - 4$ f. $y = 6$
2. The graphs of $y = \frac{1}{4}x - 6$ and $f(x) =$ __?__ are identical
3. For the relation graphed at the right, the domain is {__?__}.
4. The value of the function $\{(-3, 3), (-2, 2), (-1, 1), (0, 0)\}$ at -2 is __?__

1 Introduction to functions (See pages 249–252.)

GETTING READY

5. A relation is a set of __?__ The set of first coordinates of the ordered pairs is called the __?__ of the relation. The set of second coordinates is called the __?__ of the relation.
6. a. The symbol $f(x)$ is read "f __?__ x" or "the value of f __?__ x." It is a symbol for the number that the function f pairs with __?__.
   b. The symbol $f(3)$ is a symbol for the number that the function f pairs with __?__.
   c. If $f(x) = 4x - 1$, then $f(3) = 4($__?__$) - 1 =$ __?__. This means that (__?__, __?__) is an ordered pair of the function f.

7. **Marathons** See the news clipping at the right. The table below shows the ages and finishing times of the top eight finishers in the Manhattan Island Swim. Write a relation in which the first coordinate is the age of a swimmer and the second coordinate is the swimmer's finishing time. Is the relation a function?

[illegible]	35	45	38	24	47	51	35	48
[illegible]	7.50	7.58	7.63	7.78	7.80	7.86	7.89	7.92

In the News

Swimmers Go the Distance

Twenty-three swimmers completed NYC Swim's annual Manhattan Island Swim. Swimmers begin at Battery Park City-South Cove and swim a 28.5-mile course around Manhattan Island. The 35-year-old first-place finisher swam the distance in 7 h, 30 min, and 15 s.

Source: www.nycswim.org

8. **Health** The table at the right shows the United States Department of Agriculture limit on saturated fat intake, in grams. Write a relation in which the first coordinate is a person's daily Calorie intake and the second coordinate is the limit on saturated fat intake. Is the relation a function?

[illegible]	[illegible]
1600	18
2000	20
2200	24
2500	25
2800	31

For Exercises 9 and 10, use the following sets. Set A is the set of all dates of the year ({January 1, January 2, January 3, ...}). Set B is the set of all the people in the world.

9. A relation has domain A and range B. Each ordered pair in the relation is of the form (date, person born on that date). Is this relation a function?
10. A relation has domain B and range A. Each ordered pair in the relation is of the form (person, birth date of that person). Is this relation a function?

NEW! TRY EXERCISE icons are used to link exercises back to Examples from the section.

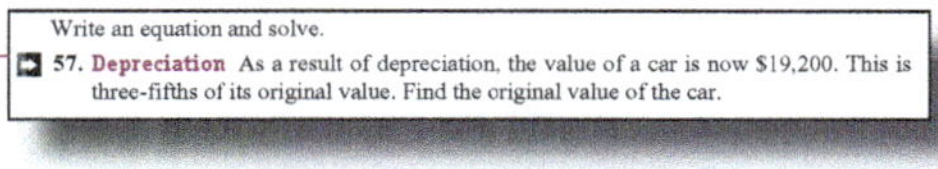
Write an equation and solve.

57. **Depreciation** As a result of depreciation, the value of a car is now $19,200. This is three-fifths of its original value. Find the original value of the car.

APPLYING CONCEPTS exercises may involve further exploration and analysis of topics, or they may integrate concepts introduced earlier in the text. **Optional** graphing calculator exercises are included, denoted by .

Working through the application exercises that contain REAL DATA will prepare you to use real-world information to answer questions and solve problems.

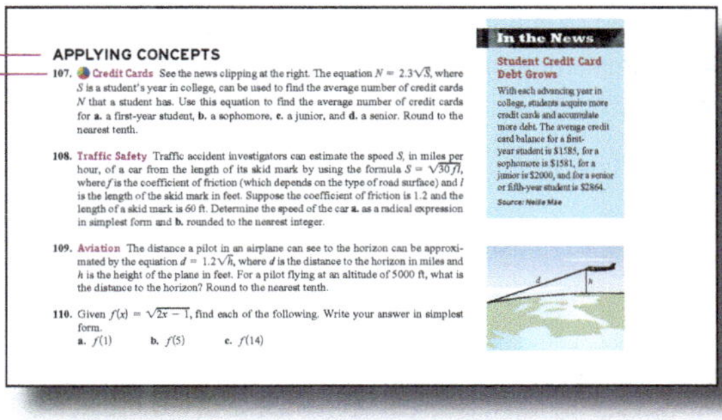
APPLYING CONCEPTS

107. **Credit Cards** See the news clipping at the right. The equation $N = 2.3\sqrt{S}$, where S is a student's year in college, can be used to find the average number of credit cards N that a student has. Use this equation to find the average number of credit cards for **a.** a first-year student, **b.** a sophomore, **c.** a junior, and **d.** a senior. Round to the nearest tenth.

108. **Traffic Safety** Traffic accident investigators can estimate the speed S, in miles per hour, of a car from the length of its skid mark by using the formula $S = \sqrt{30fl}$, where f is the coefficient of friction (which depends on the type of road surface) and l is the length of the skid mark in feet. Suppose the coefficient of friction is 1.2 and the length of a skid mark is 60 ft. Determine the speed of the car **a.** as a radical expression in simplest form and **b.** rounded to the nearest integer.

109. **Aviation** The distance a pilot in an airplane can see to the horizon can be approximated by the equation $d = 1.2\sqrt{h}$, where d is the distance to the horizon in miles and h is the height of the plane in feet. For a pilot flying at an altitude of 5000 ft, what is the distance to the horizon? Round to the nearest tenth.

110. Given $f(x) = \sqrt{2x - 1}$, find each of the following. Write your answer in simplest form.
a. $f(1)$ **b.** $f(5)$ **c.** $f(14)$

In the News

Student Credit Card Debt Grows

With each advancing year in college, students acquire more credit cards and accumulate more debt. The average credit card balance for a first-year student is $1585, for a sophomore is $1581, for a junior is $2000, and for a senior or fifth-year student is $2864.

Source: Nellie Mae

By completing the WRITING EXERCISES, you will improve your communication skills while increasing your understanding of mathematical concepts.

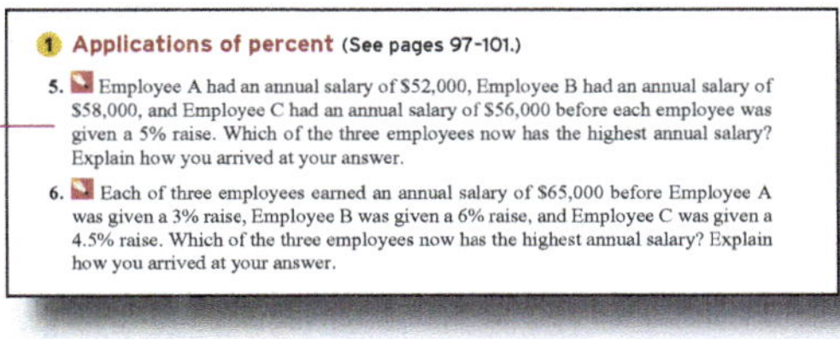
1 **Applications of percent** (See pages 97–101.)

5. Employee A had an annual salary of $52,000, Employee B had an annual salary of $58,000, and Employee C had an annual salary of $56,000 before each employee was given a 5% raise. Which of the three employees now has the highest annual salary? Explain how you arrived at your answer.

6. Each of three employees earned an annual salary of $65,000 before Employee A was given a 3% raise, Employee B was given a 6% raise, and Employee C was given a 4.5% raise. Which of the three employees now has the highest annual salary? Explain how you arrived at your answer.

NEW! PROJECTS OR GROUP ACTIVITIES appear at the end of each set of exercises. Your instructor may assign these individually, or you may be asked to work through the activities in groups.

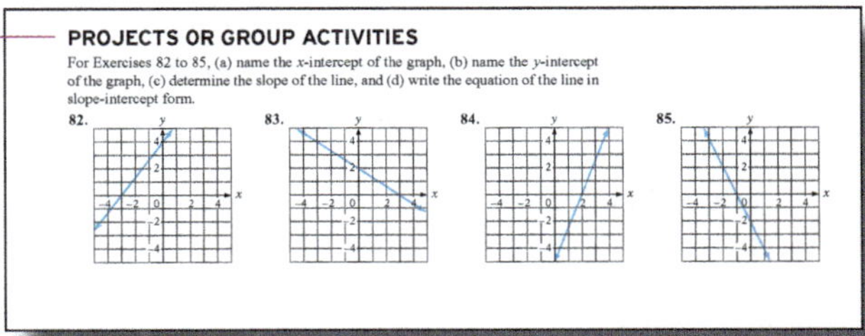
PROJECTS OR GROUP ACTIVITIES

For Exercises 82 to 85, (a) name the x-intercept of the graph, (b) name the y-intercept of the graph, (c) determine the slope of the line, and (d) write the equation of the line in slope-intercept form.

82. 83. 84. 85.

Beginning Algebra addresses a broad range of study styles by offering a WIDE VARIETY OF TOOLS FOR REVIEW.

At the end of each chapter, you will find a SUMMARY outlining KEY WORDS and ESSENTIAL RULES AND PROCEDURES presented in the chapter. Each entry includes an objective-level reference and a page reference to show you where in the chapter the concept was introduced. An example demonstrating the concept is also included.

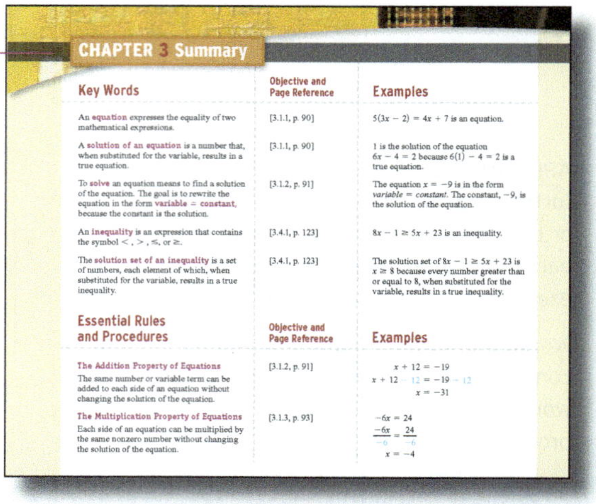

CHAPTER 3 Summary

Key Words	Objective and Page Reference	Examples
An **equation** expresses the equality of two mathematical expressions.	[3.1.1, p. 90]	$5(3x-2)=4x+7$ is an equation.
A **solution of an equation** is a number that, when substituted for the variable, results in a true equation.	[3.1.1, p. 90]	1 is the solution of the equation $6x-4=2$ because $6(1)-4=2$ is a true equation.
To **solve** an equation means to find a solution of the equation. The goal is to rewrite the equation in the form **variable = constant**, because the constant is the solution.	[3.1.2, p. 91]	The equation $x=-9$ is in the form *variable = constant*. The constant, -9, is the solution of the equation.
An **inequality** is an expression that contains the symbol $<$, $>$, $\le$, or $\ge$.	[3.4.1, p. 123]	$8x-1\ge 5x+23$ is an inequality.
The **solution set of an inequality** is a set of numbers, each element of which, when substituted for the variable, results in a true inequality.	[3.4.1, p. 123]	The solution set of $8x-1\ge 5x+23$ is $x\ge 8$ because every number greater than or equal to 8, when substituted for the variable, results in a true inequality.

Essential Rules and Procedures	Objective and Page Reference	Examples
The Addition Property of Equations The same number or variable term can be added to each side of an equation without changing the solution of the equation.	[3.1.2, p. 91]	$x+12=-19$; $x+12-12=-19-12$; $x=-31$
The Multiplication Property of Equations Each side of an equation can be multiplied by the same nonzero number without changing the solution of the equation.	[3.1.3, p. 93]	$-6x=24$; $\frac{-6x}{-6}=\frac{24}{-6}$; $x=-4$

In the CHAPTER REVIEW EXERCISES, the order in which different types of problems appear is different from the order in which the topics were presented in the chapter. The ANSWERS to these exercises include references to the section objectives upon which the exercises are based. This will help you to quickly identify where to go to review a concept if you need more practice.

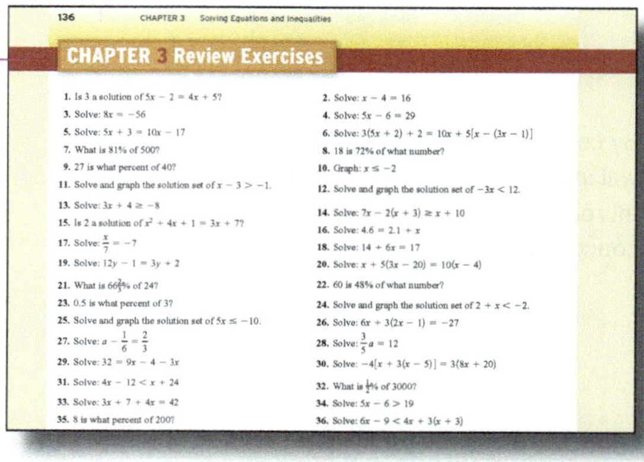

136 CHAPTER 3 Solving Equations and Inequalities

CHAPTER 3 Review Exercises

1. Is 3 a solution of $5x-2=4x+5$?
2. Solve: $x-4=16$
3. Solve: $8x=-56$
4. Solve: $5x-6=29$
5. Solve: $5x+3=10x-17$
6. Solve: $3(5x+2)+2=10x+5[x-(3x-1)]$
7. What is 81% of 500?
8. 18 is 72% of what number?
9. 27 is what percent of 40?
10. Graph: $x\le -2$
11. Solve and graph the solution set of $x-3>-1$.
12. Solve and graph the solution set of $-3x<12$.
13. Solve: $3x+4\ge -8$
14. Solve: $7x-2(x+3)\ge x+10$
15. Is 2 a solution of $x^2+4x+1=3x+7$?
16. Solve: $4.6=2.1+x$
17. Solve: $\frac{x}{7}=-7$
18. Solve: $14+6x=17$
19. Solve: $12y-1=3y+2$
20. Solve: $x+5(3x-20)=10(x-4)$
21. What is $66\frac{2}{3}\%$ of 24?
22. 60 is 48% of what number?
23. 0.5 is what percent of 3?
24. Solve and graph the solution set of $2+x<-2$.
25. Solve and graph the solution set of $5x\le -10$.
26. Solve: $6x+3(2x-1)=-27$
27. Solve: $a-\frac{1}{6}=\frac{2}{3}$
28. Solve: $\frac{3}{5}a=12$
29. Solve: $32=9x-4-3x$
30. Solve: $-4[x+3(x-5)]=3(8x+20)$
31. Solve: $4x-12<x+24$
32. What is $\frac{1}{2}\%$ of 3000?
33. Solve: $3x+7+4x=42$
34. Solve: $5x-6>19$
35. 8 is what percent of 200?
36. Solve: $6x-9<4x+3(x+3)$

Each CHAPTER TEST is designed to simulate a typical test of the concepts covered in the chapter. The ANSWERS include references to section objectives. Also provided is a reference to an Example, Problem, or Focus On, which refers students to a worked example in the text that is similar to the given test question.

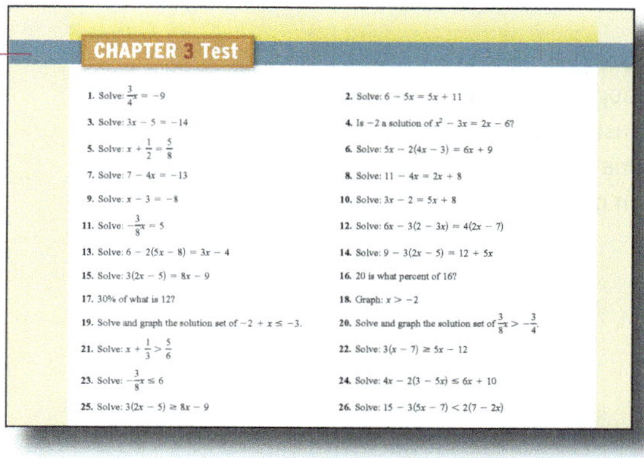

CHAPTER 3 Test

1. Solve: $\frac{3}{4}x=-9$
2. Solve: $6-5x=5x+11$
3. Solve: $3x-5=-14$
4. Is -2 a solution of $x^2-3x=2x-6$?
5. Solve: $x+\frac{1}{2}=\frac{5}{8}$
6. Solve: $5x-2(4x-3)=6x+9$
7. Solve: $7-4x=-13$
8. Solve: $11-4x=2x+8$
9. Solve: $x-3=-8$
10. Solve: $3x-2=5x+8$
11. Solve: $-\frac{3}{8}x=5$
12. Solve: $6x-3(2-3x)=4(2x-7)$
13. Solve: $6-2(5x-8)=3x-4$
14. Solve: $9-3(2x-5)=12+5x$
15. Solve: $3(2x-5)=8x-9$
16. 20 is what percent of 16?
17. 30% of what is 12?
18. Graph: $x>-2$
19. Solve and graph the solution set of $-2+x\le -3$.
20. Solve and graph the solution set of $\frac{3}{8}x>-\frac{3}{4}$.
21. Solve: $x+\frac{1}{3}>\frac{5}{6}$
22. Solve: $3(x-7)\ge 5x-12$
23. Solve: $-\frac{3}{8}x\le 6$
24. Solve: $4x-2(3-5x)\le 6x+10$
25. Solve: $3(2x-5)\ge 8x-9$
26. Solve: $15-3(5x-7)<2(7-2x)$

CUMULATIVE REVIEW EXERCISES, which appear at the end of each chapter (beginning with Chapter 2), help you maintain the skills you learned previously. The ANSWERS include references to the section objectives upon which the exercises are based.

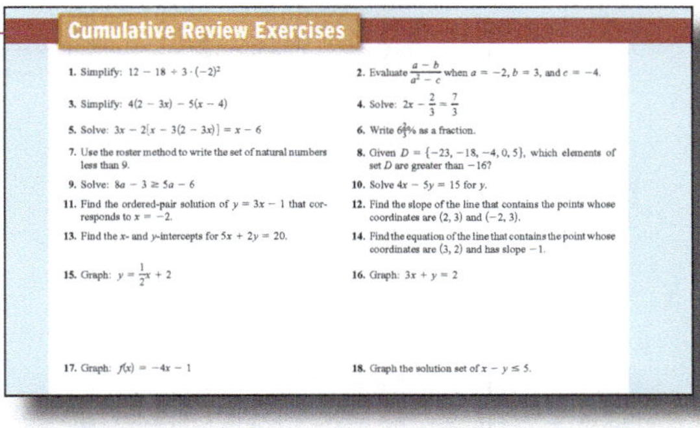

Cumulative Review Exercises

1. Simplify: $12 - 18 \div 3 \cdot (-2)^2$
2. Evaluate $\frac{a-b}{a^2-c}$ when $a = -2$, $b = 3$, and $c = -4$.
3. Simplify: $4(2 - 3x) - 5(x - 4)$
4. Solve: $2x - \frac{2}{3} = \frac{7}{3}$
5. Solve: $3x - 2[x - 3(2 - 3x)] = x - 6$
6. Write $6\frac{2}{3}\%$ as a fraction.
7. Use the roster method to write the set of natural numbers less than 9.
8. Given $D = \{-23, -18, -4, 0, 5\}$, which elements of set D are greater than -16?
9. Solve: $8a - 3 \geq 5a - 6$
10. Solve $4x - 5y = 15$ for y.
11. Find the ordered-pair solution of $y = 3x - 1$ that corresponds to $x = -2$.
12. Find the slope of the line that contains the points whose coordinates are $(2, 3)$ and $(-2, 3)$.
13. Find the x- and y-intercepts for $5x + 2y = 20$.
14. Find the equation of the line that contains the point whose coordinates are $(3, 2)$ and has slope -1.
15. Graph: $y = \frac{1}{2}x + 2$
16. Graph: $3x + y = 2$
17. Graph: $f(x) = -4x - 1$
18. Graph the solution set of $x - y \leq 5$.

A FINAL EXAM is included after the last chapter of the text. It is designed to simulate a comprehensive exam covering all the concepts presented in the text. The ANSWERS to the final exam questions are provided in the appendix at the back of the text and include references to the section objectives upon which the questions are based.

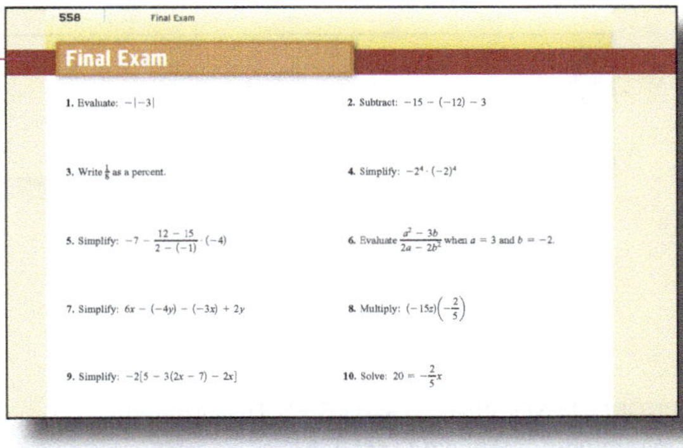

558 Final Exam

Final Exam

1. Evaluate: $-|-3|$
2. Subtract: $-15 - (-12) - 3$
3. Write $\frac{1}{8}$ as a percent.
4. Simplify: $-2^4 \cdot (-2)^4$
5. Simplify: $-7 - \frac{12 - 15}{2 - (-1)} \cdot (-4)$
6. Evaluate $\frac{a^2 - 3b}{2a - 2b^2}$ when $a = 3$ and $b = -2$.
7. Simplify: $6x - (-4y) - (-3x) + 2y$
8. Multiply: $(-15z)\left(-\frac{2}{5}\right)$
9. Simplify: $-2[5 - 3(2x - 7) - 2x]$
10. Solve: $20 = -\frac{2}{5}x$

Other Key Features

MARGINS Within the margins, you can find the following features.

> **Take Note**
>
> The expression $n(6 + n^3)$ must have parentheses. If we write $n \cdot 6 + n^3$, then by the Order of Operations Agreement, only the 6 is multiplied by n. We want n to be multiplied by the *total* of 6 and n^3.

TAKE NOTE boxes alert you to concepts that require special attention.

> **Point of Interest**
>
> The Alexandrian astronomer Ptolemy began using *omicron*, *o*, the first letter of the Greek word that means "nothing," as the symbol for zero in A.D. 150. It was not until the 13th century, however, that Fibonacci introduced 0 to the Western world as a placeholder so that we could distinguish, for example, 45 from 405.

POINT OF INTEREST boxes, which relate to the topic under discussion, may be historical in nature or may be of general interest.

How It's Used

Addition of positive and negative decimals is used in optometry. *Diopters*, which are used to measure the strength of lenses, are given as positive or negative decimals: a negative diopter lens corrects nearsightedness and a positive diopter lens corrects farsightedness. To correct more than one aspect of a person's vision, an optometrist designs an eyeglass lens that combines two or more diopter strengths.

NEW! HOW IT'S USED boxes relate to the topic under discussion. These boxes present real-world scenarios that demonstrate the utility of selected concepts from the text.

A graphing calculator can be used to evaluate variable expressions. When the value of each variable is stored in the calculator's memory and a variable expression is then entered into the calculator, the calculator evaluates that variable expression for the values of the variables stored in its memory. See the Appendix for a description of keystroking procedures.

While the text is not dependent on the use of a calculator, TECHNOLOGY boxes that focus on calculator instruction are included for selected topics. The boxes contain tips for using a graphing calculator.

PROBLEM-SOLVING STRATEGIES The problem-solving approach used throughout the text emphasizes the importance of a well-defined strategy. Model strategies are presented as guides for you to follow as you attempt the parallel Problem that accompanies each numbered Example.

EXAMPLE 2 During a recent year, nearly 1.2 million dogs or litters were registered with the American Kennel Club. The most popular breed was the Labrador retriever, with 172,841 registered. What percent of the registrations were Labrador retrievers? Round to the nearest tenth of a percent. (*Source:* American Kennel Club)

Strategy To find the percent, solve the basic percent equation using $B = 1.2$ million $= 1{,}200{,}000$ and $A = 172{,}841$. The percent is unknown.

Solution

$$PB = A$$

$$P(1{,}200{,}000) = 172{,}841 \quad \bullet\ B = 1{,}200{,}000;\ A = 172{,}841$$

$$\frac{P(1{,}200{,}000)}{1{,}200{,}000} = \frac{172{,}841}{1{,}200{,}000} \quad \bullet\ \text{Divide each side of the equation by } 1{,}200{,}000.$$

$$P \approx 0.144$$

$$P \approx 14.4\% \quad \bullet\ \text{Rewrite the decimal as a percent.}$$

Approximately 14.4% of the registrations were Labrador retrievers.

Problem 2 A student correctly answered 72 of the 80 questions on an exam. What percent of the questions were answered correctly?

Solution See page S5.

Try Exercise 43, page 105.

Instructor Resources

PRINT SUPPLEMENTS

Annotated Instructor's Edition (ISBN: 978-1-111-98944-6)
The Annotated Instructor's Edition features answers to all of the problems in the text, as well as an appendix denoting those problems that can be found in Enhanced WebAssign.

Instructor's Solutions Manual (ISBN: 978-1-133-11223-5)
Author: Rhoda Oden, *Gadsden State Community College*
The Instructor's Solutions Manual provides worked-out solutions to all of the problems in the text.

Instructor's Resource Binder with Appendix
(ISBN: 978-1-133-11248-8)
Author: Maria H. Andersen, *Muskegon Community College,* with Appendices by Richard N. Aufmann, *Palomar College,* and Joanne S. Lockwood, *Nashua Community College*
Each section of the main text is discussed in uniquely designed Teaching Guides that contain tips, examples, activities, worksheets, overheads, assessments, and solutions to all worksheets and activities.

ELECTRONIC SUPPLEMENTS

Text-Specific Videos
Author: Dana Mosely
These text-specific instructional videos provide students with visual reinforcement of concepts and explanations. The videos contain easy-to-understand language along with detailed examples and sample problems. A flexible format offers versatility. Topics can be accessed quickly, and lectures can be catered to self-paced, online, or hybrid courses. Closed captioning is provided for the hearing impaired. These videos are available through Enhanced WebAssign and CourseMate.

PowerLecture with Diploma® (ISBN: 978-1-133-11367-6)
This CD-ROM provides you with dynamic media tools for teaching. You can create, deliver, and customize tests (both print and online) in minutes with Diploma's Computerized Testing featuring algorithmic equations. The Solution Builder's online solutions manual easily builds solution sets for homework or exams. Practice Sheets, First-Day-of-Class PowerPoint® lecture slides, art and figures from the book, and a test bank in electronic format are also included on this CD-ROM.

Syllabus Creator (Included on the PowerLecture)
Authors: Richard N. Aufmann and Joanne S. Lockwood
NEW! Easily write, edit, and update your syllabus with the Aufmann/Lockwood Syllabus Creator. This software program enables you to create your new syllabus in several easy steps: first select the required course objectives; then add your contact information, course information, student expectations, grading policy, dates and location of your course, and course outline. You now have your syllabus!

Solution Builder
This online instructor database offers complete worked solutions to all exercises in the text, allowing you to create customized, secure solutions printouts (in PDF format) matched exactly to the problems you assign in class. For more information, visit www.cengage.com/solutionbuilder.

Enhanced WebAssign® (ISBN: 978-0-538-73810-1)
Exclusively from Cengage Learning, Enhanced WebAssign combines the exceptional mathematics content that you know and love with the most powerful online homework solution, WebAssign. Enhanced WebAssign engages students with immediate feedback and rich tutorial content. Interactive eBooks help students develop a deeper conceptual understanding of their subject matter. Online assignments can be built by selecting from thousands of text-specific problems. Assignments can be supplemented with problems from any Cengage Learning textbook.

Enhanced WebAssign: Start Smart Guide for Students
(ISBN: 978-0-495-38479-3)
Author: Brooks/Cole
The Enhanced WebAssign Student Start Smart Guide helps students get up and running quickly with Enhanced WebAssign so that they can study smarter and improve their performance in class.

Printed Access Card for CourseMate with eBook
(ISBN: 978-1-133-51019-2)

Instant Access Card for CourseMate with eBook
(ISBN: 978-1-133-51018-5)
Complement your text and course content with study and practice materials. Cengage Learning's Developmental Mathematics CourseMate brings course concepts to life with interactive learning, study, and exam preparation tools that support the printed textbook. Watch student comprehension soar as your class works with the printed textbook and the textbook-specific website. Developmental Mathematics CourseMate goes beyond the book to deliver what you need!

Student Resources

PRINT SUPPLEMENTS

To get access, visit CengageBrain.com

Student Solutions Manual
(ISBN: 978-1-133-11224-2)
Author: Rhoda Oden, *Gadsden State Community College*
Go beyond the answers—and improve your grade! This manual provides worked-out, step-by-step solutions to the odd-numbered problems in the text. The Student Solutions Manual gives you the information you need to truly understand how the problems are solved.

Student Workbook (ISBN: 978-1-133-11227-3)
Author: Maria H. Andersen, *Muskegon Community College*
Get a head start. The Student Workbook contains assessments, activities, and worksheets for classroom discussions, in-class activities, and group work.

AIM for Success Student Practice Sheets (ISBN: 978-1-133-11226-6)
Author: Christine S. Verity
AIM for Success Student Practice Sheets provide additional practice problems to help you learn the material.

ELECTRONIC SUPPLEMENTS

Text-Specific Videos
Author: Dana Mosely
These text-specific instructional videos provide you with visual reinforcement of concepts and explanations. The videos contain easy-to-understand language along with detailed examples and sample problems. A flexible format offers versatility. Topics can be accessed quickly, and lectures can be catered to self-paced, online, or hybrid courses. Closed captioning is provided for the hearing impaired. These videos are available through Enhanced WebAssign and CourseMate.

Enhanced WebAssign (ISBN: 978-0-538-73810-1)
Enhanced WebAssign (assigned by the instructor) provides instant feedback on homework assignments. This online homework system is easy to use and includes helpful links to textbook sections, video examples, and problem-specific tutorials.

Chapter Test Videos
(Available through Enhanced WebAssign)
Available through Enhanced WebAssign, the chapter test videos provide step-by-step solutions that follow the problem-solving methods used in the text for every end-of-chapter text question. Some solution videos feature interactive questions that provide immediate feedback on your answers.

Enhanced WebAssign: Start Smart Guide for Students
(ISBN: 978-0-495-38479-3)
Author: Brooks/Cole
If your instructor has chosen to package Enhanced WebAssign with your text, this manual will help you get up and running quickly with the Enhanced WebAssign system so that you can study smarter and improve your performance in class.

Printed Access Card for CourseMate with eBook
(ISBN: 978-1-133-51019-2)

Instant Access Card for CourseMate with eBook
(ISBN: 978-1-133-51018-5)
The more you study, the greater your success. You can make the most of your study time by accessing everything you need to succeed in one place—online with CourseMate. You can use CourseMate to read the textbook, take notes, review flashcards, watch videos, and take practice quizzes.

Acknowledgments

The authors would like to thank the people who have reviewed the seventh edition and provided many valuable suggestions.

Maria T. Alzugaray Rodriguez, *Suffolk County Community College*
Sheila Anderson, *Housatonic Community College*
Edie Carter, *Amarillo College*
Kamesh Casukhela, *The Ohio State University at Lima*
Jacqui Fields, *Wake Technical Community College*
Julie Fisher, *Austin Community College*
Shelly Hansen, *Mesa State College, Western Colorado Community College*
Gayathri Kambhampati, *Cloud County Community College–Geary Campus*
Brian Karasek, *South Mountain Community College*
Linda Kuroski, *Erie Community College City Campus*
Larry Musolino, *Lehigh Carbon Community College*
Angela Stabley, *Portland Community College*
Rose Toering, *Kilian Community College*
Edward Watkins, *Florida State College at Jacksonville*
Annette Wiesner, *University of Wisconsin–Parkside*

Special thanks go to Jean Bermingham for copyediting the manuscript and proofreading the pages, to Rhoda Oden for preparing the solutions manuals, and to Lauri Semarne for her work in ensuring the accuracy of the text. We would also like to thank the many people at Cengage Learning who worked to guide the manuscript for the eighth edition from development through production.

CHAPTER A

AIM for Success

Focus on Success

This important chapter describes study skills that are used by students who have been successful in this course. Chapter A covers a wide range of topics that focus on what you need to do to succeed in this class. It includes a complete guide to the textbook and how to use its features to become a successful student.

OBJECTIVES

A.1
- Get ready
- Motivate yourself
- Develop a "can do" attitude toward math
- Strategies for success
- Time management
- Habits of successful students

A.2
- Get the big picture
- Understand the organization
- Use the interactive method
- Use a strategy to solve word problems
- Ace the test

PREP TEST

Are you ready to succeed in this course?

1. Read this chapter. Answer all of the questions. Write down your answers on paper.
2. Write down your instructor's name.
3. Write down the classroom number.
4. Write down the days and times the class meets.
5. Bring your textbook, a notebook, and a pen or pencil to every class.
6. Be an active participant, not a passive observer.

A.1 How to Succeed in This Course

GET READY

We are committed to your success in learning mathematics and have developed many tools and resources to support you along the way.

DO YOU WANT TO EXCEL IN THIS COURSE?

Read on to learn about the skills you'll need and how best to use this book to get the results you want.

We have written this text in an *interactive* style. More about this later but, in short, this means that you are supposed to interact with the text. Do not just read the text! Work along with it. Ready? Let's begin!

WHY ARE YOU TAKING THIS COURSE?

Did you interact with the text, or did you just read the last question? Get some paper and a pencil or pen and answer the question. Really—you will have more success in math and other courses you take if you **actively participate.** Now, **interact.** Write down one reason you are taking this course.

____

____

Of course, we have no idea what you just wrote, but experience has shown us that many of you wrote something along the lines of "I have to take it to graduate" or "It is a prerequisite to another course I have to take" or "It is required for my major." Those reasons are perfectly fine. Every teacher has had to take courses that were not directly related to his or her major.

WHY DO YOU WANT TO SUCCEED IN THIS COURSE?

Think about why you want to succeed in this course. List the reasons here (not in your head . . . on the paper!):

____

____

One reason you may have listed is that math skills are important in order to be successful in your chosen career. That is certainly an important reason. Here are some other reasons.

- Math is a skill that applies across careers, which is certainly a benefit in our world of changing job requirements. A good foundation in math may enable you to more easily make a career change.
- Math can help you learn critical thinking skills, an attribute all employers want.
- Math can help you see relationships between ideas and identify patterns.

MOTIVATE YOURSELF

> **Take Note**
> Motivation alone won't lead to success. For example, suppose a person who cannot swim is rowed out to the middle of a lake and thrown overboard. That person has a lot of motivation to swim, but most likely will drown without some help. You'll need motivation *and* learning in order to succeed.

You'll find many real-life problems in this book, relating to sports, money, cars, music, and more. We hope that these topics will help you understand how mathematics is used in everyday life. To learn all of the necessary skills and to understand how you can apply them to your life outside of this course, **motivate yourself to learn.**

One of the reasons we asked you why you are taking this course was to provide motivation for you to succeed. When there is a reason to do something, that task is easier to accomplish. We understand that you may not want to be taking this course but, to achieve your career goal, this is a necessary step. Let your career goal be your motivation for success.

MAKE THE COMMITMENT TO SUCCEED!

With practice, you will improve your math skills. Skeptical? Think about when you first learned to drive a car, ride a skateboard, dance, paint, surf, or any other talent that you now have. You may have felt self-conscious or concerned that you might fail. But with time and practice, you learned the skill.

List a situation in which you accomplished your goal by spending time practicing and perfecting your skills (such as learning to play the piano or to play basketball):

____

____

You do not get "good" at something by doing it once a week. **Practice** is the backbone of any successful endeavor—including math!

DEVELOP A "CAN DO" ATTITUDE TOWARD MATH

You can do math! When you first learned the skills you just listed above, you may not have done them well. With practice, you got better. With practice, you will get better at math. Stay focused, motivated, and committed to success.

We cannot emphasize enough how important it is to overcome the "I Can't Do Math" syndrome. If you listen to interviews of very successful athletes after a particularly bad performance, you will note that they focus on the positive aspects of what they did, not the negative. Sports psychologists encourage athletes always to be positive—to have a "can do" attitude. Develop this attitude toward math and you will succeed.

Change your conversation about mathematics. Do not say "I can't do math," "I hate math," or "Math is too hard." These comments just give you an excuse to fail. You don't want to fail, and we don't want you to fail. Write it down now: **I can do math!**

____

____

STRATEGIES FOR SUCCESS

PREPARE TO SUCCEED

There are a number of things that may be worrisome to you as you begin a new semester. List some of those things now.

Here are some of the concerns expressed by our students.

- **Tuition**
  Will I be able to afford school?
- **Job**
  I must work. Will my employer give me a schedule that will allow me to go to school?
- **Anxiety**
  Will I succeed?
- **Child care**
  What will I do with my kids while I'm in class or when I need to study?
- **Time**
  Will I be able to find the time to attend class and study?
- **Degree goals**
  How long will it take me to finish school and earn my degree?

These are all important and valid concerns. Whatever your concerns, acknowledge them. Choose an education path that allows you to accommodate your concerns. Make sure they don't prevent you from succeeding.

SELECT A COURSE

Many schools offer math assessment tests. These tests evaluate your present math skills. They don't evaluate how smart you are, so don't worry about your score on the test. If you are unsure about where you should start in the math curriculum, these tests can show you where to begin. You are better off starting at a level that is appropriate for you than starting a more advanced class and then dropping it because you can't keep up. Dropping a class is a waste of time and money.

If you have difficulty with math, avoid short courses that compress the class into a few weeks. If you have struggled with math in the past, this environment does not give you the time to process math concepts. Similarly, avoid classes that meet once a week. The time delay between classes makes it difficult to make connections between concepts.

Some career goals require a number of math courses. If that is true of your major, try to take a math course every semester until you complete the requirements. Think about it this way. If you take, say, French I, and then wait two semesters before taking French II, you may forget a lot of material. Math is much the same. You must keep the concepts fresh in your mind.

TIME MANAGEMENT

One of the most important requirements in completing any task is to acknowledge the amount of time it will take to finish the job successfully. Before a construction company starts to build a skyscraper, the company spends months looking at how much time each of the phases of construction will take. This is done so that resources can be allocated when appropriate. For instance, it would not make sense to schedule the electricians to run wiring until the walls are up.

MANAGE YOUR TIME!

We know how busy you are outside of school. Do you have a full-time or a part-time job? Do you have children? Do you visit your family often? Do you play school sports or participate in the school orchestra or theater company? It can be stressful to balance all of the important activities and responsibilities in your life. Creating a time management plan will help you schedule enough time to do everything you need to do. Let's get started.

First, you need a calendar. You can use a daily planner, a calendar for a smartphone, or an online calendar, such as the ones offered by Google, MSN, or Yahoo. It is best to have a calendar on which you can fill in daily activities and be able to see a weekly or monthly view as well.

Start filling in your calendar now, even if it means stopping right here and finding a calendar. Some of the things you might include are:

- The hours each class meets
- Time for driving to and from work or school
- Leisure time, an important aspect of a healthy lifestyle
- Time for study. Plan at least one hour of study for each hour in class. This is a *minimum!*
- Time to eat
- Your work schedule
- Time for extracurricular activities such as sports, music lessons, or volunteer work
- Time for family and friends
- Time for sleep
- Time for exercise

Take Note

Be realistic about how much time you have. One gauge is that working 10 hours per week is approximately equivalent to taking one three-unit course. If your college considers 15 units a full load and you are working 10 hours per week, you should consider taking 12 units. The more you work, the fewer units you should take.

We really hope you did this. If not, please reconsider. One of the best pathways to success is understanding how much time it takes to succeed. When you finish your calendar, if it does not allow you enough time to stay physically and emotionally healthy, rethink some of your school or work activities. We don't want you to lose your job because you have to study math. On the other hand, we don't want you to fail in math because of your job.

If math is particularly difficult for you, consider taking fewer course units during the semesters you take math. This applies equally to any other subject that you may find difficult. There is no rule that you must finish college in four years. It is a myth—discard it now.

Now extend your calendar for the entire semester. Many of the entries will repeat, such as the time a class meets. In your extended calendar, include significant events that may disrupt your normal routine. These might include holidays, family outings, birthdays, anniversaries, or special events such as a concert or a football game. In addition to these events, be sure to include the dates of tests, the date of the final exam, and dates that projects or papers are due. These are all important semester events. Having them on your calendar will remind you that you need to make time for them.

CLASS TIME

To be successful, **attend class.** You should consider your commitment to attend class as serious as your commitment to your job or to keeping an appointment with a dear friend. It is difficult to overstate the importance of attending class. If you miss work, you don't get paid. If you miss class, you are not getting the full benefit of your tuition dollar. You are losing money.

If, by some unavoidable situation, you cannot attend class, find out as soon as possible what was covered in class. You might:

- Ask a friend for notes and the assignment.
- Contact your instructor and get the assignment. Missing class is no excuse for not being prepared for the next class.
- Determine whether there are online resources that you can use to help you with the topics and concepts that were discussed in the class you missed.

Going to class is important. Once you are there, **participate in class.** Stay involved and active. When your instructor asks a question, try to at least mentally answer the question. If you have a question, ask. Your instructor expects questions and wants you to understand the concept being discussed.

HOMEWORK TIME

In addition to attending class, you must **do homework.** Homework is the best way to reinforce the ideas presented in class. You should plan on at least one to two hours of homework and study for each hour you are in class. We've had many students tell us that one to two hours seems like a lot of time. That may be true, but if you want to attain your goals, you must be willing to devote the time to being successful in this math course.

You should schedule study time just as if it were class time. To do this, write down where and when you study best. For instance, do you study best at home, in the library, at the math center, under a tree, or somewhere else? Some psychologists who research successful study strategies suggest that just by varying where you study, you can increase the effectiveness of a study session. While you are considering where you prefer to study, also think about the time of day during which your study period will be most productive. Write down your thoughts.

Look at what you have written, and be sure that you can consistently be in your favorite study environment at the time you have selected. Study and homework are extremely important. Just as you should not miss class, **do not miss study time.**

Before we leave this important topic, we have a few suggestions. If at all possible, create a study hour right after class. The material will be fresh in your mind, and the immediate review, along with your homework, will help reinforce the concepts you are studying.

If you can't study right after class, make sure that you set some time *on the day of the class* to review notes and begin the homework. The longer you wait, the more difficult it is to recall some of the important points covered during class. Studying math in small chunks—one hour a day (perhaps not enough for most of us), every day, is better than seven hours in one sitting. If you are studying for an extended period of time, break up your study session by studying one subject for a while and then moving on to another subject. Try to alternate between similar or related courses. For instance, study math for a while, then science, and then back to math. Or study history for a while, then political science, and then back to history.

Meet some of the people in your class and try to **put together a study group.** The group could meet two or three times a week. During those meetings, you could quiz each other, prepare for a test, try to explain a concept to someone else in the group, or get help on a topic that is difficult for you.

After reading these suggestions, you may want to rethink where and when you study best. If so, do that now. Remember, however, that it is your individual style that is important. Choose what works for *you,* and stick to it.

HABITS OF SUCCESSFUL STUDENTS

There are a number of habits that successful students use. Think about what these might be, and write them down.

What you have written is very important. The habits you have listed are probably the things you know you must do to succeed. Here is a list of some responses from successful students we have known.

- **Set priorities.** You will encounter many distractions during the semester. Do not allow them to prevent you from reaching your goal.

- **Take responsibility.** Your instructor, this textbook, tutors, math centers, and other resources are there to help you succeed. Ultimately, however, you must choose to learn. You must choose success.
- **Hang out with successful students.** Success breeds success. When you work and study with successful students, you are in an environment that will help you succeed. Seek out people who are committed to their goals.
- **Study regularly.** We have mentioned this before, but it is too important not to be repeated.
- **Self test.** Once every few days, select homework exercises from previous assignments and use them to test your understanding. Try to do these exercises without getting help from examples in the text. These self tests will help you gain confidence that you can do these types of problems on a test given in class.
- **Try different strategies.** If you read the text and are still having difficulty understanding a concept, consider going a step further. Contact the instructor or find a tutor. Many campuses have some free tutorial services. Go to the math or learning center. Consult another textbook. Be active and get the help you need.
- **Make flash cards.** This is one of the strategies that some math students do not think to try. Flash cards are a very important part of learning math. For instance, your instructor may use words or phrases such as *linear, quadratic, exponent, base, rational,* and many others. If you don't know the meanings of these words, you will not know what is being discussed.
- **Plod along.** Your education is not a race. The primary goal is to finish. Taking too many classes and then dropping some does not get you to the end any faster. Take only the classes that you can manage.

A.2 How to Use This Text to Succeed in This Course

GET THE BIG PICTURE

One of the major resources that you will have access to the entire semester is this textbook. We have written this text with you and your success in mind. The following is a guide to the features of this text that will help you succeed.

Actually, we want you to get the *really* big picture. Take a few minutes to read the table of contents. You may feel some anxiety about all the new concepts you will be learning. Try to think of this as an exciting opportunity to learn math. Now look through the entire book. Move quickly. Don't spend more than a few seconds on each page. Scan titles, look at pictures, and notice diagrams.

Getting this "big picture" view will help you see where this course is going. To reach your goal, it's important to get an idea of the steps you will need to take along the way.

As you look through the book, find topics that interest you. What's your preference? Racing? Sailing? TV? Amusement parks? Find the Index of Applications at the back of the book, and pull out three subjects that interest you. Write those topics here.

____

____

____

UNDERSTAND THE ORGANIZATION

Look again at the Table of Contents. There are 11 chapters in this book. You'll see that every chapter is divided into sections, and each section contains a number of learning objectives. Each learning objective is labeled with a number from 1 to 5. Knowing how this book is organized will help you locate important topics and concepts as you're studying.

Before you start a new objective, take a few minutes to read the Objective Statement for that objective. Then, browse through the objective material. Especially note the words or phrases in bold type—these are important concepts that you'll need to know as you move along in the course. These words are good candidates for flash cards. If possible, include an example of the concept on the flash card, as shown at the left.

Flash Card

Rule for Multiplying Exponential Expressions

If m and n are integers, then $x^m \cdot x^n = x^{m+n}$.

Examples

$x^4 \cdot x^7 = x^{4+7} = x^{11}$

$y \cdot y^5 = y^{1+5} = y^6$

$a^2 \cdot a^6 \cdot a = a^{2+6+1} = a^9$

You will also see important concepts and rules set off in boxes. Here is one about exponents. These rules are also good candidates for flash cards.

RULE FOR MULTIPLYING EXPONENTIAL EXPRESSIONS

If m and n are integers, then $\boldsymbol{x^m \cdot x^n = x^{m+n}}$.

EXAMPLES

In each example below, we are multiplying two exponential expressions with the same base. Simplify the expression by adding the exponents.

1. $x^4 \cdot x^7 = x^{4+7} = x^{11}$

2. $y \cdot y^5 = y^{1+5} = y^6$

3. $a^2 \cdot a^6 \cdot a = a^{2+6+1} = a^9$

Leaf through Section 3.1 of Chapter 3. Write down the words in bold and any concepts or rules that are displayed in boxes.

USE THE INTERACTIVE METHOD

As we mentioned earlier, this textbook is based on an interactive approach. We want you to be actively involved in learning mathematics, and have given you many suggestions for getting "hands-on" with this book.

Focus On Look on page 93. See the Focus On? The Focus On introduces a concept (in this case, solving an equation of the form $ax = b$) and includes a step-by-step solution of the type of exercise you will find in the homework.

Focus on solving an equation of the form $ax = b$

Solve: $8x = 16$

The goal is to rewrite the equation in the form *variable* = *constant*.

$$8x = 16$$

Multiply each side of the equation by the reciprocal of 8. This is equivalent to dividing each side by 8.

$$\frac{8x}{8} = \frac{16}{8}$$

$$x = 2$$

The solution 2 checks.

The solution is 2.

Grab paper and a pencil and work along as you're reading through the Focus On. When you're done, get a clean sheet of paper. Write down the problem and try to complete the solution without looking at your notes or at the book. When you're done, check your answer. If you got it right, you're ready to move on.

Look through the text and find three instances of a Focus On. Write the concepts mentioned in each Focus On here.

____

____

____

Example/Problem Pair You'll need hands-on practice to succeed in mathematics. When we show you an example, work it out right beside our solution. Use the Example/Problem pairs to get the practice you need.

Take a look at page 94. Example 6 and Problem 6 are shown here.

EXAMPLE 6 Solve and check: $4x = 6$

Solution

$$4x = 6$$

$$\frac{4x}{4} = \frac{6}{4}$$

- **Divide each side of the equation by 4, the coefficient of *x*.**

$$x = \frac{3}{2}$$

- **Simplify each side of the equation.**

Check

$$4x = 6$$

$$4\left(\frac{3}{2}\right) \mid 6$$

$$6 = 6$$

- **This is a true equation. The solution checks.**

The solution is $\frac{3}{2}$.

Problem 6 Solve and check: $6x = 10$

Solution See page S5.

Try Exercise 93, page 96.

You'll see that each Example is fully worked out. Study the Example by carefully working through each step. Then, try to complete the Problem. Use the solution to the Example as a model for solving the Problem. If you get stuck, the solutions to the Problems are provided in the back of the book. There is a page number directly following the Problem that shows you where you can find the completely-worked-out solution. Use the solution to get a hint for the step on which you are stuck. Then, try again!

When you've arrived at your solution, check your work against the solution in the back of the book. Turn to page S5 to see the solution for Problem 6.

Remember that sometimes there is more than one way to solve a problem. But your answer should always match the answer we've given in the back of the book. If you have any questions about whether your method will always work, check with your instructor.

Now note the line that says, "Try Exercise 93, page 96." You should do the Try Exercise from the exercise set to test your understanding of the concepts that have been discussed. When you have finished the exercise, check your answer with the one given in the Answer Section. If you got the answer wrong, try again. If you continue to have difficulty, seek help from a friend, a tutor, or your instructor.

USE A STRATEGY TO SOLVE WORD PROBLEMS

Learning to solve word problems is one of the reasons you are studying math. This is where you combine all of the critical thinking skills you have learned to solve practical problems.

Try not to be intimidated by word problems. Basically, what you need is a strategy that will help you come up with the equation you will need to solve the problem. When you are looking at a word problem, try the following:

- **Read the problem.** This may seem pretty obvious, but we mean really **read** it. Don't just scan it. Read the problem slowly and carefully.
- **Write down what is known and what is unknown.** Now that you have read the problem, go back and write down everything that is known. Next, write down what it is you are trying to find. Write this—don't just think it! Be as specific as you can. For instance, if you are asked to find a distance, don't just write "I need to find the distance." Be specific and write "I need to find the distance between Earth and the moon."
- **Think of a method to find the unknown.** For instance, is there a formula that relates the known and unknown quantities? This is certainly the most difficult step. Eventually, you must write an equation to be solved.
- **Solve the equation.** Be careful as you solve the equation. There is no sense in getting to this point and then making a careless mistake. The unknown in most word problems will include a unit such as feet, dollars, or miles per hour. When you write your answer, include the unit. An answer such as 20 doesn't mean much. Is it 20 feet, 20 dollars, 20 miles per hour, or something else?
- **Check your solution.** Now that you have an answer, go back to the problem and ask yourself whether it makes sense. This is an important step. For instance, if, according to your answer, the cost of a car is $2.51, you know that something went wrong.

In this text, the solution of every word problem is broken down into two steps, **Strategy** and **Solution.** The Strategy consists of the first three steps discussed above. The Solution is the last two steps. Here is an Example from page 99 of the text. Because you have not yet studied the concepts involved in the problem, you may not be able to solve it. However, note the details given in the Strategy. When you do the Problem following an Example, be sure to include your own Strategy.

EXAMPLE 2 During a recent year, nearly 1.2 million dogs or litters were registered with the American Kennel Club. The most popular breed was the Labrador retriever, with 172,841 registered. What percent of the registrations were Labrador retrievers? Round to the nearest tenth of a percent. (*Source:* American Kennel Club)

Strategy To find the percent, solve the basic percent equation using $B = 1.2$ million $= 1{,}200{,}000$ and $A = 172{,}841$. The percent is unknown.

Solution

$$PB = A$$

$$P(1{,}200{,}000) = 172{,}841 \qquad \bullet\ B = 1{,}200{,}000;\ A = 172{,}841$$

$$\frac{P(1{,}200{,}000)}{1{,}200{,}000} = \frac{172{,}841}{1{,}200{,}000} \qquad \bullet\ \text{Divide each side of the equation by } 1{,}200{,}000.$$

$$P \approx 0.144$$

$$P \approx 14.4\% \qquad \bullet\ \text{Rewrite the decimal as a percent.}$$

Approximately 14.4% of the registrations were Labrador retrievers.

Problem 2 A student correctly answered 72 of the 80 questions on an exam. What percent of the questions were answered correctly?

Solution See page S5.

Try Exercise 43, page 105.

When you have finished studying a section, do the exercises your instructor has selected. Math is not a spectator sport. You must practice every day. Do the homework and do not get behind.

ACE THE TEST

There are a number of features in this text that will help you prepare for a test. These features will help you even more if you do just one simple thing: When you are doing your homework, go back to each previous homework assignment for the current chapter and rework two exercises. That's right—just *two* exercises. You will be surprised at how much better prepared you will be for a test by doing this.

Here are some additional aides to help you ace the test.

Chapter Summary Once you've completed a chapter, look at the Chapter Summary. The Chapter Summary is divided into two sections: Key Words and Essential Rules and Procedures. Flip to page 133 to see the Chapter Summary for Chapter 3. The summary shows all of the important topics covered in the chapter. Do you see the reference following each topic? This reference shows you the objective and page in the text where you can find more information on the concept.

Write down one Key Word and one Essential Rule or Procedure. Explain the meaning of the reference 3.1.1, page 90.

____

____

Chapter Review Exercises Turn to page 136 to see the Chapter Review Exercises for Chapter 3. When you do the review exercises, you're giving yourself an important opportunity to test your understanding of the chapter. The answer to each review exercise is given at the back of the book, along with the objective the question relates to. When you're done with the Chapter Review Exercises, check your answers. If you had trouble with any of the questions, you can restudy the objectives and retry some of the exercises in those objectives for extra help.

Go to the Answer Section at the back of the text. Find the answers for the Chapter Review Exercises for Chapter 3. Write down the answer to Exercise 34. What is the meaning of the reference 3.4.3?

____

____

Chapter Test The Chapter Test for each chapter can be found after the Chapter Review Exercises and can be used to help you prepare for your exam. The answer to each test question is given at the back of the book, along with both an objective reference and a reference to a Focus On, Example, or Problem that the question relates to. Think of these tests as "practice runs" for your in-class tests. Take the test in a quiet place, and try to work through it in the same amount of time that will be allowed for your actual exam.

The aids we have mentioned above will help you prepare for a test. You should begin your review *at least* two days before the test—three days is better. These aids will get you ready for the test.

Here are some suggestions to try while you are actually taking the test.

- **Try to relax.** We know that test situations make some students quite nervous or anxious. These feelings are normal. Try to stay calm and focused on what you know. If you have prepared as we have suggested, the answers will begin to come to you.
- **Scan the test.** Get a feeling for the big picture.
- **Read the directions carefully.** Make sure you answer each question fully.
- **Work the problems that are easiest for you first.** This will help you with your confidence and help reduce the nervous feeling you may have.

READY, SET, SUCCEED!

It takes hard work and commitment to succeed, but we know you can do it! Doing well in mathematics is just one step you'll take on your path to success. Good luck. We wish you success.

Prealgebra Review

CHAPTER 1

Focus on Success

Digital Vision

Have you read AIM for Success? It describes study skills used by students who have been successful in their math courses. This feature gives you tips on how to stay motivated, how to manage your time, and how to prepare for exams. AIM for Success also includes a complete guide to the textbook and how to use its features to be successful in this course. AIM for Success starts on page AIM-1.

OBJECTIVES

1.1
1. Order relations
2. Opposites and absolute value

1.2
1. Add integers
2. Subtract integers
3. Multiply integers
4. Divide integers
5. Application problems

1.3
1. Write rational numbers as decimals
2. Multiply and divide rational numbers
3. Add and subtract rational numbers
4. Convert among percents, fractions, and decimals

1.4
1. Exponential expressions
2. The Order of Operations Agreement

1.5
1. Measures of angles
2. Perimeter problems
3. Area problems

PREP TEST

Are you ready to succeed in this chapter?
Take the Prep Test below to find out if you are ready to learn the new material.

1. What is 127.1649 rounded to the nearest hundredth?
2. Add: $49{,}147 + 596$
3. Subtract: $5004 - 487$
4. Multiply: 407×28
5. Divide: $456 \div 19$
6. What is the smallest number into which both 8 and 12 divide evenly?
7. What is the greatest number that divides evenly into both 16 and 20?
8. Without using 1, write 21 as a product of two whole numbers.
9. Represent the shaded portion of the figure as a fraction.

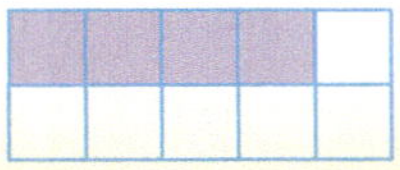

1.1 Introduction to Integers

OBJECTIVE 1 Order relations

It seems to be a human characteristic to group similar items. For instance, a botanist places plants with similar characteristics in groups called species.

Mathematicians place objects with similar properties in groups called sets. A **set** is a collection of objects. The objects in a set are called **elements** of the set.

The **roster method** of writing sets encloses a list of the elements in braces. The set of sections within an orchestra is written {brass, percussion, strings, woodwinds}.

When the elements of a set are listed, each element is listed only once. For instance, if the list of numbers 1, 2, 3, 2, 3 were placed in a set, the set would be {1, 2, 3}.

The numbers that we use to count objects, such as the number of students in a classroom or the number of people living in an apartment house, are the natural numbers.

Natural numbers = {1, 2, 3, 4, 5, 6, 7, 8, 9, 10, ...}

The three dots mean that the list of natural numbers continues on and on and that there is no largest natural number.

Point of Interest

The Alexandrian astronomer Ptolemy began using *omicron*, *o*, the first letter of the Greek word that means "nothing," as the symbol for zero in A.D. 150. It was not until the 13th century, however, that Fibonacci introduced 0 to the Western world as a placeholder so that we could distinguish, for example, 45 from 405.

The natural numbers do not include zero. The number zero is used for describing data such as the number of people who have run a two-minute mile and the number of college students at Providence College who are under the age of 10. The set of whole numbers includes the natural numbers and zero.

Whole numbers = {0, 1, 2, 3, 4, 5, 6, 7, ...}

The whole numbers do not provide all the numbers that are useful in applications. For instance, a meteorologist also needs numbers below zero.

Integers = {..., −5, −4, −3, −2, −1, 0, 1, 2, 3, 4, 5, ...}

Each integer can be shown on a number line. The integers to the left of zero on the number line are called **negative integers.** The integers to the right of zero are called **positive integers** or natural numbers. Zero is neither a positive nor a negative integer.

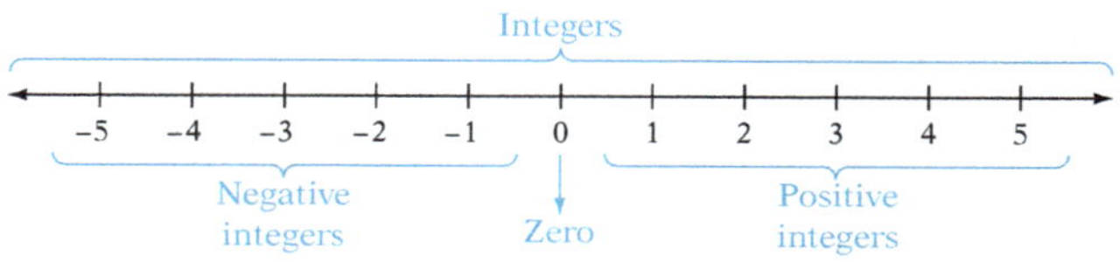

The **graph of an integer** is shown by placing a heavy dot on the number line directly above the number. The graphs of −3 and 4 are shown on the number line below.

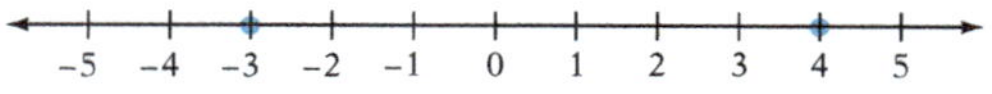

Consider the following sentences.

The quarterback threw the football, and the receiver caught *it*.

A student purchased a computer and used *it* to write English and history papers.

In the first sentence, *it* is used to mean the football; in the second sentence, *it* means the computer. In language, the word *it* can stand for many different objects. Similarly, in mathematics, a letter of the alphabet can be used to stand for a number. Such a letter is called a **variable.** Variables are used in the following definition of inequality symbols.

DEFINITION OF INEQUALITY SYMBOLS

If a and b are two numbers, and a is to the left of b on the number line, then a **is less than** b. This is written $a < b$.

EXAMPLES

1. $-4 < -1$ Negative 4 is less than negative 1.

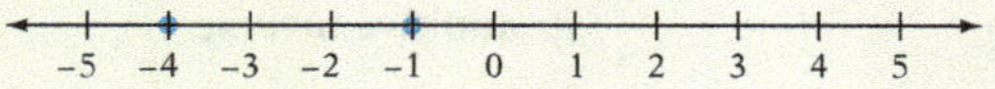

2. $-16 < -6$ Negative 16 is less than negative 6.

If a and b are two numbers, and a is to the right of b on the number line, then a **is greater than** b. This is written $a > b$.

EXAMPLES

3. $5 > 0$ Five is greater than 0.

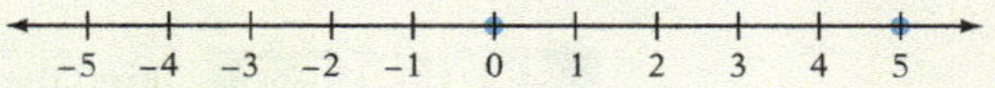

4. $-3 > -8$ Negative 3 is greater than negative 8.

There are also inequality symbols for **is less than or equal to** ($\leq$) and **is greater than or equal to** ($\geq$).

$7 \leq 15$ 7 is less than or equal to 15. This is true because $7 < 15$.

$6 \leq 6$ 6 is less than or equal to 6. This is true because $6 = 6$.

EXAMPLE 1 Use the roster method to write the set of negative integers greater than or equal to -6.

Solution $A = \{-6, -5, -4, -3, -2, -1\}$

• **A set is designated by a capital letter. The roster method encloses a list of elements in braces.**

Problem 1 Use the roster method to write the set of positive integers less than 5.

Solution See page S1.

Try Exercise 33, page 6.

EXAMPLE 2 Given $A = \{-6, -2, 0\}$, which elements of set A are less than or equal to -2?

Solution

$-6 < -2$

$-2 = -2$

$0 > -2$

• **Find the order relation between each element of set A and -2.**

The elements -6 and -2 are less than or equal to -2.

Problem 2 Given $B = \{-5, -1, 5\}$, which elements of set B are greater than -1?

Solution See page S1.

Try Exercise 45, page 6.

OBJECTIVE 2 Opposites and absolute value

Two numbers that are the same distance from zero on the number line but are on opposite sides of zero are **opposite numbers** or **opposites.** The opposite of a number is also called its **additive inverse.**

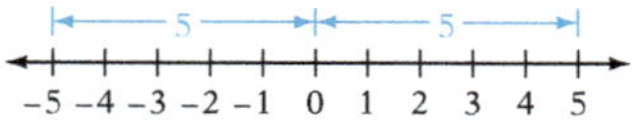

The opposite or additive inverse of 5 is -5.

The opposite or additive inverse of -5 is 5.

The negative sign can be read "the opposite of."

$-(2) = -2$ The opposite of 2 is -2.

$-(-2) = 2$ The opposite of -2 is 2.

EXAMPLE 3 Find the opposite number. **A.** 6 **B.** -51

Solution **A.** The opposite of 6 is -6.

B. The opposite of -51 is 51.

Problem 3 Find the opposite number. **A.** -9 **B.** 62

Solution See page S1.

Try Exercise 53, page 7.

The **absolute value** of a number is its distance from zero on the number line. Therefore, the absolute value of a number is a positive number or zero. The symbol for absolute value is two vertical bars, | |.

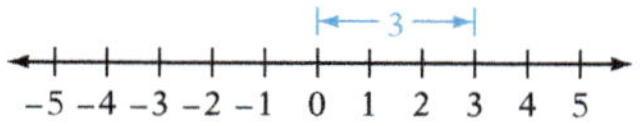

The distance from 0 to 3 is 3. Therefore, the absolute value of 3 is 3.

$|3| = 3$

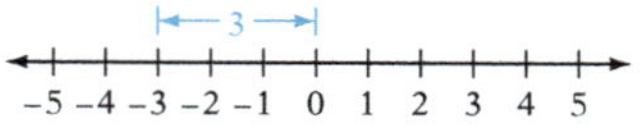

The distance from 0 to -3 is 3. Therefore, the absolute value of -3 is 3.

$|-3| = 3$

Point of Interest

The definition of *absolute value* given in the box is written in what is called rhetorical style. That is, it is written without the use of variables. This is how *all* mathematics was written prior to the Renaissance. During that period, from the 14th to the 16th century, the idea of expressing a variable symbolically was developed. In terms of that symbolism, the definition of absolute value is

$$|x| = \begin{cases} x, & x > 0 \\ 0, & x = 0 \\ -x, & x < 0 \end{cases}$$

ABSOLUTE VALUE

The absolute value of a positive number is the number itself. The absolute value of zero is zero. The absolute value of a negative number is the opposite of the negative number.

EXAMPLES

1. $|6| = 6$ **2.** $|0| = 0$ **3.** $|-6| = 6$

EXAMPLE 4 Evaluate. **A.** $|-4|$ **B.** $-|-10|$

Solution **A.** $|-4| = 4$

B. $-|-10| = -10$

- **The absolute value symbol does not affect the negative sign in front of the absolute value symbol. You can read $-|-10|$ as "the opposite of the absolute value of negative 10."**

Problem 4 Evaluate. **A.** $|-5|$ **B.** $-|-9|$

Solution See page S1.

Try Exercise 71, page 7.

1.1 Exercises

CONCEPT CHECK

Determine whether the statement is always true, sometimes true, or never true.

1. The absolute value of a number is positive.

2. The absolute value of a number is negative.

3. If x is an integer, then $|x| > -2$.

4. The opposite of a number is a positive number.

5. Classify each number as a positive integer, a negative integer, or neither.
a. -12
b. 18
c. -7
d. 0
e. $\frac{3}{4}$
f. 365

6. Place the correct symbol, $<$ or $>$, in the space provided.
a. 0 ______ any positive number.
b. 0 ______ any negative number.

1 Order relations (See pages 2–4.)

7. How do the whole numbers differ from the natural numbers?

8. Explain the difference between the symbols $<$ and $\leq$.

GETTING READY

9. The inequality $-5 < -1$ is read "negative five __?__ negative one."

10. The inequality $0 \geq -4$ is read "zero __?__ negative four."

Place the correct symbol, $<$ or $>$, between the two numbers.

11. $-2 \quad -5$
12. $-6 \quad -1$
13. $-16 \quad 1$
14. $-2 \quad 13$
15. $3 \quad -7$
16. $5 \quad -6$
17. $0 \quad -3$
18. $8 \quad 0$
19. $-42 \quad 27$
20. $-36 \quad 49$
21. $21 \quad -34$
22. $53 \quad -46$
23. $-27 \quad -39$
24. $-51 \quad -20$
25. $-131 \quad 101$
26. $127 \quad -150$

27. A number n is to the right of the number 5 on the number line. Which of the following is true?
(i) n is positive. (ii) n is negative. (iii) n is 0.
(iv) n can be positive, negative, or 0.

28. A number n is to the left of the number 5 on the number line. Which of the following is true?
(i) n is positive. (ii) n is negative. (iii) n is 0.
(iv) n can be positive, negative, or 0.

29. Do the inequalities $6 \geq 1$ and $1 \leq 6$ express the same order relation?

30. Use the "$\leq$" inequality symbol to rewrite the order relation expressed by the inequality $-2 \geq -5$.

Use the roster method to write the set.

31. the natural numbers less than 9

32. the natural numbers less than or equal to 6

33. the positive integers less than or equal to 8

34. the positive integers less than 4

35. the negative integers greater than -7

36. the negative integers greater than or equal to -5

Solve.

37. Given $A = \{-7, 0, 2, 5\}$, which elements of set A are greater than 2?

38. Given $B = \{-8, 0, 7, 15\}$, which elements of set B are greater than 7?

39. Given $D = \{-23, -18, -8, 0\}$, which elements of set D are less than -8?

40. Given $C = \{-33, -24, -10, 0\}$, which elements of set C are less than -10?

41. Given $E = \{-35, -13, 21, 37\}$, which elements of set E are greater than -10?

42. Given $F = \{-27, -14, 14, 27\}$, which elements of set F are greater than -15?

43. Given $B = \{-52, -46, 0, 39, 58\}$, which elements of set B are less than or equal to 0?

44. Given $A = \{-12, -9, 0, 12, 34\}$, which elements of set A are greater than or equal to 0?

45. Given $C = \{-23, -17, 0, 4, 29\}$, which elements of set C are greater than or equal to -17?

46. Given $D = \{-31, -12, 0, 11, 45\}$, which elements of set D are less than or equal to -12?

47. Given that set A is the positive integers less than 10, which elements of set A are greater than or equal to 5?

48. Given that set B is the positive integers less than or equal to 12, which elements of set B are greater than 6?

49. Given that set D is the negative integers greater than or equal to -10, which elements of set D are less than -4?

50. Given that set C is the negative integers greater than -8, which elements of set C are less than or equal to -3?

2 Opposites and absolute value (See pages 4–5.)

GETTING READY

51. The equation $|-5| = 5$ is read "the __?__ of negative five is five."

52. Write the statement "the opposite of negative nine is nine" in symbols.

Find the opposite number.

53. 22 **54.** 45 **55.** -31 **56.** -88

57. -168 **58.** -97 **59.** 630 **60.** 450

Evaluate.

61. $-(-18)$ **62.** $-(-30)$ **63.** $-(49)$ **64.** $-(67)$

65. $|16|$ **66.** $|19|$ **67.** $|-12|$ **68.** $|-22|$

69. $-|29|$ **70.** $-|20|$ **71.** $-|-14|$ **72.** $-|-18|$

73. $-|0|$ **74.** $|-30|$ **75.** $-|34|$ **76.** $-|-45|$

Solve.

77. Given $A = \{-8, -5, -2, 1, 3\}$, find
  a. the opposite of each element of set A.
  b. the absolute value of each element of set A.

78. Given $B = \{-11, -7, -3, 1, 5\}$, find
  a. the opposite of each element of set B.
  b. the absolute value of each element of set B.

79. True or false? The absolute value of a negative number n is greater than n.

80. A number n is positive. In order to make the statement "$|n|$ ___?___ n" a true statement, which symbol must replace the question mark?
(i) $>$ (ii) $<$ (iii) $\leq$ (iv) $=$

Place the correct symbol, $<$ or $>$, between the two numbers.

81. $|-83|$ $|58|$ **82.** $|22|$ $|-19|$ **83.** $|43|$ $|-52|$ **84.** $|-71|$ $|-92|$

85. $|-68|$ $|-42|$ **86.** $|12|$ $|-31|$ **87.** $|-45|$ $|-61|$ **88.** $|-28|$ $|43|$

APPLYING CONCEPTS

Write the given numbers in order from least to greatest.

89. $|-5|, 6, -|-8|, -19$

90. $-4, |-15|, -|-7|, 0$

91. $-(-3), -22, |-25|, |-14|$

92. $-|-26|, -(-8), |-17|, -(5)$

Meteorology A meteorologist may report a wind-chill temperature. This is the equivalent temperature, including the effects of wind and temperature, that a person would feel in calm air conditions. The table below gives the wind-chill temperature for various wind speeds and temperatures. For instance, when the temperature is 5°F and the wind is blowing at 15 mph, the wind-chill temperature is −13°F. Use this table for Exercises 93 and 94.

Wind Speed (in mph)	Wind Chill Factors Thermometer Reading (in degrees Fahrenheit)														
	25	20	15	10	5	0	−5	−10	−15	−20	−25	−30	−35	−40	−45
5	19	13	7	1	−5	−11	−16	−22	−28	−34	−40	−46	−52	−57	−63
10	15	9	3	−4	−10	−16	−22	−28	−35	−41	−47	−53	−59	−66	−72
15	13	6	0	−7	−13	−19	−26	−32	−39	−45	−51	−58	−64	−71	−77
20	11	4	−2	−9	−15	−22	−29	−35	−42	−48	−55	−61	−68	−74	−81
25	9	3	−4	−11	−17	−24	−31	−37	−44	−51	−58	−64	−71	−78	−84
30	8	1	−5	−12	−19	−26	−33	−39	−46	−53	−60	−67	−73	−80	−87
35	7	0	−7	−14	−21	−27	−34	−41	−48	−55	−62	−69	−76	−82	−89
40	6	−1	−8	−15	−22	−29	−36	−43	−50	−57	−64	−71	−78	−84	−91
45	5	−2	−9	−16	−23	−30	−37	−44	−51	−58	−65	−72	−79	−86	−93

93. Which of the following weather conditions feels colder?
- **a.** a temperature of 5°F with a 20 mph wind or a temperature of 10°F with a 15 mph wind
- **b.** a temperature of −25°F with a 10 mph wind or a temperature of −15°F with a 20 mph wind

94. Which of the following weather conditions feels warmer?
- **a.** a temperature of 5°F with a 25 mph wind or a temperature of 10°F with a 10 mph wind
- **b.** a temperature of −5°F with a 10 mph wind or a temperature of −15°F with a 5 mph wind

Complete.

95. On the number line, the two points that are four units from 0 are ___?___ and ___?___.

96. On the number line, the two points that are six units from 0 are ___?___ and ___?___.

97. On the number line, the two points that are seven units from 4 are ___?___ and ___?___.

98. On the number line, the two points that are five units from −3 are ___?___ and ___?___.

99. If a is a positive number, then $-a$ is a ___?___ number.

100. If a is a negative number, then $-a$ is a ___?___ number.

PROJECTS OR GROUP ACTIVITIES

101. Graph the numbers −5 and 3 on a number line. Use the graph to explain why −5 is less than 3 and why 3 is greater than −5.

102. Graph the numbers 1 and −2 on a number line. Use the graph to explain why 1 is greater than −2 and why −2 is less than 1.

Translate and evaluate.

103. The opposite of the additive inverse of 7

104. The absolute value of the opposite of −8

105. The opposite of the absolute value of 8

106. The absolute value of the additive inverse of −6

1.2 Operations with Integers

OBJECTIVE 1 Add integers

A number can be represented anywhere along the number line by an arrow. A positive number is represented by an arrow pointing to the right, and a negative number is represented by an arrow pointing to the left. The size of the number is represented by the length of the arrow.

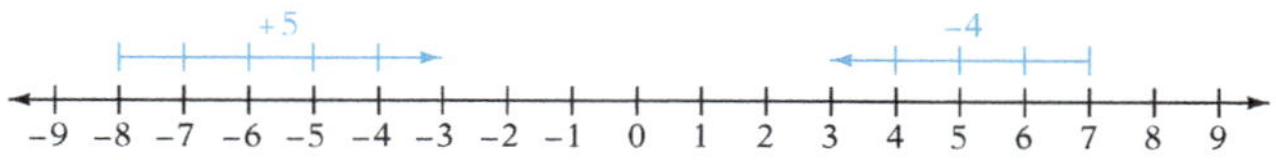

Addition is the process of finding the total of two numbers. The numbers being added are called **addends.** The total is called the **sum.** Addition of integers can be shown on the number line. To add integers, find the point on the number line corresponding to the first addend. From that point, draw an arrow representing the second addend. The sum is the number directly below the tip of the arrow.

$4 + 2 = 6$

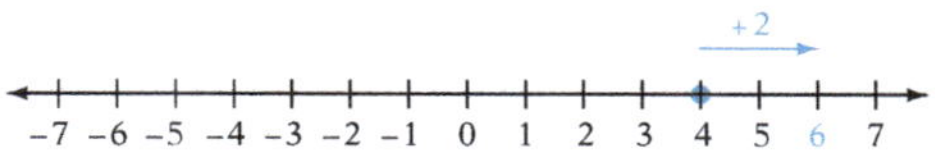

$-4 + (-2) = -6$

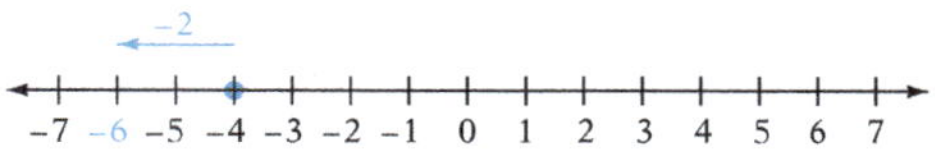

$-4 + 2 = -2$

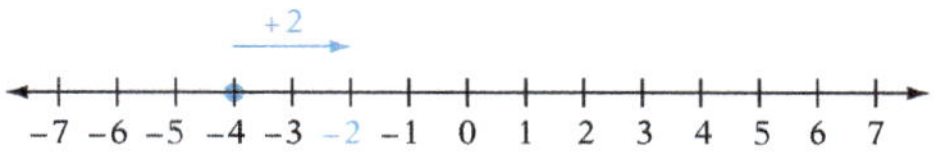

$4 + (-2) = 2$

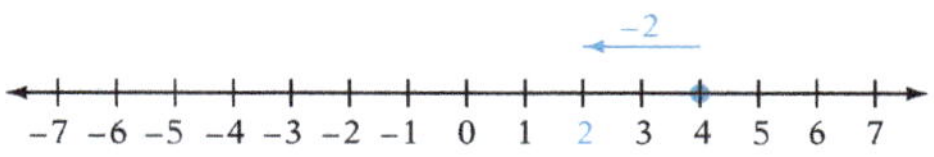

The pattern for addition shown on the preceding number lines is summarized in the following rules for adding integers.

ADDITION OF INTEGERS

Integers with the same sign

To add two numbers with the same sign, add the absolute values of the numbers. Then attach the sign of the addends.

EXAMPLES

1. $2 + 8 = 10$ **2.** $-2 + (-8) = -10$

Integers with different signs

To add two numbers with different signs, find the absolute value of each number. Then subtract the smaller of these absolute values from the larger one. Attach the sign of the number with the larger absolute value.

EXAMPLES

3. $-2 + 8 = 6$ **4.** $2 + (-8) = -6$

EXAMPLE 1 Add. **A.** $162 + (-247)$ **B.** $-14 + (-47)$ **C.** $-4 + (-6) + (-8) + 9$

Solution **A.** $162 + (-247) = -85$

- The signs are different. Subtract the absolute values of the numbers (247 − 162). Attach the sign of the number with the larger absolute value.

B. $-14 + (-47) = -61$

- The signs are the same. Add the absolute values of the numbers (14 + 47). Attach the sign of the addends.

C. $-4 + (-6) + (-8) + 9$
$= -10 + (-8) + 9$
$= -18 + 9$
$= -9$

• **To add more than two numbers, add the first two numbers. Then add the sum to the third number. Continue until all the numbers have been added.**

Problem 1 Add.

A. $-162 + 98$ **B.** $-154 + (-37)$
C. $-36 + 17 + (-21)$

Solution See page S1.

Try Exercise 25, page 15.

OBJECTIVE 2 Subtract integers

Subtraction is the process of finding the difference between two numbers. Subtraction of an integer is defined as addition of the opposite integer.

Subtract 8 − 3 by using addition of the opposite.

Subtraction → Addition of the Opposite

$$8 - (+3) = 8 + (-3) = 5$$

(+3) and (−3): Opposites

SUBTRACTION OF INTEGERS

To subtract one integer from another, add the opposite of the second integer to the first integer.

EXAMPLES

	first number	−	second number	=	first number	+	the opposite of the second number		
1.	40	−	60	=	40	+	(−60)	=	−20
2.	−40	−	60	=	−40	+	(−60)	=	−100
3.	−40	−	(−60)	=	−40	+	60	=	20
4.	40	−	(−60)	=	40	+	60	=	100

EXAMPLE 2 Subtract: $-12 - 8$

Solution $-12 - 8 = -12 + (-8)$ • **Rewrite subtraction as addition of the opposite.**

$= -20$ • **Add.**

Problem 2 Subtract: $-8 - 14$

Solution See page S1.

Try Exercise 43, page 15.

EXAMPLE 3 Subtract: $-8 - 30 - (-12) - 7 - (-14)$

Solution

$$\begin{aligned} &-8 - 30 - (-12) - 7 - (-14) \\ &= -8 + (-30) + 12 + (-7) + 14 \\ &= -38 + 12 + (-7) + 14 \\ &= -26 + (-7) + 14 \\ &= -33 + 14 \\ &= -19 \end{aligned}$$

- Rewrite each subtraction as addition of the opposite.
- Add the first two numbers. Then add the sum to the third number. Continue until all the numbers have been added.

Problem 3 Subtract: $4 - (-3) - 12 - (-7) - 20$

Solution See page S1.

Try Exercise 53, page 15.

OBJECTIVE 3

Multiply integers

> **Point of Interest**
>
> The cross × was first used as a symbol for multiplication in 1631 in a book titled *The Key to Mathematics.* In that same year, another book, *Practice of the Analytical Art*, advocated the use of a dot to indicate multiplication.

Multiplication is the process of finding the product of two numbers.

Several different symbols are used to indicate multiplication. The numbers being multiplied are called **factors;** for instance, 3 and 2 are factors in each of the examples at the right. The result is called the **product.** Note that when parentheses are used and there is no arithmetic operation symbol, the operation is multiplication.

$$3 \times 2 = 6$$
$$3 \cdot 2 = 6$$
$$(3)(2) = 6$$
$$3(2) = 6$$
$$(3)2 = 6$$

When 5 is multiplied by a sequence of decreasing integers, each product decreases by 5.

$$\begin{aligned} (5)(3) &= 15 \\ (5)(2) &= 10 \\ (5)(1) &= 5 \\ (5)(0) &= 0 \end{aligned}$$

The pattern developed can be continued so that 5 is multiplied by a sequence of negative numbers. The resulting products must be negative in order to maintain the pattern of decreasing by 5.

$$\begin{aligned} (5)(-1) &= -5 \\ (5)(-2) &= -10 \\ (5)(-3) &= -15 \\ (5)(-4) &= -20 \end{aligned}$$

This illustrates that the product of a positive number and a negative number is negative.

When -5 is multiplied by a sequence of decreasing integers, each product increases by 5.

$$\begin{aligned} (-5)(3) &= -15 \\ (-5)(2) &= -10 \\ (-5)(1) &= -5 \\ (-5)(0) &= 0 \end{aligned}$$

The pattern developed can be continued so that -5 is multiplied by a sequence of negative numbers. The resulting products must be positive in order to maintain the pattern of increasing by 5.

$$\begin{aligned} (-5)(-1) &= 5 \\ (-5)(-2) &= 10 \\ (-5)(-3) &= 15 \\ (-5)(-4) &= 20 \end{aligned}$$

This illustrates that the product of two negative numbers is positive.

This pattern for multiplication is summarized in the following rules for multiplying integers.

MULTIPLICATION OF INTEGERS

Integers with the same sign

To multiply two numbers with the same sign, multiply the absolute values of the numbers. The product is positive.

EXAMPLES

1. $4 \cdot 8 = 32$ **2.** $(-4)(-8) = 32$

Integers with different signs

To multiply two numbers with different signs, multiply the absolute values of the numbers. The product is negative.

EXAMPLES

3. $-4 \cdot 8 = -32$ **4.** $(4)(-8) = -32$

EXAMPLE 4 Multiply. **A.** $-42 \cdot 62$ **B.** $2(-3)(-5)(-7)$

Solution **A.** $-42 \cdot 62$
$= -2604$
• **The signs are different. The product is negative.**

B. $2(-3)(-5)(-7)$
$= -6(-5)(-7)$
$= 30(-7)$
$= -210$
• **To multiply more than two numbers, multiply the first two numbers. Then multiply the product by the third number. Continue until all the numbers have been multiplied.**

Problem 4 Multiply. **A.** $-38 \cdot 51$ **B.** $-7(-8)(9)(-2)$

Solution See page S1.

Try Exercise 83, page 16.

OBJECTIVE 4 Divide integers

For every division problem there is a related multiplication problem.

$$\text{Division: } \frac{8}{2} = 4 \qquad \text{Related multiplication: } 4 \cdot 2 = 8$$

This fact can be used to illustrate the rules for dividing signed numbers.

The quotient of two numbers with the same sign is positive.

$\frac{12}{3} = 4$ because $4 \cdot 3 = 12$.

$\frac{-12}{-3} = 4$ because $4(-3) = -12$.

The quotient of two numbers with different signs is negative.

$\frac{12}{-3} = -4$ because $-4(-3) = 12$.

$\frac{-12}{3} = -4$ because $-4 \cdot 3 = -12$.

DIVISION OF INTEGERS

Integers with the same sign

To divide two numbers with the same sign, divide the absolute values of the numbers. The quotient is positive.

EXAMPLES

1. $30 \div 6 = 5$ **2.** $(-30) \div (-6) = 5$

Integers with different signs

To divide two numbers with different signs, divide the absolute values of the numbers. The quotient is negative.

EXAMPLES

3. $(-30) \div 6 = -5$ **4.** $30 \div (-6) = -5$

Note that $\frac{-12}{3} = -4$, $\frac{12}{-3} = -4$, and $-\frac{12}{3} = -4$. This suggests the following rule.

If a and b are two integers, and $b \neq 0$, then $\frac{a}{-b} = \frac{-a}{b} = -\frac{a}{b}$.

Read $b \neq 0$ as "b is not equal to 0." The reason why the denominator must not be equal to 0 is explained in the following discussion of 0 and 1 in division.

ZERO AND ONE IN DIVISION

Zero divided by any number other than zero is zero.	$\frac{0}{a} = 0, a \neq 0$	because $0 \cdot a = 0$.
Any number other than zero divided by itself is 1.	$\frac{a}{a} = 1, a \neq 0$	because $1 \cdot a = a$.
Any number divided by 1 is the number.	$\frac{a}{1} = a$	because $a \cdot 1 = a$.
Division by zero is not defined.	$\frac{4}{0} = ?$	$? \times 0 = 4$ There is no number whose product with zero is 4.

EXAMPLES

1. $\frac{0}{7} = 0$ **2.** $\frac{-2}{-2} = 1$

3. $\frac{-9}{1} = -9$ **4.** $\frac{8}{0}$ is undefined.

EXAMPLE 5 Divide. **A.** $(-120) \div (-8)$ **B.** $\frac{95}{-5}$ **C.** $-\frac{-81}{3}$

Solution **A.** $(-120) \div (-8) = 15$
- **The two numbers have the same sign. The quotient is positive.**

B. $\frac{95}{-5} = -19$
- **The two numbers have different signs. The quotient is negative.**

C. $-\frac{-81}{3} = -(-27) = 27$

Problem 5 Divide. **A.** $(-135) \div (-9)$ **B.** $\frac{84}{-6}$ **C.** $-\frac{36}{-12}$

Solution See page S1.

Try Exercise 109, page 17.

OBJECTIVE 5 Application problems

In many courses, your course grade depends on the *average* of all your test scores. You compute the average by calculating the sum of all your test scores and then dividing that result by the number of tests. Statisticians call this average an **arithmetic mean.** Besides its application to finding the average of your test scores, the arithmetic mean is used in many other situations.

EXAMPLE 6 The daily low temperatures, in degrees Celsius, during one week were recorded as follows: $-8°, 2°, 0°, -7°, 1°, 6°, -1°$. Find the average daily low temperature for the week.

Strategy To find the average daily low temperature:

- ▶ Add the seven temperature readings.
- ▶ Divide the sum by 7.

Solution

$$\begin{aligned}&-8 + 2 + 0 + (-7) + 1 + 6 + (-1)\\&= -6 + 0 + (-7) + 1 + 6 + (-1)\\&= -6 + (-7) + 1 + 6 + (-1)\\&= -13 + 1 + 6 + (-1)\\&= -12 + 6 + (-1)\\&= -6 + (-1)\\&= -7\end{aligned}$$

$-7 \div 7 = -1$

The average daily low temperature was $-1°C$.

Problem 6 The daily high temperatures, in degrees Celsius, during one week were recorded as follows: $-5°, -6°, 3°, 0°, -4°, -7°, -2°$. Find the average daily high temperature for the week.

Solution See page S1.

Try Exercise 135, page 18.

1.2 Exercises

CONCEPT CHECK

Determine whether the statement is always true, sometimes true, or never true.

1. The sum of two integers is larger than either of the integers being added.

2. The sum of two nonzero integers with the same sign is positive.

3. The quotient of two integers with different signs is negative.

4. To find the opposite of a number, multiply the number by -1.

5. If x is an integer and $4x = 0$, then $x = 0$.

State whether each "−" sign is a minus sign or a negative sign.

6. $2 - (-7)$
7. $-6 - 1$
8. $-4 - (-3)$

1 Add integers (See pages 8–10.)

9. Explain how to add two integers with the same sign.
10. Explain how to add two integers with different signs.

GETTING READY

11. In the addition equation $8 + (-3) = 5$, the addends are ___?___ and ___?___, and the sum is ___?___.
12. Use the diagram at the right to complete this addition equation:
___?___ + ___?___ = ___?___.

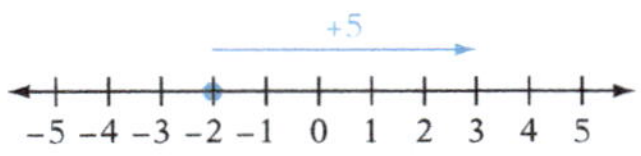

Add.

13. $-3 + (-8)$
14. $-12 + (-1)$
15. $-4 + (-5)$
16. $-12 + (-12)$
17. $6 + (-9)$
18. $4 + (-9)$
19. $-6 + 7$
20. $-12 + 6$
21. $2 + (-3) + (-4)$
22. $7 + (-2) + (-8)$
23. $-3 + (-12) + (-15)$
24. $9 + (-6) + (-16)$
25. $-17 + (-3) + 29$
26. $13 + 62 + (-38)$
27. $-3 + (-8) + 12$
28. $-27 + (-42) + (-18)$
29. $13 + (-22) + 4 + (-5)$
30. $-14 + (-3) + 7 + (-6)$

Complete Exercises 31 and 32 without actually finding the sums.

31. Is the sum $812 + (-537)$ positive or negative?
32. Is the sum of -57 and -31 positive or negative?

2 Subtract integers (See pages 10–11.)

33. Explain the meaning of the words *minus* and *negative*.
34. Explain how to rewrite $6 - (-9)$ as addition of the opposite.

GETTING READY

35. $-10 - 4 = -10 +$ ___?___ • **Rewrite subtraction as addition of the opposite.**
$=$ ___?___ • **Add.**

36. $8 - (-5) = 8 +$ ___?___ • **Rewrite subtraction as addition of the opposite.**
$=$ ___?___ • **Add.**

Subtract.

37. $16 - 8$
38. $12 - 3$
39. $7 - 14$
40. $7 - (-2)$
41. $3 - (-4)$
42. $-6 - (-3)$
43. $-4 - (-2)$
44. $6 - (-12)$
45. $-12 - 16$
46. $-4 - 3 - 2$
47. $4 - 5 - 12$
48. $12 - (-7) - 8$
49. $-12 - (-3) - (-15)$
50. $4 - 12 - (-8)$
51. $13 - 7 - 15$
52. $-6 + 19 - (-31)$
53. $-30 - (-65) - 29 - 4$
54. $42 - (-82) - 65 - 7$

Complete Exercises 55 and 56 without actually finding the difference.

55. Is the difference $-25 - 52$ positive or negative?

56. Is the difference 8 minus -5 positive or negative?

3 Multiply integers (See pages 11–12.)

57. Name the operation in each expression. Justify your answer.
a. $8(-7)$ **b.** $8 - 7$ **c.** $8 - (-7)$
d. $-xy$ **e.** $x(-y)$ **f.** $-x - y$

58. Name the operation in each expression. Justify your answer.
a. $(4)(-6)$ **b.** $4 - (6)$ **c.** $4 - (-6)$
d. $-ab$ **e.** $a(-b)$ **f.** $-a - b$

GETTING READY

59. In the equation $(-10)(7) = -70$, the factors are __?__ and __?__, and the product is __?__.

60. In the equation $15(-3) = -45$, the 15 and -3 are called the __?__, and -45 is called the __?__.

61. For the product $(-4)(-12)$, the signs of the factors are the same. The sign of the product is __?__. The product is __?__.

62. For the product $(10)(-10)$, the signs of the factors are different. The sign of the product is __?__. The product is __?__.

Multiply.

63. $14 \cdot 3$

64. $62 \cdot 9$

65. $5(-4)$

66. $4(-7)$

67. $-8(2)$

68. $-9(3)$

69. $(-5)(-5)$

70. $(-3)(-6)$

71. $(-7)(0)$

72. $-32 \cdot 4$

73. $-24 \cdot 3$

74. $19(-7)$

75. $6(-17)$

76. $-8(-26)$

77. $-4(-35)$

78. $-5(23)$

79. $5 \cdot 7(-2)$

80. $8(-6)(-1)$

81. $(-9)(-9)(2)$

82. $-8(-7)(-4)$

83. $-5(8)(-3)$

84. $(-6)(5)(7)$

85. $-1(4)(-9)$

86. $6(-3)(-2)$

87. Is the product of three negative integers positive or negative?

88. Is the product of four positive numbers and three negative numbers positive or negative?

4 Divide integers (See pages 12–13.)

GETTING READY

89. Write the division expression $\frac{-15}{3}$ using the division symbol:

__?__ ÷ __?__.

90. Write the division expression $8 \div (-4)$ as a fraction: $\frac{?}{?}$.

The quotient is __?__.

Write the related multiplication problem.

91. $\frac{-36}{-12} = 3$ **92.** $\frac{28}{-7} = -4$ **93.** $\frac{-55}{11} = -5$ **94.** $\frac{-20}{-10} = 2$

Divide.

95. $12 \div (-6)$ **96.** $18 \div (-3)$ **97.** $(-72) \div (-9)$

98. $(-64) \div (-8)$ **99.** $0 \div (-6)$ **100.** $-49 \div 0$

101. $45 \div (-5)$ **102.** $-24 \div 4$ **103.** $-36 \div 4$

104. $-56 \div 7$ **105.** $-81 \div (-9)$ **106.** $-40 \div (-5)$

107. $72 \div (-3)$ **108.** $44 \div (-4)$ **109.** $-60 \div 5$

110. $144 \div 9$ **111.** $78 \div (-6)$ **112.** $84 \div (-7)$

113. $-72 \div 4$ **114.** $-80 \div 5$ **115.** $-114 \div (-6)$

116. $-128 \div 4$ **117.** $-130 \div (-5)$ **118.** $(-280) \div 8$

Complete Exercises 119 and 120 without using a calculator.

119. State whether the quotient $-\frac{520}{-13}$ is positive or negative.

120. State whether each quotient is positive, negative, zero, or undefined.
a. $-61 \div 0$ **b.** $0 \div 85$ **c.** $-172 \div (-4)$ **d.** $-96 \div 4$

5 Application problems (See page 14.)

121. At 2:00 P.M., the temperature was 85°F. By 10:00 P.M., the temperature had dropped by 20°F. Which expression can be used to find the temperature, in degrees Fahrenheit, at 10:00 P.M.?
(i) $85 + 20$ (ii) $85 - 20$ (iii) $20 - 85$ (iv) $85 \div 20$

122. After three tests, a student's test average was 82. After the fourth test, the student's test average was 84. Was the student's grade on the fourth test higher or lower than 82?

123. Temperature Find the temperature after a rise of 9°C from −6°C.

124. Temperature Find the temperature after a rise of 7°C from −18°C.

125. Temperature The high temperature for the day was 10°C. The low temperature was −4°C. Find the difference between the high and low temperatures for the day.

126. Temperature The low temperature for the day was −2°C. The high temperature was 11°C. Find the difference between the high and low temperatures for the day.

127. Chemistry The temperature at which mercury boils is 360°C. Mercury freezes at −39°C. Find the difference between the temperature at which mercury boils and the temperature at which it freezes.

128. Chemistry The temperature at which radon boils is −62°C. Radon freezes at −71°C. Find the difference between the temperature at which radon boils and the temperature at which it freezes.

Geography The elevation, or height, of places on Earth is measured in relation to sea level, or the average level of the ocean's surface. The following table shows height above sea level as a positive number and depth below sea level as a negative number. (*Source: Information Please Almanac*)

Continent	Highest Elevation (in meters)		Lowest Elevation (in meters)	
Africa	Mt. Kilimanjaro	5895	Lake Assal	−156
Asia	Mt. Everest	8850	Dead Sea	−411
Europe	Mt. Elbrus	5642	Caspian Sea	−28
North America	Mt. Denali	6194	Death Valley	−86
South America	Mt. Aconcagua	6960	Valdes Peninsula	−40

© Peter Arnold, Inc./Alamy

Mt. Aconcagua

129. Use the table to find the difference in elevation between Mt. Elbrus and the Caspian Sea.

130. Use the table to find the difference in elevation between Mt. Aconcagua and the Valdes Peninsula.

131. Use the table to find the difference in elevation between Mt. Kilimanjaro and Lake Assal.

132. Use the table to find the difference in elevation between Mt. Denali and Death Valley.

133. Use the table to find the difference in elevation between Mt. Everest and the Dead Sea.

134. **Temperature** The date of the news clipping at the right is April 2, 2010.
- **a.** Find the difference between the high and low temperatures in the United States on that day.
- **b.** What was the difference between the high and low temperatures in the contiguous 48 states on that day?

In the News

U.S. Experiences Extreme Temperatures

The high temperature in the United States today was 93°F, recorded in Laredo, Texas. At the other extreme was Buckland, Alaska, which recorded the lowest temperature in the entire United States at −14°F. The lowest temperature in the contiguous United States was −7°F, recorded at Lake Yellowstone, Wyoming.

Source: National Weather Service

135. **Temperature** The daily low temperatures, in degrees Celsius, during one week were recorded as follows: 4°, −5°, 8°, 0°, −9°, −11°, −8°. Find the average daily low temperature for the week.

136. **Temperature** The daily high temperatures, in degrees Celsius, during one week were recorded as follows: −8°, −9°, 6°, 7°, −2°, −14°, −1°. Find the average daily high temperature for the week.

137. **Temperature** On January 22, 1943, the temperature at Spearfish, South Dakota, rose from −4°F to 45°F in two minutes. How many degrees did the temperature rise during those two minutes?

138. **Temperature** In a 24-hour period in January of 1916, the temperature in Browning, Montana, dropped from 44°F to −56°F. How many degrees did the temperature drop during that time?

Aviation The table at the right shows the average temperatures at different cruising altitudes for airplanes. Use the table for Exercises 139 and 140.

139. What is the difference between the average temperature at 12,000 ft and the average temperature at 40,000 ft?

140. How much colder is the average temperature at 30,000 ft than at 20,000 ft?

Cruising Altitude	Average Temperature
12,000 ft	16°F
20,000 ft	−12°F
30,000 ft	−48°F
40,000 ft	−70°F
50,000 ft	−70°F

141. **Golf Scores** In golf, a player's score on a hole is 0 if he or she completes the hole in *par.* **Par** is the number of strokes in which a golfer should complete a hole. In a golf match, scores are given both as a total number of strokes taken on all holes and as a value relative to par, such as -4 ("4 under par") or $+2$ ("2 over par").

In 2010, Phil Mickelson's daily scores in the Masters Tournament were -5, -1, -5, and -5. His total of -16 is found by adding the four numbers. Use the table below to find the totals for other players in the same tournament.

Player	Day 1	Day 2	Day 3	Day 4	Total
Lee Westwood	-5	-3	-4	-1	
Anthony Kim	-4	-2	$+1$	-7	
K.J. Choi	-5	-1	-2	-3	

In the News

Mickelson Earns Green Blazer

In this year's Masters Tournament, Phil Mickelson won his third Masters title by three shots over runner-up Lee Westwood.

Source: www.masters.com

David W. Leindecker/Shutterstock.com

Phil Mickelson

APPLYING CONCEPTS

Simplify.

142. $|-7 + 12|$

143. $|13 - (-4)|$

144. $|-13 - (-2)|$

145. $|18 - 21|$

Assuming the pattern is continued, find the next three numbers in the pattern.

146. $-7, -11, -15, -19, \ldots$

147. $16, 11, 6, 1, \ldots$

148. $7, -14, 28, -56, \ldots$

149. $1024, -256, 64, \ldots$

Solve.

150. 32,844 is divisible by 3. By rearranging the digits, find the largest possible number that is still divisible by 3.

151. 4563 is not divisible by 4. By rearranging the digits, find the largest possible number that is divisible by 4.

152. How many three-digit numbers of the form 8__4 are divisible by 3?

In each exercise, determine which statement is false.

153. **a.** $|3 + 4| = |3| + |4|$ **b.** $|3 - 4| = |3| - |4|$ **c.** $|4 + 3| = |4| + |3|$ **d.** $|4 - 3| = |4| - |3|$

154. **a.** $|5 + 2| = |5| + |2|$ **b.** $|5 - 2| = |5| - |2|$ **c.** $|2 + 5| = |2| + |5|$ **d.** $|2 - 5| = |2| - |5|$

Determine which statement is true for all real numbers.

155. **a.** $|x + y| \le |x| + |y|$ **b.** $|x + y| = |x| + |y|$ **c.** $|x + y| \ge |x| + |y|$

156. **a.** $||x| - |y|| \le |x| - |y|$ **b.** $||x| - |y|| = |x| - |y|$ **c.** $||x| - |y|| \ge |x| - |y|$

157. If $-4x$ equals a positive integer, is x a positive or a negative integer? Explain your answer.

PROJECTS OR GROUP ACTIVITIES

158. Is the difference between two integers always smaller than either one of the integers? If not, give an example for which the difference between two integers is greater than either integer.

On the number line, illustrate each of the following sums.

159. $-4 + 3 = -1$ 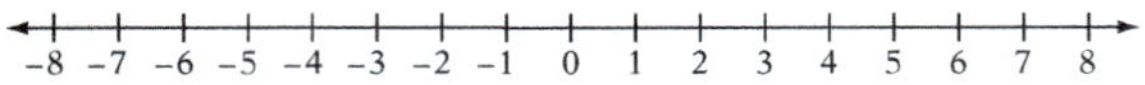

160. $-5 + 8 = 3$ 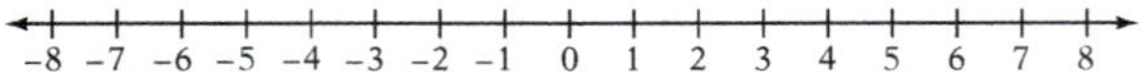

161. $2 + (-7) = -5$ 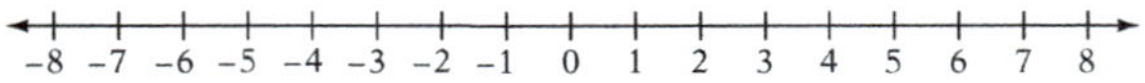

162. $1 + (-6) = -5$ 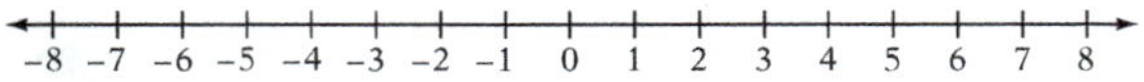

163. $-3 + (-4) = -7$ 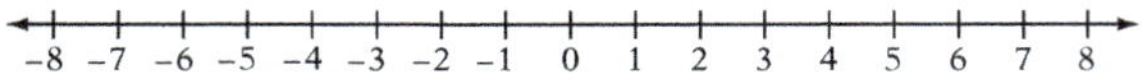

164. $-2 + (-5) = -7$ 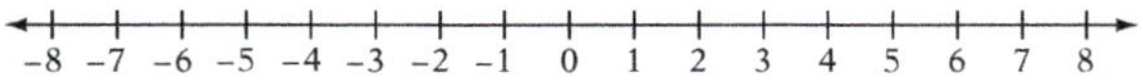

165. There are a number of models for the addition of integers. Using arrows on the number line is just one of them. Another model is a checking account. If there is a balance of \$25 in a checking account, and a check is written for \$30, the account will be overdrawn by \$5 (-5).

An alternative model uses two colors of plastic chips—say, blue for positive and red for negative—and the idea that a blue/red pair is equal to zero. To add $-8 + 3$, place 8 red chips and 3 blue chips in a circle. Pair as many red and blue chips as possible and remove the pairs from the region. There are 5 red chips remaining, or -5.

To model $(-8) + (-3)$, place 8 red chips in the region and then 3 more red chips in the region. There are no pairs of red and blue chips, so there are 11 red chips. Therefore, the answer is -11.

Use the method outlined above to model $-7 + 4$, $-2 + 6$, and $-5 + (-3)$.

166. Make up three addition problems such that each problem involves one positive and one negative addend and each problem has the sum -3. Then describe a strategy for writing these problems.

167. Make up three subtraction problems such that each problem involves a negative number minus a negative number and each problem has a difference of -8. Then describe a strategy for writing these problems.

1.3 Rational Numbers

OBJECTIVE 1 Write rational numbers as decimals

Point of Interest

As early as A.D. 630, the Hindu mathematician Brahmagupta wrote a fraction as one number over another separated by a space. The Arab mathematician al Hassar (around A.D. 1050) was the first to show a fraction with the horizontal bar separating the numerator and denominator.

A **rational number** is the quotient of two integers. Therefore, a rational number is a number that can be written in the form $\frac{a}{b}$, where a and b are integers, and b is not zero. A rational number written in this way is commonly called a **fraction.**

$\frac{a}{b}$ ← a an integer, b ← a nonzero integer

$\left.\frac{2}{3}, \frac{-4}{9}, \frac{18}{-5}, \frac{4}{1}\right\}$ Rational numbers

Because an integer can be written as the quotient of the integer and 1, every integer is a rational number.

$5 = \frac{5}{1} \qquad -3 = \frac{-3}{1}$

A number written in **decimal notation** is also a rational number.

three-tenths $0.3 = \frac{3}{10}$

thirty-five hundredths $0.35 = \frac{35}{100}$

negative four-tenths $-0.4 = -\frac{4}{10}$

A rational number written as a fraction can be written in decimal notation.

Take Note

The fraction bar can be read "divided by."

$\frac{5}{8} = 5 \div 8$

Note that dividing the numerator by the denominator results in a remainder of 0. The decimal 0.625 is a terminating decimal.

EXAMPLE 1 Write $\frac{5}{8}$ as a decimal.

Solution

$$\begin{array}{r} 0.625 \\ 8\overline{)5.000} \\ -4\,8 \\ \hline 20 \\ -16 \\ \hline 40 \\ -40 \\ \hline 0 \end{array}$$

0.625 ← This is called a **terminating decimal.**

0 ← The remainder is zero.

$\frac{5}{8} = 0.625$

Problem 1 Write $\frac{4}{25}$ as a decimal.

Solution See page S1.

Try Exercise 13, page 29.

EXAMPLE 2 Write $\frac{4}{11}$ as a decimal.

Solution

$$\begin{array}{r} 0.3636... \\ 11\overline{)4.0000} \\ \underline{-3\ 3} \\ 70 \\ \underline{-66} \\ 40 \\ \underline{-33} \\ 70 \\ \underline{-66} \\ 4 \end{array}$$

← This is called a **repeating decimal.**

← The remainder is never zero.

$\frac{4}{11} = 0.\overline{36}$ ← The bar over the digits 3 and 6 is used to show that these digits repeat.

Take Note

No matter how far we carry out the division, the remainder is never zero. The decimal $0.\overline{36}$ is a repeating decimal.

Problem 2 Write $\frac{4}{9}$ as a decimal. Place a bar over the repeating digits of the decimal.

Solution See page S1.

Try Exercise 21, page 29.

Rational numbers can be written as fractions, such as $-\frac{6}{7}$ or $\frac{8}{3}$, in which the numerator and denominator are integers. But every rational number also can be written as a repeating decimal (such as 0.25767676...) or a terminating decimal (such as 1.73). This was illustrated in Examples 1 and 2.

Numbers that cannot be written as either a repeating decimal or a terminating decimal are called **irrational numbers.** For example, 2.45445444544445... is an irrational number. Two other examples are $\sqrt{2}$ and π.

$$\sqrt{2} = 1.414213562... \qquad \pi = 3.141592654...$$

The three dots mean that the digits continue on and on without ever repeating or terminating. Although we cannot write a decimal that is exactly equal to $\sqrt{2}$ or to π, we can give approximations of these numbers. The symbol $\approx$ is read "is approximately equal to." Shown below are $\sqrt{2}$ rounded to the nearest thousandth and π rounded to the nearest hundredth.

$$\sqrt{2} \approx 1.414 \qquad \pi \approx 3.14$$

The rational numbers and the irrational numbers taken together are called the **real numbers.**

OBJECTIVE 2 Multiply and divide rational numbers

The sign rules for multiplying and dividing integers apply to multiplication and division of rational numbers.

The product of two fractions is the product of the numerators divided by the product of the denominators.

A fraction is in **simplest form** when the numerator and denominator have no common factors other than 1. The fraction $\frac{3}{8}$ is in simplest form because 3 and 8 have no common factors other than 1. The fraction $\frac{15}{50}$ is not in simplest form because the numerator and denominator have a common factor of 5. To write $\frac{15}{50}$ in simplest form, divide the numerator and denominator by the common factor 5.

$$\frac{15}{50} = \frac{\overset{1}{\cancel{5}} \cdot 3}{\underset{1}{\cancel{5}} \cdot 5 \cdot 2} = \frac{3}{10}$$

After multiplying two fractions, write the product in simplest form, as shown in Example 3.

EXAMPLE 3 Multiply: $\frac{3}{8} \cdot \frac{12}{17}$

Solution

$$\frac{3}{8} \cdot \frac{12}{17} = \frac{3 \cdot 12}{8 \cdot 17}$$

- Multiply the numerators. Multiply the denominators.

$$= \frac{3 \cdot \overset{1}{\cancel{2}} \cdot \overset{1}{\cancel{2}} \cdot 3}{2 \cdot \underset{1}{\cancel{2}} \cdot \underset{1}{\cancel{2}} \cdot 17}$$

- Write the prime factorization of each factor. Divide by the common factors.

$$= \frac{9}{34}$$

- Multiply the numbers remaining in the numerator. Multiply the numbers remaining in the denominator.

Problem 3 Multiply: $-\frac{7}{12} \cdot \frac{9}{14}$

Solution See page S2.

Try Exercise 39, page 29.

> **Take Note**
>
> The method for dividing fractions is sometimes stated "To divide fractions, invert the divisor and multiply." Inverting the divisor means writing its reciprocal.

The **reciprocal** of a fraction is the fraction with the numerator and denominator interchanged. For instance, the reciprocal of $\frac{2}{3}$ is $\frac{3}{2}$, and the reciprocal of $-\frac{5}{4}$ is $-\frac{4}{5}$. To divide fractions, multiply the dividend by the reciprocal of the divisor.

EXAMPLE 4 Divide: $\frac{3}{10} \div \left(-\frac{18}{25}\right)$

Solution

$$\frac{3}{10} \div \left(-\frac{18}{25}\right) = -\left(\frac{3}{10} \div \frac{18}{25}\right)$$

- The signs are different. The quotient is negative.

$$= -\left(\frac{3}{10} \cdot \frac{25}{18}\right)$$

- Change division to multiplication and invert the divisor.

$$= -\left(\frac{3 \cdot 25}{10 \cdot 18}\right)$$

- Multiply the numerators. Multiply the denominators.

$$= -\left(\frac{\overset{1}{\cancel{3}} \cdot \overset{1}{\cancel{5}} \cdot 5}{2 \cdot \underset{1}{\cancel{5}} \cdot 2 \cdot \underset{1}{\cancel{3}} \cdot 3}\right)$$

$$= -\frac{5}{12}$$

Problem 4 Divide: $-\frac{3}{8} \div \left(-\frac{5}{12}\right)$

Solution See page S2.

Try Exercise 45, page 29.

To multiply decimals, multiply as in the multiplication of whole numbers. Write the decimal point in the product so that the number of decimal places in the product equals the sum of the decimal places in the factors.

EXAMPLE 5 Multiply: $(-6.89)(0.00035)$

Solution

$$\begin{array}{rl} 6.89 & \text{2 decimal places} \\ \times\ 0.00035 & \text{5 decimal places} \\ \hline 3445 & \\ 2067 & \\ \hline 0.0024115 & \text{7 decimal places} \end{array}$$

• **Multiply the absolute values.**

$(-6.89)(0.00035) = -0.0024115$

• **The signs are different. The product is negative.**

Problem 5 Multiply: $(-5.44)(3.8)$

Solution See page S2.

Try Exercise 51, page 29.

To divide decimals, move the decimal point in the divisor to make it a whole number. Move the decimal point in the dividend the same number of places to the right. Place the decimal point in the quotient directly over the decimal point in the dividend. Then divide as in the division of whole numbers.

Take Note

Moving the decimal point in the numerator and denominator is the same as multiplying the numerator and the denominator by the same number. For the problem at the right, we have

$$-0.394 \div 1.7 = -\frac{0.394}{1.7} = -\frac{0.394}{1.7}\cdot\frac{10}{10} = -\frac{3.94}{17}$$

EXAMPLE 6 Divide: $-0.394 \div 1.7$. Round to the nearest hundredth.

Solution $1.7\overline{)0.3.940}$

• **Move the decimal point one place to the right in the divisor and in the dividend. Place the decimal point in the quotient.**

$$\begin{array}{r} 0.231 \approx 0.23 \\ 17.\overline{)03.940} \\ -3\,4 \\ \hline 54 \\ -51 \\ \hline 30 \\ -17 \\ \hline 13 \end{array}$$

• **The symbol $\approx$ is used to indicate that the quotient is an approximate value that has been rounded off.**

$-0.394 \div 1.7 \approx -0.23$

• **The signs are different. The quotient is negative.**

Problem 6 Divide $1.32 \div 0.27$. Round to the nearest tenth.

Solution See page S2.

Try Exercise 57, page 30.

OBJECTIVE 3 Add and subtract rational numbers

The sign rules for adding integers apply to addition of rational numbers.

To add or subtract rational numbers written as fractions, first rewrite the fractions as equivalent fractions with a common denominator. A common denominator is the **least common multiple (LCM)** of the denominators. The LCM of the denominators is also called the **lowest common denominator (LCD).**

EXAMPLE 7 Add: $-\frac{5}{6} + \frac{3}{10}$

Solution Prime factorizations of 6 and 10:

$6 = 2 \cdot 3 \quad 10 = 2 \cdot 5$

$\text{LCM} = 2 \cdot 3 \cdot 5 = 30$

• **Find the LCM of the denominators 6 and 10.**

$$-\frac{5}{6} + \frac{3}{10} = -\frac{25}{30} + \frac{9}{30}$$

• **Rewrite the fractions as equivalent fractions, using the LCM of the denominators as the common denominator.**

$$= \frac{-25 + 9}{30}$$

• **Add the numerators, and place the sum over the common denominator.**

$$= \frac{-16}{30}$$

$$= -\frac{8}{15}$$

• **Write the answer in simplest form.**

Problem 7 Subtract: $\frac{5}{9} - \frac{11}{12}$

Solution See page S2.

Try Exercise 67, page 30.

Take Note

You can find the LCM by multiplying the denominators and then dividing by the *greatest common factor* of the two denominators. In the case of 6 and 10, $6 \cdot 10 = 60$. Now divide by 2, the greatest common factor of 6 and 10.

$60 \div 2 = 30$

Alternatively, you can use as the common denominator the product of the denominators, which in this case is 60. Write each fraction with a denominator of 60. Add the fractions. Then simplify the sum.

$$-\frac{5}{6} + \frac{3}{10} = -\frac{50}{60} + \frac{18}{60}$$
$$= \frac{-50 + 18}{60}$$
$$= \frac{-32}{60}$$
$$= -\frac{8}{15}$$

The numbers $-\frac{8}{15}$, $\frac{-8}{15}$, and $\frac{8}{-15}$ all represent the same rational number. Note that in Example 7, we wrote the answer as $-\frac{8}{15}$, with the negative sign in front of the fraction. In this textbook, this is the form in which we will write answers that are negative fractions.

EXAMPLE 8 Simplify: $-\frac{3}{4} + \frac{1}{6} - \frac{5}{8}$

Solution

$$-\frac{3}{4} + \frac{1}{6} - \frac{5}{8} = -\frac{18}{24} + \frac{4}{24} - \frac{15}{24}$$

• **The LCM of 4, 6, and 8 is 24.**

$$= \frac{-18}{24} + \frac{4}{24} + \frac{-15}{24}$$

$$= \frac{-18 + 4 + (-15)}{24}$$

• **Add the numerators and place the sum over the common denominator.**

$$= \frac{-29}{24}$$

$$= -\frac{29}{24}$$

Problem 8 Simplify: $-\frac{7}{8} - \frac{5}{6} + \frac{1}{2}$

Solution See page S2.

Try Exercise 81, page 30.

How It's Used

Addition of positive and negative decimals is used in optometry. *Diopters,* which are used to measure the strength of lenses, are given as positive or negative decimals: a negative diopter lens corrects nearsightedness and a positive diopter lens corrects farsightedness. To correct more than one aspect of a person's vision, an optometrist designs an eyeglass lens that combines two or more diopter strengths.

Note that we left the answer to Example 8 as the improper fraction $-\frac{29}{24}$ rather than writing it as the mixed number $-1\frac{5}{24}$. In this text, we will normally leave answers as improper fractions and not change them to mixed numbers.

To add or subtract decimals, write the numbers so that the decimal points are in a vertical line. Then proceed as in the addition or subtraction of integers. Write the decimal point in the answer directly below the decimal points in the problem.

EXAMPLE 9 Add: $14.02 + 137.6 + 9.852$

Solution

$$\begin{array}{r} 14.02 \\ 137.6 \\ +\ \ 9.852 \\ \hline 161.472 \end{array}$$

- **Write the decimals so that the decimal points are in a vertical line.**
- **Write the decimal point in the sum directly below the decimal points in the problem.**

Problem 9 Add: $3.097 + 4.9 + 3.09$

Solution See page S2.

Try Exercise 93, page 30.

EXAMPLE 10 Add: $-114.039 + 84.76$

Solution

$$\begin{array}{r} 114.039 \\ -\ \ 84.76 \\ \hline 29.279 \end{array}$$

$$-114.039 + 84.76 = -29.279$$

- **The signs are different. Subtract the absolute value of the number with the smaller absolute value from the absolute value of the number with the larger absolute value.**
- **Attach the sign of the number with the larger absolute value.**

Problem 10 Subtract: $16.127 - 67.91$

Solution See page S2.

Try Exercise 91, page 30.

OBJECTIVE 4

Convert among percents, fractions, and decimals

"A population growth rate of 3%," "a manufacturer's discount of 25%," and "an 8% increase in pay" are typical examples of the many ways in which percent is used in applied problems. **Percent** means "parts of 100." Thus 27% means 27 parts of 100.

In applied problems involving a percent, it is usually necessary either to rewrite the percent as a fraction or a decimal, or to rewrite a fraction or a decimal as a percent.

To write 27% as a fraction, remove the percent sign and multiply by $\frac{1}{100}$.

$$27\% = 27\left(\frac{1}{100}\right) = \frac{27}{100}$$

To write a percent as a decimal, remove the percent sign and multiply by 0.01.

To write 33% as a decimal, remove the percent sign and multiply by 0.01.

$$33\% = 33(0.01) = 0.33$$

Move the decimal point two places to the left and remove the percent sign.

Note that 100% = 1.

$$100\% = 100(0.01) = 1$$

EXAMPLE 11 Write 130% as a fraction and as a decimal.

Solution $130\% = 130\left(\frac{1}{100}\right) = \frac{130}{100} = 1\frac{3}{10}$

• To write a percent as a fraction, remove the percent sign and multiply by $\frac{1}{100}$.

$130\% = 130(0.01) = 1.30$

• To write a percent as a decimal, remove the percent sign and multiply by 0.01.

Problem 11 Write 125% as a fraction and as a decimal.

Solution See page S2.

Try Exercise 111, page 31.

EXAMPLE 12 Write $33\frac{1}{3}\%$ as a fraction.

Solution $33\frac{1}{3}\% = 33\frac{1}{3}\left(\frac{1}{100}\right) = \frac{100}{3}\left(\frac{1}{100}\right)$

$= \frac{1}{3}$

• Write the mixed number $33\frac{1}{3}$ as the improper fraction $\frac{100}{3}$.

Problem 12 Write $16\frac{2}{3}\%$ as a fraction.

Solution See page S2.

Try Exercise 123, page 31.

EXAMPLE 13 Write 0.25% as a decimal.

Solution $0.25\% = 0.25(0.01) = 0.0025$

• Remove the percent sign and multiply by 0.01.

Problem 13 Write 6.08% as a decimal.

Solution See page S2.

Try Exercise 135, page 31.

A fraction or decimal can be written as a percent by multiplying by 100%. Recall that 100% = 1, and multiplying a number by 1 does not change the value of the number.

To write $\frac{5}{8}$ as a percent, multiply by 100%.

$\frac{5}{8} = \frac{5}{8}(100\%) = \frac{500}{8}\% = 62.5\%$ or $62\frac{1}{2}\%$

To write 0.82 as a percent, multiply by 100%.

$0.82 = 0.82(100\%) = 82\%$

Move the decimal point two places to the right. Then write the percent sign.

EXAMPLE 14 Write as a percent. **A.** 0.027 **B.** 1.34

Solution **A.** $0.027 = 0.027(100\%) = 2.7\%$ • To write a decimal as a percent, multiply by 100%.

B. $1.34 = 1.34(100\%) = 134\%$

Problem 14 Write as a percent. **A.** 0.043 **B.** 2.57

Solution See page S2.

Try Exercise 143, page 31.

EXAMPLE 15 Write $\frac{5}{6}$ as a percent. Round to the nearest tenth of a percent.

Solution $\frac{5}{6} = \frac{5}{6}(100\%) = \frac{500}{6}\% \approx 83.3\%$ • To write a fraction as a percent, multiply by 100%.

Problem 15 Write $\frac{5}{9}$ as a percent. Round to the nearest tenth of a percent.

Solution See page S2.

Try Exercise 155, page 31.

EXAMPLE 16 Write $\frac{7}{16}$ as a percent. Write the remainder in fractional form.

Solution $\frac{7}{16} = \frac{7}{16}(100\%) = \frac{700}{16}\% = 43\frac{3}{4}\%$ • Multiply the fraction by 100%.

Problem 16 Write $\frac{9}{16}$ as a percent. Write the remainder in fractional form.

Solution See page S2.

Try Exercise 161, page 31.

1.3 Exercises

CONCEPT CHECK

Determine whether the statement is always true, sometimes true, or never true.

1. To multiply two fractions, you must first rewrite the fractions as equivalent fractions with a common denominator.
2. A rational number can be written as a terminating decimal.
3. An irrational number is a real number.
4. 37%, 0.37, and $\frac{37}{100}$ are three numbers that have the same value.
5. To write a decimal as a percent, multiply the decimal by $\frac{1}{100}$.
6. -12 is an example of a number that is both an integer and a rational number.

1 Write rational numbers as decimals (See pages 21–22.)

GETTING READY

7. To write $\frac{2}{3}$ as a decimal, divide __?__ by __?__. The quotient is 0.6666... , which is a __?__ decimal.

8. A number such as 0.74744744474444... , whose decimal representation neither ends nor repeats, is an example of a(n) __?__ number.

Write as a decimal. Place a bar over the repeating digits of a repeating decimal.

9. $\frac{1}{3}$ **10.** $\frac{2}{3}$ **11.** $\frac{1}{4}$ **12.** $\frac{3}{4}$ **13.** $\frac{2}{5}$

14. $\frac{4}{5}$ **15.** $\frac{1}{6}$ **16.** $\frac{5}{6}$ **17.** $\frac{1}{8}$ **18.** $\frac{7}{8}$

19. $\frac{2}{9}$ **20.** $\frac{8}{9}$ **21.** $\frac{5}{11}$ **22.** $\frac{10}{11}$ **23.** $\frac{7}{12}$

24. $\frac{11}{12}$ **25.** $\frac{4}{15}$ **26.** $\frac{8}{15}$ **27.** $\frac{7}{16}$ **28.** $\frac{15}{16}$

29. $\frac{6}{25}$ **30.** $\frac{14}{25}$ **31.** $\frac{9}{40}$ **32.** $\frac{21}{40}$ **33.** $\frac{15}{22}$

34. Is $\frac{\sqrt{2}}{2}$ a rational or an irrational number?

2 Multiply and divide rational numbers (See pages 22–24.)

GETTING READY

35. The product of 1.762 and −8.4 will have __?__ decimal places.

36. The reciprocal of $\frac{4}{9}$ is __?__. To find the quotient $-\frac{2}{3} \div \frac{4}{9}$, find the product $-\frac{2}{3} \cdot$ __?__. The quotient $-\frac{2}{3} \div \frac{4}{9}$ is __?__.

Simplify.

37. $\frac{1}{2}\left(-\frac{3}{4}\right)$ **38.** $-\frac{2}{9}\left(-\frac{3}{14}\right)$ **39.** $\left(-\frac{3}{8}\right)\left(-\frac{4}{15}\right)$

40. $\frac{5}{8}\left(-\frac{7}{12}\right)\frac{16}{25}$ **41.** $\left(\frac{1}{2}\right)\left(-\frac{3}{4}\right)\left(-\frac{5}{8}\right)$ **42.** $\left(\frac{5}{12}\right)\left(-\frac{8}{15}\right)\left(-\frac{1}{3}\right)$

43. $\frac{3}{8} \div \frac{1}{4}$ **44.** $\frac{5}{6} \div \left(-\frac{3}{4}\right)$ **45.** $-\frac{5}{12} \div \frac{15}{32}$

46. $\frac{1}{8} \div \left(-\frac{5}{12}\right)$ **47.** $-\frac{4}{9} \div \left(-\frac{2}{3}\right)$ **48.** $-\frac{6}{11} \div \frac{4}{9}$

49. $(1.2)(3.47)$ **50.** $(-0.8)(6.2)$ **51.** $(-1.89)(-2.3)$

52. $(6.9)(-4.2)$ **53.** $(1.06)(-3.8)$ **54.** $(-2.7)(-3.5)$

55. State whether each product or quotient is positive or negative. Do not simplify.

a. $\left(-\frac{11}{12}\right)\left(-\frac{5}{4}\right)\left(-\frac{1}{2}\right)$

b. $-1.572 \div -8.4$

Simplify. Round to the nearest hundredth.

56. $-24.7 \div 0.09$

57. $-1.27 \div (-1.7)$

58. $9.07 \div (-3.5)$

59. $-354.2086 \div 0.1719$

3 Add and subtract rational numbers (See pages 24–26.)

GETTING READY

60. The least common multiple of the denominators of the fractions $\frac{5}{8}$, $-\frac{1}{6}$, and $\frac{2}{9}$ is ___?___.

61. Write the fraction $\frac{3}{14}$ as an equivalent fraction with the denominator 28: $\frac{3}{14} = \frac{?}{28}$.

Simplify.

62. $\frac{3}{8} + \frac{5}{8}$

63. $-\frac{1}{4} + \frac{3}{4}$

64. $\frac{7}{8} - \frac{3}{8}$

65. $-\frac{5}{6} - \frac{1}{6}$

66. $-\frac{5}{12} - \frac{3}{8}$

67. $-\frac{5}{6} - \frac{5}{9}$

68. $-\frac{6}{13} + \frac{17}{26}$

69. $-\frac{7}{12} + \frac{5}{8}$

70. $-\frac{5}{8} - \left(-\frac{11}{12}\right)$

71. $\frac{1}{3} + \frac{5}{6} - \frac{2}{9}$

72. $\frac{1}{2} - \frac{2}{3} + \frac{1}{6}$

73. $-\frac{3}{8} - \frac{5}{12} - \frac{3}{16}$

74. $-\frac{5}{16} + \frac{3}{4} - \frac{7}{8}$

75. $\frac{1}{2} - \frac{3}{8} - \left(-\frac{1}{4}\right)$

76. $\frac{3}{4} - \left(-\frac{7}{12}\right) - \frac{7}{8}$

77. $\frac{1}{3} - \frac{1}{4} - \frac{1}{5}$

78. $\frac{2}{3} - \frac{1}{2} + \frac{5}{6}$

79. $\frac{5}{16} + \frac{1}{8} - \frac{1}{2}$

80. $\frac{5}{8} - \left(-\frac{5}{12}\right) + \frac{1}{3}$

81. $\frac{1}{8} - \frac{11}{12} + \frac{1}{2}$

82. $-\frac{7}{9} + \frac{14}{15} + \frac{8}{21}$

83. $1.09 + 6.2$

84. $-32.1 - 6.7$

85. $5.13 - 8.179$

86. $-13.092 + 6.9$

87. $2.54 - 3.6$

88. $5.43 + 7.925$

89. $-16.92 - 6.925$

90. $-3.87 + 8.546$

91. $6.9027 - 17.692$

92. $2.09 - 6.72 - 5.4$

93. $16.4 + 3.09 - 7.93$

94. $-18.39 + 4.9 - 23.7$

95. $19 - (-3.72) - 82.75$

96. $-3.07 - (-2.97) - 17.4$

97. $-3.09 - 4.6 - 27.3$

Complete Exercises 98 and 99 without actually finding the sums and differences.

98. State whether each sum or difference is positive or negative.

a. $\frac{1}{5} - \frac{1}{2}$ **b.** $-21.765 + 15.1$ **c.** $0.837 + (-0.24)$ **d.** $-\frac{3}{4} + \frac{9}{10}$

99. Estimate each sum to the nearest integer.

a. $\frac{7}{8} + \frac{4}{5}$ **b.** $\frac{1}{3} + \left(-\frac{1}{2}\right)$ **c.** $-0.125 + 1.25$ **d.** $-1.3 + 0.2$

4 Convert among percents, fractions, and decimals (See pages 26–28.)

100. **a.** Explain how to convert a fraction to a percent.

b. Explain how to convert a percent to a fraction.

101. **a.** Explain how to convert a decimal to a percent.
b. Explain how to convert a percent to a decimal.

102. Explain why multiplying a number by 100% does not change the value of the number.

GETTING READY

103. To write 80% as a fraction, remove the percent sign and multiply by __?__:
$80\% = 80 \cdot$ __?__ = __?__.

104. To write 68% as a decimal, remove the percent sign and multiply by __?__:
$68\% = 68 \cdot$ __?__ = __?__.

105. To write $\frac{3}{10}$ as a percent, multiply by __?__:
$\frac{3}{10} = \frac{3}{10} \cdot$ __?__ = __?__.

106. To write 1.25 as a percent, multiply by __?__:
$1.25 = 1.25 \cdot$ __?__ = __?__.

Write as a fraction and as a decimal.

107. 75% **108.** 40% **109.** 50% **110.** 10%

111. 64% **112.** 88% **113.** 175% **114.** 160%

115. 19% **116.** 87% **117.** 5% **118.** 2%

119. 450% **120.** 380% **121.** 8% **122.** 4%

Write as a fraction.

123. $11\frac{1}{9}\%$ **124.** $37\frac{1}{2}\%$ **125.** $31\frac{1}{4}\%$ **126.** $66\frac{2}{3}\%$

127. $\frac{1}{2}\%$ **128.** $5\frac{3}{4}\%$ **129.** $6\frac{1}{4}\%$ **130.** $83\frac{1}{3}\%$

Write as a decimal.

131. 7.3% **132.** 9.1% **133.** 15.8% **134.** 0.3%

135. 9.15% **136.** 121.2% **137.** 18.23% **138.** 0.15%

Write as a percent.

139. 0.15 **140.** 0.37 **141.** 0.05 **142.** 0.02

143. 0.175 **144.** 0.125 **145.** 1.15 **146.** 2.142

147. 0.008 **148.** 0.004 **149.** 0.065 **150.** 0.083

Write as a percent. Round to the nearest tenth of a percent.

151. $\frac{27}{50}$ **152.** $\frac{83}{100}$ **153.** $\frac{1}{3}$ **154.** $\frac{3}{8}$

155. $\frac{4}{9}$ **156.** $\frac{9}{20}$ **157.** $2\frac{1}{2}$ **158.** $1\frac{2}{7}$

Write as a percent. Write the remainder in fractional form.

159. $\frac{3}{8}$ **160.** $\frac{3}{16}$ **161.** $\frac{5}{14}$ **162.** $\frac{4}{7}$

163. $1\frac{1}{4}$ **164.** $2\frac{5}{8}$ **165.** $1\frac{5}{9}$ **166.** $1\frac{13}{16}$

Complete Exercises 167 and 168 without actually finding the percents.

167. Does $\frac{4}{3}$ represent a number greater than 100% or less than 100%?

168. Does 0.055 represent a number greater than 1% or less than 1%?

Employment The graph at the right shows the responses to a survey that asked respondents, "How did you find your most recent job?" Use the graph for Exercises 169 to 171.

169. What fraction of the respondents found their most recent jobs on the Internet?

170. What fraction of the respondents found their most recent jobs through a referral?

171. Did more or less than one-quarter of the respondents find their most recent jobs through a newspaper ad?

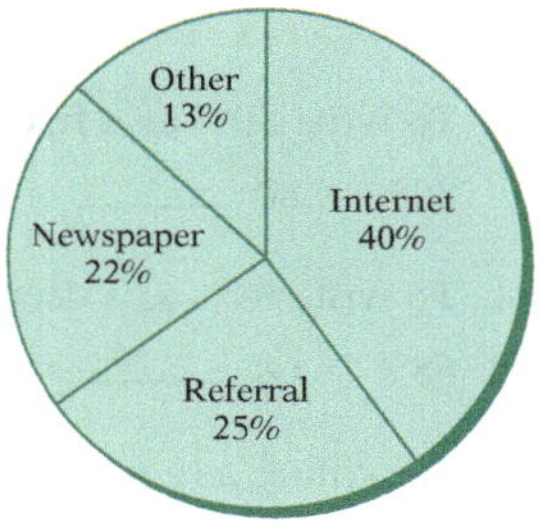

How Did You Find Your Most Recent Job?

APPLYING CONCEPTS

Classify each of the following numbers as a natural number, an integer, a positive integer, a negative integer, a rational number, an irrational number, or a real number. List all that apply.

172. -1

173. 28

174. $-\frac{9}{34}$

175. -7.707

176. $5.2\overline{6}$

177. 0.171771777...

Solve.

178. Find the average of $\frac{5}{8}$ and $\frac{3}{4}$.

179. **Temperature** The date of the news clipping at the right is March 26, 2010.

a. Find the difference between the extreme Fahrenheit temperatures.

b. Find the difference between the extreme Celsius temperatures.

In the News

The World's Hot and Cold Spots

This week's hottest temperature was 112.1°F (44.5°C), recorded at Nawabshah, Pakistan, while the coldest temperature was −87.9°F (−66.6°C), recorded at Russia's Vostok Antarctic research station.

Source: www.earthweek.com

Government The table at the right shows the surplus or deficit, in billions of dollars, for the federal budget for selected years from 1955 through 2010. A negative sign (−) indicates a deficit. Use this table for Exercises 180 to 184. (*Source:* U.S. Office of Management and Budget)

180. In which year listed was the deficit greatest?

181. Find the difference between the deficits in the years 1980 and 1985.

182. Calculate the difference between the surplus in 1960 and the deficit in 1955.

183. How many times greater was the deficit in 1985 than in 1975? Round to the nearest whole number.

184. What was the average deficit per quarter, in millions of dollars, for the year 1970?

Year	Federal Budget Surplus or Deficit (in billions of dollars)	Year	Federal Budget Surplus or Deficit (in billions of dollars)
1955	−2.993	1995	−163.952
1960	0.301	2000	236.241
1965	−1.411	2005	−318.346
1970	−2.842	2006	−248.181
1975	−53.242	2007	−160.701
1980	−73.830	2008	−458.555
1985	−212.308	2009	−1412.686
1990	−221.036	2010	−1294.131

185. **Temperature** See the news clipping at the right. What is the average normal temperature in the Northeast in February?

186. Let x represent the price of a car. If the sales tax is 6% of the price, express the total of the price of the car and the sales tax in terms of x.

187. Let x represent the price of a suit. If the suit is on sale at a discount rate of 30%, express the price of the suit after the discount in terms of x.

188. In your own words, define **a.** a rational number, **b.** an irrational number, and **c.** a real number.

189. Explain why you need a common denominator when adding two fractions and why you don't need a common denominator when multiplying two fractions.

In the News

Temps Near Normal in Northeast

This year temperatures in February averaged near normal in the Northeast. The region's average temperature was −3.2°C, which is 0.4°C above normal.

Source: National Climatic Data Center

PROJECTS OR GROUP ACTIVITIES

190. Use a calculator to determine the decimal representations of $\frac{17}{99}$, $\frac{45}{99}$, and $\frac{73}{99}$. Make a conjecture as to the decimal representation of $\frac{83}{99}$. Does your conjecture work for $\frac{33}{99}$? What about $\frac{1}{99}$?

191. A magic square is one in which the numbers in every row, column, and diagonal sum to the same number. Complete the magic square at the right.

$\frac{2}{3}$		
	$\frac{1}{6}$	$\frac{5}{6}$
		$-\frac{1}{3}$

192. Find three natural numbers a, b, and c such that $\frac{1}{a} + \frac{1}{b} + \frac{1}{c}$ is a natural number.

193. When two rational numbers are added, it is possible for the sum to be less than either addend, greater than either addend, or a number between the two addends. Give examples of each of these occurrences.

1.4 Exponents and the Order of Operations Agreement

OBJECTIVE 1

Exponential expressions

Repeated multiplication of the same factor can be written using an exponent.

$$2 \cdot 2 \cdot 2 \cdot 2 \cdot 2 = 2^5 \leftarrow \text{exponent (2 is the base)} \qquad a \cdot a \cdot a \cdot a = a^4 \leftarrow \text{exponent (} a \text{ is the base)}$$

The **exponent** indicates how many times the factor, called the **base,** occurs in the multiplication. The multiplication $2 \cdot 2 \cdot 2 \cdot 2 \cdot 2$ is in **factored form.** The exponential expression 2^5 is in **exponential form.**

Point of Interest

René Descartes (1596–1650) was the first mathematician to use exponential notation extensively as it is used today. However, for some unknown reason, he always used xx for x^2.

2^1 is read "the first power of two" or just "two." ⟶ Usually the exponent 1 is not written.

2^2 is read "the second power of two" or "two squared."

2^3 is read "the third power of two" or "two cubed."

2^4 is read "the fourth power of two."

2^5 is read "the fifth power of two."

a^5 is read "the fifth power of a."

There is a geometric interpretation of the first three natural-number powers.

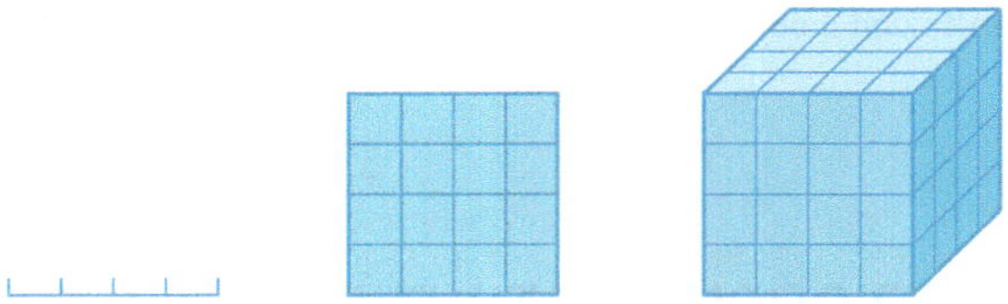

$4^1 = 4$ Length: 4 ft

$4^2 = 16$ Area: 16 ft^2

$4^3 = 64$ Volume: 64 ft^3

To evaluate an exponential expression, write each factor as many times as indicated by the exponent. Then multiply.

$$3^5 = 3 \cdot 3 \cdot 3 \cdot 3 \cdot 3 = 243$$
$$2^3 \cdot 3^2 = (2 \cdot 2 \cdot 2) \cdot (3 \cdot 3) = 8 \cdot 9 = 72$$

EXAMPLE 1 Evaluate $(-4)^2$ and -4^2.

Solution $(-4)^2 = (-4)(-4) = 16$
$-4^2 = -(4 \cdot 4) = -16$

- **The -4 is squared only when the negative sign is *inside* the parentheses. In $(-4)^2$, we are squaring -4; in -4^2, we are finding the opposite of 4^2.**

Problem 1 Evaluate $(-5)^3$ and -5^3.

Solution See page S2.

Try Exercise 13, page 37.

EXAMPLE 2 Evaluate $(-2)^4$ and $(-2)^5$.

Solution

$$\begin{aligned}(-2)^4 &= (-2)(-2)(-2)(-2)\\ &= 4(-2)(-2)\\ &= -8(-2)\\ &= 16\end{aligned}$$

$$\begin{aligned}(-2)^5 &= (-2)(-2)(-2)(-2)(-2)\\ &= 4(-2)(-2)(-2)\\ &= -8(-2)(-2)\\ &= 16(-2)\\ &= -32\end{aligned}$$

Take Note

The product of an even number of negative factors is positive. The product of an odd number of negative factors is negative.

Problem 2 Evaluate $(-3)^3$ and $(-3)^4$.

Solution See page S2.

Try Exercise 15, page 37.

EXAMPLE 3 Evaluate $(-3)^2 \cdot 2^3$ and $\left(-\frac{2}{3}\right)^3$.

Solution $(-3)^2 \cdot 2^3 = (-3)(-3) \cdot (2)(2)(2) = 9 \cdot 8 = 72$

$$\left(-\frac{2}{3}\right)^3 = \left(-\frac{2}{3}\right)\left(-\frac{2}{3}\right)\left(-\frac{2}{3}\right) = -\frac{2 \cdot 2 \cdot 2}{3 \cdot 3 \cdot 3} = -\frac{8}{27}$$

Problem 3 Evaluate $(3^3)(-2)^3$ and $\left(-\frac{2}{5}\right)^2$.

Solution See page S2.

Try Exercise 31, page 37.

OBJECTIVE 2

The Order of Operations Agreement

Evaluate $2 + 3 \cdot 5$.

There are two arithmetic operations, addition and multiplication, in this problem. The operations could be performed in different orders.

Add first.	$\underbrace{2 + 3} \cdot 5$	Multiply first.	$2 + \underbrace{3 \cdot 5}$
Then multiply.	$\underbrace{5 \quad \cdot 5}_{25}$	Then add.	$\underbrace{2 + 15}_{17}$

In order to prevent there being more than one answer to the same problem, an Order of Operations Agreement has been established.

THE ORDER OF OPERATIONS AGREEMENT

Step 1 Perform operations inside grouping symbols. **Grouping symbols** include parentheses ( ), brackets [ ], absolute value symbols | |, and the fraction bar.

Step 2 Simplify exponential expressions.

Step 3 Do multiplication and division as they occur from left to right.

Step 4 Do addition and subtraction as they occur from left to right.

EXAMPLE

	$6 + 5(1 - 2)^4$
Perform operations inside grouping symbols (Step 1).	$= 6 + 5(-1)^4$
Simplify exponential expressions (Step 2).	$= 6 + 5(1)$
Do multiplication and division from left to right (Step 3).	$= 6 + 5$
Do addition and subtraction from left to right (Step 4).	$= 11$

EXAMPLE 4 Simplify: $12 - 24(8 - 5) \div 2^2$

Solution $12 - 24(8 - 5) \div 2^2$

$= 12 - 24(3) \div 2^2$	• Perform operations inside grouping symbols.
$= 12 - 24(3) \div 4$	• Simplify exponential expressions.
$= 12 - 72 \div 4$	• Do multiplication and division as they occur from left to right.
$= 12 - 18$	
$= -6$	• Do addition and subtraction as they occur from left to right.

Problem 4 Simplify: $36 \div (8 - 5)^2 - (-3)^2 \cdot 2$

Solution See page S2.

Try Exercise 41, page 38.

One or more of the steps shown in Example 4 may not be needed to simplify an expression. In that case, proceed to the next step in the Order of Operations Agreement.

EXAMPLE 5 Simplify: $\frac{4+8}{2+1} - |3 - 1| + 2$

Solution

$\frac{4+8}{2+1} - |3 - 1| + 2$

$= \frac{12}{3} - |2| + 2$ • **Perform operations inside grouping symbols (above and below the fraction bar and inside the absolute value symbol).**

$= \frac{12}{3} - 2 + 2$ • **Find the absolute value of 2.**

$= 4 - 2 + 2$ • **Do multiplication and division as they occur from left to right.**

$= 2 + 2$ • **Do addition and subtraction as they occur from left to right.**

$= 4$

Problem 5 Simplify: $27 \div 3^2 + (-3)^2 \cdot 4$

Solution See page S2.

Try Exercise 49, page 38.

When an expression has grouping symbols inside grouping symbols, first perform the operations inside the *inner* grouping symbols by following Steps 2, 3, and 4 of the Order of Operations Agreement. Then perform the operations inside the *outer* grouping symbols by following Steps 2, 3, and 4 in sequence.

EXAMPLE 6 Simplify: $6 \div [4 - (6 - 8)] + 2^2$

Solution

$6 \div [4 - (6 - 8)] + 2^2$

$= 6 \div [4 - (-2)] + 2^2$ • **Perform operations inside the inner grouping symbols.**

$= 6 \div 6 + 2^2$ • **Perform operations inside the outer grouping symbols.**

$= 6 \div 6 + 4$ • **Simplify exponential expressions.**

$= 1 + 4$ • **Do multiplication and division.**

$= 5$ • **Do addition and subtraction.**

Problem 6 Simplify: $4 - 3[4 - 2(6 - 3)] \div 2$

Solution See page S3.

Try Exercise 51, page 38.

1.4 Exercises

CONCEPT CHECK

Rewrite each expression as an exponential expression.

1. nine to the fifth power

2. y to the fourth power

3. seven to the nth power

4. $b \cdot b \cdot b \cdot b \cdot b \cdot b \cdot b \cdot b$

Determine whether the statement is true or false.

5. $(-5)^2$, -5^2, and $-(5)^2$ all represent the same number.

6. The expression 9^4 is in exponential form.

7. To evaluate the expression $6 + 7 \cdot 10$ means to determine what one number it is equal to.

8. The Order of Operations Agreement is used for natural numbers, integers, rational numbers, and real numbers.

1 Exponential expressions (See pages 33–35.)

GETTING READY

9. In the expression $(-5)^2$, -5 is called the ___?___ and 2 is called the ___?___. To evaluate $(-5)^2$, find the product (___?___)(___?___) = ___?___.

10. The exponential expression 4^3 is read "the ___?___ power of ___?___" or "four ___?___." To evaluate 4^3, find the product (___?___)(___?___)(___?___) = ___?___.

Evaluate.

11. 6^2

12. 7^4

13. -7^2

14. -4^3

15. $(-3)^2$

16. $(-2)^3$

17. $(-3)^4$

18. $(-5)^3$

19. $\left(\frac{1}{2}\right)^2$

20. $\left(-\frac{3}{4}\right)^3$

21. $(0.3)^2$

22. $(1.5)^3$

23. $2^2 \cdot (-3)$

24. $3^4 \cdot (-5)$

25. $2^3 \cdot 3^3 \cdot (-4)$

26. $4^2 \cdot 3^2 \cdot (-7)$

27. $\left(\frac{2}{3}\right)^2 \cdot 3^3$

28. $\left(-\frac{1}{2}\right)^3 \cdot 8$

29. $(0.3)^3 \cdot 2^3$

30. $(0.5)^2 \cdot 3^3$

31. $\left(\frac{2}{3}\right)^2 \cdot \frac{1}{4} \cdot 3^3$

32. $\left(\frac{3}{4}\right)^2 \cdot 2^3 \cdot (-4)$

Complete Exercises 33 and 34 without actually finding the products.

33. Is the fifth power of negative eighteen positive or negative?

34. Is the product $-(3^2)(-5^3)$ positive or negative?

2 The Order of Operations Agreement (See pages 35–36.)

35. Why do we need an Order of Operations Agreement?

36. Describe each step in the Order of Operations Agreement.

GETTING READY

37. Simplify: $2(3^3)$

$2(3^3) = 2(\underline{\ ?\ })$ • Simplify exponential expressions.

$= \underline{\ ?\ }$ • Do multiplication and division.

38. Simplify: $3 - 5(6 - 8)^2$

$3 - 5(6 - 8)^2 = 3 - 5(\underline{\ ?\ })^2$ • Perform operations inside grouping symbols.

$= 3 - 5(\underline{\ ?\ })$ • Simplify exponential expressions.

$= 3 - \underline{\ ?\ }$ • Do multiplication and division.

$= \underline{\ ?\ }$ • Do addition and subtraction.

Simplify by using the Order of Operations Agreement.

39. $4 - 8 \div 2$

40. $3 \cdot 2^2 - 3$

41. $2(3 - 4) - (-3)^2$

42. $16 - 32 \div 2^3$

43. $24 - 18 \div 3 + 2$

44. $8 - (-3)^2 - (-2)$

45. $16 + 15 \div (-5) - 2$

46. $14 - 2^2 - |4 - 7|$

47. $3 - 2[8 - (3 - 2)]$

48. $-2^2 + 4[16 \div (3 - 5)]$

49. $6 + \dfrac{16 - 4}{2^2 + 2} - 2$

50. $24 \div \dfrac{3^2}{8 - 5} - (-5)$

51. $96 \div 2[12 + (6 - 2)] - 3^3$

52. $4 \cdot [16 - (7 - 1)] \div 10$

53. $16 \div 2 - 4^2 - (-3)^2$

54. $18 \div |9 - 2^3| + (-3)$

55. $16 - 3(8 - 3)^2 \div 5$

56. $4(-8) \div [2(7 - 3)^2]$

57. $\dfrac{(-10) + (-2)}{6^2 - 30} \div |2 - 4|$

58. $16 - 4 \cdot \dfrac{3^3 - 7}{2^3 + 2} - (-2)^2$

59. $0.3(1.7 - 4.8) + (1.2)^2$

60. $(1.65 - 1.05)^2 \div 0.4 + 0.8$

61. $\dfrac{3}{8} \div \left|\dfrac{5}{6} + \dfrac{2}{3}\right|$

62. $\left(\dfrac{3}{4}\right)^2 - \left(\dfrac{1}{2}\right)^3 \div \dfrac{3}{5}$

63. Which expression is equivalent to $9 - 2^2(1 - 5)$?

(i) $7^2(-4)$ (ii) $5(-4)$ (iii) $9 - 4(-4)$ (iv) $9 + 4(-4)$

64. Which expression is equivalent to $15 + 15 \div 3 - 4^2$?

(i) $30 \div 3 - 16$ (ii) $15 + 5 - 16$ (iii) $15 + 5 + 16$ (iv) $15 + 15 \div (-1)^2$

APPLYING CONCEPTS

65. Using the Order of Operations Agreement, describe how to simplify Exercise 55.

Place the correct symbol, $<$ or $>$, between the two numbers.

66. $(0.9)^3 \quad 1^5$

67. $(-3)^3 \quad (-2)^5$

68. $(-1.1)^2 \quad (0.9)^2$

69. **Computers** A computer with an Intel Core i7 Extreme processor can perform approximately 76.4 billion instructions per second. To the nearest second, how many seconds would it take this computer to perform 10^{12} operations?

PROJECTS OR GROUP ACTIVITIES

70. In which column is the number one million, column A, B, or C?

A	B	C
1	8	27
64	125	216
.	.	.
.	.	.
.	.	.

71. Find a rational number, r, that satisfies the condition.

a. $r^2 < r$ **b.** $r^2 = r$ **c.** $r^2 > r$

72. The sum of two natural numbers is 41. Each of the two numbers is the square of a natural number. Find the two numbers.

Determine the ones digit when the expression is evaluated.

73. 34^{202} **74.** 23^{502} **75.** 27^{622}

76. **Calculators** Does your calculator use the Order of Operations Agreement? To find out, try this problem:

$$2 + 4 \cdot 7$$

If your answer is 30, then the calculator uses the Order of Operations Agreement. If your answer is 42, it does not use that agreement.

Even if your calculator does not use the Order of Operations Agreement, you can still correctly evaluate numerical expressions. The parentheses keys, (and), are used for this purpose.

Remember that $2 + 4 \cdot 7$ means $2 + (4 \cdot 7)$ because the multiplication must be completed before the addition.

Evaluate.

a. $3 \cdot (15 - 2 \cdot 3) - 36 \div 3$ **b.** $4 \cdot 2^2 - (12 + 24 \div 6) - 5$

1.5 Concepts from Geometry

OBJECTIVE 1 Measures of angles

The word *geometry* comes from the Greek words for "earth" (*geo*) and "measure." The original purpose of geometry was to measure land. Today, geometry is used in many disciplines, such as physics, biology, geology, architecture, art, and astronomy.

Here are some basic geometric concepts.

A **plane** is a flat surface such as a table top that extends indefinitely. Figures that lie entirely in a plane are called **plane figures.**

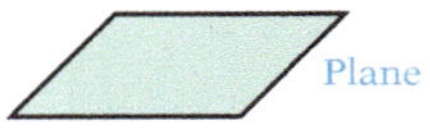

Space extends in all directions. Objects in space, such as a baseball, a house, or a tree, are called **solids.**

A **line** extends indefinitely in two directions in a plane. A line has no width.

A **ray** starts at a point and extends indefinitely in one direction. By placing a point on the ray at the right, we can name the ray AB.

A **line segment** is part of a line and has two endpoints. The line segment AB is designated by its two endpoints.

Take Note

When three letters are used to name an angle, the vertex is always the middle letter. We could also refer to the angle at the right as $\angle CAB$.

Point of Interest

The Babylonians chose 360° for the measure of one full rotation, probably because they knew there are 365 days in a year and the closest number to 365 with many divisors is 360.

How It's Used

In geography, the latitude of a point on Earth is actually the measure of an angle with its vertex at the center of Earth. The complement of this angle is called the *colatitude* of the point. Astronomers can use colatitudes to determine which stars are always visible at a given latitude and which are never visible at that latitude.

Lines in a plane can be parallel or intersect. **Parallel lines** never meet. The distance between parallel lines in a plane is always the same. We write $p \parallel q$ to indicate line p is parallel to line q. **Interecting lines** cross at a point in the plane.

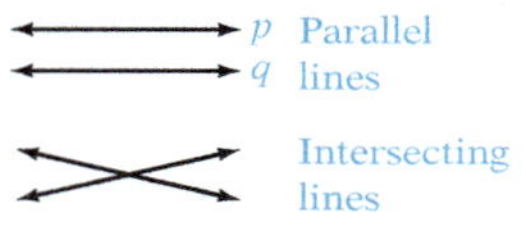

An **angle** is formed when two rays start from the same point. Rays AB and AC start from the same point A. The point at which the rays meet is called the **vertex** of the angle. The symbol $\angle$ is read "angle" and is used to name an angle. We can refer to the angle at the right as $\angle A$, $\angle BAC$, or $\angle x$.

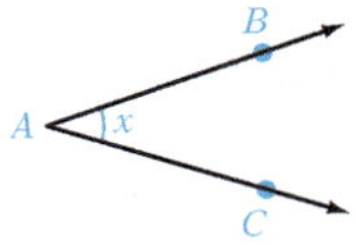

An angle can be measured in **degrees.** The symbol for degree is °. A ray rotated one revolution about its beginning point creates an angle of 360°.

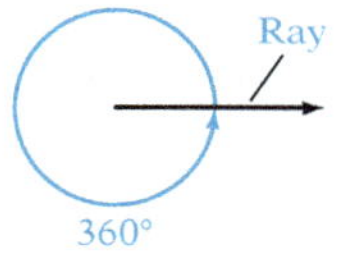

The measure of an angle is symbolized by $m\angle$. For instance, $m\angle C = 40°$. Read this as "the measure of angle C is 40°."

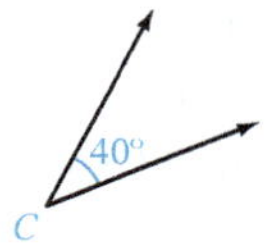

One-fourth of a revolution is one-fourth of 360°, or 90°. A 90° angle is called a **right angle.** The symbol ⊾ is used to represent a right angle. **Perpendicular lines** are intersecting lines that form right angles. We write $p \perp q$ to indicate that line p is perpendicular to line q.

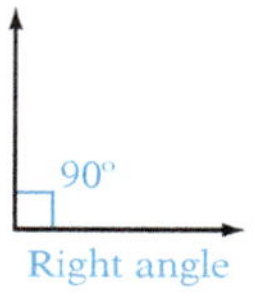

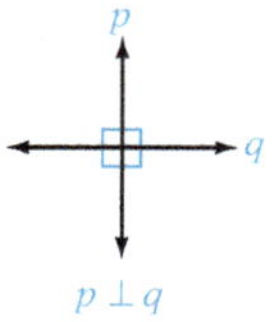

Complementary angles are two angles whose sum is 90°.

$$m\angle A + m\angle B = 35° + 55° = 90°$$

$\angle A$ and $\angle B$ are complementary angles.

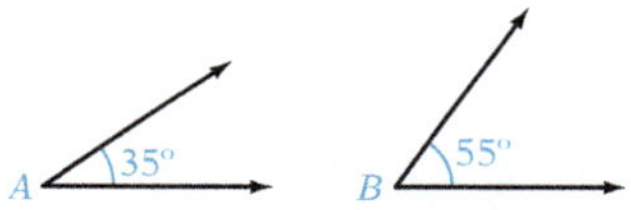

One-half of a revolution is one-half of 360°, or 180°. A 180° angle is called a **straight angle.**

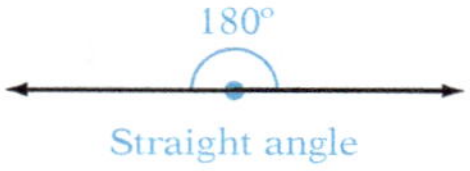

Supplementary angles are two angles whose sum is 180°.

$$m\angle A + m\angle B = 123° + 57° = 180°$$

$\angle A$ and $\angle B$ are supplementary angles.

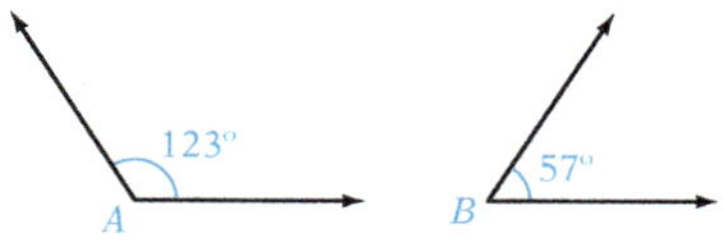

EXAMPLE 1 Find the complement of 39°.

Solution To find the complement of 39°, subtract 39° from 90°.

$90° - 39° = 51°$

51° is the complement of 39°.

Problem 1 Find the complement of 87°.

Solution See page S3.

Try Exercise 13, page 45.

EXAMPLE 2 Find the supplement of 122°.

Solution To find the supplement of 122°, subtract 122° from 180°.

$180° - 122° = 58°$

58° is the supplement of 122°.

Problem 2 Find the supplement of 87°.

Solution See page S3.

Try Exercise 19, page 45.

EXAMPLE 3 For the figure at the right, find $m\angle AOB$.

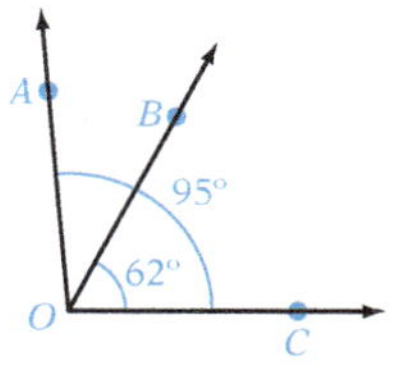

Solution $m\angle AOB$ is the difference between $m\angle AOC$ and $m\angle BOC$.

$m\angle AOB = 95° - 62° = 33°$

$m\angle AOB = 33°$

Problem 3 For the figure at the right, find $m\angle x$.

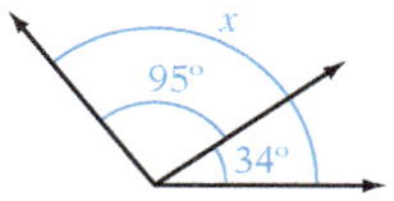

Solution See page S3.

Try Exercise 27, page 46.

OBJECTIVE 2 Perimeter problems

Perimeter is the distance around a plane figure. Perimeter is used in buying fencing for a yard, wood for the frame of a painting, and rain gutters for around a house. The perimeter of a plane figure is the sum of the lengths of the sides of the figure. Formulas for four common geometric figures are given below.

A **triangle** is a three-sided plane figure.

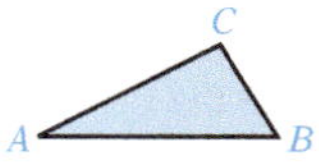

Perimeter = side 1 + side 2 + side 3

An **isosceles triangle** has two sides of the same length. An **equilateral triangle** has all three sides the same length.

A **parallelogram** is a four-sided plane figure with opposite sides parallel.

A **rectangle** is a parallelogram that has four right angles.

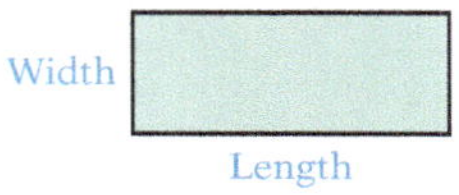

Perimeter = 2 · length + 2 · width

A **square** is a rectangle with four equal sides.

Perimeter = 4 · side

A **circle** is a plane figure in which all points are the same distance from point O, the **center of the circle.** The **diameter** of a circle is a line segment across the circle passing through the center. AB is a diameter of the circle at the right. The **radius** of a circle is a line segment from the center of the circle to a point on the circle. OC is a radius of the circle at the right. The perimeter of a circle is called its **circumference.**

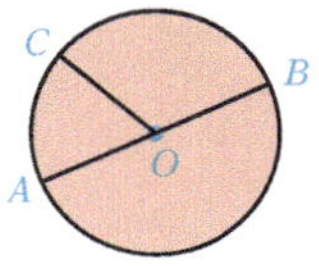

Diameter = 2 · radius or **Radius = $\frac{1}{2}$ · diameter**

Circumference = 2 · π · radius or **Circumference = π · diameter**

where $\pi \approx 3.14$ or $\frac{22}{7}$.

To find the radius of a circle given that the diameter of the circle is 25 cm, use the equation above that gives the radius of a circle in terms of the diameter.

$$\text{Radius} = \frac{1}{2} \cdot \text{diameter}$$
$$= \frac{1}{2} \cdot 25 = 12.5$$

The radius is 12.5 cm.

EXAMPLE 4 Find the perimeter of a rectangle with a width of 6 ft and a length of 18 feet.

Solution

$$\begin{aligned}\text{Perimeter} &= 2 \cdot \text{length} + 2 \cdot \text{width}\\ &= 2 \cdot 18 \text{ ft} + 2 \cdot 6 \text{ ft}\\ &= 36 \text{ ft} + 12 \text{ ft} = 48 \text{ ft}\end{aligned}$$

The perimeter is 48 ft.

Problem 4 Find the perimeter of a square that has a side of length 4.2 m.

Solution See page S3.

Try Exercise 33, page 46.

EXAMPLE 5 Find the circumference of a circle with a radius of 23 cm. Use 3.14 for π.

Solution

$$\begin{aligned}\text{Circumference} &= 2 \cdot \pi \cdot \text{radius}\\ &\approx 2 \cdot 3.14 \cdot 23 \text{ cm}\\ &\approx 144.44 \text{ cm}\end{aligned}$$

The circumference is 144.44 cm.

Problem 5 Find the circumference of a circle with a diameter of 5 in. Use 3.14 for π.

Solution See page S3.

Try Exercise 41, page 46.

EXAMPLE 6 A chain-link fence costs \$6.37 per foot. How much will it cost to fence a rectangular playground that is 108 ft wide and 195 ft long?

Strategy To find the cost of the fence:

- ▶ Find the perimeter of the playground.
- ▶ Multiply the perimeter by the per-foot cost of the fencing.

Solution

$$\begin{aligned}\text{Perimeter} &= 2 \cdot \text{length} + 2 \cdot \text{width}\\ &= 2 \cdot 195 \text{ ft} + 2 \cdot 108 \text{ ft}\\ &= 390 \text{ ft} + 216 \text{ ft} = 606 \text{ ft}\end{aligned}$$

$$\text{Cost} = 606 \times \$6.37 = \$3860.22$$

The cost is \$3860.22.

Problem 6 A metal strip is being installed around a circular table that has a diameter of 36 in. If the per-foot cost of the metal strip is \$3.21, find the cost for the metal strip. Use 3.14 for π. Round to the nearest cent.

Solution See page S3.

Try Exercise 45, page 46.

OBJECTIVE 3

Area problems

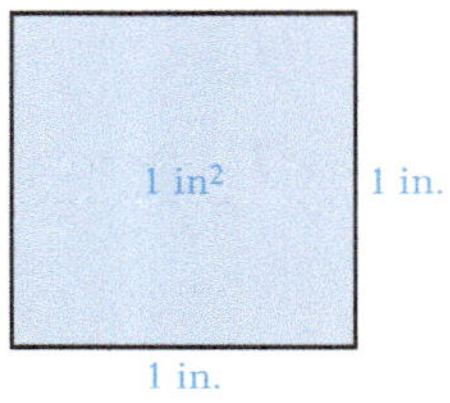

Area is a measure of the amount of surface in a region. Area is used to describe the size of a rug, a farm, a house, or a national park.

Area is measured in square units.

A square that is 1 in. on each side has an area of 1 square inch, which is written 1 in^2.

A square that is 1 cm on each side has an area of 1 square centimeter, which is written 1 cm^2.

Areas of common geometric figures are given by the following formulas.

Rectangle

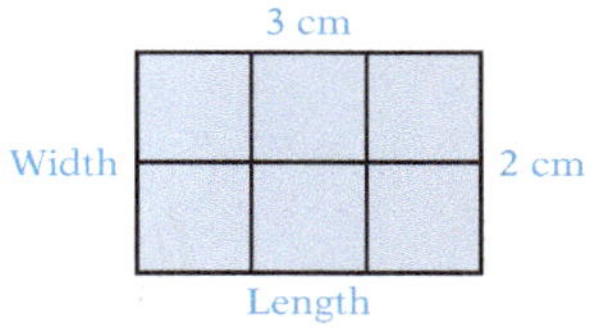

$$\textbf{Area} = \textbf{length} \cdot \textbf{width}$$
$$= 3 \text{ cm} \cdot 2 \text{ cm}$$
$$= 6 \text{ cm}^2$$

Square

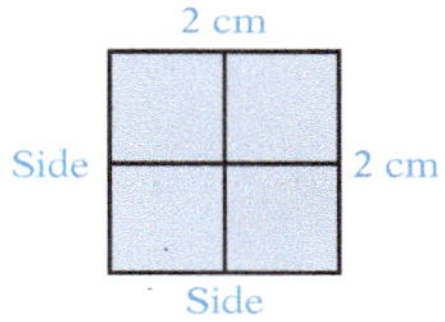

$$\textbf{Area} = \textbf{side} \cdot \textbf{side}$$
$$= 2 \text{ cm} \cdot 2 \text{ cm}$$
$$= 4 \text{ cm}^2$$

Parallelogram

The **base of a parallelogram** is one of the parallel sides. The **height of a parallelogram** is the distance between the base and the opposite parallel side. It is perpendicular to the base.

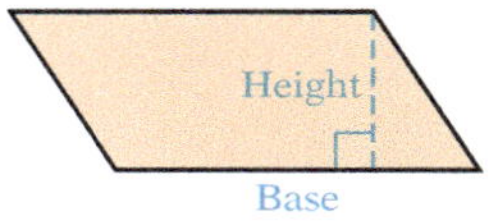

$$\textbf{Area} = \textbf{base} \cdot \textbf{height}$$
$$= 5 \text{ ft} \cdot 4 \text{ ft}$$
$$= 20 \text{ ft}^2$$

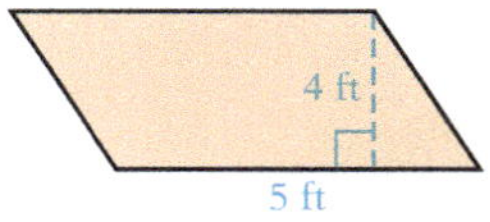

Circle

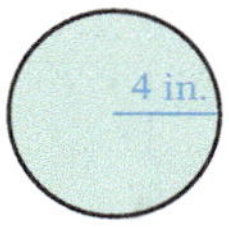

$$\textbf{Area} = \pi(\textbf{radius})^2$$
$$\approx 3.14(4 \text{ in.})^2 = 50.24 \text{ in}^2$$

Take Note

The height of a triangle is always perpendicular to the base. Sometimes it is necessary to extend the base so that a perpendicular line segment can be drawn. The extension is *not* part of the base.

Triangle

For the triangle at the right, the **base of the triangle** is AB; the **height of the triangle** is CD. Note that the height is perpendicular to the base.

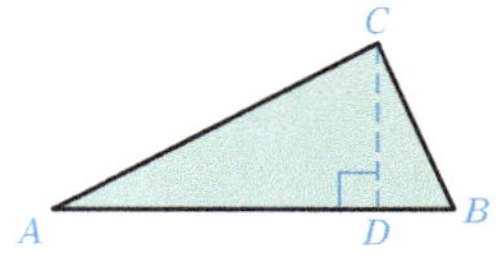

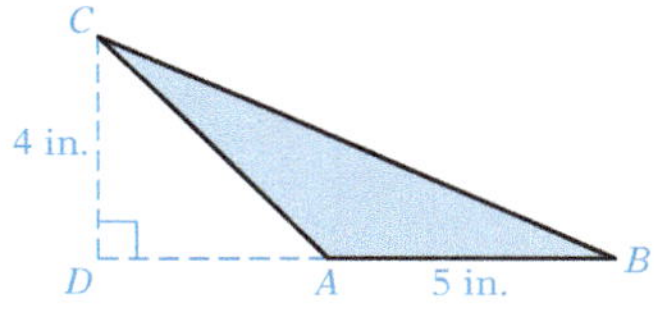

For the triangle at the left,

$$\textbf{Area} = \frac{1}{2} \cdot \textbf{base} \cdot \textbf{height}$$
$$= \frac{1}{2} \cdot 5 \text{ in.} \cdot 4 \text{ in.} = 10 \text{ in}^2$$

EXAMPLE 7 Find the area of a rectangle whose length is 8 in. and whose width is 6 in.

Solution Area = length · width

$= 8 \text{ in.} \cdot 6 \text{ in.} = 48 \text{ in}^2$

The area is 48 in^2.

Problem 7 Find the area of a triangle whose base is 5 ft and whose height is 3 ft.

Solution See page S3.

Try Exercise 51, page 47.

EXAMPLE 8 Find the area of a circle whose diameter is 5 cm. Use 3.14 for π.

Solution

$$\text{Radius} = \frac{1}{2} \cdot \text{diameter}$$

$$= \frac{1}{2} \cdot 5 \text{ cm} = 2.5 \text{ cm}$$

$$\text{Area} = \pi \cdot (\text{radius})^2$$

$$\approx 3.14(2.5 \text{ cm})^2 = 19.625 \text{ cm}^2$$

The area is 19.625 cm^2.

Problem 8 Find the area of a circle whose radius is 6 in. Use 3.14 for π.

Solution See page S3.

Try Exercise 55, page 47.

EXAMPLE 9 Find the area of the parallelogram shown at the right.

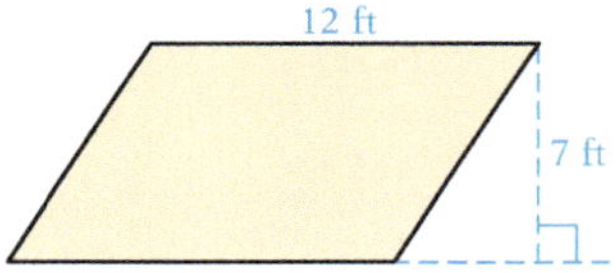

Solution Area = base · height

$= 12 \text{ ft} \cdot 7 \text{ ft} = 84 \text{ ft}^2$

The area is 84 ft^2.

Problem 9 Find the area of the parallelogram shown at the right.

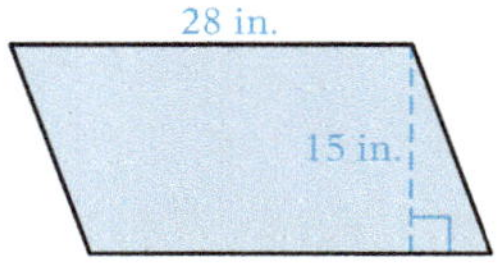

Solution See page S3.

Try Exercise 53, page 47.

EXAMPLE 10 To conserve water during a drought, a city's water department is offering homeowners a rebate on their water bill of \$1.27 per square foot of lawn that is removed from a yard and replaced with drought-resistant plants. What rebate would a homeowner receive who replaced a rectangular lawn area that is 15 ft wide and 25 ft long?

Strategy To find the amount of the rebate:

- ▶ Find the area of the replaced lawn.
- ▶ Multiply the area by the per-square-foot rebate.

Solution Area = length · width

$= 25 \text{ ft} \cdot 15 \text{ ft} = 375 \text{ ft}^2$

Rebate = 375 · \$1.27 = \$476.25

The rebate is \$476.25.

Problem 10 An interior designer is choosing between two hallway rugs. A nylon rug costs \$3.25 per square foot, and a wool rug costs \$5.93 per square foot. If the dimensions of the carpet are to be 4 ft by 15 ft, how much more expensive than the nylon rug will the wool rug be?

Solution See page S3.

Try Exercise 67, page 47.

1.5 Exercises

CONCEPT CHECK

Determine whether the statement is true or false.

1. Perpendicular lines form four 90° angles.

2. The sum of the measures of two straight angles is 360°.

3. Every square is a parallelogram.

4. Perimeter is measured in square units.

5. Is an angle whose measure is 58° less than or greater than a right angle?

6. Is an angle whose measure is 123° less than or greater than a straight angle?

1 Measures of angles (See pages 39–41.)

GETTING READY

7. If $\angle A$ and $\angle B$ are complementary angles, then $m\angle A + m\angle B =$ __?__.

8. If $\angle C$ and $\angle D$ are supplementary angles, then $m\angle C + m\angle D =$ __?__.

9. If $\angle E$ is a right angle, then $m\angle E =$ __?__.

10. If $\angle F$ is a straight angle, then $m\angle F =$ __?__.

11. When the time is 3 o'clock on an analog clock, is the measure of the smaller angle between the hands of the clock more than or less than 120°?

12. In one hour, the hour hand on an analog clock travels through how many degrees?

13. Find the complement of a 62° angle.

14. Find the complement of a 13° angle.

15. Find the supplement of a 48° angle.

16. Find the supplement of a 106° angle.

17. Find the complement of a 7° angle.

18. Find the complement of a 76° angle.

19. Find the supplement of an 89° angle.

20. Find the supplement of a 21° angle.

21. Angle AOB is a straight angle. Find $m\angle AOC$.

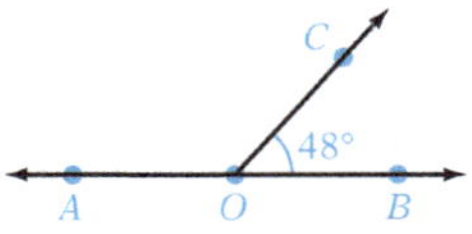

22. Angle AOB is a straight angle. Find $m\angle COB$.

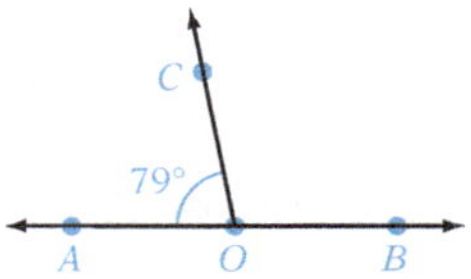

23. Find $m\angle x$.

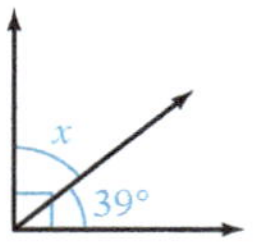

24. Find $m\angle x$.

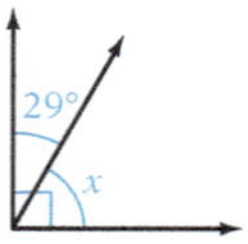

25. Find $m\angle AOB$.

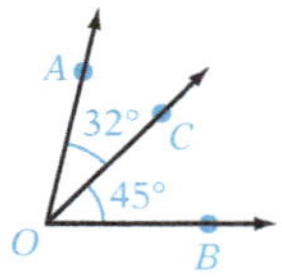

26. Find $m\angle AOB$.

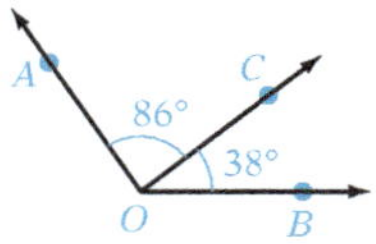

27. Find $m\angle AOC$.

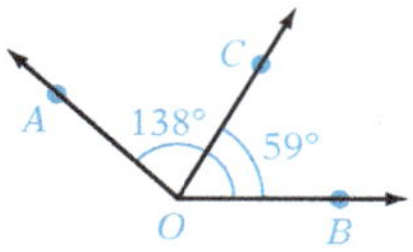

28. Find $m\angle AOC$.

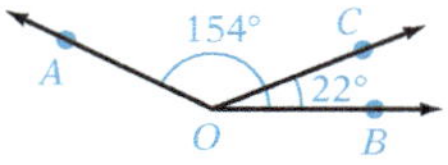

29. Find $m\angle A$.

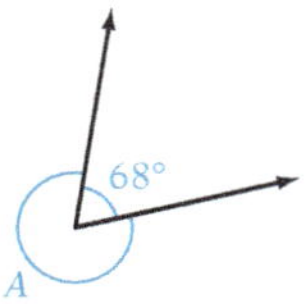

30. Find $m\angle A$.

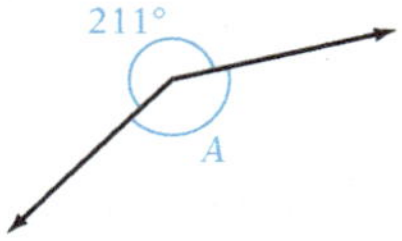

2 Perimeter problems (See pages 41–43.)

GETTING READY

31. A plane figure that has three sides is called a __?__.

32. The name of a parallelogram in which all angles are the same measure is a __?__.

33. Find the perimeter of a triangle with sides 2.51 cm, 4.08 cm, and 3.12 cm.

34. Find the perimeter of a triangle with sides 4 ft 5 in., 5 ft 3 in., and 6 ft 11 in.

35. Find the perimeter of a rectangle whose length is 4 ft 8 in. and whose width is 2 ft 5 in.

36. Find the perimeter of a rectangle whose dimensions are 5 m by 8 m.

37. Find the perimeter of a square whose side is 13 in.

38. Find the perimeter of a square whose side is 34 cm.

39. Find the circumference of a circle whose radius is 21 cm. Use 3.14 for π.

40. Find the circumference of a circle whose radius is 3.4 m. Use 3.14 for π.

41. Find the circumference of a circle whose diameter is 1.2 m. Use 3.14 for π.

42. Find the circumference of a circle whose diameter is 15 in. Use 3.14 for π.

43. Art The wood framing for an art canvas costs $4.81 per foot. How much would the wood framing cost for a rectangular picture that measures 3 ft by 5 ft?

44. Ceramics A decorative mosaic tile is being installed on the border of a square wall behind a stove. If one side of the square measures 5 ft and the cost of installing the mosaic tile is $4.86 per foot, find the cost to install the decorative border.

45. Sewing To prevent fraying, a binding is attached to the outside of a circular rug whose radius is 3 ft. If the binding costs $1.05 per foot, find the cost of the binding. Use 3.14 for π.

46. Landscaping A drip irrigation system is installed around a circular flower garden that is 4 ft in diameter. If the irrigation system costs $2.46 per foot, find the cost to place the irrigation system around the flower garden. Use 3.14 for π.

Royalty Free/Corbis

47. In the pattern at the right, the length of one side of a square is 1 unit. If the pattern were to be continued, what would the perimeter of the eighth figure in the pattern be?

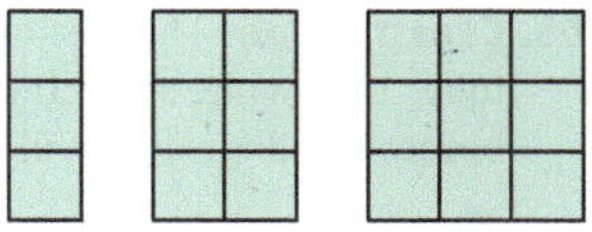

48. Which has the greater perimeter, a square whose side is 1 ft or a rectangle that has a length of 2 in. and a width of 1 in.?

3 Area problems (See pages 43–45.)

GETTING READY

49. The formula for the area of a rectangle is Area = length · __?__.

50. The formula for the area of a parallelogram is Area = base · __?__.

51. Find the area of a rectangle that measures 4 ft by 8 ft.

52. Find the area of a rectangle that measures 3.4 cm by 5.6 cm.

53. Find the area of a parallelogram whose height is 14 cm and whose base is 27 cm.

54. Find the area of a parallelogram whose height is 7 ft and whose base is 18 ft.

55. Find the area of a circle whose radius is 4 in. Use 3.14 for π.

56. Find the area of a circle whose radius is 8.2 m. Use 3.14 for π.

57. Find the area of a square whose side measures 4.1 m.

58. Find the area of a square whose side measures 5 yd.

59. Find the area of a triangle whose height is 7 cm and whose base is 15 cm.

60. Find the area of a triangle whose height is 8 in. and whose base is 13 in.

61. Find the area of a circle whose diameter is 17 in. Use 3.14 for π.

62. Find the area of a circle whose diameter is 3.6 m. Use 3.14 for π.

63. **Interior Design** See the news clipping at the right. What would be the cost of carpeting the entire living space if the cost of the carpet were $36 per square yard?

INDRANIL MUKHERJEE/AFP/Getty Images

In the News

Billion-Dollar Home Built in Mumbai

The world's first billion-dollar home is a 27-story skyscraper in downtown Mumbai, India (formerly known as Bombay). It is 550 ft high with 400,000 square feet of living space.

Source: Forbes.com

64. **Landscaping** A landscape architect recommends 0.1 gal of water per day for each square foot of lawn. How many gallons of water should be used per day on a rectangular lawn area that is 33 ft by 42 ft?

65. **Interior Design** One side of a square room measures 18 ft. How many square yards of carpet are necessary to carpet the room? *Hint*: 1 yd^2 = 9 ft^2.

66. **Carpentry** A circular, inlaid-wood design for a dining table cost $35 per square foot to build. If the radius of the design is 15 in., find the cost to build the design. Use 3.14 for π. Round to the nearest dollar. *Hint*: 144 in^2 = 1 ft^2.

67. **Interior Design** A circular stained glass window cost $48 per square foot to build. If the diameter of the window is 4 ft, find the cost to build the window. Round to the nearest dollar.

68. Construction The cost of plastering the walls of a rectangular room that is 18 ft long, 14 ft wide, and 8 ft high is \$2.56 per square foot. If 125 ft^2 is not plastered because of doors and windows, find the cost to plaster the room.

69. Conservation See the news clipping at the right. The nature reserve in Sankuru is about the size of Massachusetts. Consider Massachusetts a rectangle with a length of 150 mi and a width of 70 mi. Use these dimensions to approximate the area of the reserve in the Congo.

Ronald van der Beek/Shutterstock.com

In the News

Animal Sanctuary Established

The government of the Republic of Congo in Africa has set aside a vast expanse of land in the Sankuru Province to be used as a nature reserve. It will be a sanctuary for elephants; 11 species of primates, including the bonobos; and the okapi, a short-necked relative of the giraffe. Okapis are on the endangered species list.

Source: www.time.com

70. If both the length and the width of a rectangle are doubled, how many times larger is the area of the resulting rectangle?

71. A rectangle has a perimeter of 20 units. What dimensions will result in a rectangle with the greatest possible area? Consider only whole-number dimensions.

APPLYING CONCEPTS

72. Find the perimeter and area of the figure. Use 3.14 for π.

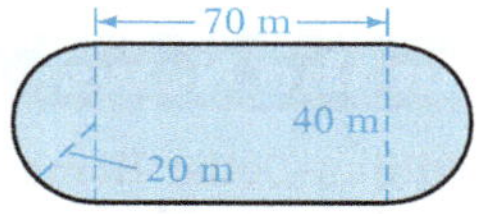

73. Find the perimeter and area of the figure.

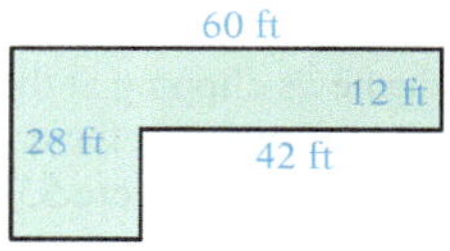

74. Find the outside perimeter and area of the shaded portion of the figure.

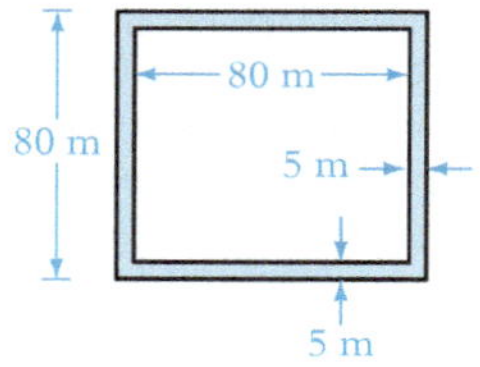

75. Find the area of the shaded portion of the figure.

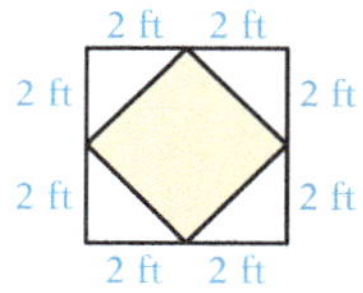

76. A trapezoid is a four-sided plane figure with two parallel sides. The area of a trapezoid is given by Area $= \frac{1}{2} \cdot$ height(base 1 + base 2). See the figure at the right.

a. Find the area of a trapezoid for which base 1 is 5 in., base 2 is 8 in., and the height is 6 in.

b. Find the area of the trapezoid shown at the right.

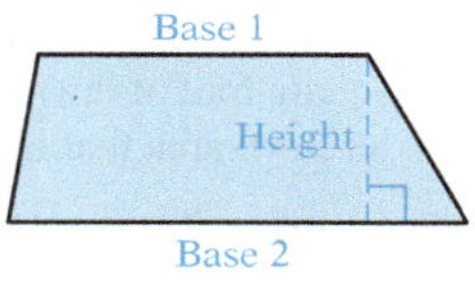

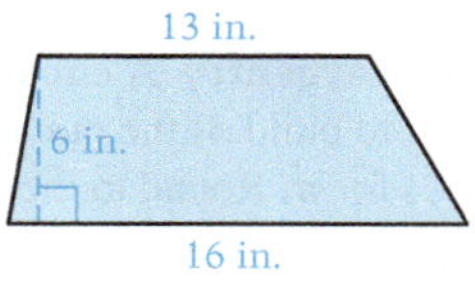

PROJECTS OR GROUP ACTIVITIES

77. Draw parallelogram *ABCD* or one similar to it and then cut it out. Cut along the dotted line to form the shaded triangle. Slide the triangle so that the slanted side corresponds to the slanted side of the parallelogram as shown. Explain how this demonstrates that the area of a parallelogram is the product of the base and the height.

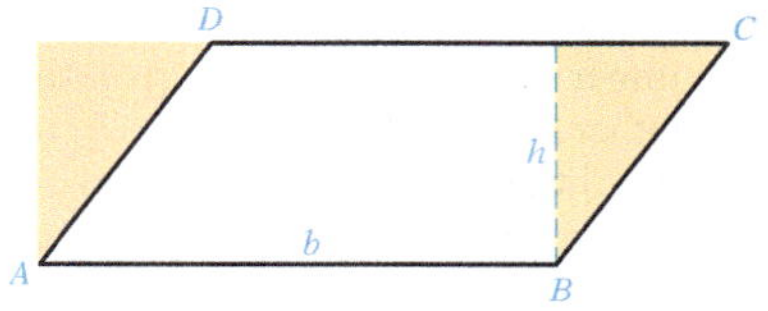

78. For each of the triangles, the base is *AB*. Draw the height.

a.
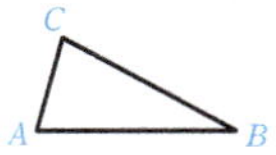

b.
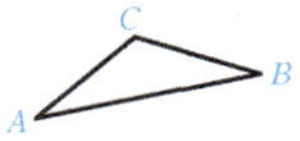

c.
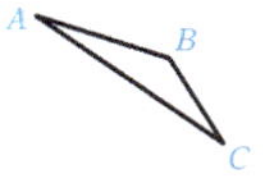

d.
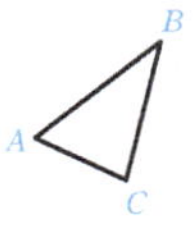

CHAPTER 1 Summary

Key Words	Objective and Page Reference	Examples
A **set** is a collection of objects. The objects in the set are called the **elements** of the set. The **roster method** of writing sets encloses a list of the elements in braces.	[1.1.1, p. 2]*	The set of even natural numbers less than 10 is written $A = \{2, 4, 6, 8\}$. The elements in this set are the numbers 2, 4, 6, and 8.
The set of **natural numbers** is $\{1, 2, 3, 4, 5, 6, 7, ...\}$. The set of **whole numbers** is $\{0, 1, 2, 3, 4, 5, 6, ...\}$. The set of **integers** is $\{..., -4, -3, -2, -1, 0, 1, 2, 3, 4, ...\}$. The integers to the left of zero on the number line are **negative integers.** The integers to the right of zero are **positive integers.**	[1.1.1, p. 2]	
The **graph of an integer** is shown by drawing a heavy dot on the number line directly above the integer.	[1.1.1, p. 2]	The graphs of -2 and 3 are shown on the number line below. 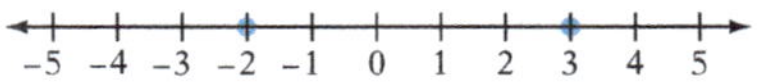

*The numbers in brackets refer to the objective where the Key Word or Essential Rule or Procedure is first introduced. For example, the reference [1.1.1] stands for Chapter 1, Section 1, Objective 1. This notation will be used in all Chapter Summaries throughout the text.

A number a **is less than** another number b, written $a < b$, if a is to the left of b on the number line. A number a **is greater than** another number b, written $a > b$, if a is to the right of b on the number line. The symbol $\leq$ means **is less than or equal to.** The symbol $\geq$ means **is greater than or equal to.**	[1.1.1, p. 3]	$-9 < 7$ $4 > -5$
Two numbers that are the same distance from zero on the number line but on opposite sides of zero are **opposite numbers,** or **opposites.** The opposite of a number is also called its **additive inverse.**	[1.1.2, p. 4]	5 and -5 are opposites.
The **absolute value** of a number is its distance from zero on the number line. The absolute value of a number is positive or zero.	[1.1.2, p. 4]	$\|-12\| = 12$ $-\|7\| = -7$
A **rational number** is a number that can be written in the form $\frac{a}{b}$, where a and b are integers and $b \neq 0$. A rational number written in this form is commonly called a **fraction.** A rational number can also be written as a repeating decimal or a terminating decimal. Numbers that cannot be written as either a repeating decimal or a terminating decimal are called **irrational numbers.** The rational numbers and the irrational numbers taken together are the **real numbers.**	[1.3.1, pp. 21–22]	$\frac{3}{8}$, $\frac{-9}{11}$, and $\frac{4}{1}$ are rational numbers. 2.636363636363..., or $2.\overline{63}$, is a repeating decimal. 3.75 is a terminating decimal. π, $\sqrt{2}$, and 2.17117111711117... are irrational numbers. $\frac{3}{8}$, $\frac{-9}{11}$, $\frac{4}{1}$, π, $\sqrt{2}$, and 2.17117111711117... are real numbers.
When rewriting fractions as equivalent fractions with a common denominator, the **lowest common denominator (LCD)** is the least common multiple of the denominators.	[1.3.3, p. 24]	The LCD of the fractions $\frac{3}{8}$ and $\frac{7}{20}$ is 40.
Percent means "parts of 100."	[1.3.4, p. 26]	63% means 63 of 100 equal parts.
An **exponent** indicates how many times the factor, called the **base,** occurs in the multiplication. For example, $3^4 = 3 \cdot 3 \cdot 3 \cdot 3$. The expression 3^4 is in **exponential form.** $3 \cdot 3 \cdot 3 \cdot 3$ is in **factored form.**	[1.4.1, p. 33]	4^3 is an exponential expression in which 4 is the base and 3 is the exponent. $4^3 = 4 \cdot 4 \cdot 4 = 64$
A **plane** is a flat surface that extends indefinitely. A **line** extends indefinitely in two directions in a plane. A **ray** starts at a point and extends indefinitely in one direction. A **line segment** is part of a line and has two endpoints.	[1.5.1, p. 39]	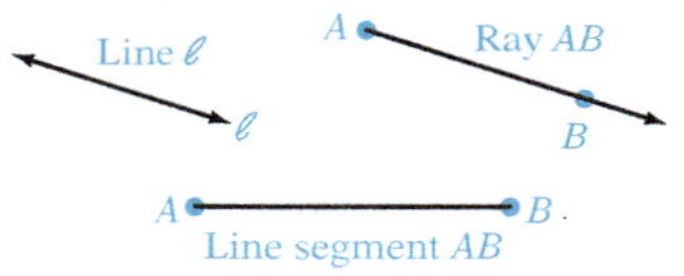
Lines in a plane can be parallel or intersect. **Parallel lines** never meet. The distance between parallel lines in a plane is always the same. **Intersecting lines** cross at a point in the plane.	[1.5.1, p. 40]	

An **angle** is formed when two rays start from the same point. The point at which the rays meet is called the **vertex** of the angle. An angle can be measured in **degrees.** The measure of an angle is symbolized by $m\angle$.

[1.5.1, p. 40]

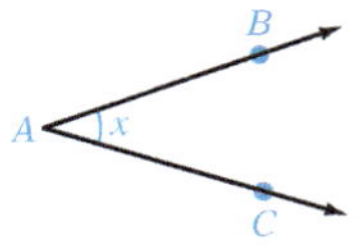

This angle can be named $\angle A$, $\angle BAC$, $\angle CAB$, or $\angle x$.

A **right angle** has a measure of 90°. **Perpendicular lines** are intersecting lines that form right angles. **Complementary angles** are two angles whose sum is 90°. A **straight angle** has a measure of 180°. **Supplementary angles** are two angles whose sum is 180°.

[1.5.1, p. 40]

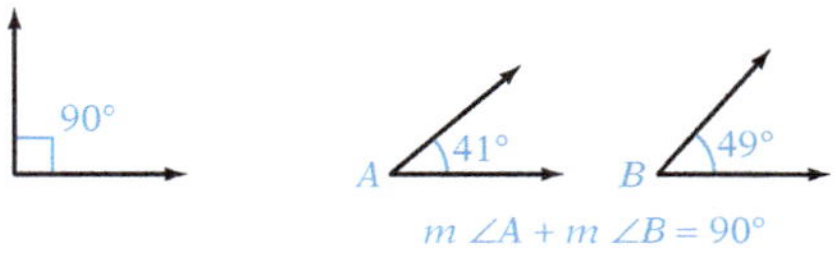

$\angle A$ and $\angle B$ are complementary angles.

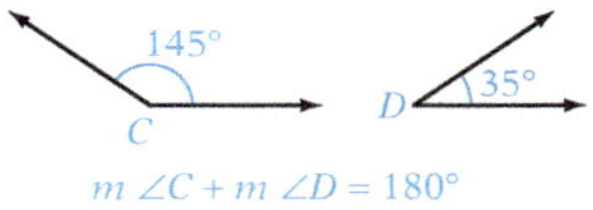

$\angle C$ and $\angle D$ are supplementary angles.

A **circle** is a plane figure in which all points are the same distance from point O, the **center** of the circle. The **diameter** of a circle is a line segment across the circle passing through the center. The **radius** of a circle is a line segment from the center of the circle to a point on the circle. The perimeter of a circle is called its **circumference.**

[1.5.2, p. 41]

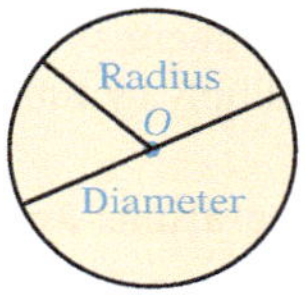

Essential Rules and Procedures	Objective and Page Reference	Examples
To add two integers with the same sign, add the absolute values of the numbers. Then attach the sign of the addends.	[1.2.1, p. 9]	$9 + 3 = 12$ $-9 + (-3) = -12$
To add two integers with different signs, find the absolute value of each number. Then subtract the smaller of these absolute values from the larger one. Attach the sign of the number with the greater absolute value.	[1.2.1, p. 9]	$9 + (-3) = 6$ $-9 + 3 = -6$
To subtract one integer from another, add the opposite of the second integer to the first integer.	[1.2.2, p. 10]	$6 - 11 = 6 + (-11) = -5$ $-2 - (-8) = -2 + 8 = 6$
To multiply two integers with the same sign, multiply the absolute values of the numbers. The product is positive.	[1.2.3, p. 12]	$3(5) = 15$ $(-3)(-5) = 15$
To multiply two integers with different signs, multiply the absolute values of the numbers. The product is negative.	[1.2.3, p. 12]	$-3(5) = -15$ $3(-5) = -15$

To divide two integers with the same sign, divide the absolute values of the numbers. The quotient is positive.	[1.2.4, p. 13]	$14 \div 2 = 7$ $-14 \div (-2) = 7$
To divide two integers with different signs, divide the absolute values of the numbers. The quotient is negative.	[1.2.4, p. 13]	$14 \div (-2) = -7$ $-14 \div 2 = -7$
Zero and One in Division $\frac{0}{a} = 0, a \neq 0$ Division by zero is undefined. $\frac{a}{a} = 1, a \neq 0$ $\frac{a}{1} = a$	[1.2.4, p. 13]	$\frac{0}{3} = 0$ $\frac{9}{0}$ is undefined. $\frac{-2}{-2} = 1$ $\frac{-6}{1} = -6$
To convert a percent to a decimal, remove the percent sign and multiply by 0.01.	[1.3.4, p. 26]	$85\% = 85(0.01) = 0.85$
To convert a percent to a fraction, remove the percent sign and multiply by $\frac{1}{100}$.	[1.3.4, p. 26]	$25\% = 25 \cdot \frac{1}{100} = \frac{25}{100} = \frac{1}{4}$
To convert a decimal or a fraction to a percent, multiply by 100%.	[1.3.4, p. 27]	$0.5 = 0.5(100\%) = 50\%$ $\frac{3}{8} = \frac{3}{8} \cdot 100\% = \frac{300}{8}\% = 37.5\%$
Order of Operations Agreement **Step 1** Perform operations inside grouping symbols. **Step 2** Simplify exponential expressions. **Step 3** Do multiplication and division as they occur from left to right. **Step 4** Do addition and subtraction as they occur from left to right.	[1.4.2, p. 35]	$16 \div (-2)^3 + (7 - 12)$ $= 16 \div (-2)^3 + (-5)$ $= 16 \div (-8) + (-5)$ $= -2 + (-5)$ $= -7$
Diameter $= 2 \cdot$ radius **Radius** $= \frac{1}{2} \cdot$ diameter	[1.5.2, p. 42]	Find the diameter of a circle whose radius is 10 in. Diameter $= 2 \cdot$ radius $= 2(10 \text{ in.}) = 20$ in.
Perimeter is the distance around a plane figure. **Triangle:** Perimeter = side 1 + side 2 + side 3 **Rectangle:** Perimeter $= 2 \cdot$ length $+ 2 \cdot$ width **Square:** Perimeter $= 4 \cdot$ side **Circle:** Circumference $= 2 \cdot \pi \cdot$ radius Circumference $= \pi \cdot$ diameter	[1.5.2, p. 41]	Find the perimeter of a rectangle whose width is 12 m and whose length is 15 m. Perimeter $= 2 \cdot 15 \text{ m} + 2 \cdot 12 \text{ m} = 54$ m Find the circumference of a circle whose radius is 3 in. Use 3.14 for π. Circumference $= 2 \cdot \pi \cdot 3 \text{ in.} \approx 18.84$ in.
Area is a measure of the amount of surface in a region. **Triangle:** Area $= \frac{1}{2} \cdot$ base $\cdot$ height **Rectangle:** Area = length $\cdot$ width **Square:** Area = side $\cdot$ side **Parallelogram:** Area = base $\cdot$ height **Circle:** Area $= \pi(\text{radius})^2$	[1.5.3, p. 43]	Find the area of a triangle whose base is 13 m and whose height is 11 m. Area $= \frac{1}{2} \cdot 13 \text{ m} \cdot 11 \text{ m} = 71.5 \text{ m}^2$ Find the area of a circle whose radius is 9 cm. Area $= \pi \cdot (9 \text{ cm})^2 \approx 254.34 \text{ cm}^2$

CHAPTER 1 Review Exercises

1. Use the roster method to write the set of natural numbers less than 7.

2. Write $\frac{5}{8}$ as a percent.

3. Evaluate: $-|-4|$

4. Subtract: $16 - (-30) - 42$

5. Find the area of a triangle whose base is 4 cm and whose height is 9 cm.

6. Write $\frac{7}{9}$ as a decimal. Place a bar over the repeating digits of the decimal.

7. Simplify: $(6.02)(-0.89)$

8. Simplify: $\dfrac{-10 + 2}{2 + (-4)} \div 2 + 6$

9. Find the opposite of -4.

10. Subtract: $16 - 30$

11. Write 0.672 as a percent.

12. Write $79\frac{1}{2}\%$ as a fraction.

13. Divide: $-72 \div 8$

14. Write $\frac{17}{20}$ as a decimal.

15. Divide: $\dfrac{5}{12} \div \left(-\dfrac{5}{6}\right)$

16. Simplify: $3^2 - 4 + 20 \div 5$

17. Find $m\angle AOB$ for the figure at the right.

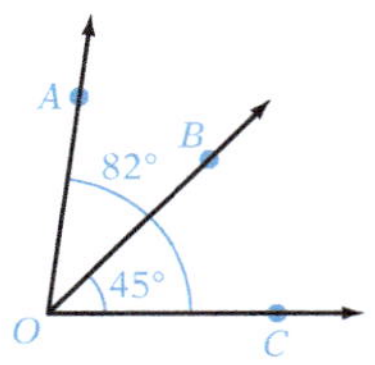

18. Add: $-22 + 14 + (-8)$

19. Multiply: $(-5)(-6)(3)$

20. Subtract: $6.039 - 12.92$

21. Given $A = \{-5, -3, 0\}$, which elements of set A are less than or equal to -3?

22. Write 7% as a decimal.

23. Evaluate: $\dfrac{3}{4} \cdot (4)^2$

24. Place the correct symbol, < or > , between the two numbers.

$-2 \quad -40$

25. Add: $13 + (-16)$

26. Find the complement of a 56° angle.

27. Add: $-\dfrac{2}{5} + \dfrac{7}{15}$

28. Evaluate: $(-3^3) \cdot 2^2$

29. Write $2\frac{7}{9}$ as a percent. Round to the nearest tenth of a percent.

30. Write 240% as a decimal.

31. Find the supplement of a 28° angle.

32. Divide: $96 \div (-12)$

33. Find the area of a circle whose diameter is 6 m. Use 3.14 for π.

34. Simplify: $2^3 \div 4 - 2(2 - 7)$

35. Evaluate: $|-3|$

36. Write $1\frac{2}{3}$ as a percent. Write the remainder in fractional form.

37. Given $C = \{-12, -8, -1, 7\}$, find
a. the opposite of each element of set C.
b. the absolute value of each element of set C.

38. Write $\frac{7}{11}$ as a decimal. Place a bar over the repeating digits of the decimal.

39. Divide: $0.2654 \div (-0.023)$. Round to the nearest tenth.

40. Simplify: $(7 - 2)^2 - 5 - 3 \cdot 4$

41. Add: $-12 + 8 + (-4)$

42. Add: $-\frac{5}{8} + \frac{1}{6}$

43. Given $D = \{-24, -17, -9, 0, 4\}$, which elements of set D are greater than -19?

44. Write 0.002 as a percent.

45. Evaluate: $-4^2 \cdot \left(\frac{1}{2}\right)^2$

46. Add: $-1.329 + 4.89$

47. Evaluate: $-|17|$

48. Subtract: $-5 - 22 - (-13) - 19 - (-6)$

49. Multiply: $\left(\frac{1}{3}\right)\left(-\frac{4}{5}\right)\left(\frac{3}{8}\right)$

50. Place the correct symbol, < or >, between the two numbers.

$$-43 \quad -34$$

51. Find the perimeter of a rectangle whose length is 12 in. and whose width is 10 in.

52. Evaluate: $(-2)^3 \cdot 4^2$

53. Write 0.075 as a percent.

54. Write $\frac{19}{35}$ as a percent. Write the remainder in fractional form.

55. Add: $14 + (-18) + 6 + (-20)$

56. Multiply: $-4(-8)(12)(0)$

57. Simplify: $2^3 - 7 + 16 \div (-3 + 5)$

58. Simplify: $\frac{3}{4} + \frac{1}{2} - \frac{3}{8}$

59. Divide: $-128 \div (-8)$

60. Place the correct symbol, < or >, between the two numbers.

$$-57 \quad 28$$

61. Evaluate: $\left(-\frac{1}{3}\right)^3 \cdot 9^2$

62. Add: $-7 + (-3) + (-12) + 16$

63. Multiply: $5(-2)(10)(-3)$

64. Use the roster method to write the set of negative integers greater than -4.

65. Landscaping A landscape company is proposing to replace a rectangular flower bed that measures 8 ft by 12 ft with sod that costs \$2.51 per square foot. Find the cost to replace the flower bed with the sod.

66. Temperature Find the temperature after a rise of 14°C from −6°C.

67. Temperature The daily low temperatures, in degrees Celsius, for a three-day period were recorded as follows: −8°, 7°, −5°. Find the average low temperature for the three-day period.

68. Temperature Use the table to find the difference between the record high temperature and the record low temperature for January in Bismarck, North Dakota.

Bismarck, North Dakota

	Jan	Feb	Mar	Apr	May	Jun	Jul	Aug	Sep	Oct	Nov	Dec
Record High	63°F (2002)	69°F (1992)	81°F (2007)	93°F (1992)	102°F (1934)	111°F (2002)	114°F (1936)	109°F (1941)	105°F (1959)	95°F (1963)	79°F (1999)	66°F (1939)
Record Low	−45°F (1916)	−45°F (1936)	−36°F (1897)	−12°F (1975)	13°F (1907)	30°F (1969)	32°F (1884)	32°F (1911)	10°F (1876)	−10°F (1991)	−30°F (1985)	−43°F (1967)

Source: www.weather.com

69. Temperature Find the temperature after a rise of 7°C from −13°C.

70. Temperature The temperature on the surface of the planet Venus is 480°C. The temperature on the surface of the dwarf planet Pluto is −234°C. Find the difference between the surface temperatures on Venus and Pluto.

CHAPTER 1 Test

1. Write 55% as a fraction.

2. Given $B = \{-8, -6, -4, -2\}$, which elements of set B are less than -5?

3. Find $m\angle x$ for the figure at the right.

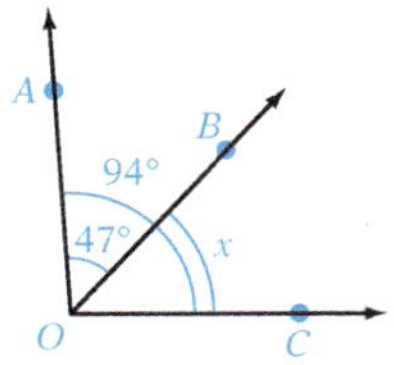

4. Write $\frac{3}{20}$ as a decimal.

5. Multiply: $\frac{3}{4}\left(-\frac{2}{21}\right)$

6. Divide: $-75 \div 5$

7. Evaluate: $\left(-\frac{2}{3}\right)^3 \cdot 3^2$

8. Add: $-7 + (-3) + 12$

9. Use the roster method to write the set of positive integers less than or equal to 6.

10. Write 1.59 as a percent.

11. Evaluate: $|-29|$

12. Place the correct symbol, $<$ or $>$, between the two numbers.

$-47 \quad -68$

13. Subtract: $-\frac{4}{9} - \frac{5}{6}$

14. Find the area of a parallelogram whose base is 10 cm and whose height is 9 cm.

15. Simplify: $8 + \frac{12 - 4}{3^2 - 1} - 6$

16. Divide: $-\frac{5}{8} \div \left(-\frac{3}{4}\right)$

17. Write $\frac{3}{13}$ as a percent. Round to the nearest tenth of a percent.

18. Write 6.2% as a decimal.

19. Subtract: $13 - (-5) - 4$

20. Write $\frac{13}{30}$ as a decimal. Place a bar over the repeating digits of the decimal.

21. Multiply: $(-0.9)(2.7)$

22. Find the complement of a 28° angle.

23. Evaluate: $2^2 \cdot (-4)^2 \cdot 10$

24. Evaluate: $-|-34|$

25. Find the circumference of a circle whose diameter is 27 in. Use 3.14 for π.

26. Given $A = \{-17, -6, 5, 9\}$, find
    a. the opposite of each element of set A.
    b. the absolute value of each element of set A.

27. Write $\frac{16}{23}$ as a percent. Write the remainder in fractional form.

28. Add: $-18.354 + 6.97$

29. Multiply: $-4(8)(-5)$

30. Simplify: $9(-4) \div [2(8 - 5)^2]$

31. **Temperature** Find the temperature after a rise of 12°C from −8°C.

32. **Temperature** Use the table to find the average of the record low temperatures in Fairbanks, Alaska, for the first four months of the year.

Marcel Jancovic/Shutterstock.com

Fairbanks, Alaska

	Jan	Feb	Mar	Apr	May	Jun	Jul	Aug	Sep	Oct	Nov	Dec
Record High	50°F (1981)	47°F (1987)	56°F (1994)	74°F (1960)	89°F (1960)	96°F (1969)	94°F (1975)	93°F (1994)	84°F (1957)	65°F (1969)	49°F (1997)	44°F (1985)
Record Low	−61°F (1969)	−58°F (1993)	−49°F (1956)	−24°F (1986)	−1°F (1964)	30°F (1950)	35°F (1959)	27°F (1987)	3°F (1992)	−27°F (1975)	−46°F (1990)	−62°F (1961)

Source: www.weather.com

33. **Recreation** The recreation department for a city is enclosing a rectangular playground that measures 150 ft by 200 ft with new fencing that costs $6.52 per foot. Find the cost of the new fencing.

CHAPTER 2

Variable Expressions

Digital Vision

Focus on Success

Have you formed or are you part of a study group? Remember that a study group can be a great way to stay focused on succeeding in this course. You can support each other, get help and offer help on homework, and prepare for tests together. (See Homework Time, page AIM-5.)

OBJECTIVES

2.1
1. Evaluate variable expressions

2.2
1. The Properties of the Real Numbers
2. Simplify variable expressions using the Properties of Addition
3. Simplify variable expressions using the Properties of Multiplication
4. Simplify variable expressions using the Distributive Property
5. Simplify general variable expressions

2.3
1. Translate a verbal expression into a variable expression
2. Translate a verbal expression into a variable expression and then simplify the resulting expression
3. Translate application problems

PREP TEST

Are you ready to succeed in this chapter?
Take the Prep Test below to find out if you are ready to learn the new material.

1. Subtract: $-12 - (-15)$
2. Divide: $-36 \div (-9)$
3. Add: $-\frac{3}{4} + \frac{5}{6}$
4. What is the reciprocal of $-\frac{9}{4}$?
5. Divide: $\left(-\frac{3}{4}\right) \div \left(-\frac{5}{2}\right)$
6. Evaluate: -2^4
7. Evaluate: $\left(\frac{2}{3}\right)^3$
8. Evaluate: $3 \cdot 4^2$
9. Evaluate: $7 - 2 \cdot 3$
10. Evaluate: $5 - 7(3 - 2^2)$

2.1 Evaluating Variable Expressions

OBJECTIVE 1

Evaluate variable expressions

> **Point of Interest**
>
> Historical manuscripts indicate that mathematics is at least 4000 years old. Yet it was only 400 years ago that mathematicians started using variables to stand for numbers. The idea that a letter can stand for some number was a critical turning point in mathematics.

Often we discuss a quantity without knowing its exact value, such as the price of gold next month, the cost of a new automobile next year, or the tuition for next semester. In algebra, a letter of the alphabet is used to stand for a quantity that is unknown or one that can change, or *vary*. The letter is called a **variable.** An expression that contains one or more variables is called a **variable expression.**

A variable expression is shown at the right. The expression can be rewritten by writing subtraction as the addition of the opposite.

$3x^2 - 5y + 2xy - x - 7$

$3x^2 + (-5y) + 2xy + (-x) + (-7)$

Note that the expression has five addends. The **terms** of a variable expression are the addends of the expression. The expression has five terms.

$$\overbrace{\underbrace{3x^2 \quad -5y \quad +2xy \quad -x}_{\text{Variable terms}} \quad \underbrace{-7}_{\text{Constant term}}}^{\text{5 terms}}$$

The terms $3x^2$, $-5y$, $2xy$, and $-x$ are **variable terms.**

The term -7 is a **constant term,** or simply a **constant.**

Each variable term is composed of a **numerical coefficient** and a **variable part** (the variable or variables and their exponents).

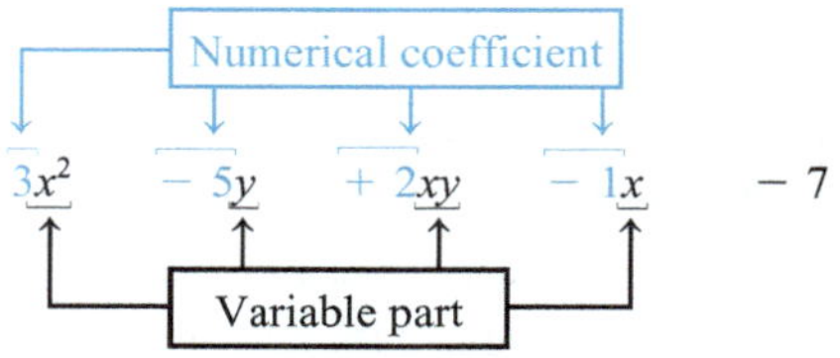

> **How It's Used**
>
> Type "convert" into an Internet browser's search box and you may get suggestions such as "convert kilograms to pounds," "convert euros to dollars," or "convert Celsius to Fahrenheit." Websites that do conversions evaluate variable expressions that are built into them. A site that converts kilograms to pounds uses a variable expression similar to 2.2*K* and evaluates it for a value of *K* that you enter.

When the numerical coefficient is 1 or -1, the 1 is usually not written ($x = 1x$ and $-x = -1x$).

EXAMPLE 1 Name the variable terms of the expression $2a^2 - 5a + 7$.

Solution $2a^2, -5a$

Problem 1 Name the constant term of the expression $6n^2 + 3n - 4$.

Solution See page S3.

Try Exercise 9, page 60.

Replacing the variable or variables in a variable expression with numbers and then simplifying the resulting numerical expression is called **evaluating the variable expression.**

EXAMPLE 2 Evaluate $ab - b^2$ when $a = 2$ and $b = -3$.

Solution

$ab - b^2$

$2(-3) - (-3)^2$ • **Replace each variable in the expression with the number it represents.**

$= 2(-3) - 9$ • **Use the Order of Operations Agreement to simplify the resulting numerical expression.**

$= -6 - 9$

$= -15$

Problem 2 Evaluate $2xy + y^2$ when $x = -4$ and $y = 2$.

Solution See page S3.

Try Exercise 35, page 61.

EXAMPLE 3 Evaluate $\frac{a^2 - b^2}{a - b}$ when $a = 3$ and $b = -4$.

Solution

$\frac{a^2 - b^2}{a - b}$

$\frac{(3)^2 - (-4)^2}{3 - (-4)}$ • **Replace each variable in the expression with the number it represents.**

$= \frac{9 - 16}{3 - (-4)}$ • **Use the Order of Operations Agreement to simplify the resulting numerical expression.**

$= \frac{-7}{7} = -1$

Problem 3 Evaluate $\frac{a^2 + b^2}{a + b}$ when $a = 5$ and $b = -3$.

Solution See page S3.

Try Exercise 37, page 61.

EXAMPLE 4 Evaluate $x^2 - 3(x - y) - z^2$ when $x = 2$, $y = -1$, and $z = 3$.

Solution

$x^2 - 3(x - y) - z^2$

$(2)^2 - 3[2 - (-1)] - (3)^2$ • **Replace each variable in the expression with the number it represents.**

$= (2)^2 - 3(3) - (3)^2$ • **Use the Order of Operations Agreement to simplify the resulting numerical expression.**

$= 4 - 3(3) - 9$

$= 4 - 9 - 9$

$= -5 - 9 = -14$

Problem 4 Evaluate $x^3 - 2(x + y) + z^2$ when $x = 2$, $y = -4$, and $z = -3$.

Solution See page S3.

Try Exercise 47, page 61.

Take Note

For your reference, geometric formulas are printed at the back of this textbook.

EXAMPLE 5 The diameter of the base of a right circular cylinder is 5 cm. The height of the cylinder is 8.5 cm. Find the volume of the cylinder. Round to the nearest tenth.

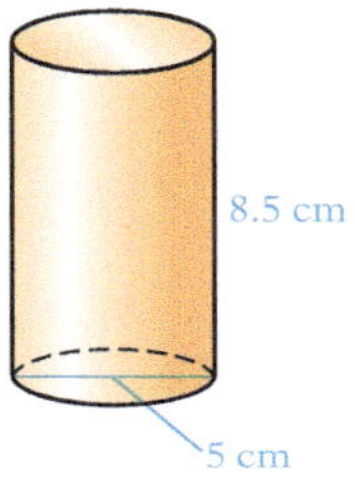

Solution

$V = \pi r^2 h$ • **Use the formula for the volume of a right circular cylinder.**

$V = \pi(2.5)^2(8.5)$ • $r = \frac{1}{2}d = \frac{1}{2}(5) = 2.5$

$V = \pi(6.25)(8.5)$ • **Use the π key on your calculator to enter the value of π.**

$V \approx 166.9$

The volume is approximately 166.9 cm^3.

Problem 5 The diameter of the base of a right circular cone is 9 cm. The height of the cone is 9.5 cm. Find the volume of the cone. Round to the nearest tenth.

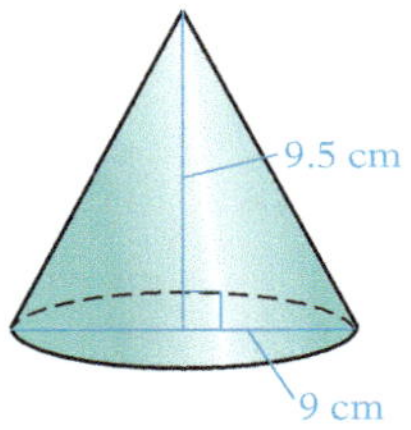

Solution See page S4.

Try Exercise 73, page 62.

A graphing calculator can be used to evaluate variable expressions. When the value of each variable is stored in the calculator's memory and a variable expression is then entered into the calculator, the calculator evaluates that variable expression for the values of the variables stored in its memory. See the Appendix for a description of keystroking procedures.

2.1 Exercises

CONCEPT CHECK

Determine whether the statement is always true, sometimes true, or never true.

1. The expression $3x^2$ is a variable expression.
2. In the expression $8y^3 - 4y$, the terms are $8y^3$ and $4y$.
3. For the expression x^5, the value of x is 1.
4. The Order of Operations Agreement is used in evaluating a variable expression.
5. The result of evaluating a variable expression is a single number.

1 Evaluate variable expressions (See pages 58–60.)

GETTING READY

6. To identify the terms of the variable expression $3x^2 - 4x - 7$, first write subtraction as addition of the opposite: $3x^2 + (\underline{\ ?\ }) + (\underline{\ ?\ })$. The terms of $3x^2 - 4x - 7$ are $\underline{\ ?\ }$, $\underline{\ ?\ }$, and $\underline{\ ?\ }$.

7. Evaluate $mn^2 - m$ when $m = -2$ and $n = 5$.

$mn^2 - m = (-2)(5)^2 - (-2)$	• **Replace m with $\underline{\ ?\ }$ and n with $\underline{\ ?\ }$.**
$= (-2)(\underline{\ ?\ }) - (-2)$	• **Simplify the exponential expression.**
$= \underline{\ ?\ } - (-2)$	• **Multiply.**
$= -50 + \underline{\ ?\ }$	• **Write subtraction as addition of the opposite.**
$= \underline{\ ?\ }$.	• **Add.**

Name the terms of the variable expression. Then underline the constant term.

8. $2x^2 + 5x - 8$
9. $-3n^2 - 4n + 7$
10. $6 - a^4$

Name the variable terms of the expression. Then underline the variable part of each term.

11. $9b^2 - 4ab + a^2$ **12.** $7x^2y + 6xy^2 + 10$ **13.** $5 - 8n - 3n^2$

Name the coefficients of the variable terms.

14. $x^2 - 9x + 2$ **15.** $12a^2 - 8ab - b^2$ **16.** $n^3 - 4n^2 - n + 9$

17. What is the meaning of the phrase "evaluate a variable expression"?

18. What is the difference between the meaning of "the value of the variable" and the meaning of "the value of the variable expression"?

Evaluate the variable expression when $a = 2$, $b = 3$, and $c = -4$.

19. $3a + 2b$ **20.** $a - 2c$ **21.** $-a^2$

22. $2c^2$ **23.** $-3a + 4b$ **24.** $3b - 3c$

25. $b^2 - 3$ **26.** $-3c + 4$ **27.** $16 \div (2c)$

28. $6b \div (-a)$ **29.** $bc \div (2a)$ **30.** $-2ab \div c$

31. $a^2 - b^2$ **32.** $b^2 - c^2$ **33.** $(a + b)^2$

34. $b^2 - 4ac$ **35.** $2a - (c + a)^2$ **36.** $(b - a)^2 + 4c$

37. $b^2 - \frac{ac}{8}$ **38.** $\frac{5ab}{6} - 3cb$ **39.** $(b - 2a)^2 + bc$

Evaluate the variable expression when $a = -2$, $b = 4$, $c = -1$, and $d = 3$.

40. $\frac{b + c}{d}$ **41.** $\frac{d - b}{c}$ **42.** $\frac{2d + b}{-a}$

43. $\frac{b + 2d}{b}$ **44.** $\frac{b - d}{c - a}$ **45.** $\frac{2c - d}{-ad}$

46. $(b + d)^2 - 4a$ **47.** $(d - a)^2 - 3c$ **48.** $(d - a)^2 \div 5$

49. $(b - c)^2 \div 5$ **50.** $b^2 - 2b + 4$ **51.** $a^2 - 5a - 6$

52. $\frac{bd}{a} \div c$ **53.** $\frac{2ac}{b} \div (-c)$ **54.** $2(b + c) - 2a$

55. $3(b - a) - bc$ **56.** $\frac{b - 2a}{bc^2 - d}$ **57.** $\frac{b^2 - a}{ad + 3c}$

58. $\frac{1}{3}d^2 - \frac{3}{8}b^2$ **59.** $\frac{5}{8}a^4 - c^2$ **60.** $\frac{-4bc}{2a - b}$

61. $\frac{abc}{b - d}$ **62.** $a^3 - 3a^2 + a$ **63.** $d^3 - 3d - 9$

64. $3dc - (4c)^2$

Evaluate the variable expression when $a = 2.7$, $b = -1.6$, and $c = -0.8$.

65. $c^2 - ab$ **66.** $(a + b)^2 - c$ **67.** $\frac{b^3}{c} - 4a$

Complete Exercises 68 and 69 without using a calculator.

68. If $\frac{b+c}{abc}$ is evaluated when $a = -25$, $b = 67$, and $c = -82$, will the result be positive or negative?

69. A sphere has a radius of 2 cm.

a. Can the exact volume of the sphere be $V\ \text{cm}^3$, where V is a whole number?

b. Can the exact volume of the sphere be $V\ \text{cm}^2$, where V is an irrational number?

Solve. Round to the nearest tenth.

70. Geometry Find the volume of a sphere that has a radius of 8.5 cm.

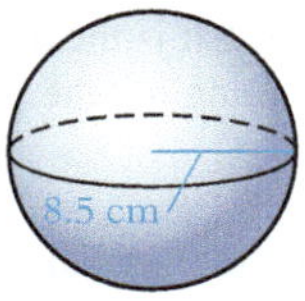

71. Geometry Find the volume of a right circular cylinder that has a radius of 1.25 in. and a height of 5.25 in.

72. Geometry The radius of the base of a right circular cylinder is 3.75 ft. The height of the cylinder is 9.5 ft. Find the surface area of the cylinder.

73. Geometry The length of one base of a trapezoid is 17.5 cm, and the length of the other base is 10.25 cm. The height is 6.75 cm. What is the area of the trapezoid?

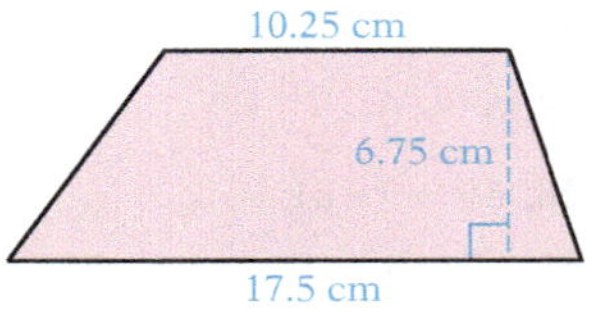

74. Geometry A right circular cone has a height of 2.75 in. The diameter of the base is 1 in. Find the volume of the cone.

75. Geometry A right circular cylinder has a height of 12.6 m. The diameter of the base is 7 m. Find the volume of the cylinder.

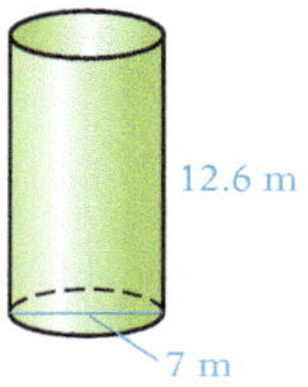

76. The value of z is the value of $a^2 - 2a$ when $a = -3$. Find the value of z^2.

77. The value of a is the value of $3x^2 - 4x - 5$ when $x = -2$. Find the value of $3a - 4$.

78. The value of c is the value of $a^2 + b^2$ when $a = 2$ and $b = -2$. Find the value of $c^2 - 4$.

79. The value of d is the value of $3w^2 - 2v$ when $w = -1$ and $v = 3$. Find the value of $d^2 - 4d$.

APPLYING CONCEPTS

Evaluate the variable expression when $a = -2$ and $b = -3$.

80. $|2a + 3b|$

81. $|-4ab|$

82. $|5a - b|$

Evaluate the following expressions for $x = 2$, $y = 3$, and $z = -2$.

83. $3^x - x^3$

84. $2^y - y^2$

85. z^y

86. z^x

87. $x^x - y^y$

88. $y^{(x^2)}$

PROJECTS OR GROUP ACTIVITIES

89. For each of the following, determine the first natural number x, greater than 2, for which the second expression is larger than the first. On the basis of your answers, make a conjecture that appears to be true about the expressions x^n and n^x, where $n = 3, 4, 5, 6, 7, \ldots$ and x is a natural number greater than 2.

a. $x^3, 3^x$

b. $x^4, 4^x$

c. $x^5, 5^x$

d. $x^6, 6^x$

Complete Exercises 90 to 93. After each exercise, evaluate the variable expression for a value of the variable that you provide.

90. A community college charges each student a \$100 student activity fee, which is added to the student's cost of tuition. Let T represent the tuition. Write a variable expression that represents the final bill after the student activity fee has been added.

91. The instructor of a sociology class is grading the first exam on a curve. The recorded grade on each exam is calculated by adding 8 points to the earned grade on the exam. Let G represent the earned grade on an exam. Write a variable expression that represents the recorded exam grade.

92. The varsity basketball coach at a high school must order jerseys for the team. The price of one jersey is \$34. Let N represent the number of players on the basketball team. Write a variable expression that represents the total cost for jerseys for the team.

93. You have a niece who is 16 years younger than you are. Let A represent your age. Write a variable expression that represents your niece's age.

2.2 Simplifying Variable Expressions

OBJECTIVE 1

The Properties of the Real Numbers

The Properties of the Real Numbers describe the ways operations on numbers can be performed. Here are some of the Properties of the Real Numbers described algebraically and in words. An example of each is provided.

THE COMMUTATIVE PROPERTY OF ADDITION

If a and b are real numbers, then $a + b = b + a$.

Two terms can be added in either order; the sum is the same.

EXAMPLE

$4 + 3 = 7$ and $3 + 4 = 7$

THE COMMUTATIVE PROPERTY OF MULTIPLICATION

If a and b are real numbers, then $a \cdot b = b \cdot a$.

Two factors can be multiplied in either order; the product is the same.

EXAMPLE

$(5)(-2) = -10$ and $(-2)(5) = -10$

THE ASSOCIATIVE PROPERTY OF ADDITION

If a, b, and c are real numbers, then $(a + b) + c = a + (b + c)$.

When three or more terms are added, the terms can be grouped (with parentheses, for example) in any order; the sum is the same.

EXAMPLE

$2 + (3 + 4) = 2 + 7 = 9$ and $(2 + 3) + 4 = 5 + 4 = 9$

THE ASSOCIATIVE PROPERTY OF MULTIPLICATION

If a, b, and c are real numbers, then $(a \cdot b) \cdot c = a \cdot (b \cdot c)$.

When three or more factors are multiplied, the factors can be grouped in any order; the product is the same.

EXAMPLE

$(-3 \cdot 4) \cdot 5 = -12 \cdot 5 = -60$ and $-3 \cdot (4 \cdot 5) = -3 \cdot 20 = -60$

THE ADDITION PROPERTY OF ZERO

If a is a real number, then $a + 0 = a$ and $0 + a = a$.

The sum of a term and zero is the term.

EXAMPLE

$4 + 0 = 4$ and $0 + 4 = 4$

THE MULTIPLICATION PROPERTY OF ZERO

If a is a real number, then $a \cdot 0 = 0$ and $0 \cdot a = 0$.

The product of a term and zero is zero.

EXAMPLE

$(5)0 = 0$ and $0(5) = 0$

THE MULTIPLICATION PROPERTY OF ONE

If a is a real number, then $a \cdot 1 = a$ and $1 \cdot a = a$.

The product of a term and 1 is the term.

EXAMPLE

$6 \cdot 1 = 6$ and $1 \cdot 6 = 6$

THE INVERSE PROPERTY OF ADDITION

If a is a real number, then $a + (-a) = 0$ and $(-a) + a = 0$.

The sum of a number and its additive inverse (or opposite) is zero.

EXAMPLE

$8 + (-8) = 0$ and $(-8) + 8 = 0$

THE INVERSE PROPERTY OF MULTIPLICATION

If a is a real number and $a \neq 0$, then $a \cdot \frac{1}{a} = 1$ and $\frac{1}{a} \cdot a = 1$.

The product of a number and its reciprocal is 1.

EXAMPLE

$7 \cdot \frac{1}{7} = 1$ and $\frac{1}{7} \cdot 7 = 1$

$\frac{1}{a}$ is the **reciprocal** of a. $\frac{1}{a}$ is also called the **multiplicative inverse** of a.

THE DISTRIBUTIVE PROPERTY

If a, b, and c are real numbers, then $a(b + c) = ab + ac$ or $(b + c)a = ba + ca$.

By the Distributive Property, the term outside the parentheses is multiplied by each term inside the parentheses.

EXAMPLE

$$2(3 + 4) = 2 \cdot 3 + 2 \cdot 4 \qquad (4 + 5)2 = 4 \cdot 2 + 5 \cdot 2$$
$$2 \cdot 7 = 6 + 8 \qquad (9)2 = 8 + 10$$
$$14 = 14 \qquad 18 = 18$$

EXAMPLE 1 Complete the statement by using the Commutative Property of Multiplication.

$(6)(5) = (?)(6)$

Solution $(6)(5) = (5)(6)$ • The Commutative Property of Multiplication states that $a \cdot b = b \cdot a$.

Problem 1 Complete the statement by using the Inverse Property of Addition.

$7 + ? = 0$

Solution See page S4.

Try Exercise 17, page 70.

EXAMPLE 2 Identify the property that justifies the statement.

$2(8 + 5) = 16 + 10$

Solution The Distributive Property
- **The Distributive Property states that $a(b + c) = ab + ac$.**

Problem 2 Identify the property that justifies the statement.

$5 + (13 + 7) = (5 + 13) + 7$

Solution See page S4.

Try Exercise 31, page 70.

OBJECTIVE 2 Simplify variable expressions using the Properties of Addition

Like terms of a variable expression are terms with the same variable parts. The terms $3x$ and $-7x$ are like terms.

Constant terms are like terms. 4 and 9 are like terms.

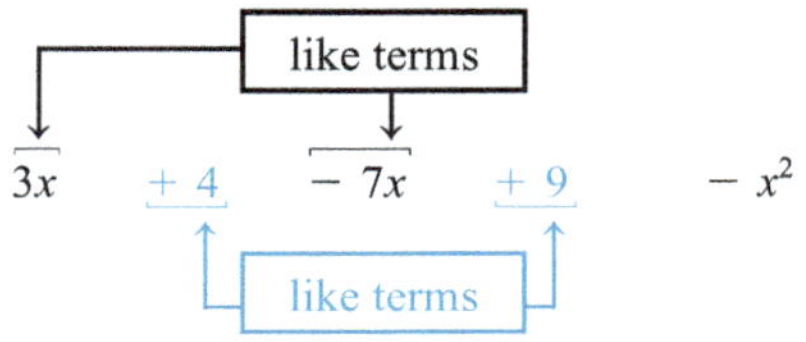

The terms $3x^2$ and $4x$ are not like terms because $x^2 = x \cdot x$ and thus the variable parts are not the same.

Here are more examples.

Not like terms	Like Terms	Like Terms
$8x$ and $3y$	$8x$ and $3x$	$8y$ and $3y$
$7x$ and $9x^2$	$7x$ and $9x$	$7x^2$ and $9x^2$
$6ab$ and $-5a$	$6a$ and $-5a$	$6ab$ and $-5ab$

To **combine like terms,** use the Distributive Property $ba + ca = (b + c)a$ to add the coefficients.

$$2x + 3x = (2 + 3)x$$
$$= 5x$$

EXAMPLE 3 Simplify. **A.** $-2y + 3y$ **B.** $5x - 11x$

Solution **A.** $-2y + 3y$

$= (-2 + 3)y$ • **Use the Distributive Property $ba + ca = (b + c)a$.**

$= 1y$ • **Add the coefficients.**

$= y$ • **Use the Multiplication Property of One.**

B. $5x - 11x$

$= [5 + (-11)]x$ • **Use the Distributive Property $ba + ca = (b + c)a$.**

$= -6x$ • **Add the coefficients.**

Problem 3 Simplify. **A.** $9x + 6x$ **B.** $-4y - 7y$

Solution See page S4.

Try Exercise 43, page 71.

In simplifying more complicated expressions, use the Properties of Addition.

The Commutative Property of Addition can be used when adding two like terms. The terms can be added in either order. The sum is the same.

$$2x + (-4x) = -4x + 2x$$
$$[2 + (-4)]x = (-4 + 2)x$$
$$-2x = -2x$$

The Associative Property of Addition is used when adding three or more terms. The terms can be grouped in any order. The sum is the same.

$$3x + 5x + 9x = (3x + 5x) + 9x = 3x + (5x + 9x)$$
$$8x + 9x = 3x + 14x$$
$$17x = 17x$$

By the Addition Property of Zero, the sum of a term and zero is the term.

$$5x + 0 = 0 + 5x = 5x$$

By the Inverse Property of Addition, the sum of a term and its additive inverse is zero.

$$7x + (-7x) = -7x + 7x = 0$$

EXAMPLE 4 Simplify. **A.** $8x + 3y - 8x$ **B.** $4x^2 + 5x - 6x^2 - 2x$

Solution **A.** $8x + 3y - 8x$

$= 3y + 8x - 8x$ • **Use the Commutative Property of Addition to rearrange the terms.**

$= 3y + (8x - 8x)$ • **Use the Associative Property of Addition to group like terms.**

$= 3y + 0$ • **Use the Inverse Property of Addition.**

$= 3y$ • **Use the Addition Property of Zero.**

B. $4x^2 + 5x - 6x^2 - 2x$

$= 4x^2 - 6x^2 + 5x - 2x$ • **Use the Commutative Property of Addition to rearrange the terms.**

$= (4x^2 - 6x^2) + (5x - 2x)$ • **Use the Associative Property of Addition to group like terms.**

$= -2x^2 + 3x$ • **Combine like terms.**

Problem 4 Simplify. **A.** $3a - 2b + 5a$ **B.** $x^2 - 7 + 9x^2 - 14$

Solution See page S4.

Try Exercise 67, page 71.

OBJECTIVE 3 Simplify variable expressions using the Properties of Multiplication

The Properties of Multiplication are used in simplifying variable expressions.

The Associative Property is used when multiplying three or more factors. $2(3x) = (2 \cdot 3)x = 6x$

The Commutative Property can be used to change the order in which factors are multiplied. $(3x) \cdot 2 = 2 \cdot (3x) = 6x$

By the Multiplication Property of One, the product of a term and 1 is the term. $(8x)(1) = (1)(8x) = 8x$

By the Inverse Property of Multiplication, the product of a term and its reciprocal is 1. $5x \cdot \frac{1}{5x} = \frac{1}{5x} \cdot 5x = 1, x \neq 0$

EXAMPLE 5 Simplify. **A.** $2(-x)$ **B.** $\frac{3}{2}\left(\frac{2x}{3}\right)$ **C.** $(16x)2$

Solution **A.** $2(-x) = 2(-1 \cdot x)$
- $-x = -1x = -1 \cdot x$

$= [2 \cdot (-1)]x$
- Use the Associative Property of Multiplication to group factors.

$= -2x$
- Multiply.

B. $\frac{3}{2}\left(\frac{2x}{3}\right) = \frac{3}{2}\left(\frac{2}{3}x\right)$
- Note that $\frac{2x}{3} = \frac{2}{3} \cdot \frac{x}{1} = \frac{2}{3}x$.

$= \left(\frac{3}{2} \cdot \frac{2}{3}\right)x$
- Use the Associative Property of Multiplication to group factors.

$= 1x$
- Use the Inverse Property of Multiplication.

$= x$
- Use the Multiplication Property of One.

C. $(16x)2 = 2(16x)$
- Use the Commutative Property of Multiplication to rearrange factors.

$= (2 \cdot 16)x$
- Use the Associative Property of Multiplication to group factors.

$= 32x$
- Multiply.

Problem 5 Simplify. **A.** $-7(-2a)$ **B.** $-\frac{5}{6}(-30y^2)$ **C.** $(-5x)(-2)$

Solution See page S4.

Try Exercise 97, page 72.

OBJECTIVE 4 Simplify variable expressions using the Distributive Property

The Distributive Property is used to remove parentheses from a variable expression.

$3(2x - 5)$
$= 3(2x) - 3(5)$
$= 6x - 15$

An extension of the Distributive Property is used when an expression contains more than two terms.

$4(x^2 + 6x - 1)$
$= 4(x^2) + 4(6x) - 4(1)$
$= 4x^2 + 24x - 4$

EXAMPLE 6 Simplify. **A.** $-3(5 + x)$ **B.** $-(2x - 4)$ **C.** $(2y - 6)2$ **D.** $5(3x + 7y - z)$

Solution **A.** $-3(5 + x)$

$= -3(5) + (-3)x$
- Use the Distributive Property.

$= -15 - 3x$
- Multiply.

B. $-(2x - 4)$

$= -1(2x - 4)$
- Just as $-x = -1x$, $-(2x - 4) = -1(2x - 4)$.

$= -1(2x) - (-1)(4)$
- Use the Distributive Property.

$= -2x + 4$
- *Note:* When a negative sign immediately precedes the parentheses, remove the parentheses and change the sign of each term inside the parentheses.

C. $(2y - 6)2$

$= (2y)(2) - (6)(2)$
- Use the Distributive Property $(b + c)a = ba + ca$.

$= 4y - 12$

D. $5(3x + 7y - z)$

$= 5(3x) + 5(7y) - 5(z)$
- Use the Distributive Property.

$= 15x + 35y - 5z$

Problem 6 Simplify. **A.** $7(4 + 2y)$ **B.** $-(5x - 12)$ **C.** $(3a - 1)5$ **D.** $-3(6a^2 - 8a + 9)$

Solution See page S4.

Try Exercise 133, page 72.

OBJECTIVE 5

Simplify general variable expressions

When simplifying variable expressions, use the Distributive Property to remove parentheses and brackets used as grouping symbols.

EXAMPLE 7 Simplify: $4(x - y) - 2(-3x + 6y)$

Solution $4(x - y) - 2(-3x + 6y)$

$= 4x - 4y + 6x - 12y$ • **Use the Distributive Property to remove parentheses.**

$= 10x - 16y$ • **Combine like terms.**

Problem 7 Simplify: $7(x - 2y) - 3(-x - 2y)$

Solution See page S4.

Try Exercise 161, page 73.

EXAMPLE 8 Simplify: $2x - 3[2x - 3(x + 7)]$

Solution $2x - 3[2x - 3(x + 7)]$

$= 2x - 3[2x - 3x - 21]$ • **Use the Distributive Property to remove the inner grouping symbols.**

$= 2x - 3[-x - 21]$ • **Combine like terms inside the grouping symbols.**

$= 2x + 3x + 63$ • **Use the Distributive Property to remove the brackets.**

$= 5x + 63$ • **Combine like terms.**

Problem 8 Simplify: $3y - 2[x - 4(2 - 3y)]$

Solution See page S4.

Try Exercise 167, page 73.

2.2 Exercises

CONCEPT CHECK

1. Which of the following pairs of terms are like terms?
   (i) $4a$ and $4b$ **(ii)** $8z^3$ and $8z^2$ **(iii)** $6ab$ and $3a$ **(iv)** $-5c^2$ and $6c^2$

2. Determine the additive inverse of each of the following.
   a. $4xy$ **b.** $-6b^2$ **c.** $-cd$

Determine whether each of the following mathematical statements is true. In each case, tell which property supports your answer.

3. $-4y \cdot 1 = -4y$

4. $(6x) \cdot 2 = 2 \cdot (6x)$

5. $5 \cdot \frac{1}{5} = 5$

6. $3(4x) = (3 \cdot 4)x$

Determine whether the statement is true or false.

7. The terms x and x^2 are like terms because both have a coefficient of 1.

8. To add like terms, add the coefficients; the variable part remains unchanged.

1 The Properties of the Real Numbers (See pages 63–66.)

GETTING READY

9. The fact that two terms can be added in either order is called the __?__ Property of Addition.

10. The fact that three or more factors can be multiplied by grouping them in any order is called the __?__ Property of Multiplication.

11. The Inverse Property of Multiplication tells us that the product of a number and its __?__ is 1.

12. The Inverse Property of Addition tells us that the sum of a number and its __?__ is zero.

Use the given property to complete the statement.

13. The Commutative Property of Multiplication: $2 \cdot 5 = 5 \cdot ?$

14. The Commutative Property of Addition: $9 + 17 = ? + 9$

15. The Associative Property of Multiplication: $(4 \cdot 5) \cdot 6 = 4 \cdot (? \cdot 6)$

16. The Associative Property of Addition: $(4 + 5) + 6 = ? + (5 + 6)$

17. The Distributive Property: $2(4 + 3) = 8 + ?$

18. The Addition Property of Zero: $? + 0 = -7$

19. The Inverse Property of Addition: $8 + ? = 0$

20. The Inverse Property of Multiplication: $\frac{1}{-5}(-5) = ?$

21. The Multiplication Property of One: $? \cdot 1 = -4$

22. The Multiplication Property of Zero: $12 \cdot ? = 0$

Identify the property that justifies the statement.

23. $-7 + 7 = 0$

24. $(-8)\left(-\frac{1}{8}\right) = 1$

25. $23 + 19 = 19 + 23$

26. $-21 + 0 = -21$

27. $2 + (6 + 14) = (2 + 6) + 14$

28. $(-3 + 9)8 = -24 + 72$

29. $3 \cdot 5 = 5 \cdot 3$

30. $-32(0) = 0$

31. $(4 \cdot 3) \cdot 5 = 4 \cdot (3 \cdot 5)$

32. $\frac{1}{4}(1) = \frac{1}{4}$

2 Simplify variable expressions using the Properties of Addition (See pages 66–67.)

33. What are *like terms*? Give an example of two like terms. Give an example of two terms that are not like terms.

34. Explain the meaning of the phrase "simplify a variable expression."

35. Which of the following are like terms? $3x, 3x^2, 5x, 5xy$

36. Which of the following are like terms? $-7a, -7b, -4a, -7a^2$

GETTING READY

37. Simplify $5a - 8a$ by combining like terms.

$5a - 8a = 5a + \underline{\ ?\ }$ • **Write subtraction as addition of the opposite.**

$= [5 + (-8)]a$ • **Use the ___?___ Property $ba + ca = (b + c)a$.**

$= \underline{\ ?\ }\ a$ • **Add the coefficients.**

38. Simplify $6xy - 6x + 7xy$ by combining like terms.

$6xy - 6x + 7xy = 6xy + 7xy - 6x$ • **Use the ___?___ Property of Addition to rearrange the terms.**

$= (6 + 7)(\underline{\ ?\ }) - 6x$ • **Use the Distributive Property $ba + ca = (b + c)a$.**

$= \underline{\ ?\ } - 6x$ • **Add the coefficients.**

Simplify.

39. $6x + 8x$ **40.** $12x + 13x$ **41.** $9a - 4a$ **42.** $12a - 3a$

43. $4y - 10y$ **44.** $8y - 6y$ **45.** $-3b - 7$ **46.** $-12y - 3$

47. $-12a + 17a$ **48.** $-3a + 12a$ **49.** $5ab - 7ab$ **50.** $9ab - 3ab$

51. $-12xy + 17xy$ **52.** $-15xy + 3xy$ **53.** $-3ab + 3ab$ **54.** $-7ab + 7ab$

55. $-\frac{1}{2}x - \frac{1}{3}x$ **56.** $-\frac{2}{5}y + \frac{3}{10}y$ **57.** $\frac{3}{8}x^2 - \frac{5}{12}x^2$ **58.** $\frac{2}{3}y^2 - \frac{4}{9}y^2$

59. $3x + 5x + 3x$ **60.** $8x + 5x + 7x$ **61.** $5a - 3a + 5a$ **62.** $10a - 17a + 3a$

63. $-5x^2 - 12x^2 + 3x^2$ **64.** $-y^2 - 8y^2 + 7y^2$ **65.** $7x - 8x + 3y$

66. $8y - 10x + 8x$ **67.** $7x - 3y + 10x$ **68.** $8y + 8x - 8y$

69. $3a - 7b - 5a + b$ **70.** $-5b + 7a - 7b + 12a$ **71.** $3x - 8y - 10x + 4x$

72. $3y - 12x - 7y + 2y$ **73.** $x^2 - 7x - 5x^2 + 5x$ **74.** $3x^2 + 5x - 10x^2 - 10x$

75. State whether the given number will be positive, negative, or zero after the variable expression $15a^2 - 12a + 7 - 21a^2 + 29a - 7$ is simplified.
a. the coefficient of a^2 **b.** the coefficient of a **c.** the constant term

76. Which expressions are equivalent to $-10x - 10y - 10y - 10x$?
(i) 0 (ii) $-20y$ (iii) $-20x$ (iv) $-20x - 20y$ (v) $-20y - 20x$

3 Simplify variable expressions using the Properties of Multiplication (See pages 67–68.)

GETTING READY

77. Simplify: $4(-12x)$

$4(-12x) = [4(-12)]x$ • **Use the ___?___ Property of Multiplication to group factors.**

$= \underline{\ ?\ }$ • **Multiply.**

78. Simplify: $\left(\frac{1}{2}a^2\right)\left(-\frac{3}{2}\right)$

$\left(\frac{1}{2}a^2\right)\left(-\frac{3}{2}\right) = (\underline{\ ?\ })\left(\frac{1}{2}a^2\right)$ • **Use the Commutative Property of Multiplication to rearrange factors.**

$= \left(-\frac{3}{2}\cdot\underline{\ ?\ }\right)a^2$ • **Use the Associative Property of Multiplication to group factors.**

$= \underline{\ ?\ }\ a^2$ • **Multiply.**

Simplify.

79. $4(3x)$ **80.** $12(5x)$ **81.** $-3(7a)$ **82.** $-2(5a)$

83. $-2(-3y)$ **84.** $-5(-6y)$ **85.** $(4x)2$ **86.** $(6x)12$

87. $(3a)(-2)$ **88.** $(7a)(-4)$ **89.** $(-3b)(-4)$ **90.** $(-12b)(-9)$

91. $-5(3x^2)$ **92.** $-8(7x^2)$ **93.** $\frac{1}{3}(3x^2)$ **94.** $\frac{1}{5}(5a)$

95. $\frac{1}{8}(8x)$ **96.** $-\frac{1}{4}(-4a)$ **97.** $-\frac{1}{7}(-7n)$ **98.** $\left(\frac{4}{3}\right)\left(\frac{3x}{4}\right)$

99. $\left(\frac{5}{12}\right)\frac{12x}{5}$ **100.** $(-6y)\left(-\frac{1}{6}\right)$ **101.** $(-10n)\left(-\frac{1}{10}\right)$ **102.** $\frac{1}{3}(9x)$

103. $\frac{1}{7}(14x)$ **104.** $-\frac{1}{5}(10x)$ **105.** $-\frac{1}{8}(16x)$ **106.** $-\frac{2}{3}(12a^2)$

107. $-\frac{5}{8}(24a^2)$ **108.** $-\frac{1}{2}(-16y)$ **109.** $-\frac{3}{4}(-8y)$ **110.** $(16y)\left(\frac{1}{4}\right)$

111. $(33y)\left(\frac{1}{11}\right)$ **112.** $(-6x)\left(\frac{1}{3}\right)$ **113.** $(-10x)\left(\frac{1}{5}\right)$ **114.** $(-8a)\left(-\frac{3}{4}\right)$

115. After $\frac{2}{7}x^2$ is multiplied by a proper fraction, is the coefficient of x^2 greater than 1 or less than 1?

116. After $-152m$ is multiplied by a whole number, is the coefficient of m greater than zero or less than zero?

4 Simplify variable expressions using the Distributive Property (See pages 68–69.)

Simplify.

117. $2(4x - 3)$ **118.** $5(2x - 7)$ **119.** $-2(a + 7)$ **120.** $-5(a + 16)$

121. $-3(2y - 8)$ **122.** $-5(3y - 7)$ **123.** $-(x + 2)$ **124.** $-(x + 7)$

125. $(5 - 3b)7$ **126.** $(10 - 7b)2$ **127.** $-3(3 - 5x)$ **128.** $-5(7 - 10x)$

129. $3(5x^2 + 2x)$ **130.** $6(3x^2 + 2x)$ **131.** $-2(-y + 9)$ **132.** $-5(-2x + 7)$

133. $(-3x - 6)5$ **134.** $(-2x + 7)7$ **135.** $2(-3x^2 - 14)$ **136.** $5(-6x^2 - 3)$

137. $-3(2y^2 - 7)$ **138.** $-8(3y^2 - 12)$ **139.** $3(x^2 + 2x - 6)$

140. $4(x^2 - 3x + 5)$ **141.** $-2(y^2 - 2y + 4)$ **142.** $-3(y^2 - 3y - 7)$

143. $2(-a^2 - 2a + 3)$ **144.** $4(-3a^2 - 5a + 7)$ **145.** $-5(-2x^2 - 3x + 7)$

146. $-3(-4x^2 + 3x - 4)$ **147.** $-(3a^2 + 5a - 4)$ **148.** $-(8b^2 - 6b + 9)$

149. After the expression $17x - 31$ is multiplied by a negative integer, is the constant term positive or negative?

150. The expression $5x - 8xy + 7y$ is multiplied by a positive integer. How many variable terms does the resulting expression have?

5 Simplify general variable expressions (See page 69.)

GETTING READY

151. When simplifying $4(3a - 7) + 2(a - 3)$, the first step is to use the Distributive Property to remove parentheses:
$4(3a - 7) + 2(a - 3) = (\underline{\ ?\ })a - \underline{\ ?\ } + (\underline{\ ?\ })a - \underline{\ ?\ }$

152. When simplifying $12m - (m - 3)$, the first step is to remove the parentheses and change the sign of each term inside the parentheses:
$12m - \underline{\ ?\ } + \underline{\ ?\ }$

Simplify.

153. $4x - 2(3x + 8)$

154. $6a - (5a + 7)$

155. $9 - 3(4y + 6)$

156. $10 - 2(11x - 3)$

157. $5n - (7 - 2n)$

158. $8 - (12 + 4y)$

159. $3(x + 2) - 5(x - 7)$

160. $2(x - 4) - 4(x + 2)$

161. $3(a - b) - 4(a + b)$

162. $2(a + 2b) - (a - 3b)$

163. $4[x - 2(x - 3)]$

164. $2[x + 2(x + 7)]$

165. $-3[2x - (x + 7)]$

166. $-2[3x - (5x - 2)]$

167. $2x - 3[x - 2(4 - x)]$

168. $-7x + 3[x - 7(3 - 2x)]$

169. $2x + 3(x - 2y) + 5(3x - 7y)$

170. $5y - 2(y - 3x) + 2(7x - y)$

171. Which expression is equivalent to $12 - 7(y - 9)$?
(i) $5(y - 9)$ (ii) $12 - 7y - 63$ (iii) $12 - 7y + 63$ (iv) $12 - 7y - 9$

172. Which expression is equivalent to $7[3b + 5(b - 6)]$?
(i) $7[3b + 5b - 6]$ (ii) $7[3b + 5b - 30]$ (iii) $7[3b + 5b + 6]$ (iv) $7[8b - 6]$

APPLYING CONCEPTS

Complete.

173. A number that has no reciprocal is $\underline{\ ?\ }$.

174. A number that is its own reciprocal is $\underline{\ ?\ }$.

175. The additive inverse of $a - b$ is $\underline{\ ?\ }$.

176. Give examples of two operations that occur in everyday experience and are not commutative (for example, putting on socks and then shoes).

PROJECTS OR GROUP ACTIVITIES

177. Determine whether the statement is true or false. If the statement is false, give an example that illustrates that it is false.
- **a.** Division is a commutative operation.
- **b.** Division is an associative operation.
- **c.** Subtraction is an associative operation.
- **d.** Subtraction is a commutative operation.
- **e.** Addition is a commutative operation.

178. Define an operation $\otimes$ as $a \otimes b = (a \cdot b) - (a + b)$.
For example, $7 \otimes 5 = (7 \cdot 5) - (7 + 5) = 35 - 12 = 23$.
- **a.** Is $\otimes$ a commutative operation? Support your answer.
- **b.** Is $\otimes$ an associative operation? Support your answer.

The chart below is an addition table. Use it to answer Exercises 179 to 185 below.

+	Δ	‡	◇
Δ	‡	◇	Δ
‡	◇	Δ	‡
◇	Δ	‡	◇

179. Find the sum of Δ and ‡.

180. What is ◇ plus ◇?

181. In our number system, zero can be added to any number without changing that number; zero is called the **additive identity.** What is the additive identity for the system in the chart above? Explain your answer.

182. Does the Commutative Property of Addition apply to this system? Explain your answer.

183. What is −Δ (the opposite of Δ) equal to? Explain your answer.

184. What is −‡ (the opposite of ‡) equal to? Explain your answer.

185. Simplify −Δ + ‡ − ◇. Explain how you arrived at your answer.

186. Which of the following expressions are equivalent?

(i) $2x + 4(2x + 1)$
(ii) $x - (4 - 9x) + 8$
(iii) $7(x - 4) - 3(2x + 6)$
(iv) $3(2x + 8) + 4(x - 5)$
(v) $6 - 2[x + (3x - 4)] + 2(9x - 5)$

2.3 Translating Verbal Expressions into Variable Expressions

OBJECTIVE 1 Translate a verbal expression into a variable expression

One of the major skills required in applied mathematics is the ability to translate a verbal expression into a variable expression. This requires recognizing the verbal phrases that translate into mathematical operations. Here is a partial list of the phrases used to indicate the different mathematical operations.

WORDS OR PHRASES FOR ADDITION

added to	6 added to y	$y + 6$
more than	8 more than x	$x + 8$
the sum of	the sum of x and z	$x + z$
increased by	t increased by 9	$t + 9$
the total of	the total of 5 and d	$5 + d$
plus	b plus 17	$b + 17$

Point of Interest

The way in which expressions are symbolized has changed over time. Here are some expressions as they may have appeared in the early 16th century.

R p. 9 for $x + 9$. The symbol **R** was used for a variable to the first power. The symbol **p.** was used for plus.

R m. 3 for $x - 3$. Again, the symbol **R** was used for the variable. The symbol **m.** was used for minus.

The square of a variable was designated by **Q,** and the cube was designated by **C.** The expression $x^3 + x^2$ was written **C p. Q.**

WORDS OR PHRASES FOR SUBTRACTION

minus	x minus 2	$x - 2$
less than	7 less than t	$t - 7$
less	7 less t	$7 - t$
subtracted from	5 subtracted from d	$d - 5$
decreased by	m decreased by 3	$m - 3$
the difference between	the difference between y and 4	$y - 4$

WORDS OR PHRASES FOR MULTIPLICATION

times	10 times t	$10t$
of	one-half of x	$\frac{1}{2}x$
the product of	the product of y and z	yz
multiplied by	b multiplied by 11	$11b$
twice	twice n	$2n$

PHRASES FOR DIVISION

divided by	x divided by 12	$\frac{x}{12}$
the quotient of	the quotient of y and z	$\frac{y}{z}$
the ratio of	the ratio of t to 9	$\frac{t}{9}$

PHRASES FOR POWER

the square of	the square of x	x^2
the cube of	the cube of a	a^3

Translating a phrase that contains the word *sum*, *difference*, *product*, or *quotient* can sometimes cause a problem. In the examples at the right, note where the operation symbol is placed.

the *sum* of x *and* y	$x + y$
the *difference* between x *and* y	$x - y$
the *product* of x *and* y	$x \cdot y$
the *quotient* of x *and* y	$\frac{x}{y}$

EXAMPLE 1 Translate into a variable expression.

A. the total of five times b and c

B. the quotient of eight less than n and fourteen

C. thirteen more than the sum of seven and the square of x

Solution **A.** the total of five times b and c

$5b + c$

- Identify words that indicate mathematical operations.
- Use the operations to write the variable expression.

B. the quotient of eight less than n and fourteen

$\frac{n - 8}{14}$

- Identify words that indicate mathematical operations.
- Use the operations to write the variable expression.

C. thirteen more than the sum of seven and the square of x

$(7 + x^2) + 13$

Problem 1 Translate into a variable expression.

A. eighteen less than the cube of x

B. y decreased by the sum of z and nine

C. the difference between q and the sum of r and t

Solution See page S4.

Try Exercise 21, page 79.

In most applications that involve translating phrases into variable expressions, the variable to be used is not given. To translate these phrases, a variable must be assigned to an unknown quantity before the variable expression can be written.

Take Note

The expression $n(6 + n^3)$ must have parentheses. If we write $n \cdot 6 + n^3$, then by the Order of Operations Agreement, only the 6 is multiplied by n. We want n to be multiplied by the *total* of 6 and n^3.

EXAMPLE 2 Translate "a number multiplied by the total of six and the cube of the number" into a variable expression.

Solution the unknown number: n

- Assign a variable to one of the unknown quantities.

the cube of the number: n^3
the total of six and the cube of the number: $6 + n^3$

- Use the assigned variable to write an expression for any other unknown quantity.

$n(6 + n^3)$

- Use the assigned variable to write the variable expression.

Problem 2 Translate "a number added to the product of five and the square of the number" into a variable expression.

Solution See page S4.

Try Exercise 47, page 79.

EXAMPLE 3 Translate "the quotient of twice a number and the difference between the number and twenty" into a variable expression.

Solution the unknown number: n
twice the number: $2n$
the difference between the number and twenty: $n - 20$

$\frac{2n}{n - 20}$

Problem 3 Translate "the product of three and the sum of seven and twice a number" into a variable expression.

Solution See page S4.

Try Exercise 49, page 79.

OBJECTIVE 2 Translate a verbal expression into a variable expression and then simplify the resulting expression

After translating a verbal expression into a variable expression, simplify the variable expression by using the Properties of the Real Numbers.

EXAMPLE 4 Translate and simplify "the total of four times an unknown number and twice the difference between the number and eight."

Solution the unknown number: n

- Assign a variable to one of the unknown quantities.

four times the unknown number: $4n$

twice the difference between the number and eight: $2(n - 8)$

- Use the assigned variable to write an expression for any other unknown quantity.

$$4n + 2(n - 8)$$

- Use the assigned variable to write the variable expression.

$$= 4n + 2n - 16$$

- Simplify the variable expression.

$$= 6n - 16$$

Problem 4 Translate and simplify "a number minus the difference between twice the number and seventeen."

Solution See page S4.

Try Exercise 65, page 80.

EXAMPLE 5 Translate and simplify "the difference between five-eighths of a number and two-thirds of the same number."

Solution the unknown number: n

- Assign a variable to one of the unknown quantities.

five-eighths of the number: $\frac{5}{8}n$

two-thirds of the number: $\frac{2}{3}n$

- Use the assigned variable to write an expression for any other unknown quantity.

$$\frac{5}{8}n - \frac{2}{3}n$$

- Use the assigned variable to write the variable expression.

$$= \frac{15}{24}n - \frac{16}{24}n$$

- Simplify the variable expression.

$$= -\frac{1}{24}n$$

Problem 5 Translate and simplify "the sum of three-fourths of a number and one-fifth of the same number."

Solution See page S4.

Try Exercise 71, page 80.

OBJECTIVE 3 Translate application problems

Many of the applications of mathematics require that you identify an unknown quantity, assign a variable to that quantity, and then attempt to express another unknown quantity in terms of that variable.

Suppose we know that the sum of two numbers is 10 and that one of the two numbers is 4. We can find the other number by subtracting 4 from 10.

one number: 4
other number: $10 - 4 = 6$
The two numbers are 4 and 6.

Now suppose we know that the sum of two numbers is 10, we don't know either number, and we want to express *both* numbers in terms of the *same* variable.

Let one number be x. Again, we can find the other number by subtracting x from 10.

one number: x
other number: $10 - x$
The two numbers are x and $10 - x$.

Note that the sum of x and $10 - x$ is 10.

$x + (10 - x) = x + 10 - x = 10$

Take Note

In Example 6, any variable can be used. For example, if the width is y, then the length is $y + 20$.

EXAMPLE 6 The length of a swimming pool is 20 ft longer than the width. Express the length of the pool in terms of the width.

Solution the width of the pool: W
• **Assign a variable to the width of the pool.**

the length is 20 more than the width: $W + 20$
• **Express the length of the pool in terms of W.**

Problem 6 An older computer takes twice as long to process a set of data as does a newer model. Express the amount of time it takes the older computer to process the data in terms of the amount of time it takes the newer model.

Solution See page S4.

Try Exercise 101, page 82.

Take Note

In Example 7, it is also correct to assign the variable to the amount in the money market fund. Then the amount in the mutual fund is $5000 - x$.

EXAMPLE 7 An investor divided \$5000 between two accounts, one a mutual fund and the other a money market fund. Use one variable to express the amounts invested in each account.

Solution the amount invested in the mutual fund: x
• **Assign a variable to the amount invested in one account.**

the amount invested in the money market fund: $5000 - x$
• **Express the amount invested in the other account in terms of x.**

Problem 7 A guitar string 6 ft long was cut into two pieces. Use one variable to express the lengths of the two pieces.

Solution See page S4.

Try Exercise 105, page 82.

2.3 Exercises

CONCEPT CHECK

Determine whether the statement is true or false.

1. "Five less than n" can be translated as "$5 - n$."

2. A variable expression contains an equals sign.

3. If the sum of two numbers is 12 and one of the two numbers is x, then the other number can be expressed as $x - 12$.

4. The words *total* and *times* both indicate multiplication.

5. The words *quotient* and *ratio* both indicate division.

1 Translate a verbal expression into a variable expression (See pages 74–76.)

GETTING READY

For each phrase, identify the words that indicate mathematical operations.

6. the sum of seven and three times m

7. twelve less than the quotient of x and negative two

8. the total of ten and fifteen divided by a number

9. twenty subtracted from the product of eight and the cube of a number

Translate into a variable expression.

10. d less than nineteen

11. the sum of six and c

12. r decreased by twelve

13. w increased by fifty-five

14. a multiplied by twenty-eight

15. y added to sixteen

16. five times the difference between n and seven

17. thirty less than the square of b

18. y less the product of three and y

19. the sum of four-fifths of m and eighteen

20. the product of negative six and b

21. nine increased by the quotient of t and five

22. four divided by the difference between p and six

23. the product of seven and the total of r and eight

24. the quotient of nine less than x and twice x

25. the product of a and the sum of a and thirteen

26. twenty-one less than the product of s and negative four

27. fourteen more than one-half of the square of z

28. the ratio of eight more than d to d

29. the total of nine times the cube of m and the square of m

30. three-eighths of the sum of t and fifteen

31. s decreased by the quotient of s and two

32. w increased by the quotient of seven and w

33. the difference between the square of c and the total of c and fourteen

34. d increased by the difference between sixteen times d and three

35. the product of eight and the total of b and five

36. a number divided by nineteen

37. thirteen less a number

38. forty more than a number

39. three-sevenths of a number

40. the square of the difference between a number and ninety

41. the quotient of twice a number and five

42. eight subtracted from the product of fifteen and a number

43. the product of a number and ten more than the number

44. fourteen added to the product of seven and a number

45. the quotient of three and the total of four and a number

46. the quotient of twelve and the sum of a number and two

47. the sum of the square of a number and three times the number

48. a number decreased by the difference between the cube of the number and ten

49. four less than the product of seven and the square of a number

50. eighty decreased by the product of thirteen and a number

51. the cube of a number decreased by the product of twelve and the number

52. Which of the following phrases translate(s) into the variable expression $25n^2 - 9$?
- (i) the difference between nine and the product of twenty-five and the square of a number
- (ii) nine subtracted from the square of twenty-five and a number
- (iii) nine less than the product of twenty-five and the square of a number

53. Which of the following phrases translate(s) into the variable expression $32 - \frac{a}{7}$?
- (i) the difference between thirty-two and the quotient of a number and seven
- (ii) thirty-two decreased by the quotient of a number and seven
- (iii) thirty-two minus the ratio of a number to seven

2 Translate a verbal expression into a variable expression and then simplify the resulting expression (See page 77.)

GETTING READY

54. The phrase "the total of one-half of a number and three-fourths of the number" can be translated as (__?__)n + (__?__)n. This expression simplifies to __?__.

55. The phrase "the difference between twelve times a number and fifteen times the number" can be translated as (__?__)n − (__?__)n. This expression simplifies to __?__.

Translate into a variable expression. Then simplify the expression.

56. a number increased by the total of the number and ten

57. a number added to the product of five and the number

58. a number decreased by the difference between nine and the number

59. eight more than the sum of a number and eleven

60. a number minus the sum of the number and fourteen

61. four more than the total of a number and nine

62. twice the sum of three times a number and forty

63. seven times the product of five and a number

64. sixteen multiplied by one-fourth of a number

65. the total of seventeen times a number and twice the number

66. the difference between nine times a number and twice the number

67. a number plus the product of the number and twelve

68. nineteen more than the difference between a number and five

69. three times the sum of the square of a number and four

70. a number subtracted from the product of the number and seven

71. three-fourths of the sum of sixteen times a number and four

72. the difference between fourteen times a number and the product of the number and seven

73. sixteen decreased by the sum of a number and nine

74. eleven subtracted from the difference between eight and a number

75. six times the total of a number and eight

76. four times the sum of a number and twenty

77. seven minus the sum of a number and two

78. three less than the sum of a number and ten

79. one-third of the sum of a number and six times the number

80. twice the quotient of four times a number and eight

81. twelve more than a number added to the difference between the number and six

82. a number plus four added to the difference between three and twice the number

83. the sum of a number and nine added to the difference between the number and twenty

84. seven increased by a number added to twice the difference between the number and two

3 Translate application problems (See pages 77–78.)

GETTING READY

85. The sum of two numbers is 25. To express both numbers in terms of the same variable, let x be one number. Then the other number is __?__.

86. The length of a rectangle is five times the width. To express the length and the width in terms of the same variable, let W be the width. Then the length is __?__.

87. The width of a rectangle is one-half the length. To express the length and the width in terms of the same variable, let L be the length. Then the width is __?__.

88. An electrician's bill is $195 for materials and $75 an hour for labor. To express the total amount of the bill in terms of the number of hours of labor, let h be the number of hours of labor. Then the cost of the labor is __?__, so the total amount for the materials and labor is __?__.

Write a variable expression.

89. The sum of two numbers is 18. Express the two numbers in terms of the same variable.

90. The sum of two numbers is 20. Express the two numbers in terms of the same variable.

91. **Astronomy** The diameter of Saturn's moon Rhea is 253 mi more than the diameter of Saturn's moon Dione. Express the diameter of Rhea in terms of the diameter of Dione. (*Source:* NASA)

Sightseeing Archive/Getty Images

92. **Noise Level** The noise level of an ambulance siren is 10 decibels louder than that of a car horn. Express the noise level of an ambulance siren in terms of the noise level of a car horn. (*Source:* League for the Hard of Hearing)

93. **Genetics** The human genome contains 11,000 more genes than the roundworm genome. Express the number of genes in the human genome in terms of the number of genes in the roundworm genome. (*Source:* Celera, USA TODAY research)

In the News

U2 Concerts Top Annual Rankings in North America

The Irish rock band U2 performed the most popular concerts on the North American circuit this year. Bruce Springsteen and the E Street Band came in second, with $28.5 million less in ticket sales.

Source: new.music.yahoo.com

94. **Rock Band Tours** See the news clipping at the right. Express Bruce Springsteen and the E Street Band's concert ticket sales in terms of U2's concert ticket sales.

95. **Space Exploration** A survey in *USA Today* reported that almost three-fourths of Americans think that money should be spent on exploration of Mars. Express the number of Americans who think that money should be spent on exploration of Mars in terms of the total number of Americans.

96. **Biology** According to the American Podiatric Medical Association, the bones in your foot account for one-fourth of all the bones in your body. Express the number of bones in your foot in terms of the number of bones in your body.

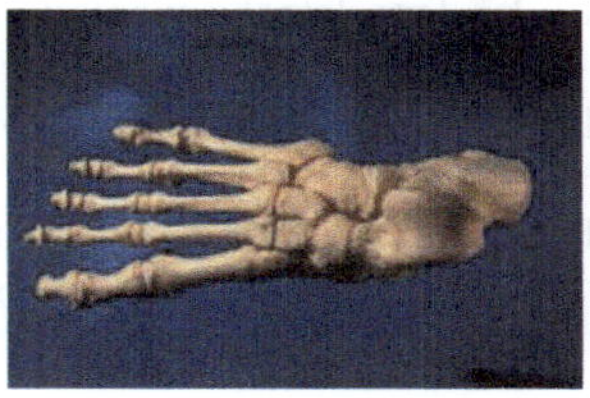

Lester V. Bergman/Corbis

97. **Sports** In football, the number of points awarded for a touchdown is three times the number of points awarded for a safety. Express the number of points awarded for a touchdown in terms of the number of points awarded for a safety.

98. **Community Colleges** According to the National Center for Education Statistics, 46% of U.S. undergraduate students attend two-year colleges. Express the number of U.S. undergraduate students who attend two-year colleges in terms of the number of U.S. undergraduate students.

99. **Major League Sports** See the news clipping at the right. Express the attendance at major league baseball games in terms of the attendance at major league basketball games.

100. **Geometry** The length of a rectangle is 5 m more than twice the width. Express the length of the rectangle in terms of the width.

101. **Geometry** In a triangle, the measure of the smallest angle is 10 degrees less than one-half the measure of the largest angle. Express the measure of the smallest angle in terms of the measure of the largest angle.

102. **Wages** An employee is paid \$1172 per week plus \$38 for each hour of overtime worked. Express the employee's weekly pay in terms of the number of hours of overtime worked.

103. **Billing** An auto repair bill is \$238 for parts and \$89 for each hour of labor. Express the amount of the repair bill in terms of the number of hours of labor.

104. **Sports** A halyard 12 ft long is cut into two pieces. Use the same variable to express the lengths of the two pieces.

105. **Coins** A coin bank contains thirty-five coins in nickels and dimes. Use the same variable to express the number of nickels and the number of dimes in the coin bank.

106. **Travel** Two cars are traveling in opposite directions and at different rates. Two hours later, the cars are 200 mi apart. Express the distance traveled by the faster car in terms of the distance traveled by the slower car.

In the News

Over 70 Million Attend Major League Baseball Games

Among major league sports, attendance at major league baseball games topped attendance at other major league sporting events. Fifty million more people went to baseball games than went to basketball games. The attendance at football games and hockey games was even less than the attendance at basketball games.

Source: Time, December 28, 2009–January 4, 2010

APPLYING CONCEPTS

107. **Chemistry** The chemical formula for glucose (sugar) is $C_6H_{12}O_6$. This formula means that there are twelve hydrogen atoms, six carbon atoms, and six oxygen atoms in each molecule of glucose. If x represents the number of atoms of oxygen in a pound of sugar, express the number of hydrogen atoms in the pound of sugar in terms of x.

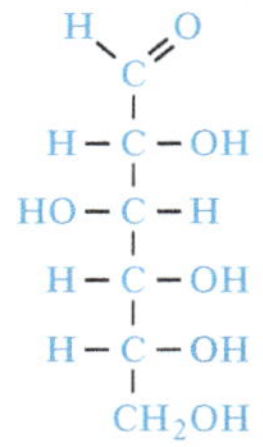

108. **Metalwork** A wire whose length is given as x inches is bent into a square. Express the length of a side of the square in terms of x.

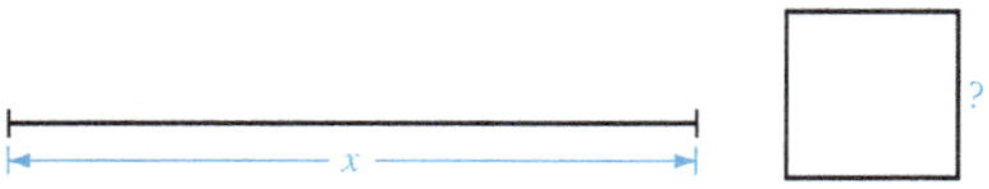

109. **Pulleys** A block-and-tackle system is designed so that pulling one end of a rope five feet will move a weight on the other end a distance of three feet. If x represents the distance the rope is pulled, express the distance the weight moves in terms of x.

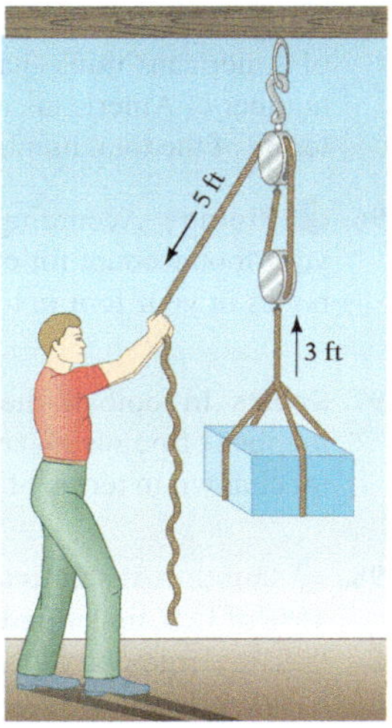

110. Translate the expressions $5x + 8$ and $5(x + 8)$ into phrases.

111. In your own words, explain how variables are used.

112. Explain the similarities and differences between the expressions "the difference between x and 5" and "5 less than x."

PROJECTS OR GROUP ACTIVITIES

113. Write five phrases that translate into the expression $p + 8$.

114. Write four phrases that translate into the expression $d - 16$.

115. Write three phrases that translate into the expression $4c$.

116. Write three phrases that translate into the expression $\frac{y}{5}$.

117. Look at the figures below.

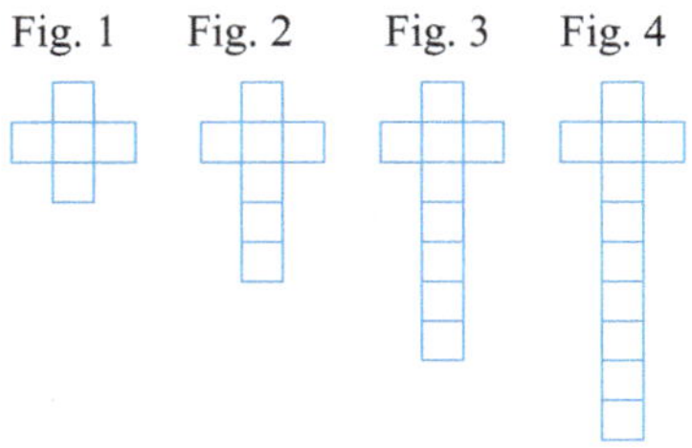

a. Look for a pattern. Fill in the table below.

Figure Number	Number of Tiles

b. Describe the number of tiles in the nth figure.

CHAPTER 2 Summary

Key Words	Objective and Page Reference	Examples
A **variable** is a letter that is used to stand for a quantity that is unknown or that can change. A **variable expression** is an expression that contains one or more variables.	[2.1.1, p. 58]	$5x - 4y + 7z$ is a variable expression. It contains the variables x, y, and z.
The **terms of a variable expression** are the addends of the expression. Each term is either a **variable term** or a **constant term.**	[2.1.1, p. 58]	The expression $4a^2 - 6b^3 + 7$ has three terms: $4a^2$, $-6b^3$, and 7. Of these, $4a^2$ and $-6b^3$ are variable terms. 7 is a constant term.

A **variable term** is composed of a **coefficient** and a **variable part.**	[2.1.1, p. 58]	For the expression $8p^4r$, 8 is the coefficient and p^4r is the variable part.
Replacing the variables in a variable expression with numbers and then simplifying the numerical expression is called **evaluating the variable expression.**	[2.1.1, p. 58]	To evaluate $a^2 - 2b$ when $a = -3$ and $b = 5$, simplify the expression $(-3)^2 - 2(5)$. $(-3)^2 - 2(5) = 9 - 10 = -1$
Like terms of a variable expression are terms with the same variable part. (Because $x^2 = x \cdot x$, x^2 and x are not like terms.) Constant terms are like terms.	[2.2.2, p. 66]	For the expression $2st - 3t + 9s - 11st$, the terms $2st$ and $-11st$ are like terms. For the expression $5x + 8 - 6x + 7 - 4x^2$, the terms $5x$ and $-6x$ are like terms. The constant terms 8 and 7 are like terms.
The **additive inverse** of a number is the opposite of the number.	[2.2.1, p. 65]	The additive inverse of 8 is -8. The additive inverse of -15 is 15.
The **multiplicative inverse** of a number is the reciprocal of the number.	[2.2.1, p. 65]	The multiplicative inverse of $\frac{3}{8}$ is $\frac{8}{3}$.

Essential Rules and Procedures

Essential Rules and Procedures	Objective and Page Reference	Examples
The Commutative Property of Addition If a and b are real numbers, then $a + b = b + a$.	[2.2.1, p. 63]	$5 + 2 = 7$ and $2 + 5 = 7$
The Commutative Property of Multiplication If a and b are real numbers, then $ab = ba$.	[2.2.1, p. 63]	$6(-3) = -18$ and $-3(6) = -18$
The Associative Property of Addition If a, b, and c are real numbers, then $(a + b) + c = a + (b + c)$.	[2.2.1, p. 64]	$-1 + (4 + 7) = -1 + 11 = 10$ $(-1 + 4) + 7 = 3 + 7 = 10$
The Associative Property of Multiplication If a, b, and c are real numbers, then $(ab)c = a(bc)$.	[2.2.1, p. 64]	$(-2 \cdot 5) \cdot 3 = -10 \cdot 3 = -30$ $-2 \cdot (5 \cdot 3) = -2 \cdot 15 = -30$
The Addition Property of Zero If a is a real number, then $a + 0 = 0 + a = a$.	[2.2.1, p. 64]	$9 + 0 = 9$ $\quad$ $0 + 9 = 9$
The Multiplication Property of Zero If a is a real number, then $a \cdot 0 = 0 \cdot a = 0$.	[2.2.1, p. 64]	$-8(0) = 0$ $\quad$ $0(-8) = 0$
The Multiplication Property of One If a is a real number, then $1 \cdot a = a \cdot 1 = a$.	[2.2.1, p. 64]	$7 \cdot 1 = 7$ $\quad$ $1 \cdot 7 = 7$
The Inverse Property of Addition If a is a real number, then $a + (-a) = (-a) + a = 0$.	[2.2.1, p. 65]	$4 + (-4) = 0$ $\quad$ $-4 + 4 = 0$

The Inverse Property of Multiplication If a is a real number and $a \neq 0$, then $a \cdot \frac{1}{a} = \frac{1}{a} \cdot a = 1$.	[2.2.1, p. 65]	$6 \cdot \frac{1}{6} = 1 \quad \frac{1}{6} \cdot 6 = 1$
The Distributive Property If a, b, and c are real numbers, then $a(b + c) = ab + ac$.	[2.2.1, p. 65]	$5(x + 3) = 5 \cdot x + 5 \cdot 3$ $= 5x + 15$
To **combine like terms,** use the Distributive Property $ba + ca = (b + c)a$ to add the coefficients.	[2.2.2, p. 66]	Simplify: $6x - 11x$ $6x - 11x = [6 + (-11)]x$ $= -5x$
The Properties of Real Numbers are used to simplify variable expressions.	[2.2.2–2.2.5, pp. 66–69]	Simplify: $5(x - y) - 3(-2x + 4y)$ $5(x - y) - 3(-2x + 4y)$ $= 5x - 5y + 6x - 12y$ $= 11x - 17y$ Simplify: $3x - 2[x + 4(x - 6)]$ $3x - 2[x + 4(x - 6)]$ $= 3x - 2[x + 4x - 24]$ $= 3x - 2[5x - 24]$ $= 3x - 10x + 48$ $= -7x + 48$
Translating verbal expressions into variable expressions requires recognizing the verbal phrases that translate into mathematical operations. See the list on pages 74–75.	[2.3.1–2.3.3, pp. 74–78]	Translate and simplify "twelve added to the sum of eight and a number." $(8 + x) + 12 = 8 + x + 12$ $= x + 20$

CHAPTER 2 Review Exercises

1. Simplify: $-7y^2 + 6y^2 - (-2y^2)$

2. Simplify: $(12x)\left(\frac{1}{4}\right)$

3. Simplify: $\frac{2}{3}(-15a)$

4. Simplify: $-2(2x - 4)$

5. Simplify: $5(2x + 4) - 3(x - 6)$

6. Evaluate $a^2 - 3b$ when $a = 2$ and $b = -4$.

7. Complete the statement by using the Inverse Property of Addition.
$-9 + ? = 0$

8. Simplify: $-4(-9y)$

9. Simplify: $-2(-3y + 9)$

10. Simplify: $3[2x - 3(x - 2y)] + 3y$

11. Simplify: $-4(2x^2 - 3y^2)$

12. Simplify: $3x - 5x + 7x$

13. Evaluate $b^2 - 3ab$ when $a = 3$ and $b = -2$.

14. Simplify: $\frac{1}{5}(10x)$

15. Simplify: $5(3 - 7b)$

16. Simplify: $2x + 3[4 - (3x - 7)]$

17. Identify the property that justifies the statement.
$-4(3) = 3(-4)$

18. Simplify: $3(8 - 2x)$

19. Simplify: $-2x^2 - (-3x^2) + 4x^2$

20. Simplify: $-3x - 2(2x - 7)$

21. Simplify: $-3(3y^2 - 3y - 7)$

22. Simplify: $-2[x - 2(x - y)] + 5y$

23. Evaluate $\frac{-2ab}{2b - a}$ when $a = -4$ and $b = 6$.

24. Simplify: $(-3)(-12y)$

25. Simplify: $4(3x - 2) - 7(x + 5)$

26. Simplify: $(16x)\left(\frac{1}{8}\right)$

27. Simplify: $-3(2x^2 - 7y^2)$

28. Evaluate $3(a - c) - 2ab$ when $a = 2$, $b = 3$, and $c = -4$.

29. Simplify: $2x - 3(x - 2)$

30. Simplify: $2a - (-3b) - 7a - 5b$

31. Simplify: $-5(2x^2 - 3x + 6)$

32. Simplify: $3x - 7y - 12x$

33. Simplify: $\frac{1}{2}(12a)$

34. Simplify: $2x + 3[x - 2(4 - 2x)]$

35. Simplify: $3x + (-12y) - 5x - (-7y)$

36. Simplify: $\left(-\frac{5}{6}\right)(-36b)$

37. Complete the statement by using the Distributive Property.
$(6 + 3)7 = 42 + ?$

38. Simplify: $4x^2 + 9x - 6x^2 - 5x$

39. Simplify: $-\frac{3}{8}(16x^2)$

40. Simplify: $-3[2x - (7x - 9)]$

41. Simplify: $-(8a^2 - 3b^2)$

42. Identify the property that justifies the statement. $-32(0) = 0$

43. Translate "b decreased by the product of seven and b" into a variable expression.

44. Translate "the sum of a number and twice the square of the number" into a variable expression.

45. Translate "three less than the quotient of six and a number" into a variable expression.

46. Translate "ten divided by the difference between y and two" into a variable expression.

47. Translate and simplify "eight times the quotient of twice a number and sixteen."

48. Translate and simplify "the product of four and the sum of two and five times a number."

49. **Geometry** The length of the base of a triangle is 15 in. more than the height of the triangle. Express the length of the base of the triangle in terms of the height of the triangle.

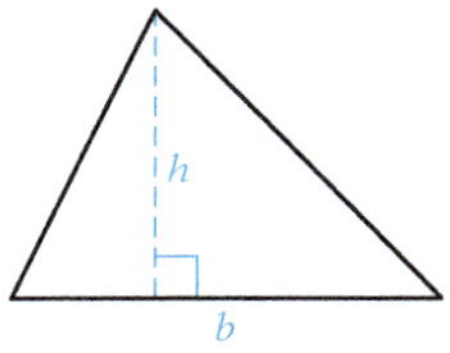

50. **Mixtures** A coffee merchant made 20 lb of a blend of coffee using only mocha java beans and espresso beans. Use the same variable to express the amounts of mocha java beans and espresso beans in the coffee blend.

CHAPTER 2 Test

1. Simplify: $(9y)4$

2. Simplify: $7x + 5y - 3x - 8y$

3. Simplify: $8n - (6 - 2n)$

4. Evaluate $3ab - (2a)^2$ when $a = -2$ and $b = -3$.

5. Identify the property that justifies the statement. $\frac{3}{8}(1) = \frac{3}{8}$

6. Simplify: $-4(-x + 10)$

7. Simplify: $\frac{2}{3}x^2 - \frac{7}{12}x^2$

8. Simplify: $(-10x)\left(-\frac{2}{5}\right)$

9. Simplify: $(-4y^2 + 8)6$

10. Complete the statement by using the Inverse Property of Addition.
$-19 + ? = 0$

11. Evaluate $\frac{-3ab}{2a + b}$ when $a = -1$ and $b = 4$.

12. Simplify: $5(x + y) - 8(x - y)$

13. Simplify: $6b - 9b + 4b$

14. Simplify: $13(6a)$

15. Simplify: $3(x^2 - 5x + 4)$

16. Evaluate $4(b - a) + bc$ when $a = 2$, $b = -3$, and $c = 4$.

17. Simplify: $6x - 3(y - 7x) + 2(5x - y)$

18. Translate "the quotient of eight more than n and seventeen" into a variable expression.

19. Translate "the difference between the sum of a and b and the square of b" into a variable expression.

20. Translate "the sum of the square of a number and the product of the number and eleven" into a variable expression.

21. Translate and simplify "twenty times the sum of a number and nine."

22. Translate and simplify "two more than a number added to the difference between the number and three."

23. Translate and simplify "a number minus the product of one-fourth and twice the number."

24. **Astronomy** The distance from Neptune to the sun is thirty times the distance from Earth to the sun. Express the distance from Neptune to the sun in terms of the distance from Earth to the sun.

25. **Carpentry** A nine-foot board is cut into two pieces of different lengths. Use the same variable to express the lengths of the two pieces.

Cumulative Review Exercises

1. Add: $-4 + 7 + (-10)$

2. Subtract: $-16 - (-25) - 4$

3. Multiply: $(-2)(3)(-4)$

4. Divide: $(-60) \div 12$

5. Write $1\frac{1}{4}$ as a decimal.

6. Write 60% as a fraction and as a decimal.

7. Use the roster method to write the set of negative integers greater than or equal to -4.

8. Write $\frac{2}{25}$ as a percent.

9. Subtract: $\frac{7}{12} - \frac{11}{16} - \left(-\frac{1}{3}\right)$

10. Divide: $\frac{5}{12} \div \left(\frac{3}{2}\right)$

11. Multiply: $\left(-\frac{9}{16}\right)\left(\frac{8}{27}\right)\left(-\frac{3}{2}\right)$

12. Simplify: $-3^2 \cdot \left(-\frac{2}{3}\right)^3$

13. Simplify: $-2^5 \div (3 - 5)^2 - (-3)$

14. Simplify: $\left(-\frac{3}{4}\right)^2 - \left(\frac{3}{8} - \frac{11}{12}\right)$

15. Evaluate $a - 3b^2$ when $a = 4$ and $b = -2$.

16. Simplify: $-2x^2 - (-3x^2) + 4x^2$

17. Simplify: $8a - 12b - 9a$

18. Simplify: $\frac{1}{3}(9a)$

19. Simplify: $\left(-\frac{5}{8}\right)(-32b)$

20. Simplify: $5(4 - 2x)$

21. Simplify: $-3(-2y + 7)$

22. Simplify: $-2(3x^2 - 4y^2)$

23. Simplify: $-4(2y^2 - 5y - 8)$

24. Simplify: $-4x - 3(2x - 5)$

25. Simplify: $3(4x - 1) - 7(x + 2)$

26. Simplify: $3x + 2[x - 4(2 - x)]$

27. Simplify: $3[4x - 2(x - 4y)] + 5y$

28. Translate "the difference between six and the product of a number and twelve" into a variable expression.

29. Translate and simplify "the total of five and the difference between a number and seven."

30. **Biology** A peregrine falcon's maximum flying speed over a quarter-mile distance is four times faster than a wildebeest's maximum running speed over the same distance. (*Source:* www.factmonster.com) Express the speed of the peregrine falcon in terms of the speed of the wildebeest.

CHAPTER

3

Solving Equations and Inequalities

Focus on Success

Are you making attending class a priority? Remember that to be successful, you must attend class. You need to be in class to hear your instructor's explanations and instructions, as well as to ask questions when something is unclear. Most students who miss a class fall behind and then find it very difficult to catch up. (See Class Time, page AIM-5.)

OBJECTIVES

3.1
1. Determine whether a given number is a solution of an equation
2. Solve equations of the form $x + a = b$
3. Solve equations of the form $ax = b$

3.2
1. Applications of percent
2. Uniform motion

3.3
1. Solve equations of the form $ax + b = c$
2. Solve equations of the form $ax + b = cx + d$
3. Solve equations containing parentheses
4. Solve application problems using formulas

3.4
1. Solve inequalities using the Addition Property of Inequalities
2. Solve inequalities using the Multiplication Property of Inequalities
3. Solve general inequalities

PREP TEST

Are you ready to succeed in this chapter?
Take the Prep Test below to find out if you are ready to learn the new material.

1. Subtract: $8 - 12$
2. Multiply: $-\frac{3}{4}\left(-\frac{4}{3}\right)$
3. Multiply: $-\frac{5}{8}(16)$
4. Write 90% as a decimal.
5. Write 0.75 as a percent.
6. Evaluate $3x^2 - 4x - 1$ when $x = -4$.
7. Simplify: $3x - 5 + 7x$
8. Simplify: $8x - 9 - 8x$
9. Simplify: $6x - 3(6 - x)$

3.1 Introduction to Equations

OBJECTIVE 1 Determine whether a given number is a solution of an equation

Point of Interest

One of the most famous equations ever stated is $E = mc^2$. This equation, stated by Albert Einstein, shows that there is a relationship between mass m and energy E.

An **equation** expresses the equality of two mathematical expressions. The expressions can be either numerical or variable expressions.

$$\left.\begin{array}{l} 9 + 3 = 12 \\ 3x - 2 = 10 \\ y^2 + 4 = 2y - 1 \\ z = 2 \end{array}\right\} \text{Equations}$$

The equation at the right is true if the variable is replaced by 5.

$x + 8 = 13$

$5 + 8 = 13$ A true equation

The equation $x + 8 = 13$ is false if the variable is replaced by 7.

$7 + 8 = 13$ A false equation

A **solution** of an equation is a number that, when substituted for the variable, results in a true equation. 5 is a solution of the equation $x + 8 = 13$. 7 is not a solution of the equation $x + 8 = 13$.

EXAMPLE 1 Is -3 a solution of the equation $4x + 16 = x^2 - 5$?

Solution

$4x + 16$	$=$	$x^2 - 5$
$4(-3) + 16$		$(-3)^2 - 5$
$-12 + 16$		$9 - 5$
4	$=$	4

- Replace the variable by the given number, -3.
- Evaluate the numerical expressions using the Order of Operations Agreement.
- Compare the results. If the results are equal, the given number is a solution. If the results are not equal, the given number is not a solution.

Yes, -3 is a solution of the equation $4x + 16 = x^2 - 5$.

Problem 1 Is $\frac{1}{4}$ a solution of $5 - 4x = 8x + 2$?

Solution See page S5.

Try Exercise 17, page 95.

EXAMPLE 2 Is -4 a solution of the equation $4 + 5x = x^2 - 2x$?

Solution

$4 + 5x$	$=$	$x^2 - 2x$
$4 + 5(-4)$		$(-4)^2 - 2(-4)$
$4 + (-20)$		$16 - (-8)$
-16	$\neq$	24

- Replace the variable by the given number, -4.
- Evaluate the numerical expressions using the Order of Operations Agreement.
- Compare the results. If the results are equal, the given number is a solution. If the results are not equal, the given number is not a solution.

No, -4 is not a solution of the equation $4 + 5x = x^2 - 2x$.

Problem 2 Is 5 a solution of $10x - x^2 = 3x - 10$?

Solution See page S5.

Try Exercise 25, page 95.

OBJECTIVE 2 Solve equations of the form $x + a = b$

To **solve an equation** means to find a solution of the equation. The simplest equation to solve is an equation of the form **variable = constant**, because the constant is the solution.

If $x = 5$, then 5 is the solution of the equation because $5 = 5$ is a true equation.

The solution of the equation shown at the right is 7. $\quad x + 2 = 9 \qquad 7 + 2 = 9$

Note that if 4 is added to each side of the equation, the solution is still 7. $\quad x + 2 + 4 = 9 + 4$, $\quad x + 6 = 13 \qquad 7 + 6 = 13$

If -5 is added to each side of the equation, the solution is still 7. $\quad x + 2 + (-5) = 9 + (-5)$, $\quad x - 3 = 4 \qquad 7 - 3 = 4$

This illustrates the Addition Property of Equations.

Take Note

Think of an equation as a balance scale. If the weights added to each side of the scale are not the same, the pans no longer balance. Similarly, the same value must be added to *each* side of the equation for the solution to remain the same.

ADDITION PROPERTY OF EQUATIONS

The same number or variable term can be added to each side of an equation without changing the solution of the equation.

This property is used in solving equations. Note the effect of adding, to each side of the equation $x + 2 = 9$, the opposite of the constant term 2. After each side of the equation is simplified, the equation is in the form variable = constant. The solution is the constant.

$$x + 2 = 9$$
$$x + 2 + (-2) = 9 + (-2)$$
$$x + 0 = 7$$
$$x = 7$$

variable = constant

The solution is 7.

In solving an equation, the goal is to rewrite the given equation in the form variable = constant. The Addition Property of Equations can be used to rewrite an equation in this form. The Addition Property of Equations is used to remove a term from one side of an equation by adding the opposite of that term to each side of the equation.

EXAMPLE 3 Solve and check: $y - 6 = 9$

Solution

$y - 6 = 9$ • The goal is to rewrite the equation in the form *variable* = *constant*.

$y - 6 + 6 = 9 + 6$ • Add 6 to each side of the equation (the Addition Property of Equations).

$y + 0 = 15$ • Simplify using the Inverse Property of Addition.

$y = 15$ • Simplify using the Addition Property of Zero. Now the equation is in the form *variable* = *constant*.

Check

$$\begin{array}{r|l} y - 6 & = 9 \\ \hline 15 - 6 & 9 \\ 9 & = 9 \end{array}$$

• This is a true equation. The solution checks.

The solution is 15. • Write the solution.

Problem 3 Solve and check: $x - \frac{1}{3} = -\frac{3}{4}$

Solution See page S5.

Try Exercise 45, page 96.

Because subtraction is defined in terms of addition, the Addition Property of Equations makes it possible to subtract the same number from each side of an equation without changing the solution of the equation.

Focus on solving an equation of the form $x + a = b$

Solve: $x + \frac{1}{2} = \frac{5}{4}$

The goal is to rewrite the equation in the form *variable* = *constant*.

$$x + \frac{1}{2} = \frac{5}{4}$$

Add the opposite of the constant term $\frac{1}{2}$ to each side of the equation. This is equivalent to subtracting $\frac{1}{2}$ from each side of the equation.

$$x + \frac{1}{2} - \frac{1}{2} = \frac{5}{4} - \frac{1}{2}$$

$$x + 0 = \frac{5}{4} - \frac{2}{4}$$

$$x = \frac{3}{4}$$

The solution $\frac{3}{4}$ checks. The solution is $\frac{3}{4}$.

EXAMPLE 4 Solve: $\frac{1}{2} = y + \frac{2}{3}$

Solution

$$\frac{1}{2} = y + \frac{2}{3}$$

$$\frac{1}{2} - \frac{2}{3} = y + \frac{2}{3} - \frac{2}{3}$$ • Subtract $\frac{2}{3}$ from each side of the equation.

$$-\frac{1}{6} = y$$ • Simplify each side of the equation.

The solution is $-\frac{1}{6}$.

Problem 4 Solve: $-8 = 5 + x$

Solution See page S5.

Try Exercise 65, page 96.

Note from the solution to Example 4 that an equation can be rewritten in the form *constant* = *variable*. Whether the equation is written in the form *variable* = *constant* or in the form *constant* = *variable*, the solution is the constant.

OBJECTIVE 3 Solve equations of the form $ax = b$

The solution of the equation shown at the right is 3.

$2x = 6$ $\qquad 2 \cdot 3 = 6$

Note that if each side of the equation is multiplied by 5, the solution is still 3.

$5 \cdot 2x = 5 \cdot 6$

$10x = 30$ $\qquad 10 \cdot 3 = 30$

If each side is multiplied by -4, the solution is still 3.

$(-4) \cdot 2x = (-4) \cdot 6$

$-8x = -24$ $\qquad -8 \cdot 3 = -24$

This illustrates the Multiplication Property of Equations.

MULTIPLICATION PROPERTY OF EQUATIONS

Each side of an equation can be multiplied by the same nonzero number without changing the solution of the equation.

This property is used in solving equations. Note the effect of multiplying each side of the equation $2x = 6$ by the reciprocal of the coefficient 2. After each side of the equation is simplified, the equation is in the form variable = constant. The solution is the constant.

$$2x = 6$$
$$\frac{1}{2} \cdot 2x = \frac{1}{2} \cdot 6$$
$$1x = 3$$
$$x = 3$$

variable = constant

The solution is 3.

In solving an equation, the goal is to rewrite the given equation in the form variable = constant. The Multiplication Property of Equations can be used to rewrite an equation in this form. The Multiplication Property of Equations is used to write the variable term with a coefficient of 1 by multiplying each side of the equation by the reciprocal of the coefficient.

EXAMPLE 5 Solve: $\frac{3x}{4} = -9$

Solution

$$\frac{3x}{4} = -9$$

- $\frac{3x}{4} = \frac{3}{4}x$

$$\frac{4}{3} \cdot \frac{3}{4}x = \frac{4}{3}(-9)$$

- Multiply each side of the equation by the reciprocal of the coefficient $\frac{3}{4}$ (the Multiplication Property of Equations).

$$1x = -12$$

- Simplify using the Inverse Property of Multiplication.

$$x = -12$$

- Simplify using the Multiplication Property of One. Now the equation is in the form *variable* = *constant*.

The solution is -12.

- Write the solution.

Problem 5 Solve: $-\frac{2x}{5} = 6$

Solution See page S5.

Try Exercise 109, page 97.

Because division is defined in terms of multiplication, the Multiplication Property of Equations makes it possible to divide each side of an equation by the same number without changing the solution of the equation.

Focus on solving an equation of the form $ax = b$

Solve: $8x = 16$

The goal is to rewrite the equation in the form *variable* = *constant*.

$$8x = 16$$

Multiply each side of the equation by the reciprocal of 8. This is equivalent to dividing each side by 8.

$$\frac{8x}{8} = \frac{16}{8}$$
$$x = 2$$

The solution 2 checks.

The solution is 2.

When using the Multiplication Property of Equations to solve an equation, multiply each side of the equation by the reciprocal of the coefficient when the coefficient is a fraction. Divide each side of the equation by the coefficient when the coefficient is an integer or a decimal.

EXAMPLE 6 Solve and check: $4x = 6$

Solution

$$4x = 6$$

$$\frac{4x}{4} = \frac{6}{4}$$

- Divide each side of the equation by 4, the coefficient of x.

$$x = \frac{3}{2}$$

- Simplify each side of the equation.

Check

$$4x = 6$$

$$4\left(\frac{3}{2}\right) \quad 6$$

$$6 = 6$$

- This is a true equation. The solution checks.

The solution is $\frac{3}{2}$.

Problem 6 Solve and check: $6x = 10$

Solution See page S5.

Try Exercise 93, page 96.

Before using one of the Properties of Equations, check to see whether one or both sides of the equation can be simplified. In Example 7, like terms appear on the left side of the equation. The first step in solving this equation is to combine the like terms so that there is only one variable term on the left side of the equation.

EXAMPLE 7 Solve: $5x - 9x = 12$

Solution

$$5x - 9x = 12$$

$$-4x = 12$$

- Combine like terms.

$$\frac{-4x}{-4} = \frac{12}{-4}$$

- Divide each side of the equation by -4.

$$x = -3$$

The solution is -3.

Problem 7 Solve: $4x - 8x = 16$

Solution See page S5.

Try Exercise 117, page 97.

3.1 Exercises

CONCEPT CHECK

1. Label each of the following as either an expression or an equation.
a. $3x + 7 = 9$ **b.** $3x + 7$ **c.** $4 - 6(y + 5)$ **d.** $a + b = 8$ **e.** $a + b - 8$

2. What is the solution of the equation $x = 8$? Use your answer to explain why the goal in solving equations is to get the variable alone on one side of the equation.

3. Which of the following are equations of the form $x + a = b$? If an equation is of the form $x + a = b$, what would you do to solve the equation?
a. $d + 7.8 = -9.2$ **b.** $0.3 = a + 1.4$ **c.** $-9 = 3y$ **d.** $-8 + c = -5.6$

4. Which of the following are equations of the form $ax = b$? If an equation is of the form $ax = b$, what would you do to solve the equation?
a. $3y = -12$ **b.** $2.4 = 0.6a$ **c.** $-5 = z - 10$ **d.** $-8c = -56$

Determine whether the statement is always true, sometimes true, or never true.

5. Both sides of an equation can be multiplied by the same number without changing the solution of the equation.

6. For an equation of the form $ax = b$, $a \neq 0$, multiplying both sides of the equation by the reciprocal of a will result in an equation of the form $x = constant$.

7. Use the Multiplication Property of Equations to remove a term from one side of an equation.

8. Adding negative 3 to each side of an equation yields the same result as subtracting 3 from each side of the equation.

1 Determine whether a given number is a solution of an equation (See page 90.)

GETTING READY

9. Determine whether -3 is a solution of the equation $9 - 5x = -3 - 9x$.

$9 - 5x$	$= -3 - 9x$	
$9 - 5(-3)$	$-3 - 9(-3)$	• **Replace the variable by the given number, ___?___.**
$9 +$ ___?___	$-3 +$ ___?___	• **Evaluate the numerical expressions.**
___?___	___?___	• **Compare the results.**

The results are ___?___; therefore, -3 is a solution of the equation $9 - 5x = -3 - 9x$.

10. Is -1 a solution of $2b - 1 = 3$?

11. Is -2 a solution of $3a - 4 = 10$?

12. Is 1 a solution of $4 - 2m = 3$?

13. Is 2 a solution of $7 - 3n = 2$?

14. Is 5 a solution of $2x + 5 = 3x$?

15. Is 4 a solution of $3y - 4 = 2y$?

16. Is 0 a solution of $4a + 5 = 3a + 5$?

17. Is 0 a solution of $4 - 3b = 4 - 5b$?

18. Is 3 a solution of $z^2 + 1 = 4 + 3z$?

19. Is 2 a solution of $2x^2 - 1 = 4x - 1$?

20. Is 4 a solution of $x(x + 1) = x^2 + 5$?

21. Is 3 a solution of $2a(a - 1) = 3a + 3$?

22. Is $-\frac{1}{4}$ a solution of $8t + 1 = -1$?

23. Is $\frac{1}{2}$ a solution of $4y + 1 = 3$?

24. Is $\frac{2}{5}$ a solution of $5m + 1 = 10m - 3$?

25. Is $\frac{3}{4}$ a solution of $8x - 1 = 12x + 3$?

26. Is 2.1 a solution of $x^2 - 4x = x + 1.89$?

27. Is 1.5 a solution of $c^2 - 3c = 4c - 8.25$?

2 Solve equations of the form $x + a = b$ (See pages 91–92.)

GETTING READY

28. To solve the equation $p - 30 = 57$, use the Addition Property of Equations to add ___?___ to each side of the equation. The solution is ___?___.

29. To solve the equation $18 + n = 25$, ___?___ 18 from each side of the equation. The solution is ___?___.

30. Without solving the equation $x + \frac{13}{15} = -\frac{21}{43}$, determine whether x is less than or greater than $-\frac{21}{43}$. Explain your answer.

31. Without solving the equation $x - \frac{11}{16} = \frac{19}{24}$, determine whether x is less than or greater than $\frac{19}{24}$. Explain your answer.

Solve and check.

32. $x + 5 = 7$ **33.** $y + 3 = 9$ **34.** $b - 4 = 11$ **35.** $z - 6 = 10$

36. $2 + a = 8$ **37.** $5 + x = 12$ **38.** $m + 9 = 3$ **39.** $t + 12 = 10$

40. $n - 5 = -2$ **41.** $x - 6 = -5$ **42.** $b + 7 = 7$ **43.** $y - 5 = -5$

44. $a - 3 = -5$ **45.** $x - 6 = -3$ **46.** $z + 9 = 2$ **47.** $n + 11 = 1$

48. $10 + m = 3$ **49.** $8 + x = 5$ **50.** $9 + x = -3$ **51.** $10 + y = -4$

52. $b - 5 = -3$ **53.** $t - 6 = -4$ **54.** $2 = x + 7$ **55.** $-8 = n + 1$

56. $4 = m - 11$ **57.** $-6 = y - 5$ **58.** $12 = 3 + w$ **59.** $-9 = 5 + x$

60. $4 = -10 + b$ **61.** $-7 = -2 + x$ **62.** $m + \frac{2}{3} = -\frac{1}{3}$ **63.** $c + \frac{3}{4} = -\frac{1}{4}$

64. $x - \frac{1}{2} = \frac{1}{2}$ **65.** $x - \frac{2}{5} = \frac{3}{5}$ **66.** $\frac{5}{8} + y = \frac{1}{8}$ **67.** $\frac{4}{9} + a = -\frac{2}{9}$

68. $-\frac{5}{6} = x - \frac{1}{4}$ **69.** $-\frac{1}{4} = c - \frac{2}{3}$ **70.** $-\frac{1}{21} = m + \frac{2}{3}$

71. $\frac{5}{9} = b - \frac{1}{3}$ **72.** $\frac{5}{12} = n + \frac{3}{4}$ **73.** $d + 1.3619 = 2.0148$

74. $w + 2.932 = 4.801$ **75.** $-0.813 + x = -1.096$ **76.** $-1.926 + t = -1.042$

3 Solve equations of the form $ax = b$ (See pages 92–94.)

77. How is the Multiplication Property of Equations used to solve an equation?

78. Why, when the Multiplication Property of Equations is used, must the number that multiplies each side of the equation not be zero?

GETTING READY

79. To solve the equation $\frac{2}{3}w = -18$, use the Multiplication Property of Equations to multiply each side of the equation by __?__. The solution is __?__.

80. To solve the equation $56 = 8n$, __?__ each side of the equation by 8. The solution is __?__.

81. Without solving the equation $-\frac{15}{41}x = -\frac{23}{25}$, determine whether x is less than or greater than zero. Explain your answer.

82. Without solving the equation $\frac{5}{28}x = -\frac{3}{44}$, determine whether x is less than or greater than zero. Explain your answer.

Solve and check.

83. $5x = 15$ **84.** $4y = 28$ **85.** $3b = -12$ **86.** $2a = -14$

87. $-3x = 6$ **88.** $-5m = 20$ **89.** $-3x = -27$ **90.** $-6n = -30$

91. $20 = 4c$ **92.** $18 = 2t$ **93.** $-32 = 8w$ **94.** $-56 = 7x$

95. $8d = 0$ **96.** $-5x = 0$ **97.** $-64 = 8a$ **98.** $-32 = -4y$

99. $\frac{x}{3} = 2$ **100.** $\frac{x}{4} = 3$ **101.** $-\frac{y}{2} = 5$ **102.** $-\frac{b}{3} = 6$

103. $\frac{n}{7} = -4$ **104.** $\frac{t}{6} = -3$ **105.** $\frac{2}{5}x = 12$ **106.** $-\frac{4}{3}c = -8$

107. $\frac{5}{6}y = -20$

108. $-\frac{2}{3}d = 8$

109. $\frac{2n}{3} = 2$

110. $\frac{5x}{6} = -10$

111. $\frac{-3z}{8} = 9$

112. $\frac{-4x}{5} = -12$

113. $-6 = -\frac{2}{3}y$

114. $-15 = -\frac{3}{5}x$

115. $3n + 2n = 20$

116. $7d - 4d = 9$

117. $10y - 3y = 21$

118. $2x - 5x = 9$

119. $\frac{x}{1.4} = 3.2$

120. $\frac{z}{2.9} = -7.8$

121. $3.4a = 7.004$

122. $-1.6m = 5.44$

123. In the equation $15x = y$, y is a positive integer. Is the value of x a negative number?

124. In the equation $-6x = y$, y is a negative integer. Is the value of x a negative number?

125. In the equation $-\frac{1}{4}x = y$, y is a positive integer. Is the value of x a negative number?

APPLYING CONCEPTS

Solve.

126. $\frac{2m + m}{5} = -9$

127. $\frac{3y - 8y}{7} = 15$

128. $\frac{1}{\frac{1}{x}} = 5$

129. $\frac{1}{\frac{1}{x}} + 8 = -19$

130. $\frac{4}{\frac{3}{b}} = 8$

131. $\frac{5}{\frac{7}{a}} - \frac{3}{\frac{7}{a}} = 6$

132. Solve for x: $x \div 28 = 1481$ remainder 25

PROJECTS OR GROUP ACTIVITIES

133. Make up an equation of the form $x + a = b$ that has 2 as a solution.

134. Make up an equation of the form $ax = b$ that has -2 as a solution.

135. Two numbers form a "two-pair" if the sum of their reciprocals equals 2. For example, $\frac{8}{15}$ and 8 are a two-pair because $\frac{15}{8} + \frac{1}{8} = 2$. If two numbers a and b form a two-pair, and $a = \frac{7}{3}$, what is the value of b?

136. Use the numbers 5, 10, and 15 to fill in the boxes in the equation $x + \square = \square - \square$.
a. What is the largest number solution possible?
b. What is the smallest number solution possible?

137. Match each lettered equation with the numbered question that can be used to solve the equation.

a. $x + 3 = 8$
b. $x - 5 = 20$
c. $4x = 16$
d. $\frac{x}{7} = 1$
e. $99 = -9x$

i. 4 times what number is equal to 16?
ii. What number minus 5 is equal to 20?
iii. 99 is equal to -9 times what number?
iv. What number plus 3 is equal to 8?
v. What number divided by 7 is equal to 1?

3.2 Applications of Equations of the Form $ax = b$

OBJECTIVE 1 Applications of percent

Solving a problem that involves a percent requires solving the basic percent equation.

THE BASIC PERCENT EQUATION

Percent · base = amount

$$P \cdot B = A$$

To translate a problem involving a percent into an equation, remember that the word *of* translates into "multiply" and the word *is* translates into "=." The base usually follows the word *of*.

Focus on solving the basic percent equation for the base

20% of what number is 30?

Given: $P = 20\% = 0.20$
$A = 30$
Unknown: Base

$$PB = A$$
$$(0.20)B = 30$$
$$\frac{0.20B}{0.20} = \frac{30}{0.20}$$
$$B = 150$$

20% of 150 is 30.

Focus on solving the basic percent equation for the percent

What percent of 40 is 30?

Given: $B = 40$
$A = 30$

Unknown: Percent

We must write the fraction as a percent in order to answer the question.

$$PB = A$$
$$P(40) = 30$$
$$40P = 30$$
$$\frac{40P}{40} = \frac{30}{40}$$
$$P = \frac{3}{4}$$
$$P = 75\%$$

30 is 75% of 40.

Focus on solving the basic percent equation for the amount

Find 25% of 200.

Given: $P = 25\% = 0.25$
$B = 200$

Unknown: Amount

$$PB = A$$
$$0.25(200) = A$$
$$50 = A$$

25% of 200 is 50.

In most cases, we write a percent as a decimal before solving the basic percent equation. However, some percents are more easily written as a fraction. For example,

$$33\frac{1}{3}\% = \frac{1}{3} \qquad 66\frac{2}{3}\% = \frac{2}{3} \qquad 16\frac{2}{3}\% = \frac{1}{6} \qquad 83\frac{1}{3}\% = \frac{5}{6}$$

EXAMPLE 1 12 is $33\frac{1}{3}\%$ of what number?

Solution

$$PB = A$$

$$\frac{1}{3}B = 12$$ • $P = 33\frac{1}{3}\% = \frac{1}{3}$; $A = 12$

$$3 \cdot \frac{1}{3}B = 3 \cdot 12$$ • Multiply each side of the equation by 3, the reciprocal of $\frac{1}{3}$.

$$B = 36$$

12 is $33\frac{1}{3}\%$ of 36.

Problem 1 27 is what percent of 60?

Solution See page S5.

Try Exercise 11, page 104.

EXAMPLE 2 During a recent year, nearly 1.2 million dogs or litters were registered with the American Kennel Club. The most popular breed was the Labrador retriever, with 172,841 registered. What percent of the registrations were Labrador retrievers? Round to the nearest tenth of a percent. (*Source:* American Kennel Club)

Strategy To find the percent, solve the basic percent equation using $B = 1.2$ million $= 1{,}200{,}000$ and $A = 172{,}841$. The percent is unknown.

Solution

$$PB = A$$

$$P(1{,}200{,}000) = 172{,}841$$ • $B = 1{,}200{,}000$; $A = 172{,}841$

$$\frac{P(1{,}200{,}000)}{1{,}200{,}000} = \frac{172{,}841}{1{,}200{,}000}$$ • Divide each side of the equation by 1,200,000.

$$P \approx 0.144$$

$$P \approx 14.4\%$$ • Rewrite the decimal as a percent.

Approximately 14.4% of the registrations were Labrador retrievers.

Problem 2 A student correctly answered 72 of the 80 questions on an exam. What percent of the questions were answered correctly?

Solution See page S5.

Try Exercise 43, page 105.

Point of Interest

As noted at the right, 172,841 of the dogs or litters registered were Labrador retrievers. Listed below are the next most popular breeds and their registrations.

Golden retrievers: 66,300
German shepherds: 57,660
Dachshunds: 54,773
Beagles: 52,026

EXAMPLE 3 According to the Centers for Disease Control and Prevention, 30.8% of the adult population of Kentucky smokes. How many adults in Kentucky smoke? Use a figure of 3,000,000 for the number of adults in Kentucky.

Strategy To find the number of adults who smoke, solve the basic percent equation using $B = 3{,}000{,}000$ and $P = 30.8\% = 0.308$. The amount is unknown.

Solution

$$PB = A$$

$$0.308(3{,}000{,}000) = A$$ • $P = 0.308$, $B = 3{,}000{,}000$

$$924{,}000 = A$$ • Multiply 0.308 by 3,000,000.

Approximately 924,000 adults in Kentucky smoke.

Problem 3 The price of a digital camcorder is \$895. A 6% sales tax is added to the price. How much is the sales tax?

Solution See page S5.

Try Exercise 39, page 105.

The simple interest that an investment earns is given by the equation

$$I = Prt$$

where I is the simple interest, P is the principal, or amount invested, r is the simple interest rate, and t is the time.

Focus on solving a problem using the simple interest equation

A \$1500 investment has an annual simple interest rate of 7%. Find the simple interest earned on the investment after 18 months.

The time is given in months, but the interest rate is an annual rate. Therefore, we must convert 18 months to years.

$$18 \text{ months} = \frac{18}{12} \text{ years} = 1.5 \text{ years}$$

To find the interest earned, solve the equation $I = Prt$ for I using $P = 1500$, $r = 0.07$, and $t = 1.5$.

$$I = Prt$$
$$I = 1500(0.07)(1.5)$$
$$I = 157.5$$

The interest earned is \$157.50.

EXAMPLE 4 In April, Marshall Wardell was charged \$8.72 in interest on an unpaid credit card balance of \$545. Find the annual interest rate for this credit card.

Strategy The interest is \$8.72. Therefore, $I = 8.72$. The unpaid balance is \$545. This is the principal on which interest is calculated. Therefore, $P = 545$. The time is 1 month. Because the *annual* interest rate must be found and the time is given as 1 month, write 1 month as $\frac{1}{12}$ year. Therefore, $t = \frac{1}{12}$. To find the interest rate, solve the equation $I = Prt$ for r.

Solution

$$I = Prt$$

$$8.72 = 545r\left(\frac{1}{12}\right)$$ • $I = 8.72, P = 545, t = \frac{1}{12}$

$$8.72 = \frac{545}{12}r$$ • Multiply 545 by $\frac{1}{12}$.

$$\frac{12}{545}(8.72) = \frac{12}{545}\left(\frac{545}{12}r\right)$$ • Multiply each side of the equation by $\frac{12}{545}$, the reciprocal of $\frac{545}{12}$.

$$0.192 = r$$

The annual interest rate is 19.2%.

Problem 4 Clarissa Adams purchased a \$1000 municipal bond that earns an annual simple interest rate of 6.4%. How much must she deposit into a bank account that earns 8% annual simple interest so that the interest earned from each account after one year is the same?

Take Note

Municipal bonds are frequently sold by government agencies to pay for roads, aqueducts, and other essential public needs. An investor who purchases one of these bonds is paid interest on the cost of the bond for a certain number of years. When that time period expires, the cost of the bond is returned to the investor.

Solution See pages S5–S6.

Try Exercise 51, page 106.

Point of Interest

In the jewelry industry, the amount of gold in a piece of jewelry is measured in *karats*. Pure gold is 24 karats. A necklace that is 18 karats is $\frac{18}{24} = 0.75 = 75\%$ gold.

The amount of a substance in a solution can be given as a percent of the total solution. For instance, if a certain fruit juice drink is advertised as containing 27% cranberry juice, then 27% of the contents of the bottle must be cranberry juice.

Problems involving mixtures are solved using the percent mixture equation

$$Q = Ar$$

where Q is the quantity of a substance in the solution, A is the amount of the solution, and r is the percent concentration of the substance.

Focus on solving a problem using the percent mixture equation

The formula for a perfume requires that the concentration of jasmine be 1.2% of the total amount of perfume. How many ounces of jasmine are in a 2-ounce bottle of this perfume?

The amount of perfume is 2 oz. Therefore, $A = 2$. The percent concentration is 1.2%, so $r = 0.012$. To find the number of ounces of jasmine, solve the equation $Q = Ar$ for Q.

$$Q = Ar$$
$$Q = 2(0.012)$$
$$Q = 0.024$$

The bottle contains 0.024 oz of jasmine.

EXAMPLE 5 To make a certain color of blue, 4 oz of cyan must be contained in 1 gal of paint. What is the percent concentration of cyan in the paint?

Strategy The amount of cyan is given in ounces and the amount of paint is given in gallons; we must convert ounces to gallons or gallons to ounces. For this problem, we will convert gallons to ounces: 1 gal = 128 oz. Therefore, $A = 128$. The quantity of cyan in the paint is 4 oz; $Q = 4$. To find the percent concentration, solve the equation $Q = Ar$ for r.

Solution

$$Q = Ar$$
$$4 = 128r$$
$$\frac{4}{128} = \frac{128r}{128}$$
$$0.03125 = r$$

The percent concentration of cyan is 3.125%.

Problem 5 The concentration of sugar in Choco-Pops cereal is 25%. If a bowl of this cereal contains 2 oz of sugar, how many ounces of cereal are in the bowl?

Solution See page S6.

Try Exercise 59, page 106.

OBJECTIVE 2 Uniform motion

> **Take Note**
> A car traveling in a *circle* at a constant speed of 45 mph is *not* in uniform motion because the direction of the car is always changing.

Any object that travels at a constant speed in a straight line is said to be in *uniform motion*. **Uniform motion** means that the speed and direction of an object do not change. For instance, a car traveling at a constant speed of 45 mph on a straight road is in uniform motion.

The solution of a uniform motion problem is based on the equation $d = rt$, where d is the distance traveled, r is the rate of travel, and t is the time spent traveling. Suppose a car travels at 50 mph for 3 h. Because the rate (50 mph) and the time (3 h) are known, we can find the distance traveled by solving the equation $d = rt$ for d.

$$d = rt$$
$$d = 50(3) \quad \bullet\ r = 50,\ t = 3$$
$$d = 150$$

The car travels a distance of 150 mi.

Focus on solving a problem using the equation $d = rt$

A jogger runs 3 mi in 45 min. What is the rate of the jogger in miles per hour?

Because the answer must be in miles per *hour* and the time is given in *minutes*, convert 45 min to hours.

$$45 \text{ min} = \frac{3}{4} \text{ h}$$

To find the rate of the jogger, solve the equation $d = rt$ for r using $d = 3$ and $t = \frac{3}{4}$.

$$d = rt$$
$$3 = r\left(\frac{3}{4}\right)$$
$$3 = \frac{3}{4}r$$
$$\left(\frac{4}{3}\right)3 = \left(\frac{4}{3}\right)\frac{3}{4}r$$
$$4 = r$$

The rate of the jogger is 4 mph.

If two objects are moving in opposite directions, then the rate at which the distance between them is increasing is the sum of the speeds of the two objects. For instance, in the diagram at the right, two cars start from the same point and travel in opposite directions. The distance between them is changing at the rate of 70 mph.

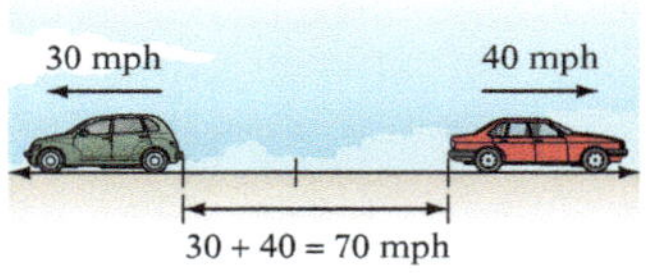

Similarly, if two objects are moving toward each other, the distance between them is decreasing at a rate that is equal to the sum of the speeds. The rate at which the two planes at the right are approaching one another is 800 mph.

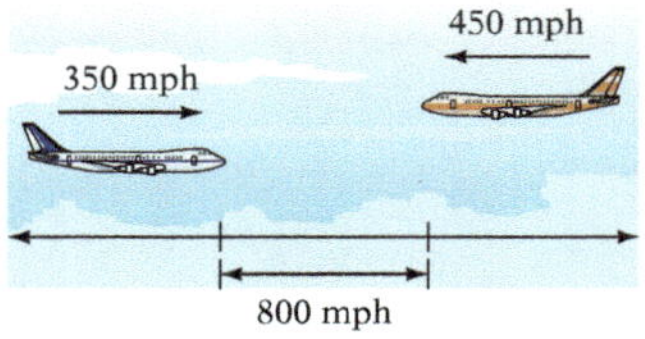

EXAMPLE 6 Two cars start from the same point and move in opposite directions. The car moving west is traveling 45 mph, and the car moving east is traveling 60 mph. In how many hours will the cars be 210 mi apart?

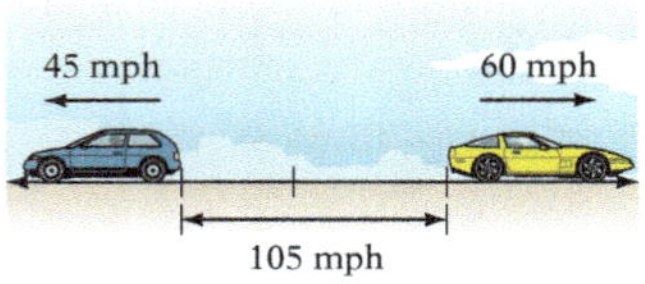

Strategy The distance is 210 mi. Therefore, $d = 210$. The cars are moving in opposite directions, so the rate at which the distance between them is changing is the sum of the rates of the two cars. The rate is 45 mph + 60 mph = 105 mph. Therefore, $r = 105$. To find the time, solve the equation $d = rt$ for t.

Solution

$$d = rt$$

$$210 = 105t \quad \bullet\ d = 210, r = 105$$

$$\frac{210}{105} = \frac{105t}{105} \quad \bullet\ \text{Divide each side of the equation by 105, the coefficient of } t.$$

$$2 = t$$

In 2 h, the cars will be 210 mi apart.

Problem 6 Two cyclists start at the same time at opposite ends of an 80-mile course. One cyclist is traveling 18 mph, and the second cyclist is traveling 14 mph. How long after they begin will they meet?

Solution See page S6.

Try Exercise 81, page 108.

If a motorboat is on a river that is flowing at a rate of 4 mph, the boat will float down the river at a speed of 4 mph even though the motor is not on. Now suppose the motor is turned on and the power adjusted so that the boat will travel at 10 mph without the aid of the current. Then, if the boat is moving *with* the current, its effective speed is the speed of the boat using power plus the speed of the current: 10 mph + 4 mph = 14 mph. However, if the boat is moving in the opposite direction of the current, the current slows the boat down, and the effective speed of the boat is the speed of the boat using power minus the speed of the current: 10 mph − 4 mph = 6 mph.

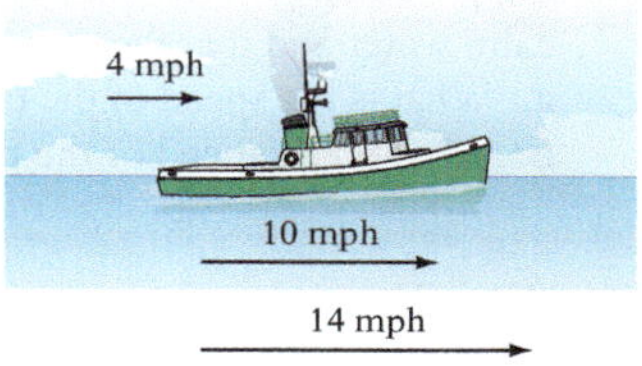

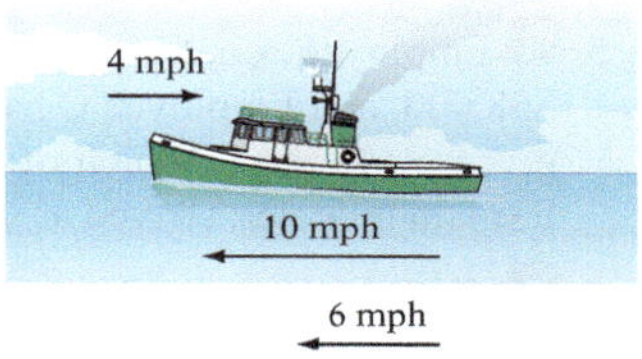

There are other situations in which these concepts may be applied.

Take Note
"ft/s" is an abbreviation for "feet per second."

EXAMPLE 7 An airline passenger is walking between two airline terminals and decides to get on a moving sidewalk that is 150 ft long. If the passenger walks at a rate of 7 ft/s and the moving sidewalk moves at a rate of 9 ft/s, how long, in seconds, will it take the passenger to walk from one end of the moving sidewalk to the other?

Strategy The distance is 150 ft. Therefore, $d = 150$. The passenger is traveling at 7 ft/s and the moving sidewalk is traveling at 9 ft/s. The rate of the passenger is the sum of the two rates, or 16 ft/s. Therefore, $r = 16$. To find the time, solve the equation $d = rt$ for t.

Solution

$$d = rt$$

$$150 = 16t \quad \bullet\ d = 150, r = 16$$

$$\frac{150}{16} = \frac{16t}{16} \quad \bullet\ \text{Divide each side of the equation by 16, the coefficient of } t.$$

$$9.375 = t$$

It will take the passenger 9.375 s to travel the length of the moving sidewalk.

Problem 7 A plane that can normally travel at 250 mph in calm air is flying into a headwind of 25 mph. How far can the plane fly in 3 h?

Solution See page S6.

Try Exercise 79, page 108.

3.2 Exercises

CONCEPT CHECK

Identify the amount and the base.

1. 30 is 75% of 40.

2. 40% of 20 is 8.

3. Keith and Jennifer started at the same time and rode toward each other on a straight road. When they met, Keith had traveled 15 mi and Jennifer had traveled 10 mi. Who had the greater average speed?

4. Suppose that you have a powerboat with the throttle set to move the boat at 8 mph in calm water and that the rate of the current of a river is 4 mph.
a. What is the speed of the boat when traveling on this river with the current?
b. What is the speed of the boat when traveling on this river against the current?

1 Applications of percent (See pages 97–101.)

5. Employee A had an annual salary of $52,000, Employee B had an annual salary of $58,000, and Employee C had an annual salary of $56,000 before each employee was given a 5% raise. Which of the three employees now has the highest annual salary? Explain how you arrived at your answer.

6. Each of three employees earned an annual salary of $65,000 before Employee A was given a 3% raise, Employee B was given a 6% raise, and Employee C was given a 4.5% raise. Which of the three employees now has the highest annual salary? Explain how you arrived at your answer.

Solve.

7. 12 is what percent of 50?

8. What percent of 125 is 50?

9. Find 18% of 40.

10. What is 25% of 60?

11. 12% of what is 48?

12. 45% of what is 9?

13. What is $33\frac{1}{3}\%$ of 27?

14. Find $16\frac{2}{3}\%$ of 30.

15. What percent of 12 is 3?

16. 10 is what percent of 15?

17. 60% of what is 3?

18. 75% of what is 6?

19. 12 is what percent of 6?

20. 20 is what percent of 16?

21. $5\frac{1}{4}\%$ of what is 21?

22. $37\frac{1}{2}\%$ of what is 15?

23. Find 15.4% of 50.

24. What is 18.5% of 46?

25. 1 is 0.5% of what?

26. 3 is 1.5% of what?

27. $\frac{3}{4}\%$ of what is 3?

28. $\frac{1}{2}\%$ of what is 3?

29. Find 125% of 16.

30. What is 250% of 12?

31. 16.4 is what percent of 20.4? Round to the nearest percent.

32. Find 18.3% of 625. Round to the nearest tenth.

33. Without solving an equation, determine whether 40% of 80 is less than, equal to, or greater than 80% of 40.

34. Without solving an equation, determine whether $\frac{1}{4}\%$ of 80 is less than, equal to, or greater than 25% of 80.

GETTING READY

Complete Exercises 35 and 36 by replacing the question marks with the correct number from the problem situation or with the word *unknown*.

35. Problem Situation: It rained on 24 of the 30 days of June. What percent of the days in June were rainy days?

Using the formula $PB = A$, $P =$ ___?___, $B =$ ___?___, and $A =$ ___?___.

36. Problem Situation: You bought a used car and made a down payment of 25% of the purchase price of \$16,000. How much was the down payment?

Using the formula $PB = A$, $P =$ ___?___, $B =$ ___?___, and $A =$ ___?___.

37. Stadiums The Arthur Ashe Tennis Stadium in New York has a seating capacity of 22,500 people. Of these seats, 1.11% are wheelchair-accessible seats. How many seats in the stadium are wheelchair accessible? Round to the nearest whole number.

38. Education The graph at the right shows the sources of revenue in a recent year for public schools in the United States. The total revenue was \$419.8 billion dollars. How many billion dollars more did public schools receive from the state than from the federal government? Round to the nearest billion dollars. (*Source:* Census Bureau)

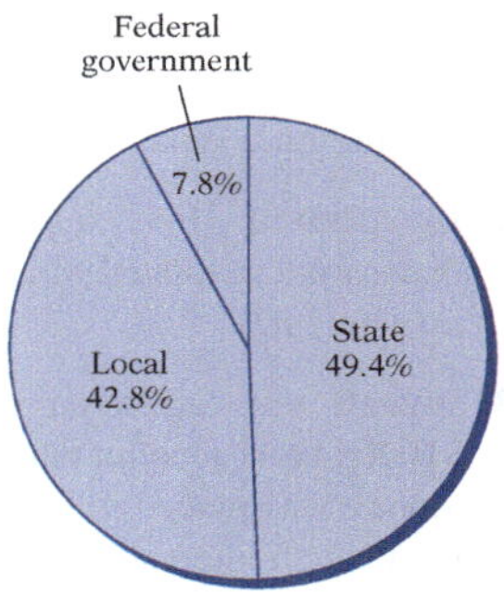

Sources of Revenue for Public Schools

39. Natural Resources On average, a person uses 13.2 gal of water per day for showering. This is 17.8% of the total amount of water used per person per day in the average single-family home. Find the total amount of water used per person per day in the average single-family home. Round to the nearest whole number. (*Source:* American Water Works Association)

40. Travel According to the annual Summer Vacation Survey conducted by Myvesta, a nonprofit consumer education organization, the average summer vacation costs \$2252. If \$1850 of this amount is charged on a credit card, what percent of the vacation cost is charged? Round to the nearest tenth of a percent.

41. The Internet The graph at the right shows the responses to a Yahoo! survey that asked people how many hours per week they typically spend online. How many people were surveyed?

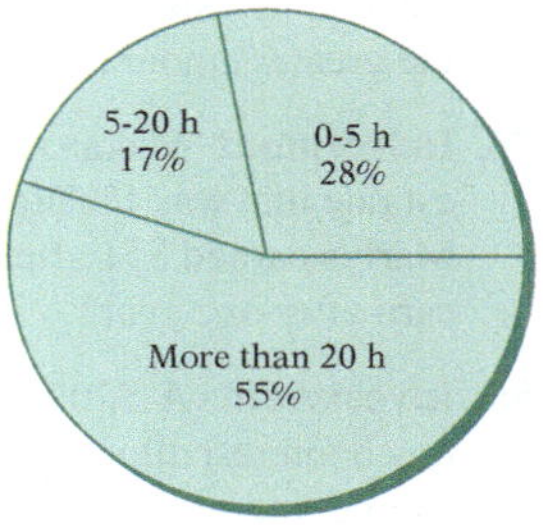

Hours per Week Spent Online

42. Health The maximum recommended daily amount of sodium for a healthy adult is 2300 mg. A McDonald's Quarter Pounder with Cheese contains approximately 1190 mg of sodium. What percentage of the maximum recommended daily allowance of sodium is contained in a McDonald's Quarter Pounder with Cheese? Round to the nearest tenth of a percent. (*Source: Consumer Reports,* August 2010)

43. Safety Recently, the National Safety Council collected data on the leading causes of accidental death. The findings revealed that for people age 20, 30 died from a fall, 47 from fire, 200 from drowning, and 1950 from motor vehicle accidents. What percent of the accidental deaths were not attributed to motor vehicle accidents? Round to the nearest percent.

44. Health According to *Health* magazine, the average American has increased his or her daily consumption of calories from 18 years ago by 11.6%. If the average daily consumption was 1970 calories 18 years ago, what is the average daily consumption today? Round to the nearest whole number.

45. Energy The Energy Information Administration reports that if every U.S. household switched 4 h of lighting per day from incandescent bulbs to compact fluorescent bulbs, we would save 31.7 billion kilowatt-hours of electricity a year, or 33% of the total electricity used for home lighting. What is the total electricity used for home lighting in this country? Round to the nearest tenth of a billion.

46. The Federal Government To override a presidential veto, at least $66\frac{2}{3}\%$ of the Senate must vote to override the veto. There are 100 senators in the Senate. What is the minimum number of votes needed to override a veto?

47. **Super Bowl** According to the Associated Press, 106.5 million people watched Super Bowl XLIV. What percent of the U.S. population watched Super Bowl XLIV? Use a figure of 310 million for the U.S. population. Round to the nearest tenth of a percent.

48. **Boston Marathon** See the news clipping at the right. What percent of the runners who started the course finished the race? Round to the nearest tenth of a percent.

49. **Consumerism** A 20-horsepower hydrostatic lawn tractor sells for $1579.99. This is 11% more than the price of the same lawn tractor with a standard transmission. What is the price of the less expensive model? Round to the nearest cent.

50. **Investment** If Miranda Perry invests $2500 in an account that earns an 8% annual simple interest rate, how much interest will Miranda have earned after 9 months?

51. **Investment** Kachina Caron invests $1200 in a simple interest account and earns $72 in 8 months. What is the annual simple interest rate?

52. **Investment** How much money must Andrea invest for 2 years in an account that earns an annual simple interest rate of 8% if she wants to earn $300 from the investment?

53. **Investment** Sal Boxer divided a gift of $3000 into two different accounts. He placed $1000 in one account that earned an annual simple interest rate of 7.5%. The remaining money was placed in an account that earned an annual simple interest rate of 8.25%. How much interest did Sal earn from the two accounts after one year?

54. **Investment** If Americo invests $2500 at an 8% annual simple interest rate and Octavia invests $3000 at a 7% annual simple interest rate, which of the two will earn the greater amount of interest in one year?

55. **Investment** Makana invested $900 in an account that earned an annual simple interest rate that was 1% higher than the rate her friend Marlys earned on her investment. If Marlys earned $51 after one year from an investment of $850, how much did Makana earn after one year?

56. **Investment** A $2000 investment at an annual simple interest rate of 6% earned as much interest after one year as another investment in an account that earned 8% annual simple interest. How much was invested at 8%?

57. **Investment** An investor places $1000 in an account that earns 9% annual simple interest and $1000 in an account that earns 6% annual simple interest. If each investment is left in the account for the same period of time, is the interest rate on the combined investment less than 6%, between 6% and 9%, or greater than 9%?

58. **Mixture Problem** The concentration of platinum in a necklace is 15%. The necklace weighs 12 g. Find the amount of platinum in the necklace.

59. **Mixture Problem** A 250-milliliter solution of a fabric dye contains 5 ml of hydrogen peroxide. What is the percent concentration of the hydrogen peroxide?

60. **Mixture Problem** A carpet is made of a blend of wool and other fibers. If the concentration of wool in the carpet is 75% and the carpet weighs 175 lb, how much wool is in the carpet?

61. **Mixture Problem** Apple Dan's 32-ounce apple-flavored fruit drink contains 8 oz of apple juice. Forty ounces of a generic brand of apple-flavored fruit drink contains 9 oz of apple juice. Which of the two brands has the greater concentration of apple juice?

62. **Mixture Problem** Bakers use simple syrup in many of their recipes. Simple syrup is made by combining 500 g of sugar with 500 g of water and mixing it well until the sugar dissolves. What is the percent concentration of sugar in simple syrup?

In the News

Thousands Complete Boston Marathon

This year, there were 26,735 entrants in the Boston Marathon, the world's oldest annual marathon. Of those registered, 23,126 people started the race, and 22,629 people finished the 26.2-mile course.

Source: www.bostonmarathon.org

63. **Mixture Problem** A pharmacist has 50 g of a topical cream that contains 75% glycerin. How many grams of the cream are not glycerin?

64. **Mixture Problem** A chemist has 100 ml of a solution that is 9% acetic acid. If the chemist adds 50 ml of pure water to this solution, what is the percent concentration of the resulting mixture?

65. **Mixture Problem** A 500-gram salt and water solution contains 50 g of salt. This mixture is left in the open air and 100 g of water evaporates from the solution. What is the percent concentration of salt in the remaining solution?

2 Uniform motion (See pages 102–103.)

GETTING READY

66. In the formula $d = rt$, d represents ___?___, r represents ___?___, and t represents ___?___.

67. A car that travels 10 mi in 30 min is traveling at a rate of ___?___ miles per hour.

68. Maria and Nathan start hiking at the same time. Maria hikes at 3 mph and Nathan hikes at 2.5 mph. After t hours, Maria has hiked ___?___ miles and Nathan has hiked ___?___ miles.

69. A plane flies at a rate of 325 mph in calm air. The wind is blowing at 30 mph. Flying with the wind, the plane travels ___?___ mph. Flying against the wind, the plane travels ___?___ mph.

70. Joe and John live 2 mi apart. They leave their houses at the same time and walk toward each other until they meet. Joe walks faster than John does.

a. Is the distance walked by Joe less than, equal to, or greater than the distance walked by John?

b. Is the time spent walking by Joe less than, equal to, or greater than the time spent walking by John?

c. What is the total distance traveled by both Joe and John?

71. Morgan and Emma ride their bikes from Morgan's house to the store. Morgan begins biking 5 min before Emma begins. Emma bikes faster than Morgan and catches up with her just as they reach the store.

a. Is the distance biked by Emma less than, equal to, or greater than the distance biked by Morgan?

b. Is the time spent biking by Emma less than, equal to, or greater than the time spent biking by Morgan?

72. **Trains** See the news clipping at the right. Find the time it will take the high-speed train to travel between the two cities. Round to the nearest tenth of an hour.

BartlomiejMagierowski/Shutterstock.com

73. A train travels at 45 mph for 3 h and then increases its speed to 55 mph for 2 more hours. How far does the train travel in the 5-hour period?

74. As part of a training program for the Boston Marathon, a runner wants to build endurance by running at a rate of 9 mph for 20 min. How far will the runner travel in that time period?

In the News

World's Fastest Train

China has unveiled the world's fastest rail link—a train that connects the cities of Guangzhou and Wuhan and can travel at speeds of up to 394.2 km/h. The distance between the two cities is 1069 km, and the train will travel that distance at an average speed of 350 km/h (217 mph). The head of the transport bureau at the Chinese railway ministry boasted, "it's the fastest train in operation in the world."

Source: news.yahoo.com

75. It takes a hospital dietician 40 min to drive from home to the hospital, a distance of 20 mi. What is the dietician's average rate of speed?

76. Marcella leaves home at 9:00 A.M. and drives to school, arriving at 9:45 A.M. If the distance between home and school is 27 mi, what is Marcella's average rate of speed?

77. The Ride for Health Bicycle Club has chosen a 36-mile course for this Saturday's ride. If the riders plan on averaging 12 mph while they are riding and they have a 1-hour lunch break planned, how long will it take them to complete the trip?

78. Palmer's average running speed is 3 km/h faster than his walking speed. If Palmer can run around a 30-kilometer course in 2 h, how many hours would it take for Palmer to walk the same course?

79. A shopping mall has a moving sidewalk that takes shoppers from the shopping area to the parking garage, a distance of 250 ft. If your normal walking rate is 5 ft/s and the moving sidewalk is traveling at 3 ft/s, how many seconds would it take you to walk on the moving sidewalk from the parking garage to the shopping area?

Denis Babenko/Shutterstock.com

80. K&B River Tours offers a river trip that takes passengers from the K&B dock to a small island that is 24 mi away. The passengers spend 1 h at the island and then return to the K&B dock. If the speed of the boat is 10 mph in calm water and the rate of the current is 2 mph, how long does the trip last?

81. Two joggers start at the same time from opposite ends of an 8-mile jogging trail and begin running toward each other. One jogger is running at a rate of 5 mph, and the other jogger is running at a rate of 7 mph. How long, in minutes, after they start will the two joggers meet?

82. **sQuba** See the news clipping at the right. Two sQubas are on opposite sides of a lake 1.6 mi wide. They start toward each other at the same time, one traveling on the surface of the water and the other traveling underwater. In how many minutes will the sQuba traveling on the surface of the water be directly above the sQuba traveling underwater? Assume they are traveling at top speed.

In the News

Underwater Driving —Not So Fast!

Swiss company Rinspeed, Inc., presented its new car, the sQuba, at the Geneva Auto Show. The sQuba can travel on land, on water, and underwater. With a new sQuba,you can expect top speeds of 77 mph when driving on land, 3 mph when driving on the surface of the water, and 1.8 mph when driving underwater!

Source: Seattle Times

83. Two cyclists start from the same point at the same time and move in opposite directions. One cyclist is traveling at 8 mph, and the other cyclist is traveling at 9 mph. After 30 min, how far apart are the two cyclists?

84. Petra, who can paddle her canoe at a rate of 10 mph in calm water, is paddling her canoe on a river whose current is 2 mph. How long will it take her to travel 4 mi upstream against the current?

85. At 8:00 A.M., a train leaves a station and travels at a rate of 45 mph. At 9:00 A.M., a second train leaves the same station on the same track and travels in the direction of the first train at a speed of 60 mph. At 10:00 A.M., how far apart are the two trains?

APPLYING CONCEPTS

86. **Consumerism** Your bill for dinner, including a 7.25% sales tax, was $92.74. You want to leave a 15% tip on the cost of the dinner before the sales tax. Find the amount of the tip to the nearest dollar.

87. Consumerism The total cost for a dinner was \$97.52. This included a 15% tip calculated on the cost of the dinner after a 6% sales tax. Find the cost of dinner before the tip and tax.

88. Business A retailer decides to increase the original price of each item in the store by 10%. After the price increase, the retailer notices a significant drop in sales and so decides to reduce the current price of each item in the store by 10%. Are the prices back to the original prices? If not, are the prices lower or higher than the original prices?

89. If a quantity increases by 100%, how many times its original value is the new value?

90. The following problem does not contain enough information: "How many hours does it take to fly from Los Angeles to New York?" What additional information do we need in order to answer the question?

PROJECTS OR GROUP ACTIVITIES

91. U.S. Population The circle graph at the right shows the population of the United States, in millions, by region. (*Source:* U.S. Census Bureau)

a. What percent of the U.S. population lives in each region? Round to the nearest tenth.

b. Which region has the largest population? In which region does the largest percent of the population live?

According to the Census Bureau, California has the highest population of all the states, with 38 million. Wyoming, with 0.1683% of the U.S. population, has the least number of residents.

c. What percent of the U.S. population lives in California? Round to the nearest tenth of a percent.

d. How many residents live in Wyoming? Round to the nearest ten thousand.

e. What percent of the U.S. population lives in the state you live in?

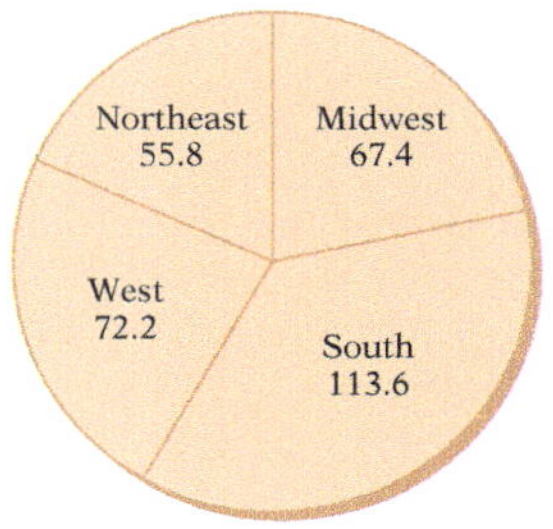

U.S. Population by Region
(in millions of residents)

92. Monthly Mortgage Payments Suppose you have a \$100,000 30-year mortgage. What is the difference between the monthly payment if the interest rate on your loan is 7.75% and the monthly payment if the interest rate is 7.25%? (*Hint:* You will need to discuss the question with an agent at a bank that provides mortgage services, discuss the question with a real estate agent, or find and use the formula for determining the monthly payment on a mortgage.)

3.3 General Equations

OBJECTIVE 1 Solve equations of the form $ax + b = c$

In solving an equation of the form $ax + b = c$, the goal is to rewrite the equation in the form *variable* = *constant*. This requires applying both the Addition and the Multiplication Properties of Equations.

Focus on solving an equation of the form $ax + b = c$

Solve: $\frac{2}{5}x - 3 = -7$

	$\frac{2}{5}x - 3 = -7$
Add 3 to each side of the equation.	$\frac{2}{5}x - 3 + 3 = -7 + 3$
Simplify.	$\frac{2}{5}x = -4$
Multiply each side of the equation by the reciprocal of the coefficient $\frac{2}{5}$.	$\frac{5}{2} \cdot \frac{2}{5}x = \frac{5}{2}(-4)$
Simplify. Now the equation is in the form *variable* = *constant*.	$x = -10$

Check

$\frac{2}{5}x - 3$	-7
$\frac{2}{5}(-10) - 3$	-7
$-4 - 3$	-7

$-7 = -7$

Write the solution. The solution is -10.

EXAMPLE 1 Solve: $3x - 7 = -5$

Solution

$3x - 7 = -5$
$3x - 7 + 7 = -5 + 7$ • Add 7 to each side of the equation.
$3x = 2$ • Simplify.
$\frac{3x}{3} = \frac{2}{3}$ • Divide each side of the equation by 3.
$x = \frac{2}{3}$ • Simplify. Now the equation is in the form *variable* = *constant*.

The solution is $\frac{2}{3}$. • Write the solution.

Problem 1 Solve: $5x + 7 = 10$

Solution See page S6.

Try Exercise 21, page 117.

EXAMPLE 2 Solve: $5 = 9 - 2x$

Solution

$5 = 9 - 2x$
$5 - 9 = 9 - 9 - 2x$ • Subtract 9 from each side of the equation.
$-4 = -2x$ • Simplify.
$\frac{-4}{-2} = \frac{-2x}{-2}$ • Divide each side of the equation by -2.
$2 = x$ • Simplify.

The solution is 2. • Write the solution.

Problem 2 Solve: $11 = 11 + 3x$

Solution See page S6.

Try Exercise 31, page 117.

Note in Example 4 in the Introduction to Equations section that the original equation, $\frac{1}{2} = y + \frac{2}{3}$, contained fractions with denominators of 2 and 3. The least common multiple of 2 and 3 is 6. The least common multiple has the property that both 2 and 3 will divide evenly into it. Therefore, if both sides of the equation are multiplied by 6, both of the denominators will divide evenly into 6. The result is an equation that does not contain any fractions. Multiplying an equation that contains fractions by the LCM of the denominators is called **clearing denominators.** It is an alternative method of solving an equation that contains fractions.

Clearing denominators is a method of solving equations. The process applies only to equations, never to expressions.

Focus on solving an equation by first clearing denominators

Solve: $\frac{1}{2} = y + \frac{2}{3}$

$$\frac{1}{2} = y + \frac{2}{3}$$

Multiply both sides of the equation by 6, the LCM of the denominators.

$$6\left(\frac{1}{2}\right) = 6\left(y + \frac{2}{3}\right)$$

Simplify each side of the equation. Use the Distributive Property on the right side of the equation.

$$3 = 6(y) + 6\left(\frac{2}{3}\right)$$

Note that multiplying both sides of the equation by the LCM of the denominators eliminates the fractions.

$$3 = 6y + 4$$

Solve the resulting equation.

$$3 - 4 = 6y + 4 - 4$$
$$-1 = 6y$$
$$\frac{-1}{6} = \frac{6y}{6}$$
$$-\frac{1}{6} = y$$

The solution is $-\frac{1}{6}$.

EXAMPLE 3 Solve: $\frac{2}{3} + \frac{1}{4}x = -\frac{1}{3}$

Solution

$$\frac{2}{3} + \frac{1}{4}x = -\frac{1}{3}$$

- **The equation contains fractions. Find the LCM of the denominators.**

$$12\left(\frac{2}{3} + \frac{1}{4}x\right) = 12\left(-\frac{1}{3}\right)$$

- **The LCM of 3 and 4 is 12. Multiply each side of the equation by 12.**

$$12\left(\frac{2}{3}\right) + 12\left(\frac{1}{4}x\right) = -4$$

- **Use the Distributive Property to multiply the left side of the equation by 12.**

$$8 + 3x = -4$$

- **The equation now contains no fractions.**

$$8 - 8 + 3x = -4 - 8$$
$$3x = -12$$
$$\frac{3x}{3} = \frac{-12}{3}$$
$$x = -4$$

The solution is -4.

Problem 3 Solve: $\frac{5}{2} - \frac{2}{3}x = \frac{1}{2}$

Solution See page S6.

Try Exercise 55, page 117.

EXAMPLE 4 Solve: $\frac{x}{2} + \frac{2}{3} = \frac{1}{6}$

Solution

$$\frac{x}{2} + \frac{2}{3} = \frac{1}{6}$$

- The equation contains fractions. Find the LCM of the denominators.

$$6\left(\frac{x}{2} + \frac{2}{3}\right) = 6\left(\frac{1}{6}\right)$$

- The LCM of 2, 3, and 6 is 6. Multiply each side of the equation by 6.

$$6\left(\frac{x}{2}\right) + 6\left(\frac{2}{3}\right) = 1$$

- Use the Distributive Property on the left side of the equation.

$$3x + 4 = 1$$

- The equation now contains no fractions.

$$3x + 4 - 4 = 1 - 4$$
$$3x = -3$$
$$\frac{3x}{3} = \frac{-3}{3}$$
$$x = -1$$

The solution is -1.

Problem 4 Solve: $\frac{x}{4} + \frac{3}{2} = \frac{3x}{8}$

Solution See page S6.

Try Exercise 63, page 117.

OBJECTIVE 2 Solve equations of the form $ax + b = cx + d$

In solving an equation of the form $ax + b = cx + d$, the goal is to rewrite the equation in the form *variable* = *constant*. Begin by rewriting the equation so that there is only one variable term in the equation. Then rewrite the equation so that there is only one constant term.

Focus on solving an equation of the form $ax + b = cx + d$

Solve: $4x - 5 = 6x + 11$

$$4x - 5 = 6x + 11$$

Subtract $6x$ from each side of the equation.

$$4x - 6x - 5 = 6x - 6x + 11$$

Simplify. Now there is only one variable term in the equation.

$$-2x - 5 = 11$$

Add 5 to each side of the equation.

$$-2x - 5 + 5 = 11 + 5$$

Simplify. Now there is only one constant term in the equation.

$$-2x = 16$$

Divide each side of the equation by -2.

$$\frac{-2x}{-2} = \frac{16}{-2}$$

Simplify. Now the equation is in the form *variable* = *constant*.

$$x = -8$$

Check

$4x - 5$	=	$6x + 11$
$4(-8) - 5$		$6(-8) + 11$
$-32 - 5$		$-48 + 11$
-37	=	-37

Write the solution.

The solution is -8.

EXAMPLE 5 Solve: $4x - 3 = 8x - 7$

Solution

$$4x - 3 = 8x - 7$$
$$4x - 8x - 3 = 8x - 8x - 7$$
• Subtract $8x$ from each side of the equation.
$$-4x - 3 = -7$$
• Simplify. Now there is only one variable term in the equation.
$$-4x - 3 + 3 = -7 + 3$$
• Add 3 to each side of the equation.
$$-4x = -4$$
• Simplify. Now there is only one constant term in the equation.
$$\frac{-4x}{-4} = \frac{-4}{-4}$$
• Divide each side of the equation by -4.
$$x = 1$$
• Simplify. Now the equation is in the form *variable* = *constant*.

The solution is 1.
• Write the solution.

Problem 5 Solve: $5x + 4 = 6 + 10x$

Solution See pages S6–S7.

Try Exercise 101, page 118.

> **Take Note**
> As shown below, we could have rewritten the equation in the form *constant* = *variable* by subtracting $4x$ from each side, adding 7 to each side, and then dividing each side by 4.
> $$4x - 3 = 8x - 7$$
> $$-3 = 4x - 7$$
> $$4 = 4x$$
> $$1 = x$$
> The solution is 1.

OBJECTIVE 3 Solve equations containing parentheses

When an equation contains parentheses, one of the steps in solving the equation requires the use of the Distributive Property. The Distributive Property is used to remove parentheses from a variable expression.

$$a(b + c) = ab + ac$$

Focus on solving an equation containing parentheses

Solve: $4 + 5(2x - 3) = 3(4x - 1)$

$$4 + 5(2x - 3) = 3(4x - 1)$$

Use the Distributive Property to remove parentheses.
$$4 + 10x - 15 = 12x - 3$$

Simplify.
$$10x - 11 = 12x - 3$$

Subtract $12x$ from each side of the equation.
$$10x - 12x - 11 = 12x - 12x - 3$$

Simplify. Now there is only one variable term in the equation.
$$-2x - 11 = -3$$

Add 11 to each side of the equation.
$$-2x - 11 + 11 = -3 + 11$$

Simplify. Now there is only one constant term in the equation.
$$-2x = 8$$

Divide each side of the equation by -2.
$$\frac{-2x}{-2} = \frac{8}{-2}$$

Simplify. Now the equation is in the form *variable* = *constant*.
$$x = -4$$

Check

$4 + 5(2x - 3)$	$= 3(4x - 1)$
$4 + 5[2(-4) - 3]$	$3[4(-4) - 1]$
$4 + 5(-8 - 3)$	$3(-16 - 1)$
$4 + 5(-11)$	$3(-17)$
$4 - 55$	-51
-51	$= -51$

Write the solution.
The solution is -4.

In Chapter 2, we discussed the use of a graphing calculator to evaluate variable expressions. The same procedure can be used to check the solution of an equation. Consider the preceding example. After we divide both sides of the equation by −2, the solution appears to be −4. To check this solution, store the value of x, −4, in the calculator. Evaluate the expression on the left side of the original equation: 4 + 5(2x − 3). The result is −51. Now evaluate the expression on the right side of the original equation: 3(4x − 1). The result is −51. Because the results are equal, the solution −4 checks. See the Appendix for a description of keystroking procedures.

EXAMPLE 6 Solve: $3x - 4(2 - x) = 3(x - 2) - 4$

Solution

$$3x - 4(2 - x) = 3(x - 2) - 4$$

$$3x - 8 + 4x = 3x - 6 - 4$$
- **Use the Distributive Property to remove parentheses.**

$$7x - 8 = 3x - 10$$
- **Simplify.**

$$7x - 3x - 8 = 3x - 3x - 10$$
- **Subtract 3x from each side of the equation.**

$$4x - 8 = -10$$

$$4x - 8 + 8 = -10 + 8$$
- **Add 8 to each side of the equation.**

$$4x = -2$$

$$\frac{4x}{4} = \frac{-2}{4}$$
- **Divide each side of the equation by 4.**

$$x = -\frac{1}{2}$$
- **The equation is in the form *variable = constant*.**

The solution is $-\frac{1}{2}$.

Problem 6 Solve: $5x - 4(3 - 2x) = 2(3x - 2) + 6$

Solution See page S7.

Try Exercise 153, page 119.

EXAMPLE 7 Solve: $3[2 - 4(2x - 1)] = 4x - 10$

Solution

$$3[2 - 4(2x - 1)] = 4x - 10$$

$$3[2 - 8x + 4] = 4x - 10$$
- **Use the Distributive Property to remove the parentheses.**

$$3[6 - 8x] = 4x - 10$$
- **Simplify inside the brackets.**

$$18 - 24x = 4x - 10$$
- **Use the Distributive Property to remove the brackets.**

$$18 - 24x - 4x = 4x - 4x - 10$$
- **Subtract 4x from each side of the equation.**

$$18 - 28x = -10$$

$$18 - 18 - 28x = -10 - 18$$
- **Subtract 18 from each side of the equation.**

$$-28x = -28$$

$$\frac{-28x}{-28} = \frac{-28}{-28}$$
- **Divide each side of the equation by −28.**

$$x = 1$$

The solution is 1.

Problem 7 Solve: $-2[3x - 5(2x - 3)] = 3x - 8$

Solution See page S7.

Try Exercise 161, page 119.

OBJECTIVE 4 Solve application problems using formulas

EXAMPLE 8 A company uses the equation $V = C - 6000t$ to determine the depreciated value V, after t years, of a milling machine that originally cost C dollars. If a milling machine originally cost \$50,000, in how many years will the depreciated value of the machine be \$38,000?

Strategy To find the number of years, replace C with 50,000 and V with 38,000 in the given equation, and solve for t.

Solution

$$V = C - 6000t$$
$$38{,}000 = 50{,}000 - 6000t \qquad \bullet\ V = 38{,}000,\ C = 50{,}000$$
$$38{,}000 - 50{,}000 = 50{,}000 - 50{,}000 - 6000t \qquad \bullet\ \text{Subtract 50,000 from each side of the equation.}$$
$$-12{,}000 = -6000t$$
$$\frac{-12{,}000}{-6000} = \frac{-6000t}{-6000} \qquad \bullet\ \text{Divide each side of the equation by } -6000\text{, the coefficient of } t.$$
$$2 = t$$

The depreciated value of the machine will be \$38,000 in 2 years.

Problem 8 The value V of an investment of \$7500 at an annual simple interest rate of 6% is given by the equation $V = 450t + 7500$, where t is the amount of time, in years, that the money is invested. In how many years will the value of a \$7500 investment be \$10,200?

Solution See page S7.

Try Exercise 175, page 120.

Take Note

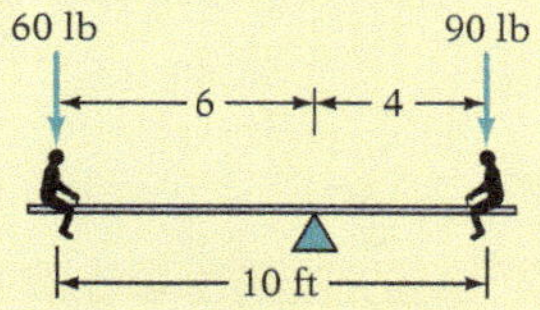

This system balances because

$$F_1x = F_2(d - x)$$
$$60(6) = 90(10 - 6)$$
$$60(6) = 90(4)$$
$$360 = 360$$

A lever system is shown below. It consists of a lever, or bar; a fulcrum; and two forces, F_1 and F_2. The distance d represents the length of the lever, x represents the distance from F_1 to the fulcrum, and $d - x$ represents the distance from F_2 to the fulcrum.

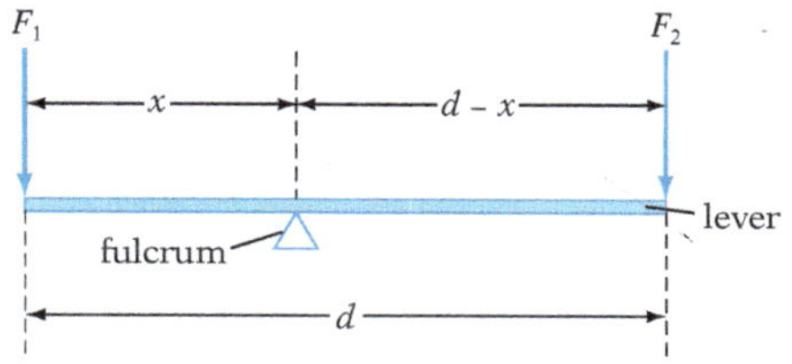

A principle of physics states that when the lever system balances,

$$F_1x = F_2(d - x)$$

EXAMPLE 9 A lever is 10 ft long. A force of 100 lb is applied to one end of the lever, and a force of 400 lb is applied to the other end. When the system balances, how far is the fulcrum from the 100-pound force?

Strategy Draw a diagram of the situation.

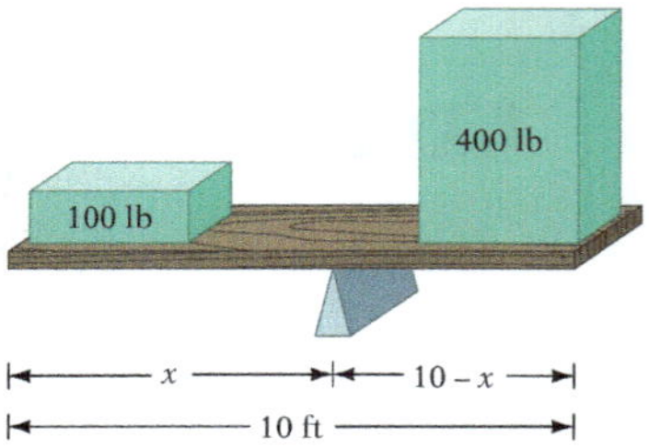

The lever is 10 ft long, so $d = 10$. One force is 100 lb, so $F_1 = 100$. The other force is 400 lb, so $F_2 = 400$. To find the distance of the fulcrum from the 100-pound force, replace the variables F_1, F_2, and d in the lever-system equation with the given values, and solve for x.

Solution

$$F_1x = F_2(d - x)$$
$$100x = 400(10 - x)$$
$$100x = 4000 - 400x$$
• Use the Distributive Property.
$$100x + 400x = 4000 - 400x + 400x$$
• Add 400x to each side of the equation.
$$500x = 4000$$
$$\frac{500x}{500} = \frac{4000}{500}$$
• Divide each side of the equation by 500, the coefficient of x.
$$x = 8$$

The fulcrum is 8 ft from the 100-pound force.

Problem 9 A lever is 14 ft long. At a distance of 6 ft from the fulcrum, a force of 40 lb is applied. How large a force must be applied to the other end of the lever so that the system will balance?

Solution See page S7.

Try Exercise 193, page 121.

3.3 Exercises

CONCEPT CHECK

1. Match each equation with the first step in solving that equation.

a. $3x - 7 = 5$	**i.** Add 7 to each side.
b. $4x + 7 = -5$	**ii.** Add 5 to each side.
c. $7x - 5 = 2$	**iii.** Subtract 7 from each side.
d. $-7x + 5 = -2$	**iv.** Subtract 5 from each side.

Determine whether the statement is true or false.

2. The same variable term can be added to both sides of an equation without changing the solution of the equation.

3. The same variable term can be subtracted from both sides of an equation without changing the solution of the equation.

4. An equation of the form $ax + b = c$ cannot be solved if a is a negative number.

5. The solution of the equation $\frac{x}{3} = 0$ is 0.

6. The solution of the equation $\frac{x}{0} = 3$ is 0.

7. In solving an equation of the form $ax + b = cx + d$, the goal is to rewrite the equation in the form *variable* = *constant*.

1 Solve equations of the form $ax + b = c$ (See pages 109–112.)

GETTING READY

8. The first step in solving the equation $5 + 8x = 29$ is to subtract ___?___ from each side of the equation. The second step is to divide each side of the equation by ___?___.

9. To clear denominators from the equation $\frac{x}{9} + 2 = \frac{1}{6}$, multiply each side of the equation by ___?___, the least common multiple of the denominators 9 and 6.

Complete Exercises 10 and 11 without actually finding the solutions.

10. Is the solution of the equation $15x + 73 = -347$ positive or negative?

11. Is the solution of the equation $17 = 25 - 40a$ positive or negative?

Solve and check.

12. $3x + 1 = 10$
13. $4y + 3 = 11$
14. $2a - 5 = 7$
15. $5m - 6 = 9$
16. $5 = 4x + 9$
17. $2 = 5b + 12$
18. $13 = 9 + 4z$
19. $9 = 7 - c$
20. $2 - x = 11$
21. $4 - 3w = -2$
22. $5 - 6x = -13$
23. $8 - 3t = 2$
24. $-5d + 3 = -12$
25. $-8x - 3 = -19$
26. $-7n - 4 = -25$
27. $-12x + 30 = -6$
28. $-13 = -11y + 9$
29. $2 = 7 - 5a$
30. $3 = 11 - 4n$
31. $-35 = -6b + 1$
32. $1 - 3x = 0$
33. $-3m - 21 = 0$
34. $7x - 3 = 3$
35. $8y + 3 = 7$
36. $6a + 5 = 9$
37. $3m + 4 = 11$
38. $9 - 4x = 6$
39. $7 - 8z = 0$
40. $5 - 6m = 2$
41. $7 - 9a = 4$
42. $2y + \frac{1}{3} = \frac{7}{3}$
43. $3x - \frac{5}{6} = \frac{13}{6}$
44. $5y + \frac{3}{7} = \frac{3}{7}$
45. $9x + \frac{4}{5} = \frac{4}{5}$
46. $4 = 7 - 2w$
47. $7 = 9 - 5a$
48. $\frac{1}{2}a - 3 = 1$
49. $\frac{1}{3}m - 1 = 5$
50. $\frac{2}{5}y + 4 = 6$
51. $\frac{3}{4}n + 7 = 13$
52. $-\frac{2}{3}x + 1 = 7$
53. $-\frac{3}{8}b + 4 = 10$
54. $\frac{x}{4} - 6 = 1$
55. $\frac{y}{5} - 2 = 3$
56. $\frac{2x}{3} - 1 = 5$
57. $\frac{3c}{7} - 1 = 8$
58. $4 - \frac{3}{4}z = -2$
59. $3 - \frac{4}{5}w = -9$
60. $17 = 7 - \frac{5}{6}t$
61. $\frac{2}{3} = y - \frac{1}{2}$
62. $\frac{3}{8} = \frac{5}{12} - \frac{1}{3}b$
63. $\frac{2}{3} = \frac{3}{4} - \frac{1}{2}y$
64. $\frac{x}{2} - \frac{1}{5} = \frac{3}{10}$
65. $\frac{2x}{15} + 3 = -\frac{1}{3}$
66. $\frac{4}{5} + \frac{3x}{10} = \frac{1}{2}$
67. $\frac{5x}{12} - \frac{1}{4} = \frac{1}{6}$
68. $7 = \frac{2x}{5} + 4$
69. $5 - \frac{4c}{7} = 8$
70. $6a + 3 + 2a = 11$
71. $5y + 9 + 2y = 23$
72. $2x - 6x + 1 = 9$
73. $b - 8b + 1 = -6$
74. $-1 = 5m + 7 - m$
75. $8 = 4n - 6 + 3n$
76. $0.15y + 0.025 = -0.074$
77. $1.2x - 3.44 = 1.3$
78. $3.5 = 3.5 + 0.076x$
79. $-6.5 = 4.3y - 3.06$

Solve.

80. If $2x - 3 = 7$, evaluate $3x + 4$.

81. If $3x + 5 = -4$, evaluate $2x - 5$.

82. If $4 - 5x = -1$, evaluate $x^2 - 3x + 1$.

83. If $2 - 3x = 11$, evaluate $x^2 + 2x - 3$.

2 Solve equations of the form $ax + b = cx + d$ (See pages 112–113.)

GETTING READY

84. To solve the equation $7x - 4 = 2x + 6$, we subtract $2x$ from each side of the equation and add 4 to each side of the equation. The resulting equation is ___?___ = ___?___. The solution of the equation is ___?___.

85. To solve $\frac{1}{6}x - 7 = \frac{3}{4}x$, begin by clearing denominators:

$(\_\_?\_\_)\left(\frac{1}{6}x - 7\right) = (\_\_?\_\_)\left(\frac{3}{4}x\right)$ • **Multiply each side of the equation by the LCM of the denominators, 6 and 4.**

$12\left(\frac{1}{6}x\right) - (12)(7) = 9x$ • **Use the __?__ Property on the left side of the equation. Simplify the right side.**

$\_\_?\_\_\,x - \_\_?\_\_ = 9x$ • **Simplify.**

$2x - \_\_?\_\_ - 84 = 9x - \_\_?\_\_$ • **Subtract $2x$ from each side of the equation.**

$-84 = \_\_?\_\_\,x$ • **Simplify.**

$-\frac{84}{7} = \frac{7x}{7}$ • **__?__ each side of the equation by __?__.**

$\_\_?\_\_ = x$ • **Simplify.**

Complete Exercises 86 and 87 without actually finding the solutions.

86. Describe the step that will allow you to rewrite the equation $2x - 3 = 7x + 12$ so that it has one variable term with a positive coefficient.

87. If you rewrite the equation $8 - y = y + 6$ so that it has one variable term on the left side of the equation, what will be the coefficient of the variable?

Solve and check.

88. $8x + 5 = 4x + 13$

89. $6y + 2 = y + 17$

90. $7m + 4 = 6m + 7$

91. $11n + 3 = 10n + 11$

92. $5x - 4 = 2x + 5$

93. $9a - 10 = 3a + 2$

94. $12y - 4 = 9y - 7$

95. $13b - 1 = 4b - 19$

96. $15x - 2 = 4x - 13$

97. $7a - 5 = 2a - 20$

98. $3x + 1 = 11 - 2x$

99. $n - 2 = 6 - 3n$

100. $2x - 3 = -11 - 2x$

101. $4y - 2 = -16 - 3y$

102. $2b + 3 = 5b + 12$

103. $m + 4 = 3m + 8$

104. $4x - 7 = 5x + 1$

105. $6d - 2 = 7d + 5$

106. $4y - 8 = y - 8$

107. $5a + 7 = 2a + 7$

108. $6 - 5x = 8 - 3x$

109. $10 - 4n = 16 - n$

110. $2x - 4 = 6x$

111. $2b - 10 = 7b$

112. $8m = 3m + 20$

113. $9y = 5y + 16$

114. $-x - 4 = -3x - 16$

115. $8 - 4x = 18 - 5x$

116. $6 - 10a = 8 - 9a$

117. $5 - 7m = 2 - 6m$

118. $8b + 5 = 5b + 7$

119. $6y - 1 = 2y + 2$

120. $7x - 8 = x - 3$

121. $10x - 3 = 3x - 1$

122. $5n + 3 = 2n + 1$

123. $8a - 2 = 4a - 5$

124. $\frac{1}{5}d = \frac{1}{2}d + 3$

125. $\frac{3}{4}x = \frac{1}{12}x + 2$

126. $\frac{2}{3}a = \frac{1}{5}a + 7$

127. $\frac{4}{5}c - 7 = \frac{1}{10}c$

128. $\frac{1}{2}y = 10 - \frac{1}{3}y$

129. $\frac{2}{3}b = 2 - \frac{5}{6}b$

130. $8.7y = 3.9y + 9.6$

131. $4.5x - 5.4 = 2.7x$

132. $5.6x = 7.2x - 6.4$

Solve.

133. If $5x = 3x - 8$, evaluate $4x + 2$.

134. If $7x + 3 = 5x - 7$, evaluate $3x - 2$.

135. If $2 - 6a = 5 - 3a$, evaluate $4a^2 - 2a + 1$.

136. If $1 - 5c = 4 - 4c$, evaluate $3c^2 - 4c + 2$.

3 Solve equations containing parentheses (See pages 113–114.)

GETTING READY

137. Use the Distributive Property to remove the parentheses from the equation $4x + 3(x + 6) = 74$: $4x +$ __?__ $+$ __?__ $= 74$.

138. Use the Distributive Property to remove the parentheses from the equation $9x - 5(4 - x) = 2(x + 2)$:

$9x -$ __?__ $+$ __?__ $=$ __?__ $+$ __?__.

139. In each of the following equations, y is the same positive integer. For which equations will x have the same solution?

(i) $5 - 2(x - 1) = y$ (ii) $3(x - 1) = y$
(iii) $5 - 2x + 2 = y$ (iv) $5 - 2x + 1 = y$

140. In solving the equation $9 - (8 - x) = 5x - 1$, how many times will you use the Distributive Property to remove parentheses?

Solve and check.

141. $5x + 2(x + 1) = 23$

142. $6y + 2(2y + 3) = 16$

143. $9n - 3(2n - 1) = 15$

144. $12 - 2(4x - 6) = 8$

145. $7 - 3(3a - 4) = 10$

146. $9m - 4(2m - 3) = 11$

147. $5(3 - 2y) + 4y = 3$

148. $4(1 - 3x) + 7x = 9$

149. $10x + 1 = 2(3x + 5) - 1$

150. $5y - 3 = 7 + 4(y - 2)$

151. $4 - 3a = 7 - 2(2a + 5)$

152. $9 - 5x = 12 - (6x + 7)$

153. $3y - 7 = 5(2y - 3) + 4$

154. $2a - 5 = 4(3a + 1) - 2$

155. $5 - (9 - 6x) = 2x - 2$

156. $7 - (5 - 8x) = 4x + 3$

157. $3[2 - 4(y - 1)] = 3(2y + 8)$

158. $5[2 - (2x - 4)] = 2(5 - 3x)$

159. $3a + 2[2 + 3(a - 1)] = 2(3a + 4)$

160. $5 + 3[1 + 2(2x - 3)] = 6(x + 5)$

161. $-2[4 - (3b + 2)] = 5 - 2(3b + 6)$

162. $-4[x - 2(2x - 3)] + 1 = 2x - 3$

163. $0.3x - 2(1.6x) - 8 = 3(1.9x - 4.1)$

164. $0.56 - 0.4(2.1y + 3) = 0.2(2y + 6.1)$

Solve.

165. If $4 - 3a = 7 - 2(2a + 5)$, evaluate $a^2 + 7a$.

166. If $9 - 5x = 12 - (6x + 7)$, evaluate $x^2 - 3x - 2$.

167. If $2z - 5 = 3(4z + 5)$, evaluate $\frac{z^2}{z - 2}$.

168. If $3n - 7 = 5(2n + 7)$, evaluate $\frac{n^2}{2n - 6}$.

4 Solve application problems using formulas (See pages 115–116.)

Physics Black ice is an ice covering on roads that is especially difficult to see and therefore extremely dangerous for motorists. The distance that a car traveling 30 mph will slide after its brakes are applied is related to the outside temperature by the formula $C = \frac{1}{4}D - 45$, where C is the Celsius temperature and D is the distance in feet that the car will slide.

169. Determine the distance a car will slide on black ice when the outside temperature is -3°C.

170. Determine the distance a car will slide on black ice when the outside temperature is -11°C.

Physics The pressure at a certain depth in the ocean can be approximated by the equation $P = \frac{1}{2}D + 15$, where P is the pressure in pounds per square inch and D is the depth in feet.

171. Find the depth of a diver when the pressure on the diver is 35 lb/in^2.

172. Find the depth of a diver when the pressure on the diver is 45 lb/in^2.

Forensic Science Forensic scientists have determined that the equation $H = 2.9L + 78.1$ can be used to approximate the height H, in centimeters, of an adult on the basis of the length L, in centimeters, of its humerus (the bone extending from the shoulder to the elbow).

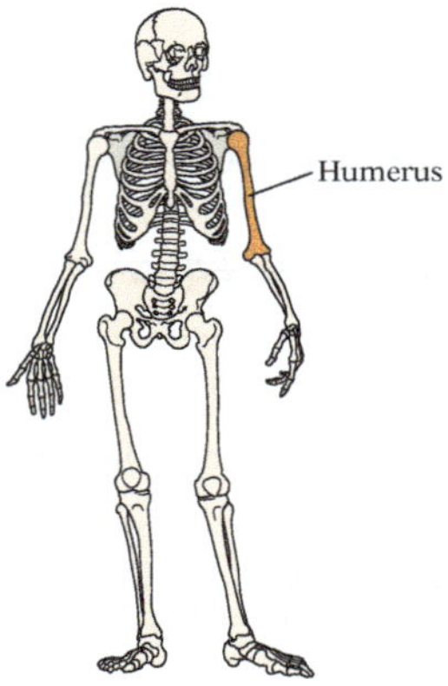

173. Use this formula to approximate the height of an adult whose humerus measures 36 cm.

174. According to this formula, what is the length of the humerus of an adult whose height is 168 cm?

Physics The distance s, in feet, that an object will fall in t seconds is given by $s = 16t^2 + vt$, where v is the initial downward velocity of the object in feet per second.

175. Find the initial velocity of an object that falls 80 ft in 2 s.

176. Find the initial velocity of an object that falls 144 ft in 3 s.

Business The fare F to be charged a customer by a taxi company is calculated using the formula $F = 2.50 + 2.30(m - 1)$, where m is the number of miles traveled.

177. A customer is charged $14.00. How many miles was the passenger driven?

178. A passenger is charged $20.90. Find the number of miles the customer was driven.

Champion Trees American Forests is an organization that maintains the National Register of Big Trees, a listing of the largest trees in the United States. The formula used to award points to a tree is $P = c + h + \frac{1}{4}s$, where P is the point total for a tree with a circumference of c inches, a height of h feet, and an average crown spread of s feet. Use this formula for Exercises 181 and 182. (*Source:* www.amfor.org)

179. Find the average crown spread of the baldcypress described in the article at the right.

180. One of the smallest trees in the United States is a Florida Crossopetalum in the Key Largo Hammocks State Botanical Site. This tree stands 11 ft tall, has a circumference of just 4.8 in., and scores 16.55 points using American Forests' formula. Find the tree's average crown spread. (*Source:* www.championtrees.org)

Employment A person's accurate typing speed can be approximated by the equation $S = \frac{W - 5e}{10}$, where S is the accurate typing speed in words per minute, W is the number of words typed in 10 min, and e is the number of errors made.

181. After taking a 10-minute typing test, a job candidate was told that he had an accurate speed of 35 words per minute. He had typed a total of 390 words. How many errors did he make?

182. A job applicant took a 10-minute typing test and was told that she had an accurate speed of 37 words per minute. If she had typed a total of 400 words, how many errors did she make?

In the News

The Senator Is a Champion

Baldcypress trees are among the most ancient of North American trees. The 3500-year-old baldcypress known as the Senator, located in Big Tree Park, Longwood, is the Florida Champion specimen of the species. With a circumference of 425 in. and a height of 118 ft, this king of the swamp forest earned a total of $557\frac{1}{4}$ points under the point system used for the National Register of Big Trees.

Source: www.championtrees.org

Seminole County Government

The Senator at Big Tree Park

Business To determine the break-even point, or the number of units that must be sold so that no profit or loss occurs, an economist uses the equation $Px = Cx + F$, where P is the selling price per unit, x is the number of units sold, C is the cost to make each unit, and F is the fixed cost.

183. An economist has determined that the selling price per unit for a television is \$550. The cost to make one television is \$325, and the fixed cost is \$78,750. Find the break-even point.

184. A manufacturing engineer determines that the cost per unit for a DVD is \$10.35 and that the fixed cost is \$13,380. The selling price for the DVD is \$21.50. Find the break-even point.

Lever Systems Use the lever-system equation $F_1x = F_2(d - x)$.

185. In the lever-system equation $F_1x = F_2(d - x)$, what does x represent? What does d represent? What does $d - x$ represent?

GETTING READY

186. Two children sit on a seesaw that is 12 ft long. One child weighs 85 lb, and the other child weighs 65 lb. If x is the distance of the 65-pound child from the fulcrum when the seesaw balances, state the value of each of the other variables in the lever-system formula $F_1x = F_2(d - x)$: $F_1 =$ __?__, $F_2 =$ __?__, and $d =$ __?__.

For Exercises 187 to 189, use the following information: Two people sit on a seesaw that is 8 ft long. The seesaw balances when the fulcrum is 3 ft from one of the people.

187. How far is the fulcrum from the other person?

188. Which person is heavier, the person who is 3 ft from the fulcrum or the other person?

189. If the two people switch places, will the seesaw still balance?

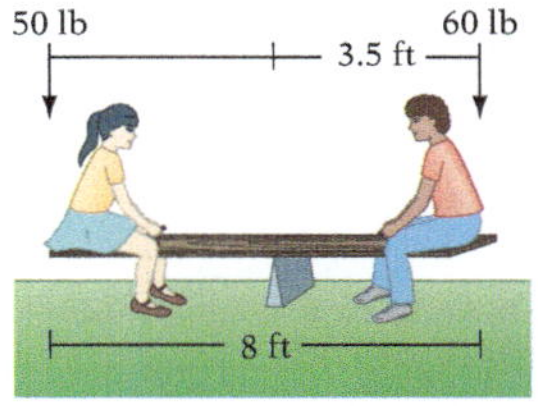

190. Suppose an 80-pound child is sitting at one end of a 7-foot seesaw, and a 60-pound child is sitting at the other end. The 80-pound child is 3 ft from the fulcrum. Use the lever-system equation to explain why the seesaw balances.

191. Two children are sitting 8 ft apart on a seesaw. One child weighs 60 lb and the second child weighs 50 lb. The fulcrum is 3.5 ft from the child weighing 60 lb. Is the seesaw balanced?

192. An adult and a child are on a seesaw that is 14 ft long. The adult weighs 175 lb and the child weighs 70 lb. How many feet from the child must the fulcrum be placed so that the seesaw balances?

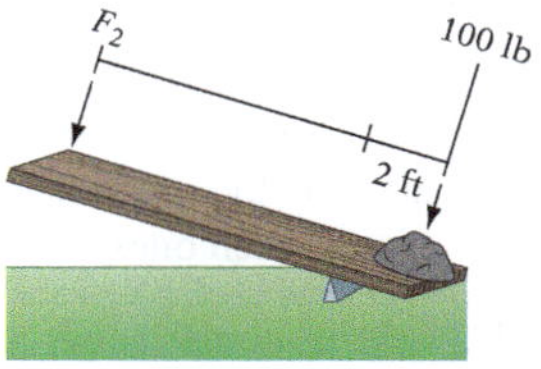

193. A lever 10 ft long is used to balance a 100-pound rock. The fulcrum is placed 2 ft from the rock. What force must be applied to the other end of the lever to balance the rock?

194. A 50-pound weight is applied to the left end of a seesaw that is 10 ft long. The fulcrum is 4 ft from the 50-pound weight. A weight of 30 lb is applied to the right end of the seesaw. Is the 30-pound weight adequate to balance the seesaw?

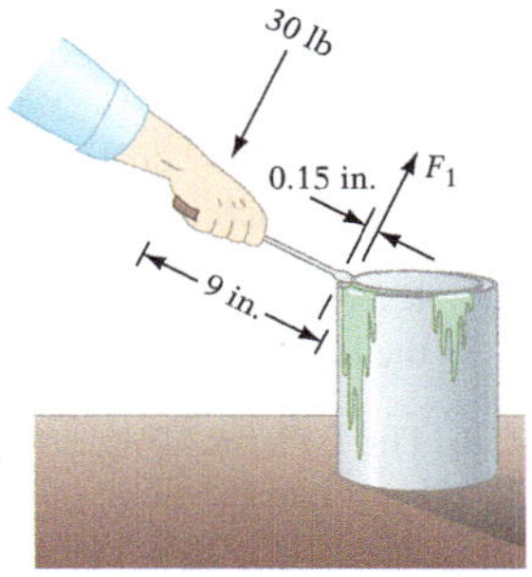

195. In preparation for a stunt, two acrobats are standing on a plank that is 18 ft long. One acrobat weighs 128 lb, and the second acrobat weighs 160 lb. How far from the 128-pound acrobat must the fulcrum be placed so that the acrobats are balanced on the plank?

196. A screwdriver 9 in. long is used as a lever to open a can of paint. The tip of the screwdriver is placed under the lip of the lid with the fulcrum 0.15 in. from the lip. A force of 30 lb is applied to the other end of the screwdriver. Find the force on the lip of the lid.

APPLYING CONCEPTS

Solve. If the equation has no solution, write "No solution."

197. $3(2x - 1) - (6x - 4) = -9$

198. $\frac{1}{5}(25 - 10b) + 4 = \frac{1}{3}(9b - 15) - 6$

199. $3[4(w + 2) - (w + 1)] = 5(2 + w)$

200. $\frac{2(5x - 6) - 3(x - 4)}{7} = x + 2$

201. One-half of a certain number equals two-thirds of the same number. Find the number.

202. Does the sentence "Solve $3x - 4(x - 1)$" make sense? Why or why not?

203. The equation $x = x + 1$ has no solution, whereas the solution of the equation $2x + 1 = 1$ is zero. Is there a difference between no solution and a solution of zero? Explain your answer.

PROJECTS OR GROUP ACTIVITIES

204. I am thinking of a number. When I subtract 4 from the number and then take 300% of the result, my new result is equal to the original number. What is the original number?

205. If $s = 5x - 3$ and $t = x + 4$, find the value of x for which $s = 3t - 1$.

206. The population of the town of Hampton increased by 10,000 people during the 1990s. In the first decade of the new millennium, the population of Hampton decreased by 10%, at which time the town had 6000 more people than at the beginning of the 1990s. Find Hampton's population at the beginning of the 1990s.

The Consumer Price Index The **consumer price index** (CPI) is a percent that is written without the percent sign. For instance, a CPI of 160.1 means 160.1%. This number means that an item that cost \$100 between 1982 and 1984 (the base years) would cost \$160.10 today. Determining the cost is an application of the basic percent equation.

$$PB = A$$
$$(\text{CPI})(\text{cost in base year}) = \text{cost today}$$
$$1.601(100) = 160.1 \qquad \bullet\ 160.1\% = 1.601$$

The table at the right gives the CPI for various products in March 2010.

Product	CPI
All items	217.6
Food and beverages	219.4
Housing	216.0
Clothing	122.1
Transportation	192.1
Medical care	387.1
Energy	210.0

207. Of the items listed, are there any items that cost more than three times as much in 2010 as they cost during the base years? If so, which ones?

208. Of the items listed, are there any items that cost more than one-and-one-half times as much in 2010 as they cost during the base years, but less than twice as much as they cost during the base years? If so, which ones?

209. If a new car cost \$30,000 in 2010, what would a comparable new car have cost during the base years? Use the transportation category.

210. The CPI for all items in the summer of 1970 was 39.0. What salary in 2010 would have had the same purchasing power as a salary of \$10,000 in 1970?

3.4 Inequalities

OBJECTIVE 1 Solve inequalities using the Addition Property of Inequalities

An expression that contains the symbol $>$, $<$, $\geq$, or $\leq$ is called an **inequality.** An inequality expresses the relative order of two mathematical expressions. The expressions can be either numerical or variable expressions.

$$\left.\begin{array}{l} 4 > 2 \\ 3x \leq 7 \\ x^2 - 2x > y + 4 \end{array}\right\} \text{Inequalities}$$

> **How It's Used**
>
> In manufacturing, the *tolerance* of a measurement is the amount by which that measurement can vary from a specified value. Inequalities are used to describe the acceptable measurements of a product within a given tolerance. For example, basketballs used in the Women's National Basketball Association have a circumference of 28.75 in. with a tolerance of 0.25 in.

The **solution set of an inequality** is a set of numbers, each element of which, when substituted for the variable, results in a true inequality. The solution set of an inequality can be graphed on the number line.

The graph of the solution set of $x > 1$ is shown at the right. The solution set is the real numbers greater than 1. The parenthesis on the graph indicates that 1 is not included in the solution set.

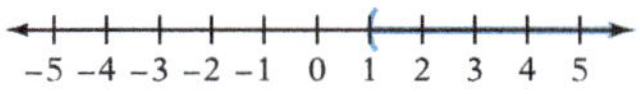

The graph of the solution set of $x \geq 1$ is shown at the right. The bracket at 1 indicates that 1 is included in the solution set.

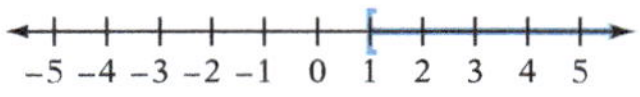

The graph of the solution set of $x < -1$ is shown at the right. The numbers less than -1 are to the left of -1 on the number line.

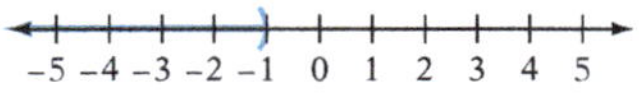

EXAMPLE 1 Graph: $x < 3$

Solution

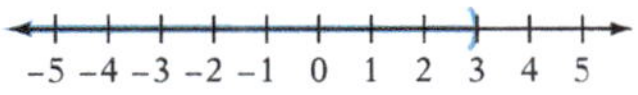

- **The solution set is the numbers less than 3.**

Problem 1 Graph: $x > -2$

Solution See page S7.

Try Exercise 9, page 129.

The inequality at the right is true if the variable is replaced by 7, 9.3, or $\frac{15}{2}$.

$$x + 3 > 8$$

$$\left.\begin{array}{r} 7 + 3 > 8 \\ 9.3 + 3 > 8 \\ \dfrac{15}{2} + 3 > 8 \end{array}\right\} \text{True inequalities}$$

The inequality $x + 3 > 8$ is false if the variable is replaced by 4, 1.5, or $-\frac{1}{2}$.

$$\left.\begin{array}{r} 4 + 3 > 8 \\ 1.5 + 3 > 8 \\ -\dfrac{1}{2} + 3 > 8 \end{array}\right\} \text{False inequalities}$$

There are many values of the variable x that will make the inequality $x + 3 > 8$ true. The solution set of $x + 3 > 8$ is any number greater than 5.

The graph of the solution set of $x + 3 > 8$ is shown at the right.

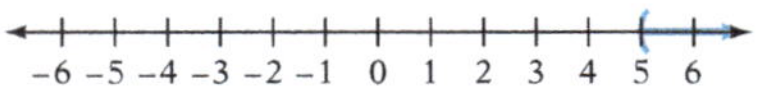

In solving an inequality, the goal is to rewrite the given inequality in the form *variable* $>$ *constant* or *variable* $<$ *constant*. The Addition Property of Inequalities is used to rewrite an inequality in this form.

ADDITION PROPERTY OF INEQUALITIES

The same number can be added to each side of an inequality without changing the solution set of the inequality.

If $a > b$, then $a + c > b + c$.

If $a < b$, then $a + c < b + c$.

EXAMPLES

1. Begin with a true inequality. $8 > 3$
   Add 5 to each side. $8 + 5 > 3 + 5$
   Simplify. The inequality is true. $13 > 8$
2. Begin with a true inequality. $-7 < -3$
   Add 2 to each side. $-7 + 2 < -3 + 2$
   Simplify. The inequality is true. $-5 < -1$

The Addition Property of Inequalities also holds true for an inequality containing the symbol $\geq$ or $\leq$.

The Addition Property of Inequalities is used when, in order to rewrite an inequality in the form *variable* $>$ *constant* or *variable* $<$ *constant*, a term must be removed from one side of the inequality. Add the opposite of the term to each side of the inequality.

Because subtraction is defined in terms of addition, the Addition Property of Inequalities makes it possible to subtract the same number from each side of an inequality without changing the solution set of the inequality.

Focus on solving an inequality using the Addition Property of Inequalities

Solve and graph. **A.** $x + 4 < 5$ **B.** $5x - 6 \leq 4x - 4$

A.

$x + 4 < 5$

Subtract 4 from each side of the inequality. $x + 4 - 4 < 5 - 4$

Simplify. $x < 1$

The solution set is $x < 1$.

The graph of the solution set of $x + 4 < 5$ is shown at the right.

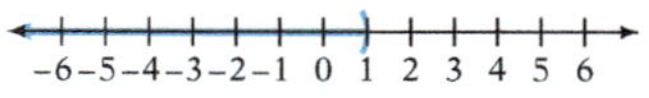

B.

$5x - 6 \leq 4x - 4$

Subtract $4x$ from each side of the inequality. $5x - 4x - 6 \leq 4x - 4x - 4$

Simplify. $x - 6 \leq -4$

Add 6 to each side of the inequality. $x - 6 + 6 \leq -4 + 6$

Simplify. $x \leq 2$

The solution set is $x \leq 2$.

The graph of the solution set of $5x - 6 \leq 4x - 4$ is shown at the right.

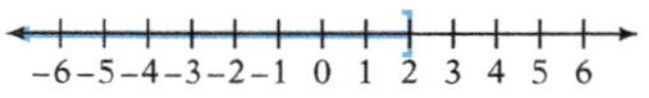

EXAMPLE 2 Solve and graph the solution set of $x + 5 > 3$.

Solution

$$x + 5 > 3$$
$$x + 5 - 5 > 3 - 5$$ • Subtract 5 from each side of the inequality.
$$x > -2$$

The solution set is $x > -2$.

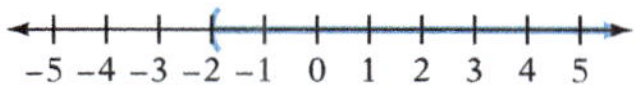

Problem 2 Solve and graph the solution set of $x + 2 < -2$.

Solution See page S7.

Try Exercise 23, page 129.

EXAMPLE 3 Solve: $7x - 14 \le 6x - 16$

Solution

$$7x - 14 \le 6x - 16$$
$$7x - 6x - 14 \le 6x - 6x - 16$$ • Subtract $6x$ from each side of the inequality.
$$x - 14 \le -16$$
$$x - 14 + 14 \le -16 + 14$$ • Add 14 to each side of the inequality.
$$x \le -2$$

The solution set is $x \le -2$.

Problem 3 Solve: $5x + 3 > 4x + 5$

Solution See page S7.

Try Exercise 33, page 130.

OBJECTIVE 2 Solve inequalities using the Multiplication Property of Inequalities

In solving an inequality, the goal is to rewrite the given inequality in the form *variable* > *constant* or *variable* < *constant*. The Multiplication Property of Inequalities is used when, in order to rewrite an inequality in this form, a coefficient must be removed from one side of the inequality.

Take Note

$c > 0$ means c is a positive number.

The inequality symbols are not changed.

MULTIPLICATION PROPERTY OF INEQUALITIES–RULE 1

Each side of an inequality can be multiplied by the same positive number without changing the solution set of the inequality.

If $a > b$ and $c > 0$, then $ac > bc$. If $a < b$ and $c > 0$, then $ac < bc$.

EXAMPLES

1. Begin with a true inequality. $5 > 4$
   Multiply each side by *positive* 2. $5(2) > 4(2)$
   Simplify. The inequality is true. $10 > 8$
2. Begin with a true inequality. $-6 < -2$
   Multiply each side by *positive* 3. $-6(3) < -2(3)$
   Simplify. The inequality is true. $-18 < -6$

Take Note

$c < 0$ means c is a negative number.

The inequality symbol is reversed in each case.

MULTIPLICATION PROPERTY OF INEQUALITIES–RULE 2

If each side of an inequality is multiplied by the same negative number and the inequality symbol is reversed, then the solution set of the inequality is not changed.

If $a > b$ and $c < 0$, then $ac < bc$. If $a < b$ and $c < 0$, then $ac > bc$.

EXAMPLES

1. Begin with a true inequality. $6 < 9$
   Multiply each side by *negative* 3 and reverse the inequality symbol. $6(-3) > 9(-3)$
   Simplify. The inequality is true. $-18 > -27$
2. Begin with a true inequality. $-1 > -5$
   Multiply each side by *negative* 2 and reverse the inequality symbol. $-1(-2) < -5(-2)$
   Simplify. The inequality is true. $2 < 10$

The Multiplication Property of Inequalities also holds true for an inequality containing the symbol $\geq$ or $\leq$.

Focus on solving an inequality using the Multiplication Property of Inequalities

Solve: $-\frac{3}{2}x \leq 6$

$$-\frac{3}{2}x \leq 6$$

Multiply each side of the inequality by the reciprocal of the coefficient $-\frac{3}{2}$. Because $-\frac{2}{3}$ is a negative number, the inequality symbol must be reversed.

$$-\frac{2}{3}\left(-\frac{3}{2}x\right) \geq -\frac{2}{3}(6)$$

Simplify.

$$x \geq -4$$

The solution set is $x \geq -4$.

The graph of the solution set of $-\frac{3}{2}x \leq 6$ is shown at the right.

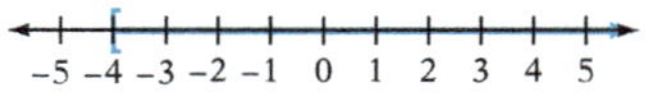

EXAMPLE 4 Solve: $-\frac{5}{8}x \leq \frac{5}{12}$

Solution

$$-\frac{5}{8}x \leq \frac{5}{12}$$

$$-\frac{8}{5}\left(-\frac{5}{8}x\right) \geq -\frac{8}{5}\left(\frac{5}{12}\right)$$

- **Multiply each side of the inequality by the reciprocal of $-\frac{5}{8}$. $-\frac{8}{5}$ is a negative number. Reverse the inequality symbol.**

$$x \geq -\frac{2}{3}$$

The solution set is $x \geq -\frac{2}{3}$.

Problem 4 Solve: $-\frac{3}{4}x \geq 18$

Solution See page S7.

Try Exercise 81, page 131.

Recall that division is defined in terms of multiplication. Therefore, the Multiplication Property of Inequalities allows each side of an inequality to be divided by the same number. When each side of an inequality is divided by a positive number, the inequality symbol remains the same. When each side of an inequality is divided by a negative number, the inequality symbol must be reversed.

Take Note

Any time an inequality is multiplied or divided by a negative number, the inequality symbol must be reversed. Compare these two examples:

$$2x < -4$$
$$\frac{2x}{2} < \frac{-4}{2}$$
$$x < -2$$

Divide each side by *positive* 2. The inequality symbol is *not* reversed.

$$-2x < 4$$
$$\frac{-2x}{-2} > \frac{4}{-2}$$
$$x > -2$$

Divide each side by *negative* 2. The inequality symbol *is* reversed.

Focus on solving an inequality using the Multiplication Property of Inequalities

Solve: $-5x > 8$

$$-5x > 8$$

Divide each side of the inequality by -5. Because -5 is a negative number, the inequality symbol must be reversed.

$$\frac{-5x}{-5} < \frac{8}{-5}$$

Simplify.

$$x < -\frac{8}{5}$$

The solution set is $x < -\frac{8}{5}$.

EXAMPLE 5 Solve and graph the solution set of $7x < -14$.

Solution

$$7x < -14$$
$$\frac{7x}{7} < \frac{-14}{7}$$
$$x < -2$$

- **Divide each side of the inequality by 7. Because 7 is a *positive* number, do not change the inequality symbol.**

The solution set is $x < -2$.

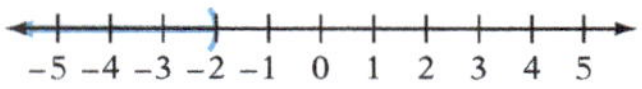

Problem 5 Solve and graph the solution set of $3x < 9$.

Solution See page S7.

Try Exercise 63, page 130.

OBJECTIVE 3 Solve general inequalities

In solving an inequality, it is often necessary to apply both the Addition and the Multiplication Properties of Inequalities.

Take Note

Solving these inequalities is similar to solving the equations in Section 3.3 *except* that when you multiply or divide the inequality by a negative number, you must reverse the inequality symbol.

Focus on solving an inequality using the Addition and Multiplication Properties of Inequalities

Solve: $3x - 2 < 5x + 4$

$$3x - 2 < 5x + 4$$

Subtract $5x$ from each side of the inequality.
Simplify.

$$3x - 5x - 2 < 5x - 5x + 4$$
$$-2x - 2 < 4$$

Add 2 to each side of the inequality.
Simplify.

$$-2x - 2 + 2 < 4 + 2$$
$$-2x < 6$$

Divide each side of the inequality by -2. Because -2 is a negative number, the inequality symbol must be reversed.

$$\frac{-2x}{-2} > \frac{6}{-2}$$

Simplify.

$$x > -3$$

The solution set is $x > -3$.

EXAMPLE 6 Solve: $7x - 3 \le 3x + 17$

Solution

$$7x - 3 \le 3x + 17$$
$$7x - 3x - 3 \le 3x - 3x + 17$$
- Subtract $3x$ from each side of the inequality.

$$4x - 3 \le 17$$
$$4x - 3 + 3 \le 17 + 3$$
- Add 3 to each side of the inequality.

$$4x \le 20$$
$$\frac{4x}{4} \le \frac{20}{4}$$
- Divide each side of the inequality by 4.

$$x \le 5$$

The solution set is $x \le 5$.

Problem 6 Solve: $5 - 4x > 9 - 8x$

Solution See page S7.

Try Exercise 103, page 132.

When an inequality contains parentheses, one of the steps in solving the inequality requires the use of the Distributive Property.

Focus on solving an inequality containing parentheses

Solve: $-2(x - 7) > 3 - 4(2x - 3)$

$$-2(x - 7) > 3 - 4(2x - 3)$$

Use the Distributive Property to remove parentheses. Simplify.

$$-2x + 14 > 3 - 8x + 12$$
$$-2x + 14 > 15 - 8x$$

Add $8x$ to each side of the inequality. Simplify.

$$-2x + 8x + 14 > 15 - 8x + 8x$$
$$6x + 14 > 15$$

Subtract 14 from each side of the inequality. Simplify.

$$6x + 14 - 14 > 15 - 14$$
$$6x > 1$$

Divide each side of the inequality by 6.

$$\frac{6x}{6} > \frac{1}{6}$$

Simplify.

$$x > \frac{1}{6}$$

The solution set is $x > \frac{1}{6}$.

EXAMPLE 7 Solve: $3(3 - 2x) \ge -5x - 2(3 - x)$

Solution

$$3(3 - 2x) \ge -5x - 2(3 - x)$$
$$9 - 6x \ge -5x - 6 + 2x$$
- Use the Distributive Property.

$$9 - 6x \ge -3x - 6$$
$$9 - 6x + 3x \ge -3x + 3x - 6$$
- Add $3x$ to each side of the inequality.

$$9 - 3x \ge -6$$
$$9 - 9 - 3x \ge -6 - 9$$
- Subtract 9 from each side of the inequality.

$$-3x \ge -15$$
$$\frac{-3x}{-3} \le \frac{-15}{-3}$$
- Divide each side of the inequality by -3. Reverse the inequality symbol.

$$x \le 5$$

The solution set is $x \le 5$.

Problem 7 Solve: $8 - 4(3x + 5) \le 6(x - 8)$

Solution See pages S7–S8.

Try Exercise 127, page 132.

3.4 Exercises

CONCEPT CHECK

1. How does the solution set of $x \le 4$ differ from the solution set of $x < 4$?

2. True or false: The solution set of $x \ge 4$ is the set $\{4, 5, 6, 7, 8, 9, ...\}$.

3. State whether or not you would reverse the inequality symbol while solving the inequality.

a. $x - 3 > 6$ **b.** $3x < 6$ **c.** $-3x > 6$

d. $3x \le -6$ **e.** $3 + x \ge -6$ **f.** $-\frac{x}{3} < 6$

Determine whether the statement is always true, sometimes true, or never true.

4. The same variable term can be added to both sides of an inequality without changing the solution set of the inequality.

5. The same variable term can be subtracted from both sides of an inequality without changing the solution set of the inequality.

6. Both sides of an inequality can be multiplied by the same number without changing the solution set of the inequality.

1 Solve inequalities using the Addition Property of Inequalities (See pages 123–125.)

GETTING READY

Complete Exercises 7 and 8 by replacing the question mark in the first statement with "includes" or "does not include." Replace the question mark in the second statement with the correct inequality symbol: $>$, $<$, $\ge$, or $\le$.

7. The graph of the solution set shown at the right __?__ the number -2. The graph is of the solution set of the inequality x __?__ -2.

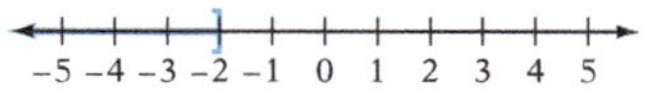

8. The graph of the solution set shown at the right __?__ the number -3. The graph is of the solution set of the inequality x __?__ -3.

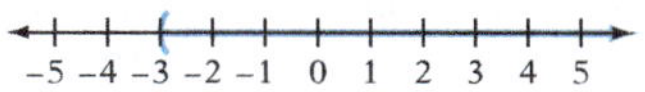

Graph.

9. $x > 2$

10. $x \ge -1$

11. $x \le 0$

12. $x < 4$

13. Which numbers are solutions of the inequality $x + 7 \le -3$?
(i) -17 **(ii)** 8 **(iii)** -10 **(iv)** 0

14. Which numbers are solutions of the inequality $x - 5 < -6$?
(i) 1 **(ii)** -1 **(iii)** 12 **(iv)** -5

Solve and graph the solution set.

15. $x + 1 < 3$

16. $y + 2 < 2$

17. $x - 5 > -2$

18. $x - 3 > -2$

19. $n + 4 \ge 7$

20. $x + 5 \ge 3$

21. $x - 6 \le -10$

22. $y - 8 \le -11$

23. $5 + x \ge 4$

24. $-2 + n \ge 0$

For Exercises 25 to 28, state whether the solution set of an inequality in the given form contains only negative numbers, only positive numbers, or both negative and positive numbers.

25. $x + n < a$, where n and a are both positive and $n > a$

26. $x + n > a$, where n and a are both positive and $n < a$

27. $x - n > -a$, where n and a are both positive and $n > a$

28. $x - n > -a$, where n and a are both positive and $n < a$

Solve.

29. $y - 3 \geq -12$

30. $x + 8 \geq -14$

31. $3x - 5 < 2x + 7$

32. $5x + 4 < 4x - 10$

33. $8x - 7 \geq 7x - 2$

34. $3n - 9 \geq 2n - 8$

35. $2x + 4 < x - 7$

36. $9x + 7 < 8x - 7$

37. $4x - 8 \leq 2 + 3x$

38. $5b - 9 < 3 + 4b$

39. $6x + 4 \geq 5x - 2$

40. $7x - 3 \geq 6x - 2$

41. $2x - 12 > x - 10$

42. $3x + 9 > 2x + 7$

43. $d + \frac{1}{2} < \frac{1}{3}$

44. $x - \frac{3}{8} < \frac{5}{6}$

45. $x + \frac{5}{8} \geq -\frac{2}{3}$

46. $y + \frac{5}{12} \geq -\frac{3}{4}$

47. $2x - \frac{1}{2} < x + \frac{3}{4}$

48. $6x - \frac{1}{3} \leq 5x - \frac{1}{2}$

49. $3x + \frac{5}{8} > 2x + \frac{5}{6}$

50. $4b - \frac{7}{12} \geq 3b - \frac{9}{16}$

51. $x + 5.8 \leq 4.6$

52. $n - 3.82 \leq 3.95$

53. $x - 0.23 \leq 0.47$

54. $3.8x < 2.8x - 3.8$

55. $1.2x < 0.2x - 7.3$

2 Solve inequalities using the Multiplication Property of Inequalities (See pages 125–127.)

56. Which numbers are solutions of the inequality $5x > 15$?
(i) 6 **(ii)** -4 **(iii)** 3 **(iv)** 5

57. Which numbers are solutions of the inequality $-4x \leq 12$?
(i) 0 **(ii)** 3 **(iii)** -3 **(iv)** -4

GETTING READY

Complete Exercises 58 and 59 by replacing the question mark with "remains the same" or "is reversed."

58. The inequality $x + 4 > -8$ can be solved by subtracting 4 from each side of the inequality. The inequality symbol __?__.

59. The inequality $-4x > 8$ can be solved by dividing each side of the inequality by -4. The inequality symbol __?__.

Solve and graph the solution set.

60. $3x < 12$

61. $8x \leq -24$

62. $5y \geq 15$

63. $24x > -48$

64. $16x \leq 16$

65. $3x > 0$

66. $-8x > 8$

67. $-2n \leq -8$

68. $-6b > 24$

69. $-4x < 8$

Solve.

70. $-5y \ge 20$

71. $3x < 5$

72. $7x > 2$

73. $-8x \le -40$

74. $-6x \le -40$

75. $10x > -25$

76. $-3x \ge \frac{6}{7}$

77. $-5x \ge \frac{10}{3}$

78. $\frac{5}{6}n < 15$

79. $\frac{2}{3}x < -12$

80. $\frac{5}{6}x < -20$

81. $-\frac{3}{8}x < 6$

82. $-\frac{3}{7}x \le 6$

83. $-\frac{2}{11}b \ge -6$

84. $-\frac{4}{7}x \ge -12$

85. $\frac{2}{3}n < \frac{1}{2}$

86. $\frac{3}{5}x > \frac{7}{10}$

87. $-\frac{2}{3}x \ge \frac{4}{7}$

88. $-\frac{3}{8}x \ge \frac{9}{14}$

89. $-\frac{3}{4}y \ge -\frac{5}{8}$

90. $-\frac{8}{9}x \ge -\frac{16}{27}$

91. $-0.27x < 0.135$

92. $-0.63x < 4.41$

93. $8.4y \ge -6.72$

94. $3.7y \ge -1.48$

95. $0.07x < -0.378$

96. $-11.7x \le 4.68$

Complete Exercises 97 and 98 without solving the inequalities or using a calculator.

97. A number n is a solution of the inequality $-0.8157n < 7.304$. Which of the following must be true?
(i) n is positive. (ii) n is negative.
(iii) n is 0. (iv) n can be positive, negative, or 0.

98. A number n is a solution of the inequality $-917n \ge -10{,}512$. Which of the following must be true?
(i) n is positive. (ii) n is negative.
(iii) n is 0. (iv) n can be positive, negative, or 0.

3 Solve general inequalities (See pages 127–128.)

99. In your own words, state the Addition Property of Inequalities and the Multiplication Property of Inequalities.

100. What differentiates solving linear equations from solving linear inequalities?

GETTING READY

101. Solve: $9 - 5x \ge -2x$

$9 - 5x \ge -2x$

$9 - 5x + \underline{\ ?\ } \ge -2x + \underline{\ ?\ }$ • Add $2x$ to each side of the inequality.

$9 - \underline{\ ?\ } \ge \underline{\ ?\ }$ • Simplify.

$9 - 9 - 3x \ge 0 - 9$ • Subtract ___?___ from each side of the inequality.

$\underline{\ ?\ } \ge \underline{\ ?\ }$ • Simplify.

$\frac{-3x}{?} \ \underline{\ ?\ } \ \frac{-9}{?}$ • Divide each side of the inequality by -3. Reverse the inequality symbol.

$x \le \underline{\ ?\ }$ • Simplify.

102. The first step of solving the inequality $8(3x - 1) < 11x + 5$ is to use the ___?___ Property to ___?___.

Solve.

103. $3x + 2 \geq 5x - 8$

104. $2n - 9 \geq 5n + 4$

105. $5x - 2 < 3x - 2$

106. $8x - 9 > 3x - 9$

107. $4x - 8 < 2x$

108. $7x - 4 < 3x$

109. $2x - 8 > 4x$

110. $3y + 2 > 7y$

111. $8 - 3x \leq 5x$

112. $10 - 3x \leq 7x$

113. $0.1(180 + x) > x$

114. $x > 0.2(50 + x)$

115. $0.15x + 55 > 0.10x + 80$

116. $-3.6b + 16 < 2.8b + 25.6$

117. $2(3x - 1) > 3x + 4$

118. $5(2x + 7) > -4x - 7$

119. $3(2x - 5) \geq 8x - 5$

120. $5x - 8 \geq 7x - 9$

121. $2(2y - 5) \leq 3(5 - 2y)$

122. $2(5x - 8) \leq 7(x - 3)$

123. $5(2 - x) > 3(2x - 5)$

124. $4(3d - 1) > 3(2 - 5d)$

125. $5(x - 2) > 9x - 3(2x - 4)$

126. $3x - 2(3x - 5) > 4(2x - 1)$

127. $4 - 3(3 - n) \leq 3(2 - 5n)$

128. $15 - 5(3 - 2x) \leq 4(x - 3)$

129. $2x - 3(x - 4) \geq 4 - 2(x - 7)$

130. $4 + 2(3 - 2y) \leq 4(3y - 5) - 6y$

131. $\frac{1}{2}(9x - 10) \leq -\frac{1}{3}(12 - 6x)$

132. $\frac{1}{4}(8 - 12d) < \frac{2}{5}(10d + 15)$

133. $\frac{2}{3}(9t - 15) + 4 < 6 + \frac{3}{4}(4 - 12t)$

134. $\frac{3}{8}(16 - 8c) - 9 \geq \frac{3}{5}(10c - 15) + 7$

135. What number is a solution of $3x - 4 \geq 5$ but not a solution of $3x - 4 > 5$?

136. What number is a solution of $8 - 2(x + 6) \leq 4$ but not a solution of $8 - 2(x + 6) < 4$?

137. Which inequalities are equivalent to the inequality $-7x - 2 > -4x + 1$?
(i) $-3 > -11x$ (ii) $3x > 3$ (iii) $-3 > 3x$ (iv) $3x < -3$

138. Which inequalities are equivalent to the inequality $10 - 7(x - 4) \geq 3$?
(i) $-7x + 38 \geq 3$ (ii) $3x - 12 \geq 3$ (iii) $35 \geq 7x$ (iv) $-18 - 7x \geq 3$

APPLYING CONCEPTS

Use the roster method to list the set of integers that are common to the solution sets of the two inequalities.

139. $5x - 12 \leq x + 8$
$3x - 4 \geq 2 + x$

140. $6x - 5 > 9x - 2$
$5x - 6 < 8x + 9$

141. $4(x - 2) \leq 3x + 5$
$7(x - 3) \geq 5x - 1$

142. $3(x + 2) < 2(x + 4)$
$4(x + 5) > 3(x + 6)$

Write an inequality that describes the graph.

143. 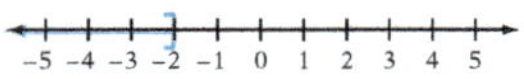

144. 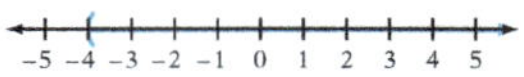

145. A theorem from geometry called the Triangle Inequality Theorem states that the sum of the lengths of two sides of a triangle must be greater than the length of the third side. Suppose two sides of a triangle measure 10 in. and 18 in. Let x be the length of the third side. What are the possible values for x?

PROJECTS OR GROUP ACTIVITIES

Graph.

146. $|x| < 3$

147. $|x| < 4$

148. $|x| > 2$

149. $|x| > 1$

Determine whether the statement is always true, sometimes true, or never true.

150. Suppose $a > 0$ and $b < 0$. Then $ab > 0$.

151. Suppose $a < 0$. Then $a^2 > 0$.

152. Suppose $a > 0$ and $b < 0$. Then $a^2 > b$.

153. Suppose $a > b$. Then $-a > -b$.

154. Suppose $a < b$. Then $ac < bc$.

155. Suppose $a \neq 0$, $b \neq 0$, and $a > b$. Then $\frac{1}{a} > \frac{1}{b}$.

CHAPTER 3 Summary

Key Words	Objective and Page Reference	Examples
An **equation** expresses the equality of two mathematical expressions.	[3.1.1, p. 90]	$5(3x - 2) = 4x + 7$ is an equation.
A **solution of an equation** is a number that, when substituted for the variable, results in a true equation.	[3.1.1, p. 90]	1 is the solution of the equation $6x - 4 = 2$ because $6(1) - 4 = 2$ is a true equation.
To **solve** an equation means to find a solution of the equation. The goal is to rewrite the equation in the form **variable = constant,** because the constant is the solution.	[3.1.2, p. 91]	The equation $x = -9$ is in the form *variable = constant*. The constant, -9, is the solution of the equation.
An **inequality** is an expression that contains the symbol $<$, $>$, $\leq$, or $\geq$.	[3.4.1, p. 123]	$8x - 1 \geq 5x + 23$ is an inequality.
The **solution set of an inequality** is a set of numbers, each element of which, when substituted for the variable, results in a true inequality.	[3.4.1, p. 123]	The solution set of $8x - 1 \geq 5x + 23$ is $x \geq 8$ because every number greater than or equal to 8, when substituted for the variable, results in a true inequality.

Essential Rules and Procedures	Objective and Page Reference	Examples
The Addition Property of Equations The same number or variable term can be added to each side of an equation without changing the solution of the equation.	[3.1.2, p. 91]	$x + 12 = -19$ $x + 12 - 12 = -19 - 12$ $x = -31$
The Multiplication Property of Equations Each side of an equation can be multiplied by the same nonzero number without changing the solution of the equation.	[3.1.3, p. 93]	$-6x = 24$ $\frac{-6x}{-6} = \frac{24}{-6}$ $x = -4$

In **solving an equation,** the goal is to rewrite the equation in the form *variable* = *constant*. This requires applying both the Addition and the Multiplication Properties of Equations.

[3.3.2, pp. 112–113]

$$7x - 4 = 3x + 16$$
$$7x - 3x - 4 = 3x - 3x + 16$$
$$4x - 4 = 16$$
$$4x - 4 + 4 = 16 + 4$$
$$4x = 20$$
$$\frac{4x}{4} = \frac{20}{4}$$
$$x = 5$$

To solve an equation containing parentheses, use the Distributive Property to remove parentheses.

[3.3.3, pp. 113–114]

$$2x + 12 = 3(x + 5)$$
$$2x + 12 = 3x + 15$$
$$2x - 3x + 12 = 3x - 3x + 15$$
$$-x + 12 = 15$$
$$-x + 12 - 12 = 15 - 12$$
$$-x = 3$$
$$-1(-x) = -1(3)$$
$$x = -3$$

The Addition Property of Inequalities

The same number or variable term can be added to each side of an inequality without changing the solution set of the inequality.

If $a > b$, then $a + c > b + c$.

If $a < b$, then $a + c < b + c$.

[3.4.1, p. 124]

$$x - 9 \le 14$$
$$x - 9 + 9 \le 14 + 9$$
$$x \le 23$$

The Multiplication Property of Inequalities

[3.4.2, pp. 125–126]

Rule 1 Each side of an inequality can be multiplied by the same **positive number** without changing the solution set of the inequality.

If $a > b$ and $c > 0$, then $ac > bc$.

If $a < b$ and $c > 0$, then $ac < bc$.

$$7x < -21$$
$$\frac{7x}{7} < \frac{-21}{7}$$
$$x < -3$$

Rule 2 If each side of an inequality is multiplied by the same **negative number** and the inequality symbol is reversed, then the solution set of the inequality is not changed.

If $a > b$ and $c < 0$, then $ac < bc$.

If $a < b$ and $c < 0$, then $ac > bc$.

$$-7x \ge 21$$
$$\frac{-7x}{-7} \le \frac{21}{-7}$$
$$x \le -3$$

In **solving an inequality,** apply both the Addition and Multiplication Properties of Inequalities.

[3.4.3, pp. 127–128]

$$4x - 1 \ge 6x + 7$$
$$4x - 6x - 1 \ge 6x - 6x + 7$$
$$-2x - 1 \ge 7$$
$$-2x - 1 + 1 \ge 7 + 1$$
$$-2x \ge 8$$
$$\frac{-2x}{-2} \le \frac{8}{-2}$$
$$x \le -4$$

The Basic Percent Equation

Percent · base = amount

$P \cdot B = A$

[3.2.1, p. 98]

40% of what number is 16?

$$PB = A$$
$$0.40B = 16$$
$$\frac{0.40B}{0.40} = \frac{16}{0.40}$$
$$B = 40$$

40% of 40 is 16.

Uniform Motion Equation

Distance = rate · time

$d = rt$

[3.2.2, p. 102]

A car travels 200 mi in 4 h. Find the rate of speed.

$$d = rt$$
$$200 = r \cdot 4$$
$$50 = r$$

The rate of speed is 50 mph.

Simple Interest Equation

Simple interest =

principal · simple interest rate · time

$I = prt$

[3.2.1, p. 100]

A $5000 investment earns an annual simple interest rate of 4%. Find the interest earned in 6 months.

$$I = prt$$
$$I = 5000(0.04)(0.5)$$
$$I = 100$$

The interest earned is $100.

Percent Mixture Equation

Quantity of a substance in a solution
= amount of the solution ·
percent concentration of the substance

$Q = Ar$

[3.2.1, p. 101]

An 8-ounce bottle of cough syrup is 2% alcohol. Find the amount of alcohol in the cough syrup.

$$Q = Ar$$
$$Q = 8(0.02)$$
$$Q = 0.16$$

The syrup contains 0.16 oz of alcohol.

Lever-System Equation

$F_1x = F_2(d - x)$, where F_1 and F_2 are two forces, d is the length of the lever, x is the distance from F_1 to the fulcrum, and $d - x$ is the distance from F_2 to the fulcrum.

[3.3.4, p. 115]

A 30-pound child is sitting at one end of an 8-foot seesaw, 5 ft from the fulcrum. A 50-pound child is sitting at the other end. The seesaw balances.

$$F_1x = F_2(d - x)$$
$$30(5) = 50(8 - 5)$$
$$30(5) = 50(3)$$
$$150 = 150$$

CHAPTER 3 Review Exercises

1. Is 3 a solution of $5x - 2 = 4x + 5$?
2. Solve: $x - 4 = 16$
3. Solve: $8x = -56$
4. Solve: $5x - 6 = 29$
5. Solve: $5x + 3 = 10x - 17$
6. Solve: $3(5x + 2) + 2 = 10x + 5[x - (3x - 1)]$
7. What is 81% of 500?
8. 18 is 72% of what number?
9. 27 is what percent of 40?
10. Graph: $x \le -2$
11. Solve and graph the solution set of $x - 3 > -1$.
12. Solve and graph the solution set of $-3x < 12$.
13. Solve: $3x + 4 \ge -8$
14. Solve: $7x - 2(x + 3) \ge x + 10$
15. Is 2 a solution of $x^2 + 4x + 1 = 3x + 7$?
16. Solve: $4.6 = 2.1 + x$
17. Solve: $\frac{x}{7} = -7$
18. Solve: $14 + 6x = 17$
19. Solve: $12y - 1 = 3y + 2$
20. Solve: $x + 5(3x - 20) = 10(x - 4)$
21. What is $66\frac{2}{3}\%$ of 24?
22. 60 is 48% of what number?
23. 0.5 is what percent of 3?
24. Solve and graph the solution set of $2 + x < -2$.
25. Solve and graph the solution set of $5x \le -10$.
26. Solve: $6x + 3(2x - 1) = -27$
27. Solve: $a - \frac{1}{6} = \frac{2}{3}$
28. Solve: $\frac{3}{5}a = 12$
29. Solve: $32 = 9x - 4 - 3x$
30. Solve: $-4[x + 3(x - 5)] = 3(8x + 20)$
31. Solve: $4x - 12 < x + 24$
32. What is $\frac{1}{2}\%$ of 3000?
33. Solve: $3x + 7 + 4x = 42$
34. Solve: $5x - 6 > 19$
35. 8 is what percent of 200?
36. Solve: $6x - 9 < 4x + 3(x + 3)$
37. Solve: $5 - 4(x + 9) > 11(12x - 9)$
38. **Geometry** Find the measure of the third angle of a triangle if the first angle is 20° and the second angle is 50°. Use the equation $A + B + C = 180°$, where A, B, and C are the measures of the angles of a triangle.
39. **Lever Systems** A lever is 12 ft long. At a distance of 2 ft from the fulcrum, a force of 120 lb is applied. How large a force must be applied to the other end of the lever so that the system will balance? Use the lever-system equation $F_1x = F_2(d - x)$.
40. **Geometry** Use the equation $P = 2L + 2W$, where P is the perimeter of a rectangle, L is its length, and W is its width, to find the width of a rectangle that has a perimeter of 49 ft and a length of 18.5 ft.
41. **Discounts** Find the discount on a 26-inch, 21-speed dual-index grip-shifting bicycle if the sale price is \$198 and the regular price is \$239.99. Use the equation $S = R - D$, where S is the sale price, R is the regular price, and D is the discount.
42. **Conservation** In Central America and Mexico, 1184 plants and animals are at risk of extinction. This represents approximately 10.7% of all the species at risk of extinction on Earth. Approximately how many plants and animals are at risk of extinction on Earth? (*Source:* World Conservation Union)

43. **Physics** The pressure at a certain depth in the ocean can be approximated by the equation $P = 15 + \frac{1}{2}D$, where P is the pressure in pounds per square inch and D is the depth in feet. Use this equation to find the depth when the pressure is 55 lb/in^2.

44. **Lever Systems** A lever is 8 ft long. A force of 25 lb is applied to one end of the lever, and a force of 15 lb is applied to the other end. Find the location of the fulcrum when the system balances. Use the lever-system equation $F_1x = F_2(d - x)$.

45. **Geometry** Find the length of a rectangle when the perimeter is 84 ft and the width is 18 ft. Use the equation $P = 2L + 2W$, where P is the perimeter of a rectangle, L is the length, and W is the width.

46. **Uniform Motion Problem** A motorboat that can travel 15 mph in calm water is traveling with the current of a river that is moving at 5 mph. How long will it take the motorboat to travel 30 mi?

47. **Investments** Cathy Serano would like to earn $125 per year from two investments. She has found one investment that will pay her $75 per year. How much must she invest in an account that earns an annual simple interest rate of 8% to reach her goal?

48. **Mixture Problem** A 150-milliliter acid solution contains 9 ml of hydrochloric acid. What is the percent concentration of hydrochloric acid?

CHAPTER 3 Test

1. Solve: $\frac{3}{4}x = -9$

2. Solve: $6 - 5x = 5x + 11$

3. Solve: $3x - 5 = -14$

4. Is -2 a solution of $x^2 - 3x = 2x - 6$?

5. Solve: $x + \frac{1}{2} = \frac{5}{8}$

6. Solve: $5x - 2(4x - 3) = 6x + 9$

7. Solve: $7 - 4x = -13$

8. Solve: $11 - 4x = 2x + 8$

9. Solve: $x - 3 = -8$

10. Solve: $3x - 2 = 5x + 8$

11. Solve: $-\frac{3}{8}x = 5$

12. Solve: $6x - 3(2 - 3x) = 4(2x - 7)$

13. Solve: $6 - 2(5x - 8) = 3x - 4$

14. Solve: $9 - 3(2x - 5) = 12 + 5x$

15. Solve: $3(2x - 5) = 8x - 9$

16. 20 is what percent of 16?

17. 30% of what is 12?

18. Graph: $x > -2$

19. Solve and graph the solution set of $-2 + x \leq -3$.

20. Solve and graph the solution set of $\frac{3}{8}x > -\frac{3}{4}$.

21. Solve: $x + \frac{1}{3} > \frac{5}{6}$

22. Solve: $3(x - 7) \geq 5x - 12$

23. Solve: $-\frac{3}{8}x \leq 6$

24. Solve: $4x - 2(3 - 5x) \leq 6x + 10$

25. Solve: $3(2x - 5) \geq 8x - 9$

26. Solve: $15 - 3(5x - 7) < 2(7 - 2x)$

27. Solve: $-6x + 16 = -2x$

28. 20 is $83\frac{1}{3}\%$ of what number?

29. Solve and graph the solution set of $\frac{2}{3}x \geq 2$.

30. Is 5 a solution of $x^2 + 2x + 1 = (x + 1)^2$?

31. **Space Travel** A person's weight on the moon is $16\frac{2}{3}\%$ of the person's weight on Earth. If an astronaut weighs 180 lb on Earth, how much would the astronaut weigh on the moon?

32. **Chemistry** A chemist mixes 100 g of water at 80°C with 50 g of water at 20°C. Use the equation $m_1 \cdot (T_1 - T) = m_2 \cdot (T - T_2)$ to find the final temperature of the water after mixing. In this equation, m_1 is the quantity of water at the hotter temperature, T_1 is the temperature of the hotter water, m_2 is the quantity of water at the cooler temperature, T_2 is the temperature of the cooler water, and T is the final temperature of the water after mixing.

33. **Manufacturing** A financial manager has determined that the cost per unit for a calculator is \$15 and that the fixed costs per month are \$2000. Find the number of calculators produced during a month in which the total cost was \$5000. Use the equation $T = U \cdot N + F$, where T is the total cost, U is the cost per unit, N is the number of units produced, and F is the fixed cost.

34. **Uniform Motion Problem** Two hikers start at the same time from opposite ends of a 15-mile course. One hiker is walking at a rate of 3.5 mph, and the second hiker is walking at a rate of 4 mph. How long after they begin will they meet?

35. **Investments** Victor Jameson has invested \$750 in a simple interest account that has an annual interest rate of 6.2%. How much must he invest in a second account that earns 5% simple interest so that the interest earned on the second account is the same as that earned on the first account?

36. **Mixture Problem** An 8-ounce glass of chocolate milk contains 2 oz of chocolate syrup. Find the percent concentration of chocolate syrup in the chocolate milk.

Cumulative Review Exercises

1. Subtract: $-6 - (-20) - 8$

2. Multiply: $(-2)(-6)(-4)$

3. Subtract: $-\frac{5}{6} - \left(-\frac{7}{16}\right)$

4. Divide: $-2\frac{1}{3} \div 1\frac{1}{6}$

5. Simplify: $-4^2 \cdot \left(-\frac{3}{2}\right)^3$

6. Simplify: $25 - 3 \cdot \frac{(5-2)^2}{2^3 + 1} - (-2)$

7. Evaluate $3(a - c) - 2ab$ when $a = 2$, $b = 3$, and $c = -4$.

8. Simplify: $3x - 8x + (-12x)$

9. Simplify: $2a - (-3b) - 7a - 5b$

10. Simplify: $(16x)\left(\frac{1}{8}\right)$

11. Simplify: $-4(-9y)$

12. Simplify: $-2(-x^2 - 3x + 2)$

13. Simplify: $-2(x - 3) + 2(4 - x)$

14. Simplify: $-3[2x - 4(x - 3)] + 2$

15. Use the roster method to write the set of negative integers greater than -8.

16. Write $\frac{7}{8}$ as a percent. Write the remainder in fractional form.

17. Write 342% as a decimal.

18. Write $62\frac{1}{2}\%$ as a fraction.

19. Is -3 a solution of $x^2 + 6x + 9 = x + 3$?

20. Solve: $x - 4 = -9$

21. Solve: $\frac{3}{5}x = -15$

22. Solve: $13 - 9x = -14$

23. Solve: $5x - 8 = 12x + 13$

24. Solve: $8x - 3(4x - 5) = -2x - 11$

25. Solve: $-\frac{3}{4}x > \frac{2}{3}$

26. Solve: $5x - 4 \geq 4x + 8$

27. Solve: $3x + 17 < 5x - 1$

28. Translate "the difference between eight and the quotient of a number and twelve" into a variable expression.

29. Translate and simplify "the sum of a number and two more than the number."

30. Dollar Bills A club treasurer has some five-dollar bills and some ten-dollar bills. The treasurer has a total of 35 bills. Use one variable to express the number of each denomination of bill.

31. Fishing A fishing wire 3 ft long is cut into two pieces, one shorter than the other. Express the length of the shorter piece in terms of the length of the longer piece.

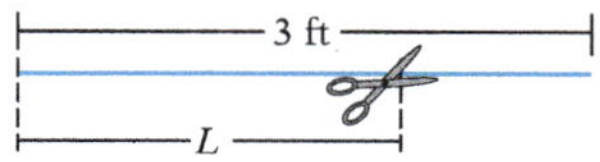

32. Taxes A computer programmer receives a weekly wage of \$1350, and \$229.50 is deducted for income tax. Find the percent of the computer programmer's salary deducted for income tax.

33. Track and Field The world record time for a 1-mile race can be approximated by the equation $t = 17.08 - 0.0067y$, where t is the time in minutes and y is the year of the race. Use this equation to predict the year in which the first "4-minute mile" was run. (The actual year was 1954.) Round to the nearest whole number.

34. Chemistry A chemist mixes 300 g of water at 75°C with 100 g of water at 15°C. Use the equation $m_1 \cdot (T_1 - T) = m_2 \cdot (T - T_2)$ to find the final temperature of the water. In this equation, m_1 is the quantity of water at the hotter temperature, T_1 is the temperature of the hotter water, m_2 is the quantity of water at the cooler temperature, T_2 is the temperature of the cooler water, and T is the final temperature of the water after mixing.

35. Lever Systems A lever is 25 ft long. At a distance of 12 ft from the fulcrum, a force of 26 lb is applied. How large a force must be applied to the other end of the lever so that the system will balance? Use the lever system equation $F_1x = F_2(d - x)$.

CHAPTER 4

Solving Equations and Inequalities: Applications

Digital Vision

Focus on Success

Do you have trouble with word problems? Word problems show the variety of ways in which math can be used. The solution of every word problem can be broken down into two steps: Strategy and Solution. The Strategy consists of reading the problem, writing down what is known and unknown, and devising a plan to find the unknown. The Solution often consists of solving an equation and then checking the solution. (See Word Problems, page AIM-10.)

OBJECTIVES

4.1
1. Translate a sentence into an equation and solve
2. Application problems

4.2
1. Perimeter problems
2. Problems involving angles formed by intersecting lines
3. Problems involving the angles of a triangle

4.3
1. Markup problems
2. Discount problems

4.4
1. Investment problems

4.5
1. Value mixture problems
2. Percent mixture problems

4.6
1. Uniform motion problems

4.7
1. Applications of inequalities

PREP TEST

Are you ready to succeed in this chapter?
Take the Prep Test below to find out if you are ready to learn the new material.

1. Simplify: $R - 0.35R$
2. Simplify: $0.08x + 0.05(400 - x)$
3. Simplify: $n + (n + 2) + (n + 4)$
4. Translate into a variable expression: "The difference between five and twice a number."
5. Write 0.4 as a percent.
6. Solve: $25x + 10(9 - x) = 120$
7. Solve: $36 = 48 - 48r$
8. Solve: $4(2x - 5) < 12$
9. A snack mixture weighs 20 oz and contains nuts and pretzels. Let n represent the number of ounces of nuts in the mixture. Express the number of ounces of pretzels in the mixture in terms of n.

4.1 Translating Sentences into Equations

OBJECTIVE 1 Translate a sentence into an equation and solve

An equation states that two mathematical expressions are equal. Therefore, translating a sentence into an equation requires recognizing the words or phrases that mean "equals." Some of the phrases that mean "equals" are *is, is equal to, amounts to,* and *represents.*

Once the sentence is translated into an equation, the equation can be solved by rewriting the equation in the form *variable* = *constant.*

EXAMPLE 1 Translate "two times the sum of a number and eight equals the sum of four times the number and six" into an equation and solve.

Solution the unknown number: n

• **Assign a variable to the unknown quantity.**

two times the sum of a number and eight	equals	the sum of four times the number and six

• **Find two verbal expressions for the same value.**

$$2(n + 8) = 4n + 6$$

• **Write a mathematical expression for each verbal expression. Write the equals sign.**

• **Solve the equation.**

$$2n + 16 = 4n + 6$$
$$2n - 4n + 16 = 4n - 4n + 6$$
$$-2n + 16 = 6$$
$$-2n + 16 - 16 = 6 - 16$$
$$-2n = -10$$
$$\frac{-2n}{-2} = \frac{-10}{-2}$$
$$n = 5$$

The number is 5.

Problem 1 Translate "nine less than twice a number is five times the sum of the number and twelve" into an equation and solve.

Solution See page S8.

Try Exercise 17, page 145.

Recall that the integers are the numbers ..., −4, −3, −2, −1, 0, 1, 2, 3, 4, ...

An **even integer** is an integer that is divisible by 2. Examples of even integers are −8, 0, and 22.

An **odd integer** is an integer that is not divisible by 2. Examples of odd integers are −17, 1, and 39.

Consecutive integers are integers that follow one another in order. Examples of consecutive integers are shown at the right. (Assume the variable n represents an integer.)

11, 12, 13
−8, −7, −6
$n, n + 1, n + 2$

Examples of **consecutive even integers** are shown at the right. (Assume the variable n represents an even integer.)

24, 26, 28
−10, −8, −6
$n, n + 2, n + 4$

Examples of **consecutive odd integers** are shown at the right. (Assume the variable n represents an odd integer.)

19, 21, 23
$-1, 1, 3$
$n, n + 2, n + 4$

Focus on solving a consecutive integer problem

The sum of three consecutive odd integers is 51. Find the integers.

Strategy

▶ Let a variable represent one of the integers. — First odd integer: n

Express each of the other integers in terms of that variable. (Remember that consecutive integers differ by 1. Consecutive even or consecutive odd integers differ by 2.) — Second odd integer: $n + 2$; Third odd integer: $n + 4$

▶ Determine the relationship among the integers. — Their sum is 51.

Solution

Write an equation. $$n + (n + 2) + (n + 4) = 51$$

Solve for n.

$$3n + 6 = 51$$
$$3n = 45$$
$$n = 15$$

The first odd integer is 15.

Substitute the value of n into the variable expressions for the second and third integers.

$$n + 2 = 15 + 2 = 17$$
$$n + 4 = 15 + 4 = 19$$

The three consecutive odd integers are 15, 17, and 19.

EXAMPLE 2 Find three consecutive even integers such that three times the second integer is six more than the sum of the first and third integers.

Strategy

▶ First even integer: n
Second even integer: $n + 2$
Third even integer: $n + 4$

▶ Three times the second integer equals six more than the sum of the first and third integers.

Solution

$$3(n + 2) = n + (n + 4) + 6$$
$$3n + 6 = 2n + 10$$
$$n + 6 = 10$$
$$n = 4$$

- **Write an equation.**
- **Solve for n.**
- **The first even integer is 4.**

$$n + 2 = 4 + 2 = 6$$
$$n + 4 = 4 + 4 = 8$$

- **Substitute the value of n into the variable expressions for the second and third integers.**

The three consecutive even integers are 4, 6, and 8.

Problem 2 Find three consecutive integers whose sum is -12.

Solution See page S8.

Try Exercise 45, page 146.

OBJECTIVE 2 Application problems

EXAMPLE 3 The temperature of the sun on the Kelvin scale is 6500 K. This is 4740° less than the temperature of the sun on the Fahrenheit scale. Find the sun's Fahrenheit temperature.

Strategy To find the Fahrenheit temperature, write and solve an equation using F to represent the Fahrenheit temperature.

Solution

6500	is	4740° less than the Fahrenheit temperature

$$\begin{aligned} 6500 &= F - 4740 \\ 6500 + 4740 &= F - 4740 + 4740 \\ 11{,}240 &= F \end{aligned}$$

The temperature of the sun is 11,240°F.

Problem 3 A molecule of octane gas has eight carbon atoms. This represents twice the number of carbon atoms in a butane gas molecule. Find the number of carbon atoms in a butane gas molecule.

Solution See page S8.

Try Exercise 57, page 147.

EXAMPLE 4 A board 10 ft long is cut into two pieces. Three times the length of the shorter piece is twice the length of the longer piece. Find the length of each piece.

Strategy Draw a diagram. To find the length of each piece, write and solve an equation using x to represent the length of the shorter piece and $10 - x$ to represent the length of the longer piece.

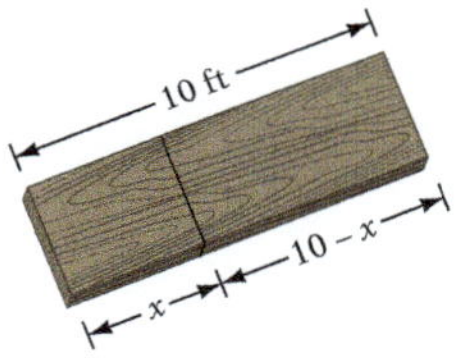

Solution

three times the length of the shorter piece	is	twice the length of the longer piece

$$\begin{aligned} 3x &= 2(10 - x) \\ 3x &= 20 - 2x \\ 3x + 2x &= 20 - 2x + 2x \\ 5x &= 20 \\ \frac{5x}{5} &= \frac{20}{5} \\ x &= 4 \\ 10 - x = 10 - 4 &= 6 \end{aligned}$$

- **The length of the shorter piece is 4 ft.**
- **Substitute the value of x into the variable expression for the longer piece and evaluate.**

The length of the shorter piece is 4 ft. The length of the longer piece is 6 ft.

Problem 4 A company manufactures 160 bicycles per day. Four times the number of 3-speed bicycles made each day equals 30 less than the number of 10-speed bicycles made each day. Find the number of 10-speed bicycles manufactured each day.

Solution See page S8.

Try Exercise 71, page 148.

Take Note

We could also let x represent the length of the longer piece and let $10 - x$ represent the length of the shorter piece. Then our equation would be

$$3(10 - x) = 2x$$

Show that this equation results in the same solutions.

4.1 Exercises

CONCEPT CHECK

For Exercises 1 to 6, determine whether the statement is true or false.

1. When two expressions represent the same value, we say that the expressions are equal to each other.
2. When translating a sentence into an equation, we can use any variable to represent an unknown number.
3. In addition to a number, the answer to an application problem must have a unit, such as meters, dollars, minutes, or miles per hour.
4. An even integer is a multiple of 2.
5. Given the consecutive odd integers -5 and -3, the next consecutive odd integer is -1.
6. If the first of three consecutive odd integers is n, then the second and third consecutive odd integers are represented as $n + 1$ and $n + 3$.
7. The sum of two numbers is 12.
   a. If x represents the larger number, represent the smaller number in terms of x.
   b. If x represents the smaller number, represent the larger number in terms of x.

1 Translate a sentence into an equation and solve (See pages 142–143.)

GETTING READY

8. When we translate a sentence into an equation, the word *is* translates into the ___?___ sign.

Translate into an equation and solve.

9. The difference between a number and fifteen is seven. Find the number.
10. The sum of five and a number is three. Find the number.
11. The product of seven and a number is negative twenty-one. Find the number.
12. The quotient of a number and four is two. Find the number.
13. Four less than three times a number is five. Find the number.
14. The difference between five and twice a number is one. Find the number.
15. Four times the sum of twice a number and three is twelve. Find the number.
16. Twenty-one is three times the difference between four times a number and five. Find the number.
17. Twelve is six times the difference between a number and three. Find the number.
18. The difference between six times a number and four times the number is negative fourteen. Find the number.
19. Twenty-two is two less than six times a number. Find the number.
20. Negative fifteen is three more than twice a number. Find the number.
21. Seven more than four times a number is three more than two times the number. Find the number.
22. The difference between three times a number and four is five times the number. Find the number.

23. Eight less than five times a number is four more than eight times the number. Find the number.

24. The sum of a number and six is four less than six times the number. Find the number.

25. Twice the difference between a number and twenty-five is three times the number. Find the number.

26. Four times a number is three times the difference between thirty-five and the number. Find the number.

27. The sum of two numbers is twenty. Three times the smaller is equal to two times the larger. Find the two numbers.

28. The sum of two numbers is fifteen. One less than three times the smaller is equal to the larger. Find the two numbers.

29. The sum of two numbers is eighteen. The total of three times the smaller and twice the larger is forty-four. Find the two numbers.

30. The sum of two numbers is two. The difference between eight and twice the smaller number is two less than four times the larger. Find the two numbers.

31. The sum of two numbers is fourteen. One number is ten more than the other number. Write two different equations that can be used to find the numbers by first writing an equation in which n represents the larger number and then writing an equation in which n represents the smaller number.

32. The sum of two numbers is seven. Twice one number is four less than the other number. Which of the following equations does *not* represent this situation?

(i) $2n = 7 - n - 4$ (ii) $2(7 - x) = x - 4$
(iii) $2x + 4 = 7 - x$ (iv) $2n - 4 = 7 - n$

GETTING READY

33. Integers that follow one another in order are called ___?___ integers.

34. Two consecutive integers differ by ___?___. Two consecutive even integers differ by ___?___. Two consecutive odd integers differ by ___?___.

35. Explain how to represent three consecutive integers using only one variable.

36. Explain why both consecutive even integers and consecutive odd integers are represented algebraically as $n, n + 2, n + 4, \ldots$.

37. The sum of three consecutive integers is 54. Find the integers.

38. The sum of three consecutive integers is 75. Find the integers.

39. The sum of three consecutive even integers is 84. Find the integers.

40. The sum of three consecutive even integers is 48. Find the integers.

41. The sum of three consecutive odd integers is 57. Find the integers.

42. The sum of three consecutive odd integers is 81. Find the integers.

43. Find two consecutive even integers such that five times the first integer is equal to four times the second integer.

44. Find two consecutive even integers such that six times the first integer equals three times the second integer.

45. Nine times the first of two consecutive odd integers equals seven times the second. Find the integers.

46. Five times the first of two consecutive odd integers is three times the second. Find the integers.

47. Find three consecutive integers whose sum is negative twenty-four.

48. Find three consecutive even integers whose sum is negative twelve.

49. Three times the smallest of three consecutive even integers is two more than twice the largest. Find the integers.

50. Twice the smallest of three consecutive odd integers is five more than the largest. Find the integers.

51. Find three consecutive odd integers such that three times the middle integer is six more than the sum of the first and third integers.

52. Find three consecutive odd integers such that four times the middle integer is two less than the sum of the first and third integers.

53. Which of the following could *not* be used to represent three consecutive even integers?
(i) $n + 2, n + 4, n + 6$ (ii) $n, n + 4, n + 6$
(iii) $n + 1, n + 3, n + 5$ (iv) $n - 2, n, n + 2$

54. If n is an odd integer, which expression represents the third consecutive odd integer after n?
(i) $n + 6$ (ii) $n + 4$ (iii) $n + 3$ (iv) $n + 2$

2 Application problems (See page 144.)

GETTING READY

55. The number of calories in a cup of low-fat milk is two-thirds the number of calories in a cup of whole milk. In this situation, let n represent the number of calories in a cup of __?__ milk, and let $\frac{2}{3}n$ represent the number of calories in a cup of __?__ milk.

56. A cup of low-fat milk has 100 calories. Use the information in Exercise 55 to write an equation that can be used to find the number of calories in a cup of whole milk: __?__ = __?__.

Write an equation and solve.

57. **Depreciation** As a result of depreciation, the value of a car is now $19,200. This is three-fifths of its original value. Find the original value of the car.

58. **Structures** The length of the Royal Gorge Bridge in Colorado is 320 m. This is one-fourth the length of the Golden Gate Bridge. Find the length of the Golden Gate Bridge.

59. **Nutrition** One slice of cheese pizza contains 290 calories. A medium-size orange has one-fifth that number of calories. How many calories are in a medium-size orange?

60. **History** John D. Rockefeller died in 1937. At the time of his death, Rockefeller had accumulated a wealth of $1,400 million, which was equal to one-sixty-fifth of the gross national product of the United States at that time. What was the U.S. gross national product in 1937? (*Source: The Wealthy 100: A Ranking of the Richest Americans, Past and Present*)

61. **Agriculture** A soil supplement that weighs 18 lb contains iron, potassium, and a mulch. The supplement contains fifteen times as much mulch as iron and twice as much potassium as iron. Find the amount of mulch in the soil supplement.

Ferenc Cegledi/Shutterstock.com
The Golden Gate Bridge

AP Photo
John D. Rockefeller

62. **Commissions** A real estate agent sold two homes and received commissions totaling \$6000. The agent's commission on one home was one and one-half times the commission on the second home. Find the agent's commission on each home.

63. **Safety** Loudness, or the intensity of sound, is measured in decibels. The sound level of a television is about 70 decibels, which is considered a safe hearing level. A food blender runs at 20 decibels higher than a TV, and a jet engine's decibel reading is 40 less than twice that of a blender. At this level, exposure can cause hearing loss. Find the intensity of the sound of a jet engine.

Point of Interest

The low-frequency pulses or whistles made by blue whales have been measured at up to 188 decibels, making them the loudest sounds produced by a living organism.

64. **Robots** Kiva Systems, Inc., builds robots that companies can use to streamline order fulfillment operations in their warehouses. Salary and other benefits for one human warehouse worker can cost a company about \$64,000 a year, an amount that is 103 times the company's yearly maintenance and operation costs for one robot. Find the yearly costs for a robot. Round to the nearest hundred. (*Source: The Boston Globe*)

65. **Geometry** Greek architects considered a rectangle whose length was approximately 1.6 times its width to be the most visually appealing. Find the length and width of a rectangle constructed in this manner if the sum of the length and width is 130 ft.

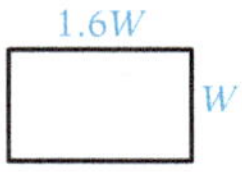

66. **Geography** Greenland, the largest island in the world, is 21 times larger than Iceland. The combined area of Greenland and Iceland is 880,000 mi^2. Find the area of Greenland.

67. **Consumerism** The cost to replace a water pump in a sports car was \$820. This included \$375 for the water pump and \$89 per hour for labor. How many hours of labor were required to replace the water pump?

68. **Utilities** The cost of electricity in a certain city is \$.09 for each of the first 300 kWh (kilowatt-hours) and \$.15 for each kWh over 300 kWh. Find the number of kilowatt-hours used by a family that receives a \$59.25 electric bill.

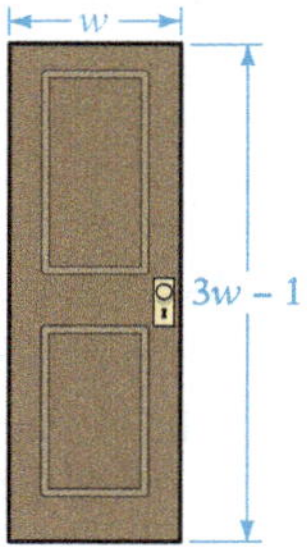

69. **Consumerism** The fee charged by a ticketing agency for a concert is \$45.50 plus \$87.50 for each ticket purchased. If your total charge for tickets is \$833, how many tickets are you purchasing?

70. **Carpentry** A carpenter is building a wood door frame. The height of the frame is 1 ft less than three times the width. What is the width of the largest door frame that can be constructed from a board 19 ft long? (*Hint:* A door frame consists of only three sides; there is no frame below a door.)

71. **Carpentry** A 20-foot board is cut into two pieces. Twice the length of the shorter piece is 4 ft more than the length of the longer piece. Find the length of the shorter piece.

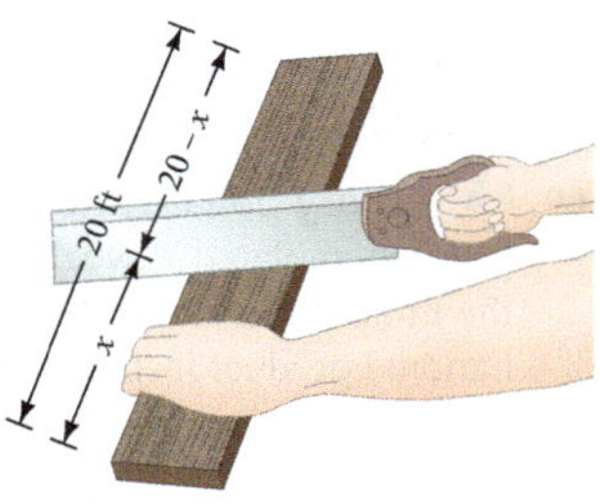

72. **Labor Unions** A union charges monthly dues of \$4.00 plus \$.25 for each hour worked during the month. A union member's dues for March were \$46.00. How many hours did the union member work during the month of March?

73. **Business** The cellular phone service for a business executive is \$80 per month plus \$.40 per minute of phone use over 900 min. In a month when the executive's cellular phone bill was \$100.40, how many minutes did the executive use the phone?

74. **Recycling** Use the information in the article below to find how many tons of plastic drink bottles were stocked for sale in U.S. stores.

In the News

Americans' Unquenchable Thirst

Despite efforts to increase recycling, the 2.16 million tons of plastic drink bottles that ended up in landfills this year represent four-fifths of the plastic drink bottles stocked for sale in U.S. stores.

And Americans can't seem to get enough of bottled water. During a recent year, stores stocked 7.5 billion gallons of bottled water, an amount that is approximately the same as the volume of water that goes over Niagara Falls every 3 h.

Source: scienceline.org

iStockphoto.com/Bilan Doty

Text Messaging For Exercises 75 and 76, use the expression $2.99 + 0.15n$, which represents the total monthly text-messaging bill for n text messages over 300 in 1 month.

75. How much does the customer pay per text message over 300 messages?

76. What is the fixed charge per month for the text-messaging service?

77. **Metalwork** A wire 12 ft long is cut into two pieces. Each piece is bent into the shape of a square. The perimeter of the larger square is twice the perimeter of the smaller square. Find the perimeter of the larger square.

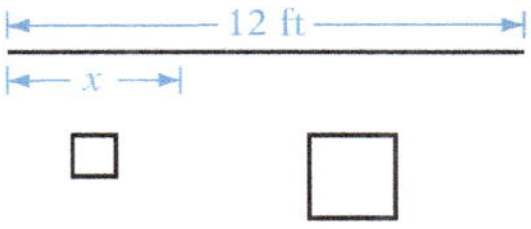

APPLYING CONCEPTS

78. The amount of liquid in a container triples every minute. The container becomes completely filled at 3:40 P.M. What fractional part of the container is filled at 3:39 P.M.?

79. **Travel** A cyclist traveling at a constant speed completes $\frac{3}{5}$ of a trip in $1\frac{1}{2}$ h. In how many additional hours will the cyclist complete the entire trip?

80. **Business** During one day at an office, one-half of the amount of money in the petty cash drawer was used in the morning, and one-third of the remaining money was used in the afternoon, leaving \$5 in the petty cash drawer at the end of the day. How much money was in the petty cash drawer at the start of the day?

81. Find four consecutive even integers whose sum is -36.

82. Find four consecutive odd integers whose sum is -48.

83. Find three consecutive odd integers such that the sum of the first and third integers is twice the second integer.

84. Find four consecutive integers such that the sum of the first and fourth integers equals the sum of the second and third integers.

85. A formula is an equation that relates variables in a known way. Find two examples of formulas that are used in your college major. Explain what each of the variables represents.

PROJECTS OR GROUP ACTIVITIES

Complete each statement with the word *even* or *odd*.

86. If k is an odd integer, then $k + 1$ is an __?__ integer.

87. If k is an odd integer, then $k - 2$ is an __?__ integer.

88. If n is an integer, then $2n$ is an __?__ integer.

89. If m and n are even integers, then $m - n$ is an __?__ integer.

90. If m and n are even integers, then mn is an __?__ integer.

91. If m and n are odd integers, then $m + n$ is an __?__ integer.

92. If m and n are odd integers, then $m - n$ is an __?__ integer.

93. If m and n are odd integers, then mn is an __?__ integer.

94. If m is an even integer and n is an odd integer, then $m - n$ is an __?__ integer.

95. If m is an even integer and n is an odd integer, then $m + n$ is an __?__ integer.

4.2 Geometry Problems

OBJECTIVE 1 Perimeter problems

Recall that the **perimeter** of a plane geometric figure is a measure of the distance around the figure. Perimeter is used, for example, in buying fencing for a lawn and in determining how much baseboard is needed for a room.

The perimeter of a triangle is the sum of the lengths of the three sides. Therefore, if a, b, and c represent the lengths of the sides of a triangle, the perimeter P of the triangle is given by $\boldsymbol{P = a + b + c}$.

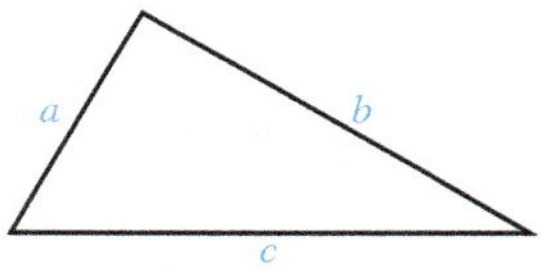

$P = a + b + c$

Two special types of triangles are shown below. An **isosceles triangle** has two sides of equal length. The two angles opposite the two sides of equal length are of equal measure. The three sides of an **equilateral triangle** are of equal length, and all three angles have the same measure.

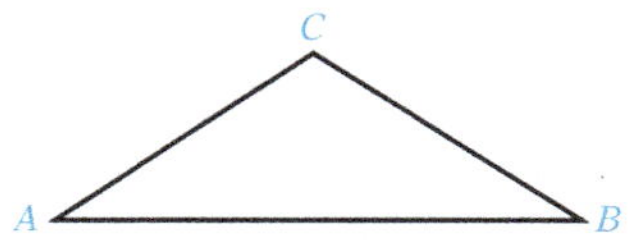

Isosceles triangle
$AC = BC$
$\angle A = \angle B$

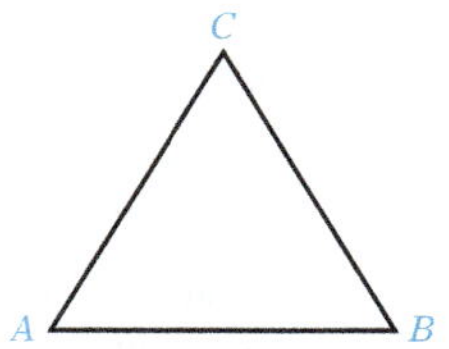

Equilateral triangle
$AB = AC = BC$
$\angle A = \angle B = \angle C$

Point of Interest

Leonardo da Vinci painted the Mona Lisa on a rectangular canvas whose height was approximately 1.6 times its width. Rectangles with these proportions, called **golden rectangles,** were used extensively in Renaissance art.

The perimeter of a rectangle is the sum of the lengths of the four sides. Let L represent the length and W represent the width of a rectangle. Then the perimeter P of the rectangle is given by $P = L + W + L + W$. Combine like terms and the formula is $\boldsymbol{P = 2L + 2W}$.

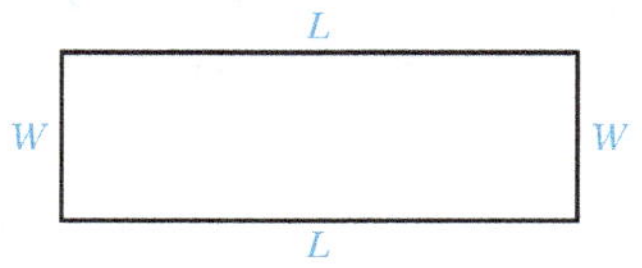

$P = 2L + 2W$

A square is a rectangle in which each side has the same length. Let s represent the length of each side of a square. Then the perimeter P of the square is given by $P = s + s + s + s$. Combine like terms and the formula is $\boldsymbol{P = 4s}$.

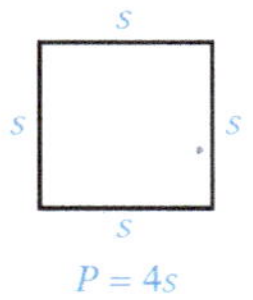

FORMULAS FOR PERIMETERS OF GEOMETRIC FIGURES

Perimeter of a triangle:	$P = a + b + c$
Perimeter of a rectangle:	$P = 2L + 2W$
Perimeter of a square:	$P = 4s$

Focus on solving a perimeter problem

The perimeter of a rectangle is 26 ft. The length of the rectangle is 1 ft more than twice the width. Find the width and length of the rectangle.

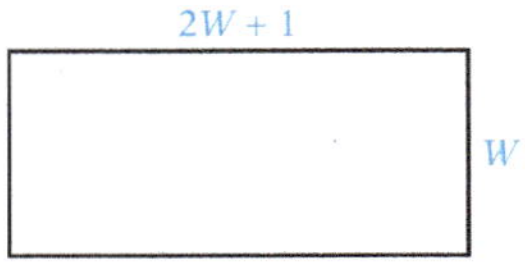

Strategy

Let a variable represent the width. Width: W

Represent the length in terms of that variable. Length: $2W + 1$

Use the formula for the perimeter of a rectangle.

Solution

$P = 2L + 2W$

$P = 26$. Substitute $2W + 1$ for L. $26 = 2(2W + 1) + 2W$

Use the Distributive Property. $26 = 4W + 2 + 2W$

Combine like terms. $26 = 6W + 2$

Subtract 2 from each side of the equation. $24 = 6W$

Divide each side of the equation by 6. $4 = W$

Find the length of the rectangle by substituting 4 for W in $2W + 1$.

$L = 2W + 1$

$= 2(4) + 1 = 8 + 1 = 9$

The width is 4 ft. The length is 9 ft.

EXAMPLE 1 The perimeter of an isosceles triangle is 25 ft. The length of the third side is 2 ft less than the length of one of the equal sides. Find the measures of the three sides of the triangle.

Strategy

- Each equal side: x
  The third side: $x - 2$
- Use the equation for the perimeter of a triangle.

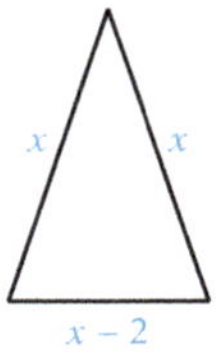

Solution

$P = a + b + c$ • **Use the formula for the perimeter of a triangle.**

$25 = x + x + (x - 2)$ • **Substitute 25 for P. Substitute the variable expressions for the three sides of the triangle.**

$25 = 3x - 2$ • **Solve the equation for x.**

$27 = 3x$

$9 = x$

$x - 2 = 9 - 2 = 7$ • **Substitute the value of *x* into the variable expression for the length of the third side.**

Each of the equal sides measures 9 ft.
The third side measures 7 ft.

Problem 1 A carpenter is designing a square patio with a perimeter of 52 ft. What is the length of each side?

Solution See page S8.

Try Exercise 17, page 158.

OBJECTIVE 2 Problems involving angles formed by intersecting lines

Recall that a unit used to measure angles is the **degree.** The symbol for degree is °. ∠ is the symbol for angle.

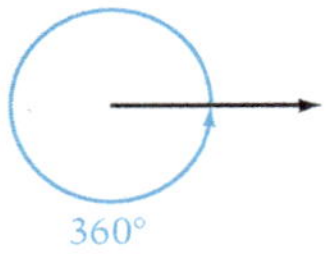

One complete revolution is 360°.

A 90° angle is called a **right angle.** The symbol ⊾ represents a right angle. Angle C ($\angle C$) is a right angle.

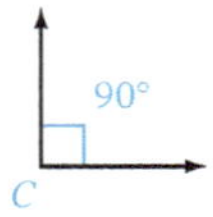

A 180° angle is called a **straight angle.** The angle at the right is a straight angle.

An **acute angle** is an angle whose measure is between 0° and 90°. $\angle A$ at the right is an acute angle.

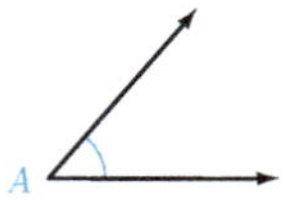

An **obtuse angle** is an angle whose measure is between 90° and 180°. $\angle B$ at the right is an obtuse angle.

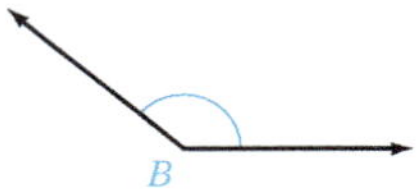

Focus on solving a problem involving angle measurement

Given the diagram at the left, find x.

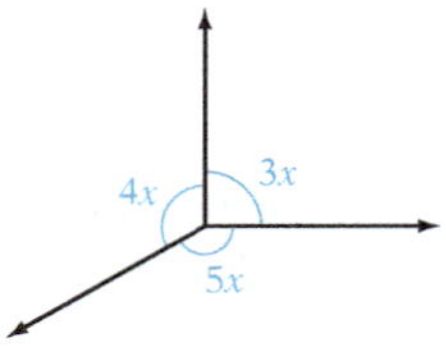

Strategy The sum of the measures of the three angles is 360°. To find x, write an equation and solve for x.

Solution

$$3x + 4x + 5x = 360°$$
$$12x = 360°$$
$$x = 30°$$

The measure of x is 30°.

Parallel lines never meet. The distance between them is always the same. The symbol ∥ means "is parallel to." In the figure at the right, $\ell_1 \parallel \ell_2$.

Perpendicular lines are intersecting lines that form right angles. The symbol ⊥ means "is perpendicular to." In the figure at the right, $p \perp q$.

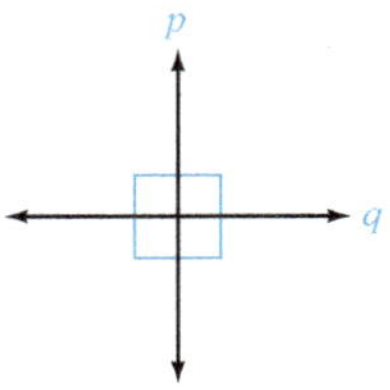

Four angles are formed by the intersection of two lines. If the two lines are perpendicular, then each of the four angles is a right angle. If the two lines are not perpendicular, then two of the angles formed are acute angles and two of the angles formed are obtuse angles. The two acute angles are always opposite each other, and the two obtuse angles are always opposite each other.

In the figure at the right, $\angle w$ and $\angle y$ are acute angles, and $\angle x$ and $\angle z$ are obtuse angles.

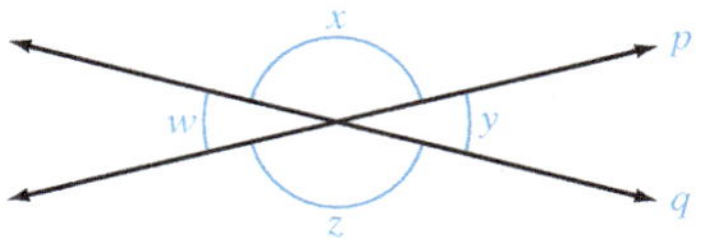

Two angles that are on opposite sides of the intersection of two lines are called **vertical angles.** Vertical angles have the same measure. $\angle w$ and $\angle y$ are vertical angles. $\angle x$ and $\angle z$ are vertical angles.

Vertical angles have the same measure.

$\angle w = \angle y$
$\angle x = \angle z$

Two angles that share a common side are called **adjacent angles.** For the figure shown above, $\angle x$ and $\angle y$ are adjacent angles, as are $\angle y$ and $\angle z$, $\angle z$ and $\angle w$, and $\angle w$ and $\angle x$. Adjacent angles of intersecting lines are supplementary angles.

Adjacent angles of intersecting lines are supplementary angles.

$\angle x + \angle y = 180°$
$\angle y + \angle z = 180°$
$\angle z + \angle w = 180°$
$\angle w + \angle x = 180°$

Focus on solving a problem involving intersecting lines

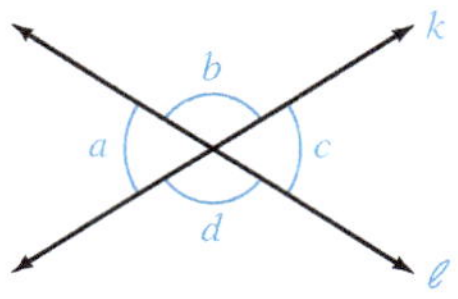

In the diagram at the left, $\angle b = 115°$. Find the measures of angles a, c, and d.

$\angle a$ is supplementary to $\angle b$ because $\angle a$ and $\angle b$ are adjacent angles of intersecting lines.

$\angle a + \angle b = 180°$
$\angle a + 115° = 180°$
$\angle a = 65°$

$\angle c = \angle a$ because $\angle c$ and $\angle a$ are vertical angles. $\angle c = 65°$

$\angle d = \angle b$ because $\angle d$ and $\angle b$ are vertical angles. $\angle d = 115°$

EXAMPLE 2 Find x.

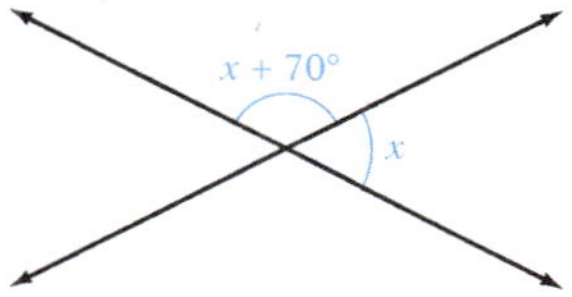

Strategy The angles labeled are adjacent angles of intersecting lines and are therefore supplementary angles. To find x, write an equation and solve for x.

Solution

$$
\begin{aligned}
x + (x + 70°) &= 180° \\
2x + 70° &= 180° \\
2x &= 110° \\
x &= 55°
\end{aligned}
$$

Problem 2 Find x.

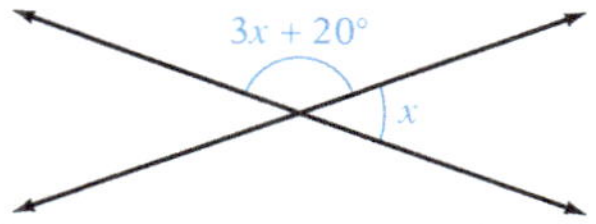

Solution See page S8.

Try Exercise 47, page 159.

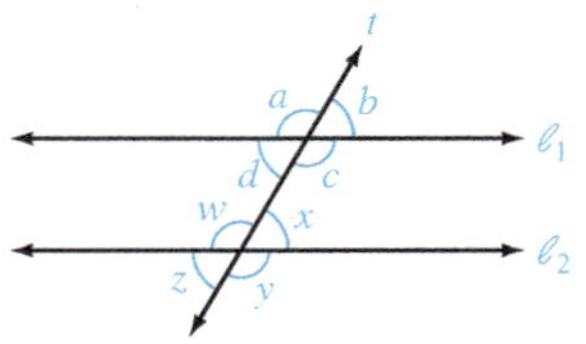

A line that intersects two other lines at different points is called a **transversal.** If the lines cut by a transversal t are parallel lines and the transversal is not perpendicular to the parallel lines, all four acute angles have the same measure and all four obtuse angles have the same measure. In the figure at the right,

$$\angle b = \angle d = \angle x = \angle z$$
$$\angle a = \angle c = \angle w = \angle y$$

Alternate interior angles are two nonadjacent angles that are on opposite sides of the transversal and between the lines. In the figure above, $\angle c$ and $\angle w$ are alternate interior angles, and $\angle d$ and $\angle x$ are alternate interior angles. Alternate interior angles have the same measure.

Alternate interior angles have the same measure.

$$\angle c = \angle w$$
$$\angle d = \angle x$$

Alternate exterior angles are two nonadjacent angles that are on opposite sides of the transversal and outside the parallel lines. In the figure above, $\angle a$ and $\angle y$ are alternate exterior angles, and $\angle b$ and $\angle z$ are alternate exterior angles. Alternate exterior angles have the same measure.

Alternate exterior angles have the same measure.

$$\angle a = \angle y$$
$$\angle b = \angle z$$

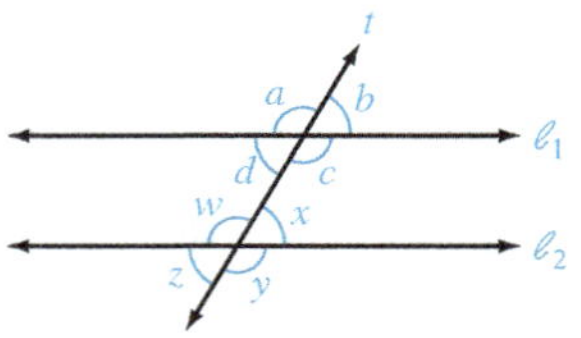

Corresponding angles are two angles that are on the same side of the transversal and are both acute angles or are both obtuse angles. In the figure at the left, the following pairs of angles are corresponding angles: $\angle a$ and $\angle w$, $\angle d$ and $\angle z$, $\angle b$ and $\angle x$, and $\angle c$ and $\angle y$. Corresponding angles have the same measure.

Corresponding angles have the same measure.

$$\angle a = \angle w$$
$$\angle d = \angle z$$
$$\angle b = \angle x$$
$$\angle c = \angle y$$

Focus on solving a problem involving parallel lines cut by a transversal

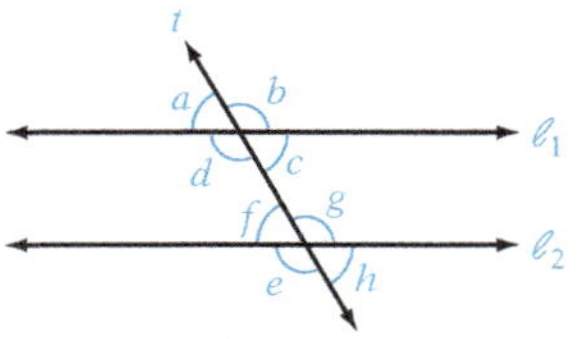

In the diagram at the left, $\ell_1 \parallel \ell_2$ and $\angle f = 58°$. Find the measures of $\angle a$, $\angle c$, and $\angle d$.

$\angle a$ and $\angle f$ are corresponding angles. $\quad \angle a = \angle f = 58°$

$\angle c$ and $\angle f$ are alternate interior angles. $\quad \angle c = \angle f = 58°$

$\angle d$ is supplementary to $\angle a$.

$$\angle d + \angle a = 180°$$
$$\angle d + 58° = 180°$$
$$\angle d = 122°$$

EXAMPLE 3 Given $\ell_1 \parallel \ell_2$, find x.

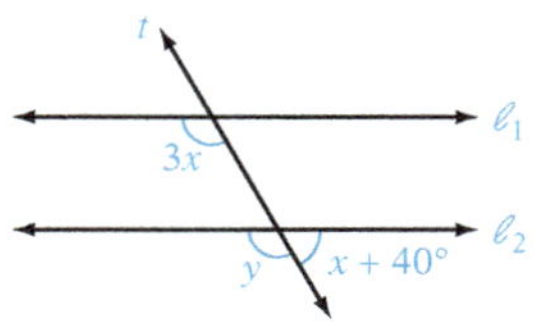

Strategy $3x = y$ because corresponding angles have the same measure.
$y + (x + 40°) = 180°$ because adjacent angles of intersecting lines are supplementary angles.

Substitute $3x$ for y and solve for x.

Solution

$$\begin{aligned} y + (x + 40°) &= 180° \\ 3x + (x + 40°) &= 180° \\ 4x + 40° &= 180° \\ 4x &= 140° \\ x &= 35° \end{aligned}$$

Problem 3 Given $\ell_1 \parallel \ell_2$, find x.

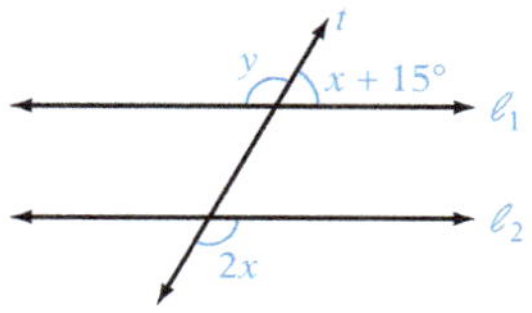

Solution See page S8.

Try Exercise 61, page 160.

OBJECTIVE 3 Problems involving the angles of a triangle

If the lines cut by a transversal are not parallel lines, the three lines will intersect at three points. In the figure at the right, the transversal t intersects lines p and q. The three lines intersect at points A, B, and C. The geometric figure formed by line segments AB, BC, and AC is a **triangle.**

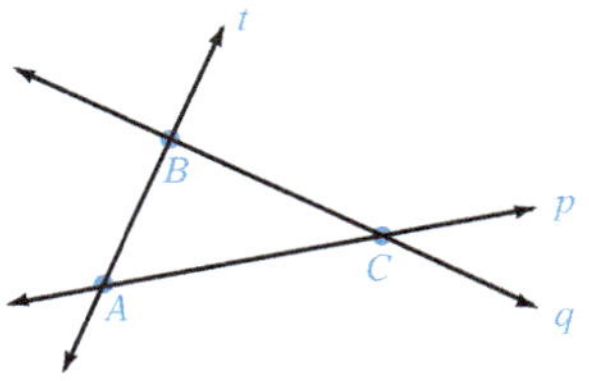

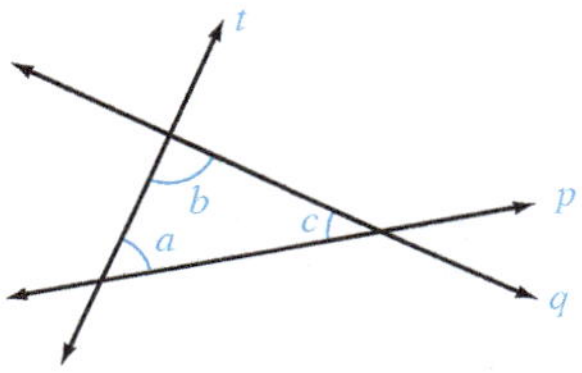

The angles within the region enclosed by the triangle are called **interior angles.** In the figure at the left, angles a, b, and c are interior angles. The sum of the measures of the interior angles is 180°.

$$\angle a + \angle b + \angle c = 180°$$

> **THE SUM OF THE MEASURES OF THE INTERIOR ANGLES OF A TRIANGLE**
>
> The sum of the measures of the interior angles of a triangle is 180°.

An angle adjacent to an interior angle is an **exterior angle.** In the figure at the right, angles m and n are exterior angles for angle a. The sum of the measures of an interior and an exterior angle is 180°.

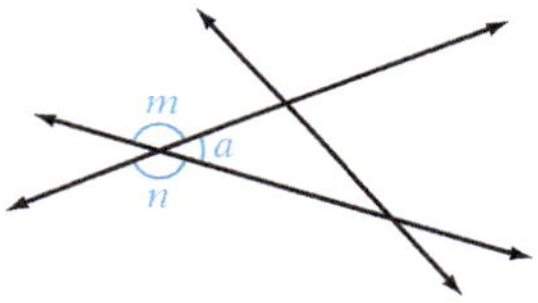

$$\angle a + \angle m = 180°$$
$$\angle a + \angle n = 180°$$

Focus on solving a problem involving the angles of a triangle

Given that $\angle c = 40°$ and $\angle e = 60°$, find the measure of $\angle d$.

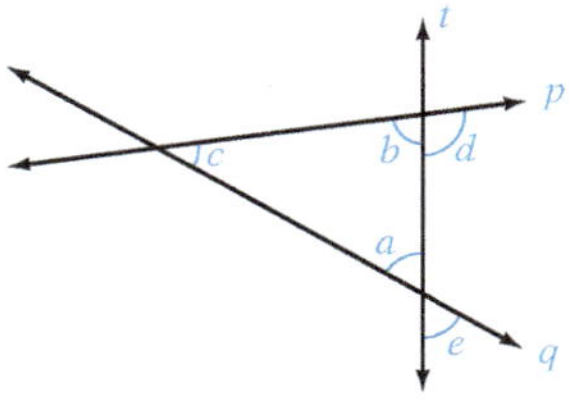

$\angle a$ and $\angle e$ are vertical angles. $\qquad \angle a = \angle e = 60°$

The sum of the interior angles is 180°.

$$\begin{aligned} \angle c + \angle a + \angle b &= 180° \\ 40° + 60° + \angle b &= 180° \\ 100° + \angle b &= 180° \\ \angle b &= 80° \end{aligned}$$

$\angle b$ and $\angle d$ are supplementary angles.

$$\angle b + \angle d = 180°$$
$$80° + \angle d = 180°$$
$$\angle d = 100°$$

EXAMPLE 4 Given that $\angle a = 45°$ and $\angle x = 100°$, find the measures of angles b, c, and y.

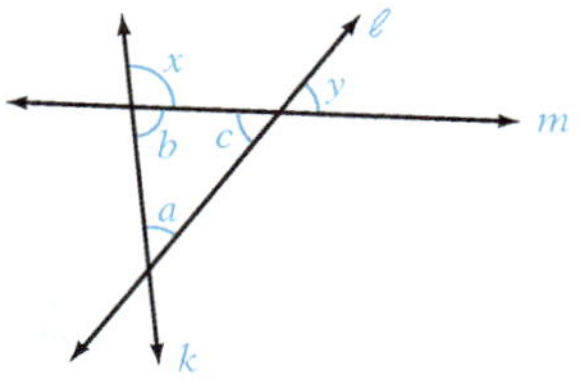

Strategy

- To find the measure of $\angle b$, use the fact that $\angle b$ and $\angle x$ are supplementary angles.
- To find the measure of $\angle c$, use the fact that the sum of the measures of the interior angles of a triangle is 180°.
- To find the measure of $\angle y$, use the fact that $\angle c$ and $\angle y$ are vertical angles.

Solution

$$\angle b + \angle x = 180°$$
$$\angle b + 100° = 180°$$
$$\angle b = 80°$$

• **$\angle b$ and $\angle x$ are supplementary angles.**

$$\angle a + \angle b + \angle c = 180°$$
$$45° + 80° + \angle c = 180°$$
$$125° + \angle c = 180°$$
$$\angle c = 55°$$

• **The sum of the measures of the interior angles of a triangle is 180°.**

$$\angle y = \angle c = 55°$$

• **$\angle c$ and $\angle y$ are vertical angles.**

Problem 4 Given that $\angle y = 55°$, find the measures of angles a, b, and d.

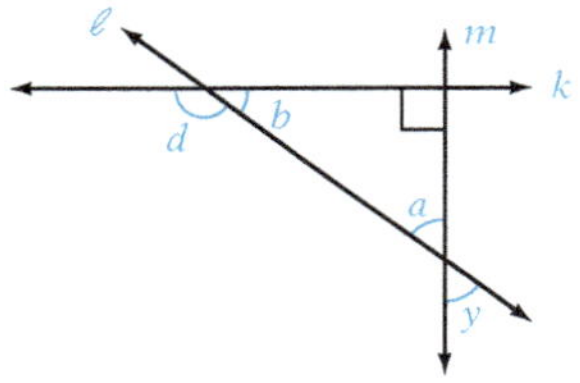

Solution See page S9.

Try Exercise 65, page 161.

EXAMPLE 5 Two angles of a triangle measure 43° and 86°. Find the measure of the third angle.

Strategy To find the measure of the third angle, use the fact that the sum of the measures of the interior angles of a triangle is 180°. Write an equation using x to represent the measure of the third angle. Solve the equation for x.

Solution

$$x + 43° + 86° = 180°$$
$$x + 129° = 180°$$
$$x = 51°$$

The measure of the third angle is 51°.

Problem 5 One angle of a triangle is a right angle, and one angle measures 27°. Find the measure of the third angle.

Solution See page S9.

Try Exercise 75, page 162.

4.2 Exercises

CONCEPT CHECK

1. Arrange the measures of an acute angle, an obtuse angle, a straight angle, and a right angle in order from smallest to largest.

2. If $\angle B$ is an acute angle, which of the following cannot be a measure of $\angle B$?

(i) 16° (ii) 89°
(iii) 90° (iv) 103°
(v) 147° (vi) 185°

3. If $\angle D$ is an obtuse angle, which of the following cannot be a measure of $\angle D$?

(i) 16° (ii) 89°
(iii) 90° (iv) 103°
(v) 147° (vi) 185°

4. Can vertical angles be acute angles?

5. Can adjacent angles be vertical angles?

6. Which is larger, the complement or the supplement of $\angle A$?

Determine whether the statement is true or false.

7. The formula for the perimeter of a rectangle is $P = 2L + 2W$, where L represents the length and W represents the width of the rectangle.

8. In the formula for the perimeter of a triangle, $P = a + b + c$, the variables a, b, and c represent the measures of the three angles of the triangle.

9. An isosceles triangle has two sides of equal measure and two angles of equal measure.

10. The perimeter of a geometric figure is a measure of the area of the figure.

1 Perimeter problems (See pages 150–152.)

11. What is the difference between an isosceles triangle and an equilateral triangle?

GETTING READY

12. Which of the following units can be used to measure perimeter?

(i) feet (ii) square miles
(iii) meters (iv) ounces
(v) cubic inches (vi) centimeters

13. The width of a rectangle is 25% of the length of the rectangle.

a. If L is the length of the rectangle, then the width of the rectangle is ___?___.

b. Using only the variable L, write an expression for the perimeter of the rectangle: 2(___?___) + 2(___?___).

c. Simplify the expression you wrote for the perimeter of the rectangle: ___?___.

14. The perimeter of an isosceles triangle is 54 ft. Let s be the length of one of the two equal sides. Is it possible for s to be 30 ft?

15. In an isosceles triangle, the third side is 50% of the length of one of the equal sides. Find the length of each side when the perimeter is 125 ft.

16. In an isosceles triangle, the length of one of the equal sides is three times the length of the third side. The perimeter is 21 m. Find the length of each side.

17. The perimeter of a rectangle is 42 m. The length of the rectangle is 3 m less than twice the width. Find the length and width of the rectangle.

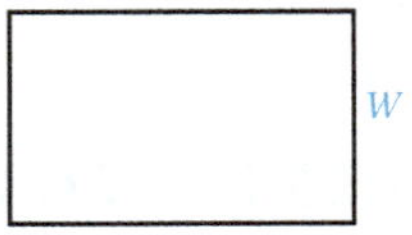

18. The width of a rectangle is 25% of the length. The perimeter is 250 cm. Find the length and width of the rectangle.

19. The perimeter of a rectangle is 120 ft. The length of the rectangle is twice the width. Find the length and width of the rectangle.

20. The perimeter of a rectangle is 50 m. The width of the rectangle is 5 m less than the length. Find the length and width of the rectangle.

21. The perimeter of a triangle is 110 cm. One side is twice the second side. The third side is 30 cm more than the second side. Find the length of each side.

22. The perimeter of a triangle is 33 ft. One side of the triangle is 1 ft longer than the second side. The third side is 2 ft longer than the second side. Find the length of each side.

23. The width of the rectangular foundation of a building is 30% of the length. The perimeter of the foundation is 338 ft. Find the length and width of the foundation.

24. The perimeter of a rectangular playground is 440 ft. If the width is 100 ft, what is the length of the playground?

25. A rectangular vegetable garden has a perimeter of 64 ft. The length of the garden is 20 ft. What is the width of the garden?

26. Each of two sides of a triangular banner measures 18 in. If the perimeter of the banner is 46 in., what is the length of the third side of the banner?

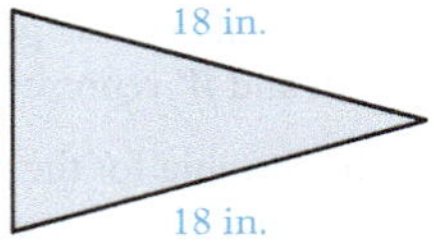

27. The perimeter of a square picture frame is 48 in. Find the length of each side of the frame.

28. A square rug has a perimeter of 32 ft. Find the length of each side of the rug.

2 Problems involving angles formed by intersecting lines (See pages 152–155.)

GETTING READY

29. Name the number of degrees in each of the following.

a. a right angle
b. a straight angle
c. one complete revolution
d. an acute angle
e. an obtuse angle

30. Find x.

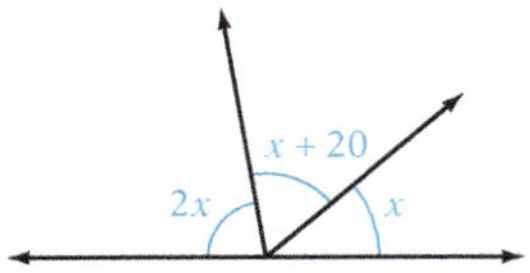

$x + x + 20 + 2x =$ ___?___ • A straight angle measures 180°.

$($___?___$)x + 20 = 180$ • Combine like terms.

$4x =$ ___?___ • Subtract 20 from each side of the equation.

$x =$ ___?___ • Divide each side of the equation by 4.

Find the measure of $\angle x$.

31.

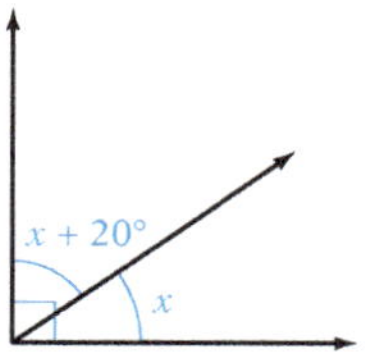

32.

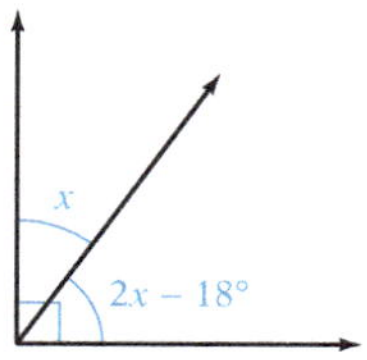

33.

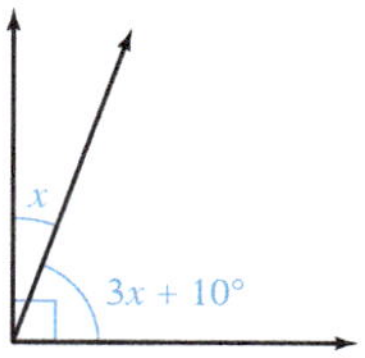

Find the measure of $\angle a$.

34.

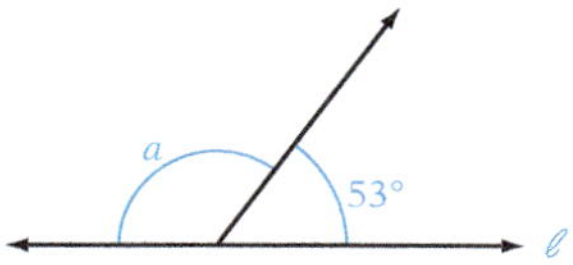

35.

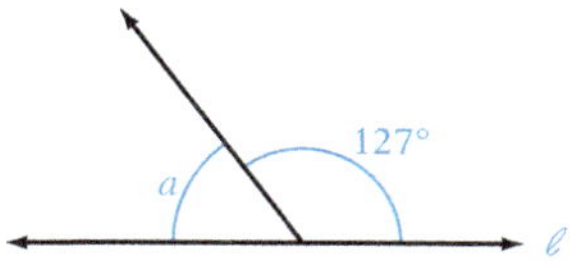

36.

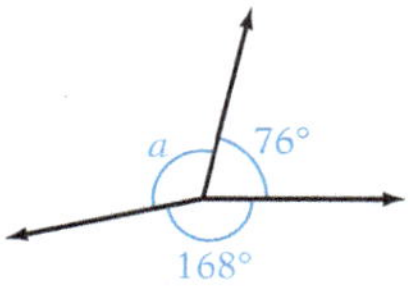

37.

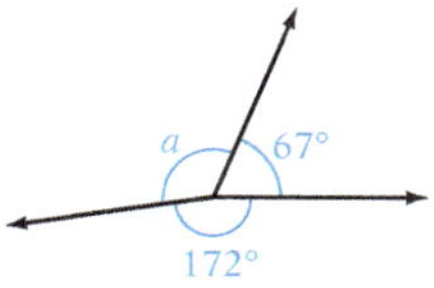

Find x.

38.

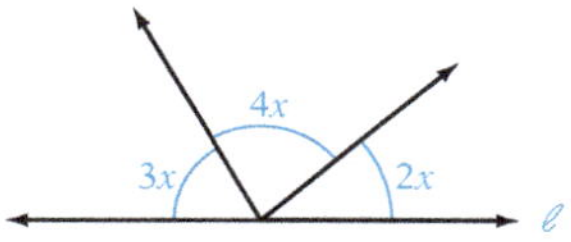

39.

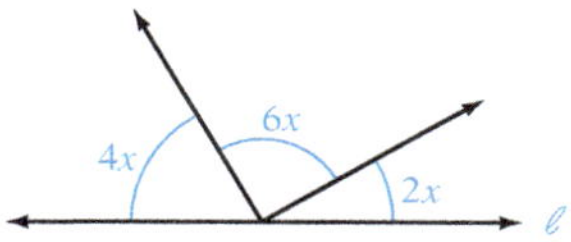

40.

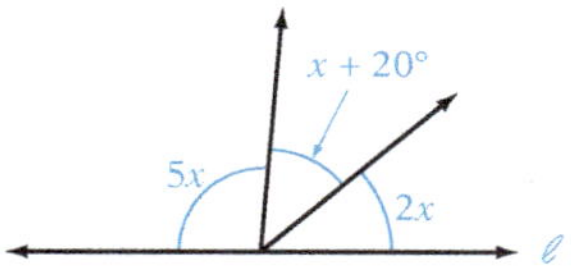

41.

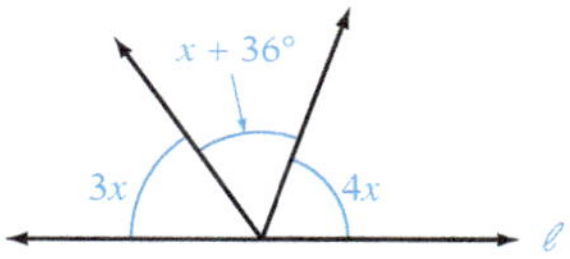

42.

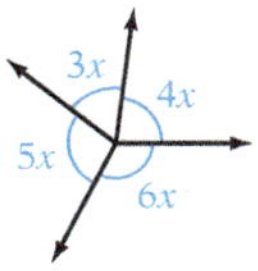

43.

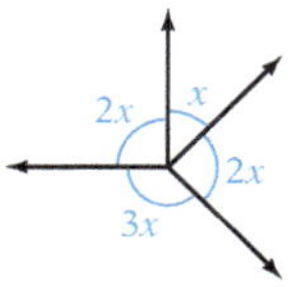

Find the measure of $\angle x$.

44.

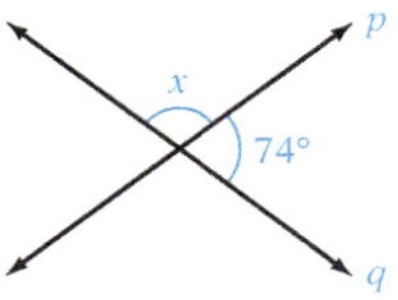

45.

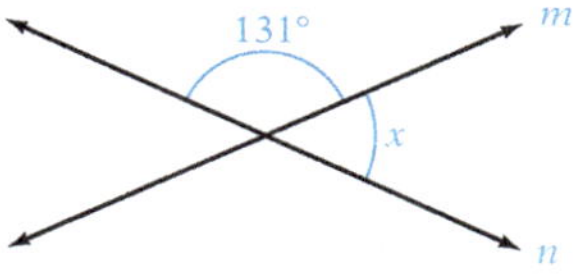

Find x.

46.

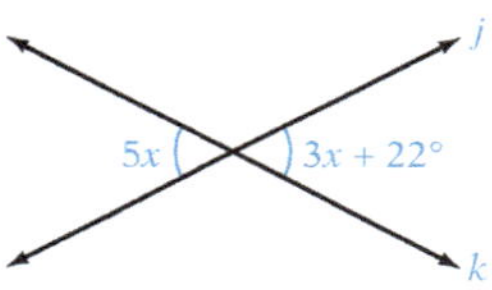

47.

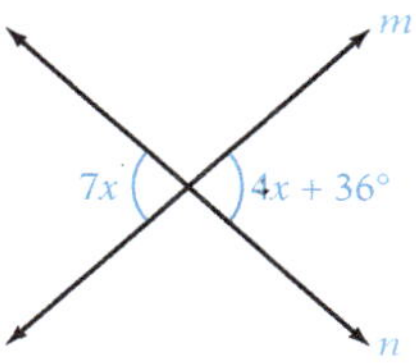

For Exercises 48 to 51, use the following diagram, in which $\ell_1 \parallel \ell_2$.

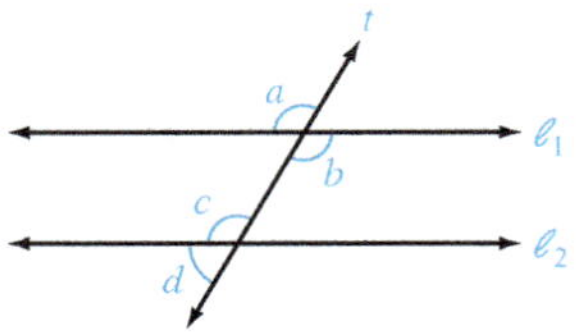

48. True or false? $\angle a$ and $\angle b$ have the same measure even if ℓ_1 and ℓ_2 are not parallel.

49. True or false? $\angle a$ and $\angle c$ have the same measure even if ℓ_1 and ℓ_2 are not parallel.

50. True or false? $\angle c$ and $\angle d$ are supplementary only if ℓ_1 and ℓ_2 are parallel.

51. True or false? $\angle c$ and $\angle d$ have the same measure only if t and ℓ_2 are perpendicular.

Given that $\ell_1 \parallel \ell_2$, find the measures of angles a and b.

52.

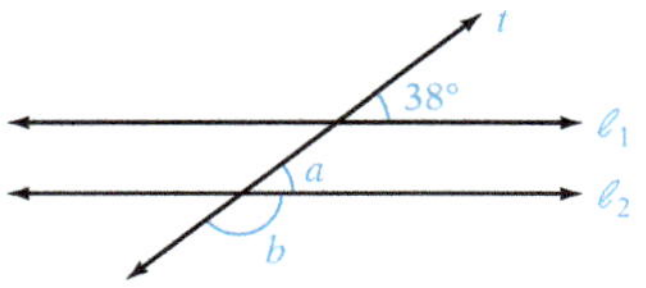

53.

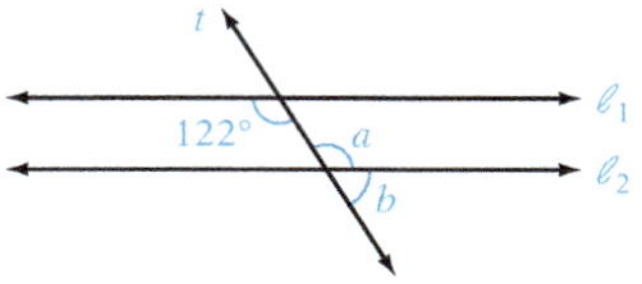

54.

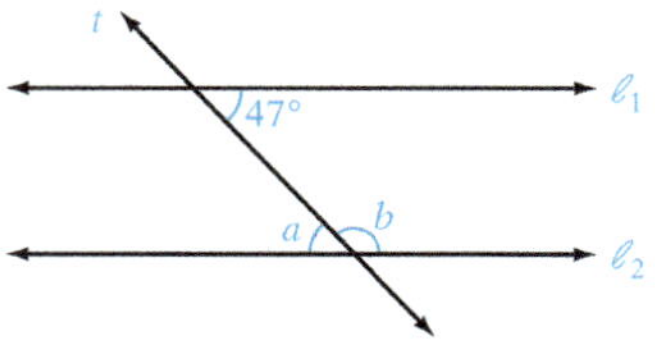

55.

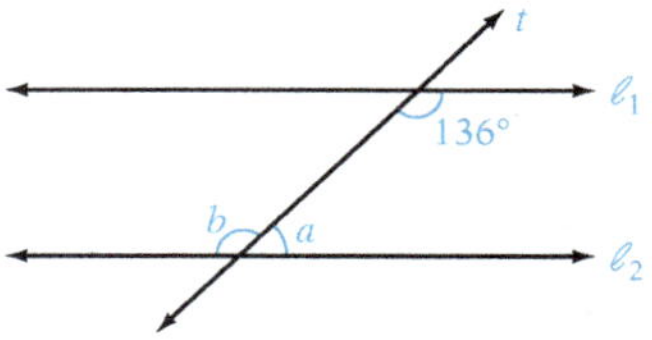

For Exercises 56 and 57, use the diagram for Exercise 54. State whether the given relationship is true even if ℓ_1 and ℓ_2 are not parallel.

56. $47° + m\angle b = 180°$

57. $m\angle a + m\angle b = 180°$

Given that $\ell_1 \parallel \ell_2$, find x.

58.

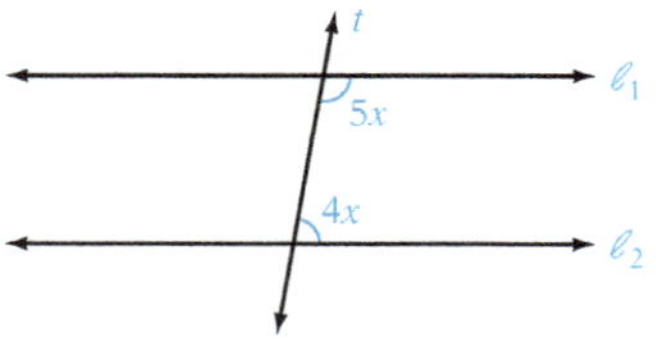

59.

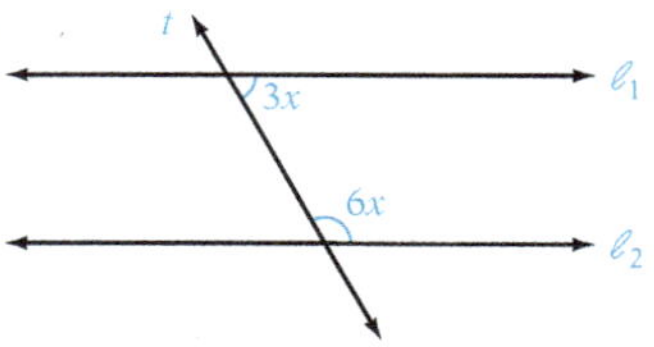

60.

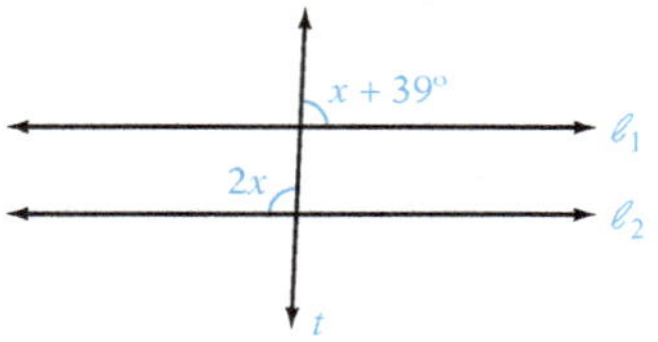

61.

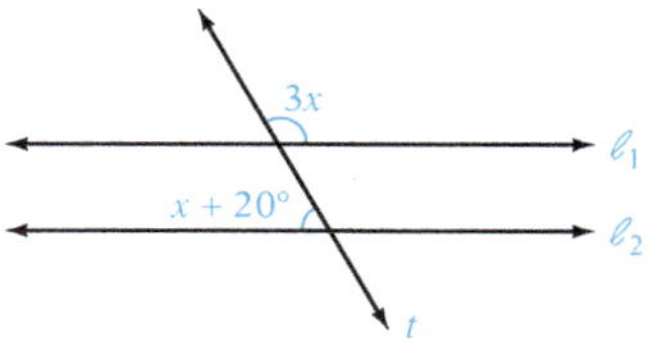

62. Given that $\angle a = 51°$, find the measure of $\angle b$.

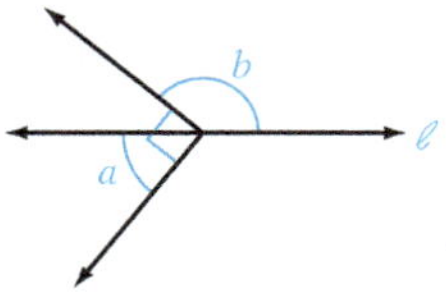

63. Given that $\angle a = 38°$, find the measure of $\angle b$.

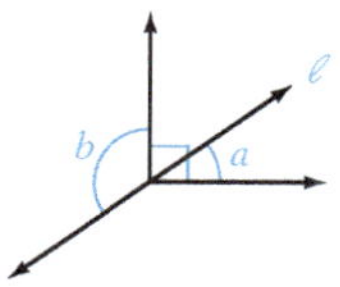

3 Problems involving the angles of a triangle (See pages 155–156.)

GETTING READY

64. Given that $\angle x = 50°$ and $\angle c = 85°$, find the measure of $\angle a$.

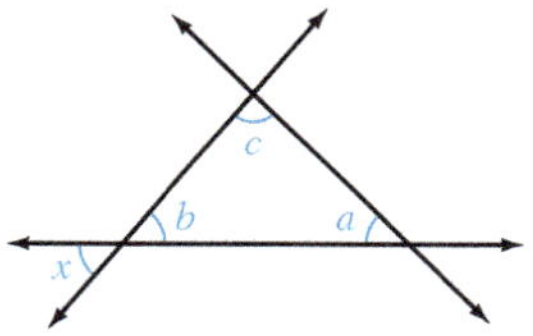

$\angle b = \angle x = 50°$	• __?__ angles have the same measure.
$\angle b + \angle c + \angle a =$ __?__	• The sum of the measures of the interior angles of a triangle is 180°.
__?__ + __?__ $+ \angle a = 180°$	• Replace $\angle b$ with 50° and $\angle c$ with 85°.
__?__ $+ \angle a = 180°$	• Add.
$\angle a =$ __?__	• Subtract 135° from each side of the equation.

65. Given that $\angle a = 95°$ and $\angle b = 70°$, find the measures of angles x and y.

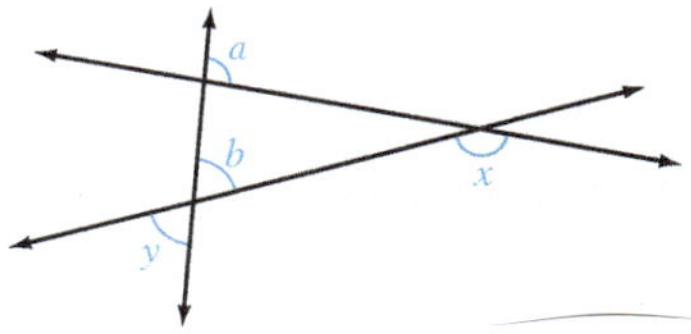

66. Given that $\angle a = 35°$ and $\angle b = 55°$, find the measures of angles x and y.

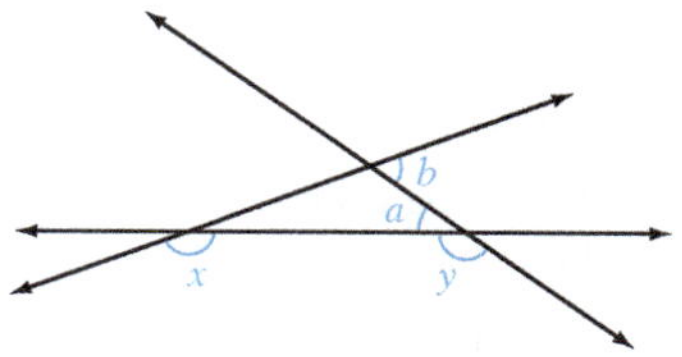

67. Given that $\angle y = 45°$, find the measures of angles a and b.

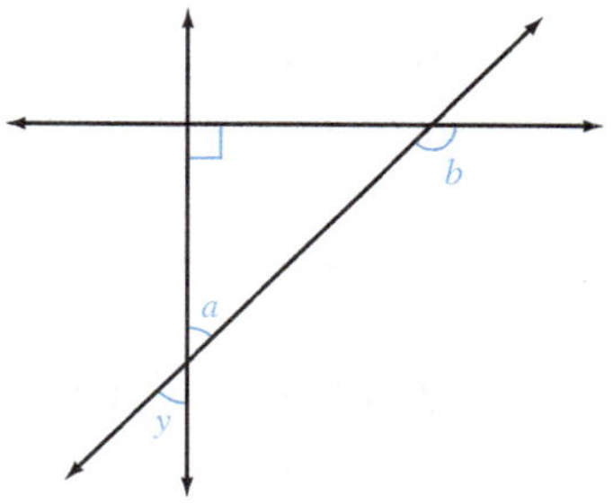

68. Given that $\angle y = 130°$, find the measures of angles a and b.

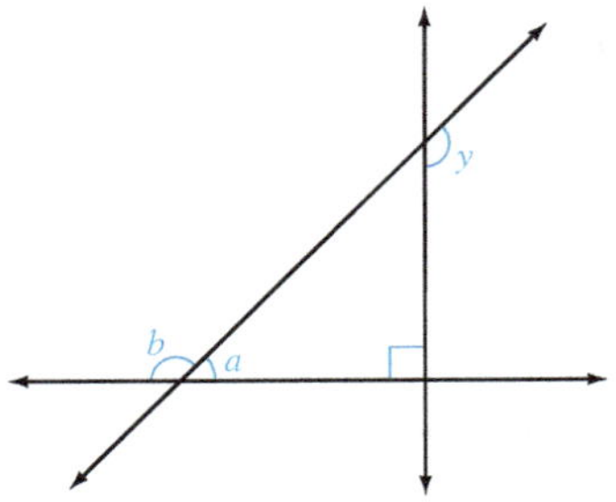

69. Given that $AO \perp OB$, express in terms of x the number of degrees in $\angle BOC$.

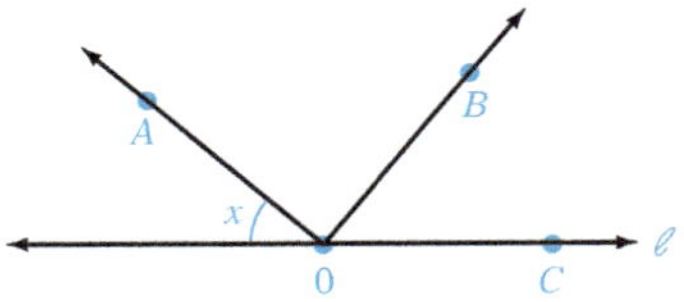

70. Given that $AO \perp OB$, express in terms of x the number of degrees in $\angle AOC$.

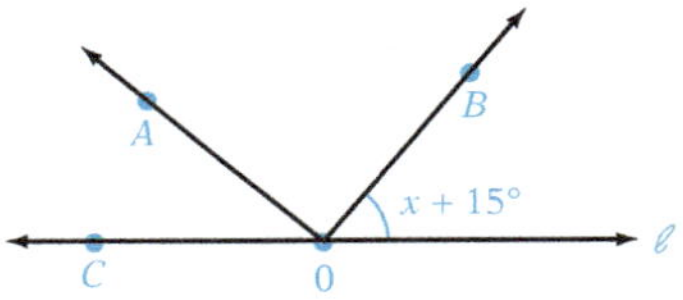

71. One angle in a triangle is a right angle, and one angle measures 30°. What is the measure of the third angle?

72. A triangle has a 45° angle and a right angle. Find the measure of the third angle.

73. Two angles of a triangle measure 42° and 103°. Find the measure of the third angle.

74. Two angles of a triangle measure 62° and 45°. Find the measure of the third angle.

75. A triangle has a 13° angle and a 65° angle. What is the measure of the third angle?

76. A triangle has a 105° angle and a 32° angle. What is the measure of the third angle?

77. In an isosceles triangle, one angle is three times the measure of one of the equal angles. Find the measure of each angle.

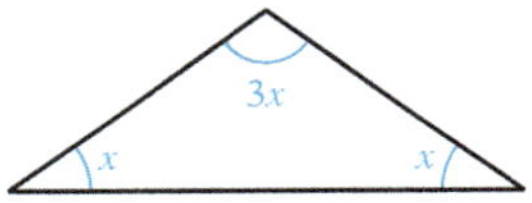

78. In an isosceles triangle, one angle is 10° less than three times the measure of one of the equal angles. Find the measure of each angle.

79. In an isosceles triangle, one angle is 16° more than twice the measure of one of the equal angles. Find the measure of each angle.

80. In a triangle, one angle is twice the measure of the second angle. The third angle is three times the measure of the second angle. Find the measure of each angle.

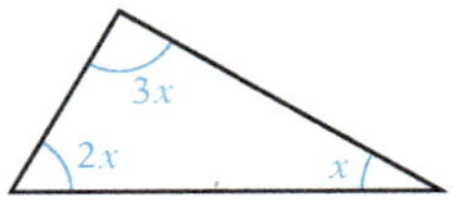

81. True or false? A triangle can have two right angles.

82. True or false? If one angle of a triangle is a right angle, then the other two angles of the triangle are complementary.

83. True or false? A triangle has nine exterior angles.

84. True or false? A triangle that has an exterior angle that is a right angle must also have an interior angle that is a right angle.

APPLYING CONCEPTS

85. A rectangle and an equilateral triangle have the same perimeter. The length of the rectangle is three times the width. Each side of the triangle is 8 cm. Find the length and width of the rectangle.

86. The length of a rectangle is 1 cm more than twice the width. If the length of the rectangle is decreased by 2 cm and the width is decreased by 1 cm, the perimeter is 20 cm. Find the length and width of the original rectangle.

87. The width of a rectangle is $8x$. The perimeter is $48x$. Find the length of the rectangle in terms of the variable x.

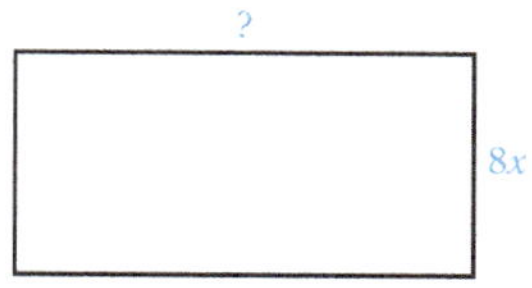

88. Prepare a report on the use of geometric form in architecture.

89. The measures of the angles of a triangle are consecutive integers. Find the measure of each angle.

PROJECTS OR GROUP ACTIVITIES

90. a. For the figure at the right, find the sum of the measures of angles x, y, and z.

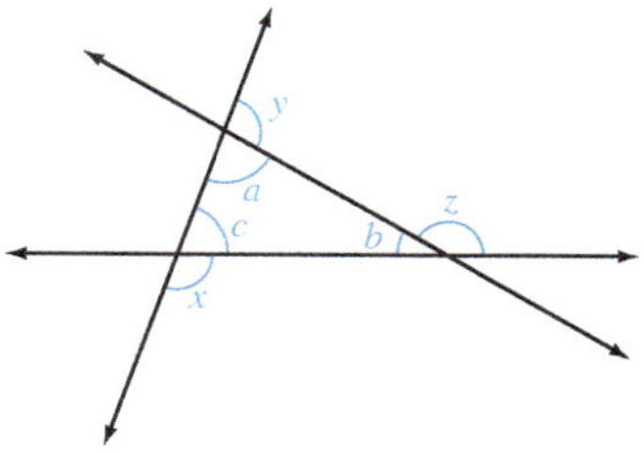

b. For the figure at the right, explain why $\angle a + \angle b = \angle x$. Write a rule that describes the relationship between an exterior angle of a triangle and the opposite interior angles. Use the rule to write an equation involving angles a, c, and z.

91. a. Draw five triangles of different sizes. For each triangle, use a protractor to find the measure of each angle. (*Note:* You will want the triangles to be fairly large so that measuring the angles is not difficult.) Then find the sum of the measures of the angles of each triangle.

b. Draw a triangle on a piece of paper as shown below. Cut out the triangle. Tear off two of the angles. Position the pieces you tore off so that angle a is adjacent to angle b and angle c is adjacent to angle b. Describe what you observe. What does this demonstrate?

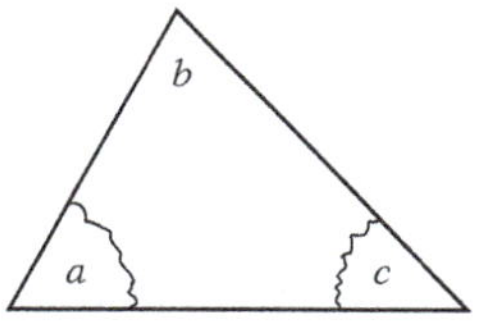

4.3 Markup and Discount Problems

OBJECTIVE 1 Markup problems

Cost is the price that a business pays for a product. **Selling price** is the price for which a business sells a product to a customer. The difference between selling price and cost is called **markup.** Markup is added to a retailer's cost to cover the expenses of operating a business and to make a profit. Markup is usually expressed as a percent of the retailer's cost. This percent is called the **markup rate.**

The basic markup equations used by a business are

$$\text{Selling price} = \text{Cost} + \text{Markup}$$
$$S = C + M$$

$$\text{Markup} = \text{Markup rate} \cdot \text{Cost}$$
$$M = r \cdot C$$

By substituting $r \cdot C$ for M in the first equation, we can also write selling price as

$$S = C + M$$
$$S = C + (r \cdot C)$$
$$S = C + rC$$

The equation $S = C + rC$ is the equation used to solve the markup problems in this section.

EXAMPLE 1 The manager of a clothing store buys a suit for \$180 and sells the suit for \$252. Find the markup rate.

Strategy Given: $C = \$180$

$S = \$252$

Unknown markup rate: r

Use the equation $S = C + rC$.

Solution

$$S = C + rC$$
$$252 = 180 + 180r$$

- **Substitute the values of *C* and *S* into the equation.**

$$252 - 180 = 180 - 180 + 180r$$
$$72 = 180r$$

- **Subtract 180 from each side of the equation.**

$$\frac{72}{180} = \frac{180r}{180}$$

- **Divide each side of the equation by 180.**

$$0.4 = r$$

- **The decimal must be changed to a percent.**

The markup rate is 40%.

Problem 1 The cost to the manager of a sporting goods store for a tennis racket is \$120. The selling price of the racket is \$180. Find the markup rate.

Solution See page S9.

Try Exercise 15, page 167.

EXAMPLE 2 The manager of a furniture store uses a markup rate of 45% on all items. The selling price of a chair is \$232. Find the cost of the chair.

Strategy Given: $r = 45\% = 0.45$

$S = \$232$

Unknown cost: C

Use the equation $S = C + rC$.

Solution

$$S = C + rC$$
$$232 = C + 0.45C$$

- **Substitute the values of *S* and *r* into the equation.**

$$232 = 1.45C$$

- $C + 0.45C = 1C + 0.45C = (1 + 0.45)C$

$$160 = C$$

The cost of the chair is \$160.

Problem 2 A hardware store employee uses a markup rate of 40% on all items. The selling price of a lawnmower is \$266. Find the cost.

Solution See page S9.

Try Exercise 21, page 167.

OBJECTIVE 2 Discount problems

Discount is the amount by which a retailer reduces the regular price of a product for a promotional sale. Discount is usually expressed as a percent of the regular price. This percent is called the **discount rate** or **markdown rate.**

The basic discount equations used by a business are

$$\text{Sale price} = \text{Regular price} - \text{Discount}$$
$$S = R - D$$

$$\text{Discount} = \text{Discount rate} \cdot \text{Regular price}$$
$$D = r \cdot R$$

By substituting $r \cdot R$ for D in the first equation, we can also write sale price as

$$S = R - D$$
$$S = R - (r \cdot R)$$
$$S = R - rR$$

The equation $S = R - rR$ is the equation used to solve the discount problems in this section.

EXAMPLE 3 In a garden supply store, the regular price of a 100-foot garden hose is \$48. During an "after-summer sale," the hose is being sold for \$36. Find the discount rate.

Strategy Given: $R = \$48$
$S = \$36$
Unknown discount rate: r
Use the equation $S = R - rR$.

Solution

$$S = R - rR$$
$$36 = 48 - 48r$$
- **Substitute the values of *R* and *S* into the equation.**

$$36 - 48 = 48 - 48 - 48r$$
$$-12 = -48r$$
- **Subtract 48 from each side of the equation.**

$$\frac{-12}{-48} = \frac{-48r}{-48}$$
- **Divide each side of the equation by −48.**

$$0.25 = r$$
- **The decimal must be changed to a percent.**

The discount rate is 25%.

Problem 3 A case of motor oil that regularly sells for \$39.80 is on sale for \$29.85. What is the discount rate?

Solution See page S9.

Try Exercise 35, page 168.

EXAMPLE 4 The sale price of a chemical sprayer is \$27.30. This price is 35% off the regular price. Find the regular price.

Strategy Given: $S = \$27.30$
$r = 35\% = 0.35$
Unknown regular price: R
Use the equation $S = R - rR$.

Solution

$$S = R - rR$$
$$27.30 = R - 0.35R$$
- **Substitute the values of *S* and *r* into the equation.**

$$27.30 = 0.65R$$
- **$R - 0.35R = 1R - 0.35R = (1 - 0.35)R$**

$$42 = R$$

The regular price is \$42.00.

Problem 4 The sale price of a cordless telephone is \$43.50. This price is 25% off the regular price. Find the regular price.

Solution See page S9.

Try Exercise 39, page 168.

4.3 Exercises

CONCEPT CHECK

1. How can you determine the markup when you know the cost and the selling price?
2. How can you determine the selling price when you know the markup and the cost?
3. How can you determine the markup when you know the cost and the markup rate?
4. How can you determine the discount when you know the sale price and the regular price?
5. How can you determine the sale price when you know the discount and the regular price?
6. How can you determine the discount when you know the regular price and the discount rate?

1 Markup problems (See pages 163–164.)

7. Explain the difference between the cost and the selling price of a product.
8. Explain the difference between markup and markup rate.

GETTING READY

Complete Exercises 9 and 10 by replacing the question marks with the correct number from the problem situation or with the word *unknown*.

9. **Problem Situation:** A pair of jeans sells for \$40. The store uses a markup rate of 25%. Find the cost of the pair of jeans.

   In the formula $S = C + rC$, $S =$ __?__, $C =$ __?__, and $r =$ __?__.

10. **Problem Situation:** A washing machine that costs \$620 is sold for \$806. Find the markup rate.

    In the formula $S = C + rC$, $S =$ __?__, $C =$ __?__, and $r =$ __?__.

11. A computer software retailer uses a markup rate of 40%. Find the selling price of a computer game that costs the retailer \$40.
12. A car dealer advertises a 5% markup over cost. Find the selling price of a car that costs the dealer \$26,000.
13. An electronics store uses a markup rate of 58%. Find the selling price of a camcorder that costs the store owner \$358.
14. The cost to a landscape architect for a 25-gallon tree is \$65. Find the selling price of the tree if the markup rate used by the architect is 30%.

15. A set of golf clubs costing \$360 is sold for \$630. Find the markup rate on the set of golf clubs.

16. A jeweler purchases a diamond ring for \$7198.50. The selling price is \$14,397. Find the markup rate.

17. A grocer purchases a bottle of fruit juice for \$1.96. The selling price of the fruit juice is \$2.45. Find the markup rate.

18. A sofa costing \$520 is sold for \$779. Find the markup rate. Round to the nearest tenth of a percent.

19. The selling price of a compact disc player is \$168. The markup rate used by the seller is 40%. Find the cost of the compact disc player.

20. A manufacturer of exercise equipment uses a markup rate of 45%. One of the manufacturer's treadmills has a selling price of \$870. Find the cost of the treadmill.

21. A markup rate of 40% was used on a basketball with a selling price of \$82.60. Find the cost of the basketball.

22. A markup rate of 25% is used on a computer that has a selling price of \$1062.50. Find the cost of the computer.

23. True or false? If a store uses a markup rate of 35%, you can find the store's cost for an item by dividing the selling price of the item by $1 + 0.35$, or 1.35.

24. If the markup rate on an item is 100%, what is the relationship between the selling price of the item and the cost of the item?

25. **Bill of Materials** Use the information in the article at the right to find the markup rate for the 8 GB iPod Touch. Round your answer to the nearest percent.

26. **Bill of Materials** Use the information in the article at the right to find the markup rate for the 32 GB iPad 3G. Round your answer to the nearest percent.

Glenn Young/Shutterstock.com

In the News

Not a Nano-Sized Markup

When you buy your latest technology gadget, do you ever wonder how much of a markup you are paying? A product's *bill of materials (BOM)* is the total cost to the manufacturer for the materials used to make the product. The rest of the price you pay is the markup. For example, the 8 GB Apple iPod Touch, with a BOM of \$149.18, sells for \$229; and the 32 GB iPad 3G, with a BOM of \$287.15, sells for \$729.

Source: www.iSuppli.com

2 Discount problems (See pages 164–166.)

27. Explain the meaning of sale price and discount.

28. Explain the difference between discount and discount rate.

GETTING READY

Complete Exercises 29 and 30 by replacing the question marks with the correct number from the problem situation or with the word *unknown*.

29. **Problem Situation:** The regular price of a digital camera is \$375. The camera is on sale for \$318.75. Find the discount rate.

In the formula $S = R - rR$, $S =$ ___?___, $R =$ ___?___, and $r =$ ___?___.

30. **Problem Situation:** The sale price of a pair of shoes is \$44, which is 20% off the regular price. Find the regular price.

In the formula $S = R + rR$, $S =$ ___?___, $R =$ ___?___, and $r =$ ___?___.

31. A tennis racket that regularly sells for \$195 is on sale for 25% off the regular price. Find the sale price.

32. A fax machine that regularly sells for \$219.99 is on sale for $33\frac{1}{3}\%$ off the regular price. Find the sale price.

33. A supplier of electrical equipment offers a 5% discount for a purchase that is paid for within 30 days. A transformer regularly sells for \$230. Find the discount price of a transformer that is paid for 10 days after the purchase.

34. A clothing wholesaler offers a discount of 10% per shirt when 10 to 20 shirts are purchased and a discount of 15% per shirt when 21 to 50 shirts are purchased. A shirt regularly sells for \$27. Find the sale price per shirt when 35 shirts are purchased.

35. A car stereo system that regularly sells for \$425 is on sale for \$318.75. Find the discount rate.

© Bill Freeman/Photo Edit

36. A pair of roller blades that regularly sells for \$99.99 is on sale for \$79.99. Find the discount rate. Round to the nearest percent.

37. A digital camera with a regular price of \$325 is on sale for \$201.50. Find the markdown rate.

38. A luggage set with a regular price of \$178 is on sale for \$103.24. Find the discount rate.

39. The sale price of a free-weight home gym is \$568, which is 20% off the regular price. Find the regular price.

40. The sale price of a toboggan is \$77, which is 30% off the regular price. Find the regular price.

41. A mechanic's tool set is on sale for \$180 after a markdown of 40% off the regular price. Find the regular price.

42. A telescope is on sale for \$285 after a markdown of 40% off the regular price. Find the regular price.

43. True or false? If an online retailer uses a discount rate of 15%, you can find the sale price for an item by multiplying the regular price of the item by $1 - 0.15$, or 0.85.

44. If the discount rate on an item is 50%, which of the following is true?
(i) $S = 2R$ (ii) $R = 2S$
(iii) $S = R$ (iv) $0.50S = R$

APPLYING CONCEPTS

45. A pair of shoes that now sells for \$63 has been marked up 40%. Find the markup on the pair of shoes.

46. The sale price of a motorcycle helmet is 25% off the regular price. The discount is \$70. Find the sale price.

47. A refrigerator selling for \$1087.50 has a markup of \$217.50. Find the markup rate.

48. The sale price of a digital copier is \$391 after a discount of \$69. Find the discount rate.

49. The manager of a camera store uses a markup rate of 30%. Find the cost of a camera selling for \$299.

50. The sale price of a high-definition television is \$450. Find the regular price if the sale price was computed by taking $\frac{1}{3}$ off the regular price followed by an additional 25% discount on the reduced price.

51. A customer buys four tires, three at the regular price and one for 20% off the regular price. The four tires cost \$608. What was the regular price of a tire?

PROJECTS OR GROUP ACTIVITIES

52. A lamp, originally priced at under \$100, was on sale for 25% off the regular price. When the regular price, a whole number of dollars, was discounted, the discounted price was also a whole number of dollars. Find the largest possible number of dollars in the regular price of the lamp.

53. A used car is on sale for 20% off the regular price of \$8500. An additional 10% discount on the sale price is offered. Is the result a 30% discount? What single discount would give the same sale price?

54. Write an application problem involving a discount rate of 35%. Choose a product and its regular price. Determine the sale price. Use complete sentences.

55. Write a report on series trade discounts. Explain how to convert a series discount to a single-discount equivalent.

4.4 Investment Problems

OBJECTIVE 1

Investment problems

Recall that the annual simple interest that an investment earns is given by the equation $\boldsymbol{I = Pr}$, where I is the simple interest, P is the principal, or the amount invested, and r is the simple interest rate.

The annual simple interest rate on a \$5000 investment is 8%.

$P = \$5000$; $r = 8\% = 0.08$; I is unknown.

$$I = Pr$$
$$I = 5000(0.08)$$
$$I = 400$$

The annual simple interest earned on the investment is \$400.

The equation $I = Pr$ is used to solve the investment problem below.

Solve: An investor has a total of \$10,000 to deposit in two simple interest accounts. On one account, the annual simple interest rate is 7%. On the second account, the annual simple interest rate is 8%. How much should be invested in each account so that the total annual interest earned is \$785?

> **Point of Interest**
>
> You may be familiar with the simple interest formula $I = Prt$. If so, you know that t represents time. In the problems in this section, time is always 1 (one year), so the formula $I = Prt$ simplifies to
>
> $$I = Pr(1)$$
> $$I = Pr$$

STRATEGY FOR SOLVING A PROBLEM INVOLVING MONEY DEPOSITED IN TWO SIMPLE INTEREST ACCOUNTS

▶ For each amount invested, use the equation $Pr = I$. Write a numerical or variable expression for the principal, the interest rate, and the interest earned. The results can be recorded in a table.

Take Note

Use the information given in the problem to fill in the principal and interest rate columns of the table. Fill in the interest earned column by multiplying the two expressions you wrote in each row.

The sum of the amounts invested is \$10,000.

Amount invested at 7%: x
Amount invested at 8%: $\$10{,}000 - x$

	Principal, P	$\cdot$	Interest rate, r	$=$	Interest earned, I
Amount at 7%	x	$\cdot$	0.07	$=$	$0.07x$
Amount at 8%	$10{,}000 - x$	$\cdot$	0.08	$=$	$0.08(10{,}000 - x)$

▶ Determine how the amounts of interest earned on the individual investments are related. For example, the total interest earned by both accounts may be known, or it may be known that the interest earned on one account is equal to the interest earned on the other account.

The sum of the interest earned by the two investments equals the total annual interest earned (\$785).

$$0.07x + 0.08(10{,}000 - x) = 785$$
$$0.07x + 800 - 0.08x = 785$$
$$-0.01x + 800 = 785$$
$$-0.01x = -15$$
$$x = 1500$$
$$10{,}000 - x = 10{,}000 - 1500 = 8500$$

- **The interest earned on the 7% account plus the interest earned on the 8% account equals the total annual interest earned.**
- **The amount invested at 7% is \$1500.**
- **Substitute the value of x into the variable expression for the amount invested at 8%.**

The amount invested at 7% is \$1500.
The amount invested at 8% is \$8500.

EXAMPLE 1 An investment counselor invested 75% of a client's money into a 9% annual simple interest money market fund. The remainder was invested in 6% annual simple interest government securities. Find the amount invested in each account if the total annual interest earned is \$3300.

Strategy ▶ Amount invested: x
Amount invested at 9%: $0.75x$
Amount invested at 6%: $0.25x$

	Principal	$\cdot$	Rate	$=$	Interest
Amount at 9%	$0.75x$	$\cdot$	0.09	$=$	$0.09(0.75x)$
Amount at 6%	$0.25x$	$\cdot$	0.06	$=$	$0.06(0.25x)$

▶ The sum of the interest earned by the two investments equals the total annual interest earned (\$3300).

Solution

$$0.09(0.75x) + 0.06(0.25x) = 3300$$
$$0.0675x + 0.015x = 3300$$
$$0.0825x = 3300$$

- **The interest earned on the 9% account plus the interest earned on the 6% account equals the total annual interest earned.**

$$x = 40{,}000$$

- **The amount invested is \$40,000.**

$0.75x = 0.75(40{,}000) = 30{,}000$ • Find the amount invested at 9%.

$0.25x = 0.25(40{,}000) = 10{,}000$ • Find the amount invested at 6%.

The amount invested at 9% is \$30,000.
The amount invested at 6% is \$10,000.

Problem 1 An investment of \$18,000 is deposited into two simple interest accounts. On one account, the annual simple interest rate is 4%. On the other, the annual simple interest rate is 6%. How much should be invested in each account so that each account earns the same interest?

Solution See page S9.

Try Exercise 15, page 172.

4.4 Exercises

CONCEPT CHECK

1. Explain what each variable in the formula $Pr = I$ represents.

2. What is the difference between interest and interest rate?

3. For the following example, name (a) the principal, (b) the interest rate, and (c) the interest earned.

The annual simple interest rate on a \$1250 investment is 5%. Find the annual simple interest earned on the investment.

4. How much interest is earned in one year on \$1000 deposited in an account that pays 5% annual simple interest?

Determine whether the statement is true or false.

5. For one year, you have x dollars deposited in an account that pays 7% annual simple interest. You will earn $0.07x$ in simple interest on this account.

6. If you have a total of \$8000 deposited in two accounts and you represent the amount you have in the first account as x, then the amount in the second account is represented as $8000 - x$.

7. The amount of interest earned on one account is $0.05x$, and the amount of interest earned on a second account is $0.08(9000 - x)$. If the two accounts earn the same amount of interest, then we can write the equation $0.05x + 0.08(9000 - x)$.

8. If the amount of interest earned on one account is $0.06x$ and the amount of interest earned on a second account is $0.09(4000 - x)$, then the total interest earned on the two accounts can be represented as $0.06x + 0.09(4000 - x)$.

1 Investment problems (See pages 169–171.)

GETTING READY

9. You invest an amount of money at an annual simple interest rate of 5.2%. You invest a second amount, \$1000 more than the first, at an annual simple interest rate of 7.2%. Let x represent the amount invested at 5.2%. Complete the following table.

	Principal, P	$\cdot$	Interest rate, r	=	Interest earned, I
Amount at 5.2%	x	$\cdot$	___?___	=	___?___
Amount at 7.2%	___?___	$\cdot$	___?___	=	___?___

10. The total annual interest earned on the investments in Exercise 9 is \$320. Use this information and the information in the table in Exercise 9 to write an equation that can be solved to find the amount of money invested at 5.2%: ___?___ + ___?___ = ___?___.

11. A dentist invested a portion of \$15,000 in a 7% annual simple interest account and the remainder in a 6.5% annual simple interest government bond. The two investments earn \$1020 in interest annually. How much was invested in each account?

12. A university alumni association invested part of \$20,000 in preferred stock that earns 8% annual simple interest and the remainder in a municipal bond that earns 7% annual simple interest. The amount of interest earned each year is \$1520. How much was invested in each account?

13. A professional athlete deposited an amount of money into a high-yield mutual fund that earns 13% annual simple interest. A second deposit, \$2500 more than the first, was placed in a certificate of deposit earning 7% annual simple interest. In one year, the total interest earned on both investments was \$475. How much money was invested in the mutual fund?

14. Jan Moser made a deposit into a 7% annual simple interest account. She made another deposit, \$1500 less than the first, in a certificate of deposit earning 9% annual simple interest. The total interest earned on both investments for one year was \$505. How much money was deposited in the certificate of deposit?

15. A team of cancer research specialists received a grant of \$300,000 to be used for cancer research. They deposited some of the money in a 10% simple interest account and the remainder in an 8.5% annual simple interest account. How much was deposited in each account if the annual interest is \$28,500?

16. Virak Ly invested part of \$30,000 in municipal bonds that earn 6.5% annual simple interest and the remainder of the money in 8.5% corporate bonds. How much is invested in each account if the total annual interest earned is \$2190?

17. To provide for retirement income, Teresa Puelo purchases a \$5000 bond that earns 7.5% annual simple interest. How much money does Teresa have invested in bonds that earn 8% annual simple interest if the total annual interest earned from the two investments is \$615?

18. After the sale of some income property, Jeremy Littlefield invested \$40,000 in a certificate of deposit that earns 7.25% annual simple interest. How much money does he have invested in certificates that earn an annual simple interest rate of 8.5% if the total annual interest earned from the two investments is \$5025?

19. Suki Hiroshi has made an investment of \$2500 at an annual simple interest rate of 7%. How much money does she have invested at an annual simple interest rate of 11% if the total interest earned is 9% of the total investment?

20. Mae Jackson has a total of \$6000 invested in two simple interest accounts. The annual simple interest rate on one account is 9%. The annual simple interest rate on the second account is 6%. How much is invested in each account if both accounts earn the same amount of interest?

21. Wayne Miller, an investment banker, invested 55% of the bank's available cash in an account that earns 8.25% annual simple interest. The remainder of the cash was placed in an account that earns 10% annual simple interest. The interest earned in one year was \$58,743.75. Find the total amount invested.

22. Mohammad Aran, a financial planner, recommended that 40% of a client's cash account be invested in preferred stock earning 9% annual simple interest. The remainder of the client's cash was placed in Treasury bonds earning 7% annual simple interest. The total annual interest earned from the two investments was \$2496. Find the total amount invested.

23. Sarah Ontkean is the manager of a mutual fund. She placed 30% of the fund's available cash in a 6% annual simple interest account, 25% in 8% corporate bonds, and the remainder in a money market fund earning 7.5% annual simple interest. The total annual interest earned from the investments was \$35,875. Find the total amount invested.

24. Joseph Abruzzio is the manager of a trust. He invested 30% of a client's cash in government bonds that earn 6.5% annual simple interest, 30% in utility stocks that earn 7% annual simple interest, and the remainder in an account that earns 8% annual simple interest. The total annual interest earned from the investments was \$5437.50. Find the total amount invested.

25. The amount of annual interest earned on the x dollars that Beth invested in one simple interest account was $0.03x$, and the amount of annual interest earned on the money that Beth invested in another simple interest account was $0.05(5000 - x)$.
a. What were the interest rates on the two accounts?
b. What was the total amount of money that Beth invested in the two accounts?

26. Refer to the investments described in Exercise 25. Which of the following could be true about the total amount T of interest earned on Beth's two accounts? There may be more than one correct answer.
(i) $T < 150$ (ii) $T > 150$ (iii) $T < 250$ (iv) $T > 250$

APPLYING CONCEPTS

27. A sales representative invests in a stock paying 9% dividends. A research consultant invests \$5000 more than the sales representative in bonds paying 8% annual simple interest. The research consultant's income from the investment is equal to the sales representative's. Find the amount of the research consultant's investment.

28. A financial manager invested 20% of a client's money in bonds paying 9% annual simple interest, 35% in an 8% simple interest account, and the remainder in 9.5% corporate bonds. Find the amount invested in each if the total annual interest earned is \$5325.

29. A plant manager invested \$3000 more in stocks than in bonds. The stocks paid 8% annual simple interest, and the bonds paid 9.5% annual simple interest. Both investments yielded the same income. Find the total annual interest received on both investments.

30. Write an essay on the topic of annual percentage rates.

PROJECTS OR GROUP ACTIVITIES

31. A bank offers a customer a 4-year certificate of deposit (CD) that earns 6.5% compound annual interest. This means that the interest earned each year is added to the principal before the interest for the next year is calculated. Find the value in 4 years of a nurse's investment of $3000 in this CD.

32. A bank offers a customer a 5-year certificate of deposit (CD) that earns 7.5% compound annual interest. This means that the interest earned each year is added to the principal before the interest for the next year is calculated. Find the value in 5 years of an accountant's investment of $2500 in this CD.

33. **Retirement Savings** Financial advisors may predict how much money one should have saved for retirement by ages 40, 50, and 60. The amount varies depending on a person's annual income. One such prediction is included in the table at the right.

Required Savings for Retirement by Age for Various Income Levels

		40	50	60
Income =	$40,000	28,000	111,000	242,000
Income =	$80,000	75,000	298,000	651,000
Income =	$120,000	131,000	524,000	1,144,000

a. According to the estimates in the table, how much should a consultant who has an annual income of $80,000 have saved for retirement by age 50?

b. Suppose a 55-year-old executive has an annual income of $120,000. The executive's retirement savings should fall between which two numbers in the table?

c. Suppose a 40-year-old manager has an annual income of $60,000. The manager's retirement savings should fall between which two numbers in the table?

d. Write an explanation of how interest and interest rates affect the level of savings required at any given age. What effect do inflation rates have on savings?

4.5 Mixture Problems

OBJECTIVE 1 Value mixture problems

A value mixture problem involves combining two ingredients that have different prices into a single blend. For example, a coffee merchant may blend two types of coffee into a single blend, or a candy manufacturer may combine two types of candy to sell as a "variety pack."

The solution of a value mixture problem is based on the equation $V = AC$, where V is the value of an ingredient, A is the amount of the ingredient, and C is the cost per unit of the ingredient.

A gold alloy costs $180 per ounce. A ring is made from 5 oz of the alloy.

$A = 5$; $C = \$180$; V is unknown.

$$V = AC$$
$$V = 5(180)$$
$$V = 900$$

The value of the ring is $900.

The equation $V = AC$ is used to solve the following value mixture problem.

Solve: A coffee merchant wants to make 9 lb of a blend of coffee costing \$6 per pound. The blend is made using a \$7 grade and a \$4 grade of coffee. How many pounds of each of these grades should be used?

STRATEGY FOR SOLVING A MIXTURE PROBLEM

► For each ingredient in the mixture, write a numerical or variable expression for the amount of the ingredient used, the unit cost of the ingredient, and the value of the amount used. For the blend, write a numerical or variable expression for the amount, the unit cost of the blend, and the value of the amount. The results can be recorded in a table.

The sum of the amounts is 9 lb.

Amount of \$7 coffee: x

Amount of \$4 coffee: $9 - x$

Take Note

Use the information given in the problem to fill in the amount and unit cost columns of the table. Fill in the value column by multiplying the two expressions you wrote in each row. Use the expressions in the last column to write the equation.

	Amount, A	·	Unit cost, C	=	Value, V
\$7 grade	x	·	7	=	$7x$
\$4 grade	$9 - x$	·	4	=	$4(9 - x)$
\$6 blend	9	·	6	=	$6(9)$

► Determine how the values of the individual ingredients are related. Use the fact that the sum of the values of the ingredients is equal to the value of the blend.

The sum of the values of the \$7 grade and the \$4 grade is equal to the value of the \$6 blend.

$$7x + 4(9 - x) = 6(9)$$
• **The value of the \$7 grade plus the value of the \$4 grade equals the value of the blend.**

$$7x + 36 - 4x = 54$$
$$3x + 36 = 54$$
$$3x = 18$$
$$x = 6$$
• **The amount of the \$7 grade is 6 lb.**

$$9 - x = 9 - 6 = 3$$
• **Substitute the value of x into the variable expression for the amount of \$4 grade.**

The merchant must use 6 lb of the \$7 coffee and 3 lb of the \$4 coffee.

EXAMPLE 1 How many ounces of a silver alloy that costs \$6 per ounce must be mixed with 10 oz of a silver alloy that costs \$8 per ounce to make a mixture that costs \$6.50 per ounce?

Strategy ► Ounces of \$6 alloy: x
Ounces of \$8 alloy: 10
Ounces of \$6.50 mixture: $x + 10$

	Amount	Cost	Value
$6 alloy	x	6	$6x$
$8 alloy	10	8	$8(10)$
$6.50 mixture	$10 + x$	6.50	$6.50(10 + x)$

▶ The sum of the values before mixing equals the value after mixing.

Solution

$$
\begin{aligned}
6x + 8(10) &= 6.50(10 + x) \\
6x + 80 &= 65 + 6.5x \\
-0.5x + 80 &= 65 \\
-0.5x &= -15 \\
x &= 30
\end{aligned}
$$

• **The value of the $6 alloy plus the value of the $8 alloy equals the value of the mixture.**

30 oz of the $6 silver alloy must be used.

Problem 1 A gardener has 20 lb of a lawn fertilizer that costs $.90 per pound. How many pounds of a fertilizer that costs $.75 per pound should be mixed with this 20 lb of lawn fertilizer to produce a mixture that costs $.85 per pound?

Solution See page S10.

Try Exercise 21, page 179.

OBJECTIVE 2

Percent mixture problems

Take Note

As stated at the right, in a 5% salt solution, 5% of the total solution is salt. However, we are referring to weight; the solution is 5% salt by weight, not by volume. For example, if the solution weighs 100 g, the salt in the solution weighs 5 g and the water weighs 95% of 100 g, or 95 g.

Recall that the amount of a substance in a solution can be given as a percent of the total solution. For example, in a 5% saltwater solution, 5% of the total solution is salt. The remaining 95% is water.

The solution of a percent mixture problem is based on the equation $Q = Ar$, where Q is the quantity of a substance in the solution, r is the percent of concentration, and A is the amount of solution.

A 500-milliliter bottle contains a 3% solution of hydrogen peroxide.

$A = 500$; $r = 3\% = 0.03$; Q is unknown.

$$
\begin{aligned}
Q &= Ar \\
Q &= 500(0.03) \\
Q &= 15
\end{aligned}
$$

The bottle contains 15 ml of hydrogen peroxide.

The equation $Q = Ar$ is used to solve the percent mixture problem below.

Solve: How many gallons of 15% salt solution must be mixed with 4 gal of a 20% salt solution to make a 17% salt solution?

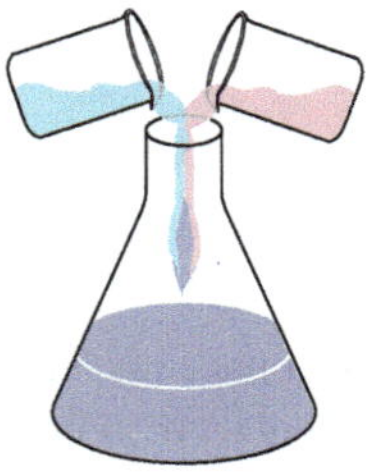

STRATEGY FOR SOLVING A PERCENT MIXTURE PROBLEM

▶ For each solution, use the equation $Ar = Q$. Write a numerical or variable expression for the amount of solution, the percent of concentration, and the quantity of the substance in the solution. The results can be recorded in a table.

The unknown amount of 15% solution: x

The amount of 20% solution: 4

The amount of 17% solution: $x + 4$

	Amount of solution, A	$\cdot$	Percent of concentration, r	=	Quantity of substance, Q
15% solution	x	$\cdot$	0.15	=	$0.15x$
20% solution	4	$\cdot$	0.20	=	$0.20(4)$
17% solution	$x + 4$	$\cdot$	0.17	=	$0.17(x + 4)$

Take Note
Use the information given in the problem to fill in the amount and percent columns of the table. Fill in the quantity column by multiplying the two expressions you wrote in each row. Use the expressions in the last column to write the equation.

► Determine how the quantities of the substance in the individual solutions are related. Use the fact that the sum of the quantities of the substances being mixed is equal to the quantity of the substance after mixing.

The sum of the quantities of salt in the 15% solution and in the 20% solution is equal to the quantity of salt in the 17% solution.

$$0.15x + 0.20(4) = 0.17(x + 4)$$
$$0.15x + 0.8 = 0.17x + 0.68$$
$$-0.02x + 0.8 = 0.68$$
$$-0.02x = -0.12$$
$$x = 6$$

• **The amount of salt in the 15% solution plus the amount of salt in the 20% solution equals the amount of salt in the 17% solution.**

6 gal of the 15% solution are required.

EXAMPLE 2 A chemist wishes to make 3 L of a 7% acid solution by mixing a 9% acid solution and a 4% acid solution. How many liters of each solution should the chemist use?

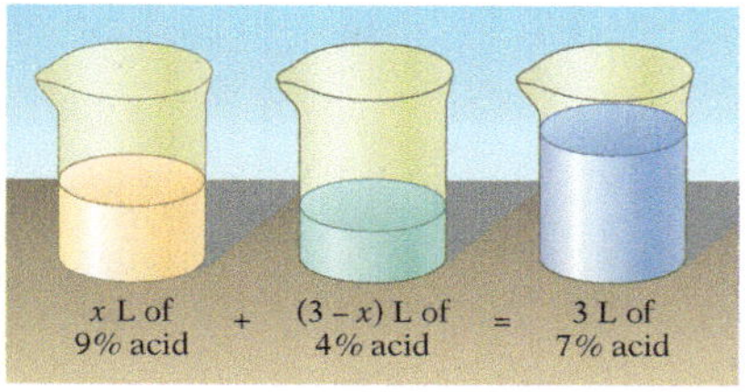

Strategy ► Liters of 9% solution: x
Liters of 4% solution: $3 - x$

	Amount	Percent	Quantity
9%	x	0.09	$0.09x$
4%	$3 - x$	0.04	$0.04(3 - x)$
7%	3	0.07	$0.07(3)$

► The sum of the quantities before mixing is equal to the quantity after mixing.

Solution

$$0.09x + 0.04(3 - x) = 0.07(3)$$
$$0.09x + 0.12 - 0.04x = 0.21$$
$$0.05x + 0.12 = 0.21$$
$$0.05x = 0.09$$

• **The amount of acid in the 9% solution plus the amount of acid in the 4% solution equals the amount of acid in the 7% solution.**

$$x = 1.8$$

• **1.8 L of the 9% solution are needed.**

$$3 - x = 3 - 1.8 = 1.2$$

• **Find the amount of the 4% solution needed.**

The chemist needs 1.8 L of the 9% solution and 1.2 L of the 4% solution.

Problem 2 How many quarts of pure orange juice are added to 5 qt of fruit drink that is 10% orange juice to make an orange drink that is 25% orange juice?

Solution See page S10.

Try Exercise 45, page 181.

4.5 Exercises

CONCEPT CHECK

For Exercises 1 to 5, determine whether the statement is true or false.

1. Both value mixture and percent mixture problems involve combining two or more ingredients into a single substance.
2. In the value mixture equation $V = AC$, the variable A represents the quantity of an ingredient.
3. Suppose we are mixing two salt solutions. Then the variable Q in the percent mixture equation $Q = Ar$ represents the amount of salt in a solution.
4. If we combine an alloy that costs \$8 an ounce with an alloy that costs \$5 an ounce, the cost of the resulting mixture will be greater than \$8 an ounce.
5. If we combine a 9% acid solution with a solution that is 4% acid, the resulting solution will be less than 4% acid.

6. Is the amount of a mixture always less than, equal to, or more than the amount of either ingredient?
7. Suppose sugar is added to a solution of sugar and water. Does the percent concentration of the sugar increase or decrease?

1 Value mixture problems (See pages 174–176.)

GETTING READY

8. The total value of a 7-pound bag of cat food that costs \$1.50 per pound is ___?___.
9. If 8 L of a solvent costs \$75 per liter, then the value of the 8 L of solvent is ___?___.
10. The cost per pound of a 5-pound bag of sugar that has a total value of \$3.80 is ___?___.
11. To make bread, Maya uses a flour mixture that combines organic whole-wheat flour that costs \$1.70 per pound with unbleached flour that costs \$.50 per pound. Ten pounds of her mixture costs \$1.30 per pound. Let x be the amount of whole-wheat flour in the mixture.

a. Complete the following table.

	Amount, A	$\cdot$	Unit cost, C	$=$	Value, V
Whole-wheat flour	x	$\cdot$	?	$=$	?
Unbleached flour	?	$\cdot$	?	$=$	?
Mixture	?	$\cdot$	?	$=$	?

b. Use the expressions in the last column of the table in part (a) to write an equation that can be solved to find the number of pounds of whole-wheat flour in Maya's mixture: ___?___ + ___?___ = ___?___.

12. Explain the meaning of each variable in the equation $V = AC$. Give an example of how this equation is used.

13. Suppose you are mixing peanuts that cost \$4 per pound with raisins that cost \$3 per pound. In the context of this situation, explain the meaning of the statement "The sum of the values of the ingredients is equal to the value of the blend."

14. At a veterinary clinic, a special high-protein dog food that costs \$6.75 per pound is mixed with a vitamin supplement that costs \$3.25 per pound. How many pounds of each should be used to make 5 lb of a mixture that costs \$4.65 per pound?

Steve Mason/Getty Images

15. A goldsmith combined an alloy that costs \$4.30 per ounce with an alloy that costs \$1.80 per ounce. How many ounces of each were used to make a mixture of 200 oz costing \$2.50 per ounce?

16. A wild bird seed mix is made by combining 100 lb of millet seed costing \$.60 per pound with sunflower seeds costing \$1.10 per pound. How many pounds of sunflower seeds are needed to make a mixture that costs \$.70 per pound?

17. How many pounds of chamomile tea that costs \$18.20 per pound must be mixed with 12 lb of orange tea that costs \$12.25 per pound to make a mixture that costs \$14.63 per pound?

18. Find the cost per pound of a coffee mixture made from 8 lb of coffee that costs \$9.20 per pound and 12 lb of coffee that costs \$5.50 per pound.

19. Find the cost per ounce of a mixture of 200 oz of a cologne that costs \$7.50 per ounce and 500 oz of a cologne that costs \$4.00 per ounce.

20. An herbalist has 30 oz of herbs costing \$2 per ounce. How many ounces of herbs costing \$1 per ounce should be mixed with these 30 oz of herbs to produce a mixture costing \$1.60 per ounce?

21. A snack food is made by mixing 5 lb of popcorn that costs \$.80 per pound with caramel that costs \$2.40 per pound. How much caramel is needed to make a mixture that costs \$1.40 per pound?

22. A grocery store offers a cheese sampler that includes a pepper cheddar cheese that costs \$16 per kilogram and Pennsylvania Jack that costs \$12 per kilogram. How many kilograms of each were used to make a 5-kilogram mixture that costs \$13.20 per kilogram?

23. A lumber company combined oak wood chips that cost \$3.10 per pound with pine wood chips that cost \$2.50 per pound. How many pounds of each were used to make an 80-pound mixture costing \$2.65 per pound?

24. The manager of a farmer's market has 500 lb of grain that costs \$1.20 per pound. How many pounds of meal costing \$.80 per pound should be mixed with the 500 lb of grain to produce a mixture that costs \$1.05 per pound?

25. A caterer made an ice cream punch by combining fruit juice that cost \$4.50 per gallon with ice cream that cost \$8.50 per gallon. How many gallons of each were used to make 100 gal of punch costing \$5.50 per gallon?

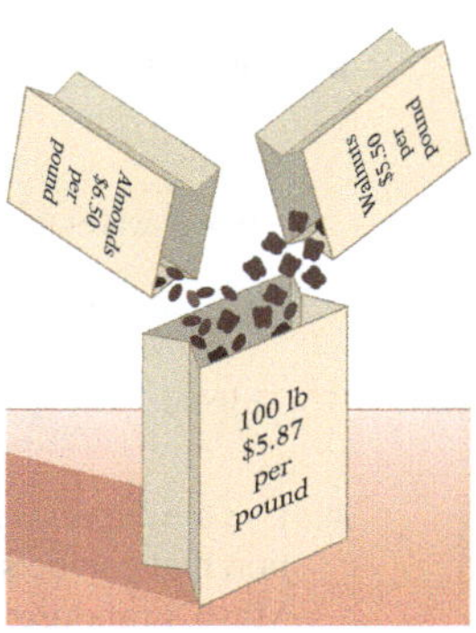

26. The manager of a specialty food store combined almonds that cost \$6.50 per pound with walnuts that cost \$5.50 per pound. How many pounds of each were used to make a 100-pound mixture that costs \$5.87 per pound?

27. Find the cost per pound of a "house blend" of coffee that is made from 12 lb of Central American coffee that costs \$8 per pound and 30 lb of South American coffee that costs \$4.50 per pound.

28. Find the cost per pound of a sugar-coated breakfast cereal made from 40 lb of sugar that costs \$2.00 per pound and 120 lb of corn flakes that cost \$1.20 per pound.

29. **Tree Conservation** A town's parks department buys trees from the tree conservation program described in the news clipping at the right. The department spends \$406 on 14 bundles of trees. How many bundles of seedlings and how many bundles of container-grown plants did the parks department buy?

In the News

Conservation Tree Planting Program Underway

The Kansas Forest Service is again offering its Conservation Tree Planting Program. Trees are sold in bundles of 25, in two sizes—seedlings cost \$17 a bundle and larger container-grown plants cost \$45 a bundle.

Source: Kansas Canopy

30. How many liters of a blue dye that costs \$1.60 per liter must be mixed with 18 L of anil that costs \$2.50 per liter to make a mixture that costs \$1.90 per liter?

31. Find the cost per ounce of a gold alloy made from 25 oz of pure gold that costs \$1282 per ounce and 40 oz of an alloy that costs \$900 per ounce.

32. Find the cost per ounce of a sunscreen made from 100 oz of a lotion that costs \$2.50 per ounce and 50 oz of a lotion that costs \$4.00 per ounce.

33. A grocer mixes peanuts that cost \$3 per pound with almonds that cost \$7 per pound. Which of the following could be true about the cost C per pound of the mixture? There may be more than one correct answer.

(i) $C = \$10$ (ii) $C > \$7$ (iii) $C < \$7$
(iv) $C < \$3$ (v) $C > \$3$ (vi) $C = \$4$

34. A snack mix is made from 3 lb of sunflower seeds that cost S dollars per pound and 4 lb of raisins that cost R dollars per pound. Which expression gives the cost C per pound of the mixture?

(i) $7(S + R)$ (ii) $3S + 4R$ (iii) $S + R$ (iv) $\dfrac{3S + 4R}{7}$

2 Percent mixture problems (See pages 176–178.)

GETTING READY

35. A 250-milliliter bottle contains a solution that is 90% isopropyl alcohol. The amount of isopropyl alcohol in the solution is 250(____?____) = ____?____ ml.

36. A 500-gram can of legumes contains 20% navy beans. The can contains ____?____ g of navy beans.

37. A 10-pound box of chocolate-covered cherries is 15% chocolate. There are ____?____ lb of chocolate in the box. There are ____?____ lb of cherries in the box.

38. A juice drink that is 60% real fruit juice is made by mixing some juice that is 20% real fruit juice with 32 oz of a juice that is 75% real fruit juice. Let x be the amount of the 20% real fruit juice that is in the mixture.

a. Complete the following table.

	Amount, A	$\cdot$	Percent of concentration, r	$=$	Quantity, Q
75% real fruit juice	?	$\cdot$	?	$=$	?
20% real fruit juice	x	$\cdot$	?	$=$	?
Mixture	?	$\cdot$	?	$=$	?

b. Use the expressions in the last column of the table in part (a) to write an equation that can be solved to find the amount of the 20% real fruit juice that is in the mixture: __?__ + __?__ = __?__.

39. Explain the meaning of each variable in the equation $Q = Ar$. Give an example of how this equation is used.

40. Suppose you are mixing a 5% acid solution with a 10% acid solution. In the context of this situation, explain the meaning of the statement "The sum of the quantities of the substances being mixed is equal to the quantity of the substance after mixing."

41. Forty ounces of a 30% gold alloy are mixed with 60 oz of a 20% gold alloy. Find the percent concentration of the resulting gold alloy.

42. One hundred ounces of juice that is 50% tomato juice are added to 200 oz of a vegetable juice that is 25% tomato juice. What is the percent concentration of tomato juice in the resulting mixture?

43. How many gallons of a 15% acid solution must be mixed with 5 gal of a 20% acid solution to make a 16% acid solution?

44. How many pounds of a chicken feed that is 50% corn must be mixed with 400 lb of a feed that is 80% corn to make a chicken feed that is 75% corn?

45. A rug is made by weaving 20 lb of yarn that is 50% wool with a yarn that is 25% wool. How many pounds of the yarn that is 25% wool must be used if the finished rug is to be 35% wool?

46. Five gallons of a dark green latex paint that is 20% yellow paint are combined with a lighter green latex paint that is 40% yellow paint. How many gallons of the lighter green paint must be used to create a green paint that is 25% yellow paint?

47. How many gallons of a plant food that is 9% nitrogen must be combined with another plant food that is 25% nitrogen to make 10 gal of a plant food that is 15% nitrogen?

48. A chemist wants to make 50 ml of a 16% acid solution by mixing a 13% acid solution and an 18% acid solution. How many milliliters of each solution should the chemist use?

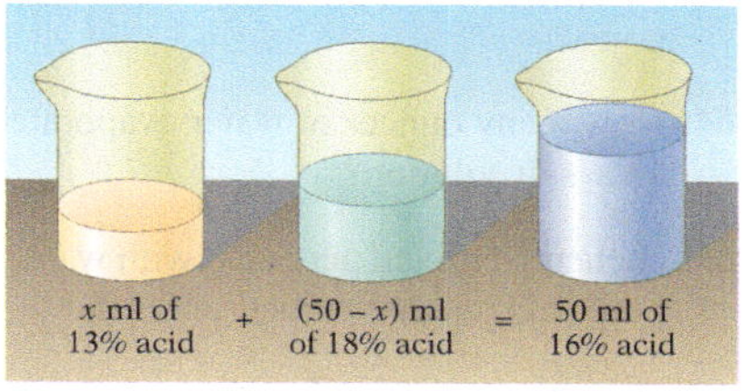

49. Five grams of sugar are added to a 45-gram serving of a breakfast cereal that is 10% sugar. What is the percent concentration of sugar in the resulting mixture?

50. Thirty ounces of pure silver are added to 50 oz of a silver alloy that is 20% silver. What is the percent concentration of the resulting alloy?

51. To make the potpourri mixture sold at a florist shop, 70 oz of a potpourri that is 80% lavender are combined with a potpourri that is 60% lavender. The resulting potpourri is 74% lavender. How much of the potpourri that is 60% lavender is used?

52. The manager of a garden shop mixes grass seed that is 40% rye grass with 40 lb of grass seed that is 60% rye grass to make a mixture that is 56% rye grass. How much of the 40% rye grass is used?

53. A hair dye is made by blending a 7% hydrogen peroxide solution and a 4% hydrogen peroxide solution. How many milliliters of each are used to make a 300-milliliter solution that is 5% hydrogen peroxide?

54. A clothing manufacturer has some pure silk thread and some thread that is 85% silk. How many kilograms of each must be woven together to make 75 kg of cloth that is 96% silk?

55. At a cosmetics company, 40 L of pure aloe cream are mixed with 50 L of a moisturizer that is 64% aloe. What is the percent concentration of aloe in the resulting mixture?

56. A hair stylist combines 12 oz of shampoo that is 20% conditioner with an 8-ounce bottle of pure shampoo. What is the percent concentration of conditioner in the 20-ounce mixture?

57. **Ethanol Fuel** See the news clipping at the right. *Gasohol* is a type of fuel made by mixing ethanol with gasoline. E10 is a fuel mixture of 10% ethanol and 90% gasoline. E20 contains 20% ethanol and 80% gasoline. How many gallons of ethanol must be added to 100 gal of E10 to make E20?

In the News

Gasohol Reduces Harmful Emissions

A new study indicates that using E20 fuel reduces carbon dioxide and hydrocarbon emissions, as compared with E10 blends or traditional gasoline.

Source: www.sciencedaily.com

58. How many ounces of pure chocolate must be added to 150 oz of chocolate topping that is 50% chocolate to make a topping that is 75% chocolate?

59. A recipe for a rice dish calls for 12 oz of a rice mixture that is 20% wild rice and 8 oz of pure wild rice. What is the percent concentration of wild rice in the 20-ounce mixture?

60. True or false? A 10% salt solution can be combined with some amount of a 20% salt solution to create a 30% salt solution.

61. True or false? When n ounces of 100% acid are mixed with $2n$ ounces of pure water, the resulting mixture is a 50% acid solution.

APPLYING CONCEPTS

62. Find the cost per ounce of a mixture of 30 oz of an alloy that costs \$4.50 per ounce, 40 oz of an alloy that costs \$3.50 per ounce, and 30 oz of an alloy that costs \$3.00 per ounce.

63. A grocer combined walnuts that cost \$5.60 per pound and cashews that cost \$7.50 per pound with 20 lb of peanuts that cost \$4.00 per pound. Find the amount of walnuts and the amount of cashews used to make the 50-pound mixture costing \$5.72 per pound.

64. How many ounces of water evaporated from 50 oz of a 12% salt solution to produce a 15% salt solution?

65. A chemist mixed pure acid with water to make 10 L of a 30% acid solution. How much pure acid and how much water did the chemist use?

66. How many grams of pure water must be added to 50 g of pure acid to make a solution that is 40% acid?

67. Tickets to a performance by a community theater company cost \$5.50 for adults and \$2.75 for children. A total of 120 tickets were sold for \$563.75. How many adults and how many children attended the performance?

PROJECTS OR GROUP ACTIVITIES

68. A radiator contains 15 gal of a 20% antifreeze solution. How many gallons must be drained from the radiator and replaced by pure antifreeze so that the radiator will contain 15 gal of a 40% antifreeze solution?

69. When 5 oz of water are added to an acid solution, the new mixture is $33\frac{1}{3}\%$ acid. When 5 oz of pure acid are added to this new mixture, the resulting mixture is 50% acid. What was the percent concentration of acid in the original mixture?

70. Explain why we look for patterns and relationships in mathematics. Include a discussion of the relationship between value mixture problems and percent mixture problems and how understanding one of these can make it easier to understand the other. Also discuss why understanding how to solve the value mixture problems in this section can be helpful in solving Exercise 67.

4.6 Uniform Motion Problems

OBJECTIVE 1

Uniform motion problems

How It's Used

A surveyor's Electronic Distance Measurer, or EDM, uses the equation $d = rt$ to find the distance between two points. A prism is placed at one point and the EDM at the other. Using the time a laser takes to travel from the EDM to the prism and back, and the speed of the laser, the EDM calculates the distance between the two points.

A train that travels constantly in a straight line at 50 mph is in *uniform motion*. Recall that **uniform motion** means the speed of an object does not change.

The solution of a uniform motion problem is based on the equation $d = rt$, where d is the distance traveled, r is the rate of travel, and t is the time spent traveling.

A car travels at 50 mph for 3 h.

$r = 50$; $t = 3$; d is unknown.

$$d = rt$$
$$d = 50(3)$$
$$d = 150$$

The car travels a distance of 150 mi.

The equation $d = rt$ is used to solve the uniform motion problem below.

Solve: A car leaves a town traveling at 35 mph. Two hours later, a second car leaves the same town, on the same road, traveling at 55 mph. How many hours after the second car leaves will the second car pass the first car?

STRATEGY FOR SOLVING A UNIFORM MOTION PROBLEM

► For each object, use the equation $rt = d$. Write a numerical or variable expression for the rate, time, and distance. The results can be recorded in a table.

Take Note

Use the information given in the problem to fill in the rate and time columns of the table. Fill in the distance column by multiplying the two expressions you wrote in each row.

The first car traveled 2 h longer than the second car.

Unknown time for the second car: t
Time for the first car: $t + 2$

	Rate, r	·	Time, t	=	Distance, d
First car	35	·	$t + 2$	=	$35(t + 2)$
Second car	55	·	t	=	$55t$

- Determine how the distances traveled by the individual objects are related. For example, the total distance traveled by both objects may be known, or it may be known that the two objects traveled the same distance.

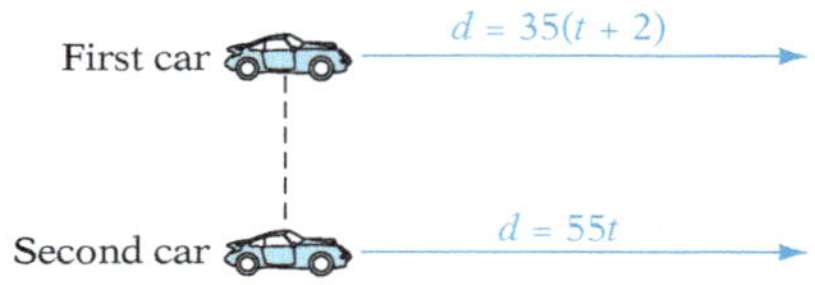

The two cars travel the same distance.

$$\begin{aligned} 35(t+2) &= 55t \\ 35t + 70 &= 55t \\ 70 &= 20t \\ 3.5 &= t \end{aligned}$$

- **The distance traveled by the first car equals the distance traveled by the second car.**

The second car will pass the first car in 3.5 h.

EXAMPLE 1 Two cars, the first traveling 10 mph faster than the second, start at the same time from the same point and travel in opposite directions. In 3 h, they are 288 mi apart. Find the rate of the second car.

Strategy

- Rate of second car: r
  Rate of first car: $r + 10$

	Rate	Time	Distance
First car	$r + 10$	3	$3(r + 10)$
Second car	r	3	$3r$

- The total distance traveled by the two cars is 288 mi.

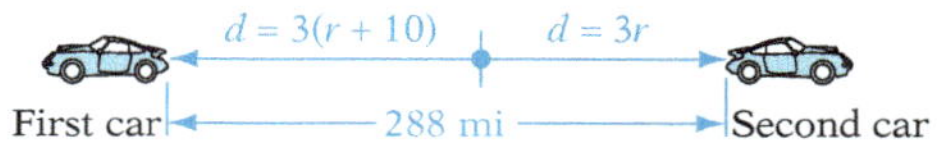

Solution

$$\begin{aligned} 3(r+10) + 3r &= 288 \\ 3r + 30 + 3r &= 288 \\ 6r + 30 &= 288 \\ 6r &= 258 \\ r &= 43 \end{aligned}$$

- **The distance traveled by the first car plus the distance traveled by the second car is 288 mi.**

The second car is traveling 43 mph.

Problem 1 Two trains, one traveling at twice the speed of the other, start at the same time from stations that are 306 mi apart and travel toward each other. In 3 h, the trains pass each other. Find the rate of each train.

Solution See page S10.

Try Exercise 7, page 186.

EXAMPLE 2 A bicycling club rides out into the country at a speed of 16 mph and returns over the same road at 12 mph. How far does the club ride out into the country if it travels a total of 7 h?

Strategy
- Time spent riding out: t
  Time spent riding back: $7 - t$

	Rate	Time	Distance
Out	16	t	$16t$
Back	12	$7 - t$	$12(7 - t)$

- The distance out equals the distance back.

Solution
$16t = 12(7 - t)$
$16t = 84 - 12t$
$28t = 84$
$t = 3$ • **The time is 3 h. Find the distance.**

The distance out $= 16t = 16(3) = 48$.

The club rides 48 mi into the country.

Problem 2 On a survey mission, a pilot flew out to a parcel of land and back in 7 h. The rate flying out was 120 mph. The rate flying back was 90 mph. How far away was the parcel of land?

Solution See page S10.

Try Exercise 13, page 186.

4.6 Exercises

CONCEPT CHECK

1. Explain what each variable in the formula $d = rt$ represents.
2. Suppose a jogger starts on a 4-mile course. Two minutes later, a second jogger starts on the same course. If both joggers arrive at the finish line at the same time, which jogger is running faster?
3. A Boeing 757 airplane leaves San Diego, California, and is flying to Dallas, Texas. One hour later, a Boeing 767 leaves San Diego, taking the same route to Dallas. If t represents the time, in hours, that the Boeing 757 is in the air, how long has the Boeing 767 been in the air?

4. If two objects are moving in opposite directions, how can the total distance between the two objects be expressed?

5. Two friends are standing 50 ft apart and begin walking toward each other on a straight sidewalk. When they meet, what is the total distance covered by the two friends?

6. Suppose two planes are heading toward each other. One plane is traveling at 450 mph, and the other plane is traveling at 375 mph. What is the rate at which the distance between the planes is changing?

1 Uniform motion problems (See pages 183–185.)

7. Two planes start from the same point and fly in opposite directions. The first plane is flying 25 mph slower than the second plane. In 2 h, the planes are 470 mi apart. Find the rate of each plane.

8. Two cyclists start from the same point and ride in opposite directions. One cyclist rides twice as fast as the other. In 3 h, they are 81 mi apart. Find the rate of each cyclist.

9. One speed skater starts across a frozen lake at an average speed of 8 m/s. Ten seconds later, a second speed skater starts from the same point and skates in the same direction at an average speed of 10 m/s. How many seconds after the second skater starts will the second skater overtake the first skater?

10. A long-distance runner started on a course running at an average speed of 6 mph. Half an hour later, a second runner began the same course at an average speed of 7 mph. How long after the second runner starts will the second runner overtake the first runner?

11. Michael Chan leaves a dock in his motorboat and travels at an average speed of 9 mph toward the Isle of Shoals, a small island off the coast of Massachusetts. Two hours later, a tour boat leaves the same dock and travels at an average speed of 18 mph toward the same island. How many hours after the tour boat leaves will Michael's boat be alongside the tour boat?

12. A jogger starts from one end of a 15-mile nature trail at 8:00 A.M. One hour later, a cyclist starts from the other end of the trail and rides toward the jogger. If the rate of the jogger is 6 mph and the rate of the cyclist is 9 mph, at what time will the two meet?

13. An executive drove from home at an average speed of 30 mph to an airport where a helicopter was waiting. The executive boarded the helicopter and flew to the corporate offices at an average speed of 60 mph. The entire distance was 150 mi. The entire trip took 3 h. Find the distance from the airport to the corporate offices.

14. A 555-mile, 5-hour plane trip was flown at two speeds. For the first part of the trip, the average speed was 105 mph. For the remainder of the trip, the average speed was 115 mph. How long did the plane fly at each speed?

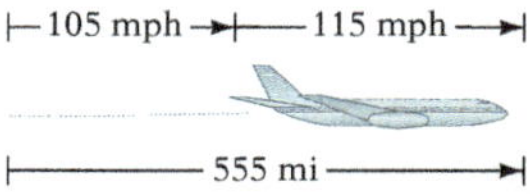

15. After a sailboat had been on the water for 3 h, a change in the wind direction reduced the average speed of the boat by 5 mph. The entire distance sailed was 57 mi. The total time spent sailing was 6 h. How far did the sailboat travel in the first 3 h?

16. A car and a bus set out at 3 P.M. from the same point, headed in the same direction. The average speed of the car is twice the average speed of the bus. In 2 h, the car is 68 mi ahead of the bus. Find the rate of the car.

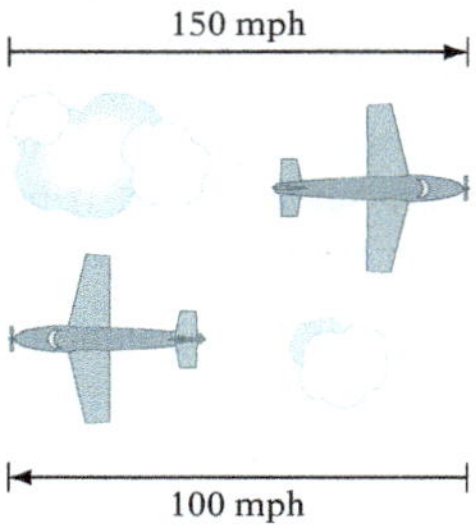

17. A passenger train leaves a train depot 2 h after a freight train leaves the same depot. The freight train is traveling 20 mph slower than the passenger train. Find the rate of each train if the passenger train overtakes the freight train in 3 h.

18. A stunt driver was needed at the production site of a Hollywood movie. The average speed of the stunt driver's flight to the site was 150 mph, and the average speed of the return trip was 100 mph. Find the distance of the round trip if the total flying time was 5 h.

19. A ship traveling east at 25 mph is 10 mi from a harbor when another ship leaves the harbor traveling east at 35 mph. How long does it take the second ship to catch up to the first ship?

20. At 10 A.M., a plane leaves Boston, Massachusetts, for Seattle, Washington, a distance of 3000 mi. One hour later, a plane leaves Seattle for Boston. Both planes are traveling at a speed of 500 mph. How many hours after the plane leaves Seattle will the planes pass each other?

21. At noon, a train leaves Washington, D.C., headed for Charleston, South Carolina, a distance of 500 mi. The train travels at a speed of 60 mph. At 1 P.M., a second train leaves Charleston headed for Washington, D.C., traveling at 50 mph. How long after the train leaves Charleston will the two trains pass each other?

22.  **Bridges** See the news clipping at the right. Two cars, the first traveling 10 km/h faster than the second, start at the same time from opposite ends of the Hangzhou Bay Bridge and travel toward each other. The cars pass each other in 12 min. Find the rate of the faster car.

© Imaginechina/Corbis

Hangzhou Bay Bridge

In the News

Longest Ocean-Crossing Bridge Opens to Public

The Hangzhou Bay Bridge is the longest ocean-crossing bridge in the world. It spans the Hangzhou Bay on the East China Sea and crosses the Qiantang River at the Yangtze River Delta. The S-shaped bridge connects Jiaxing to the north and Ningbo to the south. The bridge is 36 km long and has a speed limit of 100 kilometers per hour.

Source: www.roadtraffic-technology.com

23. A race car driver starts along a 50-mile race course traveling at an average speed of 90 mph. Fifteen minutes later, a second driver starts along the same course at an average speed of 120 mph. Will the second car overtake the first car before the drivers reach the end of the course?

24. A bus traveled on a straight road for 2 h at an average speed that was 20 mph faster than its average speed on a winding road. The time spent on the winding road was 3 h. Find the average speed on the winding road if the total trip was 210 mi.

25. A bus traveling at a rate of 60 mph overtakes a car traveling at a rate of 45 mph. If the car had a 1-hour head start, how far from the starting point does the bus overtake the car?

26. A car traveling at 48 mph overtakes a cyclist who, riding at 12 mph, had a 3-hour head start. How far from the starting point does the car overtake the cyclist?

27. A plane left Kennedy Airport Tuesday morning for a 605-mile, 5-hour trip. For the first part of the trip, the average speed was 115 mph. For the remainder of the trip, the average speed was 125 mph. How long did the plane fly at each speed?

APPLYING CONCEPTS

28. A car and a cyclist start at 10 A.M. from the same point, headed in the same direction. The average speed of the car is 5 mph more than three times the average speed of the cyclist. In 1.5 h, the car is 46.5 mi ahead of the cyclist. Find the rate of the cyclist.

29. A cyclist and a jogger set out at 11 A.M. from the same point, headed in the same direction. The average speed of the cyclist is twice the average speed of the jogger. In 1 h, the cyclist is 7 mi ahead of the jogger. Find the rate of the cyclist.

30. A car and a bus set out at 2 P.M. from the same point, headed in the same direction. The average speed of the car is 30 mph slower than twice the average speed of the bus. In 2 h, the car is 30 mi ahead of the bus. Find the rate of the car.

31. At 10 A.M., two campers left their campsite by canoe and paddled downstream at an average speed of 12 mph. They then turned around and paddled back upstream at an average rate of 4 mph. The total trip took 1 h. At what time did the campers turn around downstream?

32. At 7 A.M., two joggers start from opposite ends of an 8-mile course. One jogger is running at a rate of 4 mph, and the other is running at a rate of 6 mph. At what time will the joggers meet?

33. A truck leaves a depot at 11 A.M. and travels at a speed of 45 mph. At noon, a van leaves the same depot and travels the same route at a speed of 65 mph. At what time does the van overtake the truck?

34. Explain why the motion problems in this section are restricted to *uniform* motion.

35. Explain why 60 mph is the same as 88 ft/s.

PROJECTS OR GROUP ACTIVITIES

36. A bicyclist rides for 2 h at a speed of 10 mph and then returns at a speed of 20 mph. Find the cyclist's average speed for the trip.

37. A car travels a 1-mile track at an average speed of 30 mph. At what average speed must the car travel the next mile so that the average speed for the 2 mi is 60 mph?

38. A mountain climber ascended a mountain at 0.5 mph and descended twice as fast. The trip took 12 h. How many miles was the round trip?

39. Pat runs at 5 m/s. Chris skates at 8 m/s. How many meters can Chris skate in the time it takes Pat to run 80 m?

4.7 Inequalities

OBJECTIVE 1

Applications of inequalities

Solving application problems requires recognition of the verbal phrases that translate into mathematical symbols. Here is a partial list of the phrases used to indicate each of the four inequality symbols.

Phrase for <	**Phrases for >**
is less than	is greater than is more than exceeds
Phrases for ≤	**Phrases for ≥**
is less than or equal to maximum at most or less	is greater than or equal to minimum at least or more

EXAMPLE 1 A student must have at least 450 points out of 500 points on five tests to receive an A in a course. One student's results on the first four tests were 93, 79, 87, and 94. What scores on the last test will enable this student to receive an A in the course?

Strategy To find the scores, write and solve an inequality using N to represent the score on the last test.

Solution

total number of points on the five tests	is greater than or equal to	450

$$93 + 79 + 87 + 94 + N \geq 450$$
$$353 + N \geq 450$$
$$353 - 353 + N \geq 450 - 353$$
$$N \geq 97$$

• **Subtract 353 from each side of the equation.**

The student's score on the last test must be equal to or greater than 97.

Problem 1 An appliance dealer will make a profit on the sale of a television set if the cost of the new set is less than 70% of the selling price. What minimum selling price will enable the dealer to make a profit on a television set that costs the dealer $340?

Solution See page S10.

Try Exercise 15, page 190.

EXAMPLE 2 The base of a triangle is 8 in., and the height is $(3x - 5)$ in. Express as an integer the maximum height of the triangle when the area is less than 112 in^2.

Strategy To find the maximum height:

► Replace the variables in the area formula by the given values and solve for x.

► Replace the variable in the expression $3x - 5$ with the value found for x.

Solution

one-half the base times the height	is less than	112 in^2

$$\frac{1}{2}(8)(3x - 5) < 112$$

• **Substitute 8 for b and $(3x - 5)$ for h in the expression $\frac{1}{2}bh$.**

$$4(3x - 5) < 112$$

• **Multiply $\frac{1}{2}(8)$.**

$$12x - 20 < 112$$

• **Distributive Property**

$$12x - 20 + 20 < 112 + 20$$

• **Add 20 to each side.**

$$12x < 132$$
$$\frac{12x}{12} < \frac{132}{12}$$

• **Divide each side by 12.**

$$x < 11$$

$$3x - 5 = 3(11) - 5 = 28$$

• **Substitute the value of x into the variable expression for the height. Note that the height is less than 28 because $x < 11$.**

The maximum height of the triangle is 27 in.

Problem 2 Company A rents cars for $40 per day and 20¢ per mile driven outside a certain radius. Company B rents cars for $45 per day and 15¢ per mile driven outside a certain radius. You want to rent a car for one week. What is the maximum number of miles you can drive a Company A car outside the given radius if it is to cost you less than a Company B car?

Solution See page S10.

Try Exercise 23, page 191.

4.7 Exercises

CONCEPT CHECK

Determine whether the statement is true or false.

1. Both "is greater than" and "is more than" are represented by the inequality symbol $\geq$.
2. A minimum refers to a lower limit, whereas a maximum refers to an upper limit.
3. Given that $x > \frac{32}{6}$, the minimum integer that satisfies the inequality is 6.
4. Given that $x < \frac{25}{4}$, the maximum integer that satisfies the inequality is 7.
5. A rental car costs \$45 per day and 20¢ per mile driven outside a certain radius. If m represents the number of miles the rental car is driven outside the radius, then the expression $45 + 0.20m$ represents the cost to rent the car for one week.

1 Applications of inequalities (See pages 188–189.)

GETTING READY

For Exercises 6 to 9, translate the sentence into an inequality.

6. The maximum value of a number n is 50.
7. A number n is at least 102.
8. 500 is more than 12% of a number n.
9. The sum of a number n and 45% of the number is at most 200.

10. **Integer Problem** Three-fifths of a number is greater than two-thirds. Find the smallest integer that satisfies the inequality.
11. **Integer Problem** Four times the sum of a number and five is less than six times the number. Find the smallest integer that satisfies the inequality.
12. **Health** A health official recommends a maximum cholesterol level of 220 units. A patient has a cholesterol level of 275. By how many units must this patient's cholesterol level be reduced to satisfy the recommended maximum level?
13. **Mortgages** See the news clipping at the right. Suppose a couple's mortgage application is approved. Their monthly mortgage payment is \$2050. What is the couple's monthly household income? Round to the nearest dollar.

In the News

New Federal Standard for Mortgages

A new federal regulation states that the purchaser of a house is not to be approved for a monthly mortgage payment that is more than 38% of the purchaser's monthly household income.

Source: US News & World Report

14. **Sports** To be eligible for a basketball tournament, a basketball team must win at least 60% of its remaining games. If the team has 17 games remaining, how many games must the team win to qualify for the tournament?
15. **Recycling** A service organization will receive a bonus of \$200 for collecting more than 1850 lb of aluminum cans during its four collection drives. On the first three drives, the organization collected 505 lb, 493 lb, and 412 lb. How many pounds of cans must the organization collect on the fourth drive to receive the bonus?

JustASC/Shutterstock.com

16. **Test Scores** A professor scores all tests with a maximum of 100 points. To earn an A in this course, a student must have an average of at least 92 on four tests. One student's grades on the first three tests were 89, 86, and 90. Can this student earn an A grade?

17. Test Scores A student must have an average of at least 80 points on five tests to receive a B in a course. The student's grades on the first four tests were 75, 83, 86, and 78. What scores on the last test will enable this student to receive a B in the course?

18. Compensation A car sales representative receives a commission that is the greater of $250 or 8% of the selling price of a car. What dollar amounts in the sale price of a car will make the commission offer more attractive than the $250 flat fee?

19. Compensation A sales representative for an electronics store has the option of a monthly salary of $2000 or a 35% commission on the selling price of each item sold by the representative. What dollar amounts in sales will make the commission more attractive than the monthly salary?

Alternative Energy For Exercises 20 to 22, use the information in the article at the right.

20. a. A couple living in a town that has not changed the set-back requirement wants to install an 80-foot wind turbine on their property. How far back from the property line must the turbine be set?

b. Suppose the town lowers the 150% requirement to 125%. How far back from the property line must the turbine be set?

21. You live in a town that has not changed the set-back requirement. You want to install a wind turbine 68 ft from your property line. To the nearest foot, what is the height of the tallest wind turbine you can install?

22. You live in a town that has changed the set-back requirement to 115%. A good spot for a wind turbine on your property is 75 ft from the property line. To the nearest foot, what is the height of the tallest wind turbine you can install?

In the News

New Law Eases Restrictions on Small Wind Systems

A research project created by students in the University of New Hampshire's Environmental Politics class has led to a change in New Hampshire state law. Under the new law, a small wind turbine installed on a residential property must be set back from the property line by a distance greater than 150% of the turbine height. Individual towns may lower the 150% requirement, but they may not increase it.

Source: www.gencourt.state.nh.us

23. Compensation The sales agent for a jewelry company is offered a flat monthly salary of $3200 or a salary of $1000 plus an 11% commission on the selling price of each item sold by the agent. If the agent chooses the $3200 salary, what dollar amount does the agent expect to sell in one month?

24. Compensation A baseball player is offered an annual salary of $200,000 or a base salary of $100,000 plus a bonus of $1000 for each hit over 100 hits. How many hits must this baseball player achieve to earn more than $200,000?

25. Nutrition For a product to be labeled orange juice, a state agency requires that at least 80% of the drink be real orange juice. How many ounces of artificial flavors can be added to 32 oz of real orange juice if the drink is to be labeled orange juice?

26. Nutrition Grade A hamburger cannot contain more than 20% fat. How much fat can a butcher mix with 300 lb of lean meat to meet the 20% requirement?

27. Mixtures A silversmith combines 120 oz of an alloy that costs $4 per ounce with an alloy that costs $1.50 per ounce. How many ounces of the alloy that costs $1.50 per ounce should be used to make a mixture that costs less than $3 per ounce?

28. Mixtures The manager of a deli mixes pineapple that costs $.15 per ounce with 50 oz of cottage cheese that costs $.33 per ounce. How many ounces of pineapple should the manager use to make a mixture that costs less than $.25 per ounce?

29. Consumerism A shuttle service taking skiers to a ski area charges $8 per person each way. Four skiers are debating whether to take the shuttle bus or rent a car for $45 plus $.25 per mile. Assuming that the skiers will share the cost of the car and that they want the least expensive method of transportation, how far away is the ski area if they choose to take the shuttle service?

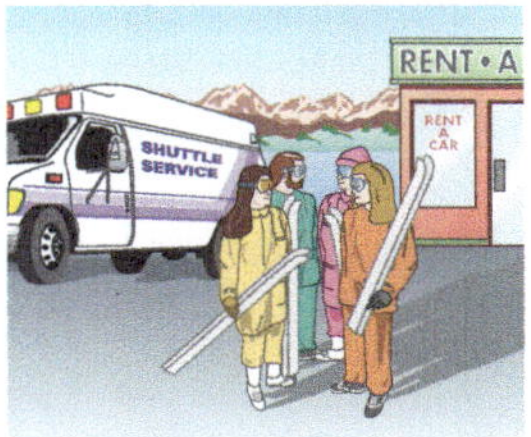

30. **Utilities** A residential water bill is based on a flat fee of \$10 plus a charge of \$.75 for each 1000 gal of water used. Find the number of gallons of water a family can use if they wish to have a monthly water bill that is less than \$55.

31. **Consumerism** Company A rents cars for \$45 per day and 8¢ per mile driven outside a particular radius. Company B rents cars for \$35 per day and 14¢ per mile driven outside the same radius. You want to rent a car for one day. Find the maximum number of miles you can drive a Company B car outside the given radius if it is to cost you less than a Company A car.

32. **Geometry** A rectangle is 8 ft wide and $(2x + 7)$ ft long. Express as an integer the maximum length of the rectangle when the perimeter is less than 54 ft.

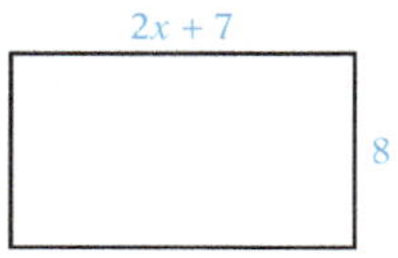

APPLYING CONCEPTS

33. **Integer Problem** Find three positive consecutive odd integers such that three times the sum of the first two integers is less than four times the third integer. (*Hint*: There is more than one solution.)

34. **Integer Problem** Find three positive consecutive even integers such that four times the sum of the first two integers is less than or equal to five times the third integer. (*Hint*: There is more than one solution.)

35. **Aviation** A maintenance crew requires between 30 min and 45 min to prepare an aircraft for its next flight. How many aircraft can this crew prepare for flight in a 6-hour period of time?

36. **Fund Raising** Your class decides to publish a calendar to raise money. The initial cost, regardless of the number of calendars printed, is \$900. After the initial cost, each calendar costs \$1.50 to produce. What is the minimum number of calendars your class must sell at \$6 per calendar to make a profit of at least \$1200?

PROJECTS OR GROUP ACTIVITIES

37. **Mixtures** A trail mix includes 100 oz of raisins costing \$.24 per ounce, 100 oz of mixed nuts costing \$.50 per ounce, and pretzels costing \$.12 per ounce. How many ounces of the pretzels should be used to make a mix costing less than \$.32 per ounce?

38. **Mixtures** An employee at a health food store mixes blueberries that cost \$.40 per ounce with 50 oz of yogurt that costs \$.24 per ounce. How many ounces of blueberries should be used to make a mixture that costs between \$.30 and \$.35 per ounce?

39. **Mixtures** A tea merchant mixes some black tea costing \$3 per pound with 10 lb of green tea costing \$4 per pound. How many pounds of the black tea should be used if the merchant wants a blend that costs between \$3.25 and \$3.75 per pound?

CHAPTER 4 Summary

Key Words | Objective and Page Reference | Examples

Consecutive integers follow one another in order.

[4.1.1, p. 142]

11, 12, 13 are consecutive integers.
$-9, -8, -7$ are consecutive integers.

The **perimeter** of a geometric figure is a measure of the distance around the figure.

[4.2.1, p. 150]

The perimeter of a triangle is the sum of the lengths of the three sides.
The perimeter of a rectangle is the sum of the lengths of the four sides.

An angle is measured in **degrees.** A 90° angle is a **right angle.** A 180° angle is a **straight angle.** One complete revolution is 360°. An **acute angle** is an angle whose measure is between 0° and 90°. An **obtuse angle** is an angle whose measure is between 90° and 180°. An **isosceles triangle** has two sides of equal measure. The three sides of an **equilateral triangle** are of equal measure.

[4.2.1, 4.2.2, pp. 150, 152]

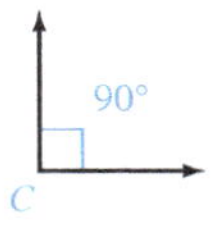

Parallel lines never meet; the distance between them is always the same. **Perpendicular lines** are intersecting lines that form right angles. Two angles that are on opposite sides of the intersection of two lines are **vertical angles;** vertical angles have the same measure. Two angles that share a common side are **adjacent angles;** adjacent angles of intersecting lines are supplementary angles.

[4.2.2, pp. 152–153]

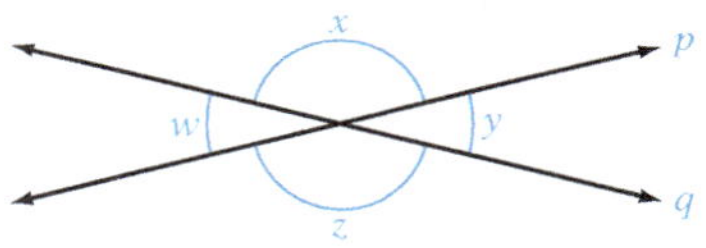

A line that intersects two other lines at two different points is a **transversal.** If the lines cut by a transversal are parallel lines, then pairs of equal angles are formed: **alternate interior angles, alternate exterior angles,** and **corresponding angles.**

[4.2.2, p. 154]

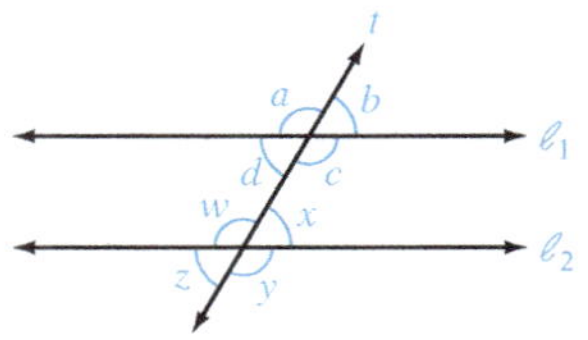

$\angle b = \angle d = \angle x = \angle z$
$\angle a = \angle c = \angle w = \angle y$

Cost is the price that a business pays for a product. **Selling price** is the price for which a business sells a product to a customer. **Markup** is the difference between selling price and cost. **Markup rate** is the markup expressed as a percent of the retailer's cost.

[4.3.1, p. 163]

If a business pays \$100 for a product and sells that product for \$140, then the cost of the product is \$100, the selling price is \$140, and the markup is \$140 − \$100 = \$40. The markup rate is 40%.

Discount is the amount by which a retailer reduces the regular price of a product. **Discount rate** is the discount expressed as a percent of the regular price.

[4.3.2, p. 164]

The regular price of a product is \$100. The product is now on sale for \$75. The discount on the product is \$100 − \$75 = \$25. The discount rate is 25%.

Essential Rules and Procedures	Objective and Page Reference	Examples
Translating a sentence into an equation requires recognizing the words or phrases that mean "equals." Some of these phrases are *is, is equal to,* and *represents.*	[4.1.1, p. 142]	Translate "eight less than the sum of a number and five is equal to twice the number" into an equation. $(x + 5) - 8 = 2x$
Consecutive Integers $n, n + 1, n + 2, \ldots$	[4.1.1, p. 142]	The sum of three consecutive integers is 135. $n + (n + 1) + (n + 2) = 135$
Consecutive Even or Consecutive Odd Integers $n, n + 2, n + 4, \ldots$	[4.1.1, pp. 142–143]	The sum of three consecutive odd integers is 135. $n + (n + 2) + (n + 4) = 135$
Formulas for Perimeter Triangle: $P = a + b + c$ Rectangle: $P = 2L + 2W$ Square: $P = 4s$	[4.2.1, p. 151]	The perimeter of a rectangle is 108 m. The length is 6 m more than three times the width. Find the width of the rectangle. $108 = 2(3W + 6) + 2W$ The perimeter of an isosceles triangle is 34 ft. The length of the third side is 5 ft less than the length of one of the equal sides. Find the length of one of the equal sides. $34 = x + x + (x - 5)$
Sum of the Angles of a Triangle The sum of the measures of the interior angles of a triangle is 180°. $\angle A + \angle B + \angle C = 180°$ The sum of an interior and a corresponding exterior angle is 180°.	[4.2.3, p. 155]	In a right triangle, the measure of one acute angle is 12° more than the measure of the smallest angle. Find the measure of the smallest angle. $x + (x + 12) + 90 = 180$
Basic Markup Equation $S = C + rC$	[4.3.1, p. 164]	The manager of a department buys a pasta maker machine for \$70 and sells the machine for \$98. Find the markup rate. $98 = 70 + 70r$
Basic Discount Equation $S = R - rR$	[4.3.2, p. 165]	The sale price of a golf putter is \$95. This price is 24% off the regular price. Find the regular price. $95 = R - 0.24R$
Annual Simple Interest Equation $I = Pr$	[4.4.1, p. 169]	You invest a portion of \$10,000 in a 7% annual simple interest account and the remainder in a 6% annual simple interest bond. The two investments earn total annual interest of \$680. How much is invested in the 7% account? $0.07x + 0.06(10{,}000 - x) = 680$

Value Mixture Equation
$V = AC$

[4.5.1, p. 174]

An herbalist has 30 oz of herbs costing \$2 per ounce. How many ounces of herbs costing \$1 per ounce should be mixed with the 30 oz to produce a mixture costing \$1.60 per ounce?

$$30(2) + 1x = 1.60(30 + x)$$

Percent Mixture Equation
$Q = Ar$

[4.5.2, p. 176]

Forty ounces of a 30% gold alloy are mixed with 60 oz of a 20% gold alloy. Find the percent concentration of the resulting gold alloy.

$$0.30(40) + 0.20(60) = x(100)$$

Uniform Motion Equation
$d = rt$

[4.6.1, p. 183]

A boat traveled from a harbor to an island at an average speed of 20 mph. The average speed on the return trip was 15 mph. The total trip took 3.5 h. How long did it take to travel to the island?

$$20t = 15(3.5 - t)$$

Translating Verbal Phrases into Inequality Symbols
Solving some application problems requires recognition of the verbal phrases that indicate the inequality symbols. See the list on page 188.

[4.7.1, p. 188]

Five times the difference between a number and eight is at least three times the number. Find the smallest number that satisfies the inequality.

$$5(x - 8) \geq 3x$$

CHAPTER 4 Review Exercises

1. Translate "seven less than twice a number is equal to the number" into an equation and solve.

2. **Music** A piano wire is 35 in. long. A note can be produced by dividing this wire into two parts so that three times the length of the shorter piece is twice the length of the longer piece. Find the length of the shorter piece.

3. **Integer Problem** The sum of two numbers is twenty-one. Three times the smaller number is two less than twice the larger number. Find the two numbers.

4. Translate "the sum of twice a number and six equals four times the difference between the number and two" into an equation and solve.

5. **Discounts** A ceiling fan that regularly sells for \$90 is on sale for \$60. Find the discount rate.

6. **Investments** A total of \$15,000 is deposited into two simple interest accounts. The annual simple interest rate on one account is 6%. The annual simple interest rate on the second account is 7%. How much should be invested in each account so that the total interest earned is \$970?

7. **Geometry** In an isosceles triangle, the measure of one of the two equal angles is 15° more than the measure of the third angle. Find the measure of each angle.

8. **Uniform Motion Problem** A motorcyclist and a bicyclist set out at 8 A.M. from the same point, headed in the same direction. The speed of the motorcyclist is three times the speed of the bicyclist. In 2 h, the motorcyclist is 60 mi ahead of the bicyclist. Find the rate of the motorcyclist.

9. **Geometry** In an isosceles triangle, the measure of one angle is 25° less than half the measure of one of the equal angles. Find the measure of each angle.

10. **Markup** The manager of a sporting goods store buys packages of bicycle cleats for $10.25 each and sells them for $18.45 each. Find the markup rate.

11. **Percent Mixture** A dairy owner mixes 5 gal of cream that is 30% butterfat with 8 gal of milk that is 4% butterfat. Find the percent concentration of butterfat in the resulting mixture.

12. **Geometry** The length of a rectangle is four times the width. The perimeter is 200 ft. Find the length and width of the rectangle.

13. **Business** The manager of an accounting firm is investigating contracts whereby large quantities of material can be photocopied at minimum cost. The contract from the Copy Center offers a fee of $50 per week and $.03 per page. The Prints 4 U Company offers a fee of $27 per week and $.05 per page. What is the minimum number of copies per week that could be ordered from the Copy Center if these copies are to cost less than copies from Prints 4 U?

14. **Discounts** The sale price of a video game is $51.46, which is 17% off the regular price. Find the regular price.

15. **Food Mixtures** The owner of a health food store combined cranberry juice that cost $7.79 per quart with apple juice that cost $7.19 per quart. How many quarts of each were used to make 10 qt of a cranapple juice mixture costing $7.61 per quart?

16. **Integer Problem** The sum of two numbers is thirty-six. The difference between the larger number and eight equals the total of four and three times the smaller number. Find the two numbers.

17. **Investments** An engineering consultant invested $14,000 in an individual retirement account paying 8.15% annual simple interest. How much money does the consultant have deposited in an account paying 12% annual simple interest if the total interest earned is 9.25% of the total investment?

18. **Percent Mixture** A pharmacist has 15 L of an 80% alcohol solution. How many liters of pure water should be added to the alcohol solution to make an alcohol solution that is 75% alcohol?

19. **Test Scores** A student's grades on five sociology exams were 68, 82, 90, 73, and 95. What is the lowest score this student can receive on the sixth exam and still have earned a total of at least 480 points?

20. Translate "the opposite of seven is equal to one-half a number less ten" into an equation and solve.

21. **Structures** The Empire State Building is 1472 ft tall. This is 654 ft less than twice the height of the Eiffel Tower. Find the height of the Eiffel Tower.

MarkS/Shutterstock.com

22. **Geometry** One angle of a triangle is 15° more than the measure of the second angle. The third angle is 15° less than the measure of the second angle. Find the measure of each angle.

23. **Carpentry** A board 10 ft long is cut into two pieces. Four times the length of the shorter piece is 2 ft less than two times the length of the longer piece. Find the length of the longer piece.

24. **Compensation** An optical engineer's consulting fee was \$600. This included \$80 for supplies and \$65 for each hour of consultation. Find the number of hours of consultation.

25. **Markup** A furniture store uses a markup rate of 60%. The store sells a solid oak curio cabinet for \$1074. Find the cost of the curio cabinet.

26. **Swimming Pools** One of the largest swimming pools in the world is located in Casablanca, Morocco, and is 480 m long. Two swimmers start at the same time from opposite ends of the pool and swim toward each other. One swimmer's rate is 65 m/min. The other swimmer's rate is 55 m/min. How many minutes after they begin swimming will they meet?

27. **Geometry** The perimeter of a triangle is 35 in. The second side is 4 in. longer than the first side. The third side is 1 in. shorter than twice the first side. Find the measure of each side.

28. **Integer Problem** The sum of three consecutive odd integers is -45. Find the integers.

29. **Geometry** A rectangle is 15 ft long and $(2x - 4)$ ft wide. Express as an integer the maximum width of the rectangle when the perimeter is less than 52 ft.

30. **Geometry** Given that $\ell_1 \parallel \ell_2$, find the measures of angles a and b.

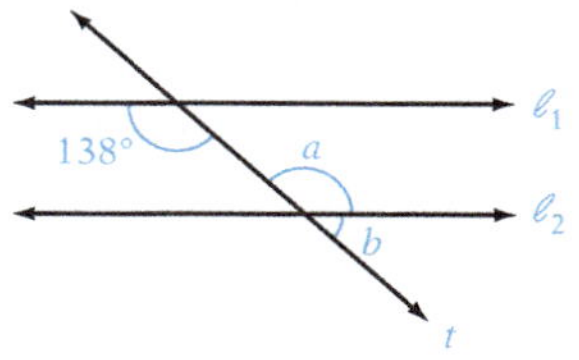

CHAPTER 4 Test

1. Translate "the sum of six times a number and thirteen is five less than the product of three and the number" into an equation and solve.

2. Translate "the difference between three times a number and fifteen is twenty-seven" into an equation and solve.

3. **Integer Problem** The sum of two numbers is 18. The difference between four times the smaller number and seven is equal to the sum of two times the larger number and five. Find the two numbers.

4. **Carpentry** A board 18 ft long is cut into two pieces. Two feet less than the product of five and the length of the shorter piece is equal to the difference between three times the length of the longer piece and eight. Find the length of each piece.

5. **Markup** The manager of a sports shop uses a markup rate of 50%. The selling price for a set of golf clubs is \$300. Find the cost of the golf clubs.

6. **Discounts** A pair of running shoes that regularly sells for \$100 is on sale for \$80. Find the discount rate.

7. **Percent Mixture** How many gallons of a 15% acid solution must be mixed with 5 gal of a 20% acid solution to make a 16% acid solution?

8. **Geometry** The perimeter of a rectangle is 38 m. The length of the rectangle is 1 m less than three times the width. Find the length and width of the rectangle.

9. **Integer Problem** Find three consecutive odd integers such that three times the first integer is one less than the sum of the second and third integers.

10. **Floral Deliveries** Florist A charges a \$3 delivery fee plus \$21 per bouquet delivered. Florist B charges a \$15 delivery fee plus \$18 per bouquet delivered. An organization wants to send each resident of a small nursing home a bouquet for Valentine's Day. How many residents are in the nursing home if it is more economical for the organization to use Florist B?

11. **Investments** A total of \$7000 is deposited into two simple interest accounts. On one account, the annual simple interest rate is 10%, and on the second account, the annual simple interest rate is 15%. How much should be invested in each account so that the total annual interest earned is \$800?

12. **Food Mixtures** A coffee merchant wants to make 12 lb of a blend of coffee costing \$6 per pound. The blend is made using a \$7 grade and a \$4 grade of coffee. How many pounds of each of these grades should be used?

13. **Uniform Motion** Two planes start at the same time from the same point and fly in opposite directions. The first plane is flying 100 mph faster than the second plane. In 3 h, the two planes are 1050 mi apart. Find the rate of each plane.

14. **Geometry** In a triangle, the first angle is 15° more than the second angle. The third angle is three times the second angle. Find the measure of each angle.

15. **Investments** A club treasurer deposited \$2400 in two simple interest accounts. The annual simple interest rate on one account is 6.75%. The annual simple interest rate on the other account is 9.45%. How much should be deposited in each account so that the same interest is earned on each account?

16. **Geometry** The width of a rectangle is 12 ft. The length is $(3x + 5)$ ft. Express as an integer the minimum length of the rectangle if the area is greater than 276 ft^2.

17. **Discounts** A file cabinet that normally sells for \$99 is on sale for 20% off. Find the sale price.

Cumulative Review Exercises

1. Given $B = \{-12, -6, -3, -1\}$, which elements of set B are less than -4?

2. Simplify: $-2 + (-8) - (-16)$

3. Simplify: $\left(-\frac{2}{3}\right)^3\left(-\frac{3}{4}\right)^2$

4. Simplify: $\frac{5}{6} - \left(\frac{2}{3}\right)^2 \div \left(\frac{1}{2} - \frac{1}{3}\right)$

5. Evaluate $-|-18|$.

6. Evaluate $b^2 - (a - b)^2$ when $a = 4$ and $b = -1$.

7. Simplify: $5x - 3y - (-4x) + 7y$

8. Simplify: $-4(3 - 2x - 5x^3)$

9. Simplify: $-2[x - 3(x - 1) - 5]$

10. Simplify: $-3x^2 - (-5x^2) + 4x^2$

11. Is 2 a solution of $4 - 2x - x^2 = 2 - 4x$?

12. Solve: $9 - x = 12$

13. Solve: $-\frac{4}{5}x = 12$

14. Solve: $8 - 5x = -7$

15. Solve: $-6x - 4(3 - 2x) = 4x + 8$

16. Write 40% as a fraction.

17. Solve and graph the solution set of $4x \geq 16$.

18. Solve: $-15x \leq 45$

19. Solve: $2x - 3 > x + 15$

20. Solve: $12 - 4(x - 1) \leq 5(x - 4)$

21. Write 0.025 as a percent.

22. Write $\frac{3}{25}$ as a percent.

23. Find $16\frac{2}{3}\%$ of 18.

24. 40% of what is 18?

25. Translate "the sum of eight times a number and twelve is equal to the product of four and the number" into an equation and solve.

26. **Construction** The area of the cement foundation of a house is 2000 ft^2. This is 200 ft^2 more than three times the area of the garage. Find the area of the garage.

27. **Repair Bills** An auto repair bill was \$563. This includes \$188 for parts and \$75 for each hour of labor. Find the number of hours of labor.

28. **Libraries** A survey of 250 librarians showed that 50 of the libraries had a particular reference book on their shelves. What percent of the libraries had the reference book?

29. **Investments** A deposit of \$4000 is made into an account that earns 11% annual simple interest. How much money is also deposited in an account that pays 14% annual simple interest if the total annual interest earned is 12% of the total investment?

30. **Markup** The manager of a department store buys a chain necklace for \$80 and sells it for \$140. Find the markup rate.

31. **Metallurgy** How many grams of a gold alloy that costs \$4 per gram must be mixed with 30 g of a gold alloy that costs \$7 per gram to make an alloy costing \$5 per gram?

32. **Percent Mixture** How many ounces of pure water must be added to 70 oz of a 10% salt solution to make a 7% salt solution?

33. **Geometry** In an isosceles triangle, the third angle is 8° less than twice the measure of one of the equal angles. Find the measure of one of the equal angles.

34. **Integer Problem** Three times the second of three consecutive even integers is 14 more than the sum of the first and third integers. Find the middle integer.

Linear Equations and Inequalities

CHAPTER 5

Focus on Success

Are you using the features of this text to learn the concepts being presented? The Focus On feature includes a step-by-step solution to the type of exercise you will be working in your homework assignment. A numbered Example provides you with a fully-worked-out solution. After studying the Example, try completing the Problem that follows. A complete solution to the Problem is given in the back of the text. Next, try the exercise listed after the Problem. You will be reinforcing the newly learned concepts at each step. (See Use the Interactive Method, page AIM-8.)

OBJECTIVES

5.1
1. Graph points in a rectangular coordinate system
2. Scatter diagrams
3. Average rate of change

5.2
1. Determine solutions of linear equations in two variables
2. Graph equations of the form $y = mx + b$
3. Graph equations of the form $Ax + By = C$

5.3
1. Find the slope of a straight line
2. Graph a line using the slope and y-intercept

5.4
1. Find the equation of a line using the equation $y = mx + b$
2. Find the equation of a line using the point-slope formula
3. Find the equation of a line given two points

5.5
1. Introduction to functions
2. Graphs of linear functions

5.6
1. Graph inequalities in two variables

PREP TEST

Are you ready to succeed in this chapter?
Take the Prep Test below to find out if you are ready to learn the new material.

1. Simplify: $\dfrac{5-(-7)}{4-8}$

2. Evaluate $\dfrac{a-b}{c-d}$ when $a = 3$, $b = -2$, $c = -3$, and $d = 2$.

3. Simplify: $-3(x - 4)$

4. Solve: $3x + 6 = 0$

5. Solve $4x + 5y = 20$ when $y = 0$.

6. Solve $3x - 7y = 11$ when $x = -1$.

7. Given $y = -4x + 5$, find the value of y when $x = -2$.

8. Simplify: $\dfrac{1}{4}(3x - 16)$

9. Which of the following are solutions of the inequality $-4 < x + 3$?
(i) 0 (ii) -3 (iii) 5 (iv) -7 (v) -10

5.1 The Rectangular Coordinate System

OBJECTIVE 1 Graph points in a rectangular coordinate system

How It's Used

Archeologists use stakes and string to overlay a rectangular coordinate system on an excavation site. The grid is reproduced on graph paper and then used to record the locations of artifacts before they are removed from the site.

A **rectangular coordinate system** is formed by two number lines, one horizontal and one vertical, that intersect at the zero point of each line. The point of intersection is called the **origin.** The two axes are called the **coordinate axes,** or simply the **axes.** Generally, the horizontal axis is labeled the ***x*-axis,** and the vertical axis is labeled the ***y*-axis.**

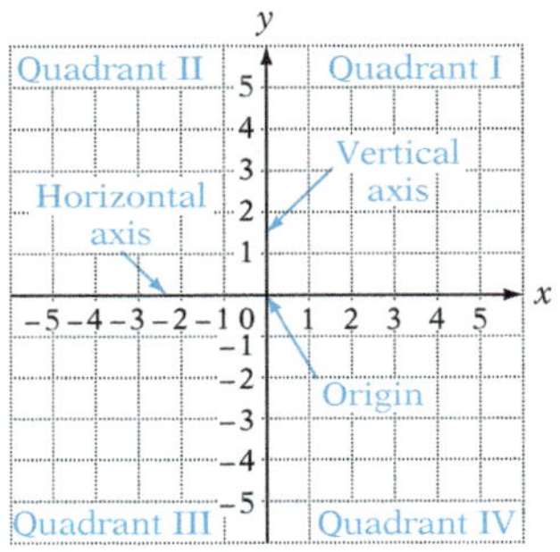

The axes determine a **plane,** which can be thought of as a large, flat sheet of paper. The two axes divide the plane into four regions called **quadrants,** which are numbered counterclockwise from I to IV starting at the upper right.

Each point in the plane can be identified by a pair of numbers called an **ordered pair.** The first number of the ordered pair measures a horizontal distance and is called the **abscissa,** or ***x*-coordinate.** The second number of the pair measures a vertical distance and is called the **ordinate,** or ***y*-coordinate.** The ordered pair (x, y) associated with a point is also called the **coordinates** of the point.

Horizontal distance → ↓ ↓ ← Vertical distance
Ordered pair → $(3, 4)$
x-coordinate → ↑ ↑ ← *y*-coordinate

To **graph,** or **plot, a point in the plane,** place a dot at the location given by the ordered pair. For example, to graph the point $(4, 1)$, start at the origin. Move 4 units to the right and then 1 unit up. Draw a dot. To graph $(-3, -2)$, start at the origin. Move 3 units left and then 2 units down. Draw a dot.

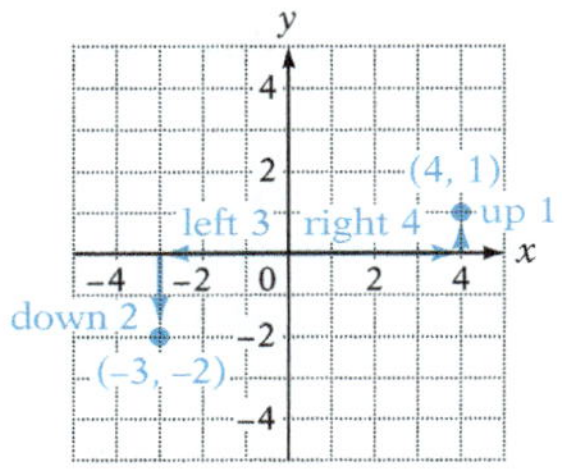

The **graph of an ordered pair** is the dot drawn at the coordinates of the point in the plane. The graphs of the ordered pairs $(4, 1)$ and $(-3, -2)$ are shown at the right.

Take Note

This concept is very important. An *ordered pair* is a *pair* of coordinates, and the *order* in which the coordinates are listed is important.

The graphs of the points whose coordinates are $(2, 3)$ and $(3, 2)$ are shown at the right. Note that they are different points. The order in which the numbers in an ordered pair are listed *is* important.

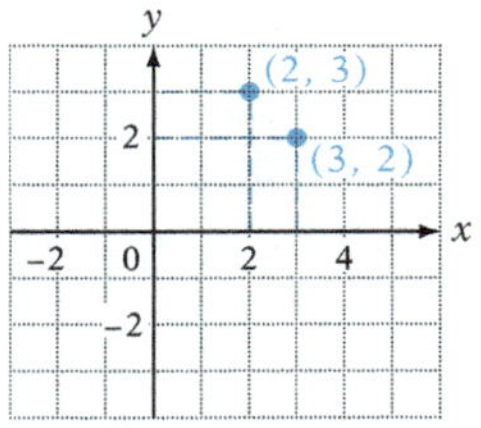

Each point in the plane is associated with an ordered pair, and each ordered pair is associated with a point in the plane. Although only integers are labeled on the coordinate grid, any ordered pair can be graphed by approximating its location. The graphs of the ordered pairs $\left(\frac{3}{2}, -\frac{4}{3}\right)$ and $(-2.4, 3.5)$ are shown at the right.

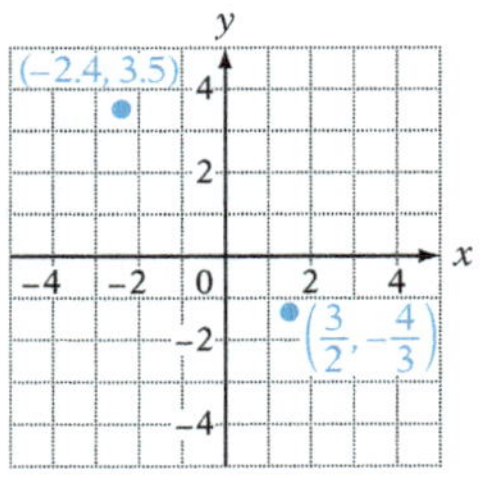

EXAMPLE 1 Graph the ordered pairs $(-2, -3)$, $(3, -2)$, $(1, 3)$, and $(4, 2)$.

Solution

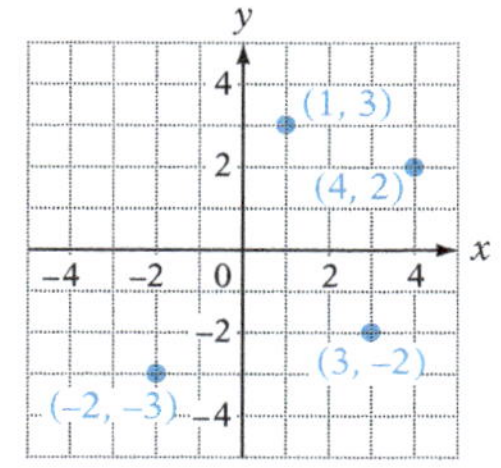

Problem 1 Graph the ordered pairs $(-1, 3)$, $(1, 4)$, $(-4, 0)$, and $(-2, -1)$.

Solution See page S11.

Try Exercise 11, page 208.

EXAMPLE 2 Find the coordinates of each of the points.

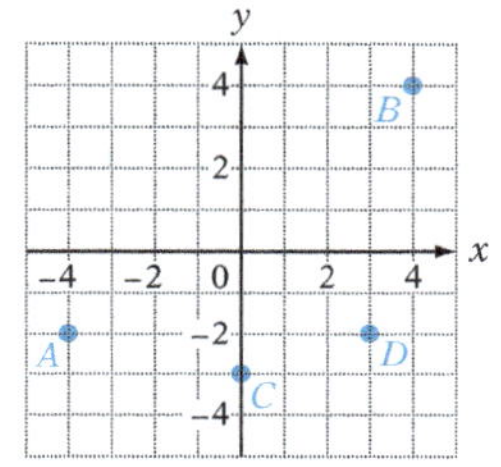

Solution $A(-4, -2)$, $B(4, 4)$, $C(0, -3)$, $D(3, -2)$

Problem 2 Find the coordinates of each of the points.

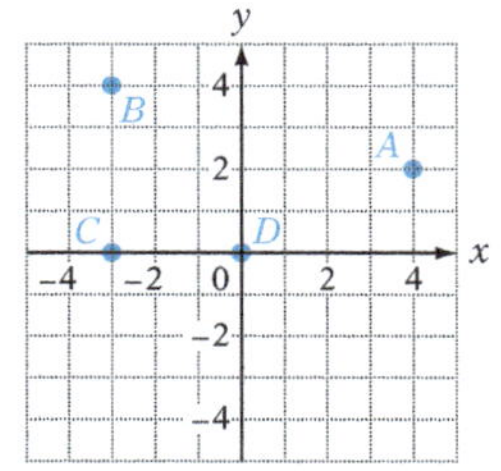

Solution See page S11.

Try Exercise 17, page 208.

OBJECTIVE 2

Scatter diagrams

There are many situations in which the relationship between two variables may be of interest. For example, a college admissions director wants to know the relationship between SAT scores (one variable) and success in college (the second variable). An employer is interested in the relationship between scores on a pre-employment test and ability to perform a job.

A researcher can investigate the relationship between two variables by means of regression analysis, which is a branch of statistics. The study of the relationship between two variables may begin with a **scatter diagram,** which is a graph of some of the known data.

The following table gives the record times for races of different lengths at a junior high track meet. Record times are rounded to the nearest second. Lengths are given in meters.

Length, x	100	200	400	800	1000	1500
Time, y	20	40	100	200	260	420

The scatter diagram for these data is shown at the right. Each ordered pair represents the length of a race and the corresponding record time. For example, the ordered pair (400, 100) indicates that the record time for the 400 m race is 100 s.

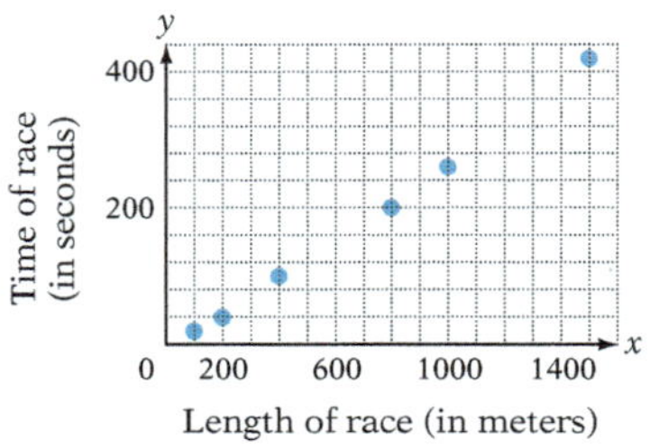

EXAMPLE 3 To test a heart medicine, a doctor measures the heart rates, in beats per minute, of five patients before and after they take the medication. The results are recorded in the following table. Graph the scatter diagram for these data.

Before medicine, x	85	80	85	75	90
After medicine, y	75	70	70	80	80

Strategy Graph the ordered pairs on a rectangular coordinate system, where the horizontal axis represents the heart rate before taking the medication and the vertical axis represents the heart rate after taking the medication.

Solution

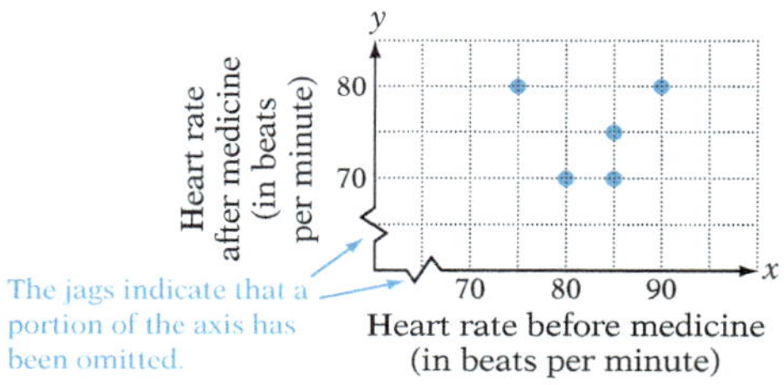

Problem 3 A study by the Federal Aviation Administration showed that narrow, over-the-wing emergency exit rows slow passenger evacuation. The table at the top of the next page shows the space between seats, in inches, and the evacuation time, in seconds, for a group of 35 passengers. The longest evacuation times are recorded in the table. Graph the scatter diagram for these data.

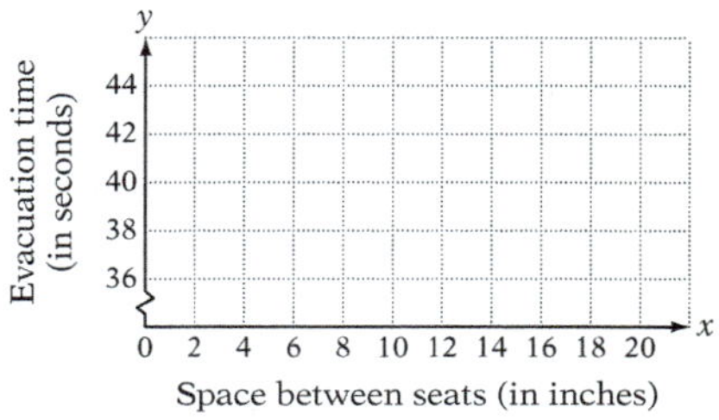

Space between seats, x	6	10	13	15	20
Evacuation time, y	44	43	38	36	37

Solution See page S11.

Try Exercise 25, page 209.

OBJECTIVE 3

Average rate of change

Population of the United States

Year	Population (in millions)
1800	5
1810	7
1820	10
1830	13
1840	17
1850	23
1860	31
1870	40
1880	50
1890	63
1900	76
1910	92
1920	106
1930	123
1940	132
1950	151
1960	179
1970	203
1980	227
1990	249
2000	281
2010	309

The table at the left shows the population of the United States for each decade from 1800 to 2010. (*Source:* U.S. Bureau of the Census) These data are graphed in the scatter diagram below.

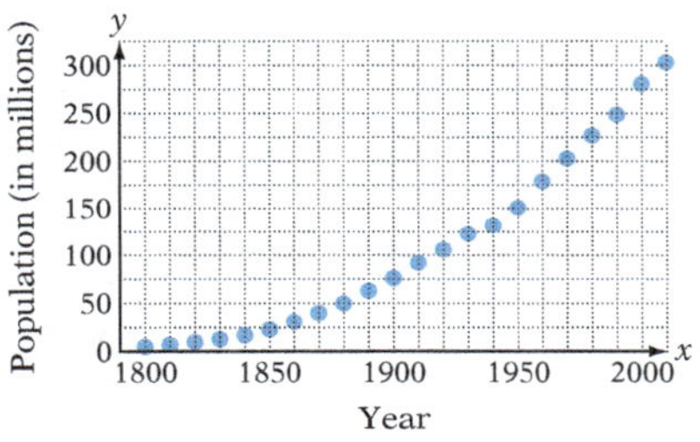

One way to describe how the population of the United States has changed over time is to calculate the change in population for a given time period. For example, we can find the change in population from 1990 to 2000.

$$\text{Population in 2000} - \text{population in 1990} = 281 \text{ million} - 249 \text{ million}$$
$$= 32 \text{ million}$$

The amount of change in the population, 32 million, does not describe how rapid the growth in population was. To describe the rate of growth, or the *rate of change,* we must consider the number of years over which that growth took place. To determine how rapidly the population grew from 1990 to 2000, divide the change in population by the number of years during which that change took place.

$$\frac{\text{Change in population}}{\text{Change in years}} = \frac{\text{population in 2000} - \text{population in 1990}}{2000 - 1990}$$
$$= \frac{281 - 249}{2000 - 1990}$$
$$= \frac{32}{10} = 3.2$$

The average rate of change was 3.2 million people per year. This means that *on average,* from 1990 to 2000, the population of the United States increased by 3.2 million people per year.

Consider the two points on the graph that correspond to these years:

(1990, 249) and (2000, 281)

Note that the average rate of change is calculated by dividing the difference between the y values of the two points by the difference between the x values of the two points.

AVERAGE RATE OF CHANGE

The average rate of change of y with respect to x is $\dfrac{\text{change in } y}{\text{change in } x}$.

EXAMPLE

Given (1990, 249) and (2000, 281),

$$\text{Average rate of change} = \frac{281 - 249}{2000 - 1990} = \frac{32}{10} = 3.2$$

To calculate the average rate of change, think

$$\frac{y \text{ value of second point} - y \text{ value of first point}}{x \text{ value of second point} - x \text{ value of first point}}$$

Focus on calculating average rate of change

Find the average rate of change per year in the U.S. population from 1980 to 1990.

In 1980, the population was 227 million: (1980, 227)
In 1990, the population was 249 million: (1990, 249)

$$\begin{aligned}\text{Average rate of change} &= \frac{\text{change in } y}{\text{change in } x}\\ &= \frac{\text{population in 1990} - \text{population in 1980}}{1990 - 1980}\\ &= \frac{249 - 227}{1990 - 1980}\\ &= \frac{22}{10} = 2.2\end{aligned}$$

The average rate of change in the population was 2.2 million people per year.

From 1990 to 2000, the average rate of change was 3.2 million people per year, whereas from 1980 to 1990, the average rate of change was 2.2 million people per year. The rate of change in the U.S. population is not constant; it varies for different intervals of time.

EXAMPLE 4 Find the average rate of change per year in the U.S. population from 1800 to 1900.

Solution In 1800, the population was 5 million: (1800, 5)
In 1900, the population was 76 million: (1900, 76)

$$\begin{aligned}\text{Average rate of change} &= \frac{\text{change in } y}{\text{change in } x}\\ &= \frac{\text{population in 1900} - \text{population in 1800}}{1900 - 1800}\\ &= \frac{76 - 5}{1900 - 1800}\\ &= \frac{71}{100} = 0.71\end{aligned}$$

The average rate of change in the population was 0.71 million people per year, or 710,000 people per year.

Note that if both the ordinates and the abscissas are subtracted in the reverse order, the result is the same.

$$\text{Average rate of change} = \frac{\text{population in 1800} - \text{population in 1900}}{1800 - 1900}$$

$$= \frac{5 - 76}{1800 - 1900} = \frac{-71}{-100} = 0.71$$

Problem 4 Find the average rate of change per year in the U.S. population from 1900 to 2000.

Solution See page S11.

Try Exercise 33, page 211.

5.1 Exercises

CONCEPT CHECK

1. In which quadrant is the graph of $(-3, 4)$?
2. In which quadrant is the graph of $(2, -5)$?
3. On which axis is the graph of $(0, -4)$?
4. On which axis is the graph of $(-6, 0)$?
5. What is the value of the y-coordinate of any point on the x-axis?
6. What is the value of the x-coordinate of any point on the y-axis?
7. Name any two points on a horizontal line that is 2 units above the x-axis.
8. Name any two points on a vertical line that is 3 units to the right of the y-axis.

1 Graph points in a rectangular coordinate system (See pages 202-203.)

GETTING READY

Complete Exercises 9 and 10 by replacing each question mark with the word *left, right, up,* or *down.*

9. To graph the point $(5, -4)$, start at the origin and move 5 units __?__ and 4 units __?__.

10. To graph the point $(-1, 7)$, start at the origin and move 1 unit __?__ and 7 units __?__.

11. Graph the ordered pairs $(-2, 1)$, $(3, -5)$, $(-2, 4)$, and $(0, 3)$.

12. Graph the ordered pairs $(5, -1)$, $(-3, -3)$, $(-1, 0)$, and $(1, -1)$.

13. Graph the ordered pairs $(0, 0)$, $(0, -5)$, $(-3, 0)$, and $(0, 2)$.

14. Graph the ordered pairs $(-4, 5)$, $(-3, 1)$, $(3, -4)$, and $(5, 0)$.

15. Graph the ordered pairs $(-1, 4)$, $(-2, -3)$, $(0, 2)$, and $(4, 0)$.

16. Graph the ordered pairs $(5, 2)$, $(-4, -1)$, $(0, 0)$, and $(0, 3)$.

17. Find the coordinates of each of the points.

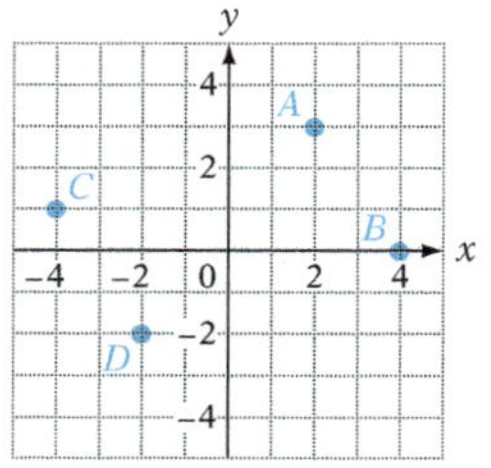

18. Find the coordinates of each of the points.

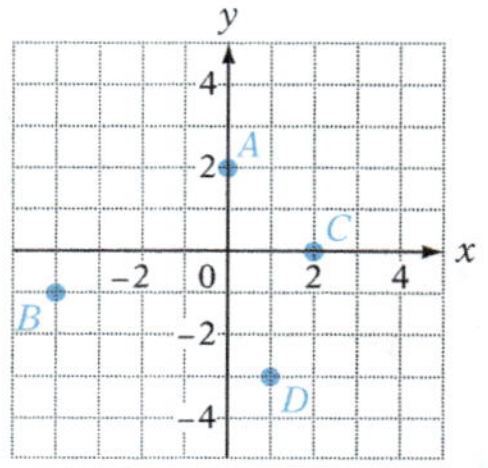

19. Find the coordinates of each of the points.

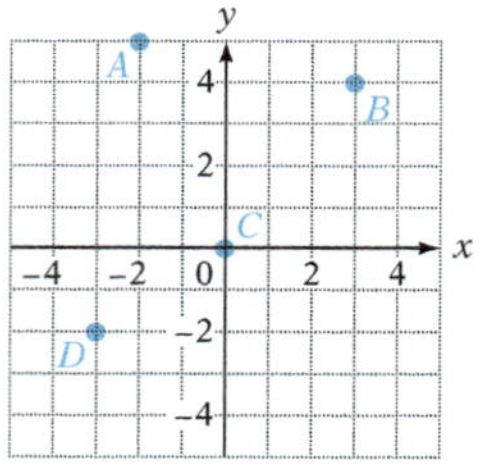

20. Find the coordinates of each of the points.

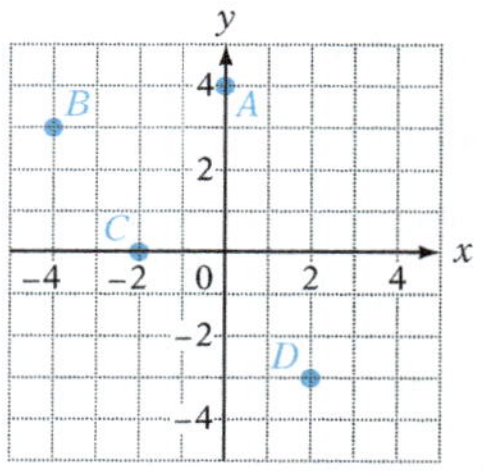

21. Let a and b be positive numbers such that $a < b$. In which quadrant is each point located?

a. (a, b) **b.** $(-a, b)$ **c.** $(b - a, -b)$ **d.** $(a - b, -b - a)$

22. Let a and b be positive numbers. State whether the two given points lie on the x-axis, the y-axis, a horizontal line other than the x-axis, or a vertical line other than the y-axis.

a. $(-a, b)$ and $(-a, 0)$ **b.** $(a, 0)$ and $(-b, 0)$

2 Scatter diagrams (See pages 203–205.)

GETTING READY

23. Look at the table of data in Exercise 25. The lowest x value is ___?___ and the highest x value is ___?___. The lowest y value is ___?___ and the highest y value is ___?___. In a scatter diagram of the data, the x-axis must show values from at least ___?___ to ___?___, and the y-axis must show values from at least ___?___ to ___?___.

24. Look at the table of data in Exercise 26. The lowest x value is ___?___ and the highest x value is ___?___. The lowest y value is ___?___ and the highest y value is ___?___. In a scatter diagram of the data, the x-axis must show values from at least ___?___ to ___?___, and the y-axis must show values from at least ___?___ to ___?___.

25. **Fuel Efficiency** The American Council for an Energy-Efficient Economy releases rankings of environmentally friendly and unfriendly cars and trucks sold in the United States. The following table shows the fuel usage, in miles per gallon, in the city and on the highway, for the twelve 2010-model vehicles ranked best for the environment. Graph the scatter diagram for these data.

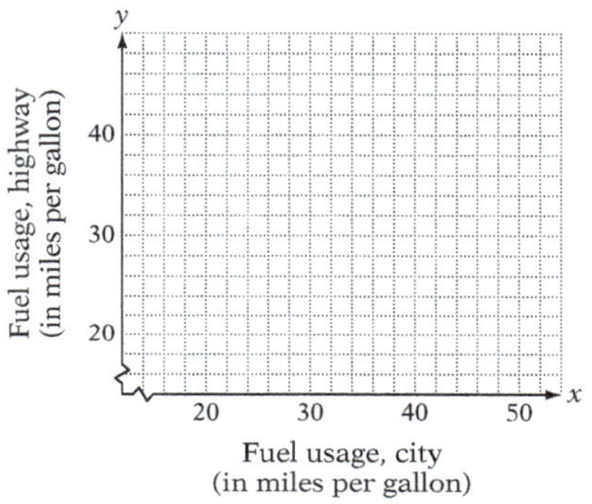

Fuel usage in mpg, city, x	51	40	29	28	34	22	22	22	21	18	21	21
Fuel usage in mpg, highway, y	48	45	35	35	31	32	31	32	27	25	22	22

26. **Fuel Efficiency** The American Council for an Energy-Efficient Economy releases rankings of environmentally friendly and unfriendly cars and trucks sold in the United States. The following table shows the fuel usage, in miles per gallon, in the city and on the highway, for the twelve 2010-model vehicles ranked worst for the environment. Graph the scatter diagram for these data.

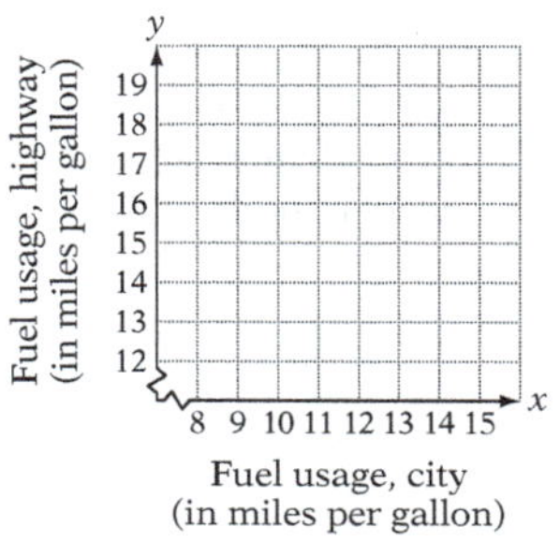

Fuel usage in mpg, city, x	8	8	9	10	12	12	10	9	11	13	13	11
Fuel usage in mpg, highway, y	13	14	15	16	17	17	17	16	15	18	19	15

27. **Olympics** The following table shows the weights, in pounds, and the heights, in inches, of some of the members of the U.S. Women's National Ice Hockey Team that won a silver medal in the 2010 Olympics. Graph the scatter diagram for these data.

AP Photo/Julie Jacobson

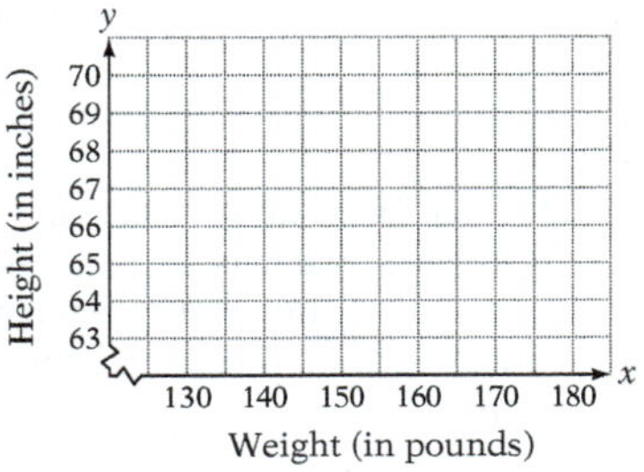

Weight, in pounds, x	140	150	140	160	170	150	160	170	150	130
Height, in inches, y	64	69	63	69	70	66	66	69	64	65

28. **Olympics** The following table shows the weights, in pounds, and the heights, in inches, of some of the members of the U.S. Men's National Ice Hockey Team that won a silver medal in the 2010 Olympics. Graph the scatter diagram for these data.

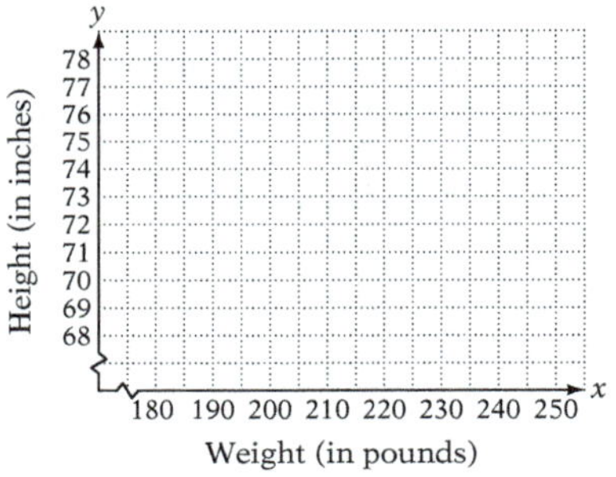

Weight, in pounds, x	220	200	200	190	210	220	210	190	190	240
Height, in inches, y	76	73	74	71	72	75	75	73	70	76

29. **Incomes** To determine trends in income levels, economists use inflation-adjusted values. These numbers take into consideration how inflation affects purchase power. The following table shows the median income, rounded to the nearest thousand, for selected years in both actual dollars (before inflation) and 2008 inflation-adjusted dollars. (*Source:* U.S. Census Bureau) Graph the scatter diagram for these data.

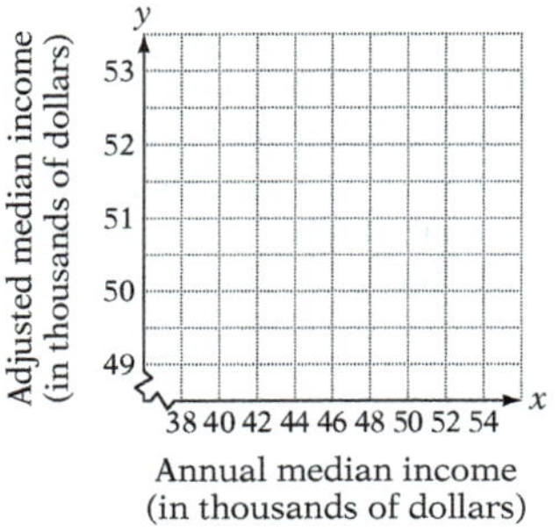

Year	Actual Median Income, x	Inflation-adjusted Median Income, y
2000	42,000	53,000
2002	42,000	51,000
2004	44,000	51,000
2006	48,000	51,000
2008	50,000	50,000

30. **World Cup Soccer** Read the news clipping at the right. The table below shows the number of shots on goal and the number of goals scored for each of the 16 teams that played for a spot in the quarter-final round of the 2010 World Cup soccer tournament. Graph the scatter diagram for these data.

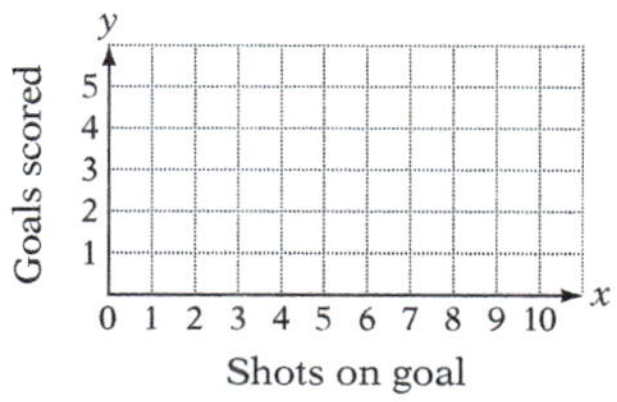

Shots on goal	8	4	6	6	8	6	7	5	9	3	6	2	7	2	8	3
Goals scored	2	1	1	2	4	1	3	1	2	1	3	0	0	0	1	0

In the News

World Cup Quarter-Final Matches Determined

Spain beat Portugal 1–0 to determine the last of the quarter-final match-ups in the 2010 World Cup. Spain's 8 shots on goal yielded just 1 goal, but that was enough to keep the Spaniards ahead of the Portuguese, who took only 3 shots on goal.

Source: www.nprstats.com

3 Average rate of change (See pages 205–207.)

GETTING READY

31. Look at the table of values in Exercise 34, in which you are asked to calculate average rates of change for pairs of data points. When finding an average rate of change between two data points in this table, will you put the values that represent time in the numerator or the denominator?

32. Look at the table of values in Exercise 36. Which of the following fractions is the correct one to use to find the average rate of change per month in the length of a fetus from the fourth month of pregnancy to the eighth month of pregnancy?

(i) $\dfrac{4-8}{16-7}$ (ii) $\dfrac{16-7}{8-4}$ (iii) $\dfrac{8-4}{16-7}$ (iv) $\dfrac{7-16}{14-8}$

33. **Temperature** On September 10 in midstate New Hampshire, the temperature at 6 A.M. was 45°F. At 2 P.M. on the same day, the temperature was 77°F. Find the average rate of change in temperature per hour.

34. **Demography** The table at the right shows the population of Colonial America for each decade from 1610 to 1780. (*Source:* infoplease.com) These figures are estimates, because this period was prior to the establishment of the U.S. Census in 1790.

a. Find the average annual rate of change in the population of the Colonies from 1650 to 1750.

b. Was the average annual rate of change in the population from 1650 to 1750 greater than or less than the average annual rate of change in the population from 1700 to 1750?

c. During which decade was the average annual rate of change the least? What was the average annual rate of change during that decade?

Population of Colonial America

Year	Population
1610	350
1620	2,300
1630	4,600
1640	26,600
1650	50,400
1660	75,100
1670	111,900
1680	151,500
1690	210,400
1700	250,900
1710	331,700
1720	466,200
1730	629,400
1740	905,600
1750	1,170,800
1760	1,593,600
1770	2,148,100
1780	2,780,400

35. **Child Development** The following table shows average weights of fetuses at various stages of prenatal development in humans. (*Source:* Surebaby.com)

Number of weeks after conception	21	24	28	30	38
Weight, in pounds	1	2	3	4	7.4

a. Find the average rate of change per week in the weight of a fetus from 21 weeks after conception to 30 weeks after conception.

b. Find the average rate of change per week in the weight of a fetus from 28 weeks after conception to 38 weeks after conception.

36. Child Development The following table shows average lengths of fetuses for selected stages of prenatal development in humans. (*Source:* Surebaby.com)

a. What is the average rate of change per month in the length of a fetus from the sixth month of pregnancy to the eighth month of pregnancy?

b. Find the average rate of change per month in the length of a fetus from the second month of pregnancy to the fourth month of pregnancy. Is this greater than or less than the average rate of change per month from the sixth month to the eighth month?

Month of pregnancy	2	4	6	8
Length, in inches	1	7	12	16

37. Education Read the news clipping below.

In the News

More Students Opt to Be Cool with Summer School

The advantage of lower costs and the chances to get ahead in their coursework lures an increasing number of students at colleges and universities to choose summer sessions. Here's a look at the summer enrollment numbers for two institutions.

Boston College Summer Enrollment

Year	2000	2005	2009
Number of students enrolled	3742	3873	6137

University of Massachusetts–Amherst Summer Enrollment

Year	2000	2005	2009
Number of students enrolled	6254	7980	9576

Source: The Boston Globe, *June 22, 2010*

a. What was the average annual rate of change in summer enrollment at Boston College between 2000 and 2005? Round to the nearest whole number.

b. What was the average annual rate of change in summer enrollment at Boston College between 2005 and 2009? Round to the nearest whole number.

c. Which was greater at the University of Massachusetts–Amherst: the average annual rate of change in summer enrollment between 2000 and 2005, or the average annual rate of change in summer enrollment between 2005 and 2009?

38. Minimum Wage The following table shows the federal minimum hourly wage for selected years. (*Source:* en.wikipedia.org)

Year	1970	1975	1980	1985	1990	1995	2000	2005	2010
Minimum wage	1.45	2.00	3.10	3.35	3.80	4.25	5.15	5.15	7.25

a. What was the average annual rate of change in the federal minimum wage from 1970 to 1990?

b. Find the average rate of change in the minimum wage from 2000 to 2010. Is this more or less than the average annual rate of change in the minimum wage from 1970 to 1990?

39. **Minimum Wage** The following table shows buying power of a worker earning the federal minimum wage for various years. This is not the *actual* minimum wage but the minimum wage adjusted for inflation. For instance, the actual minimum wage in 1970 was $1.45. Spending $1.45 in 1970 was the same as spending $8.12 in 2010. (*Source:* en.wikipedia.org)

Year	1970	1975	1980	1981	1990	1996	2007	2009
Minimum wage	8.12	8.08	8.18	8.01	6.33	6.59	6.14	7.36

a. What was the average annual rate of change in the inflation-adjusted minimum wage from 1970 to 1990? (*Note:* A negative average rate of change denotes a decrease.)

b. Prior to 1981, in which five-year period did the average annual rate of change in the inflation-adjusted minimum wage increase? What was the average annual rate of change during that period?

40. **Demography** The following table shows the past and projected populations of baby boomers for the years 2000, 2031, and 2046. (*Source:* U.S. Census Bureau)

a. Find the average rate of change per year in the population of the baby boomer generation from 2031 to 2046. (*Note:* A negative average rate of change denotes a decrease in population.)

b. Find the average rate of change per year in the population of baby boomers from 2000 to 2046. Round to the nearest hundred thousand.

Year	2000	2031	2046
Population, in millions	79	51	19

For Exercises 41 and 42, use the following table of data. The table shows the temperature, in degrees Fahrenheit, of a cup of coffee, measured every 10 min over a period of one and one-half hours.

Minutes	0	10	20	30	40	50	60	70	80	90
Temperature, °F	177.3	144.6	122.3	107.0	96.6	89.4	84.6	81.2	78.9	77.4

41. During which 10-minute interval was the average rate of change closest to -1 degree per minute?

42. During which 30-minute interval was the average rate of change closest to -1 degree per minute?

APPLYING CONCEPTS

What is the distance from the given point to the horizontal axis?

43. $(-5, 1)$ **44.** $(3, -4)$ **45.** $(-6, 0)$

What is the distance from the given point to the vertical axis?

46. $(-2, 4)$ **47.** $(1, -3)$ **48.** $(5, 0)$

49. Name the coordinates of a point plotted at the origin of the rectangular coordinate system.

50. A computer screen has a coordinate system that is different from the xy-coordinate system we have discussed. In one mode, the origin of the coordinate system is the top left point of the screen, as shown at the right. Plot the points whose coordinates are (200, 400), (0, 100), and (100, 300).

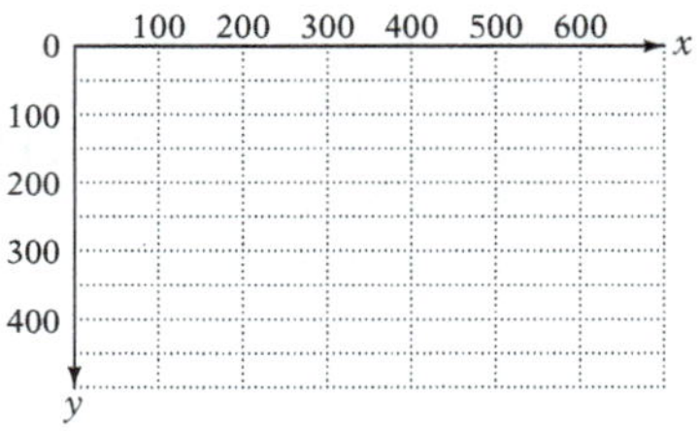

51. Write a paragraph explaining how to plot points in a rectangular coordinate system.

PROJECTS OR GROUP ACTIVITIES

52. Decide on two quantities that may be related, and collect at least ten pairs of values. Here are some examples: height and weight, time studying for a test and test grade, age of a car and its cost. Draw a scatter diagram for the data. Is there any trend? That is, as the values on the horizontal axis increase, do the values on the vertical axis increase or decrease?

53. There is an imaginary coordinate system on Earth that consists of *longitude* and *latitude.* Write a report on how location is determined on the surface of Earth.

5.2 Graphs of Straight Lines

OBJECTIVE 1 Determine solutions of linear equations in two variables

An equation of the form $y = mx + b$, where m is the coefficient of x and b is a constant, is a **linear equation in two variables.** Examples of linear equations follow.

$$y = 3x + 4 \qquad (m = 3, b = 4)$$
$$y = 2x - 3 \qquad (m = 2, b = -3)$$
$$y = -\frac{2}{3}x + 1 \qquad \left(m = -\frac{2}{3}, b = 1\right)$$
$$y = -2x \qquad (m = -2, b = 0)$$
$$y = x + 2 \qquad (m = 1, b = 2)$$

In a linear equation, the exponent of each variable is 1. The equations $y = 2x^2 - 1$ and $y = \frac{1}{x}$ are not linear equations.

A **solution of a linear equation in two variables** is an ordered pair of numbers (x, y) that makes the equation a true statement.

Focus on determining whether an ordered pair is a solution of an equation in two variables

> **Take Note**
> An ordered pair is of the form (x, y). For the ordered pair $(1, -2)$, 1 is the x value and -2 is the y value. Substitute 1 for x and -2 for y.

Is $(1, -2)$ a solution of $y = 3x - 5$?

Replace x with 1. Replace y with -2.

$$\begin{array}{c|l} y & 3x - 5 \\ \hline -2 & 3(1) - 5 \\ & 3 - 5 \end{array}$$

$$-2 = -2$$

Compare the results. If the results are equal, the given ordered pair is a solution. If the results are not equal, the given ordered pair is not a solution.

Yes, $(1, -2)$ is a solution of the equation $y = 3x - 5$.

Besides the ordered pair $(1, -2)$, there are many other ordered-pair solutions of the equation $y = 3x - 5$. For example, the method used above can be used to show that $(2, 1)$, $(-1, -8)$, $(\frac{2}{3}, -3)$, and $(0, -5)$ are also solutions.

EXAMPLE 1 Is $(-3, 2)$ a solution of $y = 2x + 2$?

Solution

$$\begin{array}{c|l} y & 2x + 2 \\ \hline 2 & 2(-3) + 2 \\ & -6 + 2 \\ & -4 \end{array}$$

• Replace x with -3 and y with 2.

$$2 \neq -4$$

No, $(-3, 2)$ is not a solution of $y = 2x + 2$.

Problem 1 Is $(2, -4)$ a solution of $y = -\frac{1}{2}x - 3$?

Solution See page S11.

Try Exercise 17, page 223.

In general, a linear equation in two variables has an infinite number of ordered-pair solutions. By choosing any value for x and substituting that value into the linear equation, we can find a corresponding value of y.

EXAMPLE 2 Find the ordered-pair solution of $y = \frac{2}{3}x - 1$ that corresponds to $x = 3$.

Solution

$y = \frac{2}{3}x - 1$

$y = \frac{2}{3}(3) - 1$ • Substitute 3 for x.

$y = 2 - 1$ • Solve for y.

$y = 1$ • When $x = 3$, $y = 1$.

The ordered-pair solution is $(3, 1)$.

Problem 2 Find the ordered-pair solution of $y = -\frac{1}{4}x + 1$ that corresponds to $x = 4$.

Solution See page S11.

Try Exercise 29, page 224.

OBJECTIVE 2 Graph equations of the form $y = mx + b$

The **graph of an equation in two variables** is a drawing of the ordered-pair solutions of the equation. For a linear equation in two variables, the graph is a straight line.

To graph a linear equation, find ordered-pair solutions of the equation. Do this by choosing any value of x and finding the corresponding value of y. Repeat this procedure, choosing different values for x, until you have found the number of solutions desired.

Because the graph of a linear equation in two variables is a straight line, and a straight line is determined by two points, it is necessary to find only two solutions. However, finding at least three points will ensure accuracy.

Focus on graphing an equation of the form $y = mx + b$

Graph $y = 2x + 1$.

Choose any values of x, and then find the corresponding values of y. The numbers 0, 2, and -1 were chosen arbitrarily for x. It is convenient to record the solutions in a table.

x	$y = 2x + 1$	y
0	$2(0) + 1$	1
2	$2(2) + 1$	5
-1	$2(-1) + 1$	-1

Take Note
If the three points you graph do not lie on a straight line, you have made an arithmetic error in calculating a point or you have plotted a point incorrectly.

Graph the ordered-pair solutions (0, 1), (2, 5), and $(-1, -1)$. Draw a line through the ordered-pair solutions.

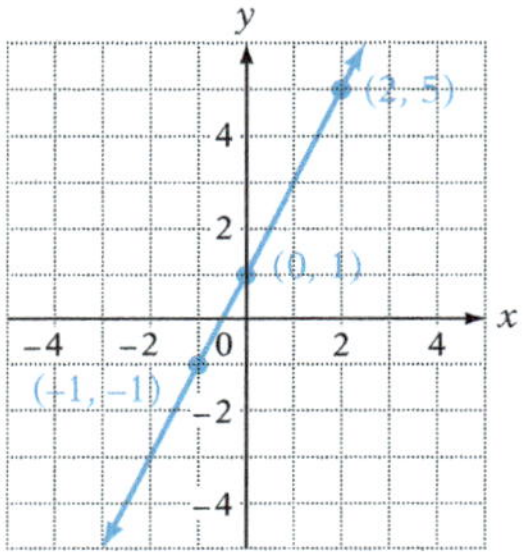

Note that the points whose coordinates are $(-2, -3)$ and $(1, 3)$ are on the graph and that these ordered pairs are solutions of the equation $y = 2x + 1$.

Remember that a graph is a drawing of the ordered-pair solutions of an equation. Therefore, every point on the graph is a solution of the equation, and every solution of the equation is a point on the graph.

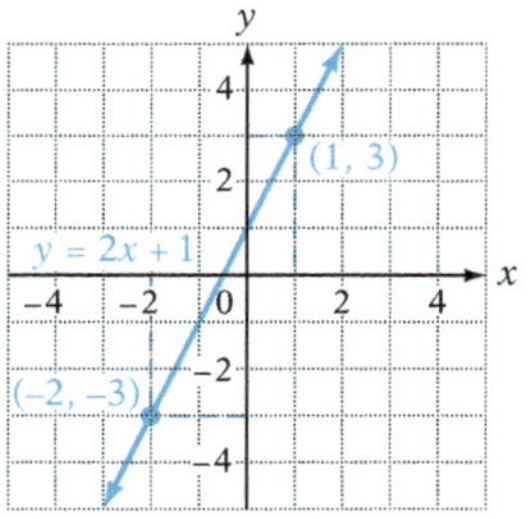

EXAMPLE 3 Graph: $y = 3x - 2$

Solution

x	$y = 3x - 2$
0	-2
2	4
-1	-5

- **Choose three values for x. Find the corresponding values of y.**

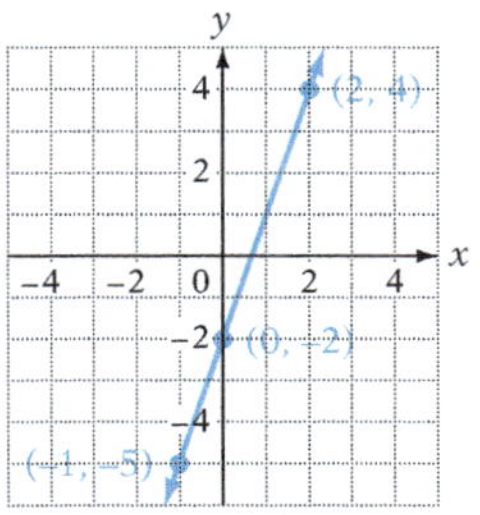

- **Graph the ordered-pair solutions. Draw a straight line through the points.**

Problem 3 Graph: $y = 3x + 1$

Solution See page S11.

Try Exercise 45, page 225.

> *Graphing calculators create graphs by plotting points and then connecting the points to form a curve. Using a graphing calculator, enter the equation $y = 3x - 2$ and verify the graph drawn in Example 3. (Refer to the Appendix for instructions on using a graphing calculator to graph a linear equation.) Trace along the graph and verify that $(0, -2)$, $(2, 4)$, and $(-1, -5)$ are coordinates of points on the graph. Now enter the equation $y = 3x + 1$ given in Problem 3. Verify that the ordered pairs you found for this equation are the coordinates of points on the graph.*

When m is a fraction in the equation $y = mx + b$, choose values of x that will simplify the evaluation. For example, to graph $y = \frac{1}{3}x - 1$, we might choose the numbers 0, 3, and -3 for x. Note that these numbers are multiples of the denominator of $\frac{1}{3}$, the coefficient of x.

x	$y = \frac{1}{3}x - 1$
0	-1
3	0
-3	-2

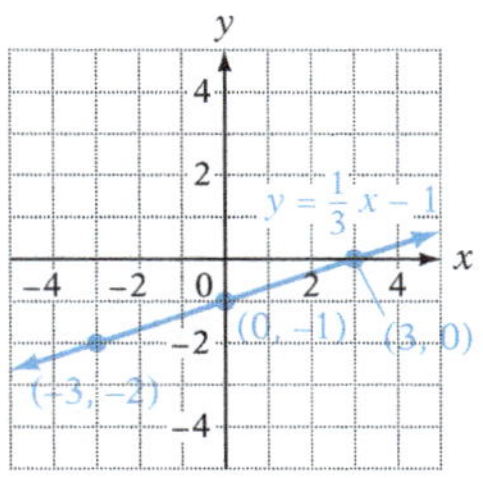

EXAMPLE 4 Graph: $y = \frac{1}{2}x - 1$

Solution

x	$y = \frac{1}{2}x - 1$
0	-1
2	0
-2	-2

- **The value of m is $\frac{1}{2}$. Choose three values for x that are multiples of the denominator 2. Find the corresponding values of y.**

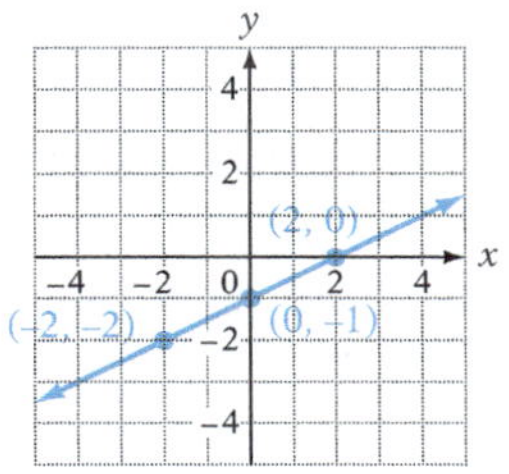

- **Graph the ordered-pair solutions. Draw a straight line through the points.**

Problem 4 Graph: $y = \frac{1}{3}x - 3$

Solution See page S11.

Try Exercise 41, page 224.

Using a graphing calculator, enter the equation $y = \frac{1}{2}x - 1$ and verify the graph drawn in Example 4. Trace along the graph and verify that (0, −1), (2, 0), and (−2, −2) are the coordinates of points on the graph. Follow the same procedure for Problem 4. (See the Appendix for instructions on entering a fractional coefficient of x.)

OBJECTIVE 3 Graph equations of the form $Ax + By = C$

An equation in the form $Ax + By = C$, where A and B are coefficients and C is a constant, is also a linear equation. Examples of these equations are shown below.

$2x + 3y = 6$	$(A = 2, B = 3, C = 6)$
$x - 2y = -4$	$(A = 1, B = -2, C = -4)$
$2x + y = 0$	$(A = 2, B = 1, C = 0)$
$4x - 5y = 2$	$(A = 4, B = -5, C = 2)$

One method of graphing an equation of the form $Ax + By = C$ involves first solving the equation for y and then following the same procedure used to graph an equation of the form $y = mx + b$. Solving the equation for y means rewriting the equation so that y is alone on one side of the equation, and the term containing x and the constant are on the other side of the equation. The Addition and Multiplication Properties of Equations are used to rewrite an equation of the form $Ax + By = C$ in the form $y = mx + b$.

Focus on solving an equation of the form $Ax + By = C$ for y

Solve the equation $3x + 2y = 4$ for y.

The equation is in the form $Ax + By = C$.

$$3x + 2y = 4$$

Use the Addition Property of Equations to subtract the term $3x$ from each side of the equation.

$$3x - 3x + 2y = -3x + 4$$

Simplify. Note that on the right side of the equation, the term containing x is first, followed by the constant.

$$2y = -3x + 4$$

Use the Multiplication Property of Equations to multiply each side of the equation by the reciprocal of the coefficient of y. (The coefficient of y is 2; the reciprocal of 2 is $\frac{1}{2}$.)

$$\frac{1}{2} \cdot 2y = \frac{1}{2}(-3x + 4)$$

Simplify. Use the Distributive Property on the right side of the equation.

$$y = \frac{1}{2}(-3x) + \frac{1}{2}(4)$$

The equation is now in the form $y = mx + b$, with $m = -\frac{3}{2}$ and $b = 2$.

$$y = -\frac{3}{2}x + 2$$

In solving the equation $3x + 2y = 4$ for y, when we multiplied both sides of the equation by $\frac{1}{2}$, we could have divided both sides of the equation by 2, as shown at the right. In simplifying the right side after dividing both sides by 2, be sure to divide *each term* by 2.

$$2y = -3x + 4$$
$$\frac{2y}{2} = \frac{-3x + 4}{2}$$
$$y = \frac{-3x}{2} + \frac{4}{2}$$
$$y = -\frac{3}{2}x + 2$$

Being able to solve an equation of the form $Ax + By = C$ for y is important because graphing calculators require that an equation be in the form $y = mx + b$ when the equation is entered for graphing.

EXAMPLE 5 Write $3x - 4y = 12$ in the form $y = mx + b$.

Solution

$$3x - 4y = 12$$
$$3x - 3x - 4y = -3x + 12$$ • **Subtract 3x from each side of the equation.**
$$-4y = -3x + 12$$ • **Simplify.**
$$\frac{-4y}{-4} = \frac{-3x + 12}{-4}$$ • **Divide each side of the equation by −4.**
$$y = \frac{-3x}{-4} + \frac{12}{-4}$$
$$y = \frac{3}{4}x - 3$$

Problem 5 Write $5x - 2y = 10$ in the form $y = mx + b$.

Solution See page S11.

Try Exercise 69, page 226.

To graph an equation of the form $Ax + By = C$, we can first solve the equation for y and then follow the same procedure used to graph an equation of the form $y = mx + b$. An example follows.

Focus on graphing an equation of the form $Ax + By = C$

Graph: $3x + 4y = 12$

Solve the equation for y.

$$3x + 4y = 12$$
$$4y = -3x + 12$$
$$y = -\frac{3}{4}x + 3$$

The value of m is $-\frac{3}{4}$.

Choose three values for x that are multiples of the denominator 4. Find the corresponding values of y.

x	$y = -\frac{3}{4}x + 3$
0	3
4	0
−4	6

Graph the ordered-pair solutions. Draw a straight line through the points.

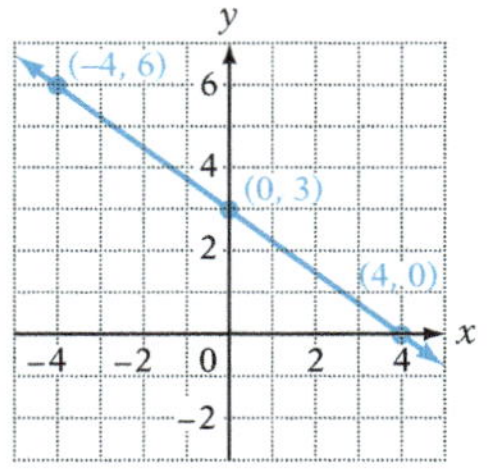

EXAMPLE 6 Graph. **A.** $2x - 5y = 10$ **B.** $x + 2y = 6$

Solution **A.** $2x - 5y = 10$

$$-5y = -2x + 10$$
$$y = \frac{2}{5}x - 2$$

• Solve the equation for y.

x	$y = \frac{2}{5}x - 2$
0	−2
5	0
−5	−4

• The value of m is $\frac{2}{5}$. Choose three values for x that are multiples of the denominator 5. Find the corresponding values of y.

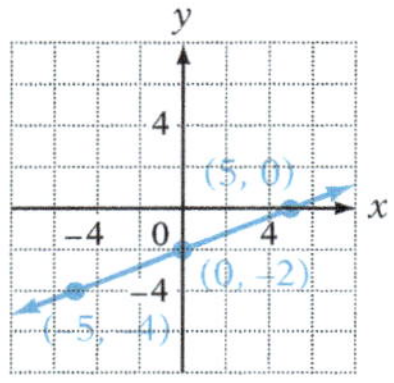

• Graph the ordered-pair solutions. Draw a straight line through the points.

B. $x + 2y = 6$

$$2y = -x + 6$$
$$y = -\frac{1}{2}x + 3$$

• Solve the equation for y.

x	$y = -\frac{1}{2}x + 3$
0	3
−2	4
4	1

• The value of m is $-\frac{1}{2}$. Choose three values for x that are multiples of the denominator 2. Find the corresponding values of y.

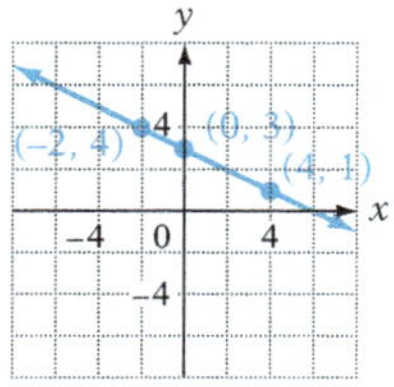

• Graph the ordered-pair solutions. Draw a straight line through the points.

Problem 6 Graph. **A.** $5x - 2y = 10$ **B.** $x - 3y = 9$

Solution See pages S11–S12.

Try Exercise 77, page 226.

The graph of the equation $2x + 3y = 6$ is shown at the right. The graph crosses the x-axis at $(3, 0)$. This point is called the ***x*-intercept.** The graph crosses the y-axis at $(0, 2)$. This point is called the ***y*-intercept.**

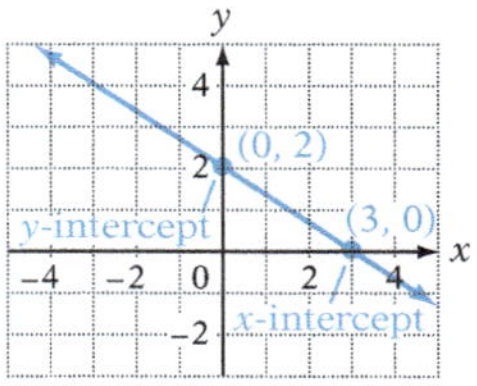

Take Note

To find the x-intercept, let $y = 0$.
To find the y-intercept, let $x = 0$.

Focus on finding the x- and y-intercepts of an equation of the form $Ax + By = C$

Find the x-intercept and the y-intercept of the graph of the equation $2x + 3y = 6$ algebraically.

To find the x-intercept, let $y = 0$.
(Any point on the x-axis has y-coordinate 0.)

The x-intercept is $(3, 0)$.

$$2x + 3y = 6$$
$$2x + 3(0) = 6$$
$$2x = 6$$
$$x = 3$$

To find the y-intercept, let $x = 0$.
(Any point on the y-axis has x-coordinate 0.)

The y-intercept is $(0, 2)$.

$$2x + 3y = 6$$
$$2(0) + 3y = 6$$
$$3y = 6$$
$$y = 2$$

Another method of graphing some equations of the form $Ax + By = C$ is to find the x- and y-intercepts, plot both intercepts, and then draw a line through the two points.

EXAMPLE 7 Find the x- and y-intercepts for $x - 2y = 4$. Graph the line.

Solution x-intercept:

$$x - 2y = 4$$
$$x - 2(0) = 4 \quad \text{• To find the } x\text{-intercept, let } y = 0.$$
$$x = 4$$

The x-intercept is $(4, 0)$.

y-intercept:

$$x - 2y = 4$$
$$0 - 2y = 4 \quad \text{• To find the } y\text{-intercept, let } x = 0.$$
$$-2y = 4$$
$$y = -2$$

The y-intercept is $(0, -2)$.

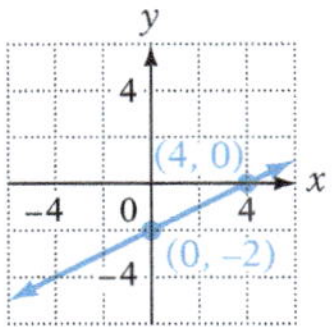

• **Graph the ordered pairs (4, 0) and (0, −2). Draw a straight line through the points.**

Problem 7 Find the x- and y-intercepts for $4x - y = 4$. Graph the line.

Solution See page S12.

Try Exercise 103, page 226.

The graph of an equation in which one of the variables is missing is either a horizontal line or a vertical line.

The equation $y = 2$ could be written $0x + y = 2$. No matter what value of x is chosen, y is always 2. Some solutions of the equation are $(3, 2)$, $(-1, 2)$, $(0, 2)$, and $(-4, 2)$. The graph is shown at the right.

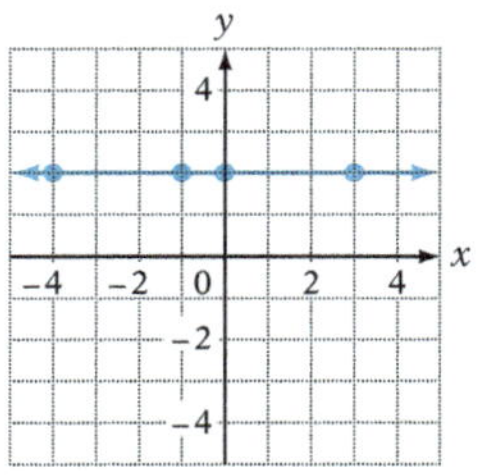

The **graph of $y = b$** is a horizontal line passing through the point whose coordinates are $(0, b)$.

Note that $(0, b)$ is the y-intercept of the graph of $y = b$. An equation of the form $y = b$ does not have an x-intercept.

The equation $x = -2$ could be written $x + 0y = -2$. No matter what value of y is chosen, x is always -2. Some solutions of the equation are $(-2, 3)$, $(-2, -2)$, $(-2, 0)$, and $(-2, 2)$. The graph is shown at the right.

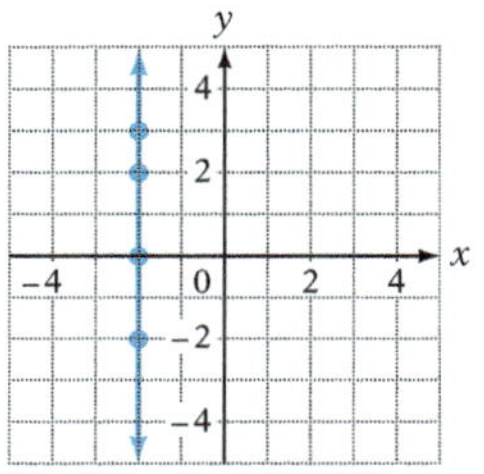

The **graph of $x = a$** is a vertical line passing through the point whose coordinates are $(a, 0)$.

Note that $(a, 0)$ is the x-intercept of the graph of $x = a$. An equation of the form $x = a$ does not have a y-intercept.

EXAMPLE 8 Graph. **A.** $y = -2$ **B.** $x = 3$

Solution

A. The graph of an equation of the form $y = b$ is a horizontal line with y-intercept $(0, b)$.

The graph of the equation $y = -2$ is a horizontal line through $(0, -2)$.

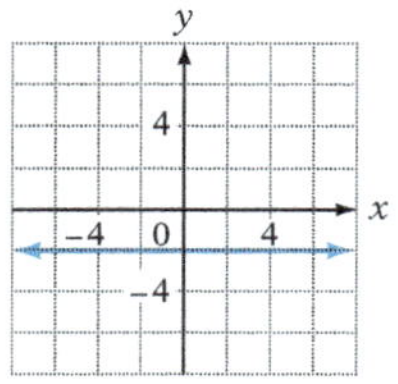

B. The graph of an equation of the form $x = a$ is a vertical line with x-intercept $(a, 0)$.

The graph of the equation $x = 3$ is a vertical line through $(3, 0)$.

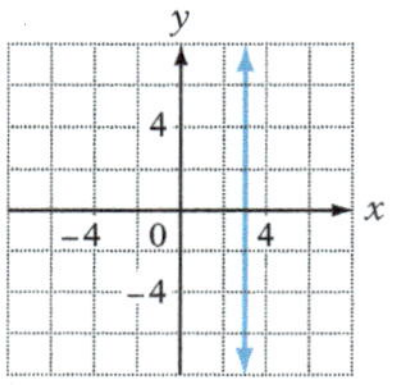

Problem 8 Graph. **A.** $y = 3$ **B.** $x = -4$

Solution See page S12.

Try Exercise 87, page 226.

5.2 Exercises

CONCEPT CHECK

1. Which of the following equations are linear equations in two variables?

(i) $y = -2x + 7$ (ii) $x - 3y = 5$ (iii) $y = -x^2 + 4$ (iv) $y^2 = x - 6$

2. Give the value of m and the value of b in each equation.

a. $y = 5x + 3$ **b.** $y = -\frac{1}{2}x - 8$ **c.** $y = x + 1$ **d.** $y = -x$

3. State whether the graph of the equation is a straight line.
a. $y = x^2 + 1$ **b.** $y = -x$ **c.** $y = \frac{1}{x}$ **d.** $y = 2 - \frac{1}{2}x$ **e.** $y = \sqrt{x} - 1$

4. Name values of x that you would choose to find integer solutions of the equation.
a. $y = \frac{3}{4}x + 2$ **b.** $y = -\frac{2}{5}x - 1$

5. Is the equation in the form $y = mx + b$, the form $Ax + By = C$, or neither?
a. $6x - 3y = 6$ **b.** $y = x - 1$ **c.** $8 - 4y = x$ **d.** $5x + 4y = 4$

6. Which coordinate of an x-intercept is 0?

7. Which coordinate of a y-intercept is 0?

8. Describe the graph of a line that has an x-intercept but no y-intercept.

9. Describe the graph of a line that has a y-intercept but no x-intercept.

10. Name the y-intercept of the graph shown at the right.

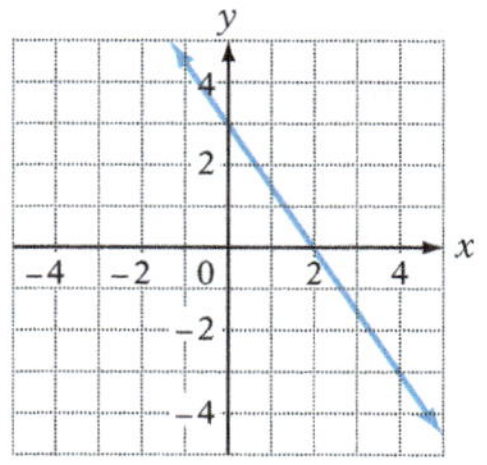

1 Determine solutions of linear equations in two variables (See pages 214–215.)

GETTING READY

11. To decide whether the ordered pair (1, 7) is a solution of the equation $y = 2x + 5$, substitute 1 for __?__ and 7 for __?__ to see whether the ordered pair (1, 7) makes the equation $y = 2x + 5$ a true statement.

12. Find the ordered-pair solution of $y = -\frac{1}{4}x + 3$ that corresponds to $x = -8$.

$y = -\frac{1}{4}x + 3$

$y = -\frac{1}{4}(-8) + 3$ • **Substitute __?__ for __?__.**

$y =$ __?__ $+ 3$ • **Multiply.**

$y =$ __?__ • **Add.**

The ordered-pair solution that corresponds to $x = -8$ is (__?__, __?__).

13. Is (3, 4) a solution of $y = -x + 7$?

14. Is (2, −3) a solution of $y = x + 5$?

15. Is (−1, 2) a solution of $y = \frac{1}{2}x - 1$?

16. Is (1, −3) a solution of $y = -2x - 1$?

17. Is (4, 1) a solution of $y = \frac{1}{4}x + 1$?

18. Is (−5, 3) a solution of $y = -\frac{2}{5}x + 1$?

19. Is (0, 4) a solution of $y = \frac{3}{4}x + 4$?

20. Is (−2, 0) a solution of $y = -\frac{1}{2}x - 1$?

21. Is (0, 0) a solution of $y = 3x + 2$?

22. Is (0, 0) a solution of $y = -\frac{3}{4}x$?

23. Find the ordered-pair solution of $y = 3x - 2$ that corresponds to $x = 3$.

24. Find the ordered-pair solution of $y = 4x + 1$ that corresponds to $x = -1$.

25. Find the ordered-pair solution of $y = \frac{2}{3}x - 1$ that corresponds to $x = 6$.

26. Find the ordered-pair solution of $y = \frac{3}{4}x - 2$ that corresponds to $x = 4$.

27. Find the ordered-pair solution of $y = -3x + 1$ that corresponds to $x = 0$.

28. Find the ordered-pair solution of $y = \frac{2}{5}x - 5$ that corresponds to $x = 0$.

29. Find the ordered-pair solution of $y = \frac{2}{5}x + 2$ that corresponds to $x = -5$.

30. Find the ordered-pair solution of $y = -\frac{1}{6}x - 2$ that corresponds to $x = 12$.

31. Use the linear equation $y = -3x + 6$.
a. For what value of x is the ordered pair $(x, 0)$ a solution of the linear equation $y = -3x + 6$?
b. Suppose (x, a) is a solution of the linear equation $y = -3x + 6$, such that $x > 2$. Is a a positive or a negative number?

32. Let $y = mx + b$ be a linear equation in which $b = -2m$. For what value of x will the ordered pair $(x, 0)$ be a solution of the equation?

2 Graph equations of the form $y = mx + b$ (See pages 216–218.)

GETTING READY

33. Find three points on the graph of $y = 3x - 4$ by finding the y values that correspond to x values of -1, 0, and 1.
a. When $x = -1$, $y =$ __?__. A point on the graph is (__?__, __?__).
b. When $x = 0$, $y =$ __?__. A point on the graph is (__?__, __?__).
c. When $x = 1$, $y =$ __?__. A point on the graph is (__?__, __?__).

34. To find points on the graph of $y = \frac{1}{5}x - 8$, it is helpful to choose x values that are divisible by __?__.

Graph.

35. $y = 2x - 3$

36. $y = -2x + 2$

37. $y = \frac{1}{3}x$

38. $y = -3x$

39. $y = \frac{2}{3}x - 1$

40. $y = \frac{3}{4}x + 2$

41. $y = -\frac{1}{4}x + 2$

42. $y = -\frac{1}{3}x + 1$

43. $y = -\frac{2}{5}x + 1$

44. $y = -\frac{1}{2}x + 3$

45. $y = 2x - 4$

46. $y = 3x - 4$

47. $y = -x + 2$

48. $y = -x - 1$

49. $y = -\frac{2}{3}x + 1$

50. $y = 5x - 4$

51. $y = -3x + 2$

52. $y = -x + 3$

Graph using a graphing calculator.

53. $y = 3x - 4$

54. $y = 2x - 3$

55. $y = -2x + 3$

56. $y = -2x - 3$

57. $y = \frac{3}{4}x + 2$

58. $y = \frac{2}{3}x - 4$

59. $y = -\frac{3}{2}x - 3$

60. $y = -\frac{2}{5}x + 2$

3 Graph equations of the form *Ax* + *By* = *C* (See pages 218–222.)

GETTING READY

61. Write the equation $x - 2y = -10$ in the form $y = mx + b$.

$x - 2y = -10$

$x - x - 2y = -x - 10$ • **Subtract __?__ from each side of the equation.**

$\underline{\quad ? \quad} = -x - 10$ • **Simplify the left side of the equation.**

$\frac{-2y}{-2} = \frac{-x - 10}{-2}$ • **Divide each side of the equation by __?__.**

$y = \underline{\quad ? \quad}\,x + \underline{\quad ? \quad}$ • **Simplify.**

62. a. The x-intercept of the graph of a linear equation is the point where the graph crosses the __?__. Its y-coordinate is __?__.

b. The y-intercept of the graph of a linear equation is the point where the graph crosses the __?__. Its x-coordinate is __?__.

Complete Exercises 63 and 64 by using the word *horizontal* or *vertical*.

63. The graph of the equation $y = 7$ is a ___?___ line passing through the point whose coordinates are (0, 7).

64. The graph of the equation $x = -10$ is a ___?___ line passing through the point whose coordinates are (−10, 0).

Write the equation in the form $y = mx + b$.

65. $3x + y = 10$
66. $2x + y = 5$
67. $4x - y = 3$
68. $5x - y = 7$
69. $3x + 2y = 6$
70. $2x + 3y = 9$
71. $2x - 5y = 10$
72. $5x - 2y = 4$
73. $2x + 7y = 14$
74. $6x - 5y = 10$
75. $x + 3y = 6$
76. $x - 4y = 12$

Graph.

77. $3x + y = 3$
78. $2x + y = 4$
79. $2x + 3y = 6$
80. $3x + 2y = 4$
81. $x - 2y = 4$
82. $x - 3y = 6$
83. $2x - 3y = 6$
84. $3x - 2y = 8$
85. $y = 4$
86. $y = -4$
87. $x = -2$
88. $x = 3$

Find the x- and y-intercepts.

89. $x - y = 3$
90. $3x + 4y = 12$
91. $y = 2x - 6$
92. $y = 2x + 10$
93. $x - 5y = 10$
94. $3x + 2y = 12$
95. $y = 3x + 12$
96. $y = 5x + 10$
97. $2x - 3y = 0$
98. $3x + 4y = 0$
99. $y = \frac{1}{2}x + 3$
100. $y = \frac{2}{3}x - 4$

Graph by using x- and y-intercepts.

101. $5x + 2y = 10$
102. $x - 3y = 6$
103. $3x - 4y = 12$

104. $2x - 3y = -6$

105. $2x + 3y = 6$

106. $x + 2y = 4$

107. $2x + 5y = 10$

108. $3x + 4y = 12$

109. $x - 3y = 6$

110. $3x - y = 6$

111. $x - 4y = 8$

112. $4x + 3y = 12$

Graph using a graphing calculator. Verify that the graph has the correct x- and y-intercepts.

113. $3x + 4y = -12$

114. $3x - 2y = -6$

115. $2x - y = 4$

116. Suppose A and B are positive and C is negative. Is the y-intercept of the graph of $Ax + By = C$ above or below the x-axis?

117. Suppose A and C are negative and B is positive. Is the x-intercept of the graph of $Ax + By = C$ to the left or to the right of the y-axis?

APPLYING CONCEPTS

118. **a.** Show that the equation $y + 3 = 2(x + 4)$ is a linear equation by writing it in the form $y = mx + b$.
b. Find the ordered-pair solution that corresponds to $x = -4$.

119. **a.** Show that the equation $y + 4 = -\frac{1}{2}(x + 2)$ is a linear equation by writing it in the form $y = mx + b$.
b. Find the ordered-pair solution that corresponds to $x = -2$.

120. For the linear equation $y = 2x - 3$, what is the increase in y when x is increased by 1?

121. For the linear equation $y = -x - 4$, what is the decrease in y when x is increased by 1?

122. Write the equation of a line that has $(0, 0)$ as both the x-intercept and the y-intercept.

123. Explain **a.** why the y-coordinate of any point on the x-axis is 0 and **b.** why the x-coordinate of any point on the y-axis is 0.

PROJECTS OR GROUP ACTIVITIES

A graphing calculator can be used to graph a linear equation. Here are the keystrokes to graph $y = \frac{2}{3}x + 1$. First the equation is entered. Then the domain (Xmin to Xmax) and the range (Ymin to Ymax) are entered. This is called the **viewing window.** Xmin and Xmax are the smallest and largest values of x that will be shown on the screen. Ymin and Ymax are the smallest and largest values of y that will be shown on the screen.

Y= CLEAR 2 X,T,θ,n ÷ 3 + 1 WINDOW (–) 10 ENTER 10 ENTER 1 ENTER (–) 10 ENTER 10 ENTER 1 ENTER GRAPH

By changing the keystrokes 2 X,T,θ,n ÷ 3 + 1, you can graph different equations.

Graph the following equations.

124. $y = 2x + 1$ For $2x$, you may enter $2 \times x$ or just $2x$. The times sign $\times$ is not necessary on many graphing calculators.

125. $y = -\frac{1}{2}x - 2$ Use the (–) key to enter a negative sign.

126. $3x + 2y = 6$ Solve for y. Then enter the equation.

127. $4x + 3y = 75$ You must adjust the viewing window. *Suggestion:* Xmin $= -25$, Xmax $= 25$, Xscl $= 5$; Ymin $= -35$, Ymax $= 35$, Yscl $= 5$. See the Appendix for assistance.

5.3 Slopes of Straight Lines

OBJECTIVE 1 Find the slope of a straight line

The graphs of $y = \frac{2}{3}x + 1$ and $y = 2x + 1$ are shown at the top of the next page. Each graph crosses the y-axis at $(0, 1)$, but the graphs have different slants. The **slope** of a line is a measure of the slant of the line. The symbol for slope is m.

The slope of a line is the ratio of the change in the y-coordinates between any two points on the line to the change in the x-coordinates.

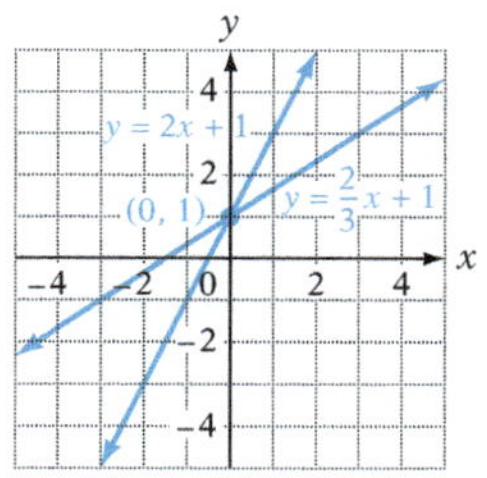

The line containing the points whose coordinates are $(-2, -3)$ and $(6, 1)$ is graphed at the right. The change in y is the difference between the two y-coordinates.

Change in $y = 1 - (-3) = 4$

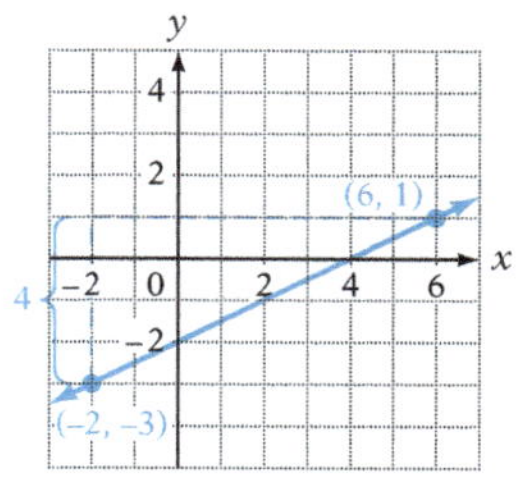

The change in x is the difference between the two x-coordinates.

Change in $x = 6 - (-2) = 8$

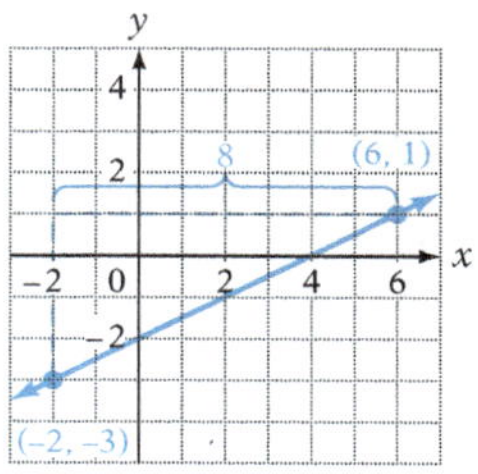

Take Note

The change in the y values can be thought of as the *rise* of the line, and the change in the x values can be thought of as the *run*. Then slope $= m = \frac{\text{rise}}{\text{run}}$.

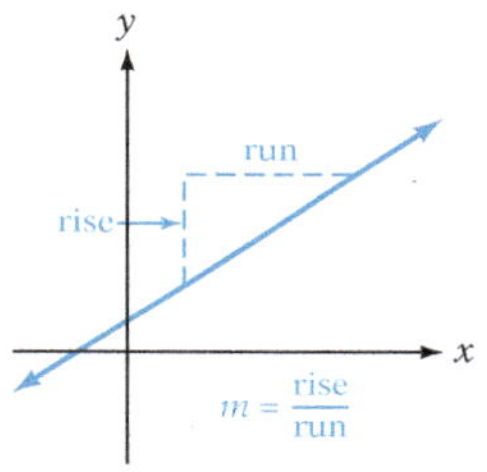

For the line containing the points whose coordinates are $(-2, -3)$ and $(6, 1)$,

$$\text{Slope} = m = \frac{\text{change in } y}{\text{change in } x} = \frac{4}{8} = \frac{1}{2}$$

The slope of the line can also be described as the ratio of the vertical change (4 units) to the horizontal change (8 units) from the point whose coordinates are $(-2, -3)$ to the point whose coordinates are $(6, 1)$.

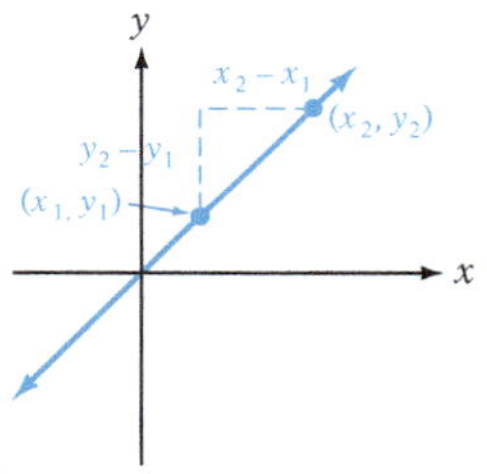

SLOPE FORMULA

The slope of a line containing two points P_1 and P_2, whose coordinates are (x_1, y_1) and (x_2, y_2), is given by

$$\textbf{Slope} = m = \frac{y_2 - y_1}{x_2 - x_1}, \; x_1 \neq x_2$$

In the slope formula, the points P_1 and P_2 are any two points on the line. The slope of a line is constant; therefore, the slope calculated using any two points on the line will be the same.

In Section 1, Objective 3 of this chapter, we calculated the average rate of change in the U.S. population for different intervals of time. In each case, the average rate of change was different. Note that the graph of the population data did not lie on a straight line.

The graph of an equation of the form $y = mx + b$ is a straight line, and the average rate of change is constant. We can calculate the average rate of change between any two points on the line and it will always be the same. For a linear equation, the rate of change is referred to as slope.

Focus on finding the slope of a line

A. Find the slope of the line containing the points whose coordinates are $(-1, 1)$ and $(2, 3)$.

Let $P_1 = (-1, 1)$ and $P_2 = (2, 3)$. Then $x_1 = -1$, $y_1 = 1$, $x_2 = 2$, and $y_2 = 3$.

$$m = \frac{y_2 - y_1}{x_2 - x_1} = \frac{3 - 1}{2 - (-1)} = \frac{2}{3}$$

It does not matter which point is named P_1 and which P_2; the slope of the line will be the same. If the points are reversed, then $P_1 = (2, 3)$ and $P_2 = (-1, 1)$.

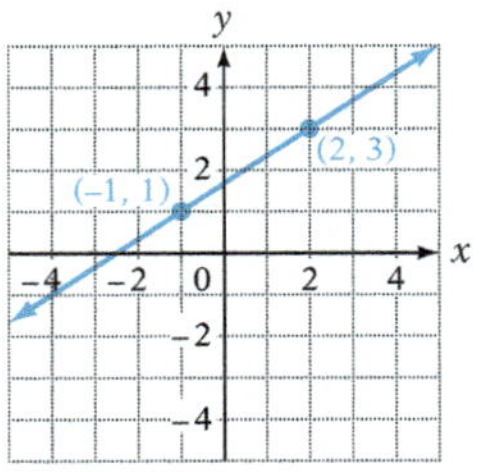

Positive slope

Take Note

Positive slope means that the value of *y* increases as the value of *x* increases.

$$m = \frac{y_2 - y_1}{x_2 - x_1} = \frac{1 - 3}{-1 - 2} = \frac{-2}{-3} = \frac{2}{3}$$

This is the same result. Here the slope is a positive number. A line that slants upward to the right has a **positive slope.**

B. Find the slope of the line containing the points whose coordinates are $(-3, 4)$ and $(2, -2)$.

Let $P_1 = (-3, 4)$ and $P_2 = (2, -2)$.

$$m = \frac{y_2 - y_1}{x_2 - x_1} = \frac{-2 - 4}{2 - (-3)} = \frac{-6}{5} = -\frac{6}{5}$$

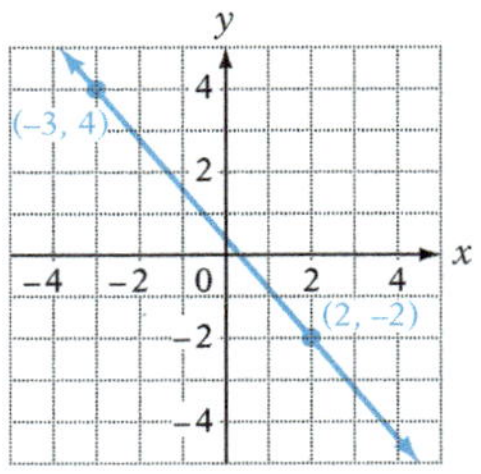

Negative slope

Take Note

Negative slope means that the value of *y* decreases as the value of *x* increases. Compare this to positive slope.

Here the slope is a negative number. A line that slants downward to the right has a **negative slope.**

C. Find the slope of the line containing the points whose coordinates are $(-1, 3)$ and $(4, 3)$.

Let $P_1 = (-1, 3)$ and $P_2 = (4, 3)$.

$$m = \frac{y_2 - y_1}{x_2 - x_1} = \frac{3 - 3}{4 - (-1)} = \frac{0}{5} = 0$$

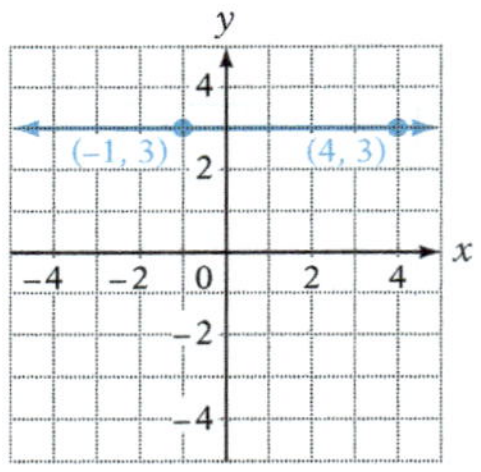

Zero slope

When $y_1 = y_2$, the graph is a horizontal line. A horizontal line has **zero slope.**

D. Find the slope of the line containing the points whose coordinates are $(2, -2)$ and $(2, 4)$.

Let $P_1 = (2, -2)$ and $P_2 = (2, 4)$.

$$m = \frac{y_2 - y_1}{x_2 - x_1} = \frac{4 - (-2)}{2 - 2} = \frac{6}{0} \leftarrow \text{Not a real number}$$

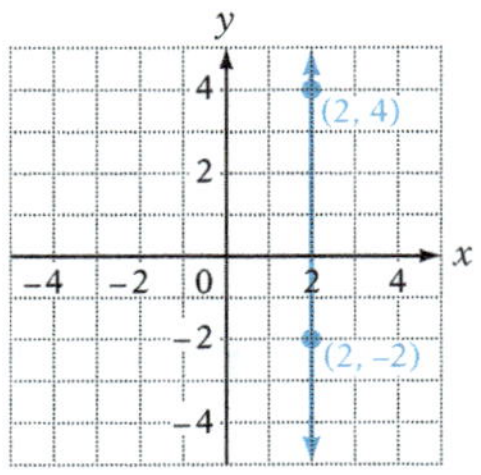

Undefined slope

When $x_1 = x_2$, the denominator of $\frac{y_2 - y_1}{x_2 - x_1}$ is 0 and the graph is a vertical line. Because division by zero is not defined, the slope of the line is not defined. The slope of a vertical line is **undefined.**

EXAMPLE 1 Find the slope of the line containing the points P_1 and P_2.

A. $P_1(-2, -1), P_2(3, 4)$ **B.** $P_1(-3, 1), P_2(2, -2)$
C. $P_1(-1, 4), P_2(-1, 0)$ **D.** $P_1(-1, 2), P_2(4, 2)$

Solution

A. $m = \frac{y_2 - y_1}{x_2 - x_1} = \frac{4 - (-1)}{3 - (-2)} = \frac{5}{5} = 1$

The slope is 1.

B. $m = \frac{y_2 - y_1}{x_2 - x_1} = \frac{-2 - 1}{2 - (-3)} = \frac{-3}{5}$

The slope is $-\frac{3}{5}$.

C. $m = \frac{y_2 - y_1}{x_2 - x_1} = \frac{0 - 4}{-1 - (-1)} = \frac{-4}{0}$

The slope is undefined.

D. $m = \frac{y_2 - y_1}{x_2 - x_1} = \frac{2 - 2}{4 - (-1)} = \frac{0}{5} = 0$

The slope is 0.

Problem 1 Find the slope of the line containing the points P_1 and P_2.

A. $P_1(-1, 2), P_2(1, 3)$ **B.** $P_1(1, 2), P_2(4, -5)$
C. $P_1(2, 3), P_2(2, 7)$ **D.** $P_1(1, -3), P_2(-5, -3)$

Solution See page S12.

Try Exercise 17, page 237.

Two lines in the rectangular coordinate system that never intersect are called **parallel lines.** The lines l_1 and l_2 graphed at the right are parallel lines. Calculating the slope of each line, we have

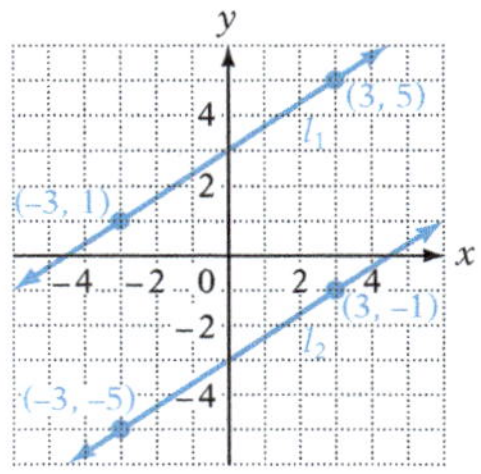

$$\text{Slope of } l_1 = \frac{y_2 - y_1}{x_2 - x_1} = \frac{5 - 1}{3 - (-3)} = \frac{4}{6} = \frac{2}{3}$$

$$\text{Slope of } l_2 = \frac{y_2 - y_1}{x_2 - x_1} = \frac{-1 - (-5)}{3 - (-3)} = \frac{4}{6} = \frac{2}{3}$$

Note that these parallel lines have the same slope. This is true of all parallel lines.

PARALLEL LINES

Two nonvertical lines in the rectangular coordinate system are parallel if and only if they have the same slope. Any two vertical lines in the rectangular coordinate system are parallel lines.

EXAMPLES

1. The graph of $y = -3x - 2$ is parallel to the graph of $y = -3x + 4$ because both lines have a slope of -3.
2. The graph of $y = -6$ is parallel to the graph of $y = 5$ because both lines have a slope of 0.
3. The graph of $x = 2$ is parallel to the graph of $x = -1$ because both lines are vertical lines.

We must distinguish between vertical and nonvertical lines in the description of parallel lines because vertical lines in the rectangular coordinate system are parallel, but their slopes are undefined.

EXAMPLE 2 Is the line containing the points $(-2, 1)$ and $(-5, -1)$ parallel to the line that contains the points $(1, 0)$ and $(4, 2)$?

Solution Find the slope of the line through $(-2, 1)$ and $(-5, -1)$.

$$m = \frac{y_2 - y_1}{x_2 - x_1} = \frac{-1 - 1}{-5 - (-2)} = \frac{-2}{-3} = \frac{2}{3}$$

Find the slope of the line through $(1, 0)$ and $(4, 2)$.

$$m = \frac{y_2 - y_1}{x_2 - x_1} = \frac{2 - 0}{4 - 1} = \frac{2}{3}$$

The slopes are equal.
The lines are parallel.

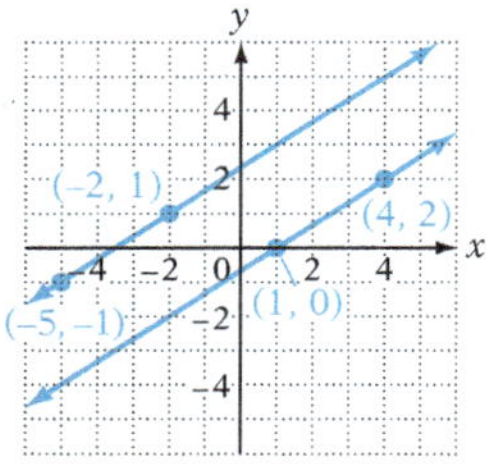

Problem 2 Is the line containing the points $(-2, -3)$ and $(7, 1)$ parallel to the line that contains the points $(1, 4)$ and $(-5, 6)$?

Solution See page S12.

Try Exercise 45, page 237.

Two lines in the rectangular coordinate system that intersect at a 90° angle (right angle) are **perpendicular lines.** The lines l_1 and l_2 graphed at the right are perpendicular lines. Calculating the slope of each line, we have

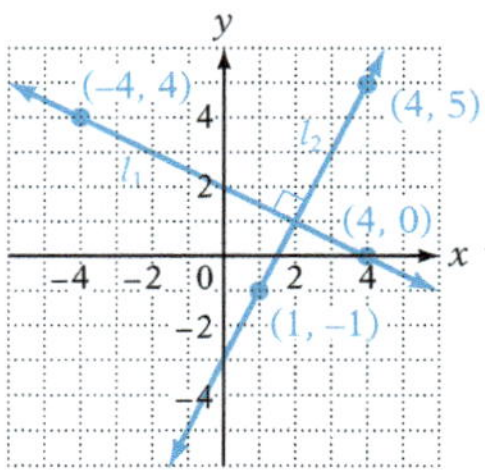

$$\text{Slope of } l_1 = \frac{y_2 - y_1}{x_2 - x_1} = \frac{0 - 4}{4 - (-4)} = \frac{-4}{8} = -\frac{1}{2}$$

$$\text{Slope of } l_2 = \frac{y_2 - y_1}{x_2 - x_1} = \frac{5 - (-1)}{4 - 1} = \frac{6}{3} = 2$$

Note that the product of the slopes of l_1 and l_2 is $\left(-\frac{1}{2}\right)(2) = \mathbf{-1}$. The product of the slopes of any two perpendicular lines is -1.

PERPENDICULAR LINES

Two nonvertical lines in the plane are perpendicular if and only if the product of their slopes is -1. A vertical and a horizontal line are perpendicular.

EXAMPLES

1. The graph of $y = -3x + 1$ is perpendicular to the graph of $y = \frac{1}{3}x + 4$ because the product of their slopes is $(-3)(\frac{1}{3}) = -1$.
2. The graph of $y = -5$ is perpendicular to the graph of $x = 2$ because a horizontal line and a vertical line are perpendicular.

EXAMPLE 3 Is the line containing the points $(3, -2)$ and $(-1, 4)$ perpendicular to the line that contains the points $(-2, 5)$ and $(1, 3)$?

Solution Find the slope of the line through $(3, -2)$ and $(-1, 4)$.

$$m = \frac{y_2 - y_1}{x_2 - x_1} = \frac{4 - (-2)}{-1 - 3} = \frac{6}{-4} = -\frac{3}{2}$$

Find the slope of the line through $(-2, 5)$ and $(1, 3)$.

$$m = \frac{y_2 - y_1}{x_2 - x_1} = \frac{3 - 5}{1 - (-2)} = \frac{-2}{3} = -\frac{2}{3}$$

The product of the slopes is $\left(-\frac{3}{2}\right)\left(-\frac{2}{3}\right) = 1$, not -1.

No, the lines are not perpendicular.

Problem 3 Is the line containing the points $(4, -5)$ and $(-2, 3)$ perpendicular to the line that contains the points $(-1, 6)$ and $(-5, 3)$?

Solution See page S12.

Try Exercise 49, page 238.

Florence Griffith-Joyner

There are many applications of the concept of slope. Here is one example.

When Florence Griffith-Joyner set the world record for the 100-meter dash, her rate of speed was approximately 9.5 m/s. The graph at the right shows the distance she ran during her record-setting run. From the graph, note that after 4 s she had traveled 38 m and that after 6 s she had traveled 57 m. The slope of the line between these two points is

$$m = \frac{57 - 38}{6 - 4} = \frac{19}{2} = 9.5$$

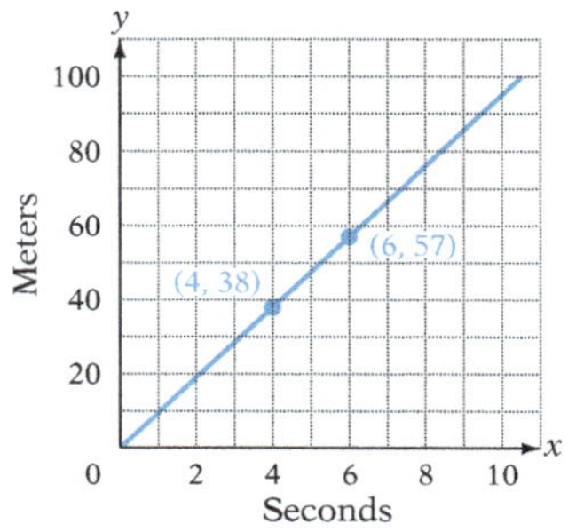

Note that the slope of the line is the same as the rate at which she was running, 9.5 m/s. The average speed of an object is related to slope.

EXAMPLE 4 The graph at the right shows the height of a plane above an airport during its 30-minute descent from cruising altitude to landing. Find the slope of the line. Write a sentence that explains the meaning of the slope.

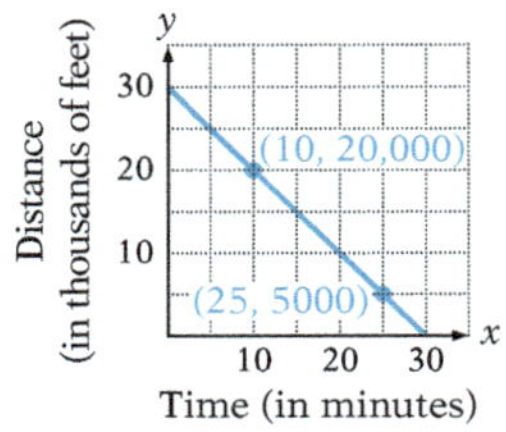

Solution $m = \dfrac{y_2 - y_1}{x_2 - x_1}$

$= \dfrac{5000 - 20{,}000}{25 - 10} = \dfrac{-15{,}000}{15} = -1000$

A slope of -1000 means that the height of the plane is *decreasing* at the rate of 1000 ft per minute.

Problem 4 The graph at the right shows the decline in the value of a used car over a five-year period. Find the slope of the line. Write a sentence that states the meaning of the slope.

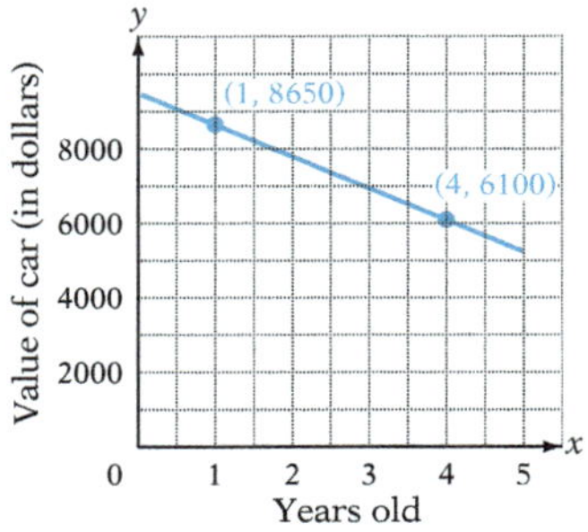

Solution See page S12.

Try Exercise 55, page 238.

OBJECTIVE 2 Graph a line using the slope and y-intercept

Recall that we can find the y-intercept of a linear equation by letting $x = 0$. To find the y-intercept of $y = 3x + 4$, let $x = 0$.

$$y = 3x + 4$$
$$y = 3(0) + 4$$
$$y = 4$$

The y-intercept is $(0, 4)$.

The constant term of $y = 3x + 4$ is the y-coordinate of the y-intercept.

In general, for any equation of the form $y = mx + b$, the y-intercept is $(0, b)$.

The graph of the equation $y = \frac{2}{3}x + 1$ is shown at the right. The points whose coordinates are $(-3, -1)$ and $(3, 3)$ are on the graph. The slope of the line is

$$m = \frac{3 - (-1)}{3 - (-3)} = \frac{4}{6} = \frac{2}{3}$$

Note that the slope of the line has the same value as the coefficient of x.

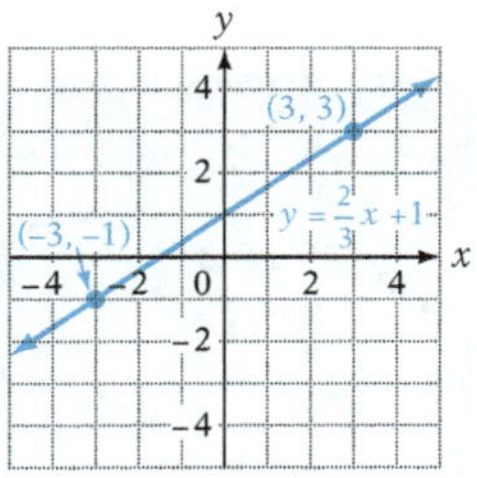

SLOPE-INTERCEPT FORM OF A STRAIGHT LINE

For any equation of the form $y = mx + b$, the slope of the line is m, the coefficient of x. The y-intercept is $(0, b)$. The equation

$$y = mx + b$$

is called the **slope-intercept form of a straight line.**

EXAMPLES

1. The slope of the graph of $y = -\frac{3}{4}x + 1$ is $-\frac{3}{4}$. The y-intercept is (0, 1).

$$y = \boxed{m}\, x + \boxed{b}$$

$$y = \boxed{-\frac{3}{4}}\, x + \boxed{1}$$

Slope $= m = -\frac{3}{4}$ ⟶ ↑ ↑ ⟵ y-intercept $= (0, b) = (0, 1)$

2. The slope of the graph of $y = 5x - 2$ is 5. The y-intercept is $(0, -2)$.
3. The slope of the graph of $y = x$ is 1. The y-intercept is (0, 0).

When the equation of a straight line is in the form $y = mx + b$, the graph can be drawn using the slope and y-intercept. First locate the y-intercept. Use the slope to find a second point on the line. Then draw a line through the two points.

Focus on graphing a line by using the slope and y-intercept

Graph $y = 2x - 3$ by using the slope and y-intercept.

y-intercept $= (0, b) = (0, -3)$

$$m = 2 = \frac{2}{1} = \frac{\text{change in } y}{\text{change in } x}$$

Beginning at the y-intercept $(0, -3)$, move up 2 units (change in y) and then right 1 unit (change in x).

$(1, -1)$ are the coordinates of a second point on the graph.

Draw a line through $(0, -3)$ and $(1, -1)$.

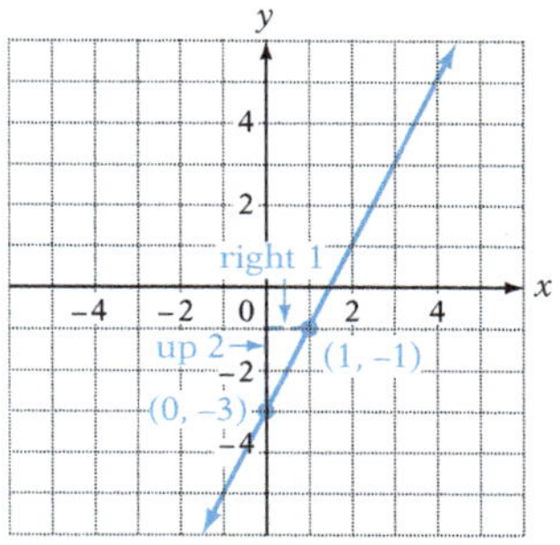

Using a graphing calculator, enter the equation $y = 2x - 3$ *and verify the graph shown above. Trace along the graph and verify that* $(0, -3)$ *is the y-intercept and that the point whose coordinates are* $(1, -1)$ *is on the graph.*

EXAMPLE 5 Graph $y = -\frac{2}{3}x + 1$ by using the slope and y-intercept.

Solution y-intercept $= (0, b) = (0, 1)$

$$m = -\frac{2}{3} = \frac{-2}{3}$$

A slope of $\frac{-2}{3}$ means to move down 2 units (change in y) and then right 3 units (change in x).

Beginning at the y-intercept (0, 1), move down 2 units and then right 3 units.

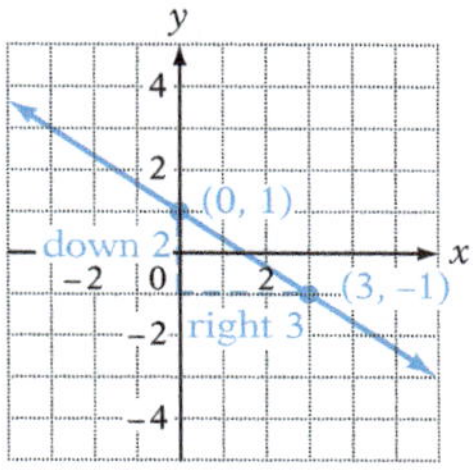

The point whose coordinates are $(3, -1)$ is a second point on the graph.

Draw a line through $(0, 1)$ and $(3, -1)$.

Problem 5 Graph $y = -\frac{1}{4}x - 1$ by using the slope and y-intercept.

Solution See page S12.

Try Exercise 65, page 239.

EXAMPLE 6 Graph $2x - 3y = 6$ by using the slope and y-intercept.

Solution Solve the equation for y.

$$2x - 3y = 6$$
$$-3y = -2x + 6$$
$$y = \frac{2}{3}x - 2$$

The slope is $m = \frac{2}{3}$.

The y-intercept is $(0, -2)$.

Beginning at the y-intercept $(0, -2)$, move up 2 units (change in y) and then right 3 units (change in x).

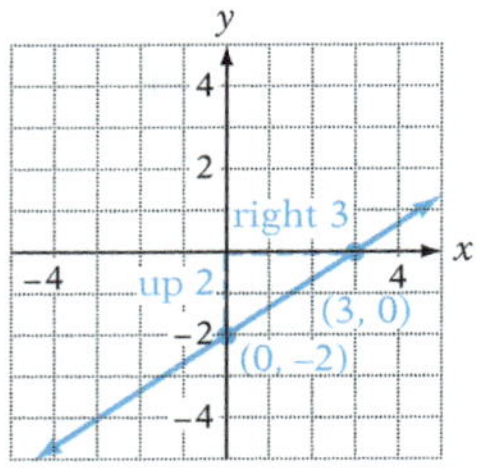

The point whose coordinates are $(3, 0)$ is a second point on the graph.

Draw a line through $(0, -2)$ and $(3, 0)$.

Problem 6 Graph $x - 2y = 4$ by using the slope and y-intercept.

Solution See page S13.

Try Exercise 75, page 240.

5.3 Exercises

CONCEPT CHECK

1. What is the symbol for slope in the equation $y = mx + b$?
2. What is the symbol for the y-coordinate of the y-intercept in the equation $y = mx + b$?
3. The formula for slope is $m =$ __?__.
4. **a.** A line that slants upward to the right has __?__ slope.
   b. A line that slants downward to the right has __?__ slope.
   c. A horizontal line has __?__ slope.
   d. The slope of a vertical line is __?__.

5. The slope of a line is -4. What is the slope of any line parallel to this line?

6. The slope of a line is $\frac{6}{5}$. What is the slope of any line parallel to this line?

7. The slope of a line is $\frac{3}{2}$. What is the slope of any line perpendicular to this line?

8. The slope of a line is -6. What is the slope of any line perpendicular to this line?

1 Find the slope of a straight line (See pages 228–234.)

9. Explain how to find the slope of a line when you know two points on the line.

10. What is the difference between a line that has zero slope and a line whose slope is undefined?

GETTING READY

11. Identify each x value and each y value to insert in the slope formula $m = \frac{y_2 - y_1}{x_2 - x_1}$ to find the slope of the line containing $P_1(1, -4)$ and $P_2(3, 2)$.

$y_2 =$ __?__; $y_1 =$ __?__; $x_2 =$ __?__; $x_1 =$ __?__

12. The slope of the line containing $P_1(-4, 3)$ and $P_2(1, -3)$ is $m = \frac{-3 - ?}{1 - (?)} =$ __?__.
Does this line slant downward to the right or upward to the right?

Find the slope of the line containing the points P_1 and P_2.

13. $P_1(4, 2), P_2(3, 4)$

14. $P_1(2, 1), P_2(3, 4)$

15. $P_1(-1, 3), P_2(2, 4)$

16. $P_1(-2, 1), P_2(2, 2)$

17. $P_1(2, 4), P_2(4, -1)$

18. $P_1(1, 3), P_2(5, -3)$

19. $P_1(-2, 3), P_2(2, 1)$

20. $P_1(5, -2), P_2(1, 0)$

21. $P_1(8, -3), P_2(4, 1)$

22. $P_1(0, 3), P_2(2, -1)$

23. $P_1(3, -4), P_2(3, 5)$

24. $P_1(-1, 2), P_2(-1, 3)$

25. $P_1(4, -2), P_2(3, -2)$

26. $P_1(5, 1), P_2(-2, 1)$

27. $P_1(0, -1), P_2(3, -2)$

28. $P_1(3, 0), P_2(2, -1)$

29. $P_1(-2, 3), P_2(1, 3)$

30. $P_1(4, -1), P_2(-3, -1)$

31. $P_1(-2, 4), P_2(-1, -1)$

32. $P_1(6, -4), P_2(4, -2)$

33. $P_1(-2, -3), P_2(-2, 1)$

34. $P_1(5, 1), P_2(5, -2)$

35. $P_1(-1, 5), P_2(5, 1)$

36. $P_1(-1, 5), P_2(7, 1)$

37. Give the slope of any line that is parallel to the line $y = 4x + 6$.

38. Give the slope of any line that is parallel to the line $y = x - 5$.

39. Give the slope of any line that is perpendicular to the line $y = -\frac{1}{3}x - 2$.

40. Give the slope of any line that is perpendicular to the line $y = 6x + 7$.

41. Are the graphs of $y = \frac{3}{8}x - 5$ and $y = \frac{3}{8}x + 2$ parallel?

42. Are the graphs of $y = -4x + 1$ and $y = 4x - 3$ parallel?

43. Are the graphs of $y = \frac{7}{2}x$ and $y = -\frac{2}{7}x + 2$ perpendicular?

44. Are the graphs of $y = 3x - 8$ and $y = -3x + 8$ perpendicular?

45. Is the line containing the points $(-4, 2)$ and $(1, 6)$ parallel to the line that contains the points $(2, -4)$ and $(7, 0)$?

46. Is the line containing the points $(-5, 0)$ and $(0, 2)$ parallel to the line that contains the points $(5, 1)$ and $(0, -1)$?

47. Is the line containing the points $(-2, -3)$ and $(7, 1)$ parallel to the line that contains the points $(6, -5)$ and $(4, 1)$?

48. Is the line containing the points $(4, -3)$ and $(2, 5)$ parallel to the line that contains the points $(-2, -3)$ and $(-4, 1)$?

49. Is the line containing the points $(1, -1)$ and $(3, -2)$ perpendicular to the line that contains the points $(-4, 1)$ and $(2, -5)$?

50. Is the line containing the points $(0, 1)$ and $(2, 4)$ perpendicular to the line that contains the points $(-4, -7)$ and $(2, 5)$?

51. Is the line containing the points $(5, 1)$ and $(3, -2)$ perpendicular to the line that contains the points $(0, -2)$ and $(3, -4)$?

52. Is the line containing the points $(5, -2)$ and $(-1, 3)$ perpendicular to the line that contains the points $(3, 4)$ and $(-2, -2)$?

53. **Postal Service** The graph at the right shows the work accomplished by an electronic mail sorter. Find the slope of the line. Write a sentence that explains the meaning of the slope.

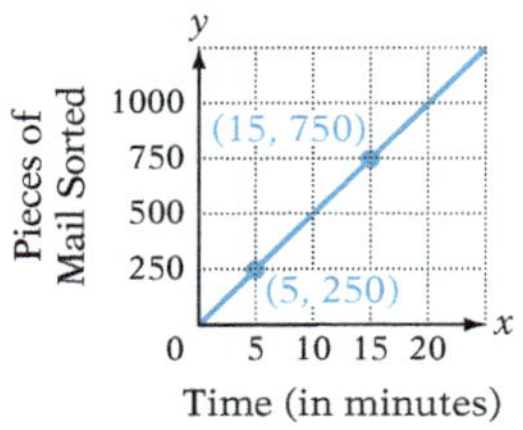

54. **Health** The graph at the right shows the relationship between distance walked and calories burned. Find the slope of the line. Write a sentence that explains the meaning of the slope.

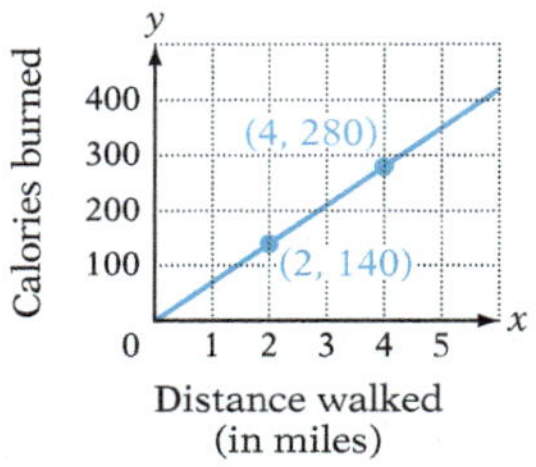

55. **Fuel Consumption** The graph at the right shows how the amount of gasoline in the tank of a car decreases as the car is driven at a constant speed of 60 mph. Find the slope of the line. Write a sentence that states the meaning of the slope.

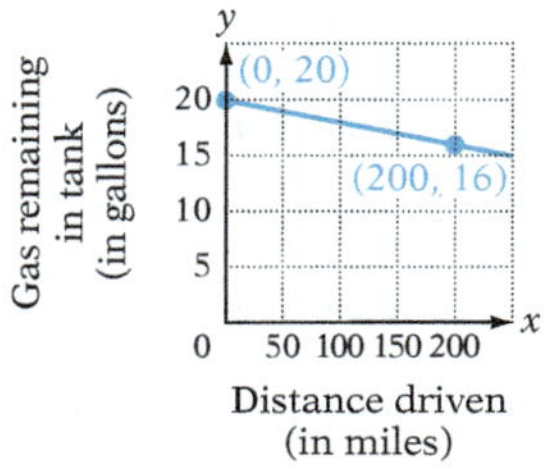

56. **Computer Science** The graph at the right shows the relationship between time and the number of kilobytes of a file remaining to be downloaded. Find the slope of the line. Write a sentence that states the meaning of the slope.

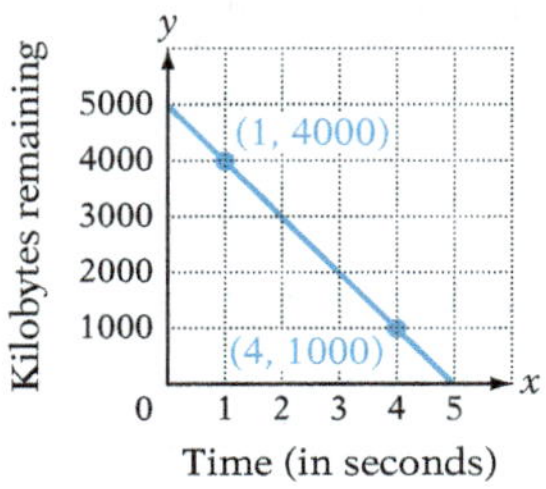

Seat Belt Use Read the news clipping below. Use the information in the news clipping for Exercises 57 and 58.

In the News

Buckling Up Saves Lives

Annual surveys conducted by the National Highway Safety Administration show that Americans' increase in seat belt use has been accompanied by a decrease in deaths due to motor vehicle accidents.

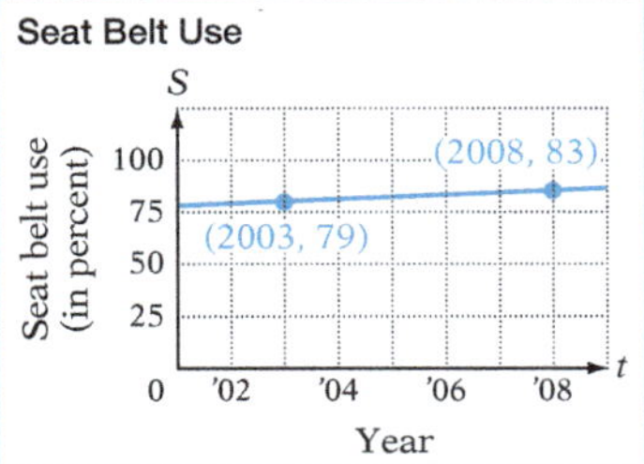

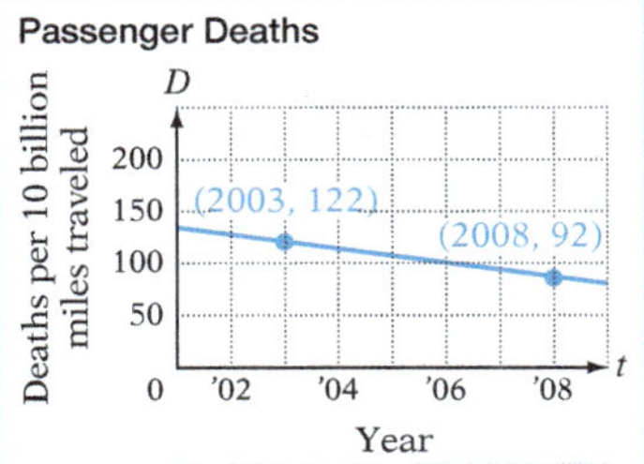

Source: National Highway Traffic Safety Association

57. Find the slope of the line in the Seat Belt Use graph. Write a sentence that states the meaning of the slope in the context of the article.

58. Find the slope of the line in the Passenger Deaths graph. Write a sentence that states the meaning of the slope in the context of the article.

59. Let l be a line passing through points (a, b) and (c, d).
   a. Describe any relationship(s) that must exist between the numbers a, b, c, and d in order for the slope of l to be undefined.
   b. Describe any relationship(s) that must exist between the numbers a, b, c, and d in order for the slope of l to be 0.
   c. Suppose a, b, c, and d are all positive with $a > c$ and $d > b$. Does l slant upward to the right or downward to the right?

60. Line l passes through the point $(-2, 4)$ and is parallel to the line that passes through the points $(-4, 1)$ and $(0, 3)$.
   a. Which quadrants does line l pass through?
   b. Are there any points (a, b) on the line l for which both a and b are negative?

2 Graph a line using the slope and y-intercept (See pages 234–236.)

GETTING READY

61. The slope of the line with equation $y = 5x - 3$ is __?__, and its y-intercept is __?__.

62. Why is an equation of the form $y = mx + b$ said to be in slope-intercept form?

Graph by using the slope and y-intercept.

63. $y = 3x + 1$

64. $y = -2x - 1$

65. $y = \frac{2}{5}x - 2$

66. $y = \frac{3}{4}x + 1$

67. $2x + y = 3$

68. $3x - y = 1$

69. $x - 2y = 4$

70. $x + 3y = 6$

71. $y = \frac{2}{3}x$

72. $y = \frac{1}{2}x$

73. $y = -x + 1$

74. $y = -x - 3$

75. $3x - 4y = 12$

76. $5x - 2y = 10$

77. $y = -4x + 2$

78. $y = 5x - 2$

79. $4x - 5y = 20$

80. $x - 3y = 6$

81. Suppose A, B, and C are all positive numbers. Does the y-intercept of the graph of $Ax + By = C$ lie above or below the x-axis? Does the graph slant upward to the right or downward to the right?

82. Suppose A is a negative number, and B and C are positive numbers. Does the y-intercept of the graph of $Ax + By = C$ lie above or below the x-axis? Does the graph slant upward to the right or downward to the right?

APPLYING CONCEPTS

83. What effect does increasing the coefficient of x have on the graph of $y = mx + b$, $m > 0$?

84. What effect does decreasing the coefficient of x have on the graph of $y = mx + b$, $m > 0$?

85. What effect does increasing the constant term have on the graph of $y = mx + b$?

86. What effect does decreasing the constant term have on the graph of $y = mx + b$?

87. Match each equation with its graph.

(i) $y = -2x + 4$

(ii) $y = 2x - 4$

(iii) $y = 2$

(iv) $2x + 4y = 0$

(v) $y = \frac{1}{2}x + 4$

(vi) $y = -\frac{1}{4}x - 2$

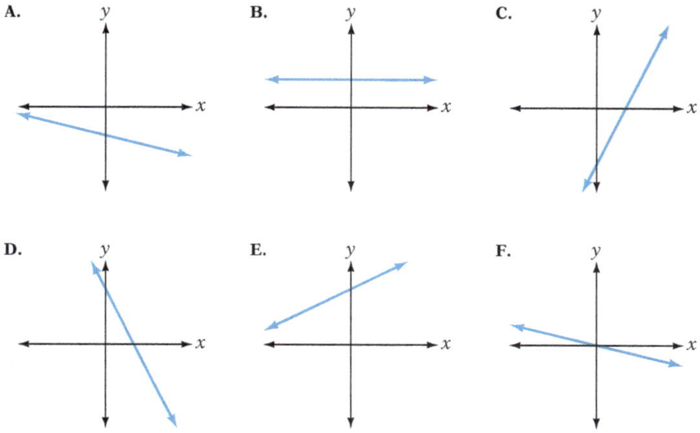

88. Which of the graphs shows a constant rate of change?

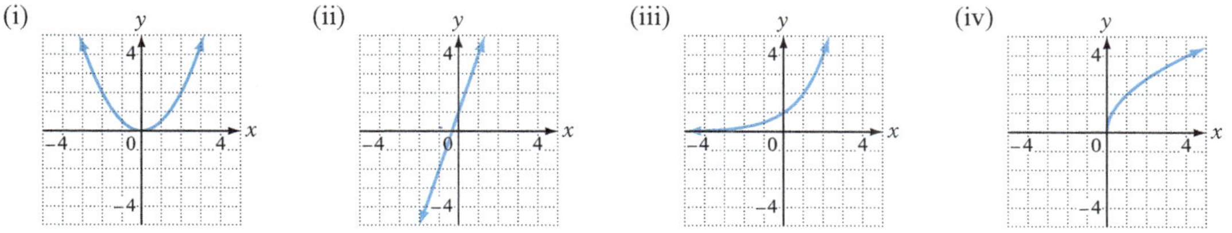

89. Do the graphs of all straight lines have a y-intercept? If not, give an example of one that does not.

90. If two lines have the same slope and the same y-intercept, must the graphs of the lines be the same? If not, give an example.

PROJECTS OR GROUP ACTIVITIES

For each pair of equations, explain how you would distinguish between their graphs.

91. a. $y = \frac{3}{5}x - 2$ **b.** $y = -\frac{3}{5}x - 2$

92. a. $y = x + 1$ **b.** $y = -x + 1$

93. a. $y = \frac{3}{4}x + 5$ **b.** $y = \frac{3}{4}x - 5$

94. a. $y = 2x - 4$ **b.** $y = 2x + 4$

95. a. $y = 6$ **b.** $x = 6$

96. a. $y = 1$ **b.** $y = -1$

5.4 Equations of Straight Lines

OBJECTIVE 1 Find the equation of a line using the equation $y = mx + b$

When the slope of a line and a point on the line are known, the equation of the line can be written using the slope-intercept form, $y = mx + b$. In Example 1 below, the known point is the y-intercept.

EXAMPLE 1 Find the equation of the line that has slope 3 and y-intercept $(0, 2)$.

Solution
$y = mx + b$
$y = 3x + b$ • The given slope, 3, is m. Replace m with 3.
$y = 3x + 2$ • The given point, $(0, 2)$, is the y-intercept. Replace b with 2.

The equation of the line that has slope 3 and y-intercept $(0, 2)$ is $y = 3x + 2$.

Problem 1 Find the equation of the line that has slope $\frac{4}{3}$ and y-intercept $(0, -1)$.

Solution See page S13.

Try Exercise 7, page 246.

In the next example, the known point is a point other than the y-intercept.

Focus on finding the equation of a line given the slope and a point on the line

Find the equation of the line that has slope $\frac{1}{2}$ and contains the point whose coordinates are $(-2, 4)$.

$y = mx + b$

The given slope, $\frac{1}{2}$, is m. Replace m with $\frac{1}{2}$.

$y = \frac{1}{2}x + b$

The given point, $(-2, 4)$, is a solution of the equation of the line. Replace x and y in the equation with the coordinates of the point.

$4 = \frac{1}{2}(-2) + b$

Solve for b, the y-intercept.

$4 = -1 + b$
$5 = b$

Write the equation of the line by replacing m and b in the equation $y = mx + b$ by their values.

$y = mx + b$
$y = \frac{1}{2}x + 5$

The equation of the line that has slope $\frac{1}{2}$ and contains the point whose coordinates are $(-2, 4)$ is $y = \frac{1}{2}x + 5$.

Take Note
Every ordered pair is of the form (x, y). For the point $(-2, 4)$, -2 is the x value and 4 is the y value. Substitute -2 for x and 4 for y.

EXAMPLE 2 Find the equation of the line that contains the point whose coordinates are $(3, -3)$ and has slope $\frac{2}{3}$.

Solution

$y = mx + b$

$y = \frac{2}{3}x + b$ • Replace *m* with the given slope.

$-3 = \frac{2}{3}(3) + b$ • Replace *x* and *y* in the equation with the coordinates of the given point.

$-3 = 2 + b$ • Solve for *b*.

$-5 = b$

$y = \frac{2}{3}x - 5$ • Write the equation of the line by replacing *m* and *b* in $y = mx + b$ by their values.

The equation of the line is $y = \frac{2}{3}x - 5$.

Problem 2 Find the equation of the line that contains the point whose coordinates are $(4, -2)$ and has slope $\frac{3}{2}$.

Solution See page S13.

Try Exercise 13, page 246.

OBJECTIVE 2 Find the equation of a line using the point-slope formula

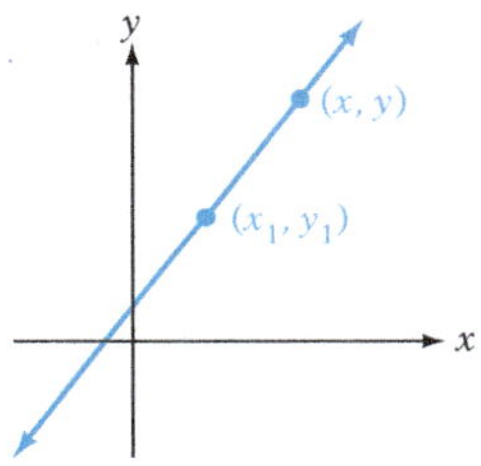

An alternative method for finding the equation of a line, given the slope and the coordinates of a point on the line, involves use of the point-slope formula. The point-slope formula is derived from the formula for slope.

Let (x_1, y_1) be the coordinates of the given point on the line, and let (x, y) be the coordinates of any other point on the line. Use the formula for slope.

$$\frac{y - y_1}{x - x_1} = m$$

Multiply both sides of the equation by $(x - x_1)$.

$$\frac{y - y_1}{x - x_1}(x - x_1) = m(x - x_1)$$

Simplify.

$$y - y_1 = m(x - x_1)$$

POINT-SLOPE FORMULA

The equation of the line that has slope *m* and contains the point whose coordinates are (x_1, y_1) can be found by the point-slope formula:

$$y - y_1 = m(x - x_1)$$

EXAMPLE 3 Use the point-slope formula to find the equation of the line that passes through the point whose coordinates are $(-2, -1)$ and has slope $\frac{3}{2}$.

Solution

$(x_1, y_1) = (-2, -1)$ • Let (x_1, y_1) be the given point.

$m = \frac{3}{2}$ • *m* is the given slope.

$y - y_1 = m(x - x_1)$ • This is the point-slope formula.

$y - (-1) = \frac{3}{2}[x - (-2)]$ • Substitute −2 for x_1, −1 for y_1, and $\frac{3}{2}$ for *m*.

$y + 1 = \frac{3}{2}(x + 2)$ • Rewrite the equation in the form $y = mx + b$.

$y + 1 = \frac{3}{2}x + 3$

$y = \frac{3}{2}x + 2$

The equation of the line is $y = \frac{3}{2}x + 2$.

Problem 3 Use the point-slope formula to find the equation of the line that passes through the point whose coordinates are (5, 4) and has slope $\frac{2}{5}$.

Solution See page S13.

Try Exercise 35, page 248.

OBJECTIVE 3 Find the equation of a line given two points

The line that passes through the points whose coordinates are $(-2, 6)$ and $(4, -3)$ is shown at the left below. In the next example, we find the equation of this line.

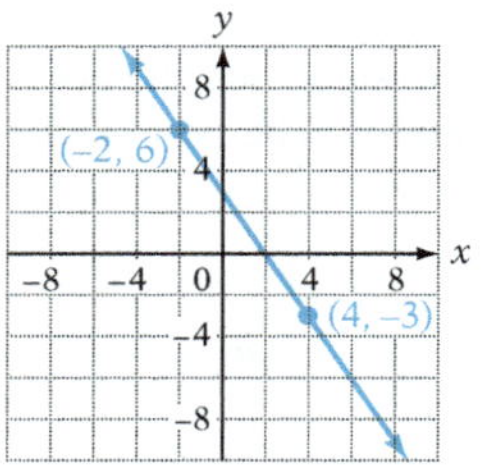

Focus on finding the equation of a line given two points

Find the equation of the line that passes through the points whose coordinates are $(-2, 6)$ and $(4, -3)$.

Use the slope formula with $(x_1, y_1) = (-2, 6)$ and $(x_2, y_2) = (4, -3)$ to find the slope of the line between the two points.

$$m = \frac{y_2 - y_1}{x_2 - x_1} = \frac{-3 - 6}{4 - (-2)} = \frac{-9}{6} = -\frac{3}{2}$$

The slope of the line is $-\frac{3}{2}$.

Now use the point-slope formula with $m = -\frac{3}{2}$ and $(x_1, y_1) = (-2, 6)$.

$$y - y_1 = m(x - x_1)$$
$$y - 6 = -\frac{3}{2}[x - (-2)]$$
$$y - 6 = -\frac{3}{2}(x + 2)$$

Use the Distributive Property.
$$y - 6 = -\frac{3}{2}x - 3$$

Write in slope-intercept form.
$$y = -\frac{3}{2}x + 3$$

Check: You can check that this is the correct equation by substituting $(-2, 6)$ and $(4, -3)$ into the equation, as shown below.

$$y = -\frac{3}{2}x + 3$$

6	$-\frac{3}{2}(-2) + 3$
6	$3 + 3$

$6 = 6$ ✓

$$y = -\frac{3}{2}x + 3$$

−3	$-\frac{3}{2}(4) + 3$
−3	$-6 + 3$

$-3 = -3$ ✓

The equation of the line is $y = -\frac{3}{2}x + 3$.

Take Note

We could have used $(4, -3)$ for (x_1, y_1) instead of $(-2, 6)$. The result would have been the same.

$$y - y_1 = m(x - x_1)$$
$$y - (-3) = -\frac{3}{2}(x - 4)$$
$$y + 3 = -\frac{3}{2}x + 6$$
$$y = -\frac{3}{2}x + 3$$

EXAMPLE 4 Find the equation of the line that passes through the points whose coordinates are $(-4, -5)$ and $(8, 4)$.

Solution Find the slope of the line between the two points.

$$m = \frac{y_2 - y_1}{x_2 - x_1} = \frac{4 - (-5)}{8 - (-4)} = \frac{9}{12} = \frac{3}{4}$$ • **Let $(x_1, y_1) = (-4, -5)$ and $(x_2, y_2) = (8, 4)$.**

$$y - y_1 = m(x - x_1)$$ • **Use the point-slope formula.**

$$y - (-5) = \frac{3}{4}[x - (-4)]$$ • **$m = \frac{3}{4}$; $(x_1, y_1) = (-4, -5)$**

$$y + 5 = \frac{3}{4}(x + 4)$$

$$y + 5 = \frac{3}{4}x + 3$$ • **Use the Distributive Property.**

$$y = \frac{3}{4}x - 2$$ • **Write in slope-intercept form.**

The equation of the line is $y = \frac{3}{4}x - 2$.

Problem 4 Find the equation of the line that passes through the points whose coordinates are $(-3, -9)$ and $(1, -1)$.

Solution See page S13.

Try Exercise 49, page 248.

If the two given points lie on a horizontal line, the procedure above can be used to find the equation of the line. However, it is quicker to just remember that the equation of a horizontal line is $y = b$, where b is the y-intercept of the graph of the line. b is also the y-coordinate of each of the two given points.

Focus on finding the equation of a horizontal line

Find the equation of the line that passes through the points whose coordinates are $(-4, -2)$ and $(2, -2)$.

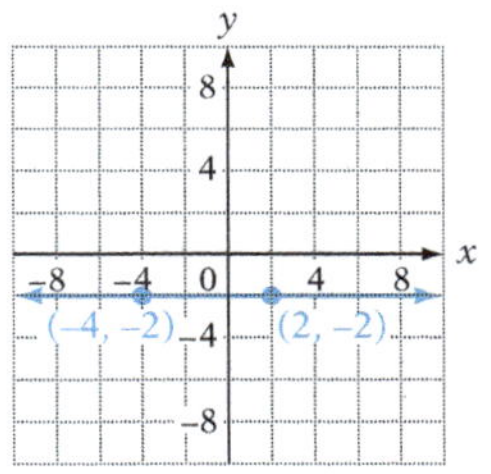

The y-coordinates of the two points are the same. The points lie on a horizontal line. The equation of the line is $y = b$, where b is the y-intercept of the graph of the line. The equation of the line is $y = -2$.

If the two given points lie on a vertical line, the procedure above cannot be used to find the equation of the line. In this case, remember that the equation of a vertical line is $x = a$, where a is the x-intercept of the graph of the line. a is also the x-coordinate of each of the two given points.

Focus on finding the equation of a vertical line

Find the equation of the line that passes through the points whose coordinates are $(3, 4)$ and $(3, -5)$.

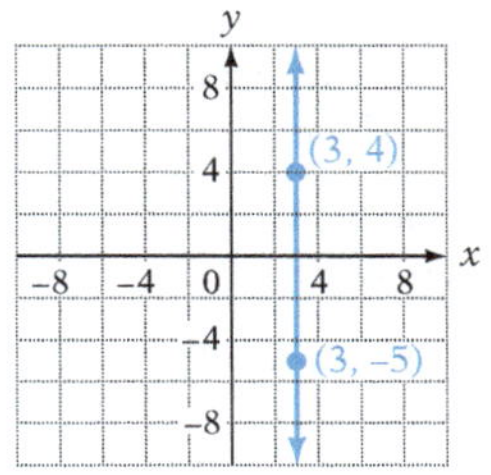

The x-coordinates of the two points are the same. The points lie on a vertical line. The equation of the line is $x = a$, where a is the x-intercept of the graph of the line. The equation of the line is $x = 3$.

5.4 Exercises

CONCEPT CHECK

1. The graph of the equation $y = 5x + 7$ has slope ___?___ and y-intercept (0, ___?___).

2. If the equation of a line has y-intercept (0, 4), then 4 can be substituted for ___?___ in the equation $y = mx + b$.

3. If it is stated that the y-intercept is 2, then the y-intercept is the point (___?___, ___?___).

4. The equation of a line contains the point $(-3, 1)$. This means that when y is ___?___, x is ___?___.

5. The point-slope formula is ___?___.

6. What properties of a line must we know in order to use the point-slope formula to find the equation of the line?

1 Find the equation of a line using the equation $y = mx + b$ (See pages 242–243.)

GETTING READY

7. In the equation of the line that has slope 3 and y-intercept (0, 1), m = ___?___ and b = ___?___. The equation is y = ___?___.

8. To find the equation of the line that contains the point whose coordinates are $(-4, 1)$ and has slope $\frac{3}{2}$, first use the given point and slope to find the y-intercept of the line:

$y = mx + b$ • **Write the slope-intercept form of the equation of a line.**

$1 = \frac{3}{2}(-4) + b$ • **Substitute ___?___ for *y*, ___?___ for *m*, and ___?___ for *x*.**

$1 =$ ___?___ $+ b$ • **Solve for *b*.**

___?___ $= b$

The y-intercept of the line through $(-4, 1)$ that has slope $\frac{3}{2}$ is (0, ___?___).

9. After you find the equation of a line given its slope and the coordinates of a point on the line, how can you determine whether you have the correct equation?

10. What point must the graph of $y = mx$ pass through?

Use the slope-intercept form.

11. Find the equation of the line that contains the point whose coordinates are (0, 2) and has slope 2.

12. Find the equation of the line that contains the point whose coordinates are $(0, -1)$ and has slope -2.

13. Find the equation of the line that contains the point whose coordinates are $(-1, 2)$ and has slope -3.

14. Find the equation of the line that contains the point whose coordinates are $(2, -3)$ and has slope 3.

15. Find the equation of the line that contains the point whose coordinates are (3, 1) and has slope $\frac{1}{3}$.

16. Find the equation of the line that contains the point whose coordinates are $(-2, 3)$ and has slope $\frac{1}{2}$.

17. Find the equation of the line that contains the point whose coordinates are $(4, -2)$ and has slope $\frac{3}{4}$.

18. Find the equation of the line that contains the point whose coordinates are $(2, 3)$ and has slope $-\frac{1}{2}$.

19. Find the equation of the line that contains the point whose coordinates are $(5, -3)$ and has slope $-\frac{3}{5}$.

20. Find the equation of the line that contains the point whose coordinates are $(5, -1)$ and has slope $\frac{1}{5}$.

21. Find the equation of the line that contains the point whose coordinates are $(2, 3)$ and has slope $\frac{1}{4}$.

22. Find the equation of the line that contains the point whose coordinates are $(-1, 2)$ and has slope $-\frac{1}{2}$.

23. Find the equation of the line that contains the point whose coordinates are $(-3, -5)$ and has slope $-\frac{2}{3}$.

24. Find the equation of the line that contains the point whose coordinates are $(-4, 0)$ and has slope $\frac{5}{2}$.

2 Find the equation of a line using the point-slope formula (See pages 243–244.)

25. When is the point-slope formula used?

26. Explain the meaning of each variable in the point-slope formula.

GETTING READY

27. In the equation of the line that has slope $-\frac{4}{5}$ and y-intercept $(0, 3)$, m is ___?___ and b is ___?___. The equation is $y =$ ___?___.

28. Make the appropriate substitutions in the point-slope formula to find the equation of the line that contains the point $(3, 2)$ and has slope 6.

$$y - y_1 = m(x - x_1)$$
$$y - (\underline{\ ?\ }) = (\underline{\ ?\ })(x - \underline{\ ?\ })$$

Use the point-slope formula.

29. Find the equation of the line that passes through the point whose coordinates are $(1, -1)$ and has slope 2.

30. Find the equation of the line that passes through the point whose coordinates are $(2, 3)$ and has slope -1.

31. Find the equation of the line that passes through the point whose coordinates are $(-2, 1)$ and has slope -2.

32. Find the equation of the line that passes through the point whose coordinates are $(-1, -3)$ and has slope -3.

33. Find the equation of the line that passes through the point whose coordinates are $(0, 0)$ and has slope $\frac{2}{3}$.

34. Find the equation of the line that passes through the point whose coordinates are $(0, 0)$ and has slope $-\frac{1}{5}$.

35. Find the equation of the line that passes through the point whose coordinates are (2, 3) and has slope $\frac{1}{2}$.

36. Find the equation of the line that passes through the point whose coordinates are (3, −1) and has slope $\frac{2}{3}$.

37. Find the equation of the line that passes through the point whose coordinates are (−4, 1) and has slope $-\frac{3}{4}$.

38. Find the equation of the line that passes through the point whose coordinates are (−5, 0) and has slope $-\frac{1}{5}$.

39. Find the equation of the line that passes through the point whose coordinates are (−2, 1) and has slope $\frac{3}{4}$.

40. Find the equation of the line that passes through the point whose coordinates are (3, −2) and has slope $\frac{1}{6}$.

41. Find the equation of the line that passes through the point whose coordinates are (−3, −5) and has slope $-\frac{4}{3}$.

42. Find the equation of the line that passes through the point whose coordinates are (3, −1) and has slope $\frac{3}{5}$.

43. **a.** Use the point-slope formula to find the equation of the line with slope m and y-intercept $(0, b)$.
b. Does your answer to part (a) simplify to the slope-intercept form of a straight line with slope m and y-intercept $(0, b)$?

44. **a.** Use the point-slope formula to find the equation of the line that goes through the point (5, 3) and has slope 0.
b. Does your answer to part (a) simplify to the equation of a horizontal line through (0, 3)?

3 Find the equation of a line given two points (See pages 244–245.)

GETTING READY

45. You are asked to find the equation of a line given the coordinates of two points on the line. Before using the point-slope formula, you must find the __?__ of the line between the two points.

46. Sketch the line described in the indicated exercise. Use your graph to determine whether the value of m in the equation of the line is positive or negative.
a. In Exercise 47, the value of m in the equation of the line is __?__.
b. In Exercise 49, the value of m in the equation of the line is __?__.

Find the equation of the line through the given points.

47. (−2, −2) and (1, 7)
48. (1, 5) and (3, 9)
49. (−5, 1) and (2, −6)
50. (−3, 9) and (1, 1)
51. (5, −1) and (−5, 11)
52. (−6, 12) and (−4, 9)
53. (−10, −3) and (5, −9)
54. (−6, −13) and (6, −1)
55. (1, 5) and (−6, 5)
56. (−3, −4) and (5, −4)
57. (5, −1) and (5, −7)
58. (−3, 6) and (−3, 0)
59. (−20, −8) and (5, 12)
60. (−6, 19) and (2, 7)
61. (0, −2) and (−6, 1)
62. (15, −9) and (−20, 5)
63. (6, −11) and (−3, 1)
64. (14, −1) and (−7, −7)
65. (3, 6) and (0, −3)
66. (5, 9) and (−5, 3)

67. $(-1, -3)$ and $(2, 6)$

68. $(-3, 6)$ and $(4, -8)$

69. $(3, -5)$ and $(3, 1)$

70. $(2, -1)$ and $(5, -1)$

71. Is it possible to find the equation of one line through three given points? Explain.

72. If $y = 2x - 3$ and (x_1, y_1) and (x_2, y_2) are the coordinates of two points on the graph of the line, what is the value of $\dfrac{y_2 - y_1}{x_2 - x_1}$?

APPLYING CONCEPTS

Is there a linear equation that contains all of the given ordered pairs? If there is, find the equation.

73. $(5, 1), (4, 2), (0, 6)$

74. $(-2, -4), (0, -3), (4, -1)$

75. $(-1, -5), (2, 4), (0, 2)$

76. $(3, -1), (12, -4), (-6, 2)$

The given ordered pairs are solutions of the same linear equation. Find n.

77. $(0, 1), (4, 9), (3, n)$

78. $(2, 2), (-1, 5), (3, n)$

79. $(2, -2), (-2, -4), (4, n)$

80. $(1, -2), (-2, 4), (4, n)$

81. The graph of a linear equation passes through the points $(2, 3)$ and $(-1, -3)$. It also passes through the point $(1, y)$. Find the value of y.

PROJECTS OR GROUP ACTIVITIES

For Exercises 82 to 85, (a) name the x-intercept of the graph, (b) name the y-intercept of the graph, (c) determine the slope of the line, and (d) write the equation of the line in slope-intercept form.

82.

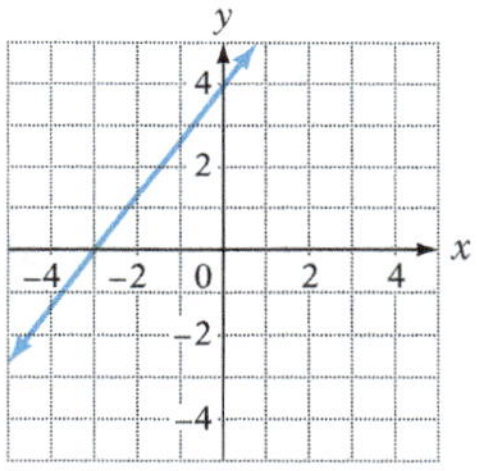

83.

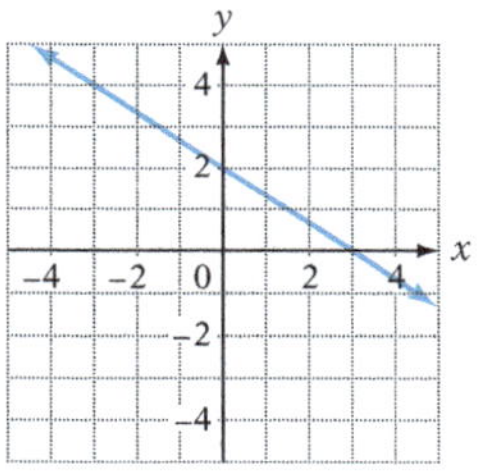

84.

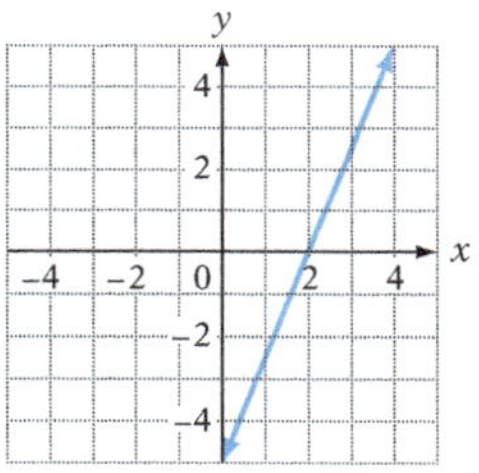

85.

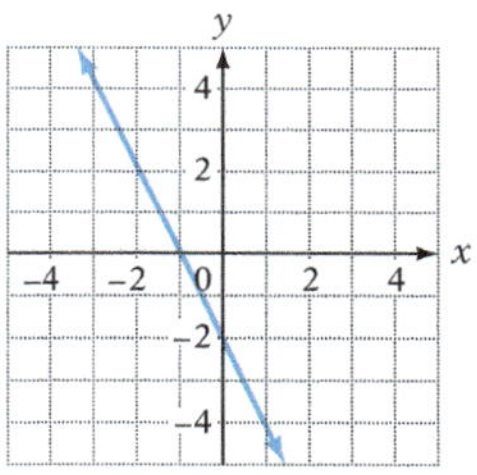

5.5 Functions

OBJECTIVE 1 Introduction to functions

The definition of *set* given in the chapter "Prealgebra Review" stated that a set is a collection of objects. Recall that the objects in a set are called the elements of the set. The elements in a set can be anything.

The set of planets in our solar system is

{Mercury, Venus, Earth, Mars, Jupiter, Saturn, Uranus, Neptune}

The objects in a set can be ordered pairs. When the elements of a set are ordered pairs, the set is called a relation. A **relation** is any set of ordered pairs.

The set $\{(1, 1), (2, 4), (3, 9), (4, 16), (5, 25)\}$ is a relation. There are five elements in the set. The elements are the ordered pairs $(1, 1)$, $(2, 4)$, $(3, 9)$, $(4, 16)$, and $(5, 25)$.

The following table shows the number of hours that each of eight students spent in the math lab during the week of the math midterm exam and the score that each student received on the math midterm.

Hours	2	3	4	4	5	6	6	7
Score	60	70	70	80	85	85	95	90

This information can be written as the relation

$$\{(2, 60), (3, 70), (4, 70), (4, 80), (5, 85), (6, 85), (6, 95), (7, 90)\}$$

where the first coordinate of each ordered pair is the hours spent in the math lab and the second coordinate is the score on the midterm exam.

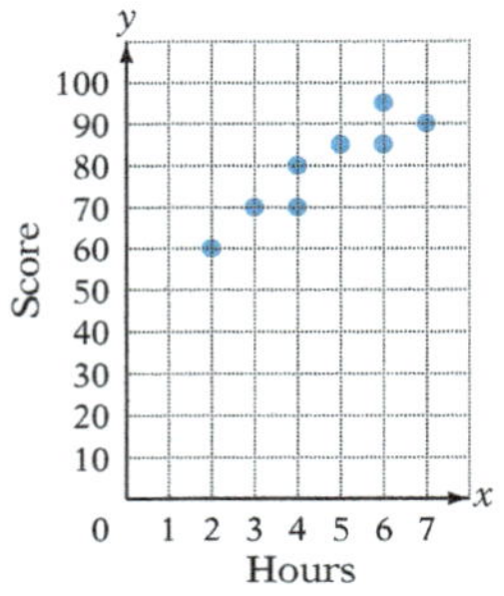

The **domain** of a relation is the set of first coordinates of the ordered pairs. The **range** is the set of second coordinates of the ordered pairs. For the relation above,

$$\text{Domain} = \{2, 3, 4, 5, 6, 7\} \qquad \text{Range} = \{60, 70, 80, 85, 90, 95\}$$

The **graph of a relation** is the graph of the ordered pairs that belong to the relation. The graph of the relation given above is shown at the left. The horizontal axis represents the domain (the hours spent in the math lab) and the vertical axis represents the range (the exam score).

A **function** is a special type of relation in which no two ordered pairs have the same first coordinate and different second coordinates. The relation above is not a function because the ordered pairs $(4, 70)$ and $(4, 80)$ have the same first coordinate and different second coordinates. The ordered pairs $(6, 85)$ and $(6, 95)$ also have the same first coordinate and different second coordinates.

The table at the right describes a grading scale that defines a relationship between a test score and a letter grade. Some of the ordered pairs in this relation are $(38, \text{F})$, $(73, \text{C})$, and $(94, \text{A})$.

Score	Letter Grade
90–100	A
80–89	B
70–79	C
60–69	D
0–59	F

This relation defines a function because no two ordered pairs can have the *same* first coordinate and *different* second coordinates. For instance, it is not possible to have an average of 73 paired with any grade other than C. Both $(73, \text{C})$ and $(73, \text{A})$ cannot be ordered pairs belonging to the function, or two students with the same score would receive different grades. Note that $(81, \text{B})$ and $(88, \text{B})$ are ordered pairs of this function. Ordered pairs of a function may have the same *second* coordinate paired with different *first* coordinates.

The domain of this function is $\{0, 1, 2, 3, \ldots, 98, 99, 100\}$.

The range of this function is $\{\text{A}, \text{B}, \text{C}, \text{D}, \text{F}\}$.

EXAMPLE 1 Find the domain and range of the relation $\{(-5, 1), (-3, 3), (-1, 5)\}$. Is the relation a function?

Solution The domain is $\{-5, -3, -1\}$.

- **The domain of the relation is the set of the first components of the ordered pairs.**

The range is $\{1, 3, 5\}$.

• **The range of the relation is the set of the second components of the ordered pairs.**

No two ordered pairs have the same first coordinate.
The relation is a function.

Problem 1 Find the domain and range of the relation $\{(1, 0), (1, 1), (1, 2), (1, 3), (1, 4)\}$. Is the relation a function?

Solution See page S13.

Try Exercise 17, page 256.

Although a function can be described in terms of ordered pairs or in a table, functions are often described by an equation. The letter f is commonly used to represent a function, but any letter can be used.

The "square" function assigns to each real number its square. The square function is described by the equation

$f(x) = x^2$ Read $f(x)$ as "f of x" or "the value of f at x."

$f(x)$ is the symbol for the number that is paired with x. In terms of ordered pairs, this is written $(x, f(x))$. $f(x)$ is the **value of the function** at x because it is the result of evaluating the variable expression. For example, $f(4)$ means to replace x by 4 and then simplify the resulting numerical expression. This process is called **evaluating the function.**

The notation $f(4)$ is used to indicate the number that is paired with 4. To evaluate $f(x) = x^2$ at 4, replace x with 4 and simplify.

$$f(x) = x^2$$
$$f(4) = 4^2$$
$$f(4) = 16$$

The square function squares a number, and when 4 is squared, the result is 16. For the square function, the number 4 is paired with 16. In other words, when x is 4, $f(x)$ is 16. The ordered pair $(4, 16)$ is an element of the function.

It is important to remember that $f(x)$ does not mean f times x. The letter f stands for the function, and $f(x)$ is the number that is paired with x.

EXAMPLE 2 Evaluate $f(x) = 2x - 4$ at $x = 3$. Write an ordered pair that is an element of the function.

Solution

$f(x) = 2x - 4$ • **Write the function.**

$f(3) = 2(3) - 4$ • **$f(3)$ is the number that is paired with 3. Replace x by 3 and evaluate.**

$f(3) = 6 - 4$

$f(3) = 2$

The ordered pair $(3, 2)$ is an element of the function.

Problem 2 Evaluate $f(x) = -5x + 1$ at $x = 2$. Write an ordered pair that is an element of the function.

Solution See page S13.

Try Exercise 25, page 256.

When a function is described by an equation and the domain is specified, the range of the function can be found by evaluating the function at each point of the domain.

EXAMPLE 3 Find the range of the function given by the equation $f(x) = -3x + 2$ if the domain is $\{-4, -2, 0, 2, 4\}$. Write five ordered pairs that belong to the function.

Solution

$$f(x) = -3x + 2$$
$$f(-4) = -3(-4) + 2 = 12 + 2 = 14$$
$$f(-2) = -3(-2) + 2 = 6 + 2 = 8$$
$$f(0) = -3(0) + 2 = 0 + 2 = 2$$
$$f(2) = -3(2) + 2 = -6 + 2 = -4$$
$$f(4) = -3(4) + 2 = -12 + 2 = -10$$

- **Write the function.**
- **Replace x by each member of the domain.**

The range is $\{-10, -4, 2, 8, 14\}$.
The ordered pairs $(-4, 14)$, $(-2, 8)$, $(0, 2)$, $(2, -4)$, and $(4, -10)$ belong to the function.

Problem 3 Find the range of the function given by the equation $f(x) = 4x - 3$ if the domain is $\{-5, -3, -1, 1\}$. Write four ordered pairs that belong to the function.

Solution See page S13.

Try Exercise 35, page 256.

OBJECTIVE 2 Graphs of linear functions

The solutions of the equation

$$y = 7x - 3$$

are ordered pairs (x, y). For example, the ordered pairs $(-1, -10)$, $(0, -3)$, and $(1, 4)$ are solutions of the equation. Therefore, this equation defines a relation.

It is not possible to substitute one value of x into the equation $y = 7x - 3$ and get two different values of y. For example, the number 1 in the domain cannot be paired with any number other than 4 in the range. (*Remember:* A function cannot have ordered pairs in which the same first coordinate is paired with different second coordinates.) Therefore, the equation defines a function.

The equation $y = 7x - 3$ is an equation of the form $y = mx + b$. In general, any equation of the form $y = mx + b$ is a function.

In the equation $y = 7x - 3$, the variable y is called the **dependent variable** because its value *depends* on the value of x. The variable x is called the **independent variable.** We choose a value for x and substitute that value into the equation to determine the value of y. We say that *y is a function of x.*

When an equation defines y as a function of x, function notation is frequently used to emphasize that the relation is a function. In this case, it is common to use the notation $f(x)$. Therefore, we can write the equation

$$y = 7x - 3$$

in function notation as

$$f(x) = 7x - 3$$

Take Note
When y is a function of x, y and $f(x)$ are interchangeable.

The **graph of a function** is a graph of the ordered pairs (x, y) of the function. Because the graph of the equation $y = mx + b$ is a straight line, a function of the form $f(x) = mx + b$ is a **linear function.**

Focus on graphing a linear function

Graph: $f(x) = \dfrac{2}{3}x - 1$

Think of the function as the equation $y = \frac{2}{3}x - 1$.

This is the equation of a straight line.

The y-intercept is $(0, -1)$. The slope is $\frac{2}{3}$.

Graph the point $(0, -1)$. From the y-intercept, go up 2 units and then right 3 units. Graph the point $(3, 1)$.

Draw a line through the two points.

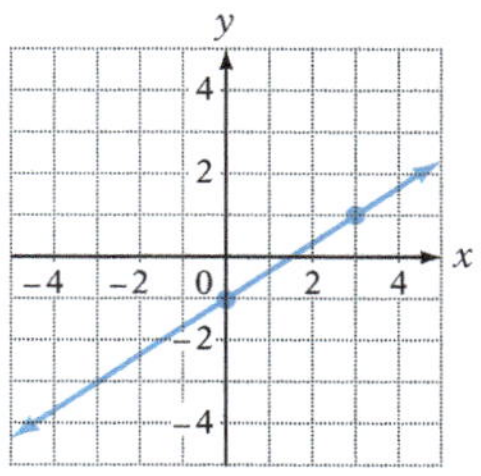

EXAMPLE 4 Graph: $f(x) = \frac{3}{4}x + 2$

Solution $f(x) = \frac{3}{4}x + 2$

$y = \frac{3}{4}x + 2$

- **Think of the function as the equation $y = \frac{3}{4}x + 2$.**
- **The graph is a straight line with y-intercept (0, 2) and slope $\frac{3}{4}$.**

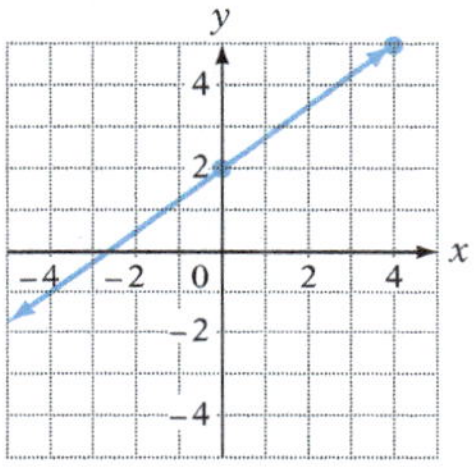

Problem 4 Graph: $f(x) = -\frac{1}{2}x - 3$

Solution See page S13.

Try Exercise 51, page 257.

Graphing calculators are used to graph functions. Using a graphing calculator, enter the equation $y = \frac{3}{4}x + 2$ and verify the graph drawn in Example 4. Trace along the graph and verify that $(-4, -1)$, $(0, 2)$, and $(4, 5)$ are coordinates of points on the graph. Now enter the equation given in Problem 4 and verify the graph you drew.

There are a variety of applications of linear functions. For example, suppose an installer of marble kitchen countertops charges \$250 plus \$180 per linear foot of countertop. The equation that describes the total cost C (in dollars) for having x feet of countertop installed is $C = 180x + 250$. In this situation, C is a function of x; the total cost depends on how many feet of countertop are installed. Therefore, we could rewrite the equation

$$C = 180x + 250$$

as the function

$$f(x) = 180x + 250$$

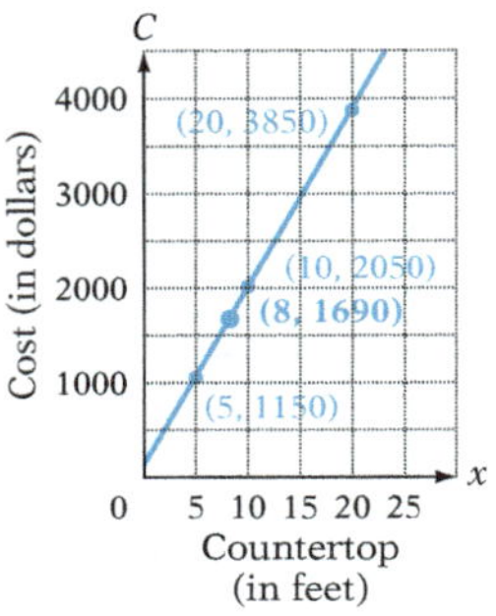

To graph $f(x) = 180x + 250$, we first choose a reasonable domain—for example, values of x between 0 and 25. It would not be reasonable to have $x \leq 0$, because no one would order the installation of 0 ft of countertop, and any amount less than 0 would be a negative amount of countertop. The upper limit of 25 is chosen because most kitchens have less than 25 ft of countertop.

Choosing $x = 5$, 10, and 20 results in the ordered pairs (5, 1150), (10, 2050), and (20, 3850). Graph these points and draw a line through them. The graph is shown at the left.

The point whose coordinates are (8, 1690) is on the graph. This ordered pair can be interpreted to mean that the installation of 8 ft of countertop costs $1690.

EXAMPLE 5 The value V of an investment of $2500 at an annual simple interest rate of 6% is given by the equation $V = 150t + 2500$, where t is the amount of time, in years, that the money is invested.

A. Write the equation in function notation.
B. Graph the equation for values of t between 0 and 10.
C. The point whose coordinates are (5, 3250) is on the graph. Write a sentence that explains the meaning of this ordered pair.

Solution

A. $V = 150t + 2500$
$f(t) = 150t + 2500$

- **The value V of the investment depends on the amount of time t it is invested. The value V is a function of the time t.**

B.

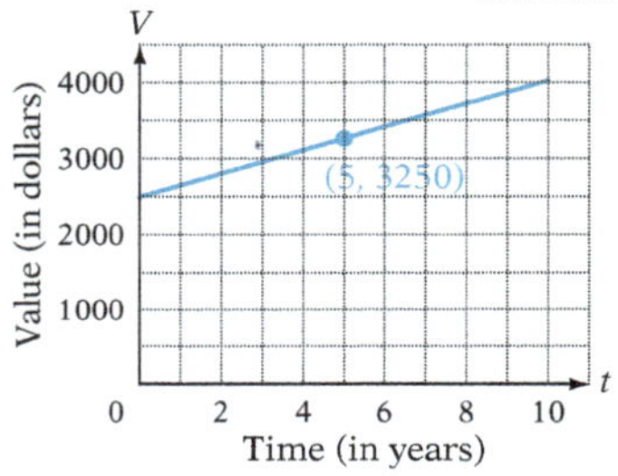

- **Some ordered pairs of the function are (2, 2800), (4, 3100), and (6, 3400).**

C. The ordered pair (5, 3250) means that in 5 years the value of the investment will be $3250.

Problem 5 A car is traveling at a uniform speed of 40 mph. The distance d (in miles) the car travels in t hours is given by the equation $d = 40t$.

A. Write the equation in function notation.

B. Use the coordinate axes at the right to graph this equation for values of t between 0 and 5.

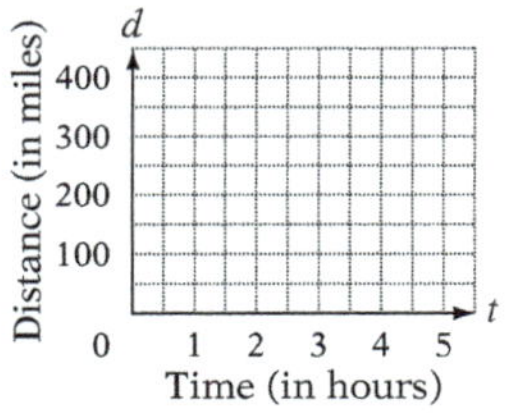

C. The point whose coordinates are (3, 120) is on the graph. Write a sentence that explains the meaning of this ordered pair.

Solution See page S13.

Try Exercise 61, page 258.

5.5 Exercises

CONCEPT CHECK

1. Determine whether the equation is a linear equation. If it is, write it in function notation.

a. $y = -\frac{3}{5}x - 2$

b. $y = x + 1$

c. $y = x^2 + 5$

d. $y^2 = 3x + 6$

e. $y = 2x - 4$

f. $y = 6$

2. The graphs of $y = \frac{1}{4}x - 6$ and $f(x) =$ ___?___ are identical

3. For the relation graphed at the right, the domain is {___?___}.

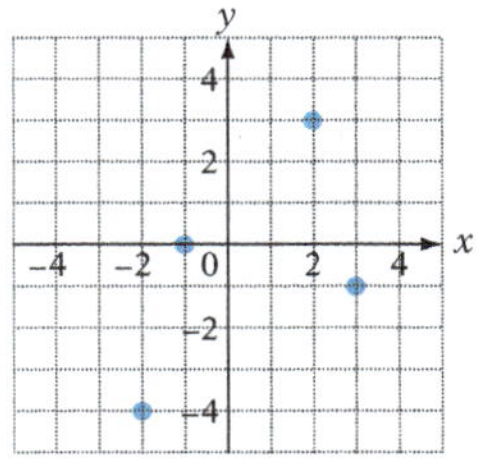

4. The value of the function $\{(-3, 3), (-2, 2), (-1, 1), (0, 0)\}$ at -2 is ___?___.

1 Introduction to functions (See pages 249–252.)

GETTING READY

5. A relation is a set of ___?___. The set of first coordinates of the ordered pairs is called the ___?___ of the relation. The set of second coordinates is called the ___?___ of the relation.

6. a. The symbol $f(x)$ is read "f ___?___ x"or "the value of f ___?___ x." It is a symbol for the number that the function f pairs with ___?___.

b. The symbol $f(3)$ is a symbol for the number that the function f pairs with ___?___.

c. If $f(x) = 4x - 1$, then $f(3) = 4($___?___$) - 1 =$ ___?___. This means that (___?___, ___?___) is an ordered pair of the function f.

7. **Marathons** See the news clipping at the right. The table below shows the ages and finishing times of the top eight finishers in the Manhattan Island Swim. Write a relation in which the first coordinate is the age of a swimmer and the second coordinate is the swimmer's finishing time. Is the relation a function?

Ages (in years)	35	45	38	24	47	51	35	48
Time (in hours)	7.50	7.58	7.63	7.78	7.80	7.86	7.89	7.92

In the News

Swimmers Go the Distance

Twenty-three swimmers completed NYC Swim's annual Manhattan Island Swim. Swimmers begin at Battery Park City–South Cove and swim a 28.5-mile course around Manhattan Island. The 35-year-old first-place finisher swam the distance in 7 h, 30 min, and 15 s.

Source: www.nycswim.org

8. **Health** The table at the right shows the United States Department of Agriculture limit on saturated fat intake, in grams. Write a relation in which the first coordinate is a person's daily Calorie intake and the second coordinate is the limit on saturated fat intake. Is the relation a function?

Daily Calories	Saturated Fat (in grams)
1600	18
2000	20
2200	24
2500	25
2800	31

For Exercises 9 and 10, use the following sets. Set A is the set of all dates of the year ({January 1, January 2, January 3, ...}). Set B is the set of all the people in the world.

9. A relation has domain A and range B. Each ordered pair in the relation is of the form (date, person born on that date). Is this relation a function?

10. A relation has domain B and range A. Each ordered pair in the relation is of the form (person, birth date of that person). Is this relation a function?

11. **Jogging** The table at the right shows the number of Calories a 150-pound person burns in one hour while running at various speeds, in miles per hour. Write a relation in which the first coordinate is the speed of the runner and the second coordinate is the number of Calories burned. Is the relation a function?

Speed (in mph)	Calories
4	411
5	514
6	618
7	720
8	823

12. **Health** The table at the right shows the birth rates of various countries, in births per thousand, and life expectancy, in years. (*Source:* www.cia.gov) Write a relation in which the first coordinate is the birth rate and the second coordinate is the life expectancy. Is the relation a function?

Country	Birth Rate	Life Expectancy
Belgium	10.48	78.6
Brazil	16.8	71.7
Martinique	14.14	79.0
Mexico	21.0	75.2
Sweden	10.36	80.4
United States	14.14	77.7

Complete the sentence using the words *domain* and *range*.

13. For the function $f(x) = 3x - 4$, $f(-1) = -7$. The number -1 is in the __?__ of the function, and the number -7 is in the __?__ of the function.

14. $f(a)$ represents a value in the __?__ of a function when a is a value in the __?__ of the function.

Find the domain and range of the relation. State whether or not the relation is a function.

15. $\{(0, 0), (2, 0), (4, 0), (6, 0)\}$

16. $\{(-2, 2), (0, 2), (1, 2), (2, 2)\}$

17. $\{(2, 2), (2, 4), (2, 6), (2, 8)\}$

18. $\{(-4, 4), (-2, 2), (0, 0), (-2, -2)\}$

19. $\{(0, 0), (1, 1), (2, 2), (3, 3)\}$

20. $\{(0, 5), (1, 4), (2, 3), (3, 2), (4, 1), (5, 0)\}$

21. $\{(-2, -3), (2, 3), (-1, 2), (1, 2), (-3, 4), (3, 4)\}$

22. $\{(-1, 0), (0, -1), (1, 0), (2, 3), (3, 5)\}$

Evaluate the function at the given value of x. Write an ordered pair that is an element of the function.

23. $f(x) = 4x; x = 10$

24. $f(x) = 8x; x = 11$

25. $f(x) = x - 5; x = -6$

26. $f(x) = x + 7; x = -9$

27. $f(x) = 3x^2; x = -2$

28. $f(x) = x^2 - 1; x = -8$

29. $f(x) = 5x + 1; x = \frac{1}{2}$

30. $f(x) = 2x - 6; x = \frac{3}{4}$

31. $f(x) = \frac{2}{5}x + 4; x = -5$

32. $f(x) = \frac{3}{2}x - 5; x = 2$

33. $f(x) = 2x^2; x = -4$

34. $f(x) = 4x^2 + 2; x = -3$

Find the range of the function defined by the given equation. Write five ordered pairs that belong to the function.

35. $f(x) = 3x - 4$; domain $= \{-5, -3, -1, 1, 3\}$

36. $f(x) = 2x + 5$; domain $= \{-10, -5, 0, 5, 10\}$

37. $f(x) = \frac{1}{2}x + 3$; domain $= \{-4, -2, 0, 2, 4\}$

38. $f(x) = \frac{3}{4}x - 1$; domain $= \{-8, -4, 0, 4, 8\}$

39. $f(x) = x^2 + 6$; domain $= \{-3, -1, 0, 1, 3\}$

40. $f(x) = 3x^2 + 6$; domain $= \{-2, -1, 0, 1, 2\}$

2 Graphs of linear functions (See pages 252-254.)

GETTING READY

41. In the linear equation $y = 5x - 9$, the independent variable is __?__ and the dependent variable is __?__. To write this equation in function notation, replace y with the symbol __?__.

42. To graph the function $f(x) = \frac{1}{2}x + 4$, replace $f(x)$ with __?__. This is the equation of a line with y-intercept __?__ and slope __?__.

Graph.

43. $f(x) = 5x$

44. $f(x) = -4x$

45. $f(x) = x + 2$

46. $f(x) = x - 3$

47. $f(x) = 6x - 1$

48. $f(x) = 3x + 4$

49. $f(x) = -2x + 3$

50. $f(x) = -5x - 2$

51. $f(x) = \frac{1}{3}x - 4$

52. $f(x) = \frac{3}{5}x + 1$

53. $f(x) = 4$

54. $f(x) = -3$

Graph using a graphing calculator.

55. $f(x) = 2x - 1$

56. $f(x) = -3x - 1$

57. $f(x) = -\frac{1}{2}x + 1$

58. $f(x) = \frac{2}{3}x + 4$

59. $f(x) = \frac{5}{2}x$

60. $f(x) = -1$

61. Taxi Fares Read the news clipping at the right. A passenger of a San Francisco taxi can use the equation $F = 2.25M + 2.65$ to calculate the fare F, in dollars, for a ride of M miles.

a. Write the equation in function notation.

b. Use the coordinate axes at the right to graph the equation for values of M between 1 and 5.

c. The point (3, 9.40) is on the graph. Write a sentence that explains the meaning of this ordered pair.

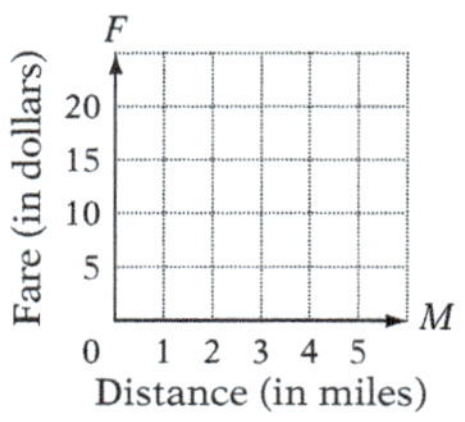

In the News

First Hybrid Taxi Cabs Retire

The nation's first hybrid taxis, 15 San Francisco vehicles, have reached the 300,000-mile retirement mark. With reduced gas costs, drivers of the hybrid cabs kept up to \$9000 more per year of their passengers' fares (\$3.10 for the first 0.2 mi plus \$.45 for each additional 0.2 mi) than drivers of conventional cabs.

Source: Los Angeles Times

62. Depreciation Depreciation is the declining value of an asset. For instance, a company purchases a truck for \$20,000. The truck is an asset worth \$20,000. In 5 years, however, the value of the truck will have declined, and it may be worth only \$4000. An equation that represents this decline is $V = 20{,}000 - 3200x$, where V is the value, in dollars, of the truck after x years.

a. Write the equation in function notation.

b. Use the coordinate axes at the right to graph the equation for values of x between 0 and 5.

c. The point (4, 7200) is on the graph. Write a sentence that explains the meaning of this ordered pair.

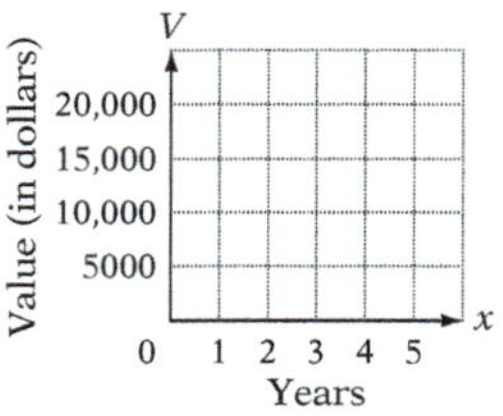

63. Depreciation A company uses the equation $V = 30{,}000 - 5000x$ to estimate the depreciated value, in dollars, of a computer. (See Exercise 62.)

a. Write the equation in function notation.

b. Use the coordinate axes at the right to graph the equation for values of x between 0 and 5.

c. The point (1, 25,000) is on the graph. Write a sentence that explains the meaning of this ordered pair.

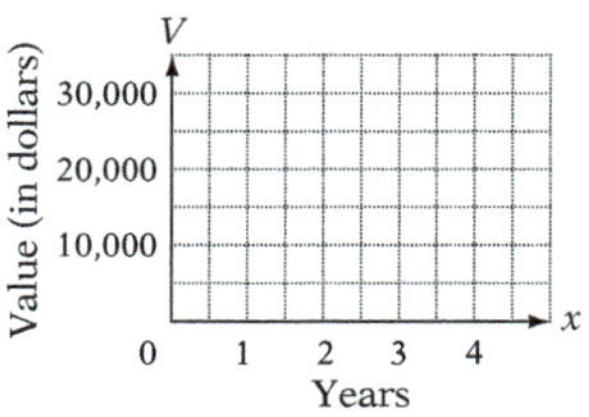

64. **Business** A rental car company charges a "drop-off" fee of \$50 to return a car to a location different than that from which it was rented. In addition, it charges a fee of \$.18 per mile the car is driven. An equation that represents the total cost to rent a car from this company is $C = 0.18m + 50$, where C is the total cost, in dollars, and m is the number of miles the car is driven.

a. Write the equation in function notation.

b. Use the coordinate axes at the right to graph the equation for values of m between 0 and 1000.

c. The point (500, 140) is on the graph. Write a sentence that explains the meaning of this ordered pair.

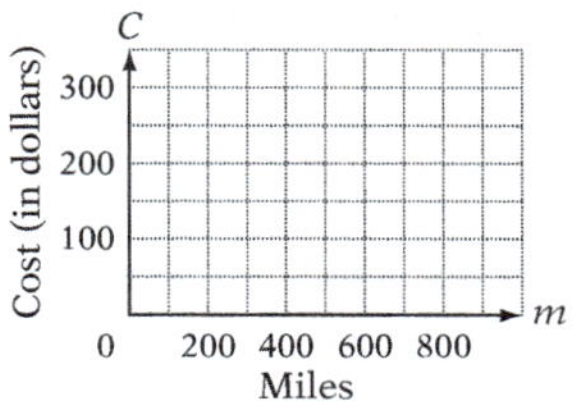

APPLYING CONCEPTS

65. **a.** A function f consists of the ordered pairs $\{(-4, -6), (-2, -2), (0, 2), (2, 6), (4, 10)\}$. Find $f(2)$.

b. A function f consists of the ordered pairs $\{(0, 9), (1, 8), (2, 7), (3, 6), (4, 5)\}$. Find $f(1)$.

66. One of the functions represented in the tables below is linear. Determine which of the functions is linear and explain why it is linear.

x	$g(x)$
1	22.50
2	25.19
3	28.20
4	31.57
5	35.35
6	39.58

x	$h(x)$
1	2.3
2	2.6
3	2.9
4	3.2
5	3.5
6	3.8

67. **Height-Time Graphs** A child's height is a function of the child's age. The graph of this function is not linear, because children go through growth spurts as they develop. However, for the graph to be reasonable, the function must be an increasing function (that is, as the age increases, the height increases), because children do not get shorter as they grow older. Match each function described below with a reasonable graph of the function.

a. The height of a plane above the ground during take-off depends on how long it has been since the plane left the gate.

b. The height of a football above the ground is related to the number of seconds that have passed since it was punted.

c. A basketball player is dribbling a basketball. The basketball's distance from the floor is related to the number of seconds that have passed since the player began dribbling the ball.

d. Two children are seated together on a roller coaster. The height of the children above the ground depends on how long they have been on the ride.

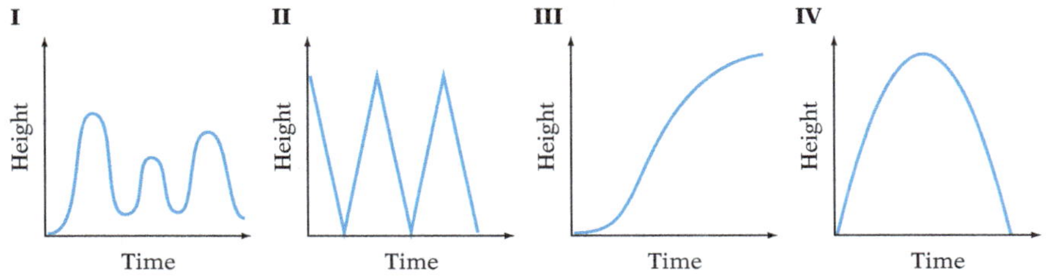

PROJECTS OR GROUP ACTIVITIES

68. Investigating a relationship between two variables is an important task in the application of mathematics. For example, botanists study the relationship between the number of bushels of wheat yielded per acre and the amount of watering per acre. Environmental scientists study the relationship between the incidence of skin cancer and the amount of ozone in the atmosphere. Business analysts study the relationship between the price of a product and the number of products that are sold at that price. Describe a relationship that is important to your major field of study.

69. Functions are a part of our everyday lives. For example, the cost to mail a package via first-class mail is a function of the weight of the package. The tuition paid by a part-time student is a function of the number of credit hours the student registers for. Provide other examples of functions.

70. Define three situations that are relations but not functions. For example, the set of ordered pairs in which the first coordinates are the runs scored by a baseball team and the second coordinates are either W for a win or L for a loss.

71. Evaluate $f(x) = x^2 + 4x + 6$ for $x = -5$ and $x = 1$. Based on the results, if $f(a) = f(b)$, does it follow that $a = b$?

5.6 Graphing Linear Inequalities

OBJECTIVE 1

Graph inequalities in two variables

Point of Interest

Linear inequalities play an important role in applied mathematics. They are used in a branch of mathematics called *linear programming,* which was developed during World War II to aid in supplying the Air Force with the machine parts necessary to keep planes flying. Today applications of linear programming have broadened to include many other disciplines.

The graph of the linear equation $y = x - 2$ separates a plane into three sets:

the set of points on the line

the set of points above the line

the set of points below the line

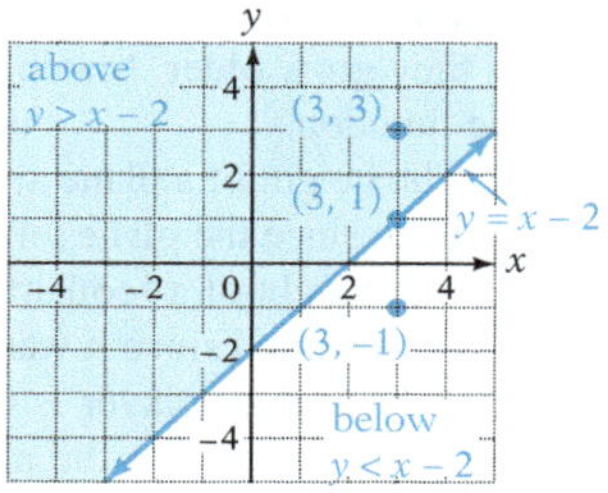

The point whose coordinates are (3, 1) is a solution of $y = x - 2$.

$$\begin{array}{c|c} y & x - 2 \\ \hline 1 & 3 - 2 \\ 1 & = 1 \end{array}$$

The point whose coordinates are (3, 3) is a solution of $y > x - 2$.

$$\begin{array}{c|c} y & x - 2 \\ \hline 3 & 3 - 2 \\ 3 & > 1 \end{array}$$

Any point above the line is a solution of $y > x - 2$.

The point whose coordinates are (3, −1) is a solution of $y < x - 2$.

$$\begin{array}{c|c} y & x - 2 \\ \hline -1 & 3 - 2 \\ -1 & < 1 \end{array}$$

Any point below the line is a solution of $y < x - 2$.

The solution set of $y = x - 2$ is all points on the line. The solution set of $y > x - 2$ is all points above the line. The solution set of $y < x - 2$ is all points below the line. The solution set of an inequality in two variables is a **half-plane.**

The following illustrates the procedure for graphing a linear inequality.

Focus on graphing a linear inequality

Graph the solution set of $2x + 3y \le 6$.

Solve the inequality for y.

$$2x + 3y \le 6$$
$$2x - 2x + 3y \le -2x + 6$$
$$3y \le -2x + 6$$
$$\frac{3y}{3} \le \frac{-2x + 6}{3}$$
$$y \le -\frac{2}{3}x + 2$$

Change the inequality to an equality and graph the line. If the inequality is $\ge$ or $\le$, the line is part of the solution set and is shown by a **solid line.** If the inequality is $>$ or $<$, the line is not part of the solution set and is shown by a **dashed line.**

$$y = -\frac{2}{3}x + 2$$

If the inequality is of the form $y > mx + b$ or $y \ge mx + b$, shade the **upper half-plane.** If the inequality is of the form $y < mx + b$ or $y \le mx + b$, shade the **lower half-plane.**

The equation $y \le -\frac{2}{3}x + 2$ is of the form $y \le mx + b$. Draw a solid line and shade the lower half-plane.

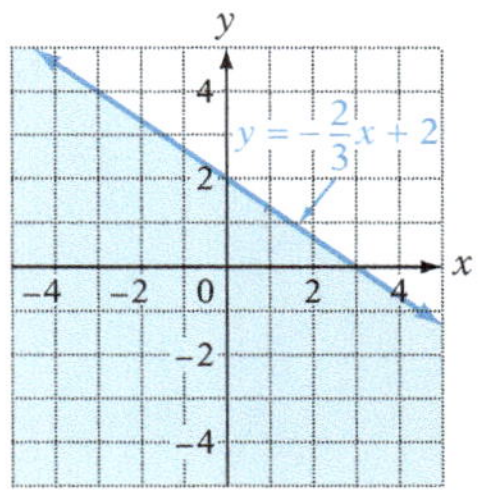

The inequality $2x + 3y \le 6$ can also be graphed as shown below.

Change the inequality to an equality.

$$2x + 3y = 6$$

Find the x- and y-intercepts of the equation. To find the x-intercept, let $y = 0$.

$$2x + 3(0) = 6$$
$$2x = 6$$
$$x = 3$$

The x-intercept is (3, 0).

To find the y-intercept, let $x = 0$.

$$2(0) + 3y = 6$$
$$3y = 6$$
$$y = 2$$

The y-intercept is (0, 2).

Graph the ordered pairs (3, 0) and (0, 2). Draw a solid line through the points because the inequality is $\le$.

Take Note

Any ordered pair is of the form (x, y). For the point (0, 0), substitute 0 for x and 0 for y in the inequality.

The point (0, 0) can be used to determine which region to shade. If (0, 0) is a solution of the inequality, then shade the region that includes the point (0, 0).

If (0, 0) is not a solution of the inequality, then shade the region that does not include the point (0, 0).

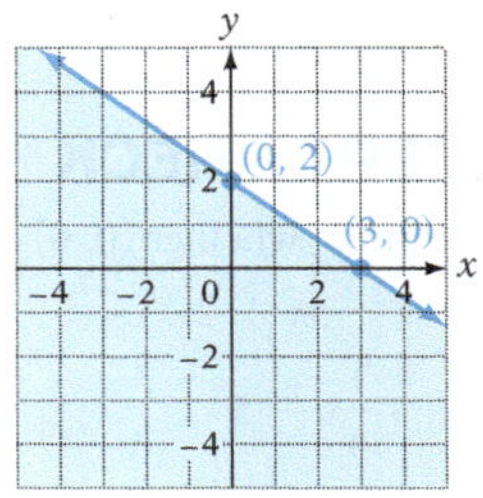

For this example, (0, 0) is a solution of the inequality. The region that contains the point (0, 0) is shaded.

$$2x + 3y \le 6$$
$$2(0) + 3(0) \le 6$$
$$0 \le 6 \quad \text{True}$$

If the line passes through the point (0, 0), another point must be used to determine which region to shade. For example, use the point (1, 0).

It is important to note that every point in the shaded region is a solution of the inequality and that every solution of the inequality is a point in the shaded region. No point outside of the shaded region is a solution of the inequality.

EXAMPLE 1 Graph the solution set of $3x + y > -2$.

Solution

$$3x + y > -2$$
$$3x - 3x + y > -3x - 2$$
$$y > -3x - 2$$

- **Solve the inequality for y.**

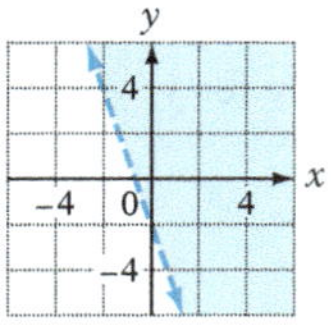

- **Graph $y = -3x - 2$ as a dashed line. Shade the upper half-plane.**

Problem 1 Graph the solution set of $x - 3y < 2$.

Solution See page S14.

Try Exercise 25, page 264.

EXAMPLE 2 Graph the solution set of $y > 3$.

Solution

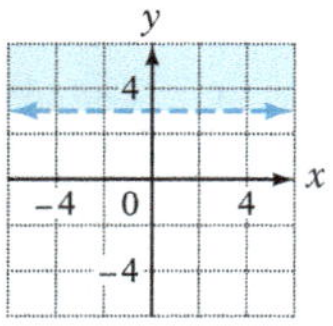

- **The inequality is solved for y. Graph $y = 3$ as a dashed line. Shade the upper half-plane.**

Problem 2 Graph the solution set of $x < 3$.

Solution See page S14.

Try Exercise 21, page 263.

5.6 Exercises

CONCEPT CHECK

1. Determine whether $(0, 0)$ is a solution of the inequality.

   a. $y < -5x + 2$ **b.** $y > x + 1$

   c. $y \leq \frac{1}{4}x - 5$ **d.** $y \geq -\frac{2}{3}x - 6$

2. When graphing the solution set of an inequality, draw a solid line when the inequality symbol is __?__ or __?__.

3. When graphing the solution set of an inequality, draw a dashed line when the inequality symbol is __?__ or __?__.

4. When graphing the solution set of an inequality of the form $y > mx + b$, should you shade the upper half-plane or the lower half-plane?

5. When graphing the solution set of an inequality of the form $y \leq mx + b$, should you shade the upper half-plane or the lower half-plane?

6. Describe the difference between the graph of $y \leq 3x + 1$ and the graph of $y < 3x + 1$.

1 Graph inequalities in two variables (See pages 260–262.)

GETTING READY

7. Fill in the blanks with "are" or "are not."
In the graph of a linear inequality, points on a solid line __?__ elements of the solution set of the inequality. Points on a dashed line __?__ elements of the solution set of the inequality.

8. In the graph of a linear inequality, the shaded portion of the graph represents the __?__ of the inequality.

Graph the solution set.

9. $y > 2x + 3$

10. $y > 3x - 9$

11. $y > \frac{3}{2}x - 4$

12. $y > -\frac{5}{4}x + 1$

13. $y \leq -\frac{3}{4}x - 1$

14. $y \leq -\frac{5}{2}x - 4$

15. $y \leq -\frac{6}{5}x - 2$

16. $y < \frac{4}{5}x + 3$

17. $x + y > 4$

18. $x - y > -3$

19. $2x + y \geq 4$

20. $3x + y \geq 6$

21. $y \leq -2$

22. $y > 3$

23. $2x + 3y \leq -6$

24. $-4x + 3y < -12$

25. $5x - 2y > 10$

26. $3x - 5y \geq -15$

27. If (0, 0) is a point on the graph of the linear inequality $Ax + By > C$, where C is not zero, is C positive or negative?

28. If $Ax + By < C$, where C is a negative number, is (0, 0) a point on the graph of this linear inequality?

APPLYING CONCEPTS

Write the inequality given its graph.

29.

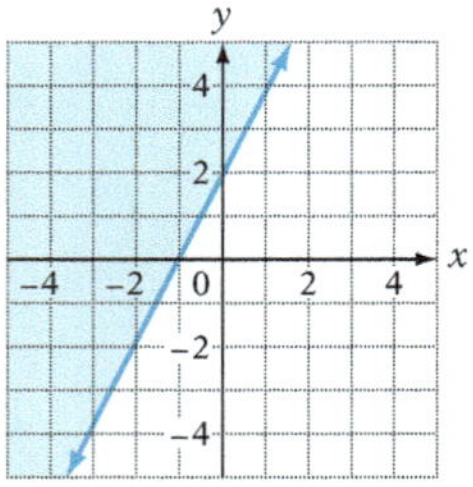

30.

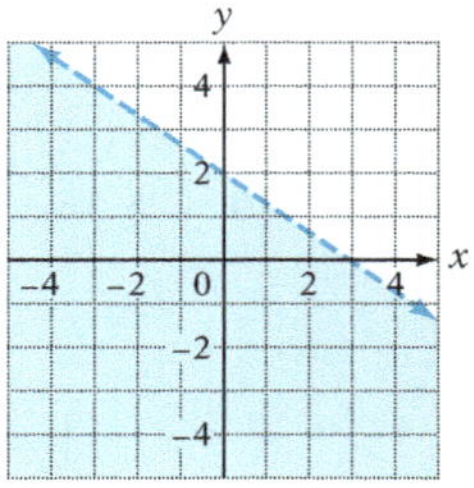

31.

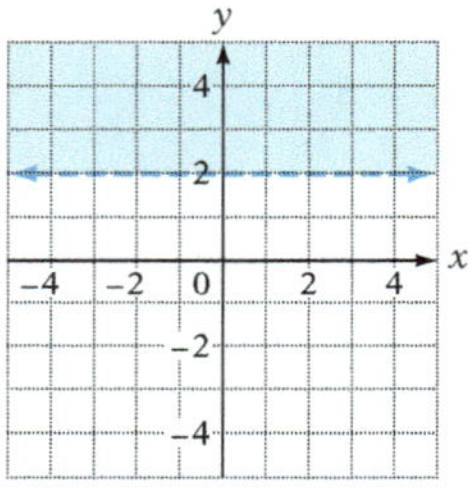

32.

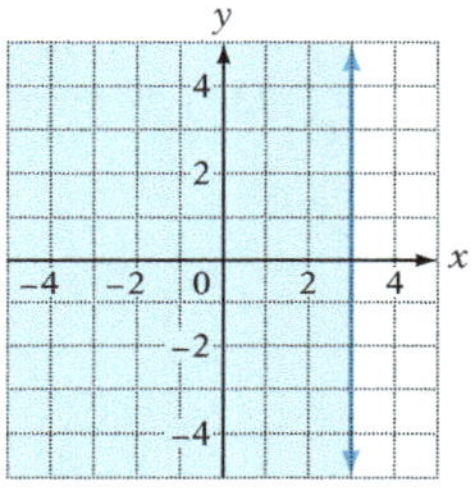

Graph the solution set.

33. $y - 5 < 4(x - 2)$

34. $y + 3 < 6(x + 1)$

35. $3x - 2(y + 1) \leq y - (5 - x)$

36. $2x - 3(y + 1) \geq y - (7 - x)$

PROJECTS OR GROUP ACTIVITIES

37. Does an inequality in two variables define a relation? Why or why not? Does an inequality in two variables define a function? Why or why not?

38. Are there any points whose coordinates satisfy both $y < 2x - 3$ and $y > -\frac{1}{4}x + 1$? If so, give the coordinates of three such points. If not, explain why not.

39. Are there any points whose coordinates satisfy both $y > 3x + 1$ and $y < 3x - 4$? If so, give the coordinates of three such points. If not, explain why not.

CHAPTER 5 Summary

Key Words	Objective and Page Reference	Examples
A **rectangular coordinate system** is formed by two number lines, one horizontal and one vertical, that intersect at the zero point of each line. The number lines that make up a rectangular coordinate system are called the **coordinate axes,** or simply the **axes.** The **origin** is the point of intersection of the two coordinate axes. Generally, the horizontal axis is labeled the **x-axis,** and the vertical axis is labeled the **y-axis.** A rectangular coordinate system divides the plane determined by the axes into four regions called **quadrants.**	[5.1.1, p. 202]	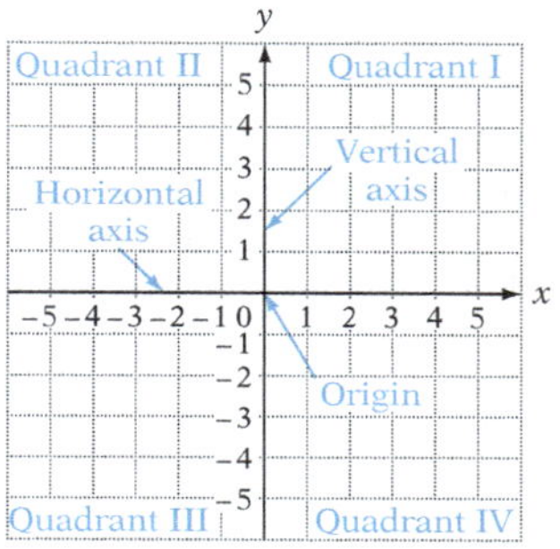
The coordinate axes determine a **plane.** Every point in the plane can be identified by an **ordered pair** (x, y). The first number in an ordered pair is called the **x-coordinate** or the **abscissa.** The second number is called the **y-coordinate** or the **ordinate.** The **coordinates** of a point are the numbers in the ordered pair associated with the point. To **graph a point in the plane,** place a dot at the location given by the ordered pair. The **graph of an ordered pair** is the dot drawn at the coordinates of the point in the plane.	[5.1.1, p. 202]	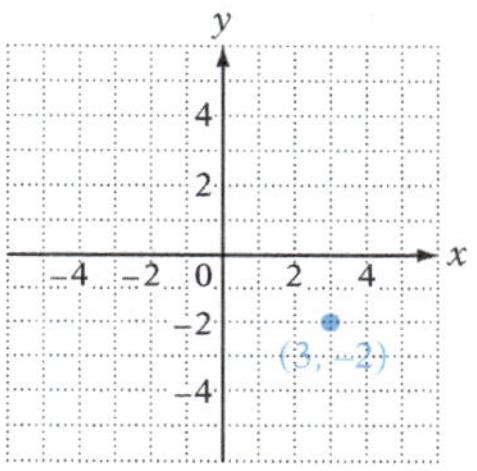 The graph of $(3, -2)$ is shown above. The ordered pair $(3, -2)$ has x-coordinate 3 and y-coordinate -2.
An equation of the form $y = mx + b$, where m is the coefficient of x and b is a constant, is a **linear equation in two variables.** A **solution of a linear equation in two variables** is an ordered pair (x, y) that makes the equation a true statement.	[5.2.1, p. 214]	$y = 2x + 3$ is an example of a linear equation in two variables. The ordered pair $(1, 5)$ is a solution of this equation because when 1 is substituted for x and 5 is substituted for y, the result is a true equation.
The **graph of an equation in two variables** is a drawing of the ordered-pair solutions of the equation. For a linear equation in two variables, the graph is a straight line.	[5.2.2, p. 216]	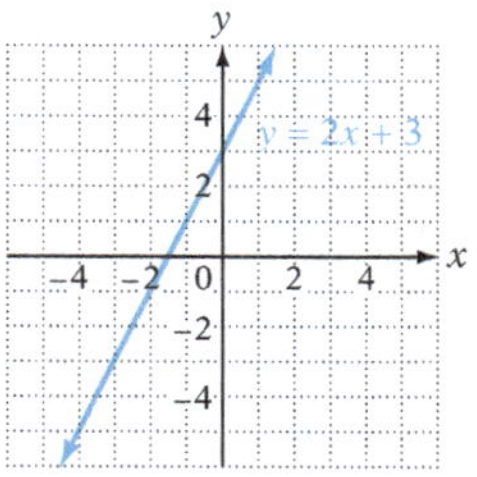 The graph of $y = 2x + 3$ is shown above.
The point at which a graph crosses the x-axis is called the **x-intercept.** At the x-intercept, the y-coordinate is 0. The point at which a graph crosses the y-axis is called the **y-intercept.** At the y-intercept, the x-coordinate is 0.	[5.2.3, p. 221]	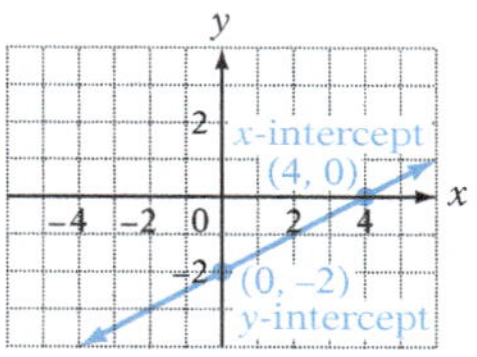

An equation of the form $Ax + By = C$ is also a **linear equation in two variables.**	[5.2.3, p. 218]	$5x - 2y = 10$ is an example of an equation in $Ax + By = C$ form.
The **graph of $y = b$** is a horizontal line with y-intercept $(0, b)$. The **graph of $x = a$** is a vertical line with x-intercept $(a, 0)$.	[5.2.3, p. 222]	The graph of $y = -3$ is a horizontal line with y-intercept $(0, -3)$. The graph of $x = 2$ is a vertical line with x-intercept $(2, 0)$.
The **slope** of a line is a measure of the slant of the line. The symbol for slope is m. A line that slants upward to the right has a **positive slope.** A line that slants downward to the right has a **negative slope.** A horizontal line has **zero slope.** The slope of a vertical line is **undefined.**	[5.3.1, pp. 228–231]	For the graph of $y = 4x - 3$, $m = 4$; the slope is positive. For the graph of $y = -6x + 1$, $m = -6$; the slope is negative. For the graph of $y = -7$, $m = 0$; the slope is 0. For the graph of $x = 8$, the slope is undefined.
Two lines in the rectangular coordinate system that never intersect are **parallel lines.** Parallel lines have the same slope.	[5.3.1, p. 231]	The lines $y = -2x + 1$ and $y = -2x - 3$ have the same slope and different y-intercepts. The lines are parallel.
A **relation** is any set of ordered pairs. The **domain** of a relation is the set of first coordinates of the ordered pairs. The **range** is the set of second coordinates of the ordered pairs. A **function** is a relation in which no two ordered pairs have the same first coordinate and different second coordinates.	[5.5.1, p. 250]	For the relation $\{(-2, -5), (0, -3), (4, -1)\}$, the domain is $\{-2, 0, 4\}$ and the range is $\{-5, -3, -1\}$. $\{(-2, -5), (0, -3), (4, -1)\}$ is a function. $\{(-2, -5), (-2, -3), (4, -1)\}$ is not a function because two ordered pairs have the same x-coordinate, -2, and different y-coordinates.
A function of the form $f(x) = mx + b$ is a **linear function.** Its graph is a straight line.	[5.5.2, p. 252]	$f(x) = -3x + 8$ is an example of a linear function.
The solution set of a linear inequality in two variables is a **half-plane.**	[5.6.1, p. 260]	The solution set of $y > x - 2$ is all points above the line $y = x - 2$. The solution set of $y < x - 2$ is all points below the line $y = x - 2$.

Essential Rules and Procedures	Objective and Page Reference	Examples
Average Rate of Change The average rate of change of y with respect to x is $\frac{\text{change in } y}{\text{change in } x}$.	[5.1.3, p. 206]	In 1979, the average price of a movie theater ticket was \$2.47. In 2009, the average price was \$7.50. (*Source:* www.natoonline.org) $\frac{7.50 - 2.47}{2009 - 1979} = \frac{5.03}{30} = 0.17$ To the nearest cent, the average annual rate of change in the price of a movie theater ticket was \$.17.

To find the *x*-intercept, let $y = 0$.
To find the *y*-intercept, let $x = 0$.

[5.2.3, p. 221]

To find the x-intercept of $4x - 3y = 12$, let $y = 0$. To find the y-intercept, let $x = 0$.

$$\begin{aligned} 4x - 3y &= 12 \\ 4x - 3(0) &= 12 \\ 4x - 0 &= 12 \\ 4x &= 12 \\ x &= 3 \end{aligned} \qquad \begin{aligned} 4x - 3y &= 12 \\ 4(0) - 3y &= 12 \\ 0 - 3y &= 12 \\ -3y &= 12 \\ y &= -4 \end{aligned}$$

The x-intercept is $(3, 0)$. The y-intercept is $(0, -4)$.

Slope of a Linear Equation

Slope $= m = \dfrac{y_2 - y_1}{x_2 - x_1}, x_2 \neq x_1$

[5.3.1, p. 229]

To find the slope of the line between the points $(2, -3)$ and $(-1, 6)$, let $P_1 = (2, -3)$ and $P_2 = (-1, 6)$. Then $(x_1, y_1) = (2, -3)$ and $(x_2, y_2) = (-1, 6)$.

$$m = \frac{y_2 - y_1}{x_2 - x_1} = \frac{6 - (-3)}{-1 - 2} = \frac{9}{-3} = -3$$

Slope-Intercept Form of a Linear Equation

$y = mx + b$

[5.3.2, p. 234]

For the equation $y = -4x + 7$, the slope is $m = -4$ and the y-intercept is $(0, b) = (0, 7)$.

Point-Slope Formula

$y - y_1 = m(x - x_1)$

[5.4.2, p. 243]

To find the equation of the line that contains the point $(6, -1)$ and has slope 2, let $(x_1, y_1) = (6, -1)$ and $m = 2$.

$$\begin{aligned} y - y_1 &= m(x - x_1) \\ y - (-1) &= 2(x - 6) \\ y + 1 &= 2x - 12 \\ y &= 2x - 13 \end{aligned}$$

Function Notation

The equation of a function is written in function notation when y is replaced by the symbol $f(x)$, where $f(x)$ is read "f of x" or "the value of f at x." To evaluate a function at a given value of x, replace x by the given value and then simplify the resulting numerical expression to find the value of $f(x)$.

[5.5.1, p. 251]

$y = x^2 + 2x - 1$ is written in function notation as $f(x) = x^2 + 2x - 1$. To evaluate $f(x) = x^2 + 2x - 1$ at $x = -3$, find $f(-3)$.

$$\begin{aligned} f(-3) &= (-3)^2 + 2(-3) - 1 \\ &= 9 - 6 - 1 \\ &= 2 \end{aligned}$$

CHAPTER 5 Review Exercises

1. Find the ordered-pair solution of $y = -\frac{2}{3}x + 2$ that corresponds to $x = 3$.
2. Find the equation of the line that contains the point whose coordinates are $(0, -1)$ and has slope 3.
3. Graph: $x = -3$
4. Graph the ordered pairs $(-3, 1)$ and $(0, 2)$.

5. Evaluate $f(x) = 3x^2 + 4$ at $x = -5$.

6. Find the slope of the line that contains the points whose coordinates are $(3, -4)$ and $(1, -4)$.

7. Graph: $y = 3x + 1$

8. Graph: $3x - 2y = 6$

9. Find the equation of the line that contains the point whose coordinates are $(-1, 2)$ and has slope $-\frac{2}{3}$.

10. Find the x- and y-intercepts for $6x - 4y = 12$.

11. Graph the line that has slope $\frac{1}{2}$ and y-intercept $(0, -1)$.

12. Graph: $f(x) = -\frac{2}{3}x + 4$

13. Find the equation of the line that contains the point whose coordinates are $(-3, 1)$ and has slope $\frac{2}{3}$.

14. Find the slope of the line that contains the points whose coordinates are $(2, -3)$ and $(4, 1)$.

15. Evaluate $f(x) = \frac{3}{5}x + 2$ at $x = -10$.

16. Find the domain and range of the relation $\{(-20, -10), (-10, -5), (0, 0), (10, 5)\}$. Is the relation a function?

17. Graph: $y = -\frac{3}{4}x + 3$

18. Graph the line that has slope 2 and y-intercept $(0, -2)$.

19. Graph the ordered pairs $(-2, -3)$ and $(2, 4)$.

20. Graph: $f(x) = 5x + 1$

21. Find the x- and y-intercepts for $2x - 3y = 12$.

22. Find the equation of the line that contains the point whose coordinates are $(-1, 0)$ and has slope 2.

23. Graph: $y = 3$

24. Graph the solution set of $3x + 2y \leq 12$.

25. Is the line containing the points $(-2, 1)$ and $(3, 5)$ parallel to the line that contains the points $(4, -5)$ and $(9, -1)$?

26. Find the equation of the line through the points whose coordinates are $(4, 14)$ and $(-8, -1)$.

27. Graph the line that has slope -1 and y-intercept $(0, 2)$.

28. Graph: $2x - 3y = 6$

29. Find the ordered-pair solution of $y = 2x - 1$ that corresponds to $x = -2$.

30. Find the equation of the line that contains the point (0, 2) and has slope -3.

31. Graph: $f(x) = 2x$

32. Evaluate $f(x) = 3x - 5$ at $x = \frac{5}{3}$.

33. Find the equation of the line through the points whose coordinates are $(-6, 0)$ and $(3, -3)$.

34. Find the equation of the line that contains the point whose coordinates are $(2, -1)$ and has slope $\frac{1}{2}$.

35. Is $(-10, 0)$ a solution of $y = \frac{1}{5}x + 2$?

36. Find the ordered-pair solution of $y = 4x - 9$ that corresponds to $x = 2$.

37. Find the x- and y-intercepts for $4x - 3y = 0$.

38. Find the equation of the line that contains the point whose coordinates are $(-2, 3)$ and has zero slope.

39. Graph the solution set of $6x - y > 6$.

40. Graph the line that has slope -3 and y-intercept (0, 1).

41. Find the equation of the line that contains the point whose coordinates are $(0, -4)$ and has slope 3.

42. Graph: $x + 2y = -4$

43. Find the domain and range of the relation $\{(-10, -5), (-5, 0), (5, 0), (-10, 0)\}$. Is the relation a function?

44. Find the range of the function given by the equation $f(x) = 3x + 7$ if the domain is $\{-20, -10, 0, 10, 20\}$.

45. Find the range of the function given by the equation $f(x) = \frac{1}{3}x + 4$ if the domain is $\{-6, -3, 0, 3, 6\}$.

46. Is the line containing the points $(-3, 6)$ and $(2, -4)$ perpendicular to the line that contains the points $(-5, -1)$ and $(7, 5)$?

47. **Airline Industry** The major airlines routinely overbook flights, but generally passengers voluntarily give up their seats when offered free airline tickets as compensation. The table below gives the number of passengers who voluntarily gave up their seats and the number who involuntarily gave up their seats during a five-year period. Numbers are rounded to the nearest ten thousand. (*Source:* USA TODAY analysis of Department of Transportation data) Graph the scatter diagram for these data.

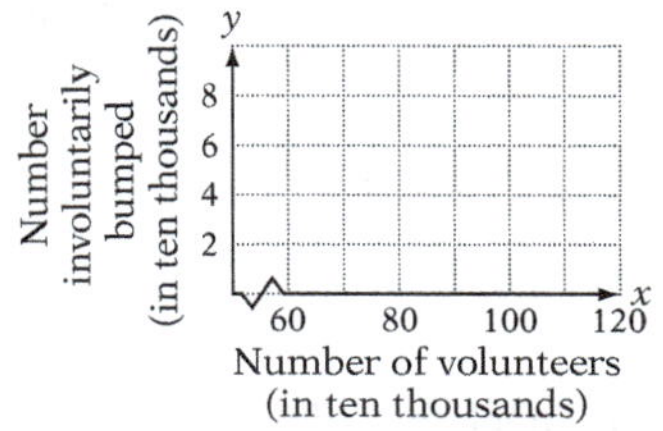

Number of volunteers, in ten thousands, x	60	77	79	90	102
Number involuntarily bumped, in ten thousands, y	4	5	5	6	5

48. Automobile Tires The following table shows the tread depth, in millimeters, of a tire and the number of miles, in thousands, that have been driven on that tire.

Miles driven	25	35	40	20	45
Tread depth	4.8	3.5	2.1	5.5	1.0

Write a relation in which the first coordinate is the number of miles driven and the second coordinate is the tread depth. Is the relation a function?

49. Building Contractors A contractor uses the equation $C = 70s + 40{,}000$ to estimate the cost of building a new house. In this equation, C is the total cost, in dollars, and s is the number of square feet in the home.

a. Write the equation in function notation.

b. Use the coordinate axes at the right to graph the equation for values of s between 0 and 5000.

c. The point (1500, 145,000) is on the graph. Write a sentence that explains the meaning of this ordered pair.

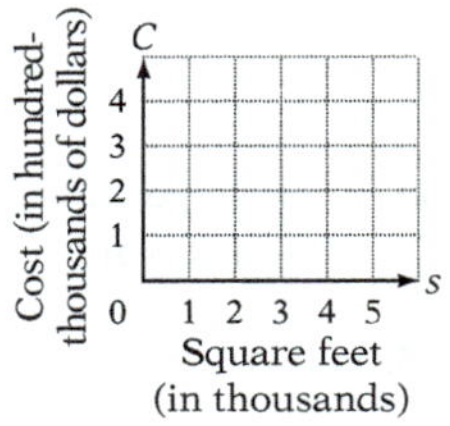

50. Hourly Wages The average hourly wage of a U.S. worker in 1980 was \$6.85. In 2010, the average hourly wage was \$18.97. (*Source:* ftp.bls.gov) Find the average annual rate of change in the average hourly wage from 1980 to 2010.

CHAPTER 5 Test

1. Find the equation of the line that contains the point whose coordinates are $(9, -3)$ and has slope $-\frac{1}{3}$.

2. Find the slope of the line that contains the points whose coordinates are $(9, 8)$ and $(-2, 1)$.

3. Find the x- and y-intercepts for $3x - 2y = 24$.

4. Find the ordered-pair solution of $y = -\frac{4}{3}x - 1$ that corresponds to $x = 9$.

5. Graph: $5x + 3y = 15$

6. Graph: $y = \frac{1}{4}x + 3$

7. Evaluate $f(x) = 4x + 7$ at $x = \frac{3}{4}$.

8. Is $(6, 3)$ a solution of $y = \frac{2}{3}x + 1$?

9. Graph the line that has slope $-\frac{2}{3}$ and y-intercept $(0, 4)$.

10. Graph the solution set of $2x - y \geq 2$.

11. Graph the ordered pairs $(3, -2)$ and $(0, 4)$.

12. Graph the line that has slope 2 and y-intercept $(0, -4)$.

13. Find the equation of the line through the points whose coordinates are $(-5, 5)$ and $(10, 14)$.

14. Find the equation of the line that contains the point whose coordinates are $(0, 7)$ and has slope $-\frac{2}{5}$.

15. Find the equation of the line that contains the point whose coordinates are $(2, 1)$ and has slope 4.

16. Evaluate $f(x) = 4x^2 - 3$ at $x = -2$.

17. Graph: $y = -2x - 1$

18. Graph the solution set of $y > 2$.

19. Graph: $x = 4$

20. Graph the line that has slope $\frac{1}{2}$ and y-intercept $(0, -3)$.

21. Graph: $f(x) = -\frac{2}{3}x + 2$

22. Graph: $f(x) = -5x$

23. Is the line containing the points $(-4, 2)$ and $(1, 5)$ perpendicular to the line that contains the points $(-1, 2)$ and $(-4, 7)$?

24. Find the range of the function given by the equation $f(x) = -3x + 5$ if the domain is $\{-7, -2, 0, 4, 9\}$.

25. Is the line containing the points $(-3, -2)$ and $(6, 0)$ parallel to the line that contains the points $(6, -4)$ and $(3, 2)$?

26. **Drivers** Drivers over age 70 number about 18 million today, up from about 13 million ten years ago. (*Source:* National Highway Traffic Safety Administration) Find the average annual rate of change in the number of drivers over age 70 for the past decade.

27. **Fire Science** The distance, in miles, from a house to a fire station and the amount, in thousands of dollars, of fire damage that house sustained in a fire are given in the following table. Graph the scatter diagram for these data.

Distance (in miles), x	3.5	4.0	5.5	6.0
Damage (in thousands of dollars), y	25	30	40	35

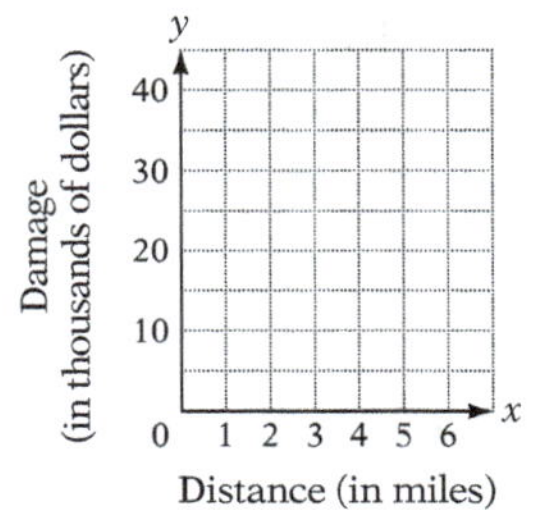

28. **Manufacturing** A company that manufactures toasters has fixed costs of \$1000 each month. The manufacturing cost per toaster is \$8. An equation that represents the total cost to manufacture the toasters is $C = 8t + 1000$, where C is the total cost, in dollars, and t is the number of toasters manufactured each month.

a. Write the equation in function notation.

b. Use the coordinate axes at the right to graph the equation for values of t between 0 and 500.

c. The point $(340, 3720)$ is on the graph. Write a sentence that explains the meaning of this ordered pair.

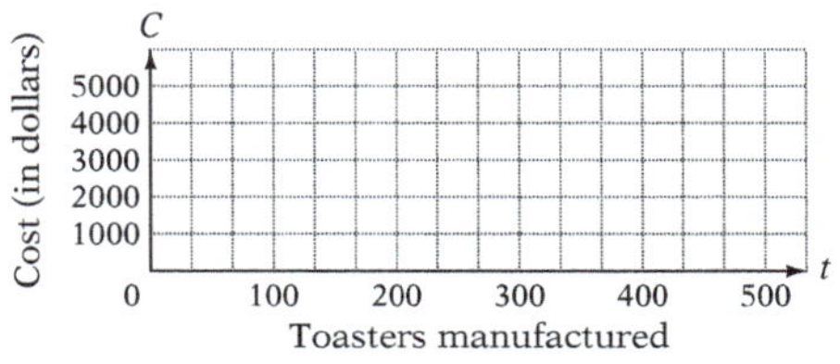

29. Test Scores The data in the following table show a reading test grade and the final exam grade in a history class.

Reading test grade	8.5	9.4	10.1	11.4	12.0
History exam	64	68	76	87	92

Write a relation in which the first coordinate is the reading test grade and the second coordinate is the score on the history final exam. Is the relation a function?

30. Lumber The graph at the right shows the cost, in dollars, per 1000 board-feet of lumber over a six-month period. Find the slope of the line. Write a sentence that states the meaning of the slope.

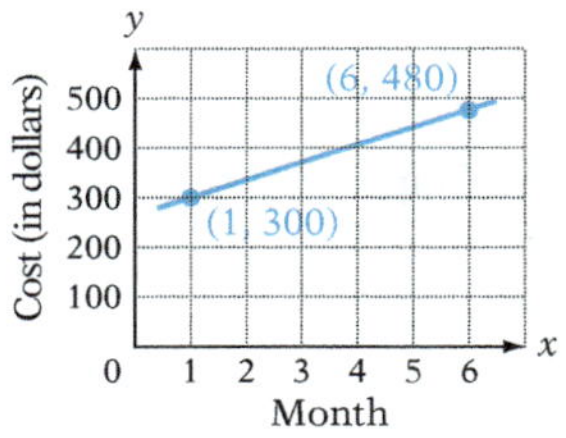

Cumulative Review Exercises

1. Simplify: $12 - 18 \div 3 \cdot (-2)^2$

2. Evaluate $\dfrac{a-b}{a^2-c}$ when $a = -2$, $b = 3$, and $c = -4$.

3. Simplify: $4(2 - 3x) - 5(x - 4)$

4. Solve: $2x - \dfrac{2}{3} = \dfrac{7}{3}$

5. Solve: $3x - 2[x - 3(2 - 3x)] = x - 6$

6. Write $6\frac{2}{3}\%$ as a fraction.

7. Use the roster method to write the set of natural numbers less than 9.

8. Given $D = \{-23, -18, -4, 0, 5\}$, which elements of set D are greater than -16?

9. Solve: $8a - 3 \geq 5a - 6$

10. Solve $4x - 5y = 15$ for y.

11. Find the ordered-pair solution of $y = 3x - 1$ that corresponds to $x = -2$.

12. Find the slope of the line that contains the points whose coordinates are $(2, 3)$ and $(-2, 3)$.

13. Find the x- and y-intercepts for $5x + 2y = 20$.

14. Find the equation of the line that contains the point whose coordinates are $(3, 2)$ and has slope -1.

15. Graph: $y = \dfrac{1}{2}x + 2$

16. Graph: $3x + y = 2$

17. Graph: $f(x) = -4x - 1$

18. Graph the solution set of $x - y \leq 5$.

19. Find the domain and range of the relation $\{(0, 4), (1, 3), (2, 2), (3, 1), (4, 0)\}$. Is the relation a function?

20. Evaluate $f(x) = -4x + 9$ at $x = 5$.

21. Find the range of the function given by the equation $f(x) = -\frac{5}{3}x + 3$ if the domain is $\{-9, -6, -3, 0, 3, 6\}$.

22. Investments An investor deposited a total of $15,000 in two simple interest accounts. On the first account, the annual simple interest rate was 4.5%. On the second account, the annual simple interest rate was 3.2%. How much was invested in the first account if the total annual interest earned on the two accounts was $584?

23. Lever Systems A lever is 8 ft long. A force of 80 lb is applied to one end of the lever, and a force of 560 lb is applied to the other end. Where is the fulcrum located when the system balances? The lever-system equation is $F_1x = F_2(d - x)$.

24. Geometry The perimeter of a triangle is 49 ft. The length of the first side is twice the length of the third side, and the length of the second side is 5 ft more than the length of the third side. Find the length of the first side.

25. Discount A dress that regularly sells for $89 is on sale for 30% off the regular price. Find the sale price.

Systems of Linear Equations

CHAPTER 6

Digital Vision

Focus on Success

Did you read Ask the Authors at the front of this text? If you did, then you know that the authors' advice is that you practice, practice, practice—and then practice some more. The more time you spend doing math outside of class, the more successful you will be in this course. (See Make the Commitment to Succeed, page AIM-3.)

OBJECTIVES

6.1 ❶ Solve systems of linear equations by graphing

6.2 ❶ Solve systems of linear equations by the substitution method

6.3 ❶ Solve systems of linear equations by the addition method

6.4 ❶ Rate-of-wind and rate-of-current problems

❷ Application problems

PREP TEST

Are you ready to succeed in this chapter?
Take the Prep Test below to find out if you are ready to learn the new material.

1. Solve $3x - 4y = 24$ for y.
2. Solve: $50 + 0.07x = 0.05(x + 1400)$
3. Simplify: $-3(2x - 7y) + 3(2x + 4y)$
4. Simplify: $4x + 2(3x - 5)$
5. Is $(-4, 2)$ a solution of $3x - 5y = -22$?
6. Find the x- and y-intercepts of $3x - 4y = 12$.
7. Are the graphs of $y = -3x + 6$ and $y = -3x - 4$ parallel?
8. Graph: $y = \frac{5}{4}x - 2$
9. One hiker starts along a trail walking at a speed of 3 mph. One-half hour later, another hiker starts on the same trail walking at a speed of 4 mph. How long after the second hiker starts will the two hikers be side by side?

6.1 Solving Systems of Linear Equations by Graphing

OBJECTIVE 1 Solve systems of linear equations by graphing

Equations considered together are called a **system of equations.** A system of equations is shown at the right.

$$2x + y = 3$$
$$x + y = 1$$

A solution of a system of linear equations can be found by graphing the lines of the system on the same coordinate axes. Three examples of **systems of linear equations in two variables** are shown below, along with the graphs of the equations of each system.

> **Take Note**
> The systems of equations at the right are *systems of linear equations* because the graph of each of the equations in the system is a line. Also, each equation has two variables. In future math courses, you will study equations that contain more than two variables.

System I

$x - 2y = -8$
$2x + 5y = 11$

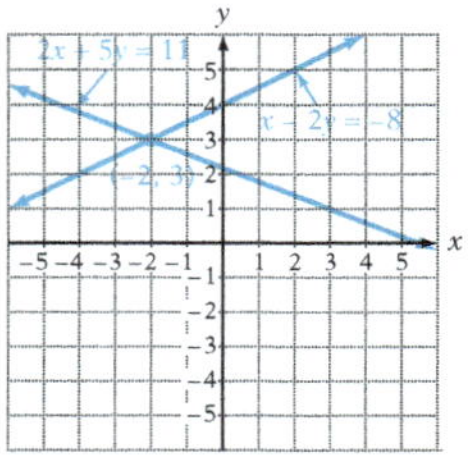

System II

$4x + 2y = 6$
$y = -2x + 3$

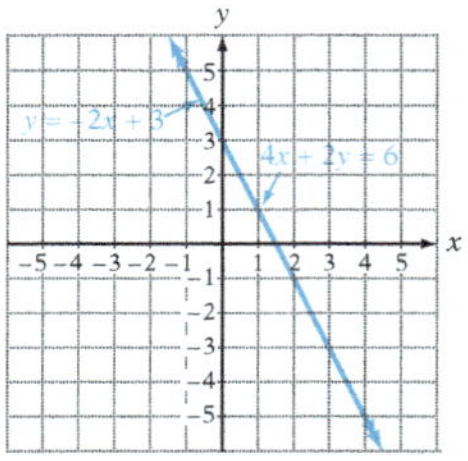

System III

$4x + 6y = 12$
$6x + 9y = -9$

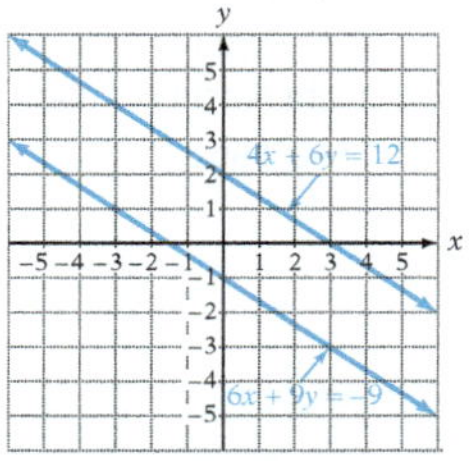

For System I, the two lines intersect at a single point whose coordinates are $(-2, 3)$. Because this point lies on both lines, it is a solution of each equation of the system of equations. We can check this by replacing x by -2 and y by 3 in each equation. The check is shown below.

$x - 2y = -8$	
$-2 - 2(3)$	-8
$-2 - 6$	-8
$-8 = -8$ ✓	

$2x + 5y = 11$	
$2(-2) + 5(3)$	11
$-4 + 15$	11
$11 = 11$ ✓	

• **Replace x by -2 and y by 3.**

A **solution of a system of equations in two variables** is an ordered pair that is a solution of each equation of the system. The ordered pair $(-2, 3)$ is a solution of System I.

Focus on determining if an ordered pair is a solution of a system of equations

Is $(-1, 4)$ a solution of the following system of equations?

$7x + 3y = 5$
$3x - 2y = 12$

Replace x by -1 and y by 4.

$7x + 3y = 5$	
$7(-1) + 3(4)$	5
$-7 + 12$	5
$5 = 5$ ✓	

$3x - 2y = 12$	
$3(-1) - 2(4)$	12
$-3 - 8$	12
$-11 \neq 12$	

Does not check.

Using the given system of equations and the graph at the right, note that the graph of the ordered pair $(-1, 4)$ lies on the graph of $7x + 3y = 5$ but *not on both lines*. The ordered pair $(-1, 4)$ is *not* a solution of the system of equations. The ordered pair $(2, -3)$, however, does lie on both lines and therefore is a solution of the system of equations.

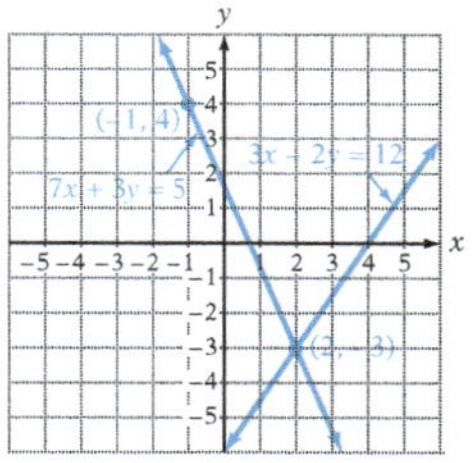

EXAMPLE 1 Is $(1, -3)$ a solution of the system $3x + 2y = -3$
$x - 3y = 6$?

Solution Substitute 1 for x and -3 for y in each equation.

$3x + 2y = -3$		$x - 3y = 6$	
$3 \cdot 1 + 2(-3)$	-3	$1 - 3(-3)$	6
$3 + (-6)$	-3	$1 - (-9)$	6
$-3 = -3$		$10 \neq 6$	

No, $(1, -3)$ is not a solution of the system of equations.

Problem 1 Is $(-1, -2)$ a solution of the system $2x - 5y = 8$
$-x + 3y = -5$?

Solution See page S14.

Try Exercise 9, page 280.

Take Note

The fact that there are an infinite number of ordered pairs that are solutions of the system of equations at the right does not mean that *every* ordered pair is a solution. For instance, $(0, 3)$, $(-2, 7)$, and $(2, -1)$ are solutions. However, $(3, 1)$, $(-1, 4)$, and $(1, 6)$ are not solutions. You should verify these statements.

System II from the preceding page and the graph of the equations of that system are shown again at the right. Note that the graph of $y = -2x + 3$ lies directly on top of the graph of $4x + 2y = 6$. Thus the two lines intersect at an infinite number of points. Because the graphs intersect at an infinite number of points, there are an infinite number of solutions of this system of equations. Since each equation represents the same set of points, the solutions of the system of equations can be stated by using the ordered pairs of either one of the equations. Therefore, we can say "The solutions are the ordered pairs that satisfy $4x + 2y = 6$," or we can say "The solutions are the ordered pairs that satisfy $y = -2x + 3$."

$$4x + 2y = 6$$
$$y = -2x + 3$$

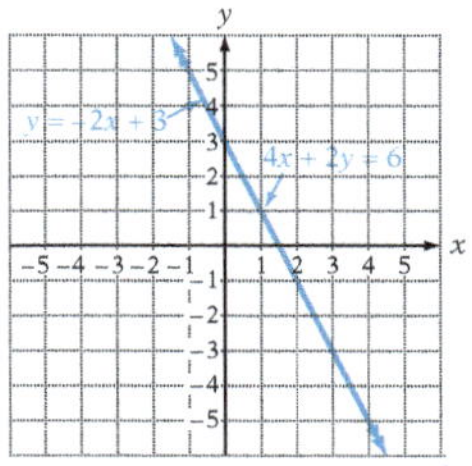

System III from the previous page and the graph of the equations of that system are shown again at the right. Note that in this case the graphs of the lines are parallel and do not intersect. Since the graphs do not intersect, there is no point that is on both lines. Therefore, the system of equations has no solution.

$$4x + 6y = 12$$
$$6x + 9y = -9$$

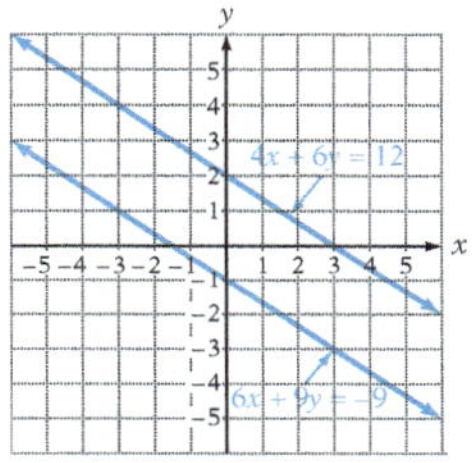

For a system of linear equations in two variables, the graphs can intersect at one point, the graphs can intersect at infinitely many points (the graphs are the same line), or the graphs can be parallel and never intersect. Such systems are called **independent, dependent,** and **inconsistent,** respectively.

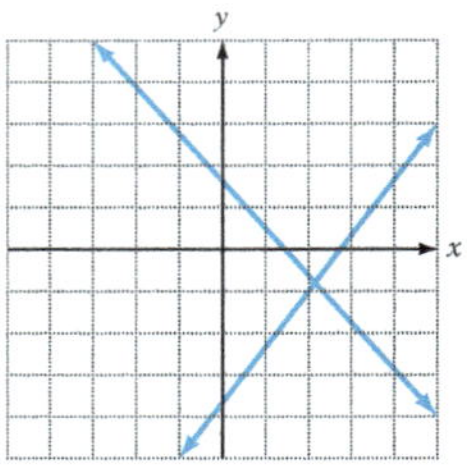

Independent

One solution

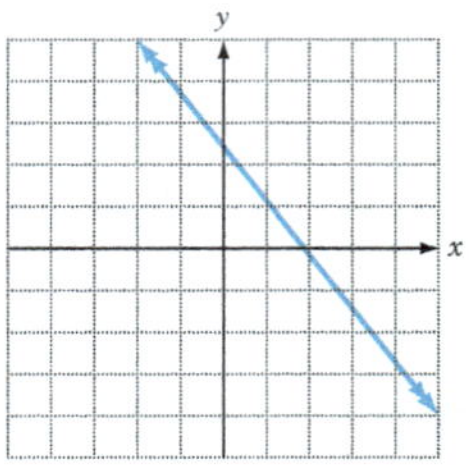

Dependent

Infinitely many solutions

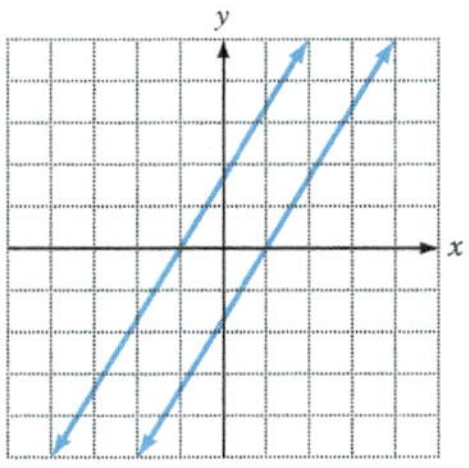

Inconsistent

No solutions

How It's Used

An economist may use two intersecting lines to represent the supply and demand for a particular product. The coordinates of the point of intersection of the lines correspond to the ideal quantity to produce and the ideal price to charge for the product in order to maximize profits.

Solving a system of equations means finding the ordered-pair solutions of the system. One way to do this is to draw the graphs of the equations in the system and determine where the graphs intersect.

Focus on solving an independent system of linear equations by graphing

Solve by graphing: $2x + 3y = 6$
$2x + y = -2$

Graph each line.

The graphs of the equations intersect at one point.

The system of equations is independent.

Find the point of intersection.

The solution is $(-3, 4)$.

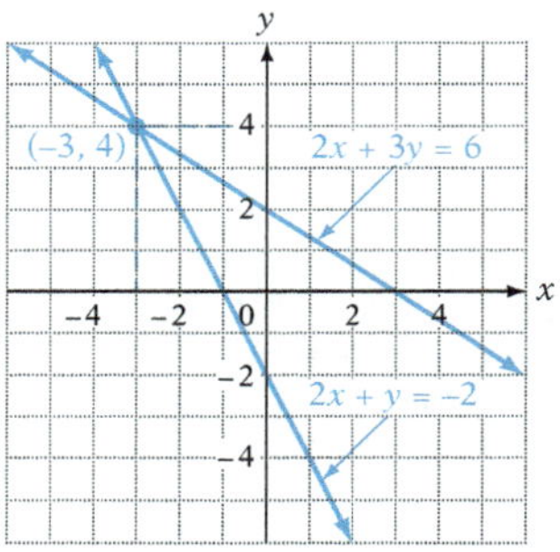

The INTERSECT feature on a graphing calculator can be used to find the solution of the system of equations given above. Refer to the Appendix for instructions on using the INTERSECT feature.

Focus on solving an inconsistent system of linear equations by graphing

Solve by graphing: $2x - y = 1$
$6x - 3y = 12$

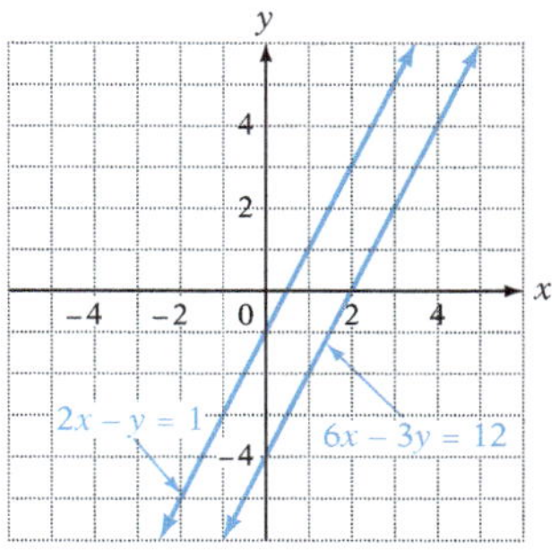

Graph each line.

The lines are parallel and therefore do not intersect.

The system of equations is inconsistent and has no solution.

Focus on solving a dependent system of linear equations by graphing

Solve by graphing: $2x + 3y = 6$
$6x + 9y = 18$

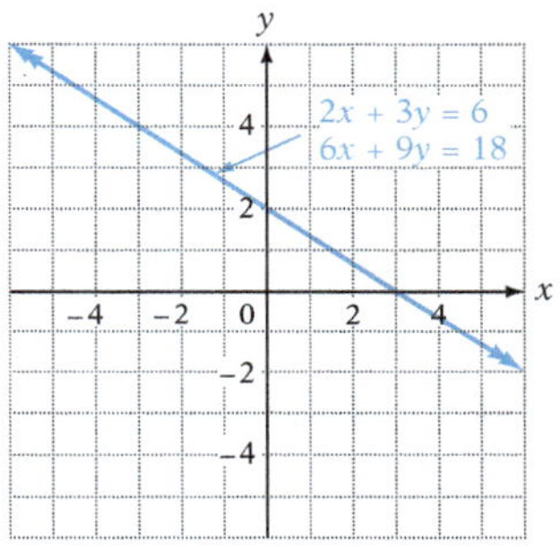

Graph each line.

The two equations represent the same line. The system of equations is dependent and therefore has an infinite number of solutions.

The solutions are the ordered pairs that are solutions of the equation $2x + 3y = 6$.

By choosing values for x, substituting these values into the equation $2x + 3y = 6$, and finding the corresponding values for y, we can find some specific ordered-pair solutions. For example, $(3, 0)$, $(0, 2)$, and $(6, -2)$ are solutions of this system of equations.

EXAMPLE 2 Solve by graphing. **A.** $x - 2y = 2$, $x + y = 5$ **B.** $4x - 2y = 6$, $y = 2x - 3$

Solution **A.**

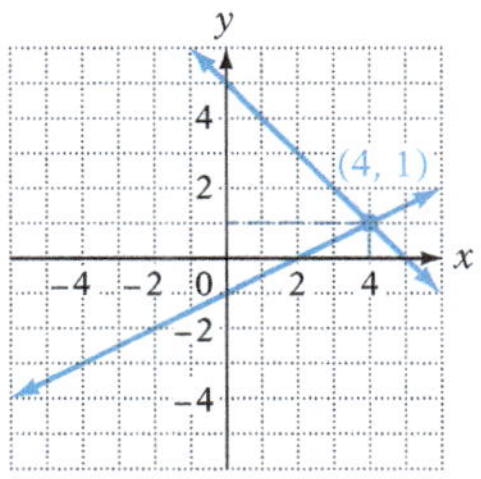

- **Graph each line. Find the point of intersection.**

The solution is (4, 1).

B.

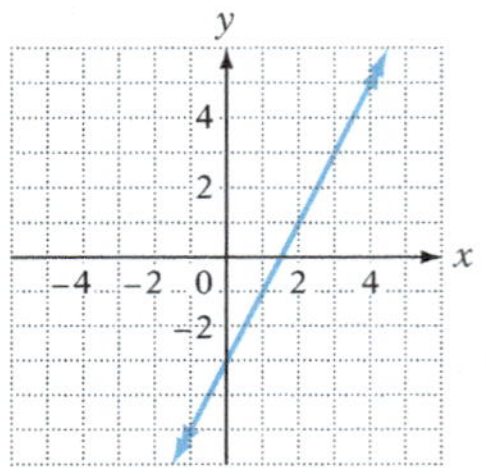

- **Graph each line. The two equations represent the same line. The system of equations has an infinite number of solutions.**

The system of equations is dependent. The solutions are the ordered pairs that satisfy the equation $y = 2x - 3$.

Problem 2 Solve by graphing. **A.** $\begin{aligned} x + 3y &= 3 \\ -x + y &= 5 \end{aligned}$ **B.** $\begin{aligned} y &= 3x - 1 \\ 6x - 2y &= -6 \end{aligned}$

Solution See page S14.

Try Exercise 49, page 282.

6.1 Exercises

CONCEPT CHECK

Determine whether the statement is always true, sometimes true, or never true.

1. A solution of a system of linear equations in two variables is an ordered pair (x, y).
2. Graphically, the solution of an independent system of linear equations in two variables is the point of intersection of the graphs of the two equations.
3. If an ordered pair is a solution of one equation in a system of linear equations but not of the other equation, then the system has no solution.
4. A system of linear equations has either one solution, represented graphically by two lines intersecting at exactly one point, or no solutions, represented graphically by two parallel lines.
5. The system of two linear equations graphed at the right has no solution.
6. An independent system of equations has no solution.

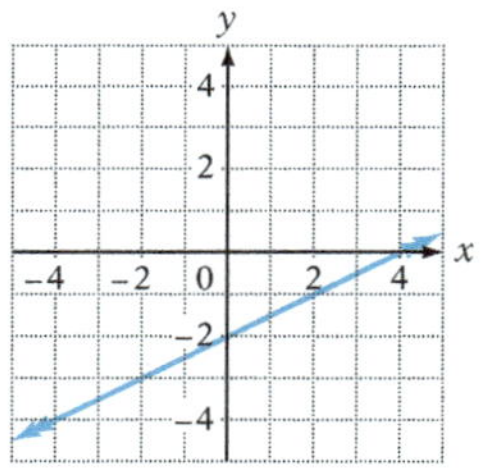

1 Solve systems of linear equations by graphing (See pages 276–280.)

7. Is (4, 3) a solution of the system $\begin{aligned} 5x - 2y &= 14 \\ x + y &= 8? \end{aligned}$

8. Is (2, 5) a solution of the system $\begin{aligned} 3x + 2y &= 16 \\ 2x - 3y &= 4? \end{aligned}$

9. Is $(-1, 3)$ a solution of the system $\begin{aligned} 4x - y &= -5 \\ 2x + 5y &= 13? \end{aligned}$

10. Is $(4, -1)$ a solution of the system $\begin{aligned} x - 4y &= 9 \\ 2x - 3y &= 11? \end{aligned}$

11. Is (0, 0) a solution of the system $\begin{aligned} 4x + 3y &= 0 \\ 2x - y &= 1? \end{aligned}$

12. Is (2, 0) a solution of the system $\begin{aligned} 3x - y &= 6 \\ x + 3y &= 2? \end{aligned}$

13. Is $(2, -3)$ a solution of the system $\begin{aligned} y &= 2x - 7 \\ 3x - y &= 9? \end{aligned}$

14. Is $(-1, -2)$ a solution of the system $\begin{aligned} 3x - 4y &= 5 \\ y &= x - 1? \end{aligned}$

15. Is (5, 2) a solution of the system $y = 2x - 8$
$y = 3x - 13$?

16. Is $(-4, 3)$ a solution of the system
$y = 2x + 11$
$y = 5x - 19$?

17. Explain how to solve a system of two equations in two variables by graphing.

GETTING READY

18. An independent system of equations has ___?___ solution(s), an inconsistent system of equations has ___?___ solution(s), and a dependent system of equations has ___?___ solution(s).

For Exercises 19 to 24, state whether the system of equations is independent, inconsistent, or dependent.

19.
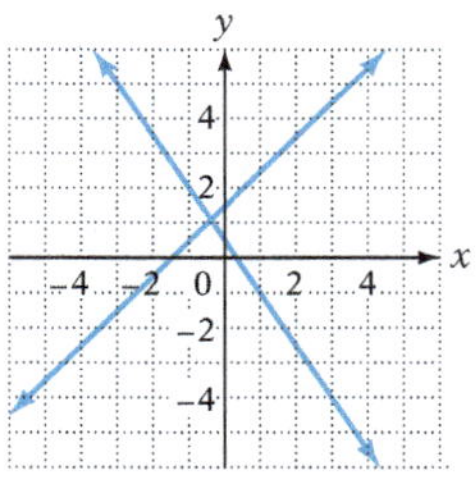

20.
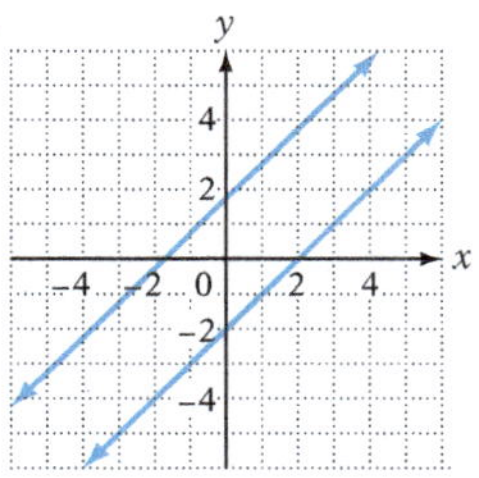

21.
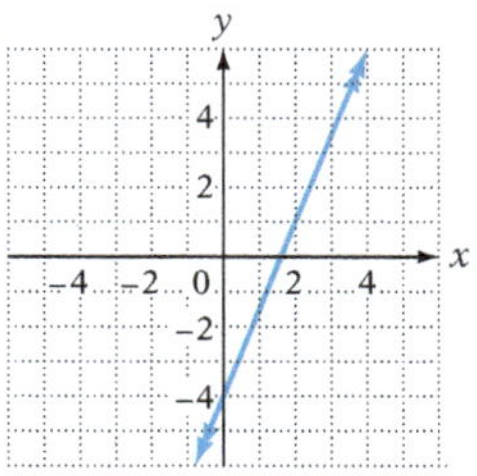

22.
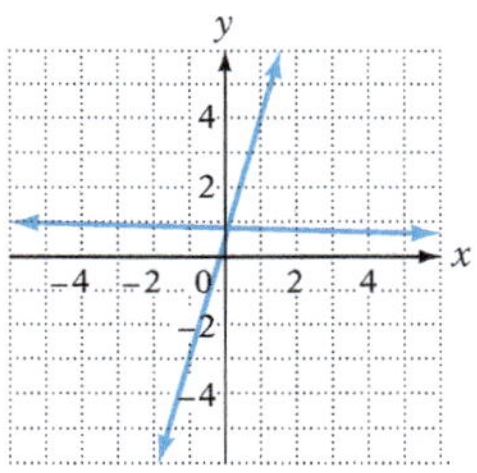

23.
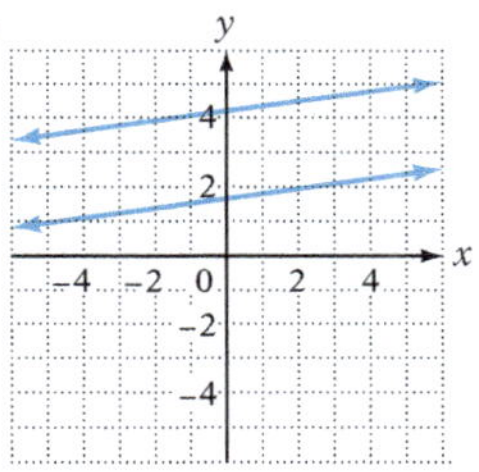

24.
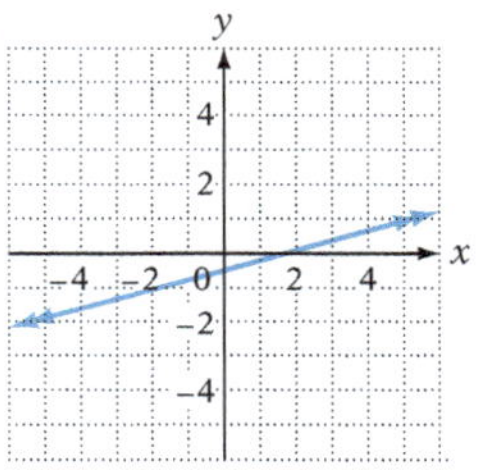

For Exercises 25 to 30, use the graph of the equations in the system to determine the solution of the system of equations.

25.
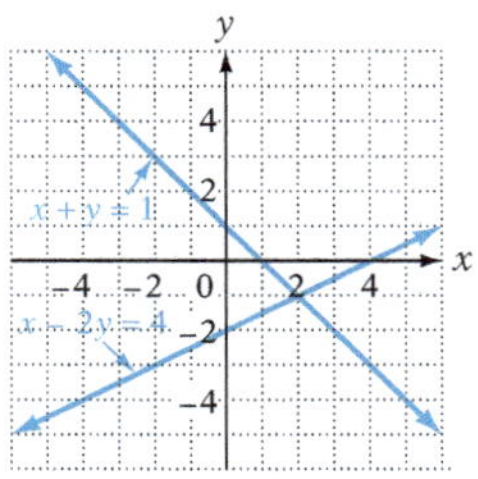

26.
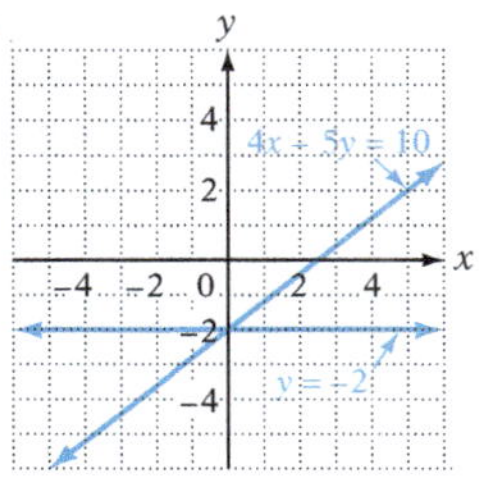

27.
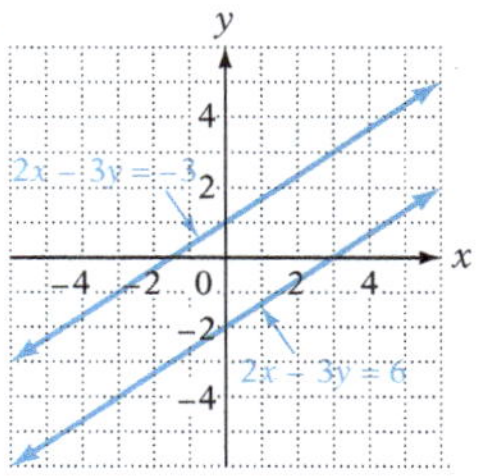

28.
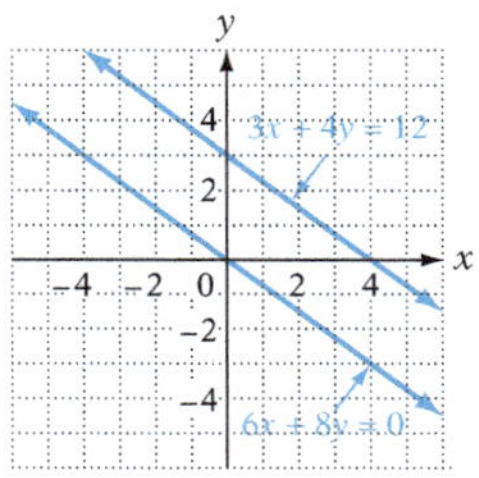

29.
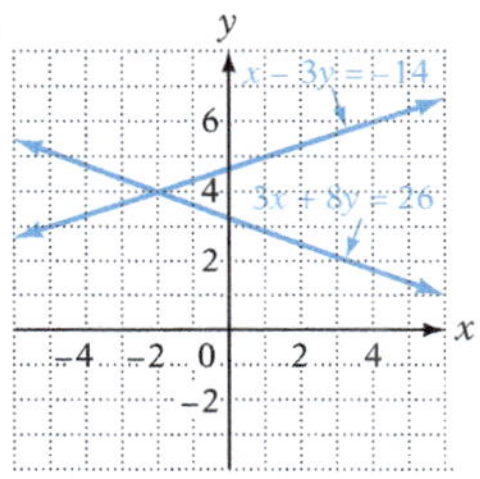

30.
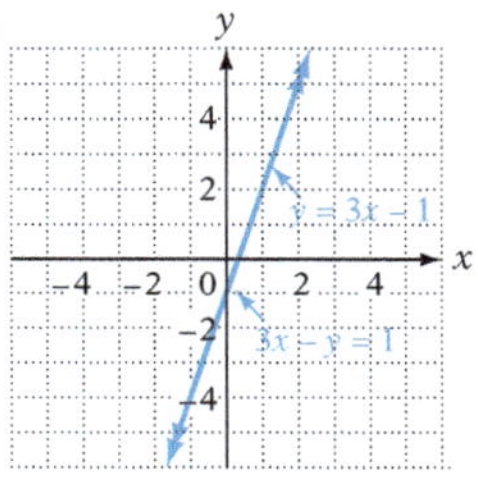

Solve by graphing.

31. $x - y = 3$
$x + y = 5$

32. $2x - y = 4$
$x + y = 5$

33. $x + 2y = 6$
$x - y = 3$

34. $3x - y = 3$
$2x + y = 2$

35. $3x - 2y = 6$
$y = 3$

36. $x = 2$
$3x + 2y = 4$

37. $x = 3$
$y = -2$

38. $x + 1 = 0$
$y - 3 = 0$

39. $y = 2x - 6$
$x + y = 0$

40. $5x - 2y = 11$
$y = 2x - 5$

41. $2x + y = -2$
$6x + 3y = 6$

42. $x + y = 5$
$3x + 3y = 6$

43. $4x - 2y = 4$
$y = 2x - 2$

44. $2x + 6y = 6$
$y = -\frac{1}{3}x + 1$

45. $x - y = 5$
$2x - y = 6$

46. $5x - 2y = 10$
$3x + 2y = 6$

47. $3x + 4y = 0$
$2x - 5y = 0$

48. $2x - 3y = 0$
$y = -\frac{1}{3}x$

49. $x - 3y = 3$
$2x - 6y = 12$

50. $4x + 6y = 12$
$6x + 9y = 18$

51. $3x + 2y = -4$
$x = 2y + 4$

52. $5x + 2y = -14$
$3x - 4y = 2$

53. $4x - y = 5$
$3x - 2y = 5$

54. $2x - 3y = 9$
$4x + 3y = -9$

Solve by graphing. Then use a graphing utility to verify your solution.

55. $5x - 2y = 10$
$3x + 2y = 6$

56. $x - y = 5$
$2x + y = 4$

57. $2x - 5y = 4$
$x - y = -1$

58. $x - 2y = -5$
$3x + 4y = -15$

59. $2x + 3y = 6$
$y = -\frac{2}{3}x + 1$

60. $2x - 5y = 10$
$y = \frac{2}{5}x - 2$

In Exercises 61 and 62, A, B, C, and D are nonzero real numbers.

61. Is the following system of equations independent, inconsistent, or dependent?
$y = Ax + B$
$y = Ax + C, B \neq C$

62. Is the following system of equations independent, inconsistent, or dependent?
$x = C$
$y = D$

APPLYING CONCEPTS

Write a system of equations given the graph.

63.

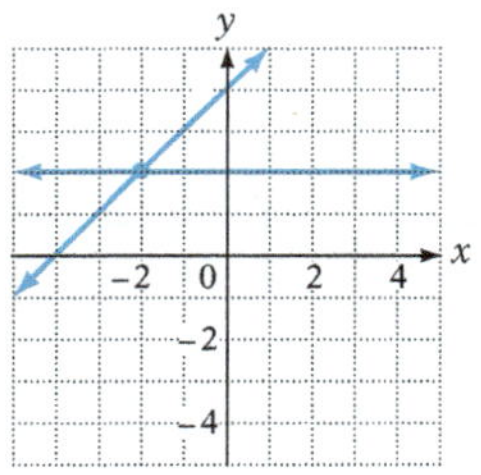

64.

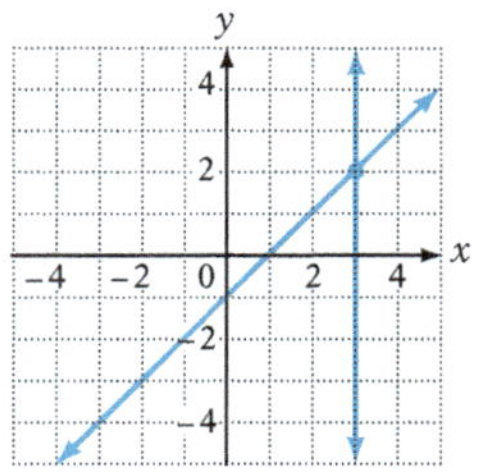

65.

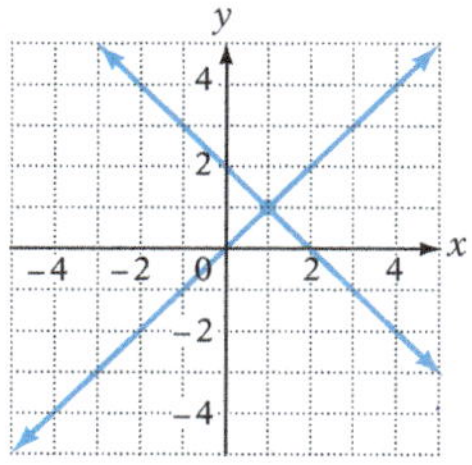

66.

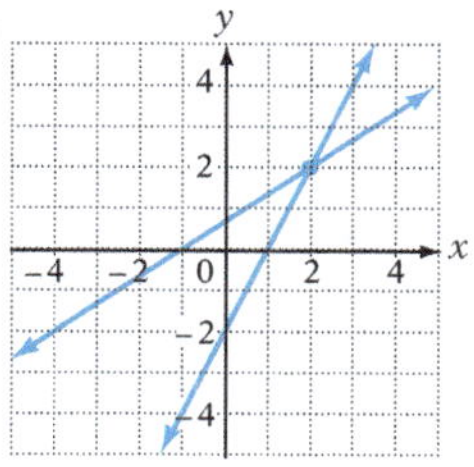

67. Determine whether the statement is always true, sometimes true, or never true.
  a. Two parallel lines have the same slope.
  b. Two different lines with the same y-intercept are parallel.
  c. Two different lines with the same slope are parallel.

68. Explain how you can determine from the graph of a system of two equations in two variables whether it is an independent system of equations. Explain how you can determine whether it is an inconsistent system of equations.

PROJECTS OR GROUP ACTIVITIES

69. Match each system of equations with its graph.

a. $2x - 3y = 6$
$2x - 5y = 10$

b. $3x - y = -5$
$x + y = 1$

c. $x + 2y = 10$
$y = x + 2$

d. $y = -3x + 5$
$y = 2x - 5$

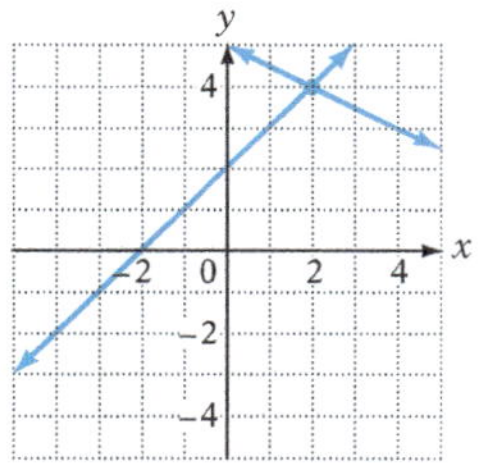

i.

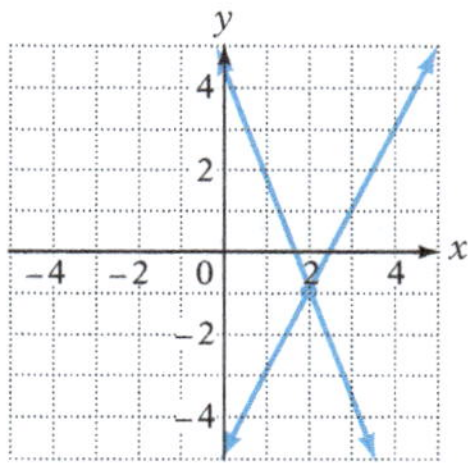

ii.

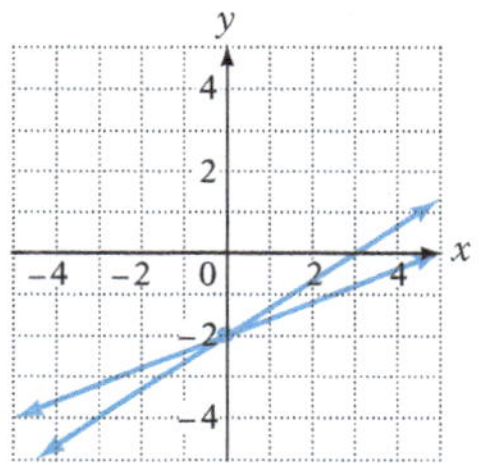

iii.

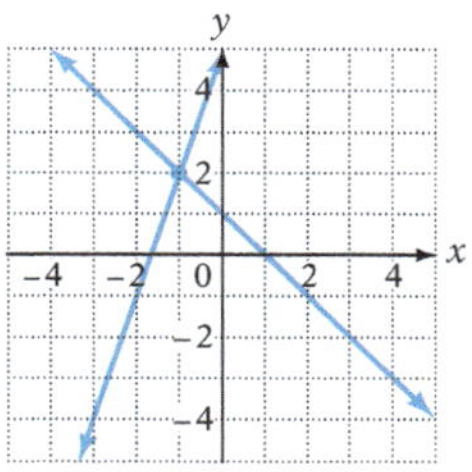

iv.

70. Write three different systems of equations: **a.** one that has $(-3, 5)$ as its only solution, **b.** one for which there is no solution, and **c.** one that is a dependent system of equations.

6.2 Solving Systems of Linear Equations by the Substitution Method

OBJECTIVE 1 Solve systems of linear equations by the substitution method

Finding a graphical solution of a system of equations is based on approximating the coordinates of a point of intersection. However, the point $\left(\frac{1}{4}, \frac{1}{2}\right)$ would be difficult to determine from a graph. An algebraic method called the **substitution method** can be used to find an exact solution of a system of equations. To use the substitution method, we must write one of the equations of the system in terms of x or in terms of y.

Focus on solving a system of linear equations by the substitution method

Solve by substitution.

A. $2x + 5y = -11$
$y = 3x - 9$

B. $5x + y = 4$
$2x - 3y = 5$

C. $y = 3x - 1$
$y = -2x - 6$

A. $2x + 5y = -11 \quad (1)$
$y = 3x - 9 \quad (2)$

Equation (2) states that $y = 3x - 9$.

Substitute $3x - 9$ for y in equation (1). $\quad 2x + 5(3x - 9) = -11$

Solve for x.

$$2x + 15x - 45 = -11$$
$$17x - 45 = -11$$
$$17x = 34$$

Note that $x = 2$. Now we must find y. $\quad x = 2$

Substitute the value of x into equation (2) and solve for y.

$$(2) \quad y = 3x - 9$$
$$y = 3 \cdot 2 - 9$$
$$y = 6 - 9$$
$$y = -3$$

The solution is $(2, -3)$.

Take Note

The graph of the system of equations at the right is shown below. Note that the lines intersect at the point whose coordinates are $(2, -3)$, which is the algebraic solution we determined by the substitution method.

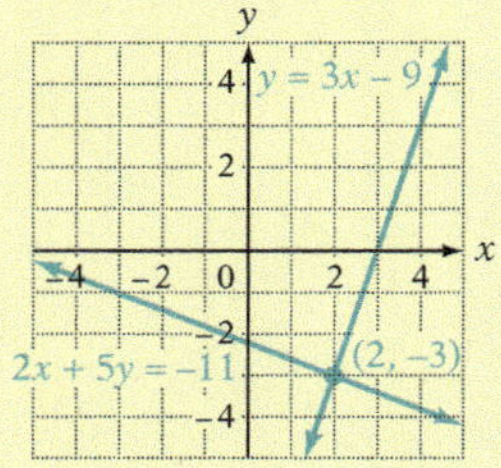

B. $5x + y = 4 \quad (1)$
$2x - 3y = 5 \quad (2)$

Equation (1) is the easier equation to solve for one variable in terms of the other.

Solve equation (1) for y.

$$5x + y = 4$$
$$y = -5x + 4$$

Substitute $-5x + 4$ for y in equation (2). Solve for x.

$$2x - 3y = 5$$
$$2x - 3(-5x + 4) = 5$$
$$2x + 15x - 12 = 5$$
$$17x - 12 = 5$$
$$17x = 17$$
$$x = 1$$

Substitute the value of x in equation (1) and solve for y.

$$(1) \quad 5x + y = 4$$
$$5(1) + y = 4$$
$$5 + y = 4$$
$$y = -1$$

The solution is $(1, -1)$.

Take Note

You can *always* check the solution of an independent system of equations. Use the skill you developed in Objective 6.1.1 to check that the ordered pair is a solution of each equation in the system.

C. $y = 3x - 1 \quad (1)$
$y = -2x - 6 \quad (2)$

Substitute $-2x - 6$ for y in equation (1).

$$y = 3x - 1$$
$$-2x - 6 = 3x - 1$$

Solve for x.

$$-5x - 6 = -1$$
$$-5x = 5$$
$$x = -1$$

Substitute the value of x into either equation and solve for y. Equation (1) is used here.

$$y = 3x - 1$$
$$y = 3(-1) - 1$$
$$y = -3 - 1$$
$$y = -4$$

The solution is $(-1, -4)$.

EXAMPLE 1 Solve by substitution: $3x + 4y = -2$ (1)
$-x + 2y = 4$ (2)

Solution

$$-x + 2y = 4$$
$$-x = -2y + 4$$
$$x = 2y - 4$$

• Solve equation (2) for x.

$$3x + 4y = -2$$
$$3(2y - 4) + 4y = -2$$
$$6y - 12 + 4y = -2$$
$$10y - 12 = -2$$
$$10y = 10$$
$$y = 1$$

• Substitute $2y - 4$ for x in equation (1).
• Solve for y.

$$-x + 2y = 4$$
$$-x + 2(1) = 4$$
$$-x + 2 = 4$$
$$-x = 2$$
$$x = -2$$

• Substitute the value of y into equation (2).
• Solve for x.

The solution is $(-2, 1)$.

Problem 1 Solve by substitution: $7x - y = 4$
$3x + 2y = 9$

Solution See page S14.

Try Exercise 31, page 287.

Take Note

In Example 2, solve equation (1) for y. The resulting equation is $y = -2x + \frac{5}{2}$. Thus both lines have the same slope, -2, but different y-intercepts. The lines are parallel.

EXAMPLE 2 Solve by substitution: $4x + 2y = 5$ (1)
$y = -2x + 1$ (2)

Solution

$$4x + 2y = 5$$
$$4x + 2(-2x + 1) = 5$$
$$4x - 4x + 2 = 5$$
$$2 = 5$$

• Substitute $-2x + 1$ for y in equation (1).

$2 = 5$ is not a true equation. The system of equations is inconsistent. The system does not have a solution.

Problem 2 Solve by substitution: $3x - y = 4$
$y = 3x + 2$

Solution See page S14.

Try Exercise 21, page 287.

Take Note

In Example 3, solve equation (1) for y. The resulting equation is $y = 3x - 2$, which is the same as equation (2).

EXAMPLE 3 Solve by substitution: $6x - 2y = 4$ (1)
$y = 3x - 2$ (2)

Solution

$$6x - 2y = 4$$
$$6x - 2(3x - 2) = 4$$
$$6x - 6x + 4 = 4$$
$$4 = 4$$

• Substitute $3x - 2$ for y in equation (1).

$4 = 4$ is a true equation. The system of equations is dependent. The solutions are the ordered pairs that satisfy the equation $y = 3x - 2$.

Problem 3 Solve by substitution: $y = -2x + 1$
$6x + 3y = 3$

Solution See page S14.

Try Exercise 45, page 287.

Note from Examples 2 and 3: If, when you are solving a system of equations, the variable is eliminated and the result is a false equation, such as $2 = 5$, the system is inconsistent and does not have a solution. If the result is a true equation, such as $4 = 4$, the system is dependent and has an infinite number of solutions.

6.2 Exercises

CONCEPT CHECK

Determine whether the statement is always true, sometimes true, or never true.

1. If one of the equations in a system of two linear equations is $y = x + 2$, then $x + 2$ can be substituted for y in the other equation of the system.
2. If a system of equations contains the equations $y = 2x + 1$ and $x + y = 5$, then $x + 2x + 1 = 5$.
3. If the equation $x = 4$ results from solving a system of equations by the substitution method, then the solution of the system of equations is 4.
4. If the true equation $6 = 6$ results from solving a system of equations by the substitution method, then the system of equations has no solutions.
5. If the false equation $0 = 7$ results from solving a system of equations by the substitution method, then the system of equations has an infinite number of solutions.
6. The ordered pair (0, 0) is a solution of a system of linear equations.

1 Solve systems of linear equations by the substitution method (See pages 283–285.)

GETTING READY

7. Use this system of equations: (1) $y = 3x - 5$
   (2) $x = 2$

   To solve the system by substitution, substitute __?__ for x in equation (1): $y = 3(2) - 5 =$ __?__.

   The solution of the system of equations is (__?__, __?__).

8. Use this system of equations: (1) $y = x + 4$
   (2) $3x + y = 12$

   To solve the system by substitution, substitute __?__ for y in equation (2): $3x + x + 4 = 12$.

   Solving the equation $3x + x + 4 = 12$ for x gives $x =$ __?__.

   To find y, substitute this value of x into equation (1): $y =$ __?__ $+ 4 =$ __?__.

   The solution of the system of equations is (__?__, __?__).

Solve by substitution.

9. $2x + 3y = 7$
   $x = 2$

10. $y = 3$
    $3x - 2y = 6$

11. $y = x - 3$
    $x + y = 5$

12. $y = x + 2$
    $x + y = 6$

13. $x = y - 2$
    $x + 3y = 2$

14. $x = y + 1$
    $x + 2y = 7$

15. $2x + 3y = 9$
$y = x - 2$

16. $3x + 2y = 11$
$y = x + 3$

17. $3x - y = 2$
$y = 2x - 1$

18. $2x - y = -5$
$y = x + 4$

19. $x = 2y - 3$
$2x - 3y = -5$

20. $x = 3y - 1$
$3x + 4y = 10$

21. $y = 4 - 3x$
$3x + y = 5$

22. $y = 2 - 3x$
$6x + 2y = 7$

23. $x = 3y + 3$
$2x - 6y = 12$

24. $x = 2 - y$
$3x + 3y = 6$

25. $3x + 5y = -6$
$x = 5y + 3$

26. $y = 2x + 3$
$4x - 3y = 1$

27. $x = 4y - 3$
$2x - 3y = 0$

28. $x = 2y$
$-2x + 4y = 6$

29. $y = 2x - 9$
$3x - y = 2$

30. $y = x + 4$
$2x - y = 6$

31. $2x - y = 4$
$3x + 2y = 6$

32. $x + y = 12$
$3x - 2y = 6$

33. $4x - 3y = 5$
$x + 2y = 4$

34. $3x - 5y = 2$
$2x - y = 4$

35. $7x - y = 4$
$5x + 2y = 1$

36. $x - 7y = 4$
$-3x + 2y = 6$

37. $7x + y = 14$
$2x - 5y = -33$

38. $3x + y = 4$
$4x - 3y = 1$

39. $x - 4y = 9$
$2x - 3y = 11$

40. $4x - y = -5$
$2x + 5y = 13$

41. $3x - y = 6$
$x + 3y = 2$

42. $3x - y = 5$
$2x + 5y = -8$

43. $4x + 3y = 0$
$2x - y = 0$

44. $5x + 2y = 0$
$x - 3y = 0$

45. $6x - 3y = 6$
$2x - y = 2$

46. $3x + y = 4$
$9x + 3y = 12$

47. $y = 2x + 11$
$y = 5x - 1$

48. $y = 2x - 8$
$y = 3x - 13$

49. $y = -4x + 2$
$y = -3x - 1$

50. $x = 3y + 7$
$x = 2y - 1$

51. $x = 4y - 2$
$x = 6y + 8$

52. $x = 3 - 2y$
$x = 5y - 10$

53. $y = 2x - 7$
$y = 4x + 5$

In Exercises 54 and 55, A, B, C, and D are nonzero real numbers.

54. Is the following system of equations independent, inconsistent, or dependent?
$x + y = A$
$x = A - y$

55. Is the following system of equations independent, inconsistent, or dependent?
$x + y = B$
$y = C - x, C \neq B$

APPLYING CONCEPTS

Rewrite each equation so that the coefficients are integers. Then solve the system of equations.

56. $0.1x - 0.6y = -0.4$
$-0.7x + 0.2y = 0.5$

57. $0.8x - 0.1y = 0.3$
$0.5x - 0.2y = -0.5$

58. $0.4x + 0.5y = 0.2$
$0.3x - 0.1y = 1.1$

59. $-0.1x + 0.3y = 1.1$
$0.4x - 0.1y = -2.2$

60. $1.2x + 0.1y = 1.9$
$0.1x + 0.3y = 2.2$

61. $1.25x - 0.01y = 1.5$
$0.24x - 0.02y = -1.52$

62. When you solve a system of equations by the substitution method, how do you determine whether the system of equations is dependent? How do you determine whether the system of equations is inconsistent?

PROJECTS OR GROUP ACTIVITIES

For what value of k does the system of equations have no solution?

63. $2x - 3y = 7$
$kx - 3y = 4$

64. $8x - 4y = 1$
$2x - ky = 3$

65. $x = 4y + 4$
$kx - 8y = 4$

66. The following was offered as a solution of the system of equations shown at the right.

(1) $y = \frac{1}{2}x + 2$
(2) $2x + 5y = 10$

$$2x + 5y = 10 \quad \bullet \text{ Equation (2)}$$
$$2x + 5\left(\frac{1}{2}x + 2\right) = 10 \quad \bullet \text{ Substitute } \tfrac{1}{2}x + 2 \text{ for } y.$$
$$2x + \frac{5}{2}x + 10 = 10 \quad \bullet \text{ Solve for } x.$$
$$\frac{9}{2}x = 0$$
$$x = 0$$

At this point the student stated that because $x = 0$, the system of equations has no solution. If this assertion is correct, is the system of equations independent, dependent, or inconsistent? If the assertion is not correct, what is the correct solution?

6.3 Solving Systems of Linear Equations by the Addition Method

OBJECTIVE 1 Solve systems of linear equations by the addition method

Another algebraic method for solving a system of equations is called the **addition method.** It is based on the Addition Property of Equations.

In the system of equations at the right, note the effect of adding equation (2) to equation (1). Because $2y$ and $-2y$ are opposites, adding the equations results in an equation with only one variable.

$$\begin{aligned} (1)\quad 3x + 2y &= 4 \\ (2)\quad 4x - 2y &= 10 \\ \hline 7x + 0y &= 14 \\ 7x &= 14 \end{aligned}$$

The solution of the resulting equation is the first coordinate of the ordered-pair solution of the system.

$$\begin{aligned} 7x &= 14 \\ x &= 2 \end{aligned}$$

The second coordinate is found by substituting the value of x into equation (1) or (2) and then solving for y. Equation (1) is used here.

$$\begin{aligned} (1)\quad 3x + 2y &= 4 \\ 3 \cdot 2 + 2y &= 4 \\ 6 + 2y &= 4 \\ 2y &= -2 \\ y &= -1 \end{aligned}$$

The solution is $(2, -1)$.

Sometimes adding the two equations does not eliminate one of the variables. In this case, use the Multiplication Property of Equations to rewrite one or both of the equations so that when the equations are added, one of the variables is eliminated.

To do this, first choose which variable to eliminate. The coefficients of that variable must be opposites. Multiply each equation by a constant that will produce coefficients that are opposites.

Focus on solving a system of linear equations by the addition method

Solve by the addition method.

A. $3x + 2y = 7$
$5x - 4y = 19$

B. $5x + 6y = 3$
$2x - 5y = 16$

C. $2x + y = 2$
$4x + 2y = -5$

A. $3x + 2y = 7 \quad (1)$
$5x - 4y = 19 \quad (2)$

To eliminate y, multiply each side of equation (1) by 2.

$$2(3x + 2y) = 2 \cdot 7$$
$$5x - 4y = 19$$

Now the coefficients of the y terms are opposites.

$$6x + 4y = 14$$
$$5x - 4y = 19$$

Add the equations.
Solve for x.

$$11x + 0y = 33$$
$$11x = 33$$
$$x = 3$$

Substitute the value of x into one of the equations and solve for y. Equation (2) is used here.

$$(2) \quad 5x - 4y = 19$$
$$5 \cdot 3 - 4y = 19$$
$$15 - 4y = 19$$
$$-4y = 4$$
$$y = -1$$

The solution is $(3, -1)$.

B. $5x + 6y = 3 \quad (1)$
$2x - 5y = 16 \quad (2)$

To eliminate x, multiply each side of equation (1) by 2 and each side of equation (2) by -5. Note how the constants are selected. The negative sign is used so that the coefficients will be opposites.

$$2(5x + 6y) = 2 \cdot 3$$
$$-5(2x - 5y) = -5 \cdot 16$$

Now the coefficients of the x terms are opposites.
Add the equations.
Solve for y.

$$10x + 12y = 6$$
$$-10x + 25y = -80$$
$$0x + 37y = -74$$
$$37y = -74$$
$$y = -2$$

Substitute the value of y into one of the equations and solve for x. Equation (1) is used here.

$$(1) \quad 5x + 6y = 3$$
$$5x + 6(-2) = 3$$
$$5x - 12 = 3$$
$$5x = 15$$
$$x = 3$$

The solution is $(3, -2)$.

C. $2x + y = 2 \quad (1)$
$4x + 2y = -5 \quad (2)$

To eliminate y, multiply each side of equation (1) by -2.

$$-4x - 2y = -4$$
$$4x + 2y = -5$$

Add the equations.
This is not a true equation.

$$0x + 0y = -9$$
$$0 = -9$$

The system of equations is inconsistent. The system does not have a solution.

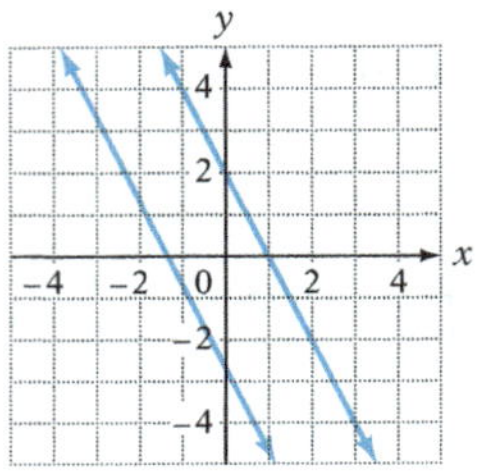

The graphs of the two equations in the preceding system of equations are shown at the left. Note that the graphs are parallel and therefore do not intersect. Thus the system of equations has no solution.

EXAMPLE 1 Solve by the addition method: $2x + 4y = 7$ (1)
$5x - 3y = -2$ (2)

Solution

$$5(2x + 4y) = 5 \cdot 7$$
$$-2(5x - 3y) = -2 \cdot (-2)$$

- **Eliminate x. Multiply each side of equation (1) by 5 and each side of equation (2) by -2.**

$$10x + 20y = 35$$
$$-10x + 6y = 4$$
$$26y = 39$$

- **Add the equations.**

$$y = \frac{39}{26} = \frac{3}{2}$$

- **Solve for y.**

$$2x + 4\left(\frac{3}{2}\right) = 7$$

- **Substitute the value of y in equation (1).**

$$2x + 6 = 7$$
$$2x = 1$$
$$x = \frac{1}{2}$$

- **Solve for x.**

The solution is $\left(\frac{1}{2}, \frac{3}{2}\right)$.

Problem 1 Solve by the addition method: $x - 2y = 1$
$2x + 4y = 0$

Solution See page S14.

Try Exercise 17, page 292.

EXAMPLE 2 Solve by the addition method: $5x = 2y - 7$ (1)
$3x + 4y = 1$ (2)

Solution

$$5x - 2y = -7$$
$$3x + 4y = 1$$

- **Write equation (1) in the form $Ax + By = C$.**

$$2(5x - 2y) = 2(-7)$$
$$3x + 4y = 1$$

- **Eliminate y. Multiply each side of equation (1) by 2.**

$$10x - 4y = -14$$
$$3x + 4y = 1$$
$$13x + 0y = -13$$
$$13x = -13$$
$$x = -1$$

- **Add the equations.**
- **Solve for x.**

$$5x = 2y - 7$$
$$5(-1) = 2y - 7$$
$$-5 = 2y - 7$$
$$2 = 2y$$
$$1 = y$$

- **Substitute the value of x into equation (1) and solve for y.**

The solution is $(-1, 1)$.

Problem 2 Solve by the addition method: $4x = y - 6$
$2x + 5y = 8$

Solution See page S15.

Try Exercise 37, page 292.

EXAMPLE 3 Solve by the addition method: $6x + 9y = 15$ (1)
$4x + 6y = 10$ (2)

Solution

$$4(6x + 9y) = 4 \cdot 15$$
$$-6(4x + 6y) = -6 \cdot 10$$

• **Eliminate x. Multiply each side of equation (1) by 4 and each side of equation (2) by −6.**

$$24x + 36y = 60$$
$$-24x - 36y = -60$$
$$0 = 0$$

• **Add the equations.**

$0 = 0$ is a true equation. The system of equations is dependent. The solutions are the ordered pairs that satisfy the equation $6x + 9y = 15$.

Problem 3 Solve by the addition method: $2x - 3y = 4$
$-4x + 6y = -8$

Solution See page S15.

Try Exercise 35, page 292.

6.3 Exercises

CONCEPT CHECK

1. How is it possible to determine whether a system of equations is independent, dependent, or inconsistent when each equation of the system is in the form $y = mx + b$? Identify each of the following systems of equations as independent, dependent, or inconsistent. Explain your answer.

a. $y = -\frac{2}{3}x + 3$
$y = -\frac{2}{3}x - 3$

b. $y = -x - 1$
$y = x + 1$

c. $y = 3x + 2$
$y = 3x + 2$

d. $y = -2x + 3$
$y = 2x + 3$

Determine whether the statement is true or false.

2. When using the addition method to solve a system of linear equations, if you multiply one side of an equation by a number, then you must multiply the other side of the equation by the same number.

3. When a system of linear equations is being solved by the addition method, in order for one of the variables to be eliminated, the coefficients of one of the variables must be opposites.

4. You are using the addition method to solve the system of equations shown at the right. The first step you will perform is to add the two equations.

$3x + 2y = 6$
$2x + 3y = -6$

1 Solve systems of linear equations by the addition method (See pages 288–291.)

GETTING READY

5. Use this system of equations: (1) $-3x - y = 5$
(2) $x - 4y = 7$

a. To eliminate x from the system of equations by using the addition method, multiply each side of equation (2) by __?__.

b. To eliminate y from the system of equations by using the addition method, multiply each side of equation (1) by __?__.

6. Use this system of equations: (1) $2x - 3y = 3$
(2) $x + 6y = 9$
   a. To eliminate x from the system of equations by using the addition method, multiply each side of equation (__?__) by (__?__).
   b. To eliminate y from the system of equations by using the addition method, multiply each side of equation (__?__) by (__?__).

Solve by the addition method.

7. $x + y = 4$, $x - y = 6$

8. $2x + y = 3$, $x - y = 3$

9. $x + y = 4$, $2x + y = 5$

10. $x - 3y = 2$, $x + 2y = -3$

11. $2x - y = 1$, $x + 3y = 4$

12. $x - 2y = 4$, $3x + 4y = 2$

13. $4x - 5y = 22$, $x + 2y = -1$

14. $3x - y = 11$, $2x + 5y = 13$

15. $2x - y = 1$, $4x - 2y = 2$

16. $x + 3y = 2$, $3x + 9y = 6$

17. $4x + 3y = 15$, $2x - 5y = 1$

18. $3x - 7y = 13$, $6x + 5y = 7$

19. $2x - 3y = 5$, $4x - 6y = 3$

20. $2x + 4y = 3$, $3x + 6y = 8$

21. $5x - 2y = -1$, $x + y = 4$

22. $4x - 3y = 1$, $8x + 5y = 13$

23. $5x + 7y = 10$, $3x - 14y = 6$

24. $7x + 10y = 13$, $4x + 5y = 6$

25. $3x - 2y = 0$, $6x + 5y = 0$

26. $5x + 2y = 0$, $3x + 5y = 0$

27. $2x - 3y = 16$, $3x + 4y = 7$

28. $3x + 4y = 10$, $4x + 3y = 11$

29. $x + 3y = 4$, $2x + 5y = 1$

30. $-2x + 7y = 9$, $3x + 2y = -1$

31. $7x - 2y = 13$, $5x + 3y = 27$

32. $3x + 5y = -11$, $2x - 7y = 3$

33. $8x - 3y = 11$, $6x - 5y = 11$

34. $4x - 8y = 36$, $3x - 6y = 27$

35. $5x + 15y = 20$, $2x + 6y = 8$

36. $2x - 3y = 4$, $-x + 4y = 3$

37. $3x = 2y + 7$, $5x - 2y = 13$

38. $2y = 4 - 9x$, $9x - y = 25$

39. $2x + 9y = 5$, $5x = 6 - 3y$

40. $3x - 4 = y + 18$, $4x + 5y = -21$

41. $2x + 3y = 7 - 2x$, $7x + 2y = 9$

42. $5x - 3y = 3y + 4$, $4x + 3y = 11$

43. $3x + y = 1$, $5x + y = 2$

44. $2x - y = 1$, $2x - 5y = -1$

45. $4x + 3y = 3$, $x + 3y = 1$

46. $2x - 5y = 4$, $x + 5y = 1$

APPLYING CONCEPTS

Solve.

47. $x - 0.2y = 0.2$, $0.2x + 0.5y = 2.2$

48. $0.5x - 1.2y = 0.3$, $0.2x + y = 1.6$

49. $1.25x - 1.5y = -1.75$, $2.5x - 1.75y = -1$

50. The point of intersection of the graphs of the equations $Ax + 2y = 2$ and $2x + By = 10$ is $(2, -2)$. Find A and B.

51. The point of intersection of the graphs of the equations $Ax - 4y = 9$ and $4x + By = -1$ is $(-1, -3)$. Find A and B.

52. Given that the graphs of the equations $2x - y = 6$, $3x - 4y = 4$, and $Ax - 2y = 0$ all intersect at the same point, find A.

53. Given that the graphs of the equations $3x - 2y = -2$, $2x - y = 0$, and $Ax + y = 8$ all intersect at the same point, find A.

54. Describe in your own words the process of solving a system of equations by the addition method.

PROJECTS OR GROUP ACTIVITIES

55. For what value of k is the system of equations dependent?

a. $2x + 3y = 7$, $4x + 6y = k$ **b.** $y = \frac{2}{3}x - 3$, $y = kx - 3$ **c.** $x = ky - 1$, $y = 2x + 2$

56. For what values of k is the system of equations independent?

a. $x + y = 7$, $kx + y = 3$ **b.** $x + 2y = 4$, $kx + 3y = 2$ **c.** $2x + ky = 1$, $x + 2y = 2$

57. For each system of equations below, graph the equations and label the point of intersection. Then add the two equations in the system, and graph the resulting equation in the same coordinate system in which you graphed the system of equations. How is the graph of the sum of the equations related to the graph of the system of equations?

a. $3x - 4y = 12$, $5x + 4y = -12$ **b.** $2x - 3y = 6$, $-2x + 5y = -10$ **c.** $2x - 3y = 2$, $5x + 4y = 5$ **d.** $2x - y = 1$, $x + 2y = 3$

58. Find an equation such that the system of equations formed by your equation and the equation $3x - 4y = 10$ has $(2, -1)$ as a solution.

6.4 Application Problems in Two Variables

OBJECTIVE 1 Rate-of-wind and rate-of-current problems

Solving motion problems that involve an object moving with or against a wind or current normally requires two variables. One variable represents the speed of the moving object in calm air or still water, and a second variable represents the rate of the wind or current.

A plane flying with the wind will travel a greater distance per hour than it would travel without the wind. The resulting rate of the plane is represented by the sum of the plane's speed and the rate of the wind.

A plane traveling against the wind, on the other hand, will travel a shorter distance per hour than it would travel without the wind. The resulting rate of the plane is represented by the difference between the plane's speed and the rate of the wind.

The same principle is used to describe the rate of a boat traveling with or against a current.

Solve: Flying with the wind, a small plane can fly 750 mi in 3 h. Against the wind, the plane can fly the same distance in 5 h. Find the rate of the plane in calm air and the rate of the wind.

STRATEGY FOR SOLVING RATE-OF-WIND AND RATE-OF-CURRENT PROBLEMS

- Choose one variable to represent the rate of the object in calm conditions and a second variable to represent the rate of the wind or current. Use these variables to express the rate of the object with and against the wind or current. Then use the time traveled with and against the wind or current, your expressions for the rate, and the fact that $rt = d$ to write expressions for the distance traveled by the object. The results can be recorded in a table.

Rate of plane in calm air: p
Rate of wind: w

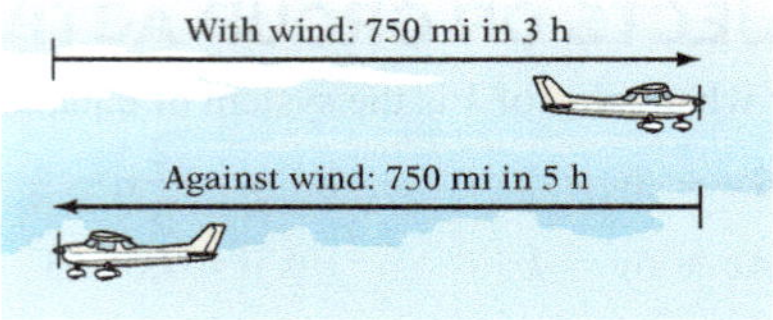

	Rate	·	Time	=	Distance
With the wind	$p + w$	·	3	=	$3(p + w)$
Against the wind	$p - w$	·	5	=	$5(p - w)$

► Determine how the expressions for distance are related.

The distance traveled with the wind is 750 mi. $3(p + w) = 750$
The distance traveled against the wind is 750 mi. $5(p - w) = 750$

Solve the system of equations.

$$3(p + w) = 750 \rightarrow \frac{3(p + w)}{3} = \frac{750}{3} \rightarrow p + w = 250$$

$$5(p - w) = 750 \rightarrow \frac{5(p - w)}{5} = \frac{750}{5} \rightarrow p - w = 150$$

$$2p = 400$$
$$p = 200$$

Substitute the value of p in the equation $p + w = 250$. Solve for w.

$$p + w = 250$$
$$200 + w = 250$$
$$w = 50$$

The rate of the plane in calm air is 200 mph.
The rate of the wind is 50 mph.

EXAMPLE 1 A 600-mile trip from one city to another takes 4 h when a plane is flying with the wind. The return trip against the wind takes 5 h. Find the rate of the plane in still air and the rate of the wind.

Strategy

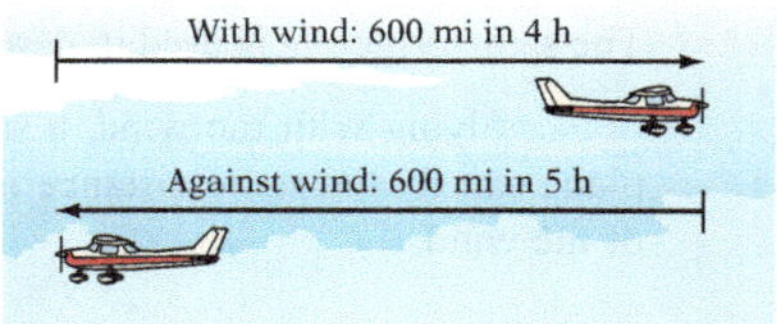

► Rate of the plane in still air: p
Rate of the wind: w

	Rate	Time	Distance
With wind	$p + w$	4	$4(p + w)$
Against wind	$p - w$	5	$5(p - w)$

► The distance traveled with the wind is 600 mi.
The distance traveled against the wind is 600 mi.

Solution

$$4(p + w) = 600 \quad (1)$$
$$5(p - w) = 600 \quad (2)$$

$$\frac{4(p + w)}{4} = \frac{600}{4}$$
• Simplify equation (1) by dividing each side of the equation by 4.

$$\frac{5(p - w)}{5} = \frac{600}{5}$$
• Simplify equation (2) by dividing each side of the equation by 5.

$$p + w = 150$$
$$p - w = 120$$
$$2p = 270$$
• Add the two equations.

$$p = 135$$
• Solve for p, the rate of the plane in still air.

$$p + w = 150$$
$$135 + w = 150$$
• Substitute the value of p into one of the equations.

$$w = 15$$
• Solve for w, the rate of the wind.

The rate of the plane in still air is 135 mph.
The rate of the wind is 15 mph.

Problem 1 A canoeist paddling with the current can travel 24 mi in 3 h. Against the current, it takes 4 h to travel the same distance. Find the rate of the current and the rate of the canoeist in calm water.

Solution See page S15.

Try Exercise 17, page 298.

OBJECTIVE 2

Application problems

Point of Interest

The Babylonians had a method for solving a system of equations. Here is an adaptation of a problem from an ancient Babylonian text (around 1500 B.C.). "There are two silver blocks. The sum of $\frac{1}{7}$ of the first block and $\frac{1}{11}$ of the second block is one sheqel (a weight). The first block diminished by $\frac{1}{7}$ of its weight equals the second diminished by $\frac{1}{11}$ of its weight. What are the weights of the two blocks?"

The application problems in this section are varieties of those problems solved earlier in the text. Each of the strategies for the problems in this section will result in a system of equations.

Solve: A jeweler purchased 5 oz of a gold alloy and 20 oz of a silver alloy for a total cost of $700. The next day, at the same prices per ounce, the jeweler purchased 4 oz of the gold alloy and 30 oz of the silver alloy for a total cost of $630. Find the cost per ounce of the silver alloy.

STRATEGY FOR SOLVING AN APPLICATION PROBLEM IN TWO VARIABLES

► Choose one variable to represent one of the unknown quantities and a second variable to represent the other unknown quantity. Write numerical or variable expressions for any remaining quantities. The results can be recorded in two tables, one for each of the conditions.

Cost per ounce of gold alloy: g Cost per ounce of silver alloy: s

First day

	Amount	·	Unit cost	=	Value
Gold alloy	5	·	g	=	$5g$
Silver alloy	20	·	s	=	$20s$

Second day

	Amount	·	Unit cost	=	Value
Gold alloy	4	·	g	=	$4g$
Silver alloy	30	·	s	=	$30s$

▶ Determine a system of equations. The strategies presented in the chapter "Solving Equations and Inequalities: Applications" can be used to determine the relationships between the expressions in the tables. Each table will give one equation of the system.

The total value of the purchase on the first day was $700. $5g + 20s = 700$
The total value of the purchase on the second day was $630. $4g + 30s = 630$

Solve the system of equations.

$$\begin{aligned} 5g + 20s &= 700 \\ 4g + 30s &= 630 \end{aligned} \rightarrow \begin{aligned} 4(5g + 20s) &= 4 \cdot 700 \\ -5(4g + 30s) &= -5 \cdot 630 \end{aligned} \rightarrow \begin{aligned} 20g + 80s &= 2800 \\ -20g - 150s &= -3150 \\ \hline -70s &= -350 \\ s &= 5 \end{aligned}$$

The cost per ounce of the silver alloy was $5.

EXAMPLE 2 A store owner purchased 20 incandescent light bulbs and 30 fluorescent bulbs for a total cost of $40. A second purchase, at the same prices, included 30 incandescent bulbs and 10 fluorescent bulbs for a total cost of $25. Find the cost of an incandescent bulb and of a fluorescent bulb.

Strategy ▶ Cost of an incandescent bulb: I
Cost of a fluorescent bulb: F

First purchase

	Amount	Unit cost	Value
Incandescent	20	I	$20I$
Fluorescent	30	F	$30F$

Second purchase

	Amount	Unit cost	Value
Incandescent	30	I	$30I$
Fluorescent	10	F	$10F$

▶ The total value of the first purchase was $40.
The total value of the second purchase was $25.

Solution

$$\begin{aligned} 20I + 30F &= 40 \quad (1) \\ 30I + 10F &= 25 \quad (2) \end{aligned}$$

$$\begin{aligned} 20I + 30F &= 40 \\ -3(30I + 10F) &= -3(25) \end{aligned}$$

• **Eliminate F. Multiply each side of equation (2) by -3.**

$$\begin{aligned} 20I + 30F &= 40 \\ -90I - 30F &= -75 \\ \hline -70I &= -35 \\ I &= 0.5 \end{aligned}$$

• **Add the two equations.**
• **Solve for I, the cost of an incandescent bulb.**

$$20I + 30F = 40$$
$$20(0.5) + 30F = 40$$
$$10 + 30F = 40$$
$$30F = 30$$
$$F = 1$$

- **Substitute the value of *I* into one of the equations.**
- **Solve for *F*.**

The cost of an incandescent bulb was $.50.
The cost of a fluorescent bulb was $1.00.

Problem 2 On Tuesday, you go into a copy center and make 85 black-and-white copies and 25 color copies for a total cost of $14.20. On Wednesday, your colleague goes to the same copy center and makes 75 black-and-white copies and 5 color copies for a total cost of $6.90. Find the cost per copy for a black-and-white copy.

Solution See page S15.

Try Exercise 29, page 299.

6.4 Exercises

CONCEPT CHECK

For Exercises 1 to 5, determine whether the statement is true or false.

1. A plane flying with the wind is traveling faster than it would be traveling without the wind.

2. The uniform motion equation $r = dt$ is used to solve rate-of-wind and rate-of-current problems.

3. If b represents the rate of a boat in calm water and c represents the rate of the current, then $b + c$ represents the rate of the boat traveling against the current.

4. If, in a system of equations, p represents the rate of a plane in calm air and w represents the rate of the wind, and the solution of the system is $p = 100$, this means that the rate of the wind is 100.

5. The system of equations at the right represents the following problem:

$$2(p + w) = 600$$
$$3(p - w) = 600$$

A plane flying with the wind flew 600 mi in 2 h. Flying against the wind, the plane could fly the same distance in 3 h. Find the rate of the plane in calm air.

6. A contractor bought 100 yd of nylon carpet for x dollars per yard and 50 yd of wool carpet for y dollars per yard. How can you represent the total cost of the carpet?

1 Rate-of-wind and rate-of-current problems (See pages 293–295.)

GETTING READY

7. A boat travels down a river for 2 h (traveling with the current), then turns around and takes 3 h to return (traveling against the current). Let b be the rate of the boat, in miles per hour, in calm water, and let c be the rate of the current, in miles per hour. Complete the following table by replacing the question marks.

	Rate, r	·	Time, t	=	Distance, d
With current	?	·	?	=	?
Against current	?	·	?	=	?

8. Suppose the distance traveled down the river by the boat in Exercise 7 is 36 mi. Use the expressions in the last column of the table in Exercise 7 to write a system of equations that can be solved to find the rate of the boat in calm water and the rate of the current.

9. Traveling with the wind, a plane flies m miles in h hours. Traveling against the wind, the plane flies n miles in h hours. Is m less than, equal to, or greater than n?

10. Traveling against the current, it takes a boat h hours to go m miles. Traveling with the current, the boat takes k hours to go m miles. Is k less than, equal to, or greater than h?

11. A whale swimming against an ocean current traveled 60 mi in 2 h. Swimming in the opposite direction, with the current, the whale was able to travel the same distance in 1.5 h. Find the speed of the whale in calm water and the rate of the ocean current.

12. A plane flying with the jet stream flew from Los Angeles to Chicago, a distance of 2250 mi, in 5 h. Flying against the jet stream, the plane could fly only 1750 mi in the same amount of time. Find the rate of the plane in calm air and the rate of the wind.

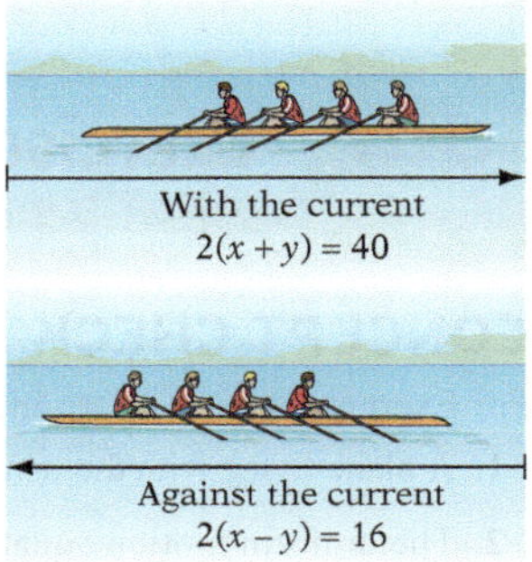

13. A rowing team rowing with the current traveled 40 km in 2 h. Rowing against the current, the team could travel only 16 km in 2 h. Find the team's rowing rate in calm water and the rate of the current.

14. Rowing with the current, a canoeist paddled 14 mi in 2 h. Against the current, the canoeist could paddle only 10 mi in the same amount of time. Find the rate of the canoeist in calm water and the rate of the current.

15. A motorboat traveling with the current went 35 mi in 3.5 h. Traveling against the current, the boat went 12 mi in 3 h. Find the rate of the boat in calm water and the rate of the current.

16. A small plane, flying into a headwind, flew 270 mi in 3 h. Flying with the wind, the plane traveled 260 mi in 2 h. Find the rate of the plane in calm air and the rate of the wind.

17. A private Learjet 31A transporting passengers was flying with a tailwind and traveled 1120 mi in 2 h. Flying against the wind on the return trip, the jet was able to travel only 980 mi in 2 h. Find the speed of the jet in calm air and the rate of the wind.

18. A rowing team rowing with the current traveled 18 mi in 2 h. Against the current, the team rowed a distance of 8 mi in the same amount of time. Find the rate of the rowing team in calm water and the rate of the current.

19. A seaplane flying with the wind flew from an ocean port to a lake, a distance of 240 mi, in 2 h. Flying against the wind, the seaplane made the trip from the lake to the ocean port in 3 h. Find the rate of the plane in calm air and the rate of the wind.

Sokolov Alexey/Shutterstock.com

20. The bird capable of the fastest flying speed is the swift. A swift flying with the wind to a favorite feeding spot traveled 26 mi in 0.2 h. On the return trip, against the wind, the swift was able to travel only 16 mi in the same amount of time. Find the rate of the swift in calm air and the rate of the wind.

21. A Boeing Apache Longbow military helicopter traveling directly into a strong headwind was able to travel 450 mi in 2.5 h. The return trip, now with a tailwind, took 1 h 40 min. Find the speed of the helicopter in calm air and the rate of the wind.

22. With the wind, a quarterback passes a football 140 ft in 2 s. Against the wind, the same pass would have traveled 80 ft in 2 s. Find the rate of the pass and the rate of the wind.

2 Application problems (See pages 295–297.)

GETTING READY

23. You purchased theater tickets for 4 adults and 2 children. For the same performance, your neighbor purchased tickets for 2 adults and 3 children. Let A be the cost per adult's ticket and C be the cost per child's ticket. Complete the following tables by replacing the question marks.

Your purchase	Amount	·	Unit cost	=	Value
Adult's ticket	?	·	?	=	?
Child's ticket	?	·	?	=	?

Neighbor's purchase	Amount	·	Unit cost	=	Value
Adult's ticket	?	·	?	=	?
Child's ticket	?	·	?	=	?

24. See Exercise 23. Suppose your tickets cost \$320 and your neighbor's tickets cost \$240. Use the expressions in the last columns of the tables in Exercise 23 to write a system of equations that can be solved to find the cost per adult's ticket and the cost per child's ticket.

© Bill Aron/PhotoEdit

25. A merchant mixes 4 lb of cinnamon tea with 1 lb of spice tea to create a mixture that costs \$12 per pound. When the merchant mixes 1 lb of the cinnamon tea with 4 lb of the spice tea, the mixture costs \$15 per pound. Is the cost per pound of the cinnamon tea less than, equal to, or greater than the cost per pound of the spice tea?

26. The total value of nickels and dimes in a bank is \$2. If the nickels were dimes and the dimes were nickels, the total value would be \$3. Is the number of nickels in the bank less than, equal to, or greater than the number of dimes in the bank?

27. Business The manager of a computer software store received two shipments of software. The cost of the first shipment, which contained 12 identical antivirus programs and 10 identical design programs, was \$1780. The second shipment, at the same prices, contained 5 copies of the antivirus program and 8 copies of the design program. The cost of the second shipment was \$1125. Find the cost for one copy of the antivirus program.

28. Business The manager of a discount clothing store received two shipments of fall clothing. The cost of the first shipment, which contained 10 identical sweaters and 20 identical jackets, was \$800. The second shipment, at the same prices, contained 5 of the same sweaters and 15 of the same jackets. The cost of the second shipment was \$550. Find the cost of one jacket.

29. Business A baker purchased 12 lb of wheat flour and 15 lb of rye flour for a total cost of \$39.87. A second purchase, at the same prices, included 15 lb of wheat flour and 10 lb of rye flour. The cost of the second purchase was \$33.30. Find the cost per pound of the wheat and rye flours.

30. Fuel Mixtures An octane number of 87 on gasoline means that the gasoline will fight engine "knock" as effectively as a reference fuel that is 87% isooctane, a type of gas. Suppose you want to fill an empty 18-gallon tank with some 87-octane gasoline and some 93-octane gasoline to produce a mixture that is 89-octane. How much of each type of gasoline must you use?

Andi Berger/Shutterstock.com

31. Food Mixtures A pastry chef created a 50-ounce sugar solution that was 34% sugar from a 20% sugar solution and a 40% sugar solution. How much of the 20% sugar solution and how much of the 40% sugar solution were used?

32. Wages Shelly Egan works as a stocker at a grocery store during her summer vacation. She gets paid a standard hourly rate for her day hours but a higher hourly rate for any hours she works during the night shift. One week she worked 17 daylight hours and 8 nighttime hours and earned \$216. The next week she earned \$246 for a total of 12 daytime and 15 nighttime hours. What rate is she being paid for daytime hours, and what is the rate for nighttime hours?

33. Sports A basketball team scored 87 points in two-point baskets and three-point baskets. If the two-point baskets had been three-point baskets and the three-point baskets had been two-point baskets, the team would have scored 93 points. Find how many two-point baskets and how many three-point baskets the team scored.

34. Consumerism The employees of a hardware store ordered lunch from a local delicatessen. The lunch consisted of 4 turkey sandwiches and 7 orders of french fries, for a total cost of \$38.30. The next day, the employees ordered 5 turkey sandwiches and 5 orders of french fries totaling \$40.75. What does the delicatessen charge for a turkey sandwich? What is the charge for an order of french fries?

Ideal Body Weight There are various formulas for calculating ideal body weight. In each of the formulas in Exercises 35 and 36, W is ideal body weight in kilograms, and x is height in inches above 60 in.

35. J. D. Robinson gave the following formula for men: $W = 52 + 1.9x$. D. R. Miller published a slightly different formula for men: $W = 56.2 + 1.41x$. At what height do both formulas give the same ideal body weight? Round to the nearest whole number.

36. J. D. Robinson gave the following formula for women: $W = 49 + 1.7x$. D. R. Miller published a slightly different formula for women: $W = 53.1 + 1.36x$. At what height do both formulas give the same ideal body weight? Round to the nearest whole number.

In the News

New Miles-per-Gallon Estimates

Beginning with model year 2008, the Environmental Protection Agency is using a new method to estimate miles-per-gallon ratings for motor vehicles. In general, estimates are lower than before. For example, under the new method, ratings for a 2010 Ford Taurus FWD are 18 mpg in the city and 25 mpg on the highway.

Source: www.fueleconomy.gov

37. Fuel Economy Read the article at the right. Suppose you use 10 gal of gas to drive a 2010 Ford Taurus FWD 208 mi. Using the new miles-per-gallon estimates given in the article, find the number of city miles and the number of highway miles you drove.

38. Investments An investment club placed a portion of its funds in a 9% annual simple interest account and the remainder in an 8% annual simple interest account. The amount of interest earned for one year was \$860. If the amounts placed in each account had been reversed, the interest earned would have been \$840. How much was invested in each account?

39. Investments An investor has two investments, one earning 5% annual simple interest and the other earning 4.5% annual simple interest. The two accounts earn \$240 in interest in one year. If the amounts in each account were reversed, the interest earned would be \$235. How much is invested in the 5% account?

APPLYING CONCEPTS

Write a system of equations. Then solve.

40. Geometry Two angles are supplementary. The larger angle is 15° more than twice the measure of the smaller angle. Find the measures of the two angles. (Supplementary angles are two angles whose sum is 180°.)

41. Geometry Two angles are complementary. The larger angle is four times the measure of the smaller angle. Find the measures of the two angles. (Complementary angles are two angles whose sum is 90°.)

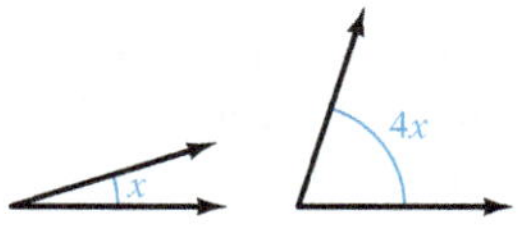

42. **Geometry** The perimeter of a rectangle is 56 cm. The length is 4 cm more than three times the width. Find the length of the rectangle.

43. **Geometry** The perimeter of a rectangle is 68 in. The length is 2 in. less than three times the width. Find the length of the rectangle.

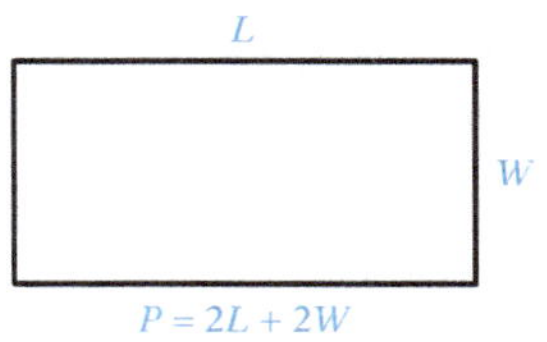

PROJECTS OR GROUP ACTIVITIES

44. Find the time t between successive alignments of the hour and minute hands on a clock. (*Hint:* Begin with the hands aligned at 12:00. Let $d°$ be the angle at which the hands next align. The time t it takes the hour hand to rotate $d°$ equals the time it takes the minute hand to rotate $(d + 360)°$. The hour hand rotates at 30° per hour, and the minute hand rotates at 360° per hour.)

CHAPTER 6 Summary

Key Words	Objective and Page Reference	Examples
Equations considered together are called a **system of equations.**	[6.1.1, p. 276]	An example of a system of equations is $4x + y = 6$, $3x + 2y = 7$
A **solution of a system of equations in two variables** is an ordered pair that is a solution of each equation of the system.	[6.1.1, p. 276]	The solution of the system of equations shown above is the ordered pair (1, 2) because it is a solution of each equation in the system.
An **independent system of equations** has one solution. The graphs of the equations in an independent system of linear equations intersect at one point.	[6.1.1, p. 278]	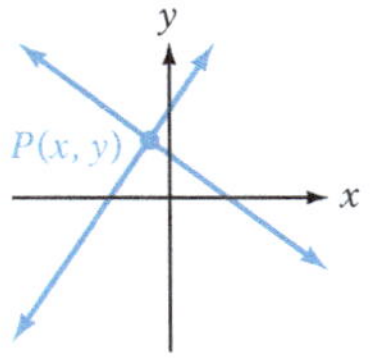
An **inconsistent system of equations** has no solution. The graphs of the equations in an inconsistent system of linear equations are parallel lines. If, when you are solving a system of equations algebraically, the variable is eliminated and the result is a false equation, such as $-3 = 8$, the system is inconsistent.	[6.1.1/6.2.1/6.3.1, pp. 278–279, 285–286, 289, 291]	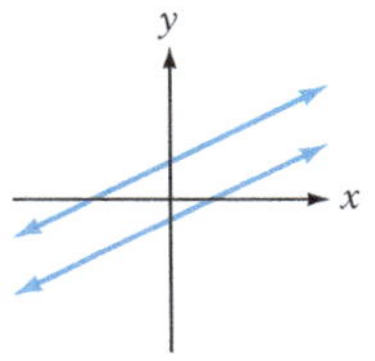

A **dependent system of equations** has an infinite number of solutions. The graphs of the equations in a dependent system of linear equations represent the same line. If, when you are solving a system of equations algebraically, the variable is eliminated and the result is a true equation, such as $1 = 1$, the system is dependent.

[6.1.1/6.2.1/6.3.1, pp. 278–280, 285–286, 291]

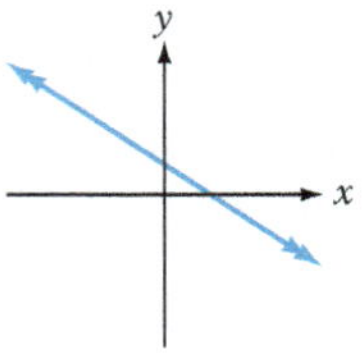

Essential Rules and Procedures

Objective and Page Reference

Examples

To solve a system of linear equations in two variables by graphing, graph each equation on the same coordinate system. If the lines intersect at one point, the point of intersection is the ordered pair that is the solution of the system. If the lines are parallel, the system is inconsistent. If the graphs represent the same line, the system is dependent.

[6.1.1, pp. 276–280]

Solve by graphing: $x + 2y = 4$
$2x + y = -1$

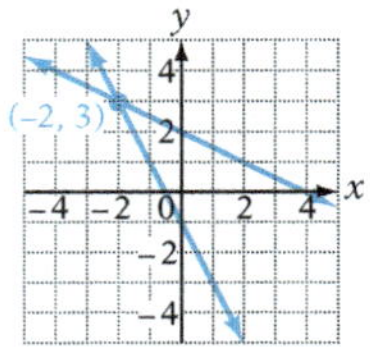

The solution is $(-2, 3)$.

To solve a system of linear equations by the substitution method, write one variable in terms of the other variable.

[6.2.1, pp. 283–286]

Solve by substitution: $2x - y = 5$ (1)
$3x + y = 5$ (2)

$3x + y = 5$
$y = -3x + 5$
• **Solve equation (2) for *y*.**

$2x - y = 5$
$2x - (-3x + 5) = 5$
$2x + 3x - 5 = 5$
$5x = 10$
$x = 2$
• **Substitute for *y* in equation (1).**

$y = -3x + 5$
$y = -3(2) + 5$
$y = -1$
• **Substitute the value of *x* to find *y*.**

The solution is $(2, -1)$.

To solve a system of linear equations by the addition method, use the Multiplication Property of Equations to rewrite one or both of the equations so that the coefficients of one variable are opposites. Then add the two equations and solve for the variables.

[6.3.1, pp. 288–291]

Solve by the addition method:

$3x + y = 4$ (1)
$x + y = 6$ (2)

$3x + y = 4$
$-x - y = -6$
• **Multiply both sides of equation (2) by −1.**

$2x = -2$
• **Add the two equations.**

$x = -1$
• **Solve for *x*.**

$x + y = 6$
$-1 + y = 6$
$y = 7$
• **Substitute the value of *x* to find *y*.**

The solution is $(-1, 7)$.

Solving motion problems that involve an object moving with or against a wind or current normally requires two variables. One variable represents the speed of the object in calm air or still water, and the second variable represents the rate of the wind or current.

[6.4.1, pp. 293–295]

Flying with the wind, a small plane can fly 1200 mi in 4 h. Against the wind, the plane can fly the same distance in 6 h. Find the rate of the plane in calm air and the rate of the wind.

$4(p + w) = 1200$
$6(p - w) = 1200$

CHAPTER 6 Review Exercises

1. Solve by substitution: $4x + 7y = 3$, $x = y - 2$

2. Solve by graphing: $3x + y = 3$, $x = 2$

3. Solve by the addition method: $3x + 8y = -1$, $x - 2y = -5$

4. Solve by substitution: $8x - y = 2$, $y = 5x + 1$

5. Solve by graphing: $x + y = 2$, $x - y = 0$

6. Solve by the addition method: $4x - y = 9$, $2x + 3y = -13$

7. Solve by substitution: $-2x + y = -4$, $x = y + 1$

8. Solve by the addition method:
$8x - y = 25$
$32x - 4y = 100$

9. Solve by the addition method: $5x - 15y = 30$, $2x + 6y = 0$

10. Solve by graphing: $3x - y = 6$, $y = -3$

11. Solve by the addition method: $7x - 2y = 0$, $2x + y = -11$

12. Solve by substitution: $x - 5y = 4$, $y = x - 4$

13. Is $(-1, -3)$ a solution of the system
$5x + 4y = -17$
$2x - y = 1$?

14. Solve by the addition method:
$6x + 4y = -3$
$12x - 10y = -15$

15. Solve by the addition method: $5x + 2y = -9$, $12x - 7y = 2$

16. Solve by graphing: $x - 3y = 12$, $y = x - 6$

17. Solve by the addition method:
$5x + 7y = 21$
$20x + 28y = 63$

18. Solve by substitution: $9x + 12y = -1$, $x - 4y = -1$

19. Solve by graphing: $4x - 2y = 8$, $y = 2x - 4$

20. Solve by the addition method:
$3x + y = -2$
$-9x - 3y = 6$

21. Solve by the addition method: $11x - 2y = 4$, $25x - 4y = 2$

22. Solve by substitution: $4x + 3y = 12$, $y = -\frac{4}{3}x + 4$

23. Solve by graphing: $y = -\frac{1}{4}x + 3$, $2x - y = 6$

24. Solve by the addition method: $2x - y = 5$, $10x - 5y = 20$

25. Solve by substitution: $6x + 5y = -2$, $y = 2x - 2$

26. Is $(-2, 0)$ a solution of the system
$-x + 9y = 2$
$6x - 4y = 12$?

27. Solve by the addition method: $6x - 18y = 7$, $9x + 24y = 2$

28. Solve by substitution: $12x - 9y = 18$, $y = \frac{4}{3}x - 3$

29. Solve by substitution: $9x - y = -3$
$18x - y = 0$

30. Solve by graphing: $x + 2y = 3$
$y = -\frac{1}{2}x + 1$

31. Solve by the addition method: $7x - 9y = 9$
$3x - y = 1$

32. Solve by substitution: $7x + 3y = -16$
$x - 2y = 5$

33. Solve by substitution: $5x - 3y = 6$
$x - y = 2$

34. Solve by the addition method: $6x + y = 12$
$9x + 2y = 18$

35. Solve by the addition method: $5x + 12y = 4$
$x + 6y = 8$

36. Solve by substitution: $6x - y = 0$
$7x - y = 1$

37. **Uniform Motion** Flying with the wind, a plane can travel 800 mi in 4 h. Against the wind, the plane requires 5 h to fly the same distance. Find the rate of the plane in calm air and the rate of the wind.

38. **Recreation** Admission to a movie theater is \$11 for adults and \$8 for children. If the receipts from 200 tickets were \$1780, how many adult tickets and how many children's tickets were sold?

39. **Uniform Motion** A canoeist traveling with the current traveled the 30 mi between two riverside campsites in 3 h. The return trip took 5 h. Find the rate of the canoeist in still water and the rate of the current.

40. **Postage** A small wood carving company mailed 190 advertisements, some requiring 46¢ in postage and others requiring 64¢ in postage. If the total cost for the mailings was \$98.20, how many mailings that required 46¢ were sent?

41. **Uniform Motion** A boat traveling with the current went 48 km in 3 h. Against the current, the boat traveled 24 km in 2 h. Find the rate of the boat in calm water and the rate of the current.

42. **Purchases** A local music shop sells some compact discs for \$15 and some for \$10. A customer spent \$120 on 10 compact discs. How many at each price did the customer purchase?

43. **Uniform Motion** With a tailwind, a flight crew flew 420 km in 3 h. Flying against the tailwind, the crew flew 440 km in 4 h. Find the rate of the plane in calm air and the rate of the wind.

44. **Uniform Motion** A paddle boat can travel 4 mi downstream in 1 h. Paddling upstream, against the current, the paddle boat travels 2 mi in 1 h. Find the rate of the boat in calm water and the rate of the current.

45. **Farming** A silo contains a mixture of lentils and corn. If 50 bushels of lentils were added to this mixture, there would be twice as many bushels of lentils as bushels of corn. If 150 bushels of corn were added to the original mixture, there would be the same amount of corn as lentils. How many bushels of each are in the silo?

46. **Uniform Motion** Flying with the wind, a small plane flew 360 mi in 3 h. Against the wind, the plane took 4 h to fly the same distance. Find the rate of the plane in calm air and the rate of the wind.

47. **Investments** An investor buys 1500 shares of stock, some costing \$6 per share and the rest costing \$25 per share. If the total cost of the stock is \$12,800, how many shares of each did the investor buy?

48. **Uniform Motion** Rowing with the current, a sculling team went 24 mi in 2 h. Rowing against the current, the team went 18 mi in 3 h. Find the rate of the sculling team in calm water and the rate of the current.

CHAPTER 6 Test

1. Solve by substitution: $4x - y = 11$
   $y = 2x - 5$

2. Solve by the addition method: $4x + 3y = 11$
   $5x - 3y = 7$

3. Is $(-2, 3)$ a solution of the system
   $2x + 5y = 11$
   $x + 3y = 7$?

4. Solve by substitution: $x = 2y + 3$
   $3x - 2y = 5$

5. Solve by the addition method: $2x - 5y = 6$
   $4x + 3y = -1$

6. Solve by graphing: $3x + 2y = 6$
   $5x + 2y = 2$

7. Solve by substitution: $4x + 2y = 3$
   $y = -2x + 1$

8. Solve by substitution: $3x + 5y = 1$
   $2x - y = 5$

9. Solve by the addition method: $7x + 3y = 11$
   $2x - 5y = 9$

10. Solve by substitution: $3x - 5y = 13$
    $x + 3y = 1$

11. Solve by the addition method: $5x + 6y = -7$
    $3x + 4y = -5$

12. Is $(2, 1)$ a solution of the system $3x - 2y = 8$
    $4x + 5y = 3$?

13. Solve by substitution: $3x - y = 5$
    $y = 2x - 3$

14. Solve by the addition method: $3x + 2y = 2$
    $5x - 2y = 14$

15. Solve by graphing: $3x + 2y = 6$
    $3x - 2y = 6$

16. Solve by substitution: $x = 3y + 1$
    $2x + 5y = 13$

17. Solve by the addition method: $5x + 4y = 7$
    $3x - 2y = 13$

18. Solve by graphing: $3x + 6y = 2$
    $y = -\frac{1}{2}x + \frac{1}{3}$

19. Solve by substitution: $4x - 3y = 1$
    $2x + y = 3$

20. Solve by the addition method: $5x - 3y = 29$
    $4x + 7y = -5$

21. Solve by substitution: $3x - 5y = -23$
    $x + 2y = -4$

22. Solve by the addition method: $9x - 2y = 17$
    $5x + 3y = -7$

23. **Uniform Motion** With the wind, a plane flies 240 mi in 2 h. Against the wind, the plane requires 3 h to fly the same distance. Find the rate of the plane in calm air and the rate of the wind.

24. **Uniform Motion** With the current, a motorboat can travel 48 mi in 3 h. Against the current, the boat requires 4 h to travel the same distance. Find the rate of the boat in calm water and the rate of the current.

Cumulative Review Exercises

1. Given $A = \{-8, -4, 0\}$, which elements of set A are less than or equal to -4?

2. Use the roster method to write the set of positive integers less than or equal to 10.

3. Simplify: $12 - 2(7 - 5)^2 \div 4$

4. Simplify: $2[5a - 3(2 - 5a) - 8]$

5. Evaluate $\frac{a^2 - b^2}{2a}$ when $a = 4$ and $b = -2$.

6. Solve: $-\frac{3}{4}x = \frac{9}{8}$

7. Solve: $4 - 3(2 - 3x) = 7x - 9$

8. Solve: $3[2 - 4(x + 1)] = 6x - 2$

9. Solve: $-7x - 5 > 4x + 50$

10. Solve: $5 + 2(x + 1) \le 13$

11. What percent of 50 is 12?

12. Find the x- and y-intercepts of the graph of $3x - 6y = 12$.

13. Find the slope of the line that contains the points whose coordinates are $(2, -3)$ and $(-3, 4)$.

14. Find the equation of the line that contains the point whose coordinates are $(-2, 3)$ and has slope $-\frac{3}{2}$.

15. Graph: $3x - 2y = 6$

16. Graph: $y = -\frac{1}{3}x + 3$

17. Graph the solution set of $y > -3x + 4$.

18. Graph: $f(x) = \frac{3}{4}x + 2$

19. Find the domain and range of the relation $\{(-5, 5), (0, 5), (1, 5), (5, 5)\}$. Is the relation a function?

20. Evaluate $f(x) = -2x - 5$ at $x = -4$.

21. Is $(2, 0)$ a solution of the system $\begin{aligned} 5x - 3y &= 10 \\ 4x + 7y &= 8 \end{aligned}$?

22. Solve by substitution: $\begin{aligned} 2x - 3y &= -7 \\ x + 4y &= 2 \end{aligned}$

23. Solve by graphing: $\begin{aligned} 2x + 3y &= 6 \\ 3x + y &= 2 \end{aligned}$

24. Solve by the addition method: $\begin{aligned} 5x - 2y &= 8 \\ 4x + 3y &= 11 \end{aligned}$

25. Find the range of the function given by the equation $f(x) = 5x - 2$ if the domain is $\{-4, -2, 0, 2, 4\}$.

26. **Manufacturing** A business manager has determined that the cost per unit for a camera is \$90 and that the fixed costs per month are \$3500. Find the number of cameras produced during a month in which the total cost was \$21,500. Use the equation $T = U \cdot N + F$, where T is the total cost, U is the cost per unit, N is the number of units produced, and F is the fixed cost.

27. **Investments** A total of \$8750 is invested in two simple interest accounts. On one account, the annual simple interest rate is 9.6%. On the second account, the annual simple interest rate is 7.2%. How much should be invested in each account so that both accounts earn the same amount of interest?

28. **Uniform Motion** Flying with the wind, a plane can fly 570 mi in 3 h. Flying against the wind, it takes the same amount of time to fly 390 mi. Find the rate of the plane in calm air and the rate of the wind.

29. **Uniform Motion** With the current, a motorboat can travel 24 mi in 2 h. Against the current, the boat requires 3 h to travel the same distance. Find the rate of the boat in calm water.

30. **Mixtures** A confectioner uses 10 lb of milk chocolate and 8 lb of dark chocolate to make a mixture costing \$98. The confectioner makes another mixture using 5 lb of milk chocolate and 12 lb of dark chocolate. The second mixture costs \$97. Find the cost per pound of the milk chocolate and the cost per pound of the dark chocolate.

Polynomials

CHAPTER

7

Digital Vision

Focus on Success

Do you get nervous before taking a math test? The more prepared you are, the less nervous you will be. There are a number of features in this text that will help you to be prepared. We suggest you start with the Chapter Summary. The Chapter Summary describes the important topics covered in the chapter. The reference following each topic shows you the objective number and the page in the text where you can find more information on the concept. Do the Chapter Review Exercises to test your understanding of the material in the chapter. If you have trouble with any of the questions, restudy the objectives the questions are taken from and retry some of the exercises in those objectives. Take the Chapter Test in a quiet place, working on it as if it were an actual exam. (See Ace the Test, page AIM-11.)

OBJECTIVES

PREP TEST

Are you ready to succeed in this chapter?
Take the Prep Test below to find out if you are ready to learn the new material.

1. Subtract: $-2 - (-3)$
2. Multiply: $-3(6)$
3. Simplify: $-\dfrac{24}{-36}$
4. Evaluate $3n^4$ when $n = -2$.
5. If $\frac{a}{b}$ is a fraction in simplest form, what number is not a possible value of b?
6. Are $2x^2$ and $2x$ like terms?
7. Simplify: $3x^2 - 4x + 1 + 2x^2 - 5x - 7$
8. Simplify: $-4y + 4y$
9. Simplify: $-3(2x - 8)$
10. Simplify: $3xy - 4y - 2(5xy - 7y)$

7.1 Addition and Subtraction of Polynomials

OBJECTIVE 1 Add polynomials

A **monomial** is a number, a variable, or a product of numbers and variables. For instance,

7	b	$\frac{2}{3}a$	$12xy^2$
A number	A variable	A product of a number and a variable	A product of a number and variables

The expression $3\sqrt{x}$ is not a monomial because $\sqrt{x}$ cannot be written as a product of variables. The expression $\frac{2x}{y^2}$ is not a monomial because it is a *quotient* of variables.

A **polynomial** is a variable expression in which the terms are monomials.

A polynomial of *one* term is a **monomial.** $-7x^2$ is a monomial.

A polynomial of *two* terms is a **binomial.** $4x + 2$ is a binomial.

A polynomial of *three* terms is a **trinomial.** $7x^2 + 5x - 7$ is a trinomial.

The terms of a polynomial in one variable are usually arranged so that the exponents on the variable decrease from left to right. This is called **descending order.**

$$4x^3 - 3x^2 + 6x - 1$$
$$5y^4 - 2y^3 + y^2 - 7y + 8$$

How It's Used

Cell phones, space probes, and modems are devices that use long strings of numbers to transmit data across great distances. Operations on polynomials are important in the design of codes that find and correct errors that occur in the transmission of the data.

The **degree of a polynomial in one variable** is the value of the largest exponent on the variable.

The degree of $4x^3 - 3x^2 + 6x - 1$ is 3.

The degree of $5y^4 - 2y^3 + y^2 - 7y + 8$ is 4.

The degree of a nonzero constant is zero.

The degree of 7 is 0.

The number zero has no degree.

Polynomials can be added, using either a vertical or a horizontal format, by combining like terms.

EXAMPLE 1 Add: $(2x^2 + x - 1) + (3x^3 + 4x^2 - 5)$
Use a vertical format.

Solution

$$\begin{array}{r} 2x^2 + x - 1 \\ \underline{3x^3 + 4x^2 \qquad - 5} \\ 3x^3 + 6x^2 + x - 6 \end{array}$$

- **Arrange the terms of each polynomial in descending order with like terms in the same column.**
- **Combine the terms in each column.**

Problem 1 Add: $(2x^2 + 4x - 3) + (5x^2 - 6x)$
Use a vertical format.

Solution See page S15.

Try Exercise 19, page 310.

EXAMPLE 2 Add: $(3x^3 - 7x + 2) + (7x^2 + 2x - 7)$
Use a horizontal format.

Solution $(3x^3 - 7x + 2) + (7x^2 + 2x - 7)$
$= 3x^3 + 7x^2 + (-7x + 2x) + (2 - 7)$ • Use the Commutative and Associative Properties of Addition to rearrange and group like terms.
$= 3x^3 + 7x^2 - 5x - 5$ • Combine like terms, and write the polynomial in descending order.

Problem 2 Add: $(-4x^2 - 3xy + 2y^2) + (3x^2 - 4y^2)$
Use a horizontal format.

Solution See page S15.

Try Exercise 25, page 311.

OBJECTIVE 2 Subtract polynomials

The opposite of the polynomial $x^2 - 2x + 3$ can be written $-(x^2 - 2x + 3)$.

The **opposite of a polynomial** is the polynomial with the sign of every term changed. $-(x^2 - 2x + 3) = -x^2 + 2x - 3$

Polynomials can be subtracted using either a vertical or a horizontal format. To subtract, add the opposite of the second polynomial to the first.

EXAMPLE 3 Subtract: $(-3x^2 - 7) - (-8x^2 + 3x - 4)$
Use a vertical format.

Solution The opposite of $-8x^2 + 3x - 4$ is $8x^2 - 3x + 4$.

$$\begin{array}{r} -3x^2 - 7 \\ 8x^2 - 3x + 4 \\ \hline 5x^2 - 3x - 3 \end{array}$$

• Write the terms of each polynomial in descending order with like terms in the same column.
• Combine the terms in each column.

Problem 3 Subtract: $(8y^2 - 4xy + x^2) - (2y^2 - xy + 5x^2)$
Use a vertical format.

Solution See page S15.

Try Exercise 43, page 311.

EXAMPLE 4 Subtract: $(5x^2 - 3x + 4) - (-3x^3 - 2x + 8)$
Use a horizontal format.

Solution $(5x^2 - 3x + 4) - (-3x^3 - 2x + 8)$
$= (5x^2 - 3x + 4) + (3x^3 + 2x - 8)$ • Rewrite subtraction as addition of the opposite.
$= 3x^3 + 5x^2 + (-3x + 2x) + (4 - 8)$ • Rearrange and group like terms.
$= 3x^3 + 5x^2 - x - 4$ • Combine like terms. Write the polynomial in descending order.

Problem 4 Subtract: $(-3a^2 - 4a + 2) - (5a^3 + 2a - 6)$
Use a horizontal format.

Solution See page S16.

Try Exercise 53, page 311.

7.1 Exercises

CONCEPT CHECK

State whether the polynomial is a monomial, a binomial, or a trinomial.

1. $8x^4 - 6x^2$

2. $4a^2b^2 + 9ab + 10$

3. $7x^3y^4$

State whether or not the expression is a monomial.

4. $3\sqrt{x}$

5. $\dfrac{4}{x}$

6. x^2y^2

State whether or not the expression is a polynomial.

7. $\dfrac{1}{5}x^3 + \dfrac{1}{2}x$

8. $\dfrac{1}{5x^2} + \dfrac{1}{2x}$

9. $x + \sqrt{5}$

1 Add polynomials (See pages 308–309.)

10. In your own words, explain the terms *monomial, binomial, trinomial,* and *polynomial.* Give an example of each.

GETTING READY

11. To use a vertical format to add polynomials, arrange the terms of each polynomial in __?__ order with __?__ terms in the same column.

12. Use a horizontal format to add $(-4y^3 + 3xy + x^3) + (2y^3 - 8xy)$.

$(-4y^3 + 3xy + x^3) + (2y^3 - 8xy)$

$= (-4y^3 + \underline{\ \ ?\ \ }) + (3xy + \underline{\ \ ?\ \ }) + x^3$ • Use the Commutative and Associative Properties of Addition to rearrange and group like terms.

$= \underline{\ \ ?\ \ }$ • Combine like terms and write the terms of the polynomial in descending order.

Add. Use a vertical format.

13. $(x^2 + 7x) + (-3x^2 - 4x)$

14. $(3y^2 - 2y) + (5y^2 + 6y)$

15. $(y^2 + 4y) + (-4y - 8)$

16. $(3x^2 + 9x) + (6x - 24)$

17. $(2x^2 + 6x + 12) + (3x^2 + x + 8)$

18. $(x^2 + x + 5) + (3x^2 - 10x + 4)$

19. $(x^3 - 7x + 4) + (2x^2 + x - 10)$

20. $(3y^3 + y^2 + 1) + (-4y^3 - 6y - 3)$

21. $(2a^3 - 7a + 1) + (-3a^2 - 4a + 1)$

22. $(5r^3 - 6r^2 + 3r) + (r^2 - 2r - 3)$

Add. Use a horizontal format.

23. $(4x^2 + 2x) + (x^2 + 6x)$

24. $(-3y^2 + y) + (4y^2 + 6y)$

25. $(4x^2 - 5xy) + (3x^2 + 6xy - 4y^2)$

26. $(2x^2 - 4y^2) + (6x^2 - 2xy + 4y^2)$

27. $(2a^2 - 7a + 10) + (a^2 + 4a + 7)$

28. $(-6x^2 + 7x + 3) + (3x^2 + x + 3)$

29. $(5x^3 + 7x - 7) + (10x^2 - 8x + 3)$

30. $(3y^3 + 4y + 9) + (2y^2 + 4y - 21)$

31. $(2r^2 - 5r + 7) + (3r^3 - 6r)$

32. $(3y^3 + 4y + 14) + (-4y^2 + 21)$

33. $(3x^2 + 7x + 10) + (-2x^3 + 3x + 1)$

34. $(7x^3 + 4x - 1) + (2x^2 - 6x + 2)$

For Exercises 35 and 36, use the following polynomials, in which a, b, c, and d are all positive numbers.

$$P = ax^3 + bx^2 - cx + d$$
$$Q = -ax^3 - bx^2 + cx - d$$
$$R = -ax^3 + bx^2 + cx + d$$

35. Which sum will be a polynomial of degree 3: $P + Q$, $Q + R$, or $P + R$?

36. Which sum will be zero: $P + Q$, $Q + R$, or $P + R$?

2 Subtract polynomials (See pages 309–310.)

GETTING READY

37. The opposite of $7x^2 + 5x - 3$ is $-(7x^2 + 5x - 3) =$ ___?___.

38. $(5x^2 + 7x - 2) - (4x^2 - 6x + 3) = (5x^2 + 7x - 2) + ($___?___$)$.

Subtract. Use a vertical format.

39. $(x^2 - 6x) - (x^2 - 10x)$

40. $(y^2 + 4y) - (y^2 + 10y)$

41. $(2y^2 - 4y) - (-y^2 + 2)$

42. $(-3a^2 - 2a) - (4a^2 - 4)$

43. $(x^2 - 2x + 1) - (x^2 + 5x + 8)$

44. $(3x^2 + 2x - 2) - (5x^2 - 5x + 6)$

45. $(4x^3 + 5x + 2) - (-3x^2 + 2x + 1)$

46. $(5y^2 - y + 2) - (-2y^3 + 3y - 3)$

47. $(2y^3 + 6y - 2) - (y^3 + y^2 + 4)$

48. $(-2x^2 - x + 4) - (-x^3 + 3x - 2)$

Subtract. Use a horizontal format.

49. $(y^2 - 10xy) - (2y^2 + 3xy)$

50. $(x^2 - 3xy) - (-2x^2 + xy)$

51. $(3x^2 + x - 3) - (x^2 + 4x - 2)$

52. $(5y^2 - 2y + 1) - (-3y^2 - y - 2)$

53. $(-2x^3 + x - 1) - (-x^2 + x - 3)$

54. $(2x^2 + 5x - 3) - (3x^3 + 2x - 5)$

55. $(4a^3 - 2a + 1) - (a^3 - 2a + 3)$

56. $(b^2 - 8b + 7) - (4b^3 - 7b - 8)$

57. $(4y^3 - y - 1) - (2y^2 - 3y + 3)$

58. $(3x^2 - 2x - 3) - (2x^3 - 2x^2 + 4)$

For Exercises 59 and 60, P and Q are polynomials such that $P = ax^2 - bx + c$ and $Q = -dx^2 - ex + f$, where $a, b, c, d, e,$ and f are all positive numbers.

59. If $a > d$, $b > e$, and $c > f$, state the sign of the coefficient of each term of $P - Q$.
 a. the x^2 term **b.** the x term **c.** the constant term

60. If $a < d$, $b > e$, and $c < f$, state the sign of the coefficient of each term of $Q - P$.
 a. the x^2 term **b.** the x term **c.** the constant term

APPLYING CONCEPTS

Simplify.

61. $\left(\frac{2}{3}a^2 + \frac{1}{2}a - \frac{3}{4}\right) - \left(\frac{5}{3}a^2 + \frac{1}{2}a + \frac{1}{4}\right)$

62. $\left(\frac{3}{5}x^2 + \frac{1}{6}x - \frac{5}{8}\right) + \left(\frac{2}{5}x^2 + \frac{5}{6}x - \frac{3}{8}\right)$

Solve.

63. What polynomial must be added to $3x^2 - 4x - 2$ so that the sum is $-x^2 + 2x + 1$?

64. What polynomial must be added to $-2x^3 + 4x - 7$ so that the sum is $x^2 - x - 1$?

65. What polynomial must be subtracted from $6x^2 - 4x - 2$ so that the difference is $2x^2 + 2x - 5$?

66. What polynomial must be subtracted from $2x^3 - x^2 + 4x - 2$ so that the difference is $x^3 + 2x - 8$?

PROJECTS OR GROUP ACTIVITIES

67. Is it possible to subtract two polynomials, each of degree 3, and have the difference be a polynomial of degree 2? If so, give an example. If not, explain why not.

68. Is it possible to add two polynomials, each of degree 3, and have the sum be a polynomial of degree 2? If so, give an example. If not, explain why not.

69. Write two polynomials, each of degree 2, whose sum is also of degree 2.

70. Write two polynomials, each of degree 2, whose sum is of degree 1.

71. Write two polynomials, each of degree 2, whose sum is of degree 0.

7.2 Multiplication of Monomials

OBJECTIVE 1

Multiply monomials

Recall that in the exponential expression x^5, x is the base and 5 is the exponent. The exponent indicates the number of times the base occurs as a factor.

The product of exponential expressions with the *same* base can be simplified by writing each expression in factored form and writing the result with an exponent.

$$x^3 \cdot x^2 = \underbrace{\overbrace{(x \cdot x \cdot x)}^{3\text{ factors}} \cdot \overbrace{(x \cdot x)}^{2\text{ factors}}}_{5\text{ factors}}$$
$$= x \cdot x \cdot x \cdot x \cdot x$$
$$= x^5$$

Adding the exponents results in the same product.

$$x^3 \cdot x^2 = x^{3+2} = x^5$$

RULE FOR MULTIPLYING EXPONENTIAL EXPRESSIONS

If m and n are integers, then $\boldsymbol{x^m \cdot x^n = x^{m+n}}$.

EXAMPLES

In each example below, we are multiplying two exponential expressions with the same base. Simplify the expression by adding the exponents.

1. $x^4 \cdot x^7 = x^{4+7} = x^{11}$
2. $y \cdot y^5 = y^{1+5} = y^6$
3. $a^2 \cdot a^6 \cdot a = a^{2+6+1} = a^9$

EXAMPLE 1 Multiply: $(2xy)(3x^2y)$

Solution $(2xy)(3x^2y)$

$= (2 \cdot 3)(x \cdot x^2)(y \cdot y)$ • Use the Commutative and Associative Properties of Multiplication to rearrange and group factors.

$= 6x^{1+2}y^{1+1}$ • Multiply variables with the same base by adding the exponents.

$= 6x^3y^2$

Problem 1 Multiply: $(3x^2)(6x^3)$

Solution See page S16.

Take Note

The Rule for Multiplying Exponential Expressions requires that the bases be the same. The expression x^3y^2 cannot be simplified.

Try Exercise 19, page 315.

EXAMPLE 2 Multiply: $(2x^2y)(-5xy^4)$

Solution $(2x^2y)(-5xy^4)$

$= [2(-5)](x^2 \cdot x)(y \cdot y^4)$ • Use the Properties of Multiplication to rearrange and group factors.

$= -10x^3y^5$ • Multiply variables with the same base by adding the exponents.

Problem 2 Multiply: $(-3xy^2)(-4x^2y^3)$

Solution See page S16.

Try Exercise 31, page 316.

OBJECTIVE 2 Simplify powers of monomials

Point of Interest

One of the first symbolic representations of powers was given by Diophantus (c. A.D. 250) in his book *Arithmetica*. He used Δ^Y for x^2 and k^Y for x^3. The symbol Δ^Y was the first two letters of the Greek word *dunamis*, meaning "power"; k^Y was from the Greek word *kubos*, meaning "cube." He also combined these symbols to denote higher powers. For instance, Δk^Y was the symbol for x^5.

A power of a monomial can be simplified by rewriting the expression in factored form and then using the Rule for Multiplying Exponential Expressions.

$$(x^2)^3 = x^2 \cdot x^2 \cdot x^2 = x^{2+2+2} = x^6$$

$$(x^4y^3)^2 = (x^4y^3)(x^4y^3) = x^4 \cdot y^3 \cdot x^4 \cdot y^3 = (x^4 \cdot x^4)(y^3 \cdot y^3) = x^{4+4}y^{3+3} = x^8y^6$$

Note that multiplying each exponent inside the parentheses by the exponent outside the parentheses gives the same result.

$$(x^2)^3 = x^{2 \cdot 3} = x^6 \qquad (x^4y^3)^2 = x^{4 \cdot 2}y^{3 \cdot 2} = x^8y^6$$

RULE FOR SIMPLIFYING POWERS OF EXPONENTIAL EXPRESSIONS

If m and n are integers, then $(x^m)^n = x^{mn}$.

EXAMPLES

Each example below is a power of an exponential expression. Simplify the expression by multiplying the exponents.

1. $(x^5)^2 = x^{5 \cdot 2} = x^{10}$
2. $(y^3)^4 = y^{3 \cdot 4} = y^{12}$

RULE FOR SIMPLIFYING POWERS OF PRODUCTS

If m, n, and p are integers, then $(x^my^n)^p = x^{mp}y^{np}$.

EXAMPLES

Each example below is a power of a product of exponential expressions. Simplify the expression by multiplying each exponent inside the parentheses by the exponent outside the parentheses.

1. $(c^5d^3)^6 = c^{5 \cdot 6}d^{3 \cdot 6} = c^{30}d^{18}$
2. $(3a^2b)^3 = 3^{1 \cdot 3}a^{2 \cdot 3}b^{1 \cdot 3} = 3^3a^6b^3 = 27a^6b^3$

EXAMPLE 3 Simplify: $(-2x)(-3xy^2)^3$

Solution $(-2x)(-3xy^2)^3$

$= (-2x)(-3)^3x^3y^6$ • Multiply each exponent in $-3xy^2$ by the exponent outside the parentheses.

$= (-2x)(-27)x^3y^6$ • Simplify $(-3)^3$.

$= [-2(-27)](x \cdot x^3)y^6$ • Use the Properties of Multiplication to rearrange and group factors.

$= 54x^4y^6$ • Multiply variable expressions with the same base by adding the exponents.

Problem 3 Simplify: $(3x)(2x^2y)^3$

Solution See page S16.

Try Exercise 73, page 316.

7.2 Exercises

CONCEPT CHECK

1. State whether the expression can be simplified using the Rule for Multiplying Exponential Expressions.
   a. $x^4 + x^5$ **b.** x^4x^5 **c.** x^4y^4 **d.** $x^4 + x^4$

2. To which of the following does the rule $(x^my^n)^p = x^{mp}y^{np}$ apply?
   (i) $(2x)^3$ (ii) $(xy^2z^3)^4$ (iii) $(2 + x^3)^3$ (iv) $(x - y)^3$ (v) $(-4xy^4z^2)^5$

State whether the expression is the product of two exponential expressions or a power of an exponential expression.

3. **a.** $b^4 \cdot b^8$ **b.** $(b^4)^8$ **c.** $(2z)^2$ **d.** $2z \cdot z$

4. **a.** $(3a^4)^5$ **b.** $(3a^4)(5a)$ **c.** $x(-xy^4)$ **d.** $(-xy)^4$

5. State whether or not the expression can be simplified using the Rule for Simplifying Powers of Products.
   a. $(xy)^3$ **b.** $(x + y)^3$ **c.** $(a^3 + b^4)^2$ **d.** $(a^3b^4)^2$

1 Multiply monomials (See page 313.)

6. Explain how to multiply two exponential expressions with the same base. Provide an example.

GETTING READY

7. Use the Rule for Multiplying Exponential Expressions to multiply:
   $(x^7)(x^2) = x^{\underline{?} + \underline{?}} = \underline{\quad ? \quad}$.

8. Multiply: $(5a^6b)(2a^5b^3)$
   $(5a^6b)(2a^5b^3)$
   $= (5 \cdot \underline{\quad ? \quad})(a^6 \cdot \underline{\quad ? \quad})(b \cdot \underline{\quad ? \quad})$ • Use the Commutative and Associative Properties of Multiplication to rearrange and group factors.
   $= (\underline{\quad ? \quad})(a^{\underline{?} + \underline{?}})(b^{\underline{?} + \underline{?}})$ • Multiply variables with the same base by ___?___ the exponents.
   $= 10a^{\underline{?}}b^{\underline{?}}$ • Simplify.

Multiply.

9. $(x)(2x)$
10. $(-3y)(y)$
11. $(3x)(4x)$
12. $(7y^3)(7y^2)$
13. $(-2a^3)(-3a^4)$
14. $(5a^6)(-2a^5)$
15. $(x^2y)(xy^4)$
16. $(x^2y^4)(xy^7)$
17. $(-2x^4)(5x^5y)$
18. $(-3a^3)(2a^2b^4)$
19. $(x^2y^4)(x^5y^4)$
20. $(a^2b^4)(ab^3)$
21. $(2xy)(-3x^2y^4)$
22. $(-3a^2b)(-2ab^3)$
23. $(x^2yz)(x^2y^4)$
24. $(-ab^2c)(a^2b^5)$
25. $(a^2b^3)(ab^2c^4)$
26. $(x^2y^3z)(x^3y^4)$
27. $(-a^2b^2)(a^3b^6)$
28. $(xy^4)(-xy^3)$
29. $(-6a^3)(a^2b)$

30. $(2a^2b^3)(-4ab^2)$
31. $(-5y^4z)(-8y^6z^5)$
32. $(3x^2y)(-4xy^2)$
33. $(10ab^2)(-2ab)$
34. $(x^2y)(yz)(xyz)$
35. $(xy^2z)(x^2y)(z^2y^2)$
36. $(-2x^2y^3)(3xy)(-5x^3y^4)$
37. $(4a^2b)(-3a^3b^4)(a^5b^2)$
38. $(3ab^2)(-2abc)(4ac^2)$

2 Simplify powers of monomials (See page 314.)

39. Explain how to simplify a power of an exponential expression. Provide an example.

GETTING READY

40. Use the Rule for Simplifying Powers of Exponential Expressions to simplify:
$(x^4)^7 = x^{(\ ?\ )(\ ?\ )} = \underline{\ \ ?\ \ }$

41. Use the Rule for Simplifying Powers of Products to simplify:
$(x^4y^2)^6 = x^{(\ ?\ )(\ ?\ )}y^{(\ ?\ )(\ ?\ )} = \underline{\ \ ?\ \ }$

42. **a.** True or false? $a^2b^5 = ab^{10}$
b. True or false? $a^2b^5 = ab^7$
c. True or false? $(a^2b^5)^2 = a^4b^{25}$
d. True or false? $(a^2 + b^5)^2 = a^4 + b^{10}$

Simplify.

43. $(x^3)^3$
44. $(y^4)^2$
45. $(x^7)^2$
46. $(y^5)^3$
47. $(2^2)^3$
48. $(3^2)^2$
49. $(-2)^2$
50. $(-3)^3$
51. $(-2^2)^3$
52. $(-2^3)^3$
53. $(-x^2)^2$
54. $(-x^2)^3$
55. $(2x)^2$
56. $(3y)^3$
57. $(-2x^2)^3$
58. $(-3y^3)^2$
59. $(x^2y^3)^2$
60. $(x^3y^4)^5$
61. $(3x^2y)^2$
62. $(-2ab^3)^4$
63. $(a^2)(3a^2)^3$
64. $(b^2)(2a^3)^4$
65. $(-2x)(2x^3)^2$
66. $(2y)(-3y^4)^3$
67. $(x^2y)(x^2y)^3$
68. $(a^3b)(ab)^3$
69. $(ab^2)^2(ab)^2$
70. $(x^2y)^2(x^3y)^3$
71. $(-2x)(-2x^3y)^3$
72. $(-3y)(-4x^2y^3)^3$
73. $(-2x)(-3xy^2)^2$
74. $(-3y)(-2x^2y)^3$
75. $(ab^2)(-2a^2b)^3$
76. $(a^2b^2)(-3ab^4)^2$
77. $(-2a^3)(3a^2b)^3$
78. $(-3b^2)(2ab^2)^3$
79. $(-3ab)^2(-2ab)^3$
80. $(-3a^2b)^3(-3ab)^3$

APPLYING CONCEPTS

Simplify.

81. $(6x)(2x^2) + (4x^2)(5x)$
82. $(2a^7)(7a^2) - (6a^3)(5a^6)$
83. $(3a^2b^2)(2ab) - (9ab^2)(a^2b)$
84. $(3x^2y^2)^2 - (2xy)^4$
85. $(5xy^3)(3x^4y^2) - (2x^3y)(x^2y^4)$
86. $a^2(ab^2)^3 - a^3(ab^3)^2$
87. $4a^2(2ab)^3 - 5b^2(a^5b)$
88. $9x^3(3x^2y)^2 - x(x^3y)^2$
89. $-2xy(x^2y)^3 - 3x^5(xy^2)^2$
90. $5a^2b(ab^2)^2 + b^3(2a^2b)^2$
91. $a^n \cdot a^n$
92. $(a^n)^2$
93. $(a^2)^n$
94. $a^2 \cdot a^n$

95. **Geometry** The length of a rectangle is $4ab$. The width is $2ab$. Find the perimeter of the rectangle in terms of ab.

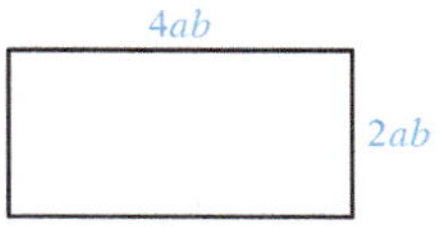

PROJECTS OR GROUP ACTIVITIES

96. Let $x_1 = -1x^1$ and, for $n > 1$, $x_n = -nx^n$. Calculate the product $(x_1)(x_2)(x_3)(x_4)(x_5)$.

97. a. Evaluate $(2^3)^2$ and $2^{(3^2)}$. Are the results the same? If not, which expression has the larger value?

b. What is the order of operations for the expression x^{m^n}?

7.3 Multiplication of Polynomials

OBJECTIVE 1 Multiply a polynomial by a monomial

To multiply a polynomial by a monomial, use the Distributive Property and the Rule for Multiplying Exponential Expressions.

EXAMPLE 1 Multiply. **A.** $-2x(x^2 - 4x - 3)$ **B.** $(5x + 4)(-2x)$ **C.** $x^3(2x^2 - 3x + 2)$

Solution **A.** $-2x(x^2 - 4x - 3)$

$= -2x(x^2) - (-2x)(4x) - (-2x)(3)$ • **Use the Distributive Property.**

$= -2x^3 + 8x^2 + 6x$ • **Use the Rule for Multiplying Exponential Expressions.**

B. $(5x + 4)(-2x)$

$= 5x(-2x) + 4(-2x)$ • **Use the Distributive Property.**

$= -10x^2 - 8x$ • **Use the Rule for Multiplying Exponential Expressions.**

C. $x^3(2x^2 - 3x + 2)$

$= 2x^5 - 3x^4 + 2x^3$ • **Use the Distributive Property and the Rule for Multiplying Exponential Expressions.**

Problem 1 Multiply. **A.** $5x(3x^2 - 2x + 4)$ **B.** $(-2y + 3)(-4y)$ **C.** $-a^2(3a^2 + 2a - 7)$

Solution See page S16.

Try Exercise 35, page 323.

OBJECTIVE 2 Multiply two polynomials

Multiplication of two polynomials requires the repeated application of the Distributive Property.

$$\begin{aligned}(y-2)(y^2+3y+1) &= (y-2)(y^2)+(y-2)(3y)+(y-2)(1)\\ &= y^3-2y^2+3y^2-6y+y-2\\ &= y^3+y^2-5y-2\end{aligned}$$

A convenient method of multiplying two polynomials is to use a vertical format similar to that used for multiplication of whole numbers.

$$\begin{array}{rrrr} & y^2 & +3y & +1\\ & & y & -2\\ \hline & -2y^2 & -6y & -2\\ y^3 & +3y^2 & +y & \\ \hline y^3 & +y^2 & -5y & -2\end{array}$$

Multiply each term in the trinomial by -2.
Multiply each term in the trinomial by y.
Like terms must be in the same column.
Add the terms in each column.

EXAMPLE 2 Multiply: $(2b^3-b+1)(2b+3)$

Solution

$$\begin{array}{rrrrr} & 2b^3 & & -b & +1\\ & & & 2b & +3\\ \hline & 6b^3 & & -3b & +3\\ 4b^4 & & -2b^2 & +2b & \\ \hline 4b^4 & +6b^3 & -2b^2 & -b & +3\end{array}$$

- **Multiply $2b^3-b+1$ by 3.**
- **Multiply $2b^3-b+1$ by $2b$. Arrange the terms in descending order.**
- **Add the terms in each column.**

Problem 2 Multiply: $(2y^3+2y^2-3)(3y-1)$

Solution See page S16.

Try Exercise 49, page 323.

EXAMPLE 3 Multiply: $(4a^3-5a-2)(3a-2)$

Solution

$$\begin{array}{rrrrr} & 4a^3 & & -5a & -2\\ & & & 3a & -2\\ \hline & -8a^3 & & +10a & +4\\ 12a^4 & & -15a^2 & -6a & \\ \hline 12a^4 & -8a^3 & -15a^2 & +4a & +4\end{array}$$

- **Multiply $4a^3-5a-2$ by -2.**
- **Multiply $4a^3-5a-2$ by $3a$.**
- **Add the terms in each column.**

Problem 3 Multiply: $(3x^3-2x^2+x-3)(2x+5)$

Solution See page S16.

Try Exercise 53, page 324.

OBJECTIVE 3 Multiply two binomials

It is often necessary to find the product of two binomials. The product can be found using a method called **FOIL,** which is based on the Distributive Property. The letters of FOIL stand for **F**irst, **O**uter, **I**nner, and **L**ast.

Focus on multiplying two binomials using the FOIL method

Multiply: $(2x + 3)(x + 5)$

Multiply the **F**irst terms.	$(2x + 3)(x + 5)$	$2x \cdot x = 2x^2$
Multiply the **O**uter terms.	$(2x + 3)(x + 5)$	$2x \cdot 5 = 10x$
Multiply the **I**nner terms.	$(2x + 3)(x + 5)$	$3 \cdot x = 3x$
Multiply the **L**ast terms.	$(2x + 3)(x + 5)$	$3 \cdot 5 = 15$
		F O I L
Add the products.	$(2x + 3)(x + 5)$	$= 2x^2 + 10x + 3x + 15$
Combine like terms.		$= 2x^2 + 13x + 15$

Take Note

FOIL is not really a different way of multiplying. It is based on the Distributive Property.

$(2x + 3)(x + 5)$
$= 2x(x + 5) + 3(x + 5)$
F O I L
$= 2x^2 + 10x + 3x + 15$
$= 2x^2 + 13x + 15$

EXAMPLE 4 Multiply: $(4x - 3)(3x - 2)$

Solution $(4x - 3)(3x - 2)$

F O I L

$= 4x(3x) + 4x(-2) + (-3)(3x) + (-3)(-2)$ • Use the FOIL method.

$= 12x^2 - 8x - 9x + 6$

$= 12x^2 - 17x + 6$ • Combine like terms.

Problem 4 Multiply: $(4y - 5)(3y - 3)$

Solution See page S16.

Try Exercise 83, page 324.

EXAMPLE 5 Multiply: $(3x - 2y)(x + 4y)$

Solution $(3x - 2y)(x + 4y)$

F O I L

$= 3x(x) + 3x(4y) + (-2y)(x) + (-2y)(4y)$ • Use the FOIL method.

$= 3x^2 + 12xy - 2xy - 8y^2$

$= 3x^2 + 10xy - 8y^2$ • Combine like terms.

Problem 5 Multiply: $(3a + 2b)(3a - 5b)$

Solution See page S16.

Try Exercise 99, page 325.

OBJECTIVE 4 Multiply binomials that have special products

The expression $(a + b)(a - b)$ is the product of the **sum and difference of two terms.** The first binomial in the expression is a sum; the second is a difference. The two terms are a and b. The first term in each binomial is a. The second term in each binomial is b.

The expression $(a + b)^2$ is the **square of a binomial.** The first term in the binomial is a. The second term in the binomial is b.

Using FOIL, it is possible to find a pattern for the product of the sum and difference of two terms and for the square of a binomial.

THE SUM AND DIFFERENCE OF TWO TERMS

$$(a + b)(a - b) = a^2 - ab + ab - b^2$$
$$= a^2 - b^2$$

Square of first term → a^2

Square of second term → b^2

EXAMPLE 6 Multiply: $(2x + 3)(2x - 3)$

Solution $(2x + 3)(2x - 3)$
- $(2x + 3)(2x - 3)$ is the product of the sum and difference of two terms.

$= (2x)^2 - 3^2$
- Square the first term. Square the second term.

$= 4x^2 - 9$
- Simplify.

Problem 6 Multiply: $(2a + 5c)(2a - 5c)$

Solution See page S16.

Try Exercise 121, page 325.

THE SQUARE OF A BINOMIAL

$$(a + b)^2 = (a + b)(a + b) = a^2 + ab + ab + b^2$$
$$= a^2 + 2ab + b^2$$

Square of first term → a^2

Twice the product of the two terms → $2ab$

Square of last term → b^2

$$(a - b)^2 = (a - b)(a - b) = a^2 - ab - ab + b^2$$
$$= a^2 - 2ab + b^2$$

Square of first term → a^2

Twice the product of the two terms → $2ab$

Square of last term → b^2

EXAMPLE 7 Multiply: $(4c + 5d)^2$

Solution $(4c + 5d)^2$
- $(4c + 5d)^2$ is the square of a binomial.

$= (4c)^2 + 2(4c)(5d) + (5d)^2$
- Square the first term. Find twice the product of the two terms. Square the second term.

$= 16c^2 + 40cd + 25d^2$
- Simplify.

Problem 7 Multiply: $(3x + 2y)^2$

Solution See page S16.

Try Exercise 117, page 325.

Note that the result in Example 7 is the same result we would get by multiplying the binomial times itself and using the FOIL method.

$$\begin{aligned}(4c + 5d)^2 &= (4c + 5d)(4c + 5d)\\ &= 16c^2 + 20cd + 20cd + 25d^2\\ &= 16c^2 + 40cd + 25d^2\end{aligned}$$

Either method can be used to square a binomial.

EXAMPLE 8 Multiply: $(3x - 2)^2$

Solution $(3x - 2)^2$ • **$(3x - 2)^2$ is the square of a binomial.**

$= (3x)^2 - 2(3x)(2) + (2)^2$ • **Square the first term. Find twice the product of the two terms. Square the last term.**

$= 9x^2 - 12x + 4$ • **Simplify.**

Problem 8 Multiply: $(6x - y)^2$

Solution See page S16.

Try Exercise 119, page 325.

OBJECTIVE 5 Application problems

EXAMPLE 9 The radius of a circle is $(x - 4)$ ft. Find the area of the circle in terms of the variable x. Leave the answer in terms of π.

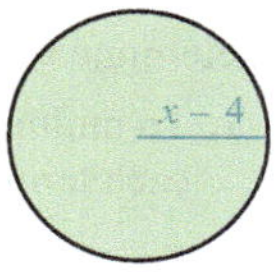

Strategy To find the area, replace the variable r in the formula $A = \pi r^2$ with the given value. Simplify the expression on the right side of the equation.

Solution

$A = \pi r^2$

$A = \pi(x - 4)^2$ • **This is the square of a binomial.**

$A = \pi(x^2 - 8x + 16)$ • **Square the binomial $x - 4$.**

$A = \pi x^2 - 8\pi x + 16\pi$ • **Use the Distributive Property.**

The area is $(\pi x^2 - 8\pi x + 16\pi)$ ft^2.

Problem 9 The length of a rectangle is $(x + 7)$ m. The width is $(x - 4)$ m. Find the area of the rectangle in terms of the variable x.

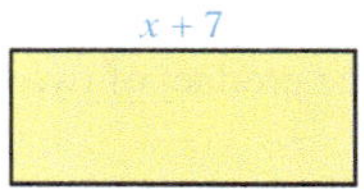

Solution See page S16.

Try Exercise 137, page 326.

EXAMPLE 10 The length of a side of a square is $(3x + 5)$ in. Find the area of the square in terms of the variable x.

Strategy To find the area of the square, replace the variable s in the formula $A = s^2$ with the given value and simplify.

Solution $A = s^2$

$A = (3x + 5)^2$ • **This is the square of a binomial.**

$A = 9x^2 + 30x + 25$ • **Square the binomial $3x + 5$.**

The area is $(9x^2 + 30x + 25)$ in^2.

Problem 10 The base of a triangle is $(x + 3)$ cm, and the height is $(4x - 6)$ cm. Find the area of the triangle in terms of the variable x.

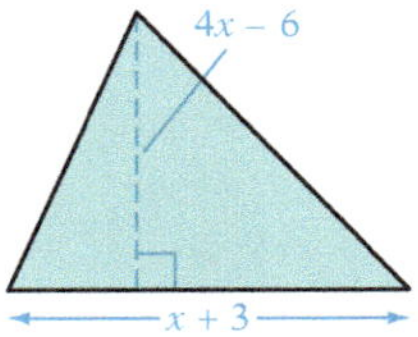

Solution See page S16.

Try Exercise 139, page 326.

7.3 Exercises

CONCEPT CHECK

Determine whether the statement is always true, sometimes true, or never true.

1. To multiply a monomial times a polynomial, use the Distributive Property to multiply each term of the polynomial by the monomial.
2. To multiply two polynomials, multiply each term of one polynomial by the other polynomial.
3. A binomial is a polynomial of degree 2.
4. $(x + 7)(x - 7)$ is the product of the sum and difference of the same two terms.
5. To square a binomial means to multiply it times itself.
6. The square of a binomial is a trinomial.
7. The FOIL method is used to multiply two polynomials.
8. Using the FOIL method, the terms $3x$ and 5 are the "First" terms in $(3x + 5)(2x + 7)$.
9. The product of two binomials is a trinomial.

1 Multiply a polynomial by a monomial (See page 317.)

10. Is the Distributive Property used to simplify the product $2(3x)$? If not, what property is used to simplify this expression?

GETTING READY

11. Multiply: $-3y(y + 7)$

$-3y(y + 7) = \underline{\ ?\ }(y) + (\underline{\ ?\ })(7)$ • **Use the ___?___ Property to multiply each term of $(y + 7)$ by ___?___.**

$= \underline{\ ?\ } - \underline{\ ?\ }$ • **In the first term, the bases are the same. Add the exponents.**

12. Multiply: $5x^3(x^2 + 2x - 10)$

$5x^3(x^2 + 2x - 10)$

$= (\underline{\ ?\ })(x^2) + (\underline{\ ?\ })(2x) - (\underline{\ ?\ })(10)$

• **Use the Distributive Property to multiply each term of ___?___ by $5x^3$.**

$= 5x^{?} + 10x^{?} - 50x^{?}$

• **Use the Rule for Multiplying Exponential Expressions.**

Multiply.

13. $x(x - 2)$

14. $y(3 - y)$

15. $-x(x + 7)$

16. $-y(7 - y)$

17. $3a^2(a - 2)$

18. $4b^2(b + 8)$

19. $-5x^2(x^2 - x)$

20. $-6y^2(y + 2y^2)$

21. $-x^3(3x^2 - 7)$

22. $-y^4(2y^2 - y^6)$

23. $2x(6x^2 - 3x)$

24. $3y(4y - y^2)$

25. $(2x - 4)3x$

26. $(2x + 1)2x$

27. $-xy(x^2 - y^2)$

28. $-x^2y(2xy - y^2)$

29. $x(2x^3 - 3x + 2)$

30. $y(-3y^2 - 2y + 6)$

31. $-a(-2a^2 - 3a - 2)$

32. $-b(5b^2 + 7b - 35)$

33. $x^2(3x^4 - 3x^2 - 2)$

34. $y^3(-4y^3 - 6y + 7)$

35. $2y^2(-3y^2 - 6y + 7)$

36. $4x^2(3x^2 - 2x + 6)$

37. $(a^2 + 3a - 4)(-2a)$

38. $(b^3 - 2b + 2)(-5b)$

39. $-3y^2(-2y^2 + y - 2)$

40. $-5x^2(3x^2 - 3x - 7)$

41. $xy(x^2 - 3xy + y^2)$

42. $ab(2a^2 - 4ab - 6b^2)$

2 Multiply two polynomials (See page 318.)

GETTING READY

43. Multiply: $(x - 3)(x^2 - 4x + 5)$

$(x - 3)(x^2 - 4x + 5) = (\underline{\ ?\ })(x^2) - (\underline{\ ?\ })(4x) + (\underline{\ ?\ })(5)$

$= x^3 - 3x^2 - 4x^2 + 12x + \underline{\ ?\ } - \underline{\ ?\ }$

$= x^3 - 7x^2 + \underline{\ ?\ } - \underline{\ ?\ }$

44. Multiply: $(x - 3)(x^2 - 4x + 5)$

$$\begin{array}{r} x^2 - 4x + 5 \\ x - 3 \\ \hline -3x^2 + 12x - \underline{\ ?\ } \\ x^3 - 4x^2 + \underline{\ ?\ } \qquad\quad \\ \hline x^3 - 7x^2 + \underline{\ ?\ } - \underline{\ ?\ } \end{array}$$

Multiply.

45. $(x^2 + 3x + 2)(x + 1)$

46. $(x^2 - 2x + 7)(x - 2)$

47. $(a - 3)(a^2 - 3a + 4)$

48. $(2x - 3)(x^2 - 3x + 5)$

49. $(-2b^2 - 3b + 4)(b - 5)$

50. $(-a^2 + 3a - 2)(2a - 1)$

51. $(3x - 5)(-2x^2 + 7x - 2)$

52. $(2a - 1)(-a^2 - 2a + 3)$

53. $(x^3 - 3x + 2)(x - 4)$

54. $(y^3 + 4y^2 - 8)(2y - 1)$

55. $(3y - 8)(5y^2 + 8y - 2)$

56. $(4y - 3)(3y^2 + 3y - 5)$

57. $(5a^3 - 15a + 2)(a - 4)$

58. $(3b^3 - 5b^2 + 7)(6b - 1)$

59. $(y + 2)(y^3 + 2y^2 - 3y + 1)$

60. $(2a - 3)(2a^3 - 3a^2 + 2a - 1)$

For Exercises 61 and 62, use the product $(ax^3 + bx + c)(dx + e)$, where a, b, c, d, and e are nonzero real numbers.

61. **a.** What is the degree of the simplified product?
b. At most, how many terms will the simplified product have?

62. Does the simplified product have an x^2 term? If so, what is its coefficient?

3 Multiply two binomials (See pages 318–319.)

GETTING READY

63. For the product $(4x - 3)(x + 5)$:

The **F**irst terms are __?__ and __?__.
The **O**uter terms are __?__ and __?__.
The **I**nner terms are __?__ and __?__.
The **L**ast terms are __?__ and __?__.

For Exercises 64 and 65, name **a.** the first terms, **b.** the outer terms, **c.** the inner terms, and **d.** the last terms of the product of the binomials.

64. $(y - 8)(2y + 3)$

65. $(3d + 4)(d - 1)$

66. Use the FOIL method to multiply $(x + 2)(8x - 3)$.
The product of the **F**irst terms is $x \cdot 8x =$ __?__.
The product of the **O**uter terms is $x \cdot (-3) =$ __?__.
The product of the **I**nner terms is $2 \cdot 8x =$ __?__.
The product of the **L**ast terms is $2 \cdot (-3) =$ __?__.
The sum of these four products is __?__.

Multiply.

67. $(x + 1)(x + 3)$

68. $(y + 2)(y + 5)$

69. $(a - 3)(a + 4)$

70. $(b - 6)(b + 3)$

71. $(y + 3)(y - 8)$

72. $(x + 10)(x - 5)$

73. $(y - 7)(y - 3)$

74. $(a - 8)(a - 9)$

75. $(2x + 1)(x + 7)$

76. $(y + 2)(5y + 1)$

77. $(3x - 1)(x + 4)$

78. $(7x - 2)(x + 4)$

79. $(4x - 3)(x - 7)$

80. $(2x - 3)(4x - 7)$

81. $(3y - 8)(y + 2)$

82. $(5y - 9)(y + 5)$

83. $(3x + 7)(3x + 11)$

84. $(5a + 6)(6a + 5)$

85. $(7a - 16)(3a - 5)$

86. $(5a - 12)(3a - 7)$

87. $(3b + 13)(5b - 6)$

88. $(x + y)(2x + y)$

89. $(2a + b)(a + 3b)$

90. $(3x - 4y)(x - 2y)$

91. $(2a - b)(3a + 2b)$

92. $(5a - 3b)(2a + 4b)$

93. $(2x + y)(x - 2y)$

94. $(3x - 7y)(3x + 5y)$

95. $(2x + 3y)(5x + 7y)$

96. $(5x + 3y)(7x + 2y)$

97. $(3a - 2b)(2a - 7b)$

98. $(5a - b)(7a - b)$

99. $(a - 9b)(2a + 7b)$

100. $(2a + 5b)(7a - 2b)$

101. $(5x + 2y)(2x - 5y)$

102. $(8x - 3y)(7x - 5y)$

103. Suppose a and b are positive numbers such that $a > b$. Will the coefficient of the x term of the product $(ax - b)(x + 1)$ be positive or negative?

104. If the constant terms of two binomials are both negative, is the constant term of the product of the two binomials also negative?

4 Multiply binomials that have special products (See pages 319–321.)

GETTING READY

105. The product $(a + b)(a - b) = a^2 - b^2$, so $(5x + 1)(5x - 1)$
$= (\underline{\ ?\ })^2 - (\underline{\ ?\ })^2 = \underline{\ ?\ } - \underline{\ ?\ }$.

106. The product $(a + b)^2 = a^2 + 2ab + b^2$, so
$(3x + 7)^2 = (\underline{\ ?\ })^2 + 2(\underline{\ ?\ })(\underline{\ ?\ }) + (\underline{\ ?\ })^2$
$= \underline{\ ?\ } + \underline{\ ?\ } + \underline{\ ?\ }$.

Multiply.

107. $(y - 5)(y + 5)$

108. $(y + 6)(y - 6)$

109. $(2x + 3)(2x - 3)$

110. $(4x - 7)(4x + 7)$

111. $(x + 1)^2$

112. $(y - 3)^2$

113. $(3a - 5)^2$

114. $(6x - 5)^2$

115. $(3x - 7)(3x + 7)$

116. $(9x - 2)(9x + 2)$

117. $(2a + b)^2$

118. $(x + 3y)^2$

119. $(x - 2y)^2$

120. $(2x - 3y)^2$

121. $(4 - 3y)(4 + 3y)$

122. $(4x - 9y)(4x + 9y)$

123. $(5x + 2y)^2$

124. $(2a - 9b)^2$

For Exercises 125 to 128, state whether the coefficient of the x term of the product is positive, negative, or zero.

125. $(ax + b)(ax - b)$, where $a > 0$ and $b > 0$

126. $(ax + b)(ax + b)$, where $a > 0$ and $b < 0$

127. $(ax + b)^2$, where $a > 0$ and $b > 0$

128. $(ax + b)^2$, where $a < 0$ and $b < 0$

5 Application problems (See pages 321-322.)

GETTING READY

129. The length of a rectangle is $(2x + 5)$ ft, and the width of the rectangle is $(2x - 5)$ ft. The expression that represents the area of the rectangle is $(\underline{\quad ? \quad})(\underline{\quad ? \quad}) = \underline{\quad ? \quad}$.

130. The length of a side of a square is $(3a - 1)$. The expression that represents the area of the square is $(\underline{\quad ? \quad})^2 = \underline{\quad ? \quad}$.

131. $(4x + 1)^2 = (\underline{\quad ? \quad})^2 + 2(\underline{\quad ? \quad})(\underline{\quad ? \quad}) + (\underline{\quad ? \quad})^2$
$= (\underline{\quad ? \quad}) + (\underline{\quad ? \quad}) + (\underline{\quad ? \quad})$

Geometry Problems

132. The length of a rectangle is $(5x)$ ft. The width is $(2x - 7)$ ft. Find the area of the rectangle in terms of the variable x.

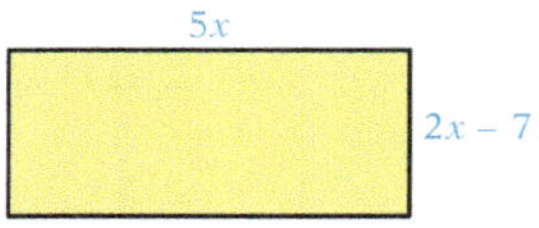

133. The width of a rectangle is $(x - 6)$ m. The length is $(2x + 3)$ m. Find the area of the rectangle in terms of the variable x.

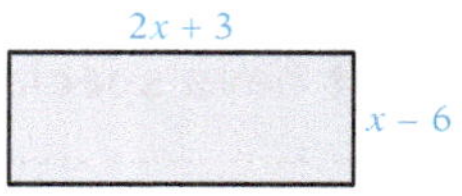

134. The width of a rectangle is $(3x + 1)$ in. The length of the rectangle is twice the width. Find the area of the rectangle in terms of the variable x.

135. The width of a rectangle is $(4x - 3)$ cm. The length of the rectangle is twice the width. Find the area of the rectangle in terms of the variable x.

136. The length of a side of a square is $(2x + 1)$ km. Find the area of the square in terms of the variable x.

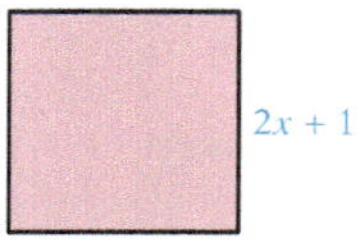

137. The length of a side of a square is $(2x - 3)$ yd. Find the area of the square in terms of the variable x.

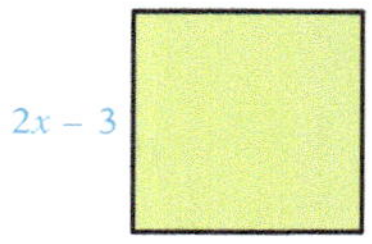

138. The base of a triangle is $(4x)$ m, and the height is $(2x + 5)$ m. Find the area of the triangle in terms of the variable x.

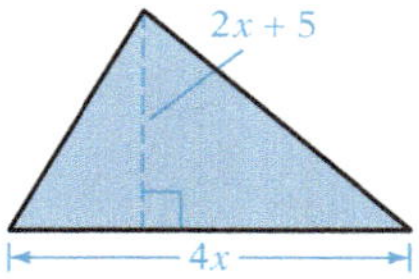

139. The base of a triangle is $(2x + 6)$ in., and the height is $(x - 8)$ in. Find the area of the triangle in terms of the variable x.

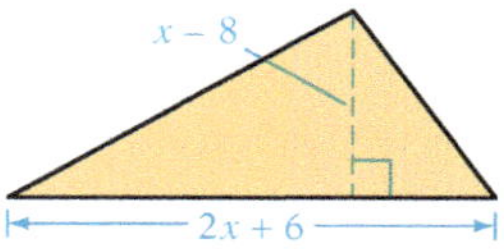

140. The radius of a circle is $(x + 4)$ cm. Find the area of the circle in terms of the variable x. Leave the answer in terms of π.

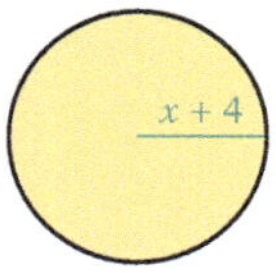

141. The radius of a circle is $(x - 3)$ ft. Find the area of the circle in terms of the variable x. Leave the answer in terms of π.

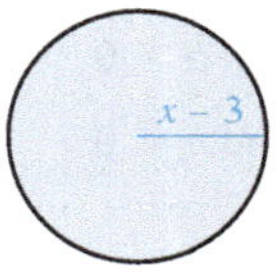

142. **The Olympics** See the news clipping at the right. The Water Cube is not actually a cube because its height is not equal to its length and width. The width of a wall of the Water Cube is 22 ft more than five times the height. (*Source:* Structurae)

a. Express the width of a wall of the Water Cube in terms of the height h.

b. Express the area of one wall of the Water Cube in terms of the height h.

The Water Cube

claudio zaccherini / Shutterstock.com

In the News

Olympic Water Cube Completed

The National Aquatics Center, also known as the Water Cube, was completed on the morning of December 26, 2006. Built in Beijing, China, for the 2008 Olympics, the Water Cube is designed to look like a "cube" of water molecules.

Source: Structurae

143. **Sports** A softball diamond has dimensions 45 ft by 45 ft. A base path border x feet wide lies on both the first-base side and the third-base side of the diamond. Express the total area of the softball diamond and the base path in terms of the variable x.

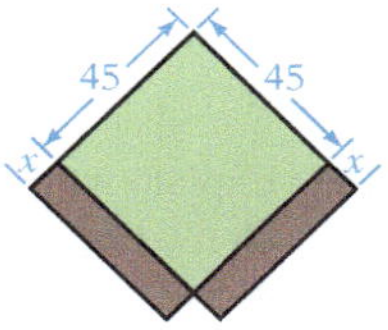

144. **Sports** An athletic field has dimensions 30 yd by 100 yd. An end zone that is w yards wide borders each end of the field. Express the total area of the field and the end zones in terms of the variable w.

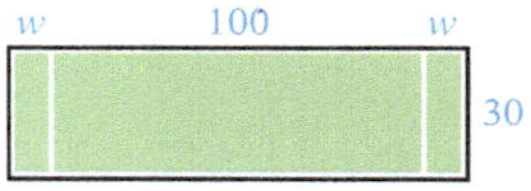

For Exercises 145 and 146, a, b, c, and d are positive integers.

145. Which rectangle has the larger area: one with a length of $(ax + b)$ ft and a width of (cx) ft, or one with a length of $(ax + b)$ ft and a width of $(cx + d)$ ft?

146. Which rectangle has the larger area: one with a length of $(ax + b)$ ft and a width of $(x + d)$ ft, or one with a length of $(ax + b)$ ft and a width of $(cx + d)$ ft?

APPLYING CONCEPTS

Simplify.

147. $(a + b)^2 - (a - b)^2$

148. $(x + 3y)^2 + (x + 3y)(x - 3y)$

149. $(3a^2 - 4a + 2)^2$

150. $(x + 4)^3$

151. $3x^2(2x^3 + 4x - 1) - 6x^3(x^2 - 2)$

152. $(3b + 2)(b - 6) + (4 + 2b)(3 - b)$

Solve.

153. Find $(4n^3)^2$ if $2n - 3 = 4n - 7$.

154. What polynomial has quotient $x^2 + 2x - 1$ when divided by $x + 3$?

155. What polynomial has quotient $3x - 4$ when divided by $4x + 5$?

156. Subtract $4x^2 - x - 5$ from the product of $x^2 + x + 3$ and $x - 4$.

157. Add $x^2 + 2x - 3$ to the product of $2x - 5$ and $3x + 1$.

158. If a polynomial of degree 3 is multiplied by a polynomial of degree 2, what is the degree of the resulting polynomial?

159. Is it possible to multiply a polynomial of degree 2 by a polynomial of degree 2 and have the product be a polynomial of degree 3? If so, give an example. If not, explain why not.

PROJECTS OR GROUP ACTIVITIES

Simplify.

160. $(x + 1)(x - 1)$

161. $(x + 1)(-x^2 + x - 1)$

162. $(x + 1)(x^3 - x^2 + x - 1)$

163. $(x + 1)(-x^4 + x^3 - x^2 + x - 1)$

Use the pattern of the answers to Exercises 160 to 163 to write the product.

164. $(x + 1)(x^5 - x^4 + x^3 - x^2 + x - 1)$

165. $(x + 1)(-x^6 + x^5 - x^4 + x^3 - x^2 + x - 1)$

166. Explain why the diagram at the right represents $(a + b)^2 = a^2 + 2ab + b^2$. Draw diagrams to represent:
a. $(x + 2)^2 = x^2 + 4x + 4$ **b.** $(y + 3)^2 = y^2 + 6y + 9$ **c.** $(z + 4)^2 = z^2 + 8z + 16$

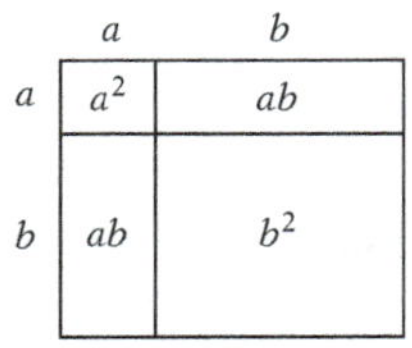

7.4 Integer Exponents and Scientific Notation

OBJECTIVE 1 Integer exponents

The quotient of two exponential expressions with the *same* base can be simplified by writing each expression in factored form, dividing by the common factors, and then writing the result with an exponent.

$$\frac{x^5}{x^2} = \frac{\overset{1}{\cancel{x}} \cdot \overset{1}{\cancel{x}} \cdot x \cdot x \cdot x}{\underset{1}{\cancel{x}} \cdot \underset{1}{\cancel{x}}} = x^3$$

Note that subtracting the exponents results in the same quotient.

$$\frac{x^5}{x^2} = x^{5-2} = x^3$$

To divide two monomials with the same base, subtract the exponents of the like bases.

Focus on dividing exponential expressions

Simplify. **A.** $\frac{a^7}{a^3}$ **B.** $\frac{r^8s^6}{r^7s}$

A. The bases are the same. Subtract the exponent in the denominator from the exponent in the numerator.

$$\frac{a^7}{a^3} = a^{7-3} = a^4$$

B. Subtract the exponents of the like bases.

$$\frac{r^8s^6}{r^7s} = r^{8-7}s^{6-1} = rs^5$$

Recall that for any number a, $a \neq 0$, $\frac{a}{a} = 1$. This property is true for exponential expressions as well. For example, for $x \neq 0$, $\frac{x^4}{x^4} = 1$.

This expression also can be simplified using the rule for dividing exponential expressions with the same base.

$$\frac{x^4}{x^4} = x^{4-4} = x^0$$

Because $\frac{x^4}{x^4} = 1$ and $\frac{x^4}{x^4} = x^0$, the following definition of zero as an exponent is used.

Take Note

In example (1) at the right, we indicated that $a \neq 0$. If we try to evaluate $(12a^3)^0$ when $a = 0$, we get $[12(0)^3]^0 = [12(0)]^0 = 0^0$. However, 0^0 is not defined. Therefore, we must assume that $a \neq 0$. To avoid stating this restriction for every example or exercise, we will assume that variables do not take on values that result in the expression 0^0.

ZERO AS AN EXPONENT

If $x \neq 0$, then $x^0 = 1$. The expression 0^0 is not defined.

EXAMPLES

1. Simplify: $(12a^3)^0$, $a \neq 0$

   Any nonzero expression to the zero power is 1.

   $(12a^3)^0 = 1$

2. Simplify: $-(y^4)^0$, $y \neq 0$

   Any nonzero expression to the zero power is 1.

   Because the negative sign is outside the parentheses, the answer is -1.

   $-(y^4)^0 = -1$

Point of Interest

In the 15th century, the expression $12^{2\bar{m}}$ was used to mean $12x^{-2}$. The use of $\bar{m}$ reflected an Italian influence. In Italy, m was used for minus and p was used for plus. It was understood that $2\bar{m}$ referred to an unnamed variable. Isaac Newton, in the 17th century, advocated the use of a negative exponent, the notation we use today.

The meaning of a negative exponent can be developed by examining the quotient $\frac{x^4}{x^6}$.

The expression can be simplified by writing the numerator and denominator in factored form, dividing by the common factors, and then writing the result with an exponent.

$$\frac{x^4}{x^6} = \frac{\overset{1}{\cancel{x}} \cdot \overset{1}{\cancel{x}} \cdot \overset{1}{\cancel{x}} \cdot \overset{1}{\cancel{x}}}{\underset{1}{\cancel{x}} \cdot \underset{1}{\cancel{x}} \cdot \underset{1}{\cancel{x}} \cdot \underset{1}{\cancel{x}} \cdot x \cdot x} = \frac{1}{x^2}$$

Now simplify the same expression by subtracting the exponents of the like bases.

$$\frac{x^4}{x^6} = x^{4-6} = x^{-2}$$

Because $\frac{x^4}{x^6} = \frac{1}{x^2}$ and $\frac{x^4}{x^6} = x^{-2}$, the expressions $\frac{1}{x^2}$ and x^{-2} must be equal. This leads to the following definition of a negative exponent.

DEFINITION OF NEGATIVE EXPONENTS

If n is a positive integer and $x \neq 0$, then $x^{-n} = \frac{1}{x^n}$ and $\frac{1}{x^{-n}} = x^n$.

EXAMPLES

In each example below, simplify the expression by writing it with a positive exponent.

1. $x^{-10} = \frac{1}{x^{10}}$ **2.** $\frac{1}{a^{-5}} = a^5$ **3.** $2^{-3} = \frac{1}{2^3} = \frac{1}{8}$

Focus on evaluating a numerical expression with a negative exponent

Evaluate 2^{-4}.

Write the expression with a positive exponent. Then simplify. $2^{-4} = \frac{1}{2^4} = \frac{1}{16}$

Take Note

Note from the example at the right that 2^{-4} is a *positive* number. A negative exponent does not indicate a negative number.

Now that negative exponents have been defined, the Rule for Dividing Exponential Expressions can be stated.

RULE FOR DIVIDING EXPONENTIAL EXPRESSIONS

If m and n are integers and $x \neq 0$, then $\frac{x^m}{x^n} = x^{m-n}$.

EXAMPLES

Simplify each expression below by using the Rule for Dividing Exponential Expressions.

1. $\frac{x^3}{x^5} = x^{3-5} = x^{-2} = \frac{1}{x^2}$ **2.** $\frac{y^6}{y^{-2}} = y^{6-(-2)} = y^8$

3. $\frac{b^{-5}}{b^{-1}} = b^{-5-(-1)} = b^{-4} = \frac{1}{b^4}$ **4.** $\frac{a^{-4}}{a^{-7}} = a^{-4-(-7)} = a^3$

EXAMPLE 1 Write $\frac{3^{-3}}{3^2}$ with a positive exponent. Then evaluate.

Solution

$\frac{3^{-3}}{3^2} = 3^{-3-2}$ • 3^{-3} and 3^2 have the same base. Subtract the exponents.

$= 3^{-5}$

$= \frac{1}{3^5}$ • Use the Definition of Negative Exponents to write the expression with a positive exponent.

$= \frac{1}{243}$ • Evaluate.

Problem 1 Write $\frac{2^{-2}}{2^3}$ with a positive exponent. Then evaluate.

Solution See page S16.

Try Exercise 33, page 335.

The rules for simplifying exponential expressions and powers of exponential expressions are true for all integers. These rules are restated here.

RULES FOR EXPONENTS

If m, n, and p are integers, then

$x^m \cdot x^n = x^{m+n}$ $\quad (x^m)^n = x^{mn}$ $\quad (x^m y^n)^p = x^{mp}y^{np}$

$\dfrac{x^m}{x^n} = x^{m-n}, x \neq 0$ $\quad x^{-n} = \dfrac{1}{x^n}, x \neq 0$ $\quad x^0 = 1, x \neq 0$

An exponential expression is in simplest form when it is written with only positive exponents.

EXAMPLE 2 Simplify: **A.** $a^{-7}b^3$ **B.** $\dfrac{x^{-4}y^6}{xy^2}$ **C.** $6d^{-4}, d \neq 0$

Solution **A.** $a^{-7}b^3 = \dfrac{b^3}{a^7}$

- **Rewrite a^{-7} with a positive exponent.**

B. $\dfrac{x^{-4}y^6}{xy^2} = x^{-4-1}y^{6-2}$

- **Divide variables with the same base by subtracting the exponents.**

$= x^{-5}y^4$

$= \dfrac{y^4}{x^5}$

- **Write the expression with only positive exponents.**

C. $6d^{-4} = 6 \cdot \dfrac{1}{d^4} = \dfrac{6}{d^4}$

- **Use the Definition of Negative Exponents to rewrite the expression with a positive exponent.**

Take Note

In Example 2C, the exponent on d is -4 (negative 4). The d^{-4} is written in the denominator as d^4. The exponent on 6 is 1 (positive 1). The 6 remains in the numerator.

Note that we indicated $d \neq 0$. This is necessary because division by zero is not defined. In this textbook, we will assume that values of the variables are chosen so that division by zero does not occur.

Problem 2 Simplify: **A.** x^5y^{-7} **B.** $\dfrac{b^8}{a^{-5}b^6}$ **C.** $4c^{-3}$

Solution See page S16.

Try Exercise 75, page 335.

EXAMPLE 3 Simplify: **A.** $\dfrac{-35a^6b^{-2}}{25a^{-2}b^5}$ **B.** $(-2x)(3x^{-2})^{-3}$

Solution **A.** $\dfrac{-35a^6b^{-2}}{25a^{-2}b^5} = -\dfrac{35a^6b^{-2}}{25a^{-2}b^5}$

- **A negative sign is placed in front of a fraction.**

$= -\dfrac{\overset{1}{\cancel{5}} \cdot 7a^{6-(-2)}b^{-2-5}}{\underset{1}{\cancel{5}} \cdot 5}$

- **Factor the coefficients. Divide by the common factors. Divide variables with the same base by subtracting the exponents.**

$= -\dfrac{7a^8b^{-7}}{5}$

$= -\dfrac{7a^8}{5b^7}$

- **Write the expression with only positive exponents.**

B. $(-2x)(3x^{-2})^{-3} = (-2x)(3^{-3}x^6)$

- **Use the Rule for Simplifying Powers of Products.**

$= \dfrac{-2x \cdot x^6}{3^3}$

- **Write the expression with positive exponents.**

$= -\dfrac{2x^7}{27}$

- **Use the Rule for Multiplying Exponential Expressions, and simplify the numerical exponential expression.**

Problem 3 Simplify: **A.** $\dfrac{12x^{-8}y^4}{-16xy^{-3}}$ **B.** $(-3ab)(2a^3b^{-2})^{-3}$

Solution See page S17.

Try Exercise 73, page 335.

OBJECTIVE 2 Scientific notation

Very large and very small numbers are encountered in the fields of science and engineering. For example, the charge of an electron is 0.000000000000000000160 coulomb. These numbers can be written more easily in scientific notation. In **scientific notation,** a number is expressed as a product of two factors, one a number between 1 and 10 and the other a power of 10.

To change a number written in decimal notation to scientific notation, write it in the form $a \times 10^n$, where a is a number between 1 and 10 and n is an integer.

Take Note
There are two steps to writing a number in scientific notation: (1) determine the number between 1 and 10, and (2) determine the exponent on 10.

For numbers greater than 10, move the decimal point to the right of the first digit. The exponent n is positive and equal to the number of places the decimal point has been moved.

$240{,}000 = 2.4 \times 10^5$

$93{,}000{,}000 = 9.3 \times 10^7$

For numbers less than 1, move the decimal point to the right of the first nonzero digit. The exponent n is negative. The absolute value of the exponent is equal to the number of places the decimal point has been moved.

$0.00030 = 3.0 \times 10^{-4}$

$0.0000832 = 8.32 \times 10^{-5}$

Look at the last example above: $0.0000832 = 8.32 \times 10^{-5}$. Using the Definition of Negative Exponents,

$$10^{-5} = \frac{1}{10^5} = \frac{1}{100{,}000} = 0.00001$$

Because $10^{-5} = 0.00001$, we can write

$$8.32 \times 10^{-5} = 8.32 \times 0.00001 = 0.0000832$$

which is the number we started with. We have not changed the value of the number; we have just written it in another form.

EXAMPLE 4 Write the number in scientific notation.
A. 824,300,000,000 **B.** 0.000000961

Solution **A.** $824{,}300{,}000{,}000 = 8.243 \times 10^{11}$
- **Move the decimal point 11 places to the left. The exponent on 10 is 11.**

B. $0.000000961 = 9.61 \times 10^{-7}$
- **Move the decimal point 7 places to the right. The exponent on 10 is −7.**

Problem 4 Write the number in scientific notation.

A. 57,000,000,000 **B.** 0.000000017

Solution See page S17.

Try Exercise 97, page 336.

Changing a number written in scientific notation to decimal notation also requires moving the decimal point.

When the exponent on 10 is positive, move the decimal point to the right the same number of places as the exponent.

$3.45 \times 10^9 = 3{,}450{,}000{,}000$

$2.3 \times 10^8 = 230{,}000{,}000$

When the exponent on 10 is negative, move the decimal point to the left the same number of places as the absolute value of the exponent.

$8.1 \times 10^{-3} = 0.0081$

$6.34 \times 10^{-6} = 0.00000634$

EXAMPLE 5 Write the number in decimal notation.

A. 7.329×10^6 **B.** 6.8×10^{-10}

Solution **A.** $7.329 \times 10^6 = 7{,}329{,}000$

- **The exponent on 10 is positive. Move the decimal point 6 places to the right.**

B. $6.8 \times 10^{-10} = 0.00000000068$

- **The exponent on 10 is negative. Move the decimal point 10 places to the left.**

Problem 5 Write the number in decimal notation.

A. 5×10^{12} **B.** 4.0162×10^{-9}

Solution See page S17.

Try Exercise 103, page 336.

The rules for multiplying and dividing with numbers in scientific notation are the same as those for calculating with algebraic expressions. The power of 10 corresponds to the variable, and the number between 1 and 10 corresponds to the coefficient of the variable.

	Algebraic Expression	Scientific Notation
Multiplication	$(4x^{-3})(2x^5) = 8x^2$	$(4 \times 10^{-3})(2 \times 10^5) = 8 \times 10^2$
Division	$\dfrac{6x^5}{3x^{-2}} = 2x^{5-(-2)} = 2x^7$	$\dfrac{6 \times 10^5}{3 \times 10^{-2}} = 2 \times 10^{5-(-2)} = 2 \times 10^7$

EXAMPLE 6 Multiply or divide.

A. $(3.0 \times 10^5)(1.1 \times 10^{-8})$ **B.** $\dfrac{7.2 \times 10^{13}}{2.4 \times 10^{-3}}$

Solution **A.** $(3.0 \times 10^5)(1.1 \times 10^{-8}) = 3.3 \times 10^{-3}$

- **Multiply 3.0 and 1.1. Add the exponents on 10.**

B. $\dfrac{7.2 \times 10^{13}}{2.4 \times 10^{-3}} = 3 \times 10^{16}$

- **Divide 7.2 by 2.4. Subtract the exponents on 10.**

Problem 6 Multiply or divide.

A. $(2.4 \times 10^{-9})(1.6 \times 10^3)$ **B.** $\dfrac{5.4 \times 10^{-2}}{1.8 \times 10^{-4}}$

Solution See page S17.

Try Exercise 121, page 337.

7.4 Exercises

CONCEPT CHECK

Determine whether the statement is true or false.

1. The expression $\frac{x^5}{y^3}$ can be simplified by subtracting the exponents.
2. The rules of exponents can be applied to expressions that contain an exponent of zero or contain negative exponents.
3. The expression 3^{-2} represents the reciprocal of 3^2.
4. $5x^0 = 0$
5. The expression 4^{-3} represents a negative number.
6. To be in simplest form, an exponential expression cannot contain any negative exponents.

1 Integer exponents (See pages 328–332.)

7. Explain how to rewrite a variable that has a negative exponent as an expression with a positive exponent.

GETTING READY

8. As long as x is not zero, x^0 is defined to be equal to __?__. Using this definition, $3^0 =$ __?__, $(7x^3)^0 =$ __?__, and $-2x^0 =$ __?__.

9. Simplify: $\frac{8x^{10}}{2x^3}$

$\frac{8x^{10}}{2x^3} = (\underline{\ ?\ })x^{\underline{?} - \underline{?}}$ • **Divide the coefficients by the common factor 2. Divide the variable parts by subtracting the exponents.**

$= \underline{\ ?\ }$ • **Simplify.**

10. Simplify: $\frac{a^3b^7}{a^5b}$

$\frac{a^3b^7}{a^5b} = a^{\underline{?} - \underline{?}}b^{\underline{?} - \underline{?}}$ • **Divide variables with the same base by subtracting the exponents.**

$= a^{\underline{?}}b^{\underline{?}}$ • **Simplify.**

$= \frac{b^6}{?}$ • **Write the expression with only positive exponents.**

Simplify.

11. $\frac{y^7}{y^3}$
12. $\frac{z^9}{z^2}$
13. $\frac{a^8}{a^5}$
14. $\frac{c^{12}}{c^5}$
15. $\frac{p^5}{p}$
16. $\frac{w^9}{w}$
17. $\frac{4x^8}{2x^5}$
18. $\frac{12z^7}{4z^3}$
19. $\frac{22k^5}{11k^4}$
20. $\frac{14m^{11}}{7m^{10}}$
21. $\frac{m^9n^7}{m^4n^5}$
22. $\frac{y^5z^6}{yz^3}$
23. $\frac{6r^4}{4r^2}$
24. $\frac{8x^9}{12x^6}$
25. $\frac{-16a^7}{24a^6}$
26. $\frac{-18b^5}{27b^4}$
27. x^{-2}
28. y^{-10}
29. $\frac{1}{a^{-6}}$
30. $\frac{1}{b^{-4}}$

Write with a positive or zero exponent. Then evaluate.

31. 5^{-2} **32.** 3^{-3} **33.** $\frac{1}{8^{-2}}$ **34.** $\frac{1}{12^{-1}}$

35. $\frac{3^{-2}}{3}$ **36.** $\frac{5^{-3}}{5}$ **37.** $\frac{2^3}{2^3}$ **38.** $\frac{3^{-2}}{3^{-2}}$

Simplify.

39. $4x^{-7}$ **40.** $-6y^{-1}$ **41.** $\frac{5}{b^{-8}}$ **42.** $\frac{-3}{y^{-3}}$

43. $\frac{1}{3x^{-2}}$ **44.** $\frac{2}{5c^{-6}}$ **45.** $(ab^5)^0$ **46.** $(32x^3y^4)^0$

47. $\frac{y^3}{y^8}$ **48.** $\frac{z^4}{z^6}$ **49.** $\frac{a^5}{a^{11}}$ **50.** $\frac{m}{m^7}$

51. $\frac{4x^2}{12x^5}$ **52.** $\frac{6y^8}{8y^9}$ **53.** $\frac{-12x}{-18x^6}$ **54.** $\frac{-24c^2}{-36c^{11}}$

55. $\frac{x^6y^5}{x^8y}$ **56.** $\frac{a^3b^2}{a^2b^3}$ **57.** $\frac{2m^6n^2}{5m^9n^{10}}$ **58.** $\frac{5r^3t^7}{6r^5t^7}$

59. $\frac{pq^3}{p^4q^4}$ **60.** $\frac{a^4b^5}{a^5b^6}$ **61.** $\frac{3x^4y^5}{6x^4y^8}$ **62.** $\frac{14a^3b^6}{21a^5b^6}$

63. $\frac{14x^4y^6z^2}{16x^3y^9z}$ **64.** $\frac{24a^2b^7c^9}{36a^7b^5c}$ **65.** $(-2xy^{-2})^3$

66. $(-3x^{-1}y^2)^2$ **67.** $(3x^{-1}y^{-2})^2$ **68.** $(5xy^{-3})^{-2}$

69. $(2x^{-1})(x^{-3})$ **70.** $(-2x^{-5})x^7$ **71.** $(-5a^2)(a^{-5})^2$

72. $(2a^{-3})(a^7b^{-1})^3$ **73.** $(-2ab^{-2})(4a^{-2}b)^{-2}$ **74.** $(3ab^{-2})(2a^{-1}b)^{-3}$

75. $\frac{a^{-3}b^{-4}}{a^2b^2}$ **76.** $\frac{3x^{-2}y^2}{6xy^2}$ **77.** $\frac{2x^{-2}y}{8xy}$

78. $\frac{3x^{-2}y}{xy}$ **79.** $\frac{2x^{-1}y^4}{x^2y^3}$ **80.** $\frac{2x^{-1}y^{-4}}{4xy^2}$

81. $\frac{12a^2b^3}{-27a^2b^2}$ **82.** $\frac{-16xy^4}{96x^4y^4}$ **83.** $\frac{-8x^2y^4}{44y^2z^5}$

84. True or false?

a. $\frac{a^{4n}}{a^n} = a^4$ b. $\frac{1}{a^{m-n}} = a^{n-m}$

85. True or false?

a. $a^{-n}a^n = 1$ b. $\frac{a^n}{b^m} = \left(\frac{a}{b}\right)^{n-m}$

2 Scientific notation (See pages 332–333.)

86. Why might a number be written in scientific notation instead of decimal notation?

GETTING READY

87. A number is written in scientific notation if it is written as the product of a number between __?__ and __?__ and an integer power of __?__.

Determine whether the number is written in scientific notation. If not, explain why not.

88. 39.4×10^3 **89.** 0.8×10^{-6} **90.** $7.1 \times 10^{2.4}$ **91.** 5.8×10^{-132}

GETTING READY

92. To write the number 354,000,000 in scientific notation, move the decimal point __?__ places to the __?__. The exponent on 10 is __?__.

93. To write the number 0.0000000086 in scientific notation, move the decimal point __?__ places to the __?__. The exponent on 10 is __?__.

Write the number in scientific notation.

94. 2,370,000 **95.** 75,000 **96.** 0.00045 **97.** 0.000076

98. 309,000 **99.** 819,000,000 **100.** 0.000000601 **101.** 0.00000000096

Write the number in decimal notation.

102. 7.1×10^5 **103.** 2.3×10^7 **104.** 4.3×10^{-5} **105.** 9.21×10^{-7}

106. 6.71×10^8 **107.** 5.75×10^9 **108.** 7.13×10^{-6} **109.** 3.54×10^{-8}

110. **Technology** See the news clipping at the right. Express in scientific notation the thickness, in meters, of the memristor.

AP Photo/Paul Sakuma

HP researchers view an image of the memristor.

In the News

HP Introduces the Memristor

Hewlett Packard has announced the design of the *memristor,* a new memory technology with the potential to be much smaller than the memory chips used in today's computers. HP has made a memristor with a thickness of 0.000000015 m (15 nanometers).

Source: The New York Times

111. **Astronomy** Astrophysicists estimate that the radius of the Milky Way galaxy is 1,000,000,000,000,000,000,000 m. Write this number in scientific notation.

112. **Geology** The mass of Earth is 5,980,000,000,000,000,000,000,000 kg. Write this number in scientific notation.

113. **Physics** Carbon nanotubes are extremely strong cylinders of carbon atoms that have remarkable properties. Some nanotubes with a diameter of 0.0000000004 m have been created. Write this number in scientific notation.

114. **Biology** The weight of a single *E. coli* bacterium is 0.000000000000665 g. Write this number in scientific notation.

115. **Archeology** The weight of the Great Pyramid of Cheops is estimated to be 12,000,000,000 lb. Write this number in scientific notation.

116. **Food Science** The frequency (in oscillations per second) of a microwave generated by a microwave oven is approximately 2,450,000,000 hertz. (1 hertz is one oscillation in 1 s.) Write this number in scientific notation.

117. **Astronomy** One light-year is the distance traveled by light in one year. One light-year is 5,880,000,000,000 mi. Write this number in scientific notation.

Will & Deni McIntyre/Getty Images

Great Pyramid of Cheops

118. **Biophysics** Biologists and physicists are working together to measure the mass of a virus. Currently, a virus with a mass of 0.0000000000000000039 g can be measured. Write this number in scientific notation.

119. **Astronomy** See the news clipping at the right. WASP-12b orbits a star that is 5.1156×10^{15} mi from Earth. (*Source:* news.yahoo.com) Write this number in decimal notation.

In the News

Hottest Planet Ever Discovered

A planet called WASP-12b is the hottest planet ever discovered, at about 4000°F. It orbits its star faster than any other known planet, completing one revolution each day.

Source: news.yahoo.com

Multiply or divide.

120. $(1.9 \times 10^{12})(3.5 \times 10^{7})$

121. $(4.2 \times 10^{7})(1.8 \times 10^{-5})$

122. $(2.3 \times 10^{-8})(1.4 \times 10^{-6})$

123. $(3 \times 10^{-20})(2.4 \times 10^{9})$

124. $\dfrac{6.12 \times 10^{14}}{1.7 \times 10^{9}}$

125. $\dfrac{6 \times 10^{-8}}{2.5 \times 10^{-2}}$

126. $\dfrac{5.58 \times 10^{-7}}{3.1 \times 10^{11}}$

127. $\dfrac{9.03 \times 10^{6}}{4.3 \times 10^{-5}}$

128. $a \times 10^{n}$ and $a \times 10^{-m}$ are numbers written in scientific notation, where n and m are positive integers such that $n < m$. Is $(a \times 10^{n})(a \times 10^{-m})$ greater than or less than 1?

129. $a \times 10^{-n}$ and $a \times 10^{-m}$ are numbers written in scientific notation, where n and m are positive integers such that $n > m$. Is $\dfrac{a \times 10^{-m}}{a \times 10^{-n}}$ greater than or less than 1?

APPLYING CONCEPTS

Evaluate.

130. $8^{-2} + 2^{-5}$

131. $9^{-2} + 3^{-3}$

132. Evaluate 2^{x} and 2^{-x} when $x = -2, -1, 0, 1,$ and 2.

133. Evaluate 3^{x} and 3^{-x} when $x = -2, -1, 0, 1,$ and 2.

Write in decimal notation.

134. 2^{-4}

135. 25^{-2}

Complete.

136. If $m = n$ and $a \neq 0$, then $\dfrac{a^{m}}{a^{n}} =$ ______.

137. If $m = n + 1$ and $a \neq 0$, then $\dfrac{a^{m}}{a^{n}} =$ ______.

Solve.

138. $(-4.8)^{x} = 1$

139. $-6.3^{x} = -1$

Determine whether each equation is true or false. If the equation is false, change the right side of the equation to make a true equation.

140. $(2a)^{-3} = \dfrac{2}{a^3}$

141. $((a^{-1})^{-1})^{-1} = \dfrac{1}{a}$

142. $(2 + 3)^{-1} = 2^{-1} + 3^{-1}$

143. If $x \neq \frac{1}{3}$, then $(3x - 1)^0 = (1 - 3x)^0$.

144. Why is the condition $x \neq \frac{1}{3}$ given in Exercise 143?

145. If x is a nonzero real number, is x^{-2} always positive, always negative, or positive or negative depending on whether x is positive or negative? Explain your answer.

146. If x is a nonzero real number, is x^{-3} always positive, always negative, or positive or negative depending on whether x is positive or negative? Explain your answer.

PROJECTS OR GROUP ACTIVITIES

147. Population and Land Allocation In this project, you are asked to determine hypothetical land allocation for the world's population today. Use the figure 7×10^9 for the current world population and the figure 3.1×10^8 for the current U.S. population. (*Source:* www.infoplease.com) One square mile is approximately 2.8×10^7 ft^2.

Rhode Island

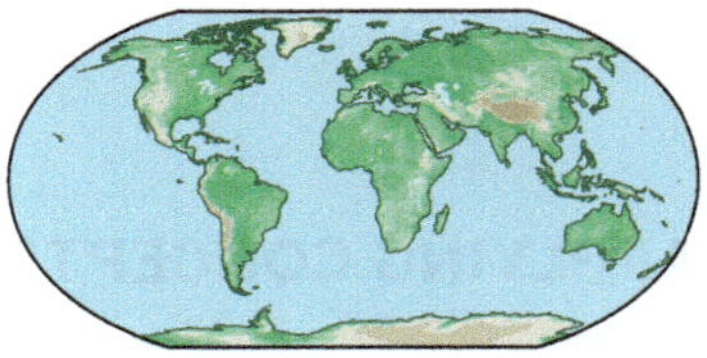

a. If every person in the world moved to Texas and each person were given an equal amount of land, how many square feet of land would each person have? The area of Texas is 2.619×10^5 mi^2.

b. If every person in the United States moved to Rhode Island and each person were given an equal amount of land, how many square feet of land would each person have? The area of Rhode Island is 1.0×10^3 mi^2. Round to the nearest whole number.

c. Suppose every person in the world were given a plot of land the size of a two-car garage (22 ft × 22 ft).

i. How many people would fit in a square mile? Round to the nearest hundred.

ii. How many square miles would be required to accommodate the entire world population? Round to the nearest hundred.

d. If the total land area of Earth were divided equally, how many acres of land would each person be allocated? Use a figure of 5.7×10^7 mi^2 for the land area of Earth. One acre is 43,560 ft^2. Round to the nearest tenth.

e. If every person on Earth were given a plot of land the size of a two-car garage, what would be the carrying capacity of Earth? Round to the nearest hundred billion.

7.5 Division of Polynomials

OBJECTIVE 1 Divide a polynomial by a monomial

Note that $\frac{8+4}{2}$ can be simplified by first adding the terms in the numerator and then dividing the result. It can also be simplified by first dividing each term in the numerator by the denominator and then adding the results.

$$\frac{8+4}{2} = \frac{12}{2} = 6$$

$$\frac{8+4}{2} = \frac{8}{2} + \frac{4}{2} = 4 + 2 = 6$$

To divide a polynomial by a monomial, divide each term in the numerator by the denominator, and write the sum of the quotients.

$$\frac{a+b}{c} = \frac{a}{c} + \frac{b}{c}$$

Focus on dividing a polynomial by a monomial

Divide: $\dfrac{6x^2 + 4x}{2x}$

Divide each term of the polynomial $6x^2 + 4x$ by the monomial $2x$. Simplify.

$$\frac{6x^2 + 4x}{2x} = \frac{6x^2}{2x} + \frac{4x}{2}$$
$$= 3x + 2$$

EXAMPLE 1 Divide: $\dfrac{6x^3 - 3x^2 + 9x}{3x}$

Solution

$$\frac{6x^3 - 3x^2 + 9x}{3x} = \frac{6x^3}{3x} - \frac{3x^2}{3x} + \frac{9x}{3x}$$
$$= 2x^2 - x + 3$$

- **Divide each term of the polynomial by the monomial 3x.**
- **Simplify each expression.**

Problem 1 Divide: $\dfrac{4x^3y + 8x^2y^2 - 4xy^3}{2xy}$

Solution See page S17.

Try Exercise 19, page 341.

EXAMPLE 2 Divide: $\dfrac{12x^2y - 6xy + 4x^2}{2xy}$

Solution

$$\frac{12x^2y - 6xy + 4x^2}{2xy}$$
$$= \frac{12x^2y}{2xy} - \frac{6xy}{2xy} + \frac{4x^2}{2xy}$$
$$= 6x - 3 + \frac{2x}{y}$$

- **Divide each term of the polynomial by the monomial 2xy.**
- **Simplify each expression.**

Problem 2 Divide: $\dfrac{24x^2y^2 - 18xy + 6y}{6xy}$

Solution See page S17.

Try Exercise 27, page 342.

OBJECTIVE 2 Divide polynomials

To divide polynomials, use a method similar to that used for division of whole numbers. The same equation used to check division of whole numbers is used to check polynomial division.

(Quotient × Divisor) + Remainder = Dividend

$$\begin{array}{r} 3 \\ 5\overline{)17} \\ \underline{-15} \\ 2 \end{array}$$

For example, for the division at the left,

(Quotient × Divisor) + Remainder = Dividend

$$(3 \times 5) \quad + \quad 2 \quad = \quad 17$$

Focus on dividing two polynomials

Divide: $(x^2 - 5x + 8) \div (x - 3)$

Step 1

$$\begin{array}{r} x \phantom{{}-5x+8} \\ x - 3\overline{)x^2 - 5x + 8} \\ \underline{x^2 - 3x} \phantom{{}+8} \\ -2x + 8 \end{array}$$

Think: $x\overline{)x^2} = \dfrac{x^2}{x} = x$

Multiply: $x(x - 3) = x^2 - 3x$

Subtract: $(x^2 - 5x) - (x^2 - 3x) = -2x$

Bring down + 8.

Step 2

$$\begin{array}{r} x - 2 \\ x - 3\overline{)x^2 - 5x + 8} \\ \underline{x^2 - 3x} \phantom{{}+8} \\ -2x + 8 \\ \underline{-2x + 6} \\ 2 \end{array}$$

Think: $x\overline{)-2x} = \dfrac{-2x}{x} = -2$

Multiply: $-2(x - 3) = -2x + 6$

Subtract: $(-2x + 8) - (-2x + 6) = 2$

The remainder is 2.

Check:

Quotient × Divisor + Remainder = Dividend

$(x - 2)(x - 3) + 2 = x^2 - 3x - 2x + 6 + 2 = x^2 - 5x + 8$

$(x^2 - 5x + 8) \div (x - 3) = x - 2 + \dfrac{2}{x - 3}$

If a term is missing in the dividend, insert the term with zero as its coefficient. This helps keep like terms in the same column. This is illustrated in Example 3.

EXAMPLE 3 Divide: $(6x + 2x^3 + 26) \div (x + 2)$

Solution

$$\begin{array}{r} 2x^2 - 4x + 14 \\ x + 2\overline{)2x^3 + 0x^2 + 6x + 26} \\ \underline{2x^3 + 4x^2} \phantom{{}+6x+26} \\ -4x^2 + 6x \phantom{{}+26} \\ \underline{-4x^2 - 8x} \phantom{{}+26} \\ 14x + 26 \\ \underline{14x + 28} \\ -2 \end{array}$$

- **Arrange the terms in descending order. There is no x^2 term in $2x^3 + 6x + 26$. Insert $0x^2$ for the missing term so that like terms will be in the same columns.**

$(6x + 2x^3 + 26) \div (x + 2) = 2x^2 - 4x + 14 - \dfrac{2}{x + 2}$

Problem 3 Divide: $(x^3 - 2x - 4) \div (x - 2)$

Solution See page S17.

Try Exercise 41, page 342.

7.5 Exercises

CONCEPT CHECK

1. Every division equation has a related multiplication equation. For instance, $\frac{16}{2} = 8$ means that $16 = 2 \cdot 8$. What is the related multiplication equation for $\frac{15x^2 + 12x}{3x} = 5x + 4$?

2. Given that $\frac{x^3 - x^2 + x - 1}{x - 1} = x^2 + 1$, name two factors of $x^3 - x^2 + x - 1$.

Determine whether the statement is true or false.

3. $5\frac{2}{3} = 5 + \frac{2}{3}$

4. For $b \neq 0$, $a \div b = \frac{a}{b}$.

5. For $c \neq 0$, $\frac{a - b}{c} = \frac{a}{c} - \frac{b}{c}$.

6. For $x \neq 0$, $\frac{9x^2 + 6x}{3x} = 3x + 6x$.

1 Divide a polynomial by a monomial (See page 339.)

GETTING READY

7. Replace each question mark to make a true statement.

$$\frac{18y^5 + 3y}{3y} = \frac{18y^5}{?} + \frac{3y}{?}$$
$$= \underline{\quad ? \quad} + \underline{\quad ? \quad}$$

8. Replace each question mark to make a true statement.

$$\frac{12x^3 - 8x^2}{4x^2} = \frac{?}{4x^2} - \frac{?}{4x^2}$$
$$= \underline{\quad ? \quad} - \underline{\quad ? \quad}$$

Divide.

9. $\frac{10a - 25}{5}$

10. $\frac{16b - 40}{8}$

11. $\frac{3a^2 + 2a}{a}$

12. $\frac{6y^2 + 4y}{y}$

13. $\frac{3x^2 - 6x}{3x}$

14. $\frac{10y^2 - 6y}{2y}$

15. $\frac{5x^2 - 10x}{-5x}$

16. $\frac{3y^2 - 27y}{-3y}$

17. $\frac{x^3 + 3x^2 - 5x}{x}$

18. $\frac{a^3 - 5a^2 + 7a}{a}$

19. $\frac{x^6 - 3x^4 - x^2}{x^2}$

20. $\frac{a^8 - 5a^5 - 3a^3}{a^2}$

21. $\frac{5x^2y^2 + 10xy}{5xy}$

22. $\frac{8x^2y^2 - 24xy}{8xy}$

23. $\frac{9y^6 - 15y^3}{-3y^3}$

24. $\frac{4x^4 - 6x^2}{-2x^2}$

25. $\frac{3x^2 - 2x + 1}{x}$

26. $\frac{8y^2 + 2y - 3}{y}$

27. $\frac{-3x^2 + 7x - 6}{x}$

28. $\frac{2y^2 - 6y + 9}{y}$

29. $\frac{16a^2b - 20ab + 24ab^2}{4ab}$

30. $\frac{22a^2b + 11ab - 33ab^2}{11ab}$

31. $\frac{9x^2y + 6xy - 3x}{xy}$

32. $\frac{18a^2b^2 + 9ab - 6}{3ab}$

33. How can multiplication be used to check that $\frac{8x^3 - 12x^2 - 4x}{4x} = 2x^2 - 3x - 1$?

2 Divide polynomials (See pages 340–341.)

Divide.

34. $(b^2 - 14b + 49) \div (b - 7)$

35. $(x^2 - x - 6) \div (x - 3)$

36. $(y^2 + 2y - 35) \div (y + 7)$

37. $(2x^2 + 5x + 2) \div (x + 2)$

38. $(2y^2 + 7) \div (y - 3)$

39. $(x^2 + 1) \div (x - 1)$

40. $(x^2 + 4) \div (x + 2)$

41. $(6x^2 - 7x) \div (3x - 2)$

42. $(a^2 + 5a + 10) \div (a + 2)$

43. $(b^2 - 8b - 9) \div (b - 3)$

44. $(2y^2 - 9y + 8) \div (2y + 3)$

45. $(3x^2 + 5x - 4) \div (x - 4)$

46. $(8x + 3 + 4x^2) \div (2x - 1)$

47. $(10 + 21y + 10y^2) \div (2y + 3)$

48. $(12a^2 - 7 - 25a) \div (3a - 7)$

49. $(5 - 23x + 12x^2) \div (4x - 1)$

50. $(24 + 6a^2 + 25a) \div (3a - 1)$

51. $(3x^2 + x^3 + 8 + 5x) \div (x + 1)$

52. $(7x - 6x^2 + x^3 - 1) \div (x - 1)$

53. $(x^4 - x^2 - 6) \div (x^2 + 2)$

54. $(x^4 + 3x^2 - 10) \div (x^2 - 2)$

55. True or false? When a sixth-degree polynomial is divided by a third-degree polynomial, the quotient is a second-degree polynomial.

56. True or false? When a polynomial of degree $3n$ is divided by a polynomial of degree n, the degree of the quotient polynomial is $2n$.

APPLYING CONCEPTS

57. Replace the question marks with expressions to make a true statement.

If $\frac{x^2 + 2x - 3}{x - 2} = x + 4 + \frac{5}{x - 2}$, then $x^2 + 2x - 3 = (?)(?) + ?$.

Solve.

58. The product of a monomial and $4b$ is $12a^2b$. Find the monomial.

59. The product of a monomial and $6x$ is $24xy^2$. Find the monomial.

60. The quotient of a polynomial and $2x + 1$ is $2x - 4 + \frac{7}{2x + 1}$. Find the polynomial.

61. The quotient of a polynomial and $x - 3$ is $x^2 - x + 8 + \frac{22}{x - 3}$. Find the polynomial.

PROJECTS OR GROUP ACTIVITIES

62. $2x - 1$ is a factor of $2x^3 - 7x^2 + 7x - 2$. The product of $2x - 1$ and what polynomial is $2x^3 - 7x^2 + 7x - 2$?

63. $4x + 1$ is a factor of $4x^3 + 9x^2 - 10x - 3$. The product of $4x + 1$ and what polynomial is $4x^3 + 9x^2 - 10x - 3$?

64. When $x^2 - x - 8$ is divided by a polynomial, the quotient is $x + 3$ and the remainder is 4. Find the polynomial.

CHAPTER 7 Summary

Key Words	Objective and Page Reference	Examples
A **monomial** is a number, a variable, or a product of numbers and variables. A **polynomial** is a variable expression in which the terms are monomials.	[7.1.1, p. 308]	5 is a number. y is a variable. $8a^2b^3$ is a product of numbers and variables. 5, y, and $8a^2b^3$ are monomials. $5 + y + 8a^2b^3$ is a polynomial.
A polynomial of one term is a **monomial.** A polynomial of two terms is a **binomial.** A polynomial of three terms is a **trinomial.**	[7.1.1, p. 308]	5, y, and $8a^2b^3$ are monomials. $x + 9$, $y^2 - 3$, and $6a + 7b$ are binomials. $x^2 + 2x - 1$ is a trinomial.
The terms of a polynomial in one variable are usually arranged so that the exponents on the variables decrease from left to right. This is called **descending order.**	[7.1.1, p. 308]	The polynomial $8x^4 + 5x^3 - 6x^2 + x - 7$ is in descending order.
The **degree of a polynomial in one variable** is the largest exponent on a variable.	[7.1.1, p. 308]	The degree of $8x^3 - 5x^2 + 4x - 12$ is 3.
The **opposite of a polynomial** is the polynomial with the sign of every term changed to its opposite.	[7.1.2, p. 309]	The opposite of the polynomial $x^2 - 3x + 4$ is $-x^2 + 3x - 4$.

Essential Rules and Procedures	Objective and Page Reference	Examples
Addition of Polynomials To add polynomials, add the coefficients of the like terms.	[7.1.1, p. 308]	$(2x^2 + 3x - 4) + (3x^3 - 4x^2 + 2x - 5)$ $= 3x^3 + (2x^2 - 4x^2) + (3x + 2x) + (-4 - 5)$ $= 3x^3 - 2x^2 + 5x - 9$
Subtraction of Polynomials To subtract polynomials, add the opposite of the second polynomial to the first.	[7.1.2, p. 309]	$(3y^2 - 8y - 9) - (5y^2 - 10y + 3)$ $= (3y^2 - 8y - 9) + (-5y^2 + 10y - 3)$ $= (3y^2 - 5y^2) + (-8y + 10y) + (-9 - 3)$ $= -2y^2 + 2y - 12$
Rule for Multiplying Exponential Expressions If m and n are integers, then $x^m \cdot x^n = x^{m+n}$.	[7.2.1, p. 313]	$b^5 \cdot b^4 = b^{5+4} = b^9$
Rule for Simplifying Powers of Exponential Expressions If m and n are integers, then $(x^m)^n = x^{mn}$.	[7.2.2, p. 314]	$(y^3)^7 = y^{3(7)} = y^{21}$
Rule for Simplifying Powers of Products If m, n, and p are integers, then $(x^m y^n)^p = x^{mp} y^{np}$.	[7.2.2, p. 314]	$(a^6 b^2)^3 = a^{6(3)} b^{2(3)} = a^{18} b^6$
To multiply a polynomial by a monomial, use the Distributive Property and the Rule for Multiplying Exponential Expressions.	[7.3.1, p. 317]	$3x(2x^2 - 8x + 5) = 6x^3 - 24x^2 + 15x$
One method of **multiplying two polynomials** is to use a vertical format similar to that used for multiplication of whole numbers.	[7.3.2, p. 318]	$\begin{array}{r} 2x^2 - 3x + 1 \\ 4x - 5 \\ \hline -10x^2 + 15x - 5 \\ 8x^3 - 12x^2 + 4x \\ \hline 8x^3 - 22x^2 + 19x - 5 \end{array}$
FOIL Method To find the product of two binomials, add the products of the **F**irst terms, the **O**uter terms, the **I**nner terms, and the **L**ast terms.	[7.3.3, pp. 318–319]	$(4x + 3)(2x - 1)$ $= (4x)(2x) + (4x)(-1) + (3)(2x) + (3)(-1)$ $= 8x^2 - 4x + 6x - 3$ $= 8x^2 + 2x - 3$
The Sum and Difference of Two Terms $(a + b)(a - b) = a^2 - b^2$	[7.3.4, p. 320]	$(3x + 4)(3x - 4) = (3x)^2 - 4^2$ $= 9x^2 - 16$
The Square of a Binomial $(a + b)^2 = a^2 + 2ab + b^2$ $(a - b)^2 = a^2 - 2ab + b^2$	[7.3.4, p. 320]	$(2x + 5)^2 = (2x)^2 + 2(2x)(5) + 5^2$ $= 4x^2 + 20x + 25$ $(2x - 5)^2 = (2x)^2 - 2(2x)(5) + (-5)^2$ $= 4x^2 - 20x + 25$
Zero as an Exponent Any nonzero expression to the zero power equals 1.	[7.4.1, p. 329]	$17^0 = 1$ $(5y)^0 = 1, y \neq 0$
Definition of Negative Exponents If n is a positive integer and $x \neq 0$, then $x^{-n} = \frac{1}{x^n}$ and $\frac{1}{x^{-n}} = x^n$.	[7.4.1, p. 330]	$x^{-6} = \frac{1}{x^6}$ and $\frac{1}{x^{-6}} = x^6$

Rule for Dividing Exponential Expressions
If m and n are integers and $x \neq 0$, then $\frac{x^m}{x^n} = x^{m-n}$.

[7.4.1, p. 330]

$$\frac{y^8}{y^3} = y^{8-3} = y^5$$

Scientific Notation
To express a number in scientific notation, write it in the form $a \times 10^n$, where a is a number between 1 and 10 and n is an integer. If the number is greater than 10, the exponent on 10 will be positive. If the number is less than 1, the exponent on 10 will be negative.

[7.4.2, pp. 332–333]

$367{,}000{,}000 = 3.67 \times 10^8$

$0.0000059 = 5.9 \times 10^{-6}$

To change a number written in scientific notation to decimal notation, move the decimal point to the right if the exponent on 10 is positive and to the left if the exponent on 10 is negative. Move the decimal point the same number of places as the absolute value of the exponent on 10.

$2.418 \times 10^7 = 24{,}180{,}000$

$9.06 \times 10^{-5} = 0.0000906$

To divide a polynomial by a monomial, divide each term in the numerator by the denominator and write the sum of the quotients.

[7.5.1, p. 339]

$$\frac{8xy^3 - 4y^2 + 12y}{4y} = \frac{8xy^3}{4y} - \frac{4y^2}{4y} + \frac{12y}{4y} = 2xy^2 - y + 3$$

To check polynomial division, use the same equation used to check division of whole numbers:

(Quotient $\times$ divisor) + remainder = dividend

[7.5.2, p. 340]

$$\begin{array}{r} x - 4 \\ x + 3 \overline{\smash{)}\, x^2 - x - 10} \\ \underline{x^2 + 3x} \\ -4x - 10 \\ \underline{-4x - 12} \\ 2 \end{array}$$

Check:

$(x - 4)(x + 3) + 2 = x^2 - x - 12 + 2$
$= x^2 - x - 10$

$(x^2 - x - 10) \div (x + 3)$
$= x - 4 + \frac{2}{x + 3}$

CHAPTER 7 Review Exercises

1. Add: $(12y^2 + 17y - 4) + (9y^2 - 13y + 3)$

2. Multiply: $(5xy^2)(-4x^2y^3)$

3. Multiply: $-2x(4x^2 + 7x - 9)$

4. Multiply: $(5a - 7)(2a + 9)$

5. Divide: $\frac{36x^2 - 42x + 60}{6}$

6. Subtract: $(5x^2 - 2x - 1) - (3x^2 - 5x + 7)$

7. Simplify: $(-3^2)^3$

8. Multiply: $(x^2 - 5x + 2)(x - 1)$

9. Multiply: $(a + 7)(a - 7)$

10. Evaluate: $\frac{6^2}{6^{-2}}$

11. Divide: $(x^2 + x - 42) \div (x + 7)$

12. Add: $(2x^3 + 7x^2 + x) + (2x^2 - 4x - 12)$

13. Multiply: $(6a^2b^5)(3a^6b)$

14. Multiply: $x^2y(3x^2 - 2x + 12)$

15. Multiply: $(2b - 3)(4b + 5)$

16. Divide: $\dfrac{16y^2 - 32y}{-4y}$

17. Subtract: $(13y^3 - 7y - 2) - (12y^2 - 2y - 1)$

18. Simplify: $(2^3)^2$

19. Multiply: $(3y^2 + 4y - 7)(2y + 3)$

20. Multiply: $(2b - 9)(2b + 9)$

21. Simplify: $(a^{-2}b^3c)^2$

22. Divide: $(6y^2 - 35y + 36) \div (3y - 4)$

23. Write 0.00000397 in scientific notation.

24. Multiply: $(xy^5z^3)(x^3y^3z)$

25. Multiply: $(6y^2 - 2y + 9)(-2y^3)$

26. Multiply: $(6x - 12)(3x - 2)$

27. Write 6.23×10^{-5} in decimal notation.

28. Subtract: $(8a^2 - a) - (15a^2 - 4)$

29. Simplify: $(-3x^2y^3)^2$

30. Multiply: $(4a^2 - 3)(3a - 2)$

31. Simplify: $(5y - 7)^2$

32. Simplify: $(-3x^{-2}y^{-3})^{-2}$

33. Divide: $(x^2 + 17x + 64) \div (x + 12)$

34. Write 2.4×10^5 in decimal notation.

35. Multiply: $(a^2b^7c^6)(ab^3c)(a^3bc^2)$

36. Multiply: $2ab^3(4a^2 - 2ab + 3b^2)$

37. Multiply: $(3x + 4y)(2x - 5y)$

38. Divide: $\dfrac{12b^7 + 36b^5 - 3b^3}{3b^3}$

39. Subtract: $(b^2 - 11b + 19) - (5b^2 + 2b - 9)$

40. Simplify: $(5a^7b^6)^2(4ab)$

41. Multiply: $(6b^3 - 2b^2 - 5)(2b^2 - 1)$

42. Multiply: $(6 - 5x)(6 + 5x)$

43. Simplify: $\dfrac{6x^{-2}y^4}{3xy}$

44. Divide: $(a^3 + a^2 + 18) \div (a + 3)$

45. Add: $(4b^3 - 7b^2 + 10) + (2b^2 - 9b - 3)$

46. Multiply: $(2a^{12}b^3)(-9b^2c^6)(3ac)$

47. Multiply: $-9x^2(2x^2 + 3x - 7)$

48. Multiply: $(10y - 3)(3y - 10)$

49. Write 9,176,000,000,000 in scientific notation.

50. Simplify: $(-3x^{-4}y)(2xy^{-3})^{-2}$

51. Simplify: $(6x^4y^7z^2)^2(-2x^3y^2z^6)^2$

52. Multiply: $(-3x^3 - 2x^2 + x - 9)(4x + 3)$

53. Simplify: $(8a + 1)^2$

54. Simplify: $\dfrac{4a^{-2}b^{-8}}{2a^{-1}b^{-2}}$

55. Divide: $(b^3 - 2b^2 - 33b - 7) \div (b - 7)$

56. **Geometry** The length of a rectangle is $5x$ m. The width is $(4x - 7)$ m. Find the area of the rectangle in terms of the variable x.

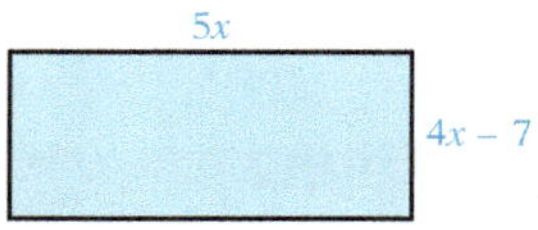

57. **Geometry** The length of a side of a square is $(5x + 4)$ in. Find the area of the square in terms of the variable x.

58. **Geometry** The base of a triangle is $(3x - 2)$ ft, and the height is $(6x + 4)$ ft. Find the area of the triangle in terms of the variable x.

59. **Geometry** The radius of a circle is $(x - 6)$ cm. Find the area of the circle in terms of the variable x. Leave the answer in terms of π.

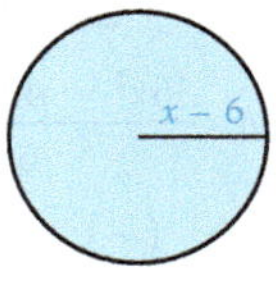

60. **Geometry** The width of a rectangle is $(3x - 8)$ mi. The length is $(5x + 4)$ mi. Find the area of the rectangle in terms of the variable x.

CHAPTER 7 Test

1. Add: $(3x^3 - 2x^2 - 4) + (8x^2 - 8x + 7)$
2. Multiply: $(-2x^3 + x^2 - 7)(2x - 3)$
3. Multiply: $2x(2x^2 - 3x)$
4. Simplify: $(-2a^2b)^3$
5. Simplify: $\dfrac{12x^2}{-3x^{-4}}$
6. Simplify: $(2ab^{-3})(3a^{-2}b^4)$
7. Subtract: $(3a^2 - 2a - 7) - (5a^3 + 2a - 10)$
8. Multiply: $(a - 2b)(a + 5b)$
9. Divide: $\dfrac{16x^5 - 8x^3 + 20x}{4x}$
10. Divide: $(4x^2 - 7) \div (2x - 3)$
11. Multiply: $(-2xy^2)(3x^2y^4)$
12. Multiply: $-3y^2(-2y^2 + 3y - 6)$
13. Simplify: $\dfrac{27xy^3}{3x^4y^3}$
14. Simplify: $(2x - 5)^2$
15. Multiply: $(2x - 7y)(5x - 4y)$
16. Multiply: $(x - 3)(x^2 - 4x + 5)$
17. Simplify: $(a^2b^{-3})^2$
18. Write 0.000029 in scientific notation.
19. Multiply: $(4y - 3)(4y + 3)$
20. Subtract: $(3y^3 - 5y + 8) - (-2y^2 + 5y + 8)$
21. Simplify: $(-3a^3b^2)^2$
22. Multiply: $(2a - 7)(5a^2 - 2a + 3)$
23. Simplify: $(3b + 2)^2$
24. Simplify: $\dfrac{-2a^2b^3}{8a^4b^8}$
25. Divide: $(8x^2 + 4x - 3) \div (2x - 3)$
26. Multiply: $(a^2b^5)(ab^2)$
27. Multiply: $(a - 3b)(a + 4b)$
28. Write 3.5×10^{-8} in decimal notation.
29. **Geometry** The length of a side of a square is $(2x + 3)$ m. Find the area of the square in terms of the variable x.

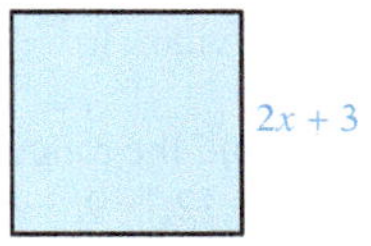

30. **Geometry** The radius of a circle is $(x - 5)$ in. Find the area of the circle in terms of the variable x. Leave the answer in terms of π.

Cumulative Review Exercises

1. Simplify: $\dfrac{3}{16} - \left(-\dfrac{3}{8}\right) - \dfrac{5}{9}$
2. Simplify: $-5^2 \cdot \left(\dfrac{2}{3}\right)^3 \cdot \left(-\dfrac{3}{8}\right)$
3. Simplify: $\left(-\dfrac{1}{2}\right)^2 \div \left(\dfrac{5}{8} - \dfrac{5}{6}\right) + 2$
4. Find the opposite of -87.
5. Write $\frac{31}{40}$ as a decimal.
6. Evaluate $\dfrac{b - (a - b)^2}{b^2}$ when $a = 3$ and $b = -2$.
7. Simplify: $-3x - (-xy) + 2x - 5xy$
8. Simplify: $(16x)\left(-\dfrac{3}{4}\right)$

9. Simplify: $-2[3x - 4(3 - 2x) + 2]$

10. Complete the statement by using the Inverse Property of Addition.
$-8 + ? = 0$

11. Solve: $12 = -\frac{2}{3}x$

12. Solve: $3x - 7 = 2x + 9$

13. Solve: $3 - 4(2 - x) = 3x + 7$

14. Solve: $-\frac{4}{5}x = 16 - x$

15. 38.4 is what percent of 160?

16. Solve: $7x - 8 \geq -29$

17. Find the slope of the line that contains the points whose coordinates are $(3, -4)$ and $(-2, 5)$.

18. Find the equation of the line that contains the point whose coordinates are $(1, -3)$ and has slope $-\frac{3}{2}$.

19. Graph: $3x - 2y = -6$

20. Graph the solution set of $y \leq \frac{4}{5}x - 3$.

21. Find the domain and range of the relation $\{(-8, -7), (-6, -5), (-4, -2), (-2, 0)\}$. Is the relation a function?

22. Evaluate $f(x) = -2x + 10$ at $x = 6$.

23. Solve by substitution: $x = 3y + 1$
$2x + 5y = 13$

24. Solve by the addition method: $9x - 2y = 17$
$5x + 3y = -7$

25. Subtract: $(5b^3 - 4b^2 - 7) - (3b^2 - 8b + 3)$

26. Multiply: $(3x - 4)(5x^2 - 2x + 1)$

27. Multiply: $(4b - 3)(5b - 8)$

28. Simplify: $(5b + 3)^2$

29. Simplify: $\frac{-3a^3b^2}{12a^4b^{-2}}$

30. Divide: $\frac{-15y^2 + 12y - 3}{-3y}$

31. Divide: $(a^2 - 3a - 28) \div (a + 4)$

32. Simplify: $(-3x^{-4}y)(-3x^{-2}y)$

33. Find the range of the function given by the equation $f(x) = -\frac{4}{3}x + 9$ if the domain is $\{-12, -9, -6, 0, 6\}$.

34. Translate and simplify "the product of five and the difference between a number and twelve."

35. Translate "the difference between eight times a number and twice the number is eighteen" into an equation and solve.

36. **Geometry** The width of a rectangle is 40% of the length. The perimeter of the rectangle is 42 m. Find the length and width of the rectangle.

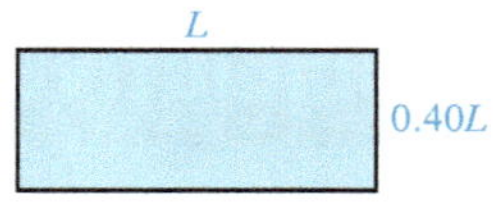

37. **Markup** A calculator costs a retailer \$24. Find the selling price when the markup rate is 80%.

38. **Food Mixtures** Fifty ounces of pure orange juice are added to 200 oz of a fruit punch that is 10% orange juice. What is the percent concentration of orange juice in the resulting mixture?

39. **Uniform Motion Problem** A car traveling at 50 mph overtakes a cyclist who, riding at 10 mph, has had a 2-hour head start. How far from the starting point does the car overtake the cyclist?

40. **Geometry** The length of a side of a square is $(3x + 2)$ ft. Find the area of the square in terms of the variable x.

CHAPTER 8

Factoring

Digital Vision

Focus on Success

Did you make a time management plan when you started this course? If not, you can still benefit from doing so. Create a schedule that gives you enough time to do everything you need to do. We want you to schedule enough time to study math each week so that you will successfully complete this course. Once you have determined the hours during which you will study, consider your study time a commitment that you cannot break. (See Time Management, page AIM-4.)

OBJECTIVES

8.1
1. Factor a monomial from a polynomial
2. Factor by grouping

8.2
1. Factor trinomials of the form $x^2 + bx + c$
2. Factor completely

8.3
1. Factor trinomials of the form $ax^2 + bx + c$ by using trial factors
2. Factor trinomials of the form $ax^2 + bx + c$ by grouping

8.4
1. Factor the difference of two squares and perfect-square trinomials

8.5
1. Factor completely

8.6
1. Solve equations by factoring
2. Application problems

PREP TEST

Are you ready to succeed in this chapter?
Take the Prep Test below to find out if you are ready to learn the new material.

1. Write 30 as a product of prime numbers.
2. Simplify: $-3(4y - 5)$
3. Simplify: $-(a - b)$
4. Simplify: $2(a - b) - 5(a - b)$
5. Solve: $4x = 0$
6. Solve: $2x + 1 = 0$
7. Multiply: $(x + 4)(x - 6)$
8. Multiply: $(2x - 5)(3x + 2)$
9. Simplify: $\dfrac{x^5}{x^2}$
10. Simplify: $\dfrac{6x^4y^3}{2xy^2}$

8.1 Common Factors

OBJECTIVE 1 Factor a monomial from a polynomial

Take Note

12 is the GCF of 24 and 60 because 12 is the largest integer that divides evenly into both 24 and 60.

The **greatest common factor (GCF)** of two or more integers is the greatest integer that is a factor of all the integers.

$24 = 2 \cdot 2 \cdot 2 \cdot 3$
$60 = 2 \cdot 2 \cdot 3 \cdot 5$
$\text{GCF} = 2 \cdot 2 \cdot 3 = 12$

The GCF of two or more monomials is the product of the GCF of the coefficients and the common variable factors.

$6x^3y = 2 \cdot 3 \cdot x \cdot x \cdot x \cdot y$
$8x^2y^2 = 2 \cdot 2 \cdot 2 \cdot x \cdot x \cdot y \cdot y$
$\text{GCF} = 2 \cdot x \cdot x \cdot y = 2x^2y$

Note that the exponent of each variable in the GCF is the same as the *smallest* exponent of that variable in any of the monomials.

The GCF of $6x^3y$ and $8x^2y^2$ is $2x^2y$.

EXAMPLE 1 Find the GCF of $12a^4b$ and $18a^2b^2c$.

Solution

$12a^4b = 2 \cdot 2 \cdot 3 \cdot a^4 \cdot b$ • Factor each monomial.

$18a^2b^2c = 2 \cdot 3 \cdot 3 \cdot a^2 \cdot b^2 \cdot c$

$\text{GCF} = 2 \cdot 3 \cdot a^2 \cdot b = 6a^2b$ • The common variable factors are a^2 and b. c is not a common factor.

The GCF of $12a^4b$ and $18a^2b^2c$ is $6a^2b$.

Problem 1 Find the GCF of $4x^6y$ and $18x^2y^6$.

Solution See page S17.

Try Exercise 21, page 354.

The Distributive Property is used to multiply factors of a polynomial. To **factor a polynomial** means to write the polynomial as a product of other polynomials.

Multiply →

Factors		**Polynomial**
$2x(x + 5)$	$=$	$2x^2 + 10x$

← Factor

In the example above, $2x$ is the GCF of the terms $2x^2$ and $10x$. It is a **common monomial factor** of the terms. $x + 5$ is a **binomial factor** of $2x^2 + 10x$.

EXAMPLE 2 Factor. **A.** $5x^3 - 35x^2 + 10x$ **B.** $16x^2y + 8x^4y^2 - 12x^4y^5$

Solution

A. The GCF is $5x$. • Find the GCF of the terms of the polynomial.

$$\frac{5x^3}{5x} = x^2, \quad \frac{-35x^2}{5x} = -7x, \quad \frac{10x}{5x} = 2$$

• Divide each term of the polynomial by the GCF.

$$5x^3 - 35x^2 + 10x = 5x(x^2) + 5x(-7x) + 5x(2)$$

• Use the quotients to rewrite the polynomial, expressing each term as a product with the GCF as one of the factors.

$$= 5x(x^2 - 7x + 2)$$

• Use the Distributive Property to write the polynomial as a product of factors.

B. $16x^2y = 2 \cdot 2 \cdot 2 \cdot 2 \cdot x^2 \cdot y$
$8x^4y^2 = 2 \cdot 2 \cdot 2 \cdot x^4 \cdot y^2$
$12x^4y^5 = 2 \cdot 2 \cdot 3 \cdot x^4 \cdot y^5$
The GCF is $4x^2y$.

• **Find the GCF of the terms of the polynomial.**

$$\frac{16x^2y}{4x^2y} = 4, \quad \frac{8x^4y^2}{4x^2y} = 2x^2y,$$
$$\frac{-12x^4y^5}{4x^2y} = -3x^2y^4$$

• **Divide each term of the polynomial by the GCF.**

$$16x^2y + 8x^4y^2 - 12x^4y^5$$
$$= 4x^2y(4) + 4x^2y(2x^2y) + 4x^2y(-3x^2y^4)$$

• **Use the quotients to rewrite the polynomial, expressing each term as a product with the GCF as one of the factors.**

$$= 4x^2y(4 + 2x^2y - 3x^2y^4)$$

• **Use the Distributive Property to write the polynomial as a product of factors.**

Problem 2 Factor. **A.** $14a^2 - 21a^4b$ **B.** $6x^4y^2 - 9x^3y^2 + 12x^2y^4$

Solution See page S17.

Try Exercise 61, page 354.

OBJECTIVE 2 Factor by grouping

In the examples below, the binomials in parentheses are called **binomial factors.**

$2a(a + b)$

$3xy(x - y)$

The Distributive Property is used to factor a common binomial factor from an expression.

In the expression at the right, the common binomial factor is $y - 3$. The Distributive Property is used to write the expression as a product of factors.

$$x(y - 3) + 4(y - 3)$$
$$= (y - 3)(x + 4)$$

EXAMPLE 3 Factor: $y(x + 2) + 3(x + 2)$

Solution $y(x + 2) + 3(x + 2)$
$= (x + 2)(y + 3)$

• **The common binomial factor is $x + 2$.**

Problem 3 Factor: $a(b - 7) + b(b - 7)$

Solution See page S17.

Try Exercise 79, page 355.

Sometimes a binomial factor must be rewritten before a common binomial factor can be found.

Focus on factoring out a common binomial factor

Factor: $a(a - b) + 5(b - a)$

$a - b$ and $b - a$ are different binomials.

Note that $(b - a) = (-a + b) = -(a - b)$.

Rewrite $(b - a)$ as $-(a - b)$ so that the terms have a common factor.

$$\begin{aligned} a(a - b) + 5(b - a) &= a(a - b) + 5[-(a - b)] \\ &= a(a - b) - 5(a - b) \\ &= (a - b)(a - 5) \end{aligned}$$

EXAMPLE 4 Factor: $2x(x - 5) + y(5 - x)$

Solution $2x(x - 5) + y(5 - x)$

$= 2x(x - 5) - y(x - 5)$ • Rewrite $5 - x$ as $-(x - 5)$ so that the terms have a common factor.

$= (x - 5)(2x - y)$ • Write the expression as a product of factors.

Problem 4 Factor: $3y(5x - 2) - 4(2 - 5x)$

Solution See page S17.

Try Exercise 89, page 355.

Some polynomials can be factored by grouping the terms so that a common binomial factor is found.

Focus on factoring by grouping

Factor: $2x^3 - 3x^2 + 4x - 6$

Group the first two terms and the last two terms (put them in parentheses).

$$2x^3 - 3x^2 + 4x - 6 = (2x^3 - 3x^2) + (4x - 6)$$

Factor out the GCF from each group.

$$= x^2(2x - 3) + 2(2x - 3)$$

Factor out the common binomial factor and write the expression as a product of factors.

$$= (2x - 3)(x^2 + 2)$$

EXAMPLE 5 Factor. **A.** $2x^3 - 3x^2 + 8x - 12$ **B.** $3y^3 - 4y^2 - 6y + 8$

Solution **A.** $2x^3 - 3x^2 + 8x - 12$

$= (2x^3 - 3x^2) + (8x - 12)$ • Group the first two terms and the last two terms.

$= x^2(2x - 3) + 4(2x - 3)$ • Factor out the GCF from each group.

$= (2x - 3)(x^2 + 4)$ • Factor out the common binomial factor and write the expression as a product of factors.

B. $3y^3 - 4y^2 - 6y + 8$
$= (3y^3 - 4y^2) - (6y - 8)$
- Group the first two terms and the last two terms. Note that $-6y + 8 = -(6y - 8)$.

$= y^2(3y - 4) - 2(3y - 4)$
- Factor out the GCF from each group.

$= (3y - 4)(y^2 - 2)$
- Factor out the common binomial factor and write the expression as a product of factors.

Problem 5 Factor. **A.** $y^5 - 5y^3 + 4y^2 - 20$ **B.** $2y^3 - 2y^2 - 3y + 3$

Solution See page S17.

Try Exercise 99, page 355.

8.1 Exercises

CONCEPT CHECK

1. Name the greatest common factor of 4, 12, and 16.

2. Name the greatest common factor of x^3, x^5, and x^6.

3. Which expression is a sum and which is a product?
a. $3x(x + 5)$ **b.** $3x^2 + 15x$

4. Name **a.** the monomial factor and **b.** the binomial factor in the expression $x(2x - 1)$.

5. Name the common binomial factor in the expression $5b(c - 6) + 8(c - 6)$.

6. Rewrite the expression $2x^3 - x^2 + 6x - 3$ by grouping the first two terms and the last two terms.

1 Factor a monomial from a polynomial (See pages 350–351.)

7. Explain the meaning of "a factor of a polynomial" and the meaning of "to factor a polynomial."

8. Explain why the statement is true.
a. The terms of the binomial $3x - 6$ have a common factor.
b. The expression $3x^2 + 15$ is not in factored form.
c. $5y - 7$ is a factor of $y(5y - 7)$.

GETTING READY

9. Use the three monomials $9x^3y^2$, $3xy^3z$, and $6x^2y^2z^2$.
a. Of the variable factors x, y, and z, the one that is *not* a factor of the GCF of $9x^3y^2$, $3xy^3z$, and $6x^2y^2z^2$ is __?__.
b. The GCF of the coefficients 9, 3, and 6 is __?__.
c. The GCF of the factors x^3, x, and x^2 is __?__.
d. The GCF of the factors y^2, y^3, and y^4 is __?__.
e. Use your answers to parts (a) through (d) to write the GCF of $9x^3y^2$, $3xy^3z$, and $6x^2y^2z^2$: __?__

10. a. Of the two expressions $25a^2 - 10a$ and $5a(5a - 2)$, the one that is written in *factored form* is __?__.

b. Use the Distributive Property to confirm that the two expressions in part (a) are equivalent:
$5a(5a - 2) = (\underline{\ ?\ })(5a) - (\underline{\ ?\ })(2) = \underline{\ ?\ } - \underline{\ ?\ }$.

c. The polynomial $25a^2 - 10a$ has a monomial factor and a binomial factor. The monomial factor is __?__ and the binomial factor is __?__.

Find the greatest common factor.

11. x^7, x^3
12. y^6, y^{12}
13. x^2y^4, xy^6
14. a^5b^3, a^3b^8
15. $x^2y^4z^6, xy^8z^2$
16. ab^2c^3, a^3b^2c
17. $14a^3, 49a^7$
18. $12y^2, 27y^4$
19. $3x^2y^2, 5ab^2$
20. $8x^2y^3, 7ab^4$
21. $9a^2b^4, 24a^4b^2$
22. $15a^4b^2, 9ab^5$
23. $ab^3, 4a^2b, 12a^2b^3$
24. $12x^2y, x^4y, 16x$
25. $2x^2y, 4xy, 8x$
26. $16x^2, 8x^4y^2, 12xy$
27. $3x^2y^2, 6x, 9x^3y^3$
28. $4a^2b^3, 8a^3, 12ab^4$

Factor.

29. $5a + 5$
30. $7b - 7$
31. $16 - 8a^2$
32. $12 + 12y^2$
33. $8x + 12$
34. $16a - 24$
35. $30a - 6$
36. $20b + 5$
37. $7x^2 - 3x$
38. $12y^2 - 5y$
39. $3a^2 + 5a^5$
40. $6b^3 - 5b^2$
41. $2x^4 - 4x$
42. $3y^4 - 9y$
43. $10x^4 - 12x^2$
44. $12a^5 - 32a^2$
45. $x^2y - xy^3$
46. $a^2b + a^4b^2$
47. $2a^5b + 3xy^3$
48. $5x^2y - 7ab^3$
49. $6a^2b^3 - 12b^2$
50. $8x^2y^3 - 4x^2$
51. $6a^2bc + 4ab^2c$
52. $10x^2yz^2 + 15xy^3z$
53. $6x^3y^3 - 12x^6y^6$
54. $3a^2b^2 - 12a^5b^5$
55. $x^3 - 3x^2 - x$
56. $a^3 + 4a^2 + 8a$
57. $2x^2 + 8x - 12$
58. $a^3 - 3a^2 + 5a$
59. $b^3 - 5b^2 - 7b$
60. $5x^2 - 15x + 35$
61. $3x^3 + 6x^2 + 9x$
62. $5y^3 - 20y^2 + 10y$
63. $2x^4 - 4x^3 + 6x^2$
64. $3y^4 - 9y^3 - 6y^2$
65. $6a^5 - 3a^3 - 2a^2$
66. $x^3y - 3x^2y^2 + 7xy^3$
67. $8x^2y^2 - 4x^2y + x^2$
68. $x^4y^4 - 3x^3y^3 + 6x^2y^2$
69. $4x^5y^5 - 8x^4y^4 + x^3y^3$
70. $16x^2y - 8x^3y^4 - 48x^2y^2$

For Exercises 71 and 72, use the polynomial $x^a + x^b + x^c$, where a, b, and c are all positive integers such that $a < b < c$.

71. What is the GCF of the terms of the polynomial $x^a + x^b + x^c$?

72. Suppose $b = a + 1$ and $c = a + 2$. Write $x^a + x^b + x^c$ in factored form.

2 Factor by grouping (See pages 351–353.)

GETTING READY

73. a. The expression $5x(3x - 2) + 4(3x - 2)$ is the sum of two terms. One term is $5x(3x - 2)$ and the other term is ___?___. The terms have the common binomial factor ___?___.

b. To write the expression $5x(3x - 2) + 4(3x - 2)$ in factored form, factor out the common binomial factor: $5x(3x - 2) + 4(3x - 2)$ $= (3x - 2)($___?___ + ___?___$)$.

74. To factor $7a^2 - 7a + 6a - 6$ by grouping, start by grouping the first two terms and the last two terms. Then factor $7a$ from the first two terms and 6 from the second two terms: $7a^2 - 7a + 6a - 6$ $= 7a($___?___ − ___?___$) + 6($___?___ − ___?___$)$. The common binomial factor is ___?___.

Factor.

75. $x(a + b) + 2(a + b)$

76. $a(x + y) + 4(x + y)$

77. $x(b + 2) - y(b + 2)$

78. $d(z - 8) + 5(z - 8)$

79. $a(y - 4) - b(y - 4)$

80. $c(x - 6) - 7(x - 6)$

Rewrite the expression so that the terms have a common binomial factor.

81. $a(x - 2) - b(2 - x)$

82. $a(x - 2) + b(2 - x)$

83. $b(a - 7) + 3(7 - a)$

84. $b(a - 7) - 3(7 - a)$

85. $x(a - 2b) - y(2b - a)$

86. $x(a - 2b) + y(2b - a)$

Factor.

87. $a(x - 2) + 5(2 - x)$

88. $a(x - 7) + b(7 - x)$

89. $b(y - 3) + 3(3 - y)$

90. $c(a - 2) - b(2 - a)$

91. $a(x - y) - 2(y - x)$

92. $3(a - b) - x(b - a)$

93. $z(c + 5) - 8(5 + c)$

94. $y(ab + 4) - 7(4 + ab)$

95. $w(3x - 4) - (4 - 3x)$

96. $d(6y - 1) - (1 - 6y)$

97. $x^3 + 4x^2 + 3x + 12$

98. $2y^3 + 4y^2 + 3y + 6$

99. $3y^3 - 12y^2 + y - 4$

100. $x^2y + 4x^2 + 3y + 12$

101. $8 + 2c + 4a^2 + a^2c$

102. $x^2 - 3x + 4ax - 12a$

103. $2y^2 - 10y + 7xy - 35x$

104. $x^3 - 4x^2 - 3x + 12$

105. $ab + 3b - 2a - 6$

106. $yz + 6z - 3y - 18$

107. $x^2a - 2x^2 - 3a + 6$

108. $3ax - 3bx - 2ay + 2by$

109. $t^2 + 4t - st - 4s$

110. $xy - 5y - 2x + 10$

111. $21x^2 + 6xy - 49x - 14y$

112. $4a^2 + 5ab - 10b - 8a$

113. $2ra + a^2 - 2r - a$

114. $2ab - 3b^2 - 3b + 2a$

115. Not all four-term expressions can be factored by grouping. Which of the following expressions can be factored by grouping?
(i) $ab - 3a - 2b - 6$ (ii) $ab + 3a - 2b - 6$ (iii) $ab + 3a + 2b - 6$

116. **a.** Which of the following expressions are equivalent to $x^2 - 5x + 6$?
(i) $x^2 - 15x + 10x + 6$ (ii) $x^2 - x - 4x + 6$ (iii) $x^2 - 2x - 3x + 6$

b. Which expression in part (a) can be factored by grouping?

APPLYING CONCEPTS

Solve.

117. **Geometry** In the equation $P = 2L + 2W$, what is the effect on P when the quantity $L + W$ doubles?

118. **Geometry** Write an expression in factored form for the shaded portion in the diagram.

a.

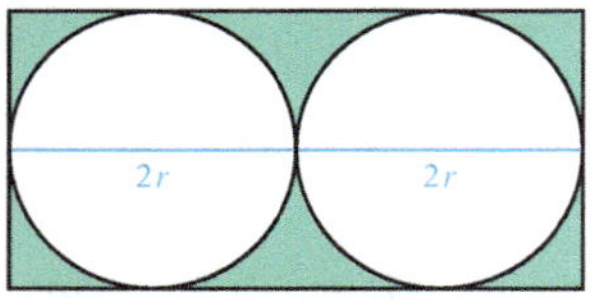

b.

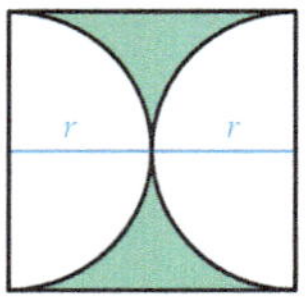

119. Replace the ? to make a true statement.

a. $a - 3 = ?(3 - a)$ **b.** $2 - (x - y) = 2 + (?)$ **c.** $4x + (3a - b) = 4x - (?)$

PROJECTS OR GROUP ACTIVITIES

120. Match equivalent expressions. There may be more than one match for an expression.

a. $3y(x - 2) - 5(x - 2)$ **i.** $3y(x - 2) + 5(x - 2)$

b. $3y(x - 2) + 5(2 - x)$ **ii.** $3y(x - 2) - 5(x + 2)$

c. $3y(x - 2) - 5(2 - x)$ **iii.** $3y(x - 2) - 5(x - 2)$

d. $3y(x - 2) + 5(x - 2)$ **iv.** $3y(x - 2) + 5(x + 2)$

e. $3y(x - 2) - 5(2 + x)$ **v.** $3xy - 6y - 5x + 10$

f. $3y(x - 2) + 5(2 + x)$ **vi.** $3xy - 6y + 5x - 10$

Factor by grouping.

121. **a.** $2x^2 + 6x + 5x + 15$ **b.** $2x^2 + 5x + 6x + 15$

122. **a.** $3x^2 + 3xy - xy - y^2$ **b.** $3x^2 - xy + 3xy - y^2$

123. **a.** $2a^2 - 2ab - 3ab + 3b^2$ **b.** $2a^2 - 3ab - 2ab + 3b^2$

Compare your answers to parts (a) and (b) of Exercises 121 to 123 in order to answer Exercise 124.

124. Do different groupings of the terms in a polynomial affect the binomial factoring?

8.2 Factoring Polynomials of the Form $x^2 + bx + c$

OBJECTIVE 1

Factor trinomials of the form $x^2 + bx + c$

Trinomials of the form $x^2 + bx + c$, where b and c are integers, are shown at the right.

$x^2 + 9x + 14, \quad b = 9, \quad c = 14$
$x^2 - x - 12, \quad b = -1, \quad c = -12$
$x^2 - 2x - 15, \quad b = -2, \quad c = -15$

Some trinomials expressed as the product of binomials are shown at the right. They are in factored form.

Trinomial		Factored Form
$x^2 + 9x + 14$	=	$(x + 2)(x + 7)$
$x^2 - x - 12$	=	$(x + 3)(x - 4)$
$x^2 - 2x - 15$	=	$(x + 3)(x - 5)$

The method by which the factors of a trinomial are found is based on FOIL. Consider the following binomial products, noting the relationship between the constant terms of the binomials and the terms of the trinomials.

Signs in the binomials are the same

$(x + 6)(x + 2) = x^2 + 2x + 6x + (6)(2) = x^2 + 8x + 12$
Sum of 6 and 2 → 8
Product of 6 and 2 → 12

$(x - 3)(x - 4) = x^2 - 4x - 3x + (-3)(-4) = x^2 - 7x + 12$
Sum of −3 and −4 → −7
Product of −3 and −4 → 12

Signs in the binomials are opposite

$(x + 3)(x - 5) = x^2 - 5x + 3x + (3)(-5) = x^2 - 2x - 15$
Sum of 3 and −5 → −2
Product of 3 and −5 → −15

$(x - 4)(x + 6) = x^2 + 6x - 4x + (-4)(6) = x^2 + 2x - 24$
Sum of −4 and 6 → 2
Product of −4 and 6 → −24

POINTS TO REMEMBER IN FACTORING $x^2 + bx + c$

1. In the trinomial, the coefficient of x is the sum of the constant terms of the binomials.
2. In the trinomial, the constant term is the product of the constant terms of the binomials.
3. When the constant term of the trinomial is positive, the constant terms of the binomials have the same sign as the coefficient of x in the trinomial.
4. When the constant term of the trinomial is negative, the constant terms of the binomials have opposite signs.

Success at factoring a trinomial depends on remembering these four points. For example, to factor $x^2 - 2x - 24$, find two numbers whose sum is -2 and whose product is -24 [Points 1 and 2]. Because the constant term of the trinomial is negative (-24), the numbers will have opposite signs [Point 4].

A systematic method of finding the correct binomial factors of $x^2 - 2x - 24$ involves listing the factors of the constant term of the trinomial and the sums of those factors.

Factors of −24	Sum of the Factors
1, −24	1 + (−24) = −23
−1, 24	−1 + 24 = 23
2, −12	2 + (−12) = −10
−2, 12	−2 + 12 = 10
3, −8	3 + (−8) = −5
−3, 8	−3 + 8 = 5
4, −6	**4 + (−6) = −2**
−4, 6	−4 + 6 = 2

Take Note

Always check your proposed factorization to ensure accuracy.

4 and −6 are two numbers whose sum is −2 and whose product is −24. Write the binomial factors of the trinomial.

$$x^2 - 2x - 24 = (x + 4)(x - 6)$$

Check: $(x + 4)(x - 6) = x^2 - 6x + 4x - 24 = x^2 - 2x - 24$

By the Commutative Property of Multiplication, the binomial factors can also be written as

$$x^2 - 2x - 24 = (x - 6)(x + 4)$$

EXAMPLE 1 Factor: $x^2 + 18x + 32$

Solution

Factors of 32	Sum
1, 32	33
2, 16	**18**
4, 8	12

- Try only positive factors of 32 [Point 3].
- Once the correct pair is found, the other factors need not be tried.

$x^2 + 18x + 32 = (x + 2)(x + 16)$

- Write the factors of the trinomial.

Check $(x + 2)(x + 16) = x^2 + 16x + 2x + 32$
$= x^2 + 18x + 32$

Problem 1 Factor: $x^2 - 8x + 15$

Solution See page S17.

Try Exercise 27, page 361.

EXAMPLE 2 Factor: $x^2 - 6x - 16$

Solution

Factors of −16	Sum
1, −16	−15
−1, 16	15
2, −8	**−6**
−2, 8	6
4, −4	0

- The factors must be of opposite signs [Point 4].

$x^2 - 6x - 16 = (x + 2)(x - 8)$

- Write the factors of the trinomial.

Check $(x + 2)(x - 8) = x^2 - 8x + 2x - 16$
$= x^2 - 6x - 16$

Problem 2 Factor: $x^2 + 3x - 18$

Solution See page S17.

Try Exercise 23, page 361.

Not all trinomials can be factored when using only integers. Consider the trinomial $x^2 - 6x - 8$.

Factors of -8	Sum
$1, -8$	-7
$-1, 8$	7
$2, -4$	-2
$-2, 4$	2

Because none of the pairs of factors of -8 has a sum of -6, the trinomial is not factorable using integers. The trinomial is said to be **nonfactorable over the integers.**

OBJECTIVE 2 Factor completely

A polynomial is **factored completely** when it is written as a product of factors that are nonfactorable over the integers.

The first step in *any* factoring problem is to determine whether the terms of the polynomial have a *common factor.* If they do, factor it out first.

EXAMPLE 3 Factor: $3x^3 + 15x^2 + 18x$

Solution The GCF of $3x^3$, $15x^2$, and $18x$ is $3x$.

- **Find the GCF of the terms of the polynomial.**

$3x^3 + 15x^2 + 18x$
$= 3x(x^2) + 3x(5x) + 3x(6)$
$= 3x(x^2 + 5x + 6)$

- **Factor out the GCF.**
- **Write the polynomial as a product of factors.**

Factors of 6	Sum
1, 6	7
2, 3	**5**

- **Factor the trinomial $x^2 + 5x + 6$. Try only positive factors of 6.**

$3x^3 + 15x^2 + 18x = 3x(x + 2)(x + 3)$

Check $3x(x + 2)(x + 3) = 3x(x^2 + 3x + 2x + 6)$
$= 3x(x^2 + 5x + 6)$
$= 3x^3 + 15x^2 + 18x$

Problem 3 Factor: $3a^2b - 18ab - 81b$

Solution See page S18.

Try Exercise 63, page 362.

EXAMPLE 4 Factor: $x^2 + 9xy + 20y^2$

Solution The terms do not have a common factor.

Factors of 20	Sum
1, 20	21
2, 10	12
4, 5	**9**

• **Try only positive factors of 20.**

$x^2 + 9xy + 20y^2 = (x + 4y)(x + 5y)$

Check $(x + 4y)(x + 5y) = x^2 + 5xy + 4xy + 20y^2$
$= x^2 + 9xy + 20y^2$

Take Note

The terms $4y$ and $5y$ are placed in the binomials. This is necessary so that when the binomials are multiplied, the middle term of the trinomial contains xy and the last term contains y^2.

Problem 4 Factor: $4x^2 - 40xy + 84y^2$

Solution See page S18.

Try Exercise 77, page 362.

8.2 Exercises

CONCEPT CHECK

1. The trinomial $x^2 - 8x + 7$ is of the form $x^2 + bx + c$. What is the value of b in the trinomial $x^2 - 8x + 7$?
2. Find two numbers whose sum is 9 and whose product is 14.
3. Find two numbers whose sum is 4 and whose product is −12.
4. When factoring a trinomial, if the constant term is positive, will the signs in the binomials be the same or different?
5. When factoring a trinomial, if the constant term is negative, will the signs in the binomials be the same or different?
6. What is the first step in factoring a trinomial?

1 Factor trinomials of the form $x^2 + bx + c$ (See pages 357–359.)

GETTING READY

Complete the table by listing the pairs of factors of the number and the sum of each pair.

7.

Factors of 18	Sum

8.

Factors of 12	Sum

9.

Factors of −21	Sum

10.

Factors of −10	Sum

11.

Factors of −28	Sum

12.

Factors of −32	Sum

Factor.

13. $x^2 + 3x + 2$

14. $x^2 + 5x + 6$

15. $x^2 - x - 2$

16. $x^2 + x - 6$

17. $a^2 + a - 12$

18. $a^2 - 2a - 35$

19. $a^2 - 3a + 2$

20. $a^2 - 5a + 4$

21. $a^2 + a - 2$

22. $a^2 - 2a - 3$

23. $b^2 + 7b - 8$

24. $y^2 - y - 6$

25. $y^2 + 6y - 55$

26. $z^2 - 4z - 45$

27. $y^2 - 5y + 6$

28. $y^2 - 8y + 15$

29. $z^2 - 14z + 45$

30. $z^2 - 12z - 160$

31. $p^2 + 2p - 35$

32. $p^2 + 12p + 27$

33. $p^2 - 6p + 8$

34. $b^2 + 9b + 20$

35. $b^2 + 13b + 40$

36. $x^2 - 11x - 42$

37. $x^2 + 9x - 70$

38. $b^2 - b - 20$

39. $b^2 + 3b - 40$

40. $y^2 - 14y - 51$

41. $y^2 - y - 72$

42. $p^2 - 4p - 21$

43. $p^2 + 16p + 39$

44. $y^2 - 8y + 32$

45. $y^2 - 9y + 81$

46. $x^2 - 15x + 56$

47. $x^2 + 21x + 38$

48. $x^2 + x - 56$

49. $x^2 + 5x - 36$

50. $a^2 - 7a - 44$

51. $a^2 - 15a + 36$

52. $c^2 - c - 90$

53. Suppose b and c are nonzero and n and m are positive constants such that $x^2 + bx + c = (x + n)(x + m)$.
  a. Is c positive or negative?
  b. Is b positive or negative?

54. Suppose b and c are nonzero and n and m are positive constants such that $x^2 + bx + c = (x - n)(x - m)$.
  a. Is c positive or negative?
  b. Is b positive or negative?

2 Factor completely (See pages 359–360.)

GETTING READY

55. Use the trinomial $5x^2 - 10x - 40$.
  a. The first step in factoring the trinomial is to look for a __?__ factor of the three terms.
  b. The GCF of the terms of the trinomial $5x^2 - 10x - 40$ is __?__.
  c. Factor the GCF from the terms of the trinomial: $5x^2 - 10x - 40 = 5($__?__$)$
  d. Factor the trinomial completely: $5x^2 - 10x - 40 = 5(x^2 - 2x - 8) = 5($__?__$)($__?__$)$

56. To factor $x^2 + 2xy - 24y^2$, find binomials of the form $(x + ny)(x + my)$, where n and m are numbers whose product is ___?___ and whose sum is ___?___. Factors of -24 whose sum is 2 are ___?___ and ___?___. The trinomial $x^2 + 2xy - 24y^2$ factors as $(x + 6y)(x - 4y)$.

Factor.

57. $2x^2 + 6x + 4$

58. $3x^2 + 15x + 18$

59. $3a^2 + 3a - 18$

60. $4x^2 - 4x - 8$

61. $ab^2 + 2ab - 15a$

62. $ab^2 + 7ab - 8a$

63. $xy^2 - 5xy + 6x$

64. $xy^2 + 8xy + 15x$

65. $2a^3 + 6a^2 + 4a$

66. $3y^3 - 15y^2 + 18y$

67. $4y^3 + 12y^2 - 72y$

68. $2x^3 - 2x^2 - 4x$

69. $5z^2 - 15z - 140$

70. $6z^2 + 12z - 90$

71. $2a^3 + 8a^2 - 64a$

72. $3a^3 - 9a^2 - 54a$

73. $x^2 - 5xy + 6y^2$

74. $x^2 + 4xy - 21y^2$

75. $a^2 - 9ab + 20b^2$

76. $a^2 - 15ab + 50b^2$

77. $x^2 - 3xy - 28y^2$

78. $s^2 + 2st - 48t^2$

79. $y^2 - 15yz - 41z^2$

80. $y^2 + 85yz + 36z^2$

81. $z^4 - 12z^3 + 35z^2$

82. $z^4 + 2z^3 - 80z^2$

83. $b^4 - 22b^3 + 120b^2$

84. $b^4 - 3b^3 - 10b^2$

85. $2y^4 - 26y^3 - 96y^2$

86. $3y^4 + 54y^3 + 135y^2$

87. $x^4 + 7x^3 - 8x^2$

88. $x^4 - 11x^3 - 12x^2$

89. $3x^3 - 36x^2 + 81x$

90. $4x^3 + 4x^2 - 24x$

91. $x^2 - 8xy + 15y^2$

92. $y^2 - 7xy - 8x^2$

93. $a^2 - 13ab + 42b^2$

94. $y^2 + 4yz - 21z^2$

95. $3x^2y + 60xy - 63y$

96. $4x^2y - 68xy - 72y$

97. $3x^3 + 3x^2 - 36x$

98. $4x^3 + 12x^2 - 160x$

99. State whether or not the trinomial has a factor of $(x + 3)$.
a. $3x^2 - 3x - 36$ **b.** $x^2y - xy - 12y$

100. State whether or not the trinomial has a factor of $(x + y)$.
a. $2x^2 - 2xy - 4y^2$ **b.** $2x^2y - 4xy - 4y$

APPLYING CONCEPTS

101. If $a(x + 3) = x^2 + 2x - 3$, find a.

102. If $-2x^3 - 6x^2 - 4x = a(x + 1)(x + 2)$, find a.

Factor.

103. $20 + c^2 + 9c$

104. $x^2y - 54y - 3xy$

105. $45a^2 + a^2b^2 - 14a^2b$

106. $12p^2 - 96p + 3p^3$

PROJECTS OR GROUP ACTIVITIES

Find all integers k such that the trinomial can be factored over the integers.

107. $x^2 + kx + 35$

108. $x^2 + kx + 18$

109. $x^2 - kx + 21$

110. $x^2 - kx + 14$

Determine the positive integer values of k for which the following polynomials are factorable over the integers.

111. $y^2 + 4y + k$

112. $z^2 + 7z + k$

113. $a^2 - 6a + k$

114. $c^2 - 7c + k$

115. $x^2 - 3x + k$

116. $y^2 + 5y + k$

117. Exercises 111 to 116 included the requirement that $k > 0$. If k is allowed to be any integer, how many different values of k are possible for each polynomial? Explain your answer.

8.3 Factoring Polynomials of the Form $ax^2 + bx + c$

OBJECTIVE 1 Factor trinomials of the form $ax^2 + bx + c$ by using trial factors

Trinomials of the form $ax^2 + bx + c$, where a, b, and c are integers and $a \neq 0$, are shown at the right.

$3x^2 - x + 4$, $a = 3$, $b = -1$, $c = 4$

$4x^2 + 5x - 8$, $a = 4$, $b = 5$, $c = -8$

These trinomials differ from those in the previous section in that the coefficient of x^2 is not 1. There are various methods of factoring these trinomials. The method described in this objective is factoring trinomials by using trial factors.

To **factor a trinomial of the form $ax^2 + bx + c$** means to express the polynomial as the product of two binomials. Factoring such polynomials by trial and error may require testing many trial factors. To reduce the number of trial factors, remember the following points.

POINTS TO REMEMBER IN FACTORING $ax^2 + bx + c$

1. If the terms of the trinomial have a common factor, factor out the common factor first.
2. If the terms of the trinomial do not have a common factor, then the terms of a binomial factor cannot have a common factor.
3. When the constant term of the trinomial is positive, the constant terms of the binomials have the same sign as the coefficient of x in the trinomial.
4. When the constant term of the trinomial is negative, the constant terms of the binomials have opposite signs.

Focus on factoring a trinomial by using trial factors

Factor. **A.** $10x^2 - x - 3$ **B.** $4x^2 - 27x + 18$

A. The terms of the trinomial $10x^2 - x - 3$ do not have a common factor; therefore, the terms of a binomial factor will not have a common factor [Point 2].

Because the constant term c of the trinomial is negative (-3), the constant terms of the binomial factors will have opposite signs [Point 4].

Find the factors of a (10) and the factors of c (-3).

Factors of 10	Factors of -3
1, 10	1, -3
2, 5	-1, 3

Using these factors, write trial factors. Use the Outer and Inner products of FOIL to check the middle term.

Trial Factors	Middle Term
$(x + 1)(10x - 3)$	$-3x + 10x = 7x$
$(x - 1)(10x + 3)$	$3x - 10x = -7x$
$\mathbf{(2x + 1)(5x - 3)}$	$\mathbf{-6x + 5x = -x}$
$(2x - 1)(5x + 3)$	$6x - 5x = x$
$(10x + 1)(x - 3)$	$-30x + x = -29x$
$(10x - 1)(x + 3)$	$30x - x = 29x$
$(5x + 1)(2x - 3)$	$-15x + 2x = -13x$
$(5x - 1)(2x + 3)$	$15x - 2x = 13x$

From the list of trial factors, $10x^2 - x - 3 = (2x + 1)(5x - 3)$.

Check: $(2x + 1)(5x - 3) = 10x^2 - 6x + 5x - 3 = 10x^2 - x - 3$

All the trial factors for this trinomial were listed in this example. However, once the correct binomial factors are found, it is not necessary to continue checking the remaining trial factors.

B. The terms of the trinomial $4x^2 - 27x + 18$ do not have a common factor; therefore, the terms of a binomial factor will not have a common factor [Point 2].

Because the constant term c of the trinomial is positive (18), the constant terms of the binomial factors will have the same sign as the coefficient of x. Because the coefficient of x is -27, both signs will be negative [Point 3].

Find the factors of a (4) and the negative factors of c (18).

Factors of 4	Factors of 18
1, 4	-1, -18
2, 2	-2, -9
	-3, -6

Using these factors, write trial factors. Use the Outer and Inner products of FOIL to check the middle term.

Trial Factors	Middle Term
$(x - 1)(4x - 18)$	Common factor
$(x - 2)(4x - 9)$	$-9x - 8x = -17x$
$(x - 3)(4x - 6)$	Common factor
$(2x - 1)(2x - 18)$	Common factor
$(2x - 2)(2x - 9)$	Common factor
$(2x - 3)(2x - 6)$	Common factor
$(4x - 1)(x - 18)$	$-72x - x = -73x$
$(4x - 2)(x - 9)$	Common factor
$\mathbf{(4x - 3)(x - 6)}$	$\mathbf{-24x - 3x = -27x}$

The correct factors have been found.

$4x^2 - 27x + 18 = (4x - 3)(x - 6)$

The last example illustrates that many of the trial factors may have common factors and thus need not be tried. For the remainder of this chapter, the trial factors with a common factor will not be listed.

EXAMPLE 1 Factor: $3x^2 + 20x + 12$

Solution

Factors of 3	Factors of 12
1, 3	1, 12 2, 6 3, 4

• Because 20 is positive, only the positive factors of 12 need be tried.

Trial Factors	Middle Term
$(x + 3)(3x + 4)$	$4x + 9x = 13x$
$(3x + 1)(x + 12)$	$36x + x = 37x$
$(3x + 2)(x + 6)$	$18x + 2x = \mathbf{20x}$

• Write the trial factors. Use FOIL to check the middle term.

$3x^2 + 20x + 12 = (3x + 2)(x + 6)$

Check $(3x + 2)(x + 6) = 3x^2 + 18x + 2x + 12$
$= 3x^2 + 20x + 12$

Problem 1 Factor: $6x^2 - 11x + 5$

Solution See page S18.

Try Exercise 25, page 370.

EXAMPLE 2 Factor: $6x^2 - 5x - 6$

Solution

Factors of 6	Factors of −6
1, 6 2, 3	1, −6 −1, 6 2, −3 −2, 3

• Find the factors of *a* (6) and the factors of *c* (−6).

Trial Factors	Middle Term
$(x - 6)(6x + 1)$	$x - 36x = -35x$
$(x + 6)(6x - 1)$	$-x + 36x = 35x$
$(2x - 3)(3x + 2)$	$4x - 9x = \mathbf{-5x}$

• Write the trial factors. Use FOIL to check the middle term.

$6x^2 - 5x - 6 = (2x - 3)(3x + 2)$

Check $(2x - 3)(3x + 2) = 6x^2 + 4x - 9x - 6$
$= 6x^2 - 5x - 6$

Problem 2 Factor: $8x^2 + 14x - 15$

Solution See page S18.

Try Exercise 31, page 370.

EXAMPLE 3 Factor: $15 - 2x - x^2$

Solution

Factors of 15	Factors of -1
1, 15 3, 5	1, -1

- The terms have no common factors. The coefficient of x^2 is -1.

Trial Factors	Middle Term
$(1 + x)(15 - x)$	$-x + 15x = 14x$
$(1 - x)(15 + x)$	$x - 15x = -14x$
$(3 + x)(5 - x)$	$-3x + 5x = 2x$
$(3 - x)(5 + x)$	$3x - 5x = \mathbf{-2x}$

- Write the trial factors. Use FOIL to check the middle term.

$15 - 2x - x^2 = (3 - x)(5 + x)$

Check $(3 - x)(5 + x) = 15 + 3x - 5x - x^2$
$= 15 - 2x - x^2$

Problem 3 Factor: $24 - 2y - y^2$

Solution See page S18.

Try Exercise 49, page 370.

The first step in factoring a trinomial is to determine whether its terms have a common factor. If so, factor out the GCF of the terms.

EXAMPLE 4 Factor: $3x^3 - 23x^2 + 14x$

Solution The GCF of $3x^3$, $23x^2$, and $14x$ is x.

- Find the GCF of the terms of the polynomial.

$3x^3 - 23x^2 + 14x = x(3x^2 - 23x + 14)$

- Factor out the GCF.

Factors of 3	Factors of 14
1, 3	$-1, -14$ $-2, -7$

- Factor the trinomial $3x^2 - 23x + 14$.

Trial Factors	Middle Term
$(x - 1)(3x - 14)$	$-14x - 3x = -17x$
$(x - 14)(3x - 1)$	$-x - 42x = -43x$
$(x - 2)(3x - 7)$	$-7x - 6x = -13x$
$(x - 7)(3x - 2)$	$-2x - 21x = \mathbf{-23x}$

$3x^3 - 23x^2 + 14x = x(x - 7)(3x - 2)$

Check $x(x - 7)(3x - 2) = x(3x^2 - 2x - 21x + 14)$
$= x(3x^2 - 23x + 14)$
$= 3x^3 - 23x^2 + 14x$

Problem 4 Factor: $4a^2b^2 - 30a^2b + 14a^2$

Solution See page S18.

Try Exercise 61, page 370.

OBJECTIVE 2

Factor trinomials of the form $ax^2 + bx + c$ by grouping

Take Note

In this objective, we are using the skills taught in Objective 2 of Section 1 of this chapter. You may want to review that material before studying this objective.

In the previous objective, trinomials of the form $ax^2 + bx + c$ were factored by using trial factors. In this objective, factoring by grouping is used.

To factor $ax^2 + bx + c$, first find two factors of $a \cdot c$ whose sum is b. Use the two factors to rewrite the middle term of the trinomial as the sum of two terms. Then use factoring by grouping to write the factorization of the trinomial.

Focus on factoring a trinomial by grouping

Factor. **A.** $2x^2 + 13x + 15$ **B.** $6x^2 - 11x - 10$ **C.** $3x^2 - 2x - 4$

A. $2x^2 + 13x + 15$

$a = 2, c = 15, a \cdot c = 2 \cdot 15 = 30$

Find two positive factors of 30 whose sum is 13.

Positive Factors of 30	Sum
1, 30	31
2, 15	17
3, 10	13
5, 6	11

The factors are 3 and 10.

Use the factors 3 and 10 to rewrite $13x$ as $3x + 10x$.

$$2x^2 + 13x + 15$$
$$= 2x^2 + 3x + 10x + 15$$

Factor by grouping.

$$= (2x^2 + 3x) + (10x + 15)$$
$$= x(2x + 3) + 5(2x + 3)$$
$$= (2x + 3)(x + 5)$$

Check: $(2x + 3)(x + 5) = 2x^2 + 10x + 3x + 15$
$= 2x^2 + 13x + 15$

B. $6x^2 - 11x - 10$

$a = 6, c = -10, a \cdot c = 6(-10) = -60$

Find two factors of -60 whose sum is -11.

Factors of −60	Sum
1, −60	−59
−1, 60	59
2, −30	−28
−2, 30	28
3, −20	−17
−3, 20	17
4, −15	−11

The required sum has been found. The remaining factors need not be checked. The factors are 4 and -15.

Use the factors 4 and -15 to rewrite $-11x$ as $4x - 15x$.

$$6x^2 - 11x - 10$$
$$= 6x^2 + 4x - 15x - 10$$

Factor by grouping.

Note: $-15x - 10 = -(15x + 10)$

$$= (6x^2 + 4x) - (15x + 10)$$
$$= 2x(3x + 2) - 5(3x + 2)$$
$$= (3x + 2)(2x - 5)$$

Check: $(3x + 2)(2x - 5) = 6x^2 - 15x + 4x - 10$
$= 6x^2 - 11x - 10$

C. $3x^2 - 2x - 4$

$a = 3, c = -4, a \cdot c = 3(-4) = -12$

Find two factors of -12 whose sum is -2.

Factors of -12	Sum
$1, -12$	-11
$-1, 12$	11
$2, -6$	-4
$-2, 6$	4
$3, -4$	-1
$-3, 4$	1

No integer factors of -12 have a sum of -2. Therefore, $3x^2 - 2x - 4$ is nonfactorable over the integers.

EXAMPLE 5 Factor: $2x^2 + 19x - 10$

Solution $a \cdot c = 2(-10) = -20$ • **Find $a \cdot c$.**

$-1(20) = -20$
$-1 + 20 = 19$ • **Find two numbers whose product is -20 and whose sum is 19.**

$2x^2 + 19x - 10$
$= 2x^2 - x + 20x - 10$ • **Rewrite $19x$ as $-x + 20x$.**
$= (2x^2 - x) + (20x - 10)$ • **Factor by grouping.**
$= x(2x - 1) + 10(2x - 1)$
$= (2x - 1)(x + 10)$

Problem 5 Factor: $2a^2 + 13a - 7$

Solution See page S18.

Try Exercise 77, page 371.

EXAMPLE 6 Factor: $8y^2 - 10y - 3$

Solution $a \cdot c = 8(-3) = -24$ • **Find $a \cdot c$.**

$2(-12) = -24$
$2 + (-12) = -10$ • **Find two numbers whose product is -24 and whose sum is -10.**

$8y^2 - 10y - 3$
$= 8y^2 + 2y - 12y - 3$ • **Rewrite $-10y$ as $2y - 12y$.**
$= (8y^2 + 2y) - (12y + 3)$ • **Factor by grouping.**
$= 2y(4y + 1) - 3(4y + 1)$
$= (4y + 1)(2y - 3)$

Problem 6 Factor: $4a^2 - 11a - 3$

Solution See page S18.

Try Exercise 71, page 371.

Remember that the first step in factoring a trinomial is to determine whether the terms have a common factor. If so, factor out the GCF of the terms.

EXAMPLE 7 Factor: $24x^2y - 76xy + 40y$

Solution

$24x^2y - 76xy + 40y$
$= 4y(6x^2 - 19x + 10)$

• The terms of the polynomial have a common factor, $4y$. Factor out the GCF.

$a \cdot c = 6(10) = 60$

• To factor $6x^2 - 19x + 10$, first find $a \cdot c$.

$-4(-15) = 60$
$-4 + (-15) = -19$

• Find two numbers whose product is 60 and whose sum is -19.

$6x^2 - 19x + 10$
$= 6x^2 - 4x - 15x + 10$
$= (6x^2 - 4x) - (15x - 10)$
$= 2x(3x - 2) - 5(3x - 2)$
$= (3x - 2)(2x - 5)$

• Rewrite $-19x$ as $-4x - 15x$.
• Factor by grouping.

$24x^2y - 76xy + 40y$
$= 4y(6x^2 - 19x + 10)$
$= 4y(3x - 2)(2x - 5)$

• Write the complete factorization of the given polynomial.

Problem 7 Factor: $15x^3 + 40x^2 - 80x$

Solution See page S19.

Try Exercise 111, page 371.

8.3 Exercises

CONCEPT CHECK

Replace the ? to make a true statement.

1. $6x^2 + 11x - 10 = (3x - 2)(?)$

2. $40x^2 + 41x + 10 = (8x + 5)(?)$

3. $20x^2 - 31x + 12 = (5x - 4)(?)$

4. $12x^2 - 4x - 21 = (6x + 7)(?)$

Fill in the blanks.

5. To factor $2x^2 - 5x + 2$ by grouping, find two numbers whose product is ______ and whose sum is ______.

6. To factor $3x^2 + 2x - 5$ by grouping, find two numbers whose product is ______ and whose sum is ______.

7. To factor $4x^2 - 8x + 3$ by grouping, $-8x$ must be written as ______ + ______.

8. To factor $6x^2 + 7x - 3$ by grouping, $7x$ must be written as ______ + ______.

9. When multiplying two binomials, which portion of the FOIL method determines the middle term of the resulting trinomial?

1 Factor trinomials of the form $ax^2 + bx + c$ by using trial factors (See pages 363-366.)

GETTING READY

10. Use trial factors to factor $2x^2 - 7x - 15$.

a. Complete the table of the factors of 2 and -15.

Factors of 2	Factors of -15
1, ?	1, ?
	-1, ?
	3, ?
	-3, ?

b. Complete the table by finding the middle term for each pair of trial factors.

Trial Factors	Middle Term
$(x - 15)(2x + 1)$	?
$(x + 15)(2x - 1)$	?
$(x - 5)(2x + 3)$	?
$(x + 5)(2x - 3)$	?

c. Use the results of part (b) to write the factored form of $2x^2 - 7x - 15$: (___?___)(___?___).

Factor by using trial factors.

11. $2x^2 + 3x + 1$
12. $5x^2 + 6x + 1$
13. $2y^2 + 7y + 3$
14. $3y^2 + 7y + 2$
15. $2a^2 - 3a + 1$
16. $3a^2 - 4a + 1$
17. $2b^2 - 11b + 5$
18. $3b^2 - 13b + 4$
19. $2x^2 + x - 1$
20. $7x^2 + 50x + 7$
21. $2x^2 - 5x - 3$
22. $3x^2 + 5x - 2$
23. $6z^2 - 7z + 3$
24. $9z^2 + 3z + 2$
25. $6t^2 - 11t + 4$
26. $10t^2 + 11t + 3$
27. $8x^2 + 33x + 4$
28. $4x^2 - 3x - 1$
29. $6b^2 - 19b + 15$
30. $4z^2 + 5z - 6$
31. $3p^2 + 22p - 16$
32. $7p^2 + 19p + 10$
33. $6x^2 - 17x + 12$
34. $15x^2 - 19x + 6$
35. $5b^2 + 33b - 14$
36. $8x^2 - 30x + 25$
37. $6a^2 + 7a - 24$
38. $14a^2 + 15a - 9$
39. $18t^2 - 9t - 5$
40. $12t^2 + 28t - 5$
41. $15a^2 + 26a - 21$
42. $6a^2 + 23a + 21$
43. $8y^2 - 26y + 15$
44. $18y^2 - 27y + 4$
45. $3z^2 + 95z + 10$
46. $8z^2 - 36z + 1$
47. $28 + 3z - z^2$
48. $15 - 2z - z^2$
49. $8 - 7x - x^2$
50. $12 + 11x - x^2$
51. $9x^2 + 33x - 60$
52. $16x^2 - 16x - 12$
53. $24x^2 - 52x + 24$
54. $60x^2 + 95x + 20$
55. $35a^4 + 9a^3 - 2a^2$
56. $15a^4 + 26a^3 + 7a^2$
57. $15b^2 - 115b + 70$
58. $25b^2 + 35b - 30$
59. $10x^3 + 12x^2 + 2x$
60. $9x^3 - 39x^2 + 12x$
61. $4yz^3 + 5yz^2 - 6yz$
62. $2yz^3 - 17yz^2 + 8yz$
63. $9x^3y + 12x^2y + 4xy$
64. $9a^3b - 9a^2b^2 - 10ab^3$

For Exercises 65 and 66, let $(nx + p)$ and $(mx + q)$ be factors of the trinomial $ax^2 + bx + c$, in which a and c are even and b is odd.

65. If n is even, which of the numbers p, m, and q must be odd?

66. If p is even, which of the numbers n, m, and q must be odd?

2 Factor trinomials of the form $ax^2 + bx + c$ by grouping (See pages 367–369.)

GETTING READY

67. Complete the table of pairs of factors of -18.

Factors of −18	Sum of the Factors
1, ?	?
−1, ?	?
2, ?	?
−2, ?	?
3, ?	?
−3, ?	?

68. To factor $2x^2 - 3x - 9$ by grouping, first find two integers whose product is $2(-9) =$ ___?___ and whose sum is the coefficient of the middle term, ___?___. These two integers are ___?___ and ___?___.

Factor by grouping.

69. $2t^2 - t - 10$

70. $2t^2 + 5t - 12$

71. $3p^2 - 16p + 5$

72. $6p^2 + 5p + 1$

73. $12y^2 - 7y + 1$

74. $6y^2 - 5y + 1$

75. $5x^2 - 62x - 7$

76. $9x^2 - 13x - 4$

77. $12y^2 + 19y + 5$

78. $5y^2 - 22y + 8$

79. $7a^2 + 47a - 14$

80. $11a^2 - 54a - 5$

81. $4z^2 + 11z + 6$

82. $6z^2 - 25z + 14$

83. $8y^2 + 17y + 9$

84. $12y^2 - 145y + 12$

85. $6b^2 - 13b + 6$

86. $15b^2 - 43b + 22$

87. $15x^2 - 82x + 24$

88. $13z^2 + 49z - 8$

89. $10z^2 - 29z + 10$

90. $15z^2 - 44z + 32$

91. $4x^2 + 6x + 2$

92. $12x^2 + 33x - 9$

93. $15y^2 - 50y + 35$

94. $30y^2 + 10y - 20$

95. $2x^3 - 11x^2 + 5x$

96. $2x^3 - 3x^2 - 5x$

97. $3a^2 + 5ab - 2b^2$

98. $2a^2 - 9ab + 9b^2$

99. $4y^2 - 11yz + 6z^2$

100. $2y^2 + 7yz + 5z^2$

101. $12 - x - x^2$

102. $18 + 17x - x^2$

103. $360y^2 + 4y - 4$

104. $10t^2 - 5t - 50$

105. $16t^2 + 40t - 96$

106. $3p^3 - 16p^2 + 5p$

107. $6p^3 + 5p^2 + p$

108. $26z^2 + 98z - 24$

109. $30z^2 - 87z + 30$

110. $12a^3 + 14a^2 - 48a$

111. $42a^3 + 45a^2 - 27a$

112. $36p^2 - 9p^3 - p^4$

113. $9x^2y - 30xy^2 + 25y^3$

114. $8x^2y - 38xy^2 + 35y^3$

115. $9x^3y - 24x^2y^2 + 16xy^3$

116. $45a^3b - 78a^2b^2 + 24ab^3$

In Exercises 117 to 120, information is given about the signs of b and c in $ax^2 + bx + c$, where $a > 0$. If you want to factor $ax^2 + bx + c$ by grouping, you look for factors of ac whose sum is b. In each case, state whether the factors of ac should be two positive numbers, two negative numbers, or one positive and one negative number.

117. $b > 0$ and $c > 0$

118. $b < 0$ and $c < 0$

119. $b < 0$ and $c > 0$

120. $b > 0$ and $c < 0$

APPLYING CONCEPTS

Factor.

121. $6y + 8y^3 - 26y^2$

122. $22p^2 - 3p^3 + 16p$

123. $a^3b - 24ab - 2a^2b$

124. $3xy^2 - 14xy + 2xy^3$

125. $25t^2 + 60t - 10t^3$

126. $3xy^3 + 2x^3y - 7x^2y^2$

127. $2(y + 2)^2 - (y + 2) - 3$

128. $3(a + 2)^2 - (a + 2) - 4$

129. $10(x + 1)^2 - 11(x + 1) - 6$

130. $4(y - 1)^2 - 7(y - 1) - 2$

131. Given that $x + 2$ is a factor of $x^3 - 2x^2 - 5x + 6$, factor $x^3 - 2x^2 - 5x + 6$ completely.

132. In your own words, explain how the signs of the last terms of the two binomial factors of a trinomial are determined.

PROJECTS OR GROUP ACTIVITIES

Find all integers k such that the trinomial can be factored over the integers.

133. $2x^2 + kx + 3$

134. $2x^2 + kx - 3$

135. $3x^2 + kx + 2$

136. $3x^2 + kx - 2$

137. $2x^2 + kx + 5$

138. $2x^2 + kx - 5$

139. **Geometry** The area of a rectangle is $(3x^2 + x - 2)$ ft^2. Find the dimensions of the rectangle in terms of the variable x. Given that $x > 0$, specify the dimension that is the length and the dimension that is the width. Can x be negative? Can $x = 0$? Explain your answers.

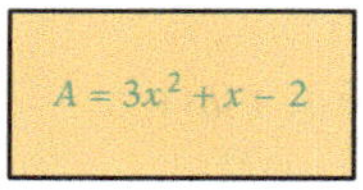

8.4 Special Factoring

OBJECTIVE 1 Factor the difference of two squares and perfect-square trinomials

Recall from Objective 4 in Section 3 of the chapter on polynomials that the product of the sum and difference of the same two terms equals the square of the first term minus the square of the second term.

$$(a + b)(a - b) = a^2 - b^2$$

How It's Used

A secure transmission of your credit card number to an Internet retailer depends on how easy it is to factor a number. One strategy hackers use to find the factors is to try to write the credit card number as the difference of two squares. For instance, one way to show that 901 factors is to write $901 = 1225 - 324 = 35^2 - 18^2 = (35 - 18)(35 + 18) = 17(53)$.

Take Note

Convince yourself that the sum of two squares is nonfactorable over the integers by trying to factor $x^2 + 4$.

The expression $a^2 - b^2$ is the **difference of two squares.** The pattern just mentioned suggests the following rule for factoring the difference of two squares.

RULE FOR FACTORING THE DIFFERENCE OF TWO SQUARES

Difference of Two Squares		Sum and Difference of Two Terms
$a^2 - b^2$	$=$	$(a + b)(a - b)$

EXAMPLES

Each expression is the difference of two squares.

Factor.

1. $x^2 - 4 = x^2 - 2^2 = (x + 2)(x - 2)$
2. $y^2 - 9 = y^2 - 3^2 = (y + 3)(y - 3)$

$a^2 + b^2$ is the sum of two squares. It is nonfactorable over the integers.

EXAMPLE 1 Factor. **A.** $x^2 - 16$ **B.** $x^2 - 10$ **C.** $z^6 - 25$

Solution

A. $x^2 - 16 = x^2 - 4^2$ • Write $x^2 - 16$ as the difference of two squares.

$= (x + 4)(x - 4)$ • The factors are the sum and difference of the terms x and 4.

B. $x^2 - 10$ is nonfactorable over the integers. • Because 10 is not the square of an integer, $x^2 - 10$ cannot be written as the difference of two squares.

C. $z^6 - 25 = (z^3)^2 - 5^2$ • Write $z^6 - 25$ as the difference of two squares.

$= (z^3 + 5)(z^3 - 5)$ • The factors are the sum and difference of the terms z^3 and 5.

Problem 1 Factor. **A.** $25a^2 - b^2$ **B.** $6x^2 - 1$ **C.** $n^8 - 36$

Solution See page S19.

Try Exercise 17, page 376.

EXAMPLE 2 Factor: $z^4 - 16$

Solution

$z^4 - 16 = (z^2)^2 - (4)^2$ • This is the difference of two squares.

$= (z^2 + 4)(z^2 - 4)$ • The factors are the sum and difference of the terms z^2 and 4.

$= (z^2 + 4)(z + 2)(z - 2)$ • Factor $z^2 - 4$, which is the difference of two squares. $z^2 + 4$ is nonfactorable over the integers.

Problem 2 Factor: $n^4 - 81$

Solution See page S19.

Try Exercise 37, page 376.

Recall from Objective 4 of Section 3 in the chapter on polynomials the pattern for finding the square of a binomial.

$$(a + b)^2 = (a + b)(a + b) = a^2 + ab + ab + b^2$$
$$= a^2 + 2ab + b^2$$

- a^2: Square of the first term
- $2ab$: Twice the product of the two terms
- b^2: Square of the last term

The square of a binomial is a **perfect-square trinomial.** The pattern above suggests the following rule for factoring a perfect-square trinomial.

RULE FOR FACTORING A PERFECT-SQUARE TRINOMIAL

Perfect-Square Trinomial				Square of a Binomial
$a^2 + 2ab + b^2$	=	$(a + b)(a + b)$	=	$(a + b)^2$
$a^2 - 2ab + b^2$	=	$(a - b)(a - b)$	=	$(a - b)^2$

Note in these patterns that the sign in the binomial is the sign of the middle term of the trinomial.

Focus on factoring a perfect square trinomial

Factor. **A.** $4x^2 - 20x + 25$ **B.** $9x^2 + 30x + 16$

A. $4x^2 - 20x + 25$

Check that the first term and the last term are squares. $4x^2 = (2x)^2, 25 = 5^2$

Use the squared terms to factor the trinomial as the square of a binomial. The sign in the binomial is the sign of the middle term of the trinomial. $(2x - 5)^2$

Check the factorization.
$$(2x - 5)^2$$
$$= (2x)^2 + 2(2x)(-5) + (-5)^2$$
$$= 4x^2 - 20x + 25$$

The factorization is correct. $4x^2 - 20x + 25 = (2x - 5)^2$

B. $9x^2 + 30x + 16$

Check that the first term and the last term are squares. $9x^2 = (3x)^2, 16 = 4^2$

Use the squared terms to factor the trinomial as the square of a binomial. The sign in the binomial is the sign of the middle term of the trinomial. $(3x + 4)^2$

Check the factorization.
$$(3x + 4)^2$$
$$= (3x)^2 + 2(3x)(4) + 4^2$$
$$= 9x^2 + 24x + 16$$

Take Note
A perfect-square trinomial can always be factored using either of the methods presented in Section 3 of this chapter. However, noticing that a trinomial is a perfect-square trinomial can save you a considerable amount of time.

$9x^2 + 24x + 16 \neq 9x^2 + 30x + 16$

The proposed factorization is not correct.

In this case, the polynomial is not a perfect-square trinomial. It may, however, still factor. In fact, $9x^2 + 30x + 16 = (3x + 2)(3x + 8)$. If a trinomial does not check as a perfect-square trinomial, try to factor it by another method.

EXAMPLE 3 Factor. **A.** $9x^2 - 30x + 25$ **B.** $4x^2 + 37x + 9$

Solution **A.** $9x^2 = (3x)^2$, $25 = 5^2$ • Check that the first and last terms are squares.

$(3x - 5)^2$ • Use the squared terms to factor the trinomial as the square of a binomial.

$(3x - 5)^2$
$= (3x)^2 + 2(3x)(-5) + (-5)^2$ • Check the factorization.
$= 9x^2 - 30x + 25$ • The factorization checks.

$9x^2 - 30x + 25 = (3x - 5)^2$

B. $4x^2 = (2x)^2$, $9 = 3^2$ • Check that the first and last terms are squares.

$(2x + 3)^2$ • Use the squared terms to factor the trinomial as the square of a binomial.

$(2x + 3)^2$ • Check the factorization.
$= (2x)^2 + 2(2x)(3) + 3^2$
$= 4x^2 + 12x + 9$ • The factorization does not check.

$4x^2 + 37x + 9$ • Use another method to factor the trinomial.
$= (4x + 1)(x + 9)$

Problem 3 Factor. **A.** $16y^2 + 8y + 1$ **B.** $x^2 + 14x + 36$

Solution See page S19.

Try Exercise 33, page 376.

8.4 Exercises

CONCEPT CHECK

Which of the expressions in the list are perfect squares?

1. 4; 8; $25x^6$; $12y^{10}$; $100x^4y^4$

2. 9; 18; $15a^8$; $49b^{12}$; $64a^{16}b^2$

3. Which of the expressions are the difference of two squares?
(i) $a^2 - 36$ (ii) $b^2 - 12$ (iii) $c^2 + 25$ (iv) $d^2 - 100$

4. Which expression is the sum and difference of two terms?
(i) $(a + 4)(a + 4)$ (ii) $(a + 4)(b - 4)$ (iii) $(a + 4)(a - 4)$

Determine whether the statement is always true, sometimes true, or never true.

5. A binomial is factorable.

6. A trinomial is factorable.

7. If a binomial is multiplied times itself, the result is a perfect-square trinomial.

8. In a perfect-square trinomial, the first and last terms are perfect squares.

1 Factor the difference of two squares and perfect-square trinomials (See pages 372–375.)

GETTING READY

9. a. The binomial $9x^2 - 4$ is in the form $a^2 - b^2$, where $a =$ ___?___ and $b =$ ___?___.

b. Use the formula $a^2 - b^2 = (a + b)(a - b)$ to factor $9x^2 - 4$: $9x^2 - 4 =$ (___?___)(___?___).

10. a. The trinomial $16y^2 - 8y + 1$ is in the form $a^2 - 2ab + b^2$, where $a =$ ___?___ and $b =$ ___?___.

b. Use the formula $a^2 - 2ab + b^2 = (a - b)^2$ to factor $16y^2 - 8y + 1$: $16y^2 - 8y + 1 =$ (___?___)^2.

11. Provide an example of each of the following.
a. the difference of two squares
b. the product of the sum and difference of two terms
c. a perfect-square trinomial
d. the square of a binomial
e. the sum of two squares

12. Explain the rule for factoring
a. the difference of two squares.
b. a perfect-square trinomial.

Factor.

13. $x^2 - 4$ **14.** $x^2 - 9$ **15.** $a^2 - 81$ **16.** $a^2 - 49$

17. $4x^2 - 9$ **18.** $9x^2 - 16$ **19.** $y^2 + 6y + 9$ **20.** $y^2 + 14y + 49$

21. $a^2 - 2a + 1$ **22.** $x^2 - 12x + 36$ **23.** $z^2 - 18z - 81$ **24.** $x^2 + 8x - 16$

25. $x^6 - 9$ **26.** $y^{12} - 121$ **27.** $25x^2 - 4$ **28.** $9x^2 - 49$

29. $1 - 49x^2$ **30.** $1 - 64x^2$ **31.** $x^2 + 2xy + y^2$ **32.** $x^2 + 6xy + 9y^2$

33. $4a^2 + 4a + 1$ **34.** $25x^2 + 10x + 1$ **35.** $t^2 + 36$ **36.** $x^2 + 64$

37. $x^4 - y^2$ **38.** $b^4 - 16a^2$ **39.** $9x^2 - 16y^2$ **40.** $25z^2 - y^2$

41. $16b^2 + 24b + 9$ **42.** $4a^2 - 20a + 25$ **43.** $4b^2 + 28b + 49$

44. $9a^2 - 42a + 49$ **45.** $25a^2 + 30ab + 9b^2$ **46.** $4a^2 - 12ab + 9b^2$

47. $x^2y^2 - 4$ **48.** $a^2b^2 - 25$ **49.** $9x^2 + 13x + 4$

50. $x^2 + 10x + 16$

51. Which expressions are equivalent to $x^4 - 81$?
(i) $(x^2 + 9)(x + 3)(x - 3)$ (ii) $(2x + 9)(2x^2 - 9)$
(iii) $(x^2 + 9)(x^2 - 9)$ (iv) $(x + 3)^2(x - 3)^2$

52. Which expressions can be factored as the square of a binomial, given that a and b are positive numbers?
(i) $a^2x^2 - 2abx + b^2$ (ii) $a^2x^2 - 2abx - b^2$
(iii) $a^2x^2 + 2abx + b^2$ (iv) $a^2x^2 + 2abx - b^2$

APPLYING CONCEPTS

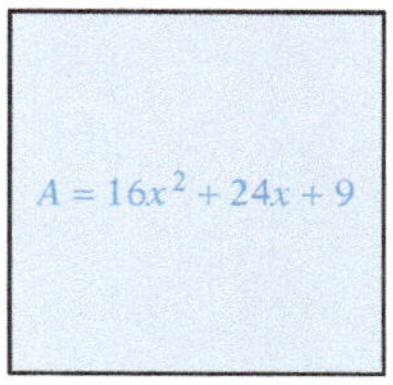

53. Geometry The area of a square is $(16x^2 + 24x + 9)$ ft^2. Find the dimensions of the square in terms of the variable x. Can $x = 0$? What are the possible values of x?

The cube of an integer is a **perfect cube.** Because $2^3 = 8$, 8 is a perfect cube. Because $4^3 = 64$, 64 is a perfect cube. A variable expression can be a perfect cube; the exponents on variables of perfect cubes are multiples of 3. Therefore, x^3, x^6, and x^9 are perfect cubes. The sum and the difference of two perfect cubes are factorable. They can be written as the product of a binomial and a trinomial. Their factoring patterns are shown below.

$a^3 + b^3$ is the sum of two cubes. $a^3 + b^3 = (a + b)(a^2 - ab + b^2)$
$a^3 - b^3$ is the difference of two cubes. $a^3 - b^3 = (a - b)(a^2 + ab + b^2)$

To factor $x^3 - 8$, write the binomial as the difference of two perfect cubes. Use the factoring pattern shown above. Replace a with x and b with 2.

$$x^3 - 8 = (x)^3 - (2)^3$$
$$= (x - 2)(x^2 + 2x + 4)$$

Factor.

54. $x^3 + 8$ **55.** $y^3 + 27$ **56.** $y^3 - 27$ **57.** $x^3 - 1$

58. $y^3 + 64$ **59.** $x^3 - 125$ **60.** $8x^3 - 1$ **61.** $27y^3 + 1$

PROJECTS OR GROUP ACTIVITIES

Find all integers k such that the trinomial is a perfect-square trinomial.

62. $4x^2 - kx + 9$ **63.** $25x^2 - kx + 1$ **64.** $36x^2 + kxy + y^2$ **65.** $64x^2 + kxy + y^2$

66. $x^2 + 6x + k$ **67.** $x^2 - 4x + k$ **68.** $x^2 - 2x + k$ **69.** $x^2 + 10x + k$

70. Select any odd integer greater than 1, square it, and then subtract 1. Is the result evenly divisible by 8? Prove that this procedure always produces a number that is divisible by 8. (*Suggestion:* Any odd integer greater than 1 can be expressed as $2n + 1$, where n is a natural number.)

8.5 Factoring Polynomials Completely

OBJECTIVE 1 Factor completely

This section is devoted to describing a strategy for factoring polynomials and reviewing the factoring techniques you have learned in this chapter.

GENERAL FACTORING STRATEGY

When factoring a polynomial completely, ask yourself the following questions about the polynomial.

1. Do the terms contain a common factor? If so, factor out the common factor.
2. Is the polynomial the difference of two squares? If so, factor.
3. Is the polynomial a perfect-square trinomial? If so, factor.
4. Is the polynomial a trinomial that is the product of two binomials? If so, factor.
5. Does the polynomial contain four terms? If so, try factoring by grouping.
6. Is each binomial factor nonfactorable over the integers? If not, factor.

When factoring a polynomial, remember that you may have to factor more than once in order to write the polynomial as a product of factors, each of which is nonfactorable over the integers.

EXAMPLE 1 Factor. **A.** $3x^2 - 48$ **B.** $x^3 - 3x^2 - 4x + 12$ **C.** $4x^2y^2 + 12xy^2 + 9y^2$

Solution

A. $3x^2 - 48$

$= 3(x^2 - 16)$ • **The GCF of the terms is 3. Factor out the common factor.**

$= 3(x + 4)(x - 4)$ • **Factor the difference of two squares.**

B. $x^3 - 3x^2 - 4x + 12$

$= (x^3 - 3x^2) - (4x - 12)$ • **The polynomial contains four terms. Factor by grouping.**

$= x^2(x - 3) - 4(x - 3)$

$= (x - 3)(x^2 - 4)$

$= (x - 3)(x + 2)(x - 2)$ • **Factor the difference of two squares.**

C. $4x^2y^2 + 12xy^2 + 9y^2$

$= y^2(4x^2 + 12x + 9)$ • **The GCF of the terms is y^2. Factor out the common factor.**

$= y^2(2x + 3)^2$ • **Factor the perfect-square trinomial.**

Problem 1 Factor. **A.** $12x^3 - 75x$ **B.** $a^2b - 7a^2 - b + 7$ **C.** $4x^3 + 28x^2 - 120x$

Solution See page S19.

Try Exercise 15, page 379.

8.5 Exercises

CONCEPT CHECK

The first step in any factoring problem is to factor out the greatest common factor. The second step depends on the number of terms the polynomial has. For each of the following cases, state what the next step in factoring could be.

1. The polynomial has two terms.
2. The polynomial has three terms.
3. The polynomial has four terms.

1 Factor completely (See page 378.)

GETTING READY

4. When factoring a polynomial, always look first for a __?__ factor.
5. When a polynomial is factored completely, each factor is __?__ over the integers.
6. Which factor in $(x^2 - 81)(x^2 + 81)$ can be factored?

Factor.

7. $2x^2 - 18$
8. $y^3 - 10y^2 + 25y$
9. $x^4 + 2x^3 - 35x^2$
10. $a^4 - 11a^3 + 24a^2$
11. $5b^2 + 75b + 180$
12. $6y^2 - 48y + 72$
13. $3a^2 + 36a + 10$
14. $5a^2 - 30a + 4$
15. $2x^2y + 16xy - 66y$
16. $3a^2b + 21ab - 54b$
17. $x^3 - 6x^2 - 5x$
18. $b^3 - 8b^2 - 7b$
19. $3y^2 - 36$
20. $3y^2 - 147$
21. $20a^2 + 12a + 1$
22. $12a^2 - 36a + 27$
23. $x^2y^2 - 7xy^2 - 8y^2$
24. $a^2b^2 + 3a^2b - 88a^2$
25. $10a^2 - 5ab - 15b^2$
26. $16x^2 - 32xy + 12y^2$
27. $50 - 2x^2$
28. $72 - 2x^2$
29. $12a^3b - a^2b^2 - ab^3$
30. $2x^3y - 7x^2y^2 + 6xy^3$
31. $2ax - 2a + 2bx - 2b$
32. $4ax - 12a - 2bx + 6b$
33. $12a^3 - 12a^2 + 3a$
34. $18a^3 + 24a^2 + 8a$
35. $243 + 3a^2$
36. $75 + 27y^2$
37. $12a^3 - 46a^2 + 40a$
38. $24x^3 - 66x^2 + 15x$
39. $x^3 - 2x^2 - x + 2$
40. $ay^2 - by^2 - a + b$
41. $4a^3 + 20a^2 + 25a$
42. $2a^3 - 8a^2b + 8ab^2$
43. $27a^2b - 18ab + 3b$
44. $a^2b^2 - 6ab^2 + 9b^2$
45. $48 - 12x - 6x^2$
46. $21x^2 - 11x^3 - 2x^4$
47. $ax^2 - 4a + bx^2 - 4b$
48. $a^2x - b^2x - a^2y + b^2y$
49. $x^4 - x^2y^2$
50. $b^4 - a^2b^2$
51. $18a^3 + 24a^2 + 8a$
52. $32xy^2 - 48xy + 18x$

53. $2b + ab - 6a^2b$

54. $20x - 11xy - 3xy^2$

55. $4x - 20 - x^3 + 5x^2$

56. $ay^2 - by^2 - 9a + 9b$

57. $72xy^2 + 48xy + 8x$

58. $4x^2y + 8xy + 4y$

59. $15y^2 - 2xy^2 - x^2y^2$

60. $4x^4 - 38x^3 + 48x^2$

61. $y^3 - 9y$

62. $a^4 - 16$

63. $2x^4y^2 - 2x^2y^2$

64. $6x^5y - 6xy^5$

65. $x^9 - x^5$

66. $8b^5 - 2b^3$

67. $24x^3y + 14x^2y - 20xy$

68. $12x^3y - 60x^2y + 63xy$

69. $4x^4y^2 - 20x^3y^2 + 25x^2y^2$

70. $9x^4y^2 + 24x^3y^2 + 16x^2y^2$

71. $m^4 - 256$

72. $81 - t^4$

73. $y^8 - 81$

For Exercises 74 and 75, the factored form of a polynomial P is $x^2(x - a)(x + b)$, where $a > 0$ and $b > 0$.

74. What is the degree of the polynomial P?

75. If the middle term of the polynomial P has a negative coefficient, is b greater than, less than, or equal to a?

APPLYING CONCEPTS

Factor.

76. $(4x - 3)^2 - y^2$

77. $(2a + 3)^2 - 25b^2$

78. $(x^2 - 4x + 4) - y^2$

79. $(4x^2 + 12x + 9) - 4y^2$

80. **Number Problems** The product of two numbers is 48. One of the two numbers is a perfect square. The other is a prime number. Find the sum of the two numbers.

PROJECTS OR GROUP ACTIVITIES

81. Show how you can use the difference of two squares to find the products $42 \cdot 38$ and $84 \cdot 76$.

82. List any three consecutive natural numbers. What is the relationship between the square of the middle number and the product of the first and third numbers? Is this relationship always true? Try to prove your answer.

83. The values of a, b, c, and d are 1, 3, 5, and 7, but not necessarily in that order. Find the largest possible value of $2ab + 2bc + 2cd + 2da$.

8.6 Solving Equations

OBJECTIVE 1 Solve equations by factoring

Recall that the Multiplication Property of Zero states that the product of a number and zero is zero.

If a is a real number, then $a \cdot 0 = 0$.

Consider the equation $a \cdot b = 0$. If this is a true equation, then either $a = 0$ or $b = 0$.

PRINCIPLE OF ZERO PRODUCTS

If the product of two factors is zero, then at least one of the factors must be zero.

If $\mathbf{a \cdot b = 0}$, then $\mathbf{a = 0}$ or $\mathbf{b = 0}$.

The Principle of Zero Products is used in solving equations.

Take Note

$x - 2$ is equal to a number. $x - 3$ is equal to a number. In $(x - 2)(x - 3)$, two numbers are being multiplied. Since their product is zero, one of the numbers must be equal to zero. The number $x - 2$ is equal to 0 or the number $x - 3$ is equal to 0.

Focus on solving an equation using the Principle of Zero Products

Solve: $(x - 2)(x - 3) = 0$

If $(x - 2)(x - 3) = 0$, then $(x - 2) = 0$ or $(x - 3) = 0$.

$$(x - 2)(x - 3) = 0$$

Solve each equation for x.

$$x - 2 = 0 \qquad x - 3 = 0$$
$$x = 2 \qquad x = 3$$

Check:

$$\begin{array}{c|c} (x - 2)(x - 3) & 0 \\ \hline (2 - 2)(2 - 3) & 0 \\ 0(-1) & 0 \\ 0 = 0 & \end{array} \qquad \begin{array}{c|c} (x - 2)(x - 3) & 0 \\ \hline (3 - 2)(3 - 3) & 0 \\ 1(0) & 0 \\ 0 = 0 & \end{array}$$

A true equation — A true equation

Write the solutions. The solutions are 2 and 3.

An equation that can be written in the form $ax^2 + bx + c = 0$, $a \neq 0$, is a **quadratic equation.** A quadratic equation is in **standard form** when the polynomial is equal to zero and its terms are in descending order.

$$3x^2 + 2x + 1 = 0$$
$$4x^2 - 3x + 2 = 0$$

A quadratic equation can be solved by using the Principle of Zero Products when the polynomial $ax^2 + bx + c$ is factorable.

EXAMPLE 1 Solve: $2x^2 + x = 6$

Solution

$2x^2 + x = 6$ • This is a quadratic equation.

$2x^2 + x - 6 = 0$ • Write it in standard form.

$(2x - 3)(x + 2) = 0$ • Factor the trinomial.

$2x - 3 = 0 \qquad x + 2 = 0$ • Set each factor equal to zero (the Principle of Zero Products).

$2x = 3 \qquad x = -2$ • Solve each equation for x.

$x = \frac{3}{2}$

Check

$2x^2 + x$	6
$2\left(\frac{3}{2}\right)^2 + \frac{3}{2}$	6
$2\left(\frac{9}{4}\right) + \frac{3}{2}$	6
$\frac{9}{2} + \frac{3}{2}$	6

$6 = 6$

$2x^2 + x$	6
$2(-2)^2 + (-2)$	6
$2 \cdot 4 - 2$	6
$8 - 2$	6

$6 = 6$

The solutions are $\frac{3}{2}$ and -2. • Write the solutions.

Problem 1 Solve: $2x^2 - 50 = 0$

Solution See page S19.

Try Exercise 25, page 385.

Example 1 illustrates the steps involved in solving a quadratic equation by factoring.

STEPS IN SOLVING A QUADRATIC EQUATION BY FACTORING

1. Write the equation in standard form.
2. Factor the polynominal.
3. Set each factor equal to zero.
4. Solve each equation for the variable.
5. Check the solutions.

EXAMPLE 2 Solve: $(x - 3)(x - 10) = -10$

Solution

$(x - 3)(x - 10) = -10$ • This is a quadratic equation. The Principle of Zero Products cannot be used unless 0 is on one side of the equation.

$x^2 - 13x + 30 = -10$ • Multiply $(x - 3)(x - 10)$.

$x^2 - 13x + 40 = 0$ • Write the equation in standard form.

$(x - 8)(x - 5) = 0$ • Factor.

$x - 8 = 0 \qquad x - 5 = 0$ • Set each factor equal to zero.

$x = 8 \qquad x = 5$ • Solve each equation for x.

The solutions are 8 and 5. • Write the solutions.

Problem 2 Solve: $(x + 2)(x - 7) = 52$

Solution See page S19.

Try Exercise 53, page 385.

OBJECTIVE 2 Application problems

EXAMPLE 3 The sum of the squares of two consecutive positive odd integers is equal to 130. Find the two integers.

Strategy

- ▶ First positive odd integer: n
  Second positive odd integer: $n + 2$
  Square of the first positive odd integer: n^2
  Square of the second positive odd integer: $(n + 2)^2$
- ▶ The sum of the square of the first positive odd integer and the square of the second positive odd integer is 130.

Solution

$$n^2 + (n + 2)^2 = 130$$
• **This is a quadratic equation.**

$$n^2 + n^2 + 4n + 4 = 130$$
• **Square $n + 2$.**

$$2n^2 + 4n - 126 = 0$$
• **Combine like terms. Subtract 130 from each side of the equation.**

$$2(n^2 + 2n - 63) = 0$$
• **Factor out the common factor of 2.**

$$n^2 + 2n - 63 = 0$$
• **Divide each side of the equation by 2.**

$$(n - 7)(n + 9) = 0$$
• **Factor the trinomial.**

$$n - 7 = 0 \qquad n + 9 = 0$$
• **Set each factor equal to zero.**

$$n = 7 \qquad n = -9$$
• **Solve for n.**

Because -9 is not a positive odd integer, it is not a solution. The first odd integer is 7.

$$n + 2 = 7 + 2 = 9$$
• **Substitute the value of n into the variable expression for the second positive odd integer and evaluate.**

The two integers are 7 and 9.

Problem 3 The sum of the squares of two consecutive positive integers is 85. Find the two integers.

Solution See page S19.

Try Exercise 69, page 386.

EXAMPLE 4 A stone is thrown into a well with an initial velocity of 8 ft/s. The well is 440 ft deep. How many seconds later will the stone hit the bottom of the well? Use the equation $d = vt + 16t^2$, where d is the distance in feet, v is the initial velocity in feet per second, and t is the time in seconds.

Strategy To find the time for the stone to drop to the bottom of the well, replace the variables d and v by their given values and solve for t.

Solution

$$\begin{aligned} d &= vt + 16t^2 \\ 440 &= 8t + 16t^2 \\ 0 &= 16t^2 + 8t - 440 \\ 0 &= 8(2t^2 + t - 55) \\ 0 &= 2t^2 + t - 55 \\ 0 &= (2t + 11)(t - 5) \end{aligned}$$

- **Replace *d* with 440 and *v* with 8.**
- **Write the equation in standard form.**
- **Factor out the common factor of 8.**
- **Divide each side of the equation by 8.**
- **Factor the trinomial.**

$$\begin{aligned} 2t + 11 &= 0 & t - 5 &= 0 \\ 2t &= -11 & t &= 5 \\ t &= -\frac{11}{2} \end{aligned}$$

- **Set each factor equal to zero.**
- **Solve each equation for *t*.**

Because time cannot be a negative number, $-\frac{11}{2}$ is not a solution.

The time is 5 s.

Problem 4 The length of a rectangle is 3 m more than twice the width. The area of the rectangle is 90 m^2. Find the length and width of the rectangle.

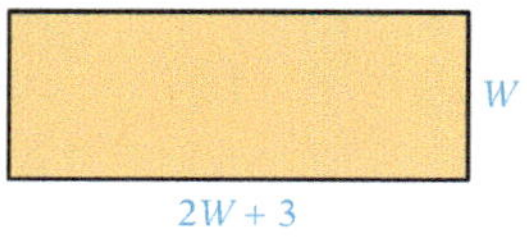

Solution See page S19.

Try Exercise 75, page 386.

8.6 Exercises

CONCEPT CHECK

1. Determine whether the equation is a quadratic equation.
a. $2x^2 - 8 = 0$ **b.** $2x - 8 = 0$ **c.** $x^2 = 8x$

2. Write the equation in standard form.
a. $x^2 + 4 = 4x$ **b.** $x + x^2 = 6$

3. Can the equation be solved by using the Principle of Zero Products without first rewriting the equation?
a. $4x(6x + 7) = 0$ **b.** $0 = (4x - 5)(3x + 8)$ **c.** $2x(x - 5) - 5 = 0$
d. $(x - 7)(y + 3) = 0$ **e.** $0 = (2x - 3)x + 3$ **f.** $0 = (2x - 3)(x + 3)$

Determine whether the statement is true or false.

4. If you multiply two numbers and the product is zero, then either one or both of the numbers must be zero.

5. $2x^2 + 5x - 7$ is a quadratic equation.

6. $3x + 1 = 0$ is a quadratic equation in standard form.

1 Solve equations by factoring (See pages 381–383.)

7. What does the Principle of Zero Products state?

8. Why is it possible to solve some quadratic equations by using the Principle of Zero Products?

GETTING READY

9. Of the two quadratic equations $0 = x^2 + x - 2$ and $5x^2 + 2x = 3$, the one that is in standard form is __?__.

10. Solve: $2x^2 - 5x - 3 = 0$

$2x^2 - 5x - 3 = 0$		• The equation is in __?__ form.
$(x - \_\_?\_\_)(2x + \_\_?\_\_) = 0$		• Factor the trinomial.
$x - 3 = \_\_?\_\_$	$2x + 1 = \_\_?\_\_$	• Use the Principle of Zero Products to set each factor equal to zero.
$x = \_\_?\_\_$	$x = \_\_?\_\_$	• Solve each equation for x.

Solve.

11. $(y + 3)(y + 2) = 0$

12. $(y - 3)(y - 5) = 0$

13. $(z - 7)(z - 3) = 0$

14. $(z + 8)(z - 9) = 0$

15. $x(x - 5) = 0$

16. $x(x + 2) = 0$

17. $a(a - 9) = 0$

18. $a(a + 12) = 0$

19. $y(2y + 3) = 0$

20. $t(4t - 7) = 0$

21. $2a(3a - 2) = 0$

22. $4b(2b + 5) = 0$

23. $9x^2 - 1 = 0$

24. $16x^2 - 49 = 0$

25. $x^2 + 6x + 8 = 0$

26. $x^2 - 8x + 15 = 0$

27. $z^2 + 5z - 14 = 0$

28. $z^2 + z - 72 = 0$

29. $x^2 - 5x + 6 = 0$

30. $2y^2 - y - 1 = 0$

31. $2a^2 - 9a - 5 = 0$

32. $3a^2 + 14a + 8 = 0$

33. $2x^2 - 6x - 20 = 0$

34. $3y^2 + 12y - 63 = 0$

35. $x^2 - 7x = 0$

36. $2a^2 - 8a = 0$

37. $a^2 + 5a = -4$

38. $a^2 - 5a = 24$

39. $y^2 - 5y = -6$

40. $y^2 - 7y = 8$

41. $2t^2 + 7t = 4$

42. $3t^2 + t = 10$

43. $3t^2 - 13t = -4$

44. $5t^2 - 16t = -12$

45. $x(x - 12) = -27$

46. $x(x - 11) = 12$

47. $y(y - 7) = 18$

48. $y(y + 8) = -15$

49. $p(p + 3) = -2$

50. $p(p - 1) = 20$

51. $y(y + 4) = 45$

52. $y(y - 8) = -15$

53. $(x + 8)(x - 3) = -30$

54. $(x + 4)(x - 1) = 14$

55. $(y + 3)(y + 10) = -10$

56. $(z - 5)(z + 4) = 52$

57. $(2x + 5)(x + 1) = -1$

58. $(y + 3)(2y + 3) = 5$

For Exercises 59 to 62, $ax^2 + bx + c = 0$, where $a > 0$, is a quadratic equation that can be solved by factoring and then using the Principle of Zero Products.

59. If $ax^2 + bx + c = 0$ has one positive solution and one negative solution, is c greater than, less than, or equal to 0?

60. If zero is one solution of $ax^2 + bx + c = 0$, is c greater than, less than, or equal to 0?

61. If $ax^2 + bx + c$ is a perfect-square trinomial, how many solutions does $0 = ax^2 + bx + c$ have?

62. If the solutions of $ax^2 + bx + c = 0$ are opposites, what kind of trinomial is $ax^2 + bx + c$?

2 Application problems (See pages 383–384.)

GETTING READY

Complete Exercises 63 and 64 using this problem situation: The sum of the squares of two consecutive positive integers is 113. Find the two integers.

63. a. Let x represent a positive integer. Then the next consecutive positive integer is ___?___, and an expression that represents the sum of the squares of the two integers is ___?___.

b. An equation that can be used to find the integers is ___?___ + ___?___ = ___?___.

64. The solutions of the equation you wrote in Exercise 63 are −8 and 7. Which solution should be eliminated and why?

65. Integer Problem The square of a positive number is 6 more than five times the positive number. Find the number.

66. Integer Problem The square of a negative number is 15 more than twice the negative number. Find the number.

67. Integer Problem The sum of two numbers is 6. The sum of the squares of the two numbers is 20. Find the two numbers.

68. Integer Problem The sum of two numbers is 8. The sum of the squares of the two numbers is 34. Find the two numbers.

69. Integer Problem The sum of the squares of two consecutive positive integers is 41. Find the two integers.

70. Integer Problem The sum of the squares of two consecutive positive even integers is 100. Find the two integers.

71. Integer Problem The product of two consecutive positive integers is 240. Find the two integers.

72. Integer Problem The product of two consecutive positive even integers is 168. Find the two integers.

73. Geometry The length of the base of a triangle is three times the height. The area of the triangle is 54 ft^2. Find the base and height of the triangle.

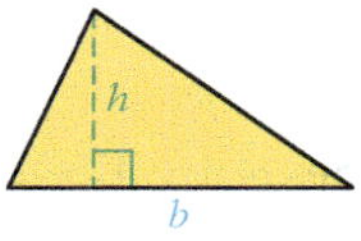

74. Geometry The height of a triangle is 4 m more than twice the length of the base. The area of the triangle is 35 m^2. Find the height of the triangle.

75. Geometry The length of a rectangle is 2 ft more than twice the width. The area is 144 ft^2. Find the length and width of the rectangle.

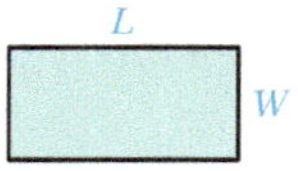

76. Geometry The width of a rectangle is 5 ft less than the length. The area of the rectangle is 176 ft^2. Find the length and width of the rectangle.

77. **Geometry** The length of each side of a square is extended 4 m. The area of the resulting square is 64 m^2. Find the length of a side of the original square.

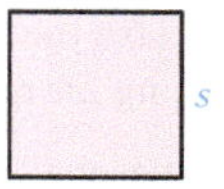

78. **Geometry** The length of each side of a square is extended 2 cm. The area of the resulting square is 64 cm^2. Find the length of a side of the original square.

79. **Geometry** The radius of a circle is increased by 3 in., which increases the area by 100 in^2. Find the radius of the original circle. Round to the nearest hundredth.

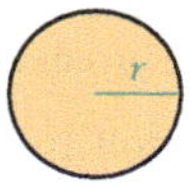

80. **Geometry** The length of a rectangle is 5 cm, and the width is 3 cm. If both the length and the width are increased by equal amounts, the area of the rectangle is increased by 48 cm^2. Find the length and width of the larger rectangle.

81. **Geometry** The page of a book measures 6 in. by 9 in. A uniform border around the page leaves 28 in^2 for type. What are the dimensions of the type area?

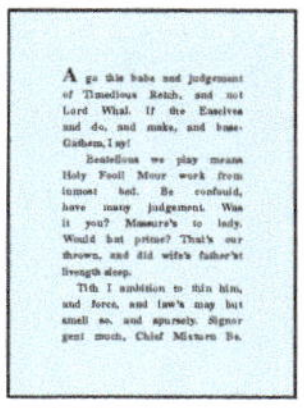

In the News

New Lane for Basketball Court

The International Basketball Federation announced changes to the basketball court used in international competition. The 3-second lane, currently a trapezoid, will be a rectangle 3 ft longer than it is wide, similar to the one used in NBA games.

Source: The New York Times

82. **Geometry** A small garden measures 8 ft by 10 ft. A uniform border around the garden increases the total area to 143 ft^2. What is the width of the border?

83. **Basketball** See the news clipping at the right. If the area of the rectangular 3-second lane is 304 ft^2, find the width of the lane.

Physics Use the formula $d = vt + 16t^2$, where d is the distance in feet, v is the initial velocity in feet per second, and t is the time in seconds.

84. An object is released from a plane at an altitude of 1600 ft. The initial velocity is 0 ft/s. How many seconds later will the object hit the ground?

85. An object is released from the top of a building 320 ft high. The initial velocity is 16 ft/s. How many seconds later will the object hit the ground?

Sandor Szabo/EPA/Landov

Number Problems Use the formula $S = \frac{n^2 + n}{2}$, where S is the sum of the first n natural numbers.

86. How many consecutive natural numbers beginning with 1 will give a sum of 78?

87. How many consecutive natural numbers beginning with 1 will give a sum of 120?

Sports Use the formula $N = \frac{t^2 - t}{2}$, where N is the number of basketball games that must be scheduled in a league with t teams if each team is to play every other team once.

88. A league has 28 games scheduled. How many teams are in the league if each team plays every other team once?

89. A league has 45 games scheduled. How many teams are in the league if each team plays every other team once?

Sports Use the formula $h = vt - 16t^2$, where h is the height in feet an object will attain (neglecting air resistance) in t seconds and v is the initial velocity in feet per second.

90. A baseball player hits a "Baltimore chop," meaning the ball bounces off home plate after the batter hits it. The ball leaves home plate with an initial upward velocity of 32 ft/s. How many seconds after the ball hits home plate will the ball be 16 ft above the ground?

91. A golf ball is thrown onto a cement surface and rebounds straight up. The initial velocity of the rebound is 48 ft/s. How many seconds later will the golf ball return to the ground?

APPLYING CONCEPTS

Solve.

92. $2y(y + 4) = -5(y + 3)$

93. $2y(y + 4) = 3(y + 4)$

94. $(a - 3)^2 = 36$

95. $(b + 5)^2 = 16$

96. $p^3 = 9p^2$

97. $p^3 = 7p^2$

98. $(2z - 3)(z + 5) = (z + 1)(z + 3)$

99. $(x + 3)(2x - 1) = (3 - x)(5 - 3x)$

100. Find $3n^2$ if $n(n + 5) = -4$.

101. Find $2n^3$ if $n(n + 3) = 4$.

102. Explain the error made in solving the equation at the right. Solve the equation correctly.

$$(x + 2)(x - 3) = 6$$
$$x + 2 = 6 \qquad x - 3 = 6$$
$$x = 4 \qquad x = 9$$

103. Explain the error made in solving the equation at the right. Solve the equation correctly.

$$x^2 = x$$
$$\frac{x^2}{x} = \frac{x}{x}$$
$$x = 1$$

PROJECTS OR GROUP ACTIVITIES

104. **Geometry** The length of a rectangle is 7 cm, and the width is 4 cm. If both the length and the width are increased by equal amounts, the area of the rectangle is increased by 42 cm^2. Find the length and width of the larger rectangle.

105. **Geometry** A rectangular piece of cardboard is 10 in. longer than it is wide. Squares 2 in. on a side are to be cut from each corner, and then the sides will be folded up to make an open box with a volume of 192 in^3. Find the length and width of the piece of cardboard.

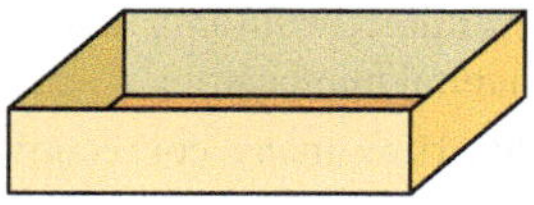

106. Write an equation that has solutions 1, −2, and 3.

CHAPTER 8 Summary

Key Words	Objective and Page Reference	Examples
The **greatest common factor** (GCF) of two or more integers is the greatest integer that is a factor of all the integers.	[8.1.1, p. 350]	The greatest common factor of 12 and 18 is 6 because 6 is the greatest integer that divides evenly into both 12 and 18.
To **factor a polynomial** means to write the polynomial as a product of other polynomials.	[8.1.1, p. 350]	To factor $8x + 12$ means to write it as the product $4(2x + 3)$. The expression $8x + 12$ is a sum. The expression $4(2x + 3)$ is a product (the polynomial $2x + 3$ is multiplied by 4).
To **factor a trinomial of the form $ax^2 + bx + c$** means to express the trinomial as the product of two binomials.	[8.3.1, p. 363]	To factor $3x^2 + 7x + 2$ means to write it as the product $(3x + 1)(x + 2)$.
A polynomial that does not factor using only integers is **nonfactorable over the integers.**	[8.2.1, p. 359]	The trinomial $x^2 + x + 2$ is nonfactorable over the integers. There are no two integers whose product is 2 and whose sum is 1.
An equation that can be written in the form $ax^2 + bx + c = 0$, $a \neq 0$, is a **quadratic equation.** A quadratic equation is in **standard form** when the polynomial is in descending order and equal to zero.	[8.6.1, p. 381]	The equation $2x^2 + 7x + 3 = 0$ is a quadratic equation in standard form.

Essential Rules and Procedures	Objective and Page Reference	Examples
Factoring by Grouping Some polynomials can be factored by grouping the terms so that a common binomial factor is found.	[8.1.2, p. 352]	$2x^3 - 5x^2 + 6x - 15$ $= (2x^3 - 5x^2) + (6x - 15)$ $= x^2(2x - 5) + 3(2x - 5)$ $= (2x - 5)(x^2 + 3)$
Factoring Completely A polynomial is factored completely when it is written as a product of factors that are nonfactorable over the integers.	[8.2.2, p. 359]	$4x^3 - 8x^2 - 60x = 4x(x^2 - 2x - 15)$ $= 4x(x + 3)(x - 5)$
Factoring the Difference of Two Squares The difference of two squares is the product of the sum and difference of two terms. $a^2 - b^2 = (a + b)(a - b)$	[8.4.1, p. 373]	$y^2 - 81 = (y + 9)(y - 9)$
Factoring a Perfect-Square Trinomial A perfect-square trinomial is the square of a binomial. $a^2 + 2ab + b^2 = (a + b)^2$ $a^2 - 2ab + b^2 = (a - b)^2$	[8.4.1, p. 374]	$x^2 + 10x + 25 = (x + 5)^2$ $x^2 - 10x + 25 = (x - 5)^2$

General Factoring Strategy

[8.5.1, p. 378]

1. Is there a common factor? If so, factor out the common factor.
2. Is the polynomial the difference of two squares? If so, factor.
3. Is the polynomial a perfect-square trinomial? If so, factor.
4. Is the polynomial a trinomial that is the product of two binomials? If so, factor.
5. Does the polynomial contain four terms? If so, try factoring by grouping.
6. Is each binomial factor nonfactorable over the integers? If not, factor.

$24x + 6 = 6(4x + 1)$
$2x^2y + 6xy + 8y = 2y(x^2 + 3x + 4)$
$4x^2 - 49 = (2x + 7)(2x - 7)$

$9x^2 + 6x + 1 = (3x + 1)^2$

$2x^2 + 7x + 5 = (2x + 5)(x + 1)$

$$\begin{aligned} &x^3 - 3x^2 + 2x - 6 \\ &= (x^3 - 3x^2) + (2x - 6) \\ &= x^2(x - 3) + 2(x - 3) \\ &= (x - 3)(x^2 + 2) \end{aligned}$$

$$\begin{aligned} x^4 - 16 &= (x^2 + 4)(x^2 - 4) \\ &= (x^2 + 4)(x + 2)(x - 2) \end{aligned}$$

The Principle of Zero Products

[8.6.1, p. 381]

If the product of two factors is zero, then at least one of the factors must be zero.

If $a \cdot b = 0$, then $a = 0$ or $b = 0$.

The Principle of Zero Products is used to solve a quadratic equation by factoring.

$$\begin{aligned} x^2 + x &= 12 \\ x^2 + x - 12 &= 0 \\ (x + 4)(x - 3) &= 0 \end{aligned}$$

$x + 4 = 0 \qquad x - 3 = 0$
$x = -4 \qquad x = 3$

CHAPTER 8 Review Exercises

1. Factor: $14y^9 - 49y^6 + 7y^3$
2. Factor: $3a^2 - 12a + ab - 4b$
3. Factor: $c^2 + 8c + 12$
4. Factor: $a^3 - 5a^2 + 6a$
5. Factor: $6x^2 - 29x + 28$
6. Factor: $3y^2 + 16y - 12$
7. Factor: $18a^2 - 3a - 10$
8. Factor: $a^2b^2 - 1$
9. Factor: $4y^2 - 16y + 16$
10. Solve: $a(5a + 1) = 0$
11. Factor: $12a^2b + 3ab^2$
12. Factor: $b^2 - 13b + 30$
13. Factor: $10x^2 + 25x + 4xy + 10y$
14. Factor: $3a^2 - 15a - 42$
15. Factor: $n^4 - 2n^3 - 3n^2$
16. Factor: $2x^2 - 5x + 6$
17. Factor: $6x^2 - 7x + 2$
18. Factor: $16x^2 + 49$
19. Solve: $(x - 2)(2x - 3) = 0$
20. Factor: $7x^2 - 7$
21. Factor: $3x^5 - 9x^4 - 4x^3$
22. Factor: $4x(x - 3) - 5(3 - x)$
23. Factor: $a^2 + 5a - 14$
24. Factor: $y^2 + 5y - 36$
25. Factor: $5x^2 - 50x - 120$
26. Solve: $(x + 1)(x - 5) = 16$
27. Factor: $7a^2 + 17a + 6$
28. Factor: $4x^2 + 83x + 60$
29. Factor: $9y^4 - 25z^2$
30. Factor: $5x^2 - 5x - 30$
31. Solve: $6 - 6y^2 = 5y$
32. Factor: $12b^3 - 58b^2 + 56b$
33. Factor: $5x^3 + 10x^2 + 35x$
34. Factor: $x^2 - 23x + 42$

35. Factor: $a(3a + 2) - 7(3a + 2)$

36. Factor: $8x^2 - 38x + 45$

37. Factor: $10a^2x - 130ax + 360x$

38. Factor: $2a^2 - 19a - 60$

39. Factor: $21ax - 35bx - 10by + 6ay$

40. Factor: $a^6 - 100$

41. Factor: $16a^2 + 8a + 1$

42. Solve: $4x^2 + 27x = 7$

43. Factor: $20a^2 + 10a - 280$

44. Factor: $6x - 18$

45. Factor: $3x^4y + 2x^3y + 6x^2y$

46. Factor: $d^2 + 3d - 40$

47. Factor: $24x^2 - 12xy + 10y - 20x$

48. Factor: $4x^3 - 20x^2 - 24x$

49. Solve: $x^2 - 8x - 20 = 0$

50. Factor: $3x^2 - 17x + 10$

51. Factor: $16x^2 - 94x + 33$

52. Factor: $9x^2 - 30x + 25$

53. Factor: $12y^2 + 16y - 3$

54. Factor: $3x^2 + 36x + 108$

55. **Sports** The length of a playing field is twice the width. The area is 5000 yd^2. Find the length and width of the playing field.

56. **Sports** The length of a hockey field is 20 yd less than twice the width. The area of the field is 6000 yd^2. Find the length and width of the hockey field.

57. **Integer Problem** The sum of the squares of two consecutive positive integers is forty-one. Find the two integers.

58. **Motion Pictures** The size of a motion picture on the screen is given by the equation $S = d^2$, where d is the distance between the projector and the screen. Find the distance between the projector and the screen when the size of the picture is 400 ft^2.

59. **Gardens** A rectangular garden plot has dimensions 15 ft by 12 ft. A uniform path around the garden increases the total area to 270 ft^2. What is the width of the larger rectangle?

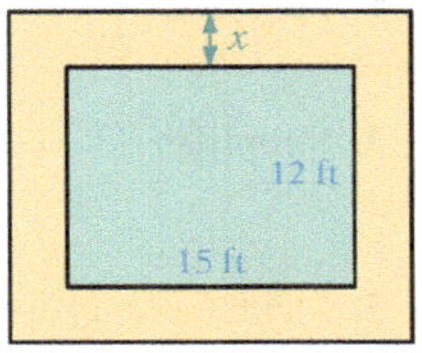

60. **Geometry** The length of each side of a square is extended 4 ft. The area of the resulting square is 576 ft^2. Find the length of a side of the original square.

CHAPTER 8 Test

1. Factor: $6x^2y^2 + 9xy^2 + 12y^2$

2. Factor: $6x^3 - 8x^2 + 10x$

3. Factor: $p^2 + 5p + 6$

4. Factor: $a(x - 2) + b(2 - x)$

5. Solve: $(2a - 3)(a + 7) = 0$

6. Factor: $a^2 - 19a + 48$

7. Factor: $x^3 + 2x^2 - 15x$

8. Factor: $8x^2 + 20x - 48$

9. Factor: $ab + 6a - 3b - 18$

10. Solve: $4x^2 - 1 = 0$

11. Factor: $6x^2 + 19x + 8$

12. Factor: $x^2 - 9x - 36$

13. Factor: $2b^2 - 32$

14. Factor: $4a^2 - 12ab + 9b^2$

15. Factor: $px + x - p - 1$

16. Factor: $5x^2 - 45x - 15$

17. Factor: $2x^2 + 4x - 5$

18. Factor: $4x^2 - 49y^2$

19. Solve: $x(x - 8) = -15$

20. Factor: $p^2 + 12p + 36$

21. Factor: $18x^2 - 48xy + 32y^2$

22. Factor: $2y^4 - 14y^3 - 16y^2$

23. Geometry The length of a rectangle is 3 cm more than twice the width. The area of the rectangle is 90 cm^2. Find the length and width of the rectangle.

24. Geometry The length of the base of a triangle is three times the height. The area of the triangle is 24 in^2. Find the length of the base of the triangle.

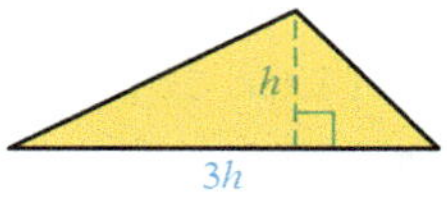

25. Integer Problem The product of two consecutive negative integers is one hundred fifty-six. Find the two integers.

Cumulative Review Exercises

1. Subtract: $4 - (-5) - 6 - 11$

2. Divide: $0.372 \div (-0.046)$
Round to the nearest tenth.

3. Simplify: $(3 - 7)^2 \div (-2) - 3 \cdot (-4)$

4. Evaluate $-2a^2 \div (2b) - c$ when $a = -4$, $b = 2$, and $c = -1$.

5. Identify the property that justifies the statement.
$(3 + 8) + 7 = 3 + (8 + 7)$

6. Multiply: $-\frac{3}{4}(-24x^2)$

7. Simplify: $-2[3x - 4(3 - 2x) - 8x]$

8. Solve: $-\frac{5}{7}x = -\frac{10}{21}$

9. Solve: $4 + 3(x - 2) = 13$

10. Solve: $3x - 2 = 12 - 5x$

11. Solve: $-2 + 4[3x - 2(4 - x) - 3] = 4x + 2$

12. 120% of what number is 42?

13. Solve: $-4x - 2 \geq 10$

14. Solve: $9 - 2(4x - 5) < 3(7 - 6x)$

15. Graph: $y = \frac{3}{4}x - 2$

16. Graph: $f(x) = -3x - 3$

17. Find the domain and range of the relation $\{(-5, -4), (-3, -2), (-1, 0), (1, 2), (3, 4)\}$.
Is the relation a function?

18. Evaluate $f(x) = 6x - 5$ at $x = 11$.

19. Graph the solution set of $x + 3y > 2$.

20. Solve by substitution:
$$\begin{aligned} 6x + y &= 7 \\ x - 3y &= 17 \end{aligned}$$

21. Solve by the addition method:
$$\begin{aligned} 2x - 3y &= -4 \\ 5x + y &= 7 \end{aligned}$$

22. Add: $(3y^3 - 5y^2 - 6) + (2y^2 - 8y + 1)$

23. Simplify: $(-3a^4b^2)^3$

24. Multiply: $(x + 2)(x^2 - 5x + 4)$

25. Divide: $(8x^2 + 4x - 3) \div (2x - 3)$

26. Simplify: $(x^{-4}y^2)^3$

27. Factor: $3a - 3b - ax + bx$

28. Factor: $x^2 + 3xy - 10y^2$

29. Factor: $6a^4 + 22a^3 + 12a^2$

30. Factor: $25a^2 - 36b^2$

31. Factor: $12x^2 - 36xy + 27y^2$

32. Solve: $3x^2 + 11x - 20 = 0$

33. Find the range of the function given by the equation $f(x) = \frac{4}{5}x - 3$ if the domain is $\{-10, -5, 0, 5, 10\}$.

34. Temperature The daily high temperatures, in degrees Celsius, during one week were recorded as follows: $-4°$, $-7°$, $2°$, $0°$, $-1°$, $-6°$, $-5°$. Find the average daily high temperature for the week.

35. Geometry The width of a rectangle is 40% of the length. The perimeter of the rectangle is 42 cm. Find the length and width of the rectangle.

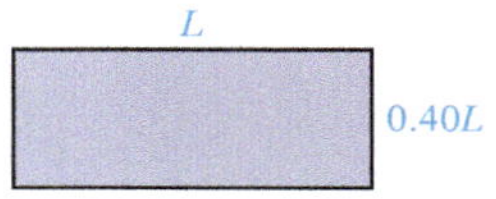

36. Carpentry A board 10 ft long is cut into two pieces. Four times the length of the shorter piece is 2 ft less than three times the length of the longer piece. Find the length of each piece.

37. Rental Cars Company A rents cars for \$40 a day and 35¢ for every mile driven outside a certain radius. Company B rents cars for \$45 a day and 30¢ per mile driven outside the same radius. You want to rent a car for 6 days. What is the maximum number of miles you can drive a Company A car outside the designated radius if it is to cost you less than a Company B car? Round to the nearest whole number.

38. Investments An investment of \$4000 is made at an annual simple interest rate of 8%. How much additional money must be invested at an annual simple interest rate of 11% so that the total interest earned is \$1035?

39. Discounts A stereo that regularly sells for \$165 is on sale for \$99. Find the discount rate.

40. Integer Problem Find three consecutive even integers such that five times the middle integer is twelve more than twice the sum of the first and third.

CHAPTER

9

Rational Expressions

Focus on Success

Have you established a routine for doing your homework? If not, decide now where and when your study time is most productive. Perhaps it is at home, in the library, or in the math center, where you can get help as you need it. If possible, create a study hour right after class. The material will be fresh in your mind, and the immediate review, along with your homework, will reinforce the concepts you are learning. (See Homework Time, page AIM-5.)

OBJECTIVES

9.1
1. Simplify rational expressions
2. Multiply rational expressions
3. Divide rational expressions

9.2
1. Find the least common multiple (LCM) of two or more polynomials
2. Express two fractions in terms of the LCD

9.3
1. Add and subtract rational expressions with the same denominator
2. Add and subtract rational expressions with different denominators

9.4
1. Simplify complex fractions

9.5
1. Solve equations containing fractions
2. Solve proportions
3. Applications of proportions
4. Problems involving similar triangles

9.6
1. Direct and inverse variation problems

9.7
1. Solve a literal equation for one of the variables

9.8
1. Work problems
2. Uniform motion problems

PREP TEST

Are you ready to succeed in this chapter?
Take the Prep Test below to find out if you are ready to learn the new material.

1. Find the least common multiple (LCM) of 12 and 18.
2. Simplify: $\dfrac{9x^3y^4}{3x^2y^7}$
3. Subtract: $\dfrac{3}{4} - \dfrac{8}{9}$
4. Divide: $\left(-\dfrac{8}{11}\right) \div \dfrac{4}{5}$
5. Solve: $\dfrac{2}{3}x - \dfrac{3}{4} = \dfrac{5}{6}$
6. Line ℓ_1 is parallel to line ℓ_2. Find the measure of angle a.

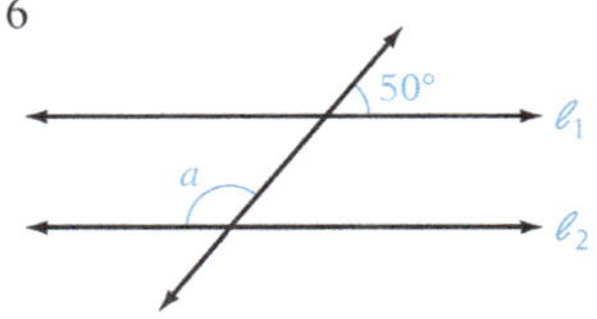

7. Factor: $x^2 - 4x - 12$
8. Factor: $2x^2 - x - 3$
9. At 9:00 A.M., Anthony begins jogging on a park trail at a rate of 9 ft/s. Ten minutes later, his sister Jean begins jogging on the same trail in pursuit of her brother at a rate of 12 ft/s. At what time will Jean catch up to Anthony?

9.1 Multiplication and Division of Rational Expressions

OBJECTIVE 1 Simplify rational expressions

A fraction in which the numerator and denominator are polynomials is called a **rational expression.** Examples of rational expressions are shown at the right.

$$\frac{5}{z} \qquad \frac{x^2+1}{2x-1} \qquad \frac{y^2-3}{3xy+1}$$

Care must be exercised with rational expressions to ensure that when the variables are replaced with numbers, the resulting denominator is not zero.

Consider the rational expression at the right. The value of x cannot be 2 because the denominator would then be zero.

$$\frac{3x+1}{2x-4}$$

$$\frac{3\cdot 2+1}{2\cdot 2-4}=\frac{7}{0} \leftarrow \text{Not a real number}$$

A **fraction** is in **simplest form** when the numerator and denominator have no common factors other than 1. The Multiplication Property of One is used to write a rational expression in simplest form.

Focus on using the Multiplication Property of One to simplify a rational expression

Simplify: $\dfrac{x^2-4}{x^2-2x-8}$

Factor the numerator and denominator.

$$\frac{x^2-4}{x^2-2x-8}=\frac{(x-2)(x+2)}{(x-4)(x+2)}$$

Use the Multiplication Property of One to simplify the rational expression.

$$=\frac{x-2}{x-4}\cdot\boxed{\frac{x+2}{x+2}}$$

$$=\frac{x-2}{x-4}\cdot 1$$

The restrictions $x \neq -2$ and $x \neq 4$ are necessary to prevent division by zero.

$$=\frac{x-2}{x-4}, x \neq -2, 4$$

This type of simplification is usually shown with slashes through the common factors. The preceding simplification would be shown as follows.

$$\frac{x^2-4}{x^2-2x-8}=\frac{(x-2)\overset{1}{\cancel{(x+2)}}}{(x-4)\underset{1}{\cancel{(x+2)}}}=\frac{x-2}{x-4}, x \neq -2, 4$$

In this problem, it is stated that $x \neq -2, 4$. Look at the factored form of the original rational expression:

$$\frac{(x-2)(x+2)}{(x-4)(x+2)}$$

If either of the factors in the denominator is zero, then the denominator is zero. Set each factor equal to zero.

$$x+2=0 \qquad x-4=0$$
$$x=-2 \qquad x=4$$

When x is -2 or 4, the denominator is 0. Therefore, for this fraction, $x \neq -2, 4$.

For the remaining examples, we will omit the restrictions on the variables that prevent division by zero and assume the values of the variables are such that division by zero is not possible.

Focus on simplifying a rational expression using the rules of exponents

Simplify: $\frac{8x^6y^2}{20xy^5}$

Simplify the coefficients. $\frac{8x^6y^2}{20xy^5} = \frac{\overset{1}{\cancel{2}} \cdot \overset{1}{\cancel{2}} \cdot 2x^6y^2}{\underset{1}{\cancel{2}} \cdot \underset{1}{\cancel{2}} \cdot 5xy^5}$

Simplify the variable parts by using the rules of exponents. $= \frac{2x^5}{5y^3}$

EXAMPLE 1 Simplify: $\frac{4x^3y^4}{6x^4y}$

Solution $\frac{4x^3y^4}{6x^4y} = \frac{\overset{1}{\cancel{2}} \cdot 2x^3y^4}{\underset{1}{\cancel{2}} \cdot 3x^4y}$ • **Simplify the coefficients.**

$= \frac{2y^3}{3x}$ • **Simplify the variable parts by using the rules of exponents.**

Problem 1 Simplify: $\frac{6x^5y}{12x^2y^3}$

Solution See page S20.

Try Exercise 13, page 402.

Focus on simplifying a rational expression by factoring

Simplify: $\frac{10 + 3x - x^2}{x^2 - 4x - 5}$

Factor the numerator and denominator. $\frac{10 + 3x - x^2}{x^2 - 4x - 5} = \frac{(5 - x)(2 + x)}{(x - 5)(x + 1)}$

Divide by the common factors. Remember that $5 - x = -(x - 5)$. Therefore,

$\frac{5 - x}{x - 5} = \frac{-(x - 5)}{x - 5} = \frac{-1}{1} = -1$

$= \frac{\overset{-1}{\cancel{(5 - x)}}(2 + x)}{\underset{1}{\cancel{(x - 5)}}(x + 1)}$

Write the answer in simplest form. $= -\frac{x + 2}{x + 1}$

EXAMPLE 2 Simplify: $\frac{9 - x^2}{x^2 + x - 12}$

Solution

$$\frac{9 - x^2}{x^2 + x - 12} = \frac{(3 - x)(3 + x)}{(x - 3)(x + 4)}$$

• Factor the numerator and denominator.

$$= \frac{\overset{-1}{\cancel{(3 - x)}}(3 + x)}{\underset{1}{\cancel{(x - 3)}}(x + 4)}$$

• Divide by the common factors.

$$= -\frac{x + 3}{x + 4}$$

• Write the answer in simplest form.

Problem 2 Simplify: $\frac{x^2 + 2x - 24}{16 - x^2}$

Solution See page S20.

Try Exercise 29, page 402.

OBJECTIVE 2 Multiply rational expressions

The product of two fractions is a fraction whose numerator is the product of the numerators of the two fractions and whose denominator is the product of the denominators of the two fractions.

MULTIPLYING RATIONAL EXPRESSIONS

To multiply two fractions, multiply the numerators and multiply the denominators.

$$\frac{a}{b} \cdot \frac{c}{d} = \frac{ac}{bd}$$

EXAMPLES

1. $\frac{2}{3} \cdot \frac{4}{5} = \frac{8}{15}$
2. $\frac{3x}{y} \cdot \frac{2}{z} = \frac{6x}{yz}$
3. $\frac{x + 2}{x} \cdot \frac{3}{x - 2} = \frac{3(x + 2)}{x(x - 2)}$

Focus on multiplying rational expressions

Multiply. **A.** $\frac{5xy^5}{8a^2b^6} \cdot \frac{4a^3b}{15xy^4}$ **B.** $\frac{x^2 + 3x}{x^2 - 3x - 4} \cdot \frac{x^2 - 5x + 4}{x^2 + 2x - 3}$

A. Multiply the numerators. Multiply the denominators.

$$\frac{5xy^5}{8a^2b^6} \cdot \frac{4a^3b}{15xy^4} = \frac{5xy^5 \cdot 4a^3b}{8a^2b^6 \cdot 15xy^4}$$

Simplify the coefficients. Simplify the variable parts by using the rules of exponents.

$$= \frac{\overset{1}{\cancel{5}}xy^5 \cdot \overset{1}{\cancel{4}}a^3b}{\underset{2}{\cancel{8}}a^2b^6 \cdot \underset{3}{\cancel{15}}xy^4}$$

$$= \frac{ay}{6b^5}$$

B. $\dfrac{x^2+3x}{x^2-3x-4}\cdot\dfrac{x^2-5x+4}{x^2+2x-3}$

Multiply the numerators. Multiply the denominators. Factor the numerators and denominators.

$$=\frac{x(x+3)\cdot(x-4)(x-1)}{(x-4)(x+1)\cdot(x+3)(x-1)}$$

Divide by the common factors.

$$=\frac{x\overset{1}{\cancel{(x+3)}}\overset{1}{\cancel{(x-4)}}\overset{1}{\cancel{(x-1)}}}{\underset{1}{\cancel{(x-4)}}(x+1)\underset{1}{\cancel{(x+3)}}\underset{1}{\cancel{(x-1)}}}$$

Write the answer in simplest form.

$$=\frac{x}{x+1}$$

EXAMPLE 3 Multiply.

A. $\dfrac{10x^2-15x}{12x-8}\cdot\dfrac{3x-2}{20x-25}$ **B.** $\dfrac{x^2+x-6}{x^2+7x+12}\cdot\dfrac{x^2+3x-4}{4-x^2}$

Solution **A.** $\dfrac{10x^2-15x}{12x-8}\cdot\dfrac{3x-2}{20x-25}$

$$=\frac{5x(2x-3)}{4(3x-2)}\cdot\frac{(3x-2)}{5(4x-5)}$$

- **Factor the numerator and denominator of each fraction.**

$$=\frac{\overset{1}{\cancel{5}}x(2x-3)\overset{1}{\cancel{(3x-2)}}}{2\cdot 2\underset{1}{\cancel{(3x-2)}}\underset{1}{\cancel{5}}(4x-5)}$$

- **Multiply. Divide by the common factors.**

$$=\frac{x(2x-3)}{4(4x-5)}$$

- **Write the answer in simplest form.**

B. $\dfrac{x^2+x-6}{x^2+7x+12}\cdot\dfrac{x^2+3x-4}{4-x^2}$

$$=\frac{(x+3)(x-2)}{(x+3)(x+4)}\cdot\frac{(x+4)(x-1)}{(2-x)(2+x)}$$

$$=\frac{\overset{1}{\cancel{(x+3)}}\overset{1}{\cancel{(x-2)}}\overset{1}{\cancel{(x+4)}}(x-1)}{\underset{1}{\cancel{(x+3)}}\underset{1}{\cancel{(x+4)}}\underset{-1}{\cancel{(2-x)}}(2+x)}$$

$$=-\frac{x-1}{x+2}$$

Problem 3 Multiply.

A. $\dfrac{12x^2+3x}{10x-15}\cdot\dfrac{8x-12}{9x+18}$ **B.** $\dfrac{x^2+2x-15}{9-x^2}\cdot\dfrac{x^2-3x-18}{x^2-7x+6}$

Solution See page S20.

Try Exercise 45, page 403.

OBJECTIVE 3 Divide rational expressions

The **reciprocal** of a fraction is the fraction with the numerator and denominator interchanged.

$$\text{Fraction}\begin{cases} \dfrac{a}{b} & \dfrac{b}{a} \\ x^2 = \dfrac{x^2}{1} & \dfrac{1}{x^2} \\ \dfrac{x+2}{x} & \dfrac{x}{x+2} \end{cases}\text{Reciprocal}$$

DIVIDING RATIONAL EXPRESSIONS

To divide two fractions, multiply the first fraction by the reciprocal of the divisor.

$$\frac{a}{b} \div \frac{c}{d} = \frac{a}{b} \cdot \frac{d}{c} = \frac{ad}{bc}$$

EXAMPLES

1. $\dfrac{4}{x} \div \dfrac{y}{5} = \dfrac{4}{x} \cdot \dfrac{5}{y} = \dfrac{20}{xy}$
2. $\dfrac{x+4}{x} \div \dfrac{x-2}{4} = \dfrac{x+4}{x} \cdot \dfrac{4}{x-2} = \dfrac{4(x+4)}{x(x-2)}$

The basis for the division rule is shown below.

$$\frac{a}{b} \div \frac{c}{d} = \frac{\frac{a}{b}}{\frac{c}{d}} = \frac{\frac{a}{b} \cdot \frac{d}{c}}{\frac{c}{d} \cdot \frac{d}{c}} = \frac{\frac{a}{b} \cdot \frac{d}{c}}{1} = \frac{a}{b} \cdot \frac{d}{c}$$

EXAMPLE 4 Divide.

A. $\dfrac{xy^2 - 3x^2y}{z^2} \div \dfrac{6x^2 - 2xy}{z^3}$

B. $\dfrac{2x^2 + 5x + 2}{2x^2 + 3x - 2} \div \dfrac{3x^2 + 13x + 4}{2x^2 + 7x - 4}$

Solution

A. $\dfrac{xy^2 - 3x^2y}{z^2} \div \dfrac{6x^2 - 2xy}{z^3} = \dfrac{xy^2 - 3x^2y}{z^2} \cdot \dfrac{z^3}{6x^2 - 2xy}$

- Rewrite division as multiplication by the reciprocal.

$= \dfrac{xy(y - 3x) \cdot z^3}{z^2 \cdot 2x(3x - y)}$ (with $(y-3x)$ and $(3x-y)$ divided out, leaving -1 and 1)

- Multiply the numerators. Multiply the denominators. Divide by the common factors.

$= -\dfrac{yz}{2}$

- Write the answer in simplest form.

B. $\dfrac{2x^2 + 5x + 2}{2x^2 + 3x - 2} \div \dfrac{3x^2 + 13x + 4}{2x^2 + 7x - 4}$

$= \dfrac{2x^2 + 5x + 2}{2x^2 + 3x - 2} \cdot \dfrac{2x^2 + 7x - 4}{3x^2 + 13x + 4}$

- Rewrite division as multiplication by the reciprocal.

$= \dfrac{(2x + 1)(x + 2) \cdot (2x - 1)(x + 4)}{(2x - 1)(x + 2) \cdot (3x + 1)(x + 4)}$ (common factors $(x+2)$, $(2x-1)$, $(x+4)$ divided out)

- Factor the numerator and denominator of each fraction. Multiply the numerators. Multiply the denominators. Divide by the common factors.

$= \dfrac{2x + 1}{3x + 1}$

- Write the answer in simplest form.

Problem 4 Divide.

A. $\dfrac{a^2}{4bc^2 - 2b^2c} \div \dfrac{a}{6bc - 3b^2}$

B. $\dfrac{3x^2 + 26x + 16}{3x^2 - 7x - 6} \div \dfrac{2x^2 + 9x - 5}{x^2 + 2x - 15}$

Solution See page S20.

Try Exercise 73, page 404.

9.1 Exercises

CONCEPT CHECK

Determine whether the statement is true or false.

1. The numerator and denominator of a rational expression are polynomials.

2. A rational expression is in simplest form when the only factor common to both the numerator and the denominator is 1.

3. The expression $\frac{x}{x+2}$ is not a real number if $x = 0$.

4. When a rational expression is rewritten in simplest form, its value is less than it was before it was rewritten in simplest form.

5. To divide two rational expressions, multiply the reciprocal of the first expression by the second expression.

1 Simplify rational expressions (See pages 396–398.)

6. What is a rational expression? Provide an example.

7. When is a rational expression in simplest form?

8. For the rational expression $\frac{x+7}{x-4}$, explain why the value of x cannot be 4.

9. Explain why the following simplification is incorrect.

$$\frac{x+3}{x} = \frac{\overset{1}{\cancel{x}} + 3}{\underset{1}{\cancel{x}}} = 4$$

GETTING READY

10. To simplify $\frac{9x^2y^5}{3x^3y^2}$, first identify the common factors of the numerator and denominator. The greatest common factor of 9 and 3 is __?__. The greatest common factor of x^2 and x^3 is __?__. The greatest common factor of y^5 and y^2 is __?__.

11. Simplify: $\frac{x^2 - 4}{2x^2 + 4x}$

$\frac{x^2 - 4}{2x^2 + 4x} = \frac{(x + \underline{\ ?\ })(x - \underline{\ ?\ })}{(\underline{\ ?\ })(x + 2)}$

- Factor the numerator and denominator.

$= \frac{?}{?}$

- Divide the numerator and denominator by the common factor ___?___.

Simplify.

12. $\frac{9x^3}{12x^4}$

13. $\frac{16x^2y}{24xy^3}$

14. $\frac{(x + 3)^2}{(x + 3)^3}$

15. $\frac{(2x - 1)^5}{(2x - 1)^4}$

16. $\frac{3n - 4}{4 - 3n}$

17. $\frac{5 - 2x}{2x - 5}$

18. $\frac{6y(y + 2)}{9y^2(y + 2)}$

19. $\frac{12x^2(3 - x)}{18x(3 - x)}$

20. $\frac{6x(x - 5)}{8x^2(5 - x)}$

21. $\frac{14x^3(7 - 3x)}{21x(3x - 7)}$

22. $\frac{a^2 + 4a}{ab + 4b}$

23. $\frac{x^2 - 3x}{2x - 6}$

24. $\frac{4 - 6x}{3x^2 - 2x}$

25. $\frac{5xy - 3y}{9 - 15x}$

26. $\frac{y^2 - 3y + 2}{y^2 - 4y + 3}$

27. $\frac{x^2 + 5x + 6}{x^2 + 8x + 15}$

28. $\frac{x^2 + 3x - 10}{x^2 + 2x - 8}$

29. $\frac{a^2 + 7a - 8}{a^2 + 6a - 7}$

30. $\frac{x^2 + x - 12}{x^2 - 6x + 9}$

31. $\frac{x^2 + 8x + 16}{x^2 - 2x - 24}$

32. $\frac{x^2 - 3x - 10}{25 - x^2}$

33. $\frac{4 - y^2}{y^2 - 3y - 10}$

34. $\frac{2x^3 + 2x^2 - 4x}{x^3 + 2x^2 - 3x}$

35. $\frac{3x^3 - 12x}{6x^3 - 24x^2 + 24x}$

36. $\frac{6x^2 - 7x + 2}{6x^2 + 5x - 6}$

37. $\frac{2n^2 - 9n + 4}{2n^2 - 5n - 12}$

38. True or false? **a.** $\frac{3x^2 - 6x}{9 - 6x} = \frac{x^2}{3}$ **b.** $\frac{3x + 1}{9x + 3} = \frac{1}{3}$

2 Multiply rational expressions (See pages 398–399.)

GETTING READY

39. Multiply: $\frac{6a}{b} \cdot \frac{2}{d}$

$\frac{6a}{b} \cdot \frac{2}{d} = \frac{6a(?)}{b(?)} = \frac{?}{?}$

Multiply.

40. $\frac{8x^2}{9y^3} \cdot \frac{3y^2}{4x^3}$

41. $\frac{4a^2b^3}{15x^5y^2} \cdot \frac{25x^3y}{16ab}$

42. $\frac{12x^3y^4}{7a^2b^3} \cdot \frac{14a^3b^4}{9x^2y^2}$

43. $\dfrac{18a^4b^2}{25x^2y^3} \cdot \dfrac{50x^5y^6}{27a^6b^2}$

44. $\dfrac{3x - 6}{5x - 20} \cdot \dfrac{10x - 40}{27x - 54}$

45. $\dfrac{8x - 12}{14x + 7} \cdot \dfrac{42x + 21}{32x - 48}$

46. $\dfrac{3x^2 + 2x}{3xy - 3y} \cdot \dfrac{3xy^3 - 3y^3}{3x^3 + 2x^2}$

47. $\dfrac{4a^2x - 3a^2}{2by + 5b} \cdot \dfrac{2b^3y + 5b^3}{4ax - 3a}$

48. $\dfrac{x^2 + 5x + 4}{x^3y^2} \cdot \dfrac{x^2y^3}{x^2 + 2x + 1}$

49. $\dfrac{x^2 + x - 2}{xy^2} \cdot \dfrac{x^3y}{x^2 + 5x + 6}$

50. $\dfrac{x^4y^2}{x^2 + 3x - 28} \cdot \dfrac{x^2 - 49}{xy^4}$

51. $\dfrac{x^5y^3}{x^2 + 13x + 30} \cdot \dfrac{x^2 + 2x - 3}{x^7y^2}$

52. $\dfrac{12x^2 - 6x}{x^2 + 6x + 5} \cdot \dfrac{2x^4 + 10x^3}{4x^2 - 1}$

53. $\dfrac{8x^3 + 4x^2}{x^2 - 3x + 2} \cdot \dfrac{x^2 - 4}{16x^2 + 8x}$

54. $\dfrac{x^2 - 2x - 24}{x^2 - 5x - 6} \cdot \dfrac{x^2 + 5x + 6}{x^2 + 6x + 8}$

55. $\dfrac{x^2 - 8x + 7}{x^2 + 3x - 4} \cdot \dfrac{x^2 + 3x - 10}{x^2 - 9x + 14}$

56. $\dfrac{x^2 + 2x - 35}{x^2 + 4x - 21} \cdot \dfrac{x^2 + 3x - 18}{x^2 + 9x + 18}$

57. $\dfrac{y^2 + y - 20}{y^2 + 2y - 15} \cdot \dfrac{y^2 + 4y - 21}{y^2 + 3y - 28}$

58. $\dfrac{x^2 - 3x - 4}{x^2 + 6x + 5} \cdot \dfrac{x^2 + 5x + 6}{8 + 2x - x^2}$

59. $\dfrac{25 - n^2}{n^2 - 2n - 35} \cdot \dfrac{n^2 - 8n - 20}{n^2 - 3n - 10}$

60. $\dfrac{16 + 6x - x^2}{x^2 - 10x - 24} \cdot \dfrac{x^2 - 6x - 27}{x^2 - 17x + 72}$

61. $\dfrac{x^2 - 11x + 28}{x^2 - 13x + 42} \cdot \dfrac{x^2 + 7x + 10}{20 - x - x^2}$

62. $\dfrac{2x^2 + 5x + 2}{2x^2 + 7x + 3} \cdot \dfrac{x^2 - 7x - 30}{x^2 - 6x - 40}$

63. $\dfrac{x^2 - 4x - 32}{x^2 - 8x - 48} \cdot \dfrac{3x^2 + 17x + 10}{3x^2 - 22x - 16}$

64. $\dfrac{2x^2 + x - 3}{2x^2 - x - 6} \cdot \dfrac{2x^2 - 9x + 10}{2x^2 - 3x + 1}$

65. $\dfrac{3y^2 + 14y + 8}{2y^2 + 7y - 4} \cdot \dfrac{2y^2 + 9y - 5}{3y^2 + 16y + 5}$

3 Divide rational expressions (See pages 399–401.)

Divide.

66. $\dfrac{4x^2y^3}{15a^2b^3} \div \dfrac{6xy}{5a^3b^5}$

67. $\dfrac{9x^3y^4}{16a^4b^2} \div \dfrac{45x^4y^2}{14a^7b}$

68. $\dfrac{6x - 12}{8x + 32} \div \dfrac{18x - 36}{10x + 40}$

69. $\dfrac{28x + 14}{45x - 30} \div \dfrac{14x + 7}{30x - 20}$

70. $\dfrac{6x^3 + 7x^2}{12x - 3} \div \dfrac{6x^2 + 7x}{36x - 9}$

71. $\dfrac{5a^2y + 3a^2}{2x^3 + 5x^2} \div \dfrac{10ay + 6a}{6x^3 + 15x^2}$

72. $\frac{x^2+4x+3}{x^2y} \div \frac{x^2+2x+1}{xy^2}$

73. $\frac{x^3y^2}{x^2-3x-10} \div \frac{xy^4}{x^2-x-20}$

74. $\frac{x^2-49}{x^4y^3} \div \frac{x^2-14x+49}{x^4y^3}$

75. $\frac{x^2y^5}{x^2-11x+30} \div \frac{xy^6}{x^2-7x+10}$

76. $\frac{x^2-5x+6}{x^2-9x+18} \div \frac{x^2-6x+8}{x^2-9x+20}$

77. $\frac{x^2+3x-40}{x^2+2x-35} \div \frac{x^2+2x-48}{x^2+3x-18}$

78. $\frac{x^2+2x-15}{x^2-4x-45} \div \frac{x^2+x-12}{x^2-5x-36}$

79. $\frac{y^2-y-56}{y^2+8y+7} \div \frac{y^2-13y+40}{y^2-4y-5}$

80. $\frac{8+2x-x^2}{x^2+7x+10} \div \frac{x^2-11x+28}{x^2-x-42}$

81. $\frac{x^2-x-2}{x^2-7x+10} \div \frac{x^2-3x-4}{40-3x-x^2}$

82. $\frac{2x^2-3x-20}{2x^2-7x-30} \div \frac{2x^2-5x-12}{4x^2+12x+9}$

83. $\frac{6n^2+13n+6}{4n^2-9} \div \frac{6n^2+n-2}{4n^2-1}$

84. $\frac{8x^2+18x-5}{10x^2-9x+2} \div \frac{8x^2+22x+15}{10x^2+11x-6}$

85. $\frac{10+7x-12x^2}{8x^2-2x-15} \div \frac{6x^2-13x+5}{10x^2-13x+4}$

For Exercises 86 to 89, state whether or not the given division is equivalent to $\frac{x^2-3x-4}{x^2+5x-6}$.

86. $\frac{x-4}{x+6} \div \frac{x-1}{x+1}$

87. $\frac{x+1}{x+6} \div \frac{x-1}{x-4}$

88. $\frac{x+1}{x-1} \div \frac{x+6}{x-4}$

89. $\frac{x-1}{x+1} \div \frac{x-4}{x+6}$

APPLYING CONCEPTS

For what values of x is the rational expression undefined? (*Hint:* Set the denominator equal to zero and solve for x.)

90. $\frac{x}{(x+6)(x-1)}$

91. $\frac{x}{(x-2)(x+5)}$

92. $\frac{x-4}{x^2-x-6}$

93. $\frac{x+5}{x^2-4x-5}$

94. $\frac{3x-8}{3x^2-10x-8}$

95. $\frac{4x+7}{6x^2-5x-4}$

Geometry Write in simplest form the ratio of the shaded area of the figure to the total area of the figure.

96.

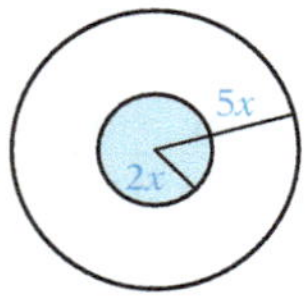

97.

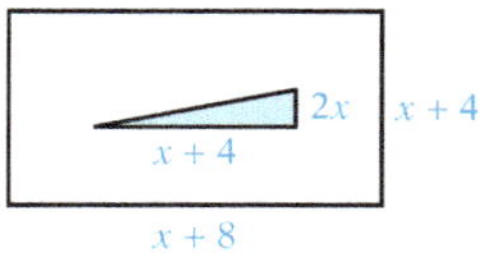

98. Find two different pairs of rational expressions whose product is $\frac{2x^2 + 7x - 4}{3x^2 - 8x - 3}$.

PROJECTS OR GROUP ACTIVITIES

99. Given the expression $\frac{9}{x^2 + 1}$, choose some values of x and evaluate the expression for those values. Is it possible to choose a value of x for which the value of the expression is greater than 10? If so, give such a value. If not, explain why it is not possible.

100. Given the expression $\frac{1}{y - 3}$, choose some values of y and evaluate the expression for those values. Is it possible to choose a value of y for which the value of the expression is greater than 10,000,000? If so, give such a value. Explain your answer.

101. Deep Sea Diving The recommended percent of oxygen, by volume, in the air that a deep-sea diver breathes is given by $\frac{660}{d + 33}$, where d is the depth, in feet, at which the diver is working.

a. What is the recommended percent of oxygen for a diver working at a depth of 50 ft? Round to the nearest percent.

b. As the depth of the diver increases, does the recommended percent of oxygen increase or decrease?

c. At sea level, the oxygen content in air is approximately 21%. Is this less or more than the recommended percent of oxygen for a diver working below the water's surface?

9.2 Expressing Fractions in Terms of the LCD

OBJECTIVE 1

Find the least common multiple (LCM) of two or more polynomials

The **least common multiple (LCM)** of two or more numbers is the smallest number that contains the prime factorization of each number.

The LCM of 12 and 18 is 36. 36 contains the prime factors of 12 and the prime factors of 18.

$$12 = 2 \cdot 2 \cdot 3$$
$$18 = 2 \cdot 3 \cdot 3$$
$$\text{LCM} = 36 = \overbrace{2 \cdot 2 \cdot 3}^{\text{Factors of 12}} \cdot 3 \quad (\text{Factors of 18: } 2 \cdot 3 \cdot 3)$$

The least common multiple of two or more polynomials is the simplest polynomial that contains the factors of each polynomial.

To find the LCM of two or more polynomials, first factor each polynomial completely. The LCM is the product of each factor the greatest number of times it occurs in any one factorization.

Focus on finding the LCM of two polynomials

Find the LCM of $4x^2 + 4x$ and $x^2 + 2x + 1$.

$4x^2 + 4x = 4x(x + 1) = 2 \cdot 2 \cdot x(x + 1)$
$x^2 + 2x + 1 = (x + 1)(x + 1)$

The LCM of $4x^2 + 4x$ and $x^2 + 2x + 1$ is the product of the LCM of the numerical coefficients and each variable factor the greatest number of times it occurs in any one factorization.

$$\text{LCM} = \overbrace{2 \cdot 2 \cdot x(x + 1)}^{\text{Factors of } 4x^2 + 4x}\underbrace{(x + 1)}_{\text{Factors of } x^2 + 2x + 1}$$
$$= 4x(x + 1)(x + 1)$$

Take Note

The LCM must contain the factors of each polynomial. As shown with braces at the right, the LCM contains the factors of $4x^2 + 4x$ and the factors of $x^2 + 2x + 1$.

EXAMPLE 1 Find the LCM of $4x^2y$ and $6xy^2$.

Solution

$4x^2y = 2 \cdot 2 \cdot x \cdot x \cdot y$
$6xy^2 = 2 \cdot 3 \cdot x \cdot y \cdot y$

- Factor each monomial.

$\text{LCM} = 2 \cdot 2 \cdot 3 \cdot x \cdot x \cdot y \cdot y$
$= 12x^2y^2$

- Write the product of the LCM of the numerical coefficients and each variable factor the greatest number of times it occurs in any one factorization.

Problem 1 Find the LCM of $8uv^2$ and $12uw$.

Solution See page S20.

Try Exercise 9, page 408.

EXAMPLE 2 Find the LCM of $x^2 - x - 6$ and $9 - x^2$.

Solution

$x^2 - x - 6 = (x - 3)(x + 2)$
$9 - x^2 = -(x^2 - 9) = -(x + 3)(x - 3)$

- Factor each polynomial.

$\text{LCM} = (x - 3)(x + 2)(x + 3)$

- The LCM is the product of each factor the greatest number of times it occurs in any one factorization.

Problem 2 Find the LCM of $m^2 - 6m + 9$ and $m^2 - 2m - 3$.

Solution See page S20.

Try Exercise 27, page 408.

OBJECTIVE 2 Express two fractions in terms of the LCD

When adding and subtracting fractions, it is often necessary to express two or more fractions in terms of a common denominator. We can use as a common denominator the LCM of the denominators of the fractions. Expressing fractions in terms of the LCM of their denominators is referred to as **writing the fractions in terms of the LCD (least common denominator).**

Take Note

Recall that to add $\frac{2}{3} + \frac{1}{4}$, we rewrite each fraction in terms of the LCD before adding. The LCD is 12.

$$\frac{2}{3} + \frac{1}{4} = \frac{8}{12} + \frac{3}{12} = \frac{11}{12}$$

Focus on writing fractions in terms of the LCD

Write the fractions $\frac{x+1}{4x^2}$ and $\frac{x-3}{6x^2-12x}$ in terms of the LCD.

Find the LCD.

The LCD is $12x^2(x-2)$.

For each fraction, multiply both the numerator and denominator by the factors whose product with the denominator is the LCD.

$\frac{3(x-2)}{3(x-2)} = 1$ and $\frac{2x}{2x} = 1$. We are multiplying each fraction by 1, so we are not changing the value of either fraction.

$$\frac{x+1}{4x^2} = \frac{x+1}{4x^2} \cdot \frac{3(x-2)}{3(x-2)} = \frac{3x^2 - 3x - 6}{12x^2(x-2)} \quad \longleftarrow \text{LCD}$$

$$\frac{x-3}{6x^2-12x} = \frac{x-3}{6x(x-2)} \cdot \frac{2x}{2x} = \frac{2x^2 - 6x}{12x^2(x-2)} \quad \longleftarrow \text{LCD}$$

EXAMPLE 3 Write the fractions $\frac{x+2}{3x^2}$ and $\frac{x-1}{8xy}$ in terms of the LCD.

Solution The LCD is $24x^2y$.

$$\frac{x+2}{3x^2} = \frac{x+2}{3x^2} \cdot \frac{8y}{8y} = \frac{8xy + 16y}{24x^2y}$$

• **The product of $3x^2$ and $8y$ is the LCD.**

$$\frac{x-1}{8xy} = \frac{x-1}{8xy} \cdot \frac{3x}{3x} = \frac{3x^2 - 3x}{24x^2y}$$

• **The product of $8xy$ and $3x$ is the LCD.**

Problem 3 Write the fractions $\frac{x-3}{4xy^2}$ and $\frac{2x+1}{9y^2z}$ in terms of the LCD.

Solution See page S20.

Try Exercise 43, page 409.

EXAMPLE 4 Write the fractions $\frac{2x-1}{2x-x^2}$ and $\frac{x}{x^2+x-6}$ in terms of the LCD.

Solution

$$\frac{2x-1}{2x-x^2} = \frac{2x-1}{-(x^2-2x)} = -\frac{2x-1}{x^2-2x}$$

• **Rewrite $\frac{2x-1}{2x-x^2}$ with a denominator of $x^2 - 2x$.**

The LCD is $x(x-2)(x+3)$.

$$\frac{2x-1}{2x-x^2} = -\frac{2x-1}{x(x-2)} \cdot \frac{x+3}{x+3} = -\frac{2x^2+5x-3}{x(x-2)(x+3)}$$

$$\frac{x}{x^2+x-6} = \frac{x}{(x-2)(x+3)} \cdot \frac{x}{x} = \frac{x^2}{x(x-2)(x+3)}$$

Problem 4 Write the fractions $\frac{x+4}{x^2-3x-10}$ and $\frac{2x}{x^2-25}$ in terms of the LCD.

Solution See page S20.

Try Exercise 59, page 409.

9.2 Exercises

CONCEPT CHECK

Determine whether the statement is true or false.

1. The least common multiple of two numbers is the smallest number that contains all the prime factors of both numbers.
2. The least common denominator is the least common multiple of the denominators of two or more fractions.
3. The LCM of x^2, x^5, and x^8 is x^2.
4. We can rewrite $\frac{x}{y}$ as $\frac{4x}{4y}$ by using the Multiplication Property of One.
5. To rewrite a rational expression in terms of a common denominator, determine the factor by which you must multiply the denominator so that the denominator will be the common denominator. Then multiply the numerator and denominator of the fraction by that factor.

1 Find the least common multiple (LCM) of two or more polynomials (See pages 405–406.)

GETTING READY

6. a. In the LCM of $12x^4y^3$ and $15x^2y^5$, the exponent on x is __?__ and the exponent on y is __?__.

   b. Because $12 = 3 \cdot 4$, and $15 = 3 \cdot 5$, the coefficient in the LCM of $12x^4y^3$ and $15x^2y^5$ is (__?__)(__?__)(__?__) = (__?__).

Find the LCM of the expressions.

7. $8x^3y$, $12xy^2$
8. $6ab^2$, $18ab^3$
9. $10x^4y^2$, $15x^3y$
10. $12a^2b$, $18ab^3$
11. $8x^2$, $4x^2 + 8x$
12. $6y^2$, $4y + 12$
13. $2x^2y$, $3x^2 + 12x$
14. $4xy^2$, $6xy^2 + 12y^2$
15. $3x + 3$, $2x^2 + 4x + 2$
16. $4x - 12$, $2x^2 - 12x + 18$
17. $(x - 1)(x + 2)$, $(x - 1)(x + 3)$
18. $(2x - 1)(x + 4)$, $(2x + 1)(x + 4)$
19. $(2x + 3)^2$, $(2x + 3)(x - 5)$
20. $(x - 7)(x + 2)$, $(x - 7)^2$
21. $x - 1$, $x - 2$, $(x - 1)(x - 2)$
22. $(x + 4)(x - 3)$, $x + 4$, $x - 3$
23. $x^2 - x - 6$, $x^2 + x - 12$
24. $x^2 + 3x - 10$, $x^2 + 5x - 14$
25. $x^2 + 5x + 4$, $x^2 - 3x - 28$
26. $x^2 - 10x + 21$, $x^2 - 8x + 15$
27. $x^2 - 2x - 24$, $x^2 - 36$
28. $x^2 + 7x + 10$, $x^2 - 25$
29. $x^2 - 7x - 30$, $x^2 - 5x - 24$
30. $2x^2 - 7x + 3$, $2x^2 + x - 1$
31. $3x^2 - 11x + 6$, $3x^2 + 4x - 4$
32. $2x^2 - 9x + 10$, $2x^2 + x - 15$
33. $15 + 2x - x^2$, $x - 5$, $x + 3$
34. $5 + 4x - x^2$, $x - 5$, $x + 1$

35. How many factors of $x - 3$ are in the LCM of each pair of expressions?
a. $x^2 + x - 12$ and $x^2 - 9$
b. $x^2 - x - 12$ and $x^2 + 6x + 9$
c. $x^2 - 6x + 9$ and $x^2 + x - 12$

36. How many factors of a are in the LCM of $(a^3b^2)^2$ and a^4b^6? How many factors of b?

2 Express two fractions in terms of the LCD (See pages 406–407.)

GETTING READY

For Exercises 37 to 40, use the fractions $\frac{x^2}{y(y-3)}$ and $\frac{x}{(y-3)^2}$.

37. The LCD of the fractions is ___?___.

38. To write the first fraction in terms of the LCD, multiply its numerator and its denominator by ___?___.

39. To write the second fraction in terms of the LCD, multiply its numerator and its denominator by ___?___.

40. Write the two fractions in terms of the LCD: ___?___ and ___?___.

Write each fraction in terms of the LCD.

41. $\frac{4}{x}; \frac{3}{x^2}$

42. $\frac{5}{ab^2}; \frac{6}{ab}$

43. $\frac{x}{3y^2}; \frac{z}{4y}$

44. $\frac{5y}{6x^2}; \frac{7}{9xy}$

45. $\frac{y}{x(x-3)}; \frac{6}{x^2}$

46. $\frac{a}{y^2}; \frac{6}{y(y+5)}$

47. $\frac{9}{(x-1)^2}; \frac{6}{x(x-1)}$

48. $\frac{a^2}{y(y+7)}; \frac{a}{(y+7)^2}$

49. $\frac{3}{x-3}; -\frac{5}{x(3-x)}$

50. $\frac{b}{y(y-4)}; \frac{b^2}{4-y}$

51. $\frac{3}{(x-5)^2}; \frac{2}{5-x}$

52. $\frac{3}{7-y}; \frac{2}{(y-7)^2}$

53. $\frac{3}{x^2+2x}; \frac{4}{x^2}$

54. $\frac{2}{y-3}; \frac{3}{y^3-3y^2}$

55. $\frac{x-2}{x+3}; \frac{x}{x-4}$

56. $\frac{x^2}{2x-1}; \frac{x+1}{x+4}$

57. $\frac{3}{x^2+x-2}; \frac{x}{x+2}$

58. $\frac{3x}{x-5}; \frac{4}{x^2-25}$

59. $\frac{5}{2x^2-9x+10}; \frac{x-1}{2x-5}$

60. $\frac{x-3}{3x^2+4x-4}; \frac{2}{x+2}$

61. $\frac{x}{x^2+x-6}; \frac{2x}{x^2-9}$

62. $\frac{x-1}{x^2+2x-15}; \frac{x}{x^2+6x+5}$

63. $\dfrac{x}{9 - x^2}; \dfrac{x - 1}{x^2 - 6x + 9}$

64. $\dfrac{2x}{10 + 3x - x^2}; \dfrac{x + 2}{x^2 - 8x + 15}$

APPLYING CONCEPTS

Write each expression in terms of the LCD.

65. $\dfrac{3}{10^2}; \dfrac{5}{10^4}$

66. $\dfrac{8}{10^3}; \dfrac{9}{10^5}$

67. $b; \dfrac{5}{b}$

68. $3; \dfrac{2}{n}$

69. $1; \dfrac{y}{y - 1}$

70. $x; \dfrac{x}{x^2 - 1}$

71. $\dfrac{x^2 + 1}{(x - 1)^3}; \dfrac{x + 1}{(x - 1)^2}; \dfrac{1}{x - 1}$

72. $\dfrac{a^2 + a}{(a + 1)^3}; \dfrac{a + 1}{(a + 1)^2}; \dfrac{1}{a + 1}$

73. $\dfrac{1}{x^2 + 2x + xy + 2y}; \dfrac{1}{x^2 + xy - 2x - 2y}$

74. $\dfrac{1}{ab + 3a - 3b - b^2}; \dfrac{1}{ab + 3a + 3b + b^2}$

75. When is the LCM of two expressions equal to their product?

PROJECTS OR GROUP ACTIVITIES

76. Match the polynomials with their LCM. An LCM may be used more than once.

a. $x^2 - 4$ and $x^2 + 3x + 2$
b. $x + 3$ and $x^2 + 5x + 6$
c. $x^2 - x - 2$ and $x^2 + 2x + 1$
d. $x - 4$ and $x^2 - 1$
e. $2 - x$ and $x^2 + 3x + 2$
f. $4 - x$ and $x^2 - 1$
g. $x - 4$ and $1 - x^2$
h. $2 + x - x^2$ and $(x + 1)^2$

i. $(x + 3)(x + 2)$
ii. $(x - 4)(x + 1)(x - 1)$
iii. $(x + 2)(x - 2)(x + 1)$
iv. $(x - 2)(x + 1)(x + 1)$

9.3 Addition and Subtraction of Rational Expressions

OBJECTIVE 1 Add and subtract rational expressions with the same denominator

When adding rational expressions in which the denominators are the same, add the numerators. The denominator of the sum is the common denominator. The sum is written in simplest form.

When subtracting rational expressions in which the denominators are the same, subtract the numerators. The denominator of the difference is the common denominator. Write the answer in simplest form.

Take Note
Be careful with signs when subtracting rational expressions. In example (4) at the right, note that we must subtract the *entire* numerator $2x + 3$.

$(3x - 1) - (2x + 3) =$
$3x - 1 - 2x - 3$

ADDING AND SUBTRACTING RATIONAL EXPRESSIONS

To add or subtract rational expressions in which the denominators are the same, add or subtract the numerators. The denominator of the sum or difference is the common denominator. Write the answer in simplest form.

$$\frac{a}{b} + \frac{c}{b} = \frac{a+c}{b} \qquad \frac{a}{b} - \frac{c}{b} = \frac{a-c}{b}$$

EXAMPLES

1. $\frac{5x}{18} + \frac{7x}{18} = \frac{12x}{18} = \frac{2x}{3}$
2. $\frac{x}{x^2-1} + \frac{1}{x^2-1} = \frac{x+1}{x^2-1} = \frac{\overset{1}{\cancel{(x+1)}}}{\underset{1}{\cancel{(x+1)}}(x-1)} = \frac{1}{x-1}$
3. $\frac{2x}{x-2} - \frac{4}{x-2} = \frac{2x-4}{x-2} = \frac{2\overset{1}{\cancel{(x-2)}}}{\underset{1}{\cancel{(x-2)}}} = 2$
4. $\frac{3x-1}{x^2-5x+4} - \frac{2x+3}{x^2-5x+4}$
$= \frac{(3x-1)-(2x+3)}{x^2-5x+4} = \frac{x-4}{x^2-5x+4}$
$= \frac{\overset{1}{\cancel{(x-4)}}}{\underset{1}{\cancel{(x-4)}}(x-1)} = \frac{1}{x-1}$

EXAMPLE 1 Add or subtract. **A.** $\frac{7}{x^2} + \frac{9}{x^2}$ **B.** $\frac{3x^2}{x^2-1} - \frac{x+4}{x^2-1}$

Solution **A.** $\frac{7}{x^2} + \frac{9}{x^2} = \frac{7+9}{x^2}$

- **The denominators are the same. Add the numerators.**

$= \frac{16}{x^2}$

B. $\frac{3x^2}{x^2-1} - \frac{x+4}{x^2-1} = \frac{3x^2-(x+4)}{x^2-1}$

- **The denominators are the same. Subtract the numerators.**

$= \frac{3x^2-x-4}{x^2-1}$

$= \frac{(3x-4)\overset{1}{\cancel{(x+1)}}}{(x-1)\underset{1}{\cancel{(x+1)}}}$

- **Write the answer in simplest form.**

$= \frac{3x-4}{x-1}$

Problem 1 Add or subtract. **A.** $\frac{3}{xy} + \frac{12}{xy}$ **B.** $\frac{2x^2}{x^2-x-12} - \frac{7x+4}{x^2-x-12}$

Solution See page S20.

Try Exercise 15, page 415.

OBJECTIVE 2

Add and subtract rational expressions with different denominators

Before two fractions with different denominators can be added or subtracted, each fraction must be expressed in terms of a common denominator. In this text, we express each fraction in terms of the LCD, which is the LCM of the denominators.

Focus on adding rational expressions with different denominators

Add: $\dfrac{x-3}{x^2-2x} + \dfrac{6}{x^2-4}$

> **Take Note**
> This objective requires use of the skills learned in the objective on writing two fractions in terms of the LCD and the objective on adding and subtracting rational expressions with the same denominator.

Find the LCD.

$x^2 - 2x = x(x-2)$
$x^2 - 4 = (x-2)(x+2)$

The LCD is $x(x-2)(x+2)$.

Write each fraction in terms of the LCD. Multiply the first fraction by $\frac{x+2}{x+2}$ and the second fraction by $\frac{x}{x}$.

$$\frac{x-3}{x^2-2x} + \frac{6}{x^2-4}$$
$$= \frac{x-3}{x(x-2)} \cdot \frac{x+2}{x+2} + \frac{6}{(x-2)(x+2)} \cdot \frac{x}{x}$$
$$= \frac{x^2-x-6}{x(x-2)(x+2)} + \frac{6x}{x(x-2)(x+2)}$$

Add the fractions.

$$= \frac{x^2-x-6+6x}{x(x-2)(x+2)}$$
$$= \frac{x^2+5x-6}{x(x-2)(x+2)}$$

Factor the numerator to determine whether there are common factors in the numerator and denominator.

$$= \frac{(x+6)(x-1)}{x(x-2)(x+2)}$$

EXAMPLE 2 Add or subtract.

A. $\dfrac{y}{x} - \dfrac{4y}{3x} + \dfrac{3y}{4x}$ **B.** $\dfrac{2x}{x-3} - \dfrac{5}{3-x}$ **C.** $x - \dfrac{3}{5x}$

Solution **A.** The LCD is $12x$. • Find the LCD.

$$\frac{y}{x} - \frac{4y}{3x} + \frac{3y}{4x}$$
$$= \frac{y}{x} \cdot \frac{12}{12} - \frac{4y}{3x} \cdot \frac{4}{4} + \frac{3y}{4x} \cdot \frac{3}{3}$$
• Rewrite each fraction so that it has a denominator of $12x$.
$$= \frac{12y}{12x} - \frac{16y}{12x} + \frac{9y}{12x}$$
$$= \frac{12y - 16y + 9y}{12x}$$
• Add and subtract the fractions.
$$= \frac{5y}{12x}$$
• Simplify the numerator.

B. The LCM of $x-3$ and $3-x$ is $x-3$. • $3 - x = -(x-3)$

$$\frac{2x}{x-3} - \frac{5}{3-x}$$
$$= \frac{2x}{x-3} - \frac{5}{-(x-3)} \cdot \frac{-1}{-1}$$
• Multiply $\dfrac{5}{-(x-3)}$ by $\dfrac{-1}{-1}$ so that the denominator will be $x-3$.
$$= \frac{2x}{x-3} - \frac{-5}{x-3}$$
$$= \frac{2x-(-5)}{x-3}$$
• Subtract the fractions.
$$= \frac{2x+5}{x-3}$$
• Simplify the numerator.

C. The LCD is $5x$.

$$x - \frac{3}{5x} = \frac{x}{1} - \frac{3}{5x}$$

• Write x as $\frac{x}{1}$ and multiply it by $\frac{5x}{5x}$.

$$= \frac{x}{1} \cdot \frac{5x}{5x} - \frac{3}{5x}$$

$$= \frac{5x^2}{5x} - \frac{3}{5x}$$

$$= \frac{5x^2 - 3}{5x}$$

• Subtract the fractions.

Problem 2 Add or subtract.

A. $\frac{z}{8y} - \frac{4z}{3y} + \frac{5z}{4y}$ **B.** $\frac{5x}{x-2} - \frac{3}{2-x}$ **C.** $y + \frac{5}{y-7}$

Solution See page S21.

Try Exercise 51, page 416.

EXAMPLE 3 Add or subtract.

A. $\frac{2x}{2x-3} - \frac{1}{x+1}$ **B.** $\frac{x+3}{x^2-2x-8} + \frac{3}{4-x}$

Solution **A.** The LCD is $(2x-3)(x+1)$.

$$\frac{2x}{2x-3} - \frac{1}{x+1}$$

$$= \frac{2x}{2x-3} \cdot \frac{x+1}{x+1} - \frac{1}{x+1} \cdot \frac{2x-3}{2x-3}$$

• Rewrite each fraction so that it has a denominator of $(2x-3)(x+1)$.

$$= \frac{2x^2+2x}{(2x-3)(x+1)} - \frac{2x-3}{(2x-3)(x+1)}$$

$$= \frac{(2x^2+2x) - (2x-3)}{(2x-3)(x+1)}$$

• Subtract the fractions.

$$= \frac{2x^2+3}{(2x-3)(x+1)}$$

• Simplify the numerator.

B. The LCD is $(x-4)(x+2)$.

$$\frac{x+3}{x^2-2x-8} + \frac{3}{4-x}$$

$$= \frac{x+3}{(x-4)(x+2)} + \frac{3}{-(x-4)} \cdot \frac{-1 \cdot (x+2)}{-1 \cdot (x+2)}$$

• Rewrite each fraction so that it has a denominator of $(x-4)(x+2)$.

$$= \frac{x+3}{(x-4)(x+2)} + \frac{-3(x+2)}{(x-4)(x+2)}$$

$$= \frac{(x+3) + (-3)(x+2)}{(x-4)(x+2)}$$

• Add the fractions.

$$= \frac{x+3-3x-6}{(x-4)(x+2)}$$

• Simplify the numerator.

$$= \frac{-2x-3}{(x-4)(x+2)}$$

Problem 3 Add or subtract. **A.** $\frac{4x}{3x - 1} - \frac{9}{x + 4}$ **B.** $\frac{2x - 1}{x^2 - 25} + \frac{2}{5 - x}$

Solution See page S21.

Try Exercise 87, page 417.

9.3 Exercises

CONCEPT CHECK

Determine whether the statement is true or false.

1. To add two fractions, add the numerators and the denominators.
2. The procedure for subtracting two rational expressions is the same as that for subtracting two arithmetic fractions.
3. To add two rational expressions, first multiply both expressions by the LCD.
4. If $x \neq -2$ and $x \neq 0$, then $\frac{x}{x + 2} + \frac{3}{x + 2} = \frac{x + 3}{x + 2} = \frac{3}{2}$.

1 Add and subtract rational expressions with the same denominator (See pages 410–411.)

GETTING READY

5. Add: $\frac{6a + 1}{a - 4} + \frac{2a - 1}{a - 4}$

$\frac{6a + 1}{a - 4} + \frac{2a - 1}{a - 4} = \frac{6a + 1 + 2a - 1}{a - 4}$ • The denominators are the same. Add the __?__.

$= \frac{?}{a - 4}$ • Simplify the numerator. The result is in simplest form because the numerator and denominator have no common factors.

6. Subtract: $\frac{5x}{5x - 14} - \frac{2x - 5}{5x - 14}$

$\frac{5x}{5x - 14} - \frac{2x - 5}{5x - 14}$ • The __?__ are the same.

$= \frac{? - (?)}{5x - 14}$ • Subtract the numerators.

$= \frac{5x - ? + ?}{5x - 14}$ • Remove the parentheses.

$= \frac{?}{5x - 14}$ • Simplify the numerator. The fraction is in simplest form.

Add or subtract.

7. $\dfrac{3}{y^2} + \dfrac{8}{y^2}$

8. $\dfrac{6}{ab} - \dfrac{2}{ab}$

9. $\dfrac{3}{x + 4} - \dfrac{10}{x + 4}$

10. $\dfrac{x}{x + 6} - \dfrac{2}{x + 6}$

11. $\dfrac{3x}{2x + 3} + \dfrac{5x}{2x + 3}$

12. $\dfrac{6y}{4y + 1} - \dfrac{11y}{4y + 1}$

13. $\dfrac{2x + 1}{x - 3} + \dfrac{3x + 6}{x - 3}$

14. $\dfrac{4x + 3}{2x - 7} + \dfrac{3x - 8}{2x - 7}$

15. $\dfrac{5x - 1}{x + 9} - \dfrac{3x + 4}{x + 9}$

16. $\dfrac{6x - 5}{x - 10} - \dfrac{3x - 4}{x - 10}$

17. $\dfrac{x - 7}{2x + 7} - \dfrac{4x - 3}{2x + 7}$

18. $\dfrac{2n}{3n + 4} - \dfrac{5n - 3}{3n + 4}$

19. $\dfrac{x}{x^2 + 2x - 15} - \dfrac{3}{x^2 + 2x - 15}$

20. $\dfrac{3x}{x^2 + 3x - 10} - \dfrac{6}{x^2 + 3x - 10}$

21. $\dfrac{2x + 3}{x^2 - x - 30} - \dfrac{x - 2}{x^2 - x - 30}$

22. $\dfrac{3x - 1}{x^2 + 5x - 6} - \dfrac{2x - 7}{x^2 + 5x - 6}$

23. $\dfrac{4y + 7}{2y^2 + 7y - 4} - \dfrac{y - 5}{2y^2 + 7y - 4}$

24. $\dfrac{x + 1}{2x^2 - 5x - 12} + \dfrac{x + 2}{2x^2 - 5x - 12}$

25. $\dfrac{2x^2 + 3x}{x^2 - 9x + 20} + \dfrac{2x^2 - 3}{x^2 - 9x + 20} - \dfrac{4x^2 + 2x + 1}{x^2 - 9x + 20}$

26. $\dfrac{2x^2 + 3x}{x^2 - 2x - 63} - \dfrac{x^2 - 3x + 21}{x^2 - 2x - 63} - \dfrac{x - 7}{x^2 - 2x - 63}$

Determine whether the fractions have been added or subtracted.

27. $\dfrac{8a}{15b} \; ? \; \dfrac{2a}{15b} = \dfrac{2a}{3b}$

28. $\dfrac{8a}{15b} \; ? \; \dfrac{2a}{15b} = \dfrac{2a}{5b}$

29. $\dfrac{x + 4}{2x + 1} \; ? \; \dfrac{x - 1}{2x + 1} = \dfrac{5}{2x + 1}$

30. $\dfrac{x + 4}{2x + 1} \; ? \; \dfrac{x - 1}{2x + 1} = \dfrac{2x + 3}{2x + 1}$

2 Add and subtract rational expressions with different denominators (See pages 411–414.)

GETTING READY

31. To add or subtract rational expressions that have different denominators, you must first express each fraction in terms of a __?__, which can be the __?__ of the denominators of the rational expressions.

32. The LCM of the denominators of two fractions is also called the __?__.

33. True or false? To add $\dfrac{5}{8x^3} + \dfrac{7}{12x^5}$, use a common denominator of $4x^3$.

34. True or false? $\dfrac{3}{x - 8} + \dfrac{3}{8 - x} = 0$.

Add or subtract.

35. $\frac{4}{x} + \frac{5}{y}$

36. $\frac{7}{a} + \frac{5}{b}$

37. $\frac{12}{x} - \frac{5}{2x}$

38. $\frac{5}{3a} - \frac{3}{4a}$

39. $\frac{1}{2x} - \frac{5}{4x} + \frac{7}{6x}$

40. $\frac{7}{4y} + \frac{11}{6y} - \frac{8}{3y}$

41. $\frac{5}{3x} - \frac{2}{x^2} + \frac{3}{2x}$

42. $\frac{6}{y^2} + \frac{3}{4y} - \frac{2}{5y}$

43. $\frac{2}{x} - \frac{3}{2y} + \frac{3}{5x} - \frac{1}{4y}$

44. $\frac{5}{2a} + \frac{7}{3b} - \frac{2}{b} - \frac{3}{4a}$

45. $\frac{2x + 1}{3x} + \frac{x - 1}{5x}$

46. $\frac{4x - 3}{6x} + \frac{2x + 3}{4x}$

47. $\frac{x - 3}{6x} + \frac{x + 4}{8x}$

48. $\frac{2x - 3}{2x} + \frac{x + 3}{3x}$

49. $\frac{2x + 9}{9x} - \frac{x - 5}{5x}$

50. $\frac{3y - 2}{12y} - \frac{y - 3}{18y}$

51. $\frac{x + 4}{2x} - \frac{x - 1}{x^2}$

52. $\frac{x - 2}{3x^2} - \frac{x + 4}{x}$

53. $\frac{x - 10}{4x^2} + \frac{x + 1}{2x}$

54. $\frac{x + 5}{3x^2} + \frac{2x + 1}{2x}$

55. $y + \frac{8}{3y}$

56. $\frac{7}{2n} - n$

57. $\frac{4}{x + 4} + x$

58. $x + \frac{3}{x + 2}$

59. $5 - \frac{x - 2}{x + 1}$

60. $3 + \frac{x - 1}{x + 1}$

61. $\frac{2x + 1}{6x^2} - \frac{x - 4}{4x}$

62. $\frac{x + 3}{6x} - \frac{x - 3}{8x^2}$

63. $\frac{x + 2}{xy} - \frac{3x - 2}{x^2y}$

64. $\frac{3x - 1}{xy^2} - \frac{2x + 3}{xy}$

65. $\frac{4x - 3}{3x^2y} + \frac{2x + 1}{4xy^2}$

66. $\frac{5x + 7}{6xy^2} - \frac{4x - 3}{8x^2y}$

67. $\frac{x - 2}{8x^2} - \frac{x + 7}{12xy}$

68. $\frac{3x - 1}{6y^2} - \frac{x + 5}{9xy}$

69. $\frac{4}{x - 2} + \frac{5}{x + 3}$

70. $\frac{2}{x - 3} + \frac{5}{x - 4}$

71. $\frac{6}{x-7} - \frac{4}{x+3}$

72. $\frac{3}{y+6} - \frac{4}{y-3}$

73. $\frac{2x}{x+1} - \frac{1}{x-3}$

74. $\frac{3x}{x-4} - \frac{2}{x+6}$

75. $\frac{4x}{2x-1} - \frac{5}{x-6}$

76. $\frac{6x}{x+5} - \frac{3}{2x+3}$

77. $\frac{2a}{a-7} + \frac{5}{7-a}$

78. $\frac{4x}{6-x} + \frac{5}{x-6}$

79. $\frac{x+1}{x-6} - \frac{x+1}{6-x}$

80. $\frac{y+5}{y-2} + \frac{y+3}{2-y}$

81. $\frac{b+1}{b-1} + \frac{b-1}{b+1}$

82. $\frac{x-2}{x+1} + \frac{x-3}{x-1}$

83. $\frac{x}{x^2-9} + \frac{3}{x-3}$

84. $\frac{y}{y^2-16} + \frac{1}{y-4}$

85. $\frac{2x}{x^2-x-6} - \frac{3}{x+2}$

86. $\frac{5x}{x^2+2x-8} - \frac{2}{x+4}$

87. $\frac{3x-1}{x^2-10x+25} - \frac{3}{x-5}$

88. $\frac{2a+3}{a^2-7a+12} - \frac{2}{a-3}$

89. $\frac{x+4}{x^2-x-42} + \frac{3}{7-x}$

90. $\frac{x+3}{x^2-3x-10} + \frac{2}{5-x}$

91. $\frac{x}{2x+4} - \frac{2}{x^2+2x}$

92. $\frac{x+2}{4x+16} - \frac{2}{x^2+4x}$

93. $\frac{x-1}{x^2-x-2} + \frac{3}{x^2-3x+2}$

94. $\frac{a+2}{a^2+a-2} + \frac{3}{a^2+2a-3}$

95. $\frac{1}{x+1} + \frac{x}{x-6} - \frac{5x-2}{x^2-5x-6}$

96. $\frac{x}{x-4} + \frac{5}{x+5} - \frac{11x-8}{x^2+x-20}$

APPLYING CONCEPTS

Simplify.

97. $\frac{a}{a-b}+\frac{b}{b-a}+1$

98. $\frac{y}{x-y}+2-\frac{x}{y-x}$

99. $b-3+\frac{5}{b+4}$

100. $2y-1+\frac{6}{y+5}$

101. $\frac{(n+1)^2}{(n-1)^2}-1$

102. $1-\frac{(y-2)^2}{(y+2)^2}$

103. $\frac{x^2+x-6}{x^2+2x-8}\cdot\frac{x^2+5x+4}{x^2+2x-3}-\frac{2}{x-1}$

104. $\frac{x^2+9x+20}{x^2+4x-5}\div\frac{x^2-49}{x^2+6x-7}-\frac{x}{x-7}$

105. $\frac{x^2-9}{x^2+6x+9}\div\frac{x^2+x-20}{x^2-x-12}+\frac{1}{x+1}$

106. $\frac{x^2-25}{x^2+10x+25}\cdot\frac{x^2-7x+10}{x^2-x-2}+\frac{1}{x+1}$

Rewrite the expression as the sum of two fractions in simplest form.

107. $\frac{5b+4a}{ab}$

108. $\frac{6x+7y}{xy}$

109. $\frac{3x^2+4xy}{x^2y^2}$

110. $\frac{2mn^2+8m^2n}{m^3n^3}$

PROJECTS OR GROUP ACTIVITIES

111. Find the rational expression in simplest form that represents the sum of the reciprocals of the consecutive integers n and $n+1$.

112. Find the rational expression in simplest form that represents the difference of the reciprocals of the consecutive integers n and $n+1$.

113. Complete the equation.

a. $\frac{x-1}{x+1}-?=\frac{1}{x}$

b. $\frac{5}{2x}-?=\frac{2}{2x+1}$

114. Let x and y be positive integers. If $A=\frac{1}{x}+\frac{1}{y}+1$ and $B=\frac{x+y}{xy}$, then which of the following is true: $A<B$, $A>B$, or $A=B$?

9.4 Complex Fractions

OBJECTIVE 1 Simplify complex fractions

A **complex fraction** is a fraction whose numerator or denominator contains one or more fractions. Examples of complex fractions are shown at the right.

$$\frac{3}{2-\frac{1}{2}},\quad \frac{4+\frac{1}{x}}{3+\frac{2}{x}},\quad \frac{\frac{1}{x-1}+x+3}{x-3+\frac{1}{x+4}}$$

Focus on simplifying a complex fraction

Simplify: $\dfrac{1 - \dfrac{4}{x^2}}{1 + \dfrac{2}{x}}$

How It's Used

There are many instances of complex fractions in application problems. The fraction $\dfrac{1}{\dfrac{1}{r_1} + \dfrac{1}{r_2}}$ is used to determine the total resistance in certain electric circuits.

Find the LCD of the fractions in the numerator and denominator.

The LCD of $\dfrac{4}{x^2}$ and $\dfrac{2}{x}$ is x^2.

Multiply the numerator and denominator of the complex fraction by the LCD. We are multiplying the complex fraction by $\dfrac{x^2}{x^2}$, which equals 1, so we are not changing the value of the fraction.

$$\frac{1 - \dfrac{4}{x^2}}{1 + \dfrac{2}{x}} = \frac{1 - \dfrac{4}{x^2}}{1 + \dfrac{2}{x}} \cdot \frac{x^2}{x^2}$$

Use the Distributive Property to multiply $\left(1 - \dfrac{4}{x^2}\right)x^2$ and $\left(1 + \dfrac{2}{x}\right)x^2$.

$$= \frac{1 \cdot x^2 - \dfrac{4}{x^2} \cdot x^2}{1 \cdot x^2 + \dfrac{2}{x} \cdot x^2}$$

$$= \frac{x^2 - 4}{x^2 + 2x}$$

Simplify.

$$= \frac{(x - 2)\overset{1}{\cancel{(x + 2)}}}{x\underset{1}{\cancel{(x + 2)}}}$$

$$= \frac{x - 2}{x}$$

The method shown above of simplifying a complex fraction by multiplying the numerator and denominator by the LCD is used in Example 1. However, a different approach is to rewrite the numerator and denominator of the complex fraction as single fractions and then divide the numerator by the denominator. The example shown above is simplified below using this alternative method.

Rewrite the numerator and denominator of the complex fraction as single fractions.

$$\frac{1 - \dfrac{4}{x^2}}{1 + \dfrac{2}{x}} = \frac{1 \cdot \dfrac{x^2}{x^2} - \dfrac{4}{x^2}}{1 \cdot \dfrac{x}{x} + \dfrac{2}{x}} = \frac{\dfrac{x^2}{x^2} - \dfrac{4}{x^2}}{\dfrac{x}{x} + \dfrac{2}{x}} = \frac{\dfrac{x^2 - 4}{x^2}}{\dfrac{x + 2}{x}}$$

Recall that the fraction bar can be read "divided by." Divide the numerator of the complex fraction by the denominator. Rewrite division as multiplication by the reciprocal.

$$= \frac{x^2 - 4}{x^2} \div \frac{x + 2}{x} = \frac{x^2 - 4}{x^2} \cdot \frac{x}{x + 2}$$

Multiply the fractions.
Factor the numerator.

$$= \frac{(x^2 - 4)x}{x^2(x + 2)} = \frac{(x + 2)(x - 2)x}{x^2(x + 2)}$$

Simplify.

$$= \frac{x - 2}{x}$$

Note that this is the same result as before.

EXAMPLE 1 Simplify. **A.** $\dfrac{\frac{1}{x}+\frac{1}{2}}{\frac{1}{x^2}-\frac{1}{4}}$ **B.** $\dfrac{1-\frac{2}{x}-\frac{15}{x^2}}{1-\frac{11}{x}+\frac{30}{x^2}}$

Solution **A.** The LCM of the denominators, x, 2, x^2, and 4, is $4x^2$.

$$\frac{\frac{1}{x}+\frac{1}{2}}{\frac{1}{x^2}-\frac{1}{4}} = \frac{\frac{1}{x}+\frac{1}{2}}{\frac{1}{x^2}-\frac{1}{4}}\cdot\frac{4x^2}{4x^2}$$

- **Multiply the numerator and denominator of the complex fraction by $4x^2$.**

$$= \frac{\frac{1}{x}\cdot 4x^2 + \frac{1}{2}\cdot 4x^2}{\frac{1}{x^2}\cdot 4x^2 - \frac{1}{4}\cdot 4x^2}$$

- **Use the Distributive Property.**

$$= \frac{4x + 2x^2}{4 - x^2}$$

$$= \frac{2x\overset{1}{\cancel{(2+x)}}}{(2-x)\underset{1}{\cancel{(2+x)}}}$$

- **Factor the numerator and denominator. Divide by the common factors.**

$$= \frac{2x}{2-x}$$

- **Write the answer in simplest form.**

B. The LCM of the denominators, x and x^2, is x^2.

$$\frac{1-\frac{2}{x}-\frac{15}{x^2}}{1-\frac{11}{x}+\frac{30}{x^2}} = \frac{1-\frac{2}{x}-\frac{15}{x^2}}{1-\frac{11}{x}+\frac{30}{x^2}}\cdot\frac{x^2}{x^2}$$

- **Multiply the numerator and denominator of the complex fraction by x^2.**

$$= \frac{1\cdot x^2 - \frac{2}{x}\cdot x^2 - \frac{15}{x^2}\cdot x^2}{1\cdot x^2 - \frac{11}{x}\cdot x^2 + \frac{30}{x^2}\cdot x^2}$$

- **Use the Distributive Property.**

$$= \frac{x^2 - 2x - 15}{x^2 - 11x + 30}$$

$$= \frac{\overset{1}{\cancel{(x-5)}}(x+3)}{\underset{1}{\cancel{(x-5)}}(x-6)}$$

- **Factor the numerator and denominator. Divide by the common factors.**

$$= \frac{x+3}{x-6}$$

- **Write the answer in simplest form.**

Problem 1 Simplify. **A.** $\dfrac{\frac{1}{3}-\frac{1}{x}}{\frac{1}{9}-\frac{1}{x^2}}$ **B.** $\dfrac{1+\frac{4}{x}+\frac{3}{x^2}}{1+\frac{10}{x}+\frac{21}{x^2}}$

Solution See page S21.

Try Exercise 19, page 422.

9.4 Exercises

CONCEPT CHECK

The following are the examples of complex fractions given at the beginning of Objective 9.4.1. By what fraction would you multiply each complex fraction in order to simplify it?

1. $\dfrac{3}{2-\dfrac{1}{2}}$

2. $\dfrac{4+\dfrac{1}{x}}{3+\dfrac{2}{x}}$

3. $\dfrac{\dfrac{1}{x-1}+x+3}{x-3+\dfrac{1}{x+4}}$

Determine whether the statement is true or false.

4. To simplify a complex fraction, multiply the complex fraction by the LCD of the fractions in the numerator and denominator of the complex fraction.

5. When we multiply the numerator and denominator of a complex fraction by the same expression, we are using the Multiplication Property of One.

6. Our goal in simplifying a complex fraction is to rewrite it so that there are no fractions in the numerator or in the denominator. We then express the fraction in simplest form.

1 Simplify complex fractions (See pages 418–420.)

GETTING READY

For Exercises 7 to 10, use the complex fraction $\dfrac{1-\dfrac{5}{x}+\dfrac{6}{x^2}}{1-\dfrac{4}{x^2}}$.

7. To simplify the complex fraction, multiply the numerator and denominator of the complex fraction by the LCD of the fractions __?__, __?__, and __?__. The LCD is __?__.

8. When you multiply the numerator of the complex fraction by x^2, the numerator of the complex fraction simplifies to __?__, which factors as (__?__)(__?__).

9. When you multiply the denominator of the complex fraction by x^2, the denominator of the complex fraction simplifies to __?__, which factors as (__?__)(__?__).

10. The numerator from Exercise 8 and the denominator from Exercise 9 have a common factor of __?__, so the simplified form of the complex fraction is __?__.

Simplify.

11. $\dfrac{\dfrac{1}{3}+\dfrac{3}{x}}{\dfrac{1}{9}-\dfrac{9}{x^2}}$

12. $\dfrac{\dfrac{1}{2}+\dfrac{2}{x}}{\dfrac{1}{4}-\dfrac{4}{x^2}}$

13. $\dfrac{2-\dfrac{8}{x+4}}{3-\dfrac{12}{x+4}}$

14. $\dfrac{5-\dfrac{25}{x+5}}{1-\dfrac{3}{x+5}}$

15. $\dfrac{1+\dfrac{5}{y-2}}{1-\dfrac{2}{y-2}}$

16. $\dfrac{2-\dfrac{11}{2x-1}}{3-\dfrac{17}{2x-1}}$

17. $\dfrac{\dfrac{3}{x-2}+3}{\dfrac{4}{x-2}+4}$

18. $\dfrac{\dfrac{3}{2x+1}-3}{2-\dfrac{4x}{2x+1}}$

19. $\dfrac{2-\dfrac{3}{x}-\dfrac{2}{x^2}}{2+\dfrac{5}{x}+\dfrac{2}{x^2}}$

20. $\dfrac{2+\dfrac{5}{x}-\dfrac{12}{x^2}}{4-\dfrac{4}{x}-\dfrac{3}{x^2}}$

21. $\dfrac{1-\dfrac{1}{x}-\dfrac{6}{x^2}}{1-\dfrac{9}{x^2}}$

22. $\dfrac{1+\dfrac{4}{x}+\dfrac{4}{x^2}}{1-\dfrac{2}{x}-\dfrac{8}{x^2}}$

23. $\dfrac{1-\dfrac{5}{x}-\dfrac{6}{x^2}}{1+\dfrac{6}{x}+\dfrac{5}{x^2}}$

24. $\dfrac{1-\dfrac{7}{a}+\dfrac{12}{a^2}}{1+\dfrac{1}{a}-\dfrac{20}{a^2}}$

25. $\dfrac{1-\dfrac{6}{x}+\dfrac{8}{x^2}}{\dfrac{4}{x^2}+\dfrac{3}{x}-1}$

26. $\dfrac{1+\dfrac{3}{x}-\dfrac{18}{x^2}}{\dfrac{21}{x^2}-\dfrac{4}{x}-1}$

27. $\dfrac{x-\dfrac{4}{x+3}}{1+\dfrac{1}{x+3}}$

28. $\dfrac{y+\dfrac{1}{y-2}}{1+\dfrac{1}{y-2}}$

29. $\dfrac{1-\dfrac{x}{2x+1}}{x-\dfrac{1}{2x+1}}$

30. $\dfrac{1-\dfrac{2x-2}{3x-1}}{x-\dfrac{4}{3x-1}}$

31. $\dfrac{x-5+\dfrac{14}{x+4}}{x+3-\dfrac{2}{x+4}}$

32. $\dfrac{a+4+\dfrac{5}{a-2}}{a+6+\dfrac{15}{a-2}}$

33. $\dfrac{x+3-\dfrac{10}{x-6}}{x+2-\dfrac{20}{x-6}}$

34. $\dfrac{x-7+\dfrac{5}{x-1}}{x-3+\dfrac{1}{x-1}}$

35. $\dfrac{1-\dfrac{2}{x+1}}{1+\dfrac{1}{x-2}}$

36. $\dfrac{1-\dfrac{1}{x+2}}{1+\dfrac{2}{x-1}}$

37. $\dfrac{1-\dfrac{2}{x+4}}{1+\dfrac{3}{x-1}}$

38. $\dfrac{1+\dfrac{1}{x-2}}{1-\dfrac{3}{x+2}}$

39. $\dfrac{\dfrac{1}{x}-\dfrac{2}{x-1}}{\dfrac{3}{x}+\dfrac{1}{x-1}}$

40. $\dfrac{\dfrac{3}{n+1}+\dfrac{1}{n}}{\dfrac{2}{n+1}+\dfrac{3}{n}}$

41. $\dfrac{\dfrac{3}{b-4}-\dfrac{2}{b+1}}{\dfrac{5}{b+1}-\dfrac{1}{b-4}}$

42. $\dfrac{\dfrac{5}{x-5}-\dfrac{3}{x-1}}{\dfrac{6}{x-1}+\dfrac{2}{x-5}}$

43. True or false? The reciprocal of the complex fraction $\dfrac{1}{1-\dfrac{1}{x}}$ is $\dfrac{x-1}{x}$.

44. True or false? If the denominator of a complex fraction is the reciprocal of its numerator, then the complex fraction is equal to the square of the numerator of the complex fraction.

APPLYING CONCEPTS

Simplify.

45. $1 + \dfrac{1}{1 + \dfrac{1}{2}}$

46. $1 + \dfrac{1}{1 + \dfrac{1}{1 + \dfrac{1}{2}}}$

47. $1 - \dfrac{1}{1 - \dfrac{1}{x}}$

48. $1 - \dfrac{1}{1 - \dfrac{1}{y + 1}}$

49. $\dfrac{a^{-1} - b^{-1}}{a^{-2} - b^{-2}}$

50. $\dfrac{x^{-2} - y^{-2}}{x^{-2}y^{-2}}$

51. How would you explain to a classmate why we multiply the numerator and denominator of a complex fraction by the LCD of the fractions in the numerator and denominator?

PROJECTS OR GROUP ACTIVITIES

The complex fraction $\dfrac{1}{\frac{1}{r_1} + \frac{1}{r_2}}$ is mentioned in How It's Used on page 419. The fraction gives the total resistance, in ohms, of an electrical circuit that contains two parallel resistors with resistances of r_1 and r_2.

52. Show that the resistance fraction can be rewritten in the form $\dfrac{r_1 r_2}{r_1 + r_2}$.

53. Suppose an electrical circuit contains two parallel resistors with resistances of $r_1 = 2$ ohms and $r_2 = 3$ ohms. Calculate the total resistance in the circuit twice, once using the complex fraction shown above and once using the fraction as rewritten in Exercise 52.

54. Repeat Exercise 53 using $r_1 = 6$ ohms and $r_2 = 8$ ohms.

55. Which form of the resistance fraction did you find easier to work with when doing the calculations in Exercises 53 and 54? Why?

9.5 Equations Containing Fractions

OBJECTIVE 1 Solve equations containing fractions

In the chapter "Solving Equations and Inequalities," equations containing fractions were solved by the method of clearing denominators. Recall that to **clear denominators,** we multiply each side of an equation by the LCD. The result is an equation that contains no fractions. In this section, we will again solve equations containing fractions by multiplying each side of the equation by the LCD. The difference between this section and the chapter "Solving Equations and Inequalities" is that the fractions in these equations contain variables in the denominators.

Focus on solving an equation containing fractions

Solve: $\frac{3x-1}{4x}+\frac{2}{3x}=\frac{7}{6x}$

> **Take Note**
>
> Note that we are now solving *equations*, not operating on *expressions*. We are not writing each fraction in terms of the LCD; we are multiplying both sides of the equation by the LCD.

Find the LCD. The LCD of the fractions is 12*x*.

$$\frac{3x-1}{4x}+\frac{2}{3x}=\frac{7}{6x}$$

Multiply each side of the equation by the LCD.

$$12x\left(\frac{3x-1}{4x}+\frac{2}{3x}\right)=12x\left(\frac{7}{6x}\right)$$

Simplify by using the Distributive Property.

$$12x\left(\frac{3x-1}{4x}\right)+12x\left(\frac{2}{3x}\right)=12x\left(\frac{7}{6x}\right)$$

$$\frac{12x}{1}\left(\frac{3x-1}{4x}\right)+\frac{12x}{1}\left(\frac{2}{3x}\right)=\frac{12x}{1}\left(\frac{7}{6x}\right)$$

$$3(3x-1)+4(2)=2(7)$$

Solve for *x*.

$$9x-3+8=14$$
$$9x+5=14$$
$$9x=9$$
$$x=1$$

1 checks as a solution.
The solution is 1.

EXAMPLE 1 Solve: $\frac{4}{x}-\frac{x}{2}=\frac{7}{2}$

Solution

$$\frac{4}{x}-\frac{x}{2}=\frac{7}{2}$$

- **The LCD of the fractions is 2*x*.**

$$2x\left(\frac{4}{x}-\frac{x}{2}\right)=2x\left(\frac{7}{2}\right)$$

- **Multiply each side of the equation by 2*x*.**

$$\frac{2x}{1}\cdot\frac{4}{x}-\frac{2x}{1}\cdot\frac{x}{2}=\frac{2x}{1}\cdot\frac{7}{2}$$

$$8-x^2=7x$$

- **This is a quadratic equation.**

$$0=x^2+7x-8$$

- **Write the quadratic equation in standard form.**

$$0=(x+8)(x-1)$$

- **Solve by factoring.**

$$x+8=0 \qquad x-1=0$$
$$x=-8 \qquad x=1$$

Both −8 and 1 check as solutions.
The solutions are −8 and 1.

Problem 1 Solve: $x+\frac{1}{3}=\frac{4}{3x}$

Solution See page S21.

Try Exercise 19, page 432.

Occasionally, a value of a variable in a fractional equation makes one of the denominators zero. In this case, that value of the variable is not a solution of the equation.

Focus on solving an equation that has no solution

Solve: $\frac{2x}{x-2} = 1 + \frac{4}{x-2}$

Find the LCD. The LCD is $x - 2$.

$$\frac{2x}{x-2} = 1 + \frac{4}{x-2}$$

Multiply each side of the equation by the LCD.

$$(x-2)\frac{2x}{x-2} = (x-2)\left(1 + \frac{4}{x-2}\right)$$

Simplify by using the Distributive Property.

$$(x-2)\left(\frac{2x}{x-2}\right) = (x-2)\cdot 1 + (x-2)\cdot\frac{4}{x-2}$$

Solve for x.

$$2x = x - 2 + 4$$
$$2x = x + 2$$
$$x = 2$$

When x is replaced by 2, the denominators of $\frac{2x}{x-2}$ and $\frac{4}{x-2}$ are zero. Therefore, 2 is not a solution of the equation.

The equation has no solution.

EXAMPLE 2 Solve: $\frac{3x}{x-4} = 5 + \frac{12}{x-4}$

Solution

$$\frac{3x}{x-4} = 5 + \frac{12}{x-4}$$

- **The LCD of the fractions is $x - 4$.**

$$(x-4)\cdot\frac{3x}{x-4} = (x-4)\left(5 + \frac{12}{x-4}\right)$$

- **Multiply each side of the equation by $x - 4$.**

$$3x = (x-4)5 + 12$$
$$3x = 5x - 20 + 12$$

- **Use the Distributive Property on the right-hand side of the equation.**

$$3x = 5x - 8$$
$$-2x = -8$$
$$x = 4$$

- **Solve for x.**

4 does not check as a solution.
The equation has no solution.

Problem 2 Solve: $\frac{5x}{x+2} = 3 - \frac{10}{x+2}$

Solution See page S22.

Try Exercise 25, page 432.

OBJECTIVE 2

Solve proportions

How It's Used

Stock market analysts use many ratios in their work. Examples include the *price/earnings ratio*, the *current ratio*, and the *quick ratio*.

Quantities such as 4 meters, 15 seconds, and 8 gallons are number quantities written with **units.** In these examples, the units are meters, seconds, and gallons.

A **ratio** is the quotient of two quantities that have the same unit.

The length of a living room is 16 ft, and the width is 12 ft. The ratio of the length to the width is written

$$\frac{16\text{ ft}}{12\text{ ft}} = \frac{16}{12} = \frac{4}{3}$$

A ratio is in simplest form when the two numbers do not have a common factor. Note that the units are not written.

A **rate** is the quotient of two quantities that have different units.

There are 2 lb of salt in 8 gal of water. The salt-to-water rate is

$$\frac{2\text{ lb}}{8\text{ gal}} = \frac{1\text{ lb}}{4\text{ gal}}$$

A rate is in simplest form when the two numbers do not have a common factor. The units are written as part of the rate.

A **proportion** is an equation that states the equality of two ratios or rates.

Examples of proportions are shown below.

$$\frac{30\text{ mi}}{4\text{ h}} = \frac{15\text{ mi}}{2\text{ h}} \qquad \frac{4}{6} = \frac{20}{30} \qquad \frac{3}{4} = \frac{x}{12}$$

Because a proportion is an equation containing fractions, the same method used to solve an equation containing fractions is used to solve a proportion. Multiply each side of the equation by the LCD of the fractions. Then solve for the variable.

Focus on solving a proportion

Solve the proportion $\frac{4}{x} = \frac{2}{3}$.

$$\frac{4}{x} = \frac{2}{3}$$

Multiply each side of the proportion by the LCD.

$$3x\left(\frac{4}{x}\right) = 3x\left(\frac{2}{3}\right)$$

$$12 = 2x$$

Solve the equation.

$$6 = x$$

The solution is 6.

EXAMPLE 3 Solve. **A.** $\frac{8}{x+3} = \frac{4}{x}$ **B.** $\frac{6}{x+4} = \frac{12}{5x-13}$

Solution **A.**

$$\frac{8}{x+3} = \frac{4}{x}$$

- **The LCD is $x(x + 3)$.**

$$x(x+3)\frac{8}{x+3} = x(x+3)\frac{4}{x}$$

- **Multiply each side of the equation by $x(x + 3)$.**

$$8x = (x+3)4$$
$$8x = 4x + 12$$

- **Use the Distributive Property.**

$$4x = 12$$

- **Solve for x.**

$$x = 3$$

- **Remember to check the solution because the original equation is a fractional equation.**

The solution is 3.

B.

$$\frac{6}{x+4} = \frac{12}{5x-13}$$

- **The LCD is $(5x - 13)(x + 4)$.**

$$(5x-13)(x+4)\frac{6}{x+4} = (5x-13)(x+4)\frac{12}{5x-13}$$

- **Multiply each side of the equation by $(5x - 13)(x + 4)$.**

$$(5x-13)6 = (x+4)12$$
$$30x - 78 = 12x + 48$$

- **Use the Distributive Property.**

$$18x - 78 = 48$$

- **Solve for x.**

$$18x = 126$$
$$x = 7$$

- **Remember to check the solution because the original equation is a fractional equation.**

The solution is 7.

Problem 3 Solve. **A.** $\frac{2}{5} = \frac{6}{5x + 5}$ **B.** $\frac{5}{2x - 3} = \frac{10}{x + 3}$

Solution See page S22.

Try Exercise 45, page 433.

OBJECTIVE 3 Applications of proportions

Take Note

It is also correct to write the proportion with the loan amounts in the numerators and the monthly payments in the denominators. The solution will be the same.

EXAMPLE 4 The monthly loan payment for a car is $29.50 for each $1000 borrowed. At this rate, find the monthly payment for a $9000 car loan.

Strategy To find the monthly payment, write and solve a proportion using P to represent the monthly car payment.

Solution

$$\frac{29.50}{1000} = \frac{P}{9000}$$

• **The monthly payments are in the numerators. The loan amounts are in the denominators.**

$$9000\left(\frac{29.50}{1000}\right) = 9000\left(\frac{P}{9000}\right)$$

• **Multiply each side of the equation by the LCD.**

$$265.50 = P$$

The monthly payment is $265.50.

Problem 4 Nine ceramic tiles are required to tile a 4-square-foot area. At this rate, how many square feet can be tiled with 270 ceramic tiles?

© Paul Almasy/Corbis

Solution See page S22.

Try Exercise 59, page 433.

EXAMPLE 5 An investment of $1200 earns $96 each year. At the same rate, how much additional money must be invested to earn $128 each year?

Strategy To find the additional amount of money that must be invested, write and solve a proportion using x to represent the additional money. Then $1200 + x$ is the total amount invested.

Solution

$$\frac{1200}{96} = \frac{1200 + x}{128}$$

• **The amounts invested are in the numerators. The amounts earned are in the denominators.**

$$\frac{25}{2} = \frac{1200 + x}{128}$$

• **Simplify $\frac{1200}{96}$.**

$$128\left(\frac{25}{2}\right) = 128\left(\frac{1200 + x}{128}\right)$$

• **Multiply each side of the equation by the LCD.**

$$1600 = 1200 + x$$
$$400 = x$$

An additional $400 must be invested.

Problem 5 Three ounces of a medication are required for a 120-pound adult. At the same rate, how many additional ounces of the medication are required for a 180-pound adult?

Solution See page S22.

Try Exercise 79, page 434.

OBJECTIVE 4 Problems involving similar triangles

Clayton Sharrard/PhotoEdit, Inc.

Similar objects have the same shape but not necessarily the same size. A tennis ball is similar to a basketball. A model ship is similar to an actual ship.

Similar objects have corresponding parts; for example, the rudder on the model ship corresponds to the rudder on the actual ship. The relationship between the sizes of the corresponding parts can be written as a ratio, and each ratio will be the same. If the rudder on the model ship is $\frac{1}{100}$ the size of the rudder on the actual ship, then the model wheelhouse is $\frac{1}{100}$ the size of the actual wheelhouse, the width of the model is $\frac{1}{100}$ the width of the actual ship, and so on.

The two triangles ABC and DEF shown at the right are similar triangles. Side $\overline{AB}$ corresponds to $\overline{DE}$, side $\overline{BC}$ corresponds to $\overline{EF}$, and side $\overline{AC}$ corresponds to $\overline{DF}$. The height $\overline{CH}$ corresponds to the height $\overline{FK}$. The ratios of the lengths of corresponding parts are equal.

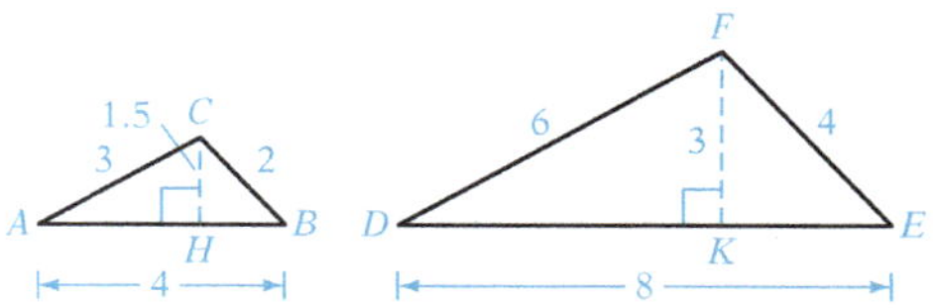

$$\frac{AB}{DE} = \frac{4}{8} = \frac{1}{2}, \quad \frac{AC}{DF} = \frac{3}{6} = \frac{1}{2}, \quad \frac{BC}{EF} = \frac{2}{4} = \frac{1}{2}, \quad \text{and} \quad \frac{CH}{FK} = \frac{1.5}{3} = \frac{1}{2}$$

Because the ratios of corresponding parts are equal, three proportions can be formed using the sides of the triangles.

$$\frac{AB}{DE} = \frac{AC}{DF}, \quad \frac{AB}{DE} = \frac{BC}{EF}, \quad \text{and} \quad \frac{AC}{DF} = \frac{BC}{EF}$$

Three proportions can also be formed by using the sides and heights of the triangles.

$$\frac{AB}{DE} = \frac{CH}{FK}, \quad \frac{AC}{DF} = \frac{CH}{FK}, \quad \text{and} \quad \frac{BC}{EF} = \frac{CH}{FK}$$

The corresponding angles in similar triangles are equal. Therefore,

$$\angle A = \angle D, \quad \angle B = \angle E, \quad \text{and} \quad \angle C = \angle F$$

Focus on using the properties of similar triangles to find the area of a triangle

Triangles ABC and DEF are similar triangles. Find the area of triangle ABC.

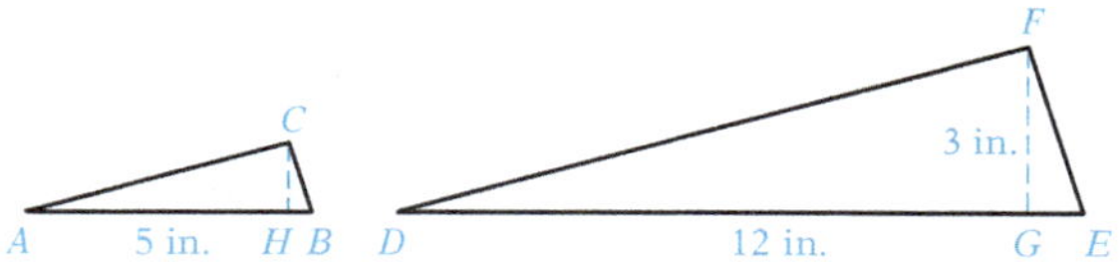

Strategy ▶ Solve a proportion to find the height of triangle ABC.

▶ Use the formula for the area of a triangle.

Solution Write a proportion using the ratios of corresponding sides. $\frac{AB}{DE} = \frac{CH}{FG}$

$AB = 5, DE = 12, FG = 3$ $\frac{5}{12} = \frac{CH}{3}$

The LCM of 12 and 3 is 12. Multiply each side of the equation by 12. $12 \cdot \frac{5}{12} = 12 \cdot \frac{CH}{3}$

Simplify. $5 = 4(CH)$

Solve for CH. $1.25 = CH$

Use the formula for the area of a triangle. The base is 5 in. The height is 1.25 in.

$$A = \frac{1}{2}bh$$

$$A = \frac{1}{2}(5)(1.25)$$

$$A = 3.125$$

The area of the triangle is 3.125 in^2.

Take Note

Vertical angles of intersecting lines, parallel lines, and angles of a triangle are discussed in the section on Geometry Problems in the chapter "Solving Equations and Inequalities: Applications."

It is also true that if the three angles of one triangle are equal, respectively, to the three angles of another triangle, then the two triangles are similar.

A line segment $\overline{DE}$ is drawn parallel to the base AB in the triangle at the right. $\angle x = \angle m$ and $\angle y = \angle n$ because corresponding angles are equal. Because $\angle C = \angle C$, the three angles of triangle DEC are equal, respectively, to the three angles of triangle ABC. Triangle DEC is similar to triangle ABC.

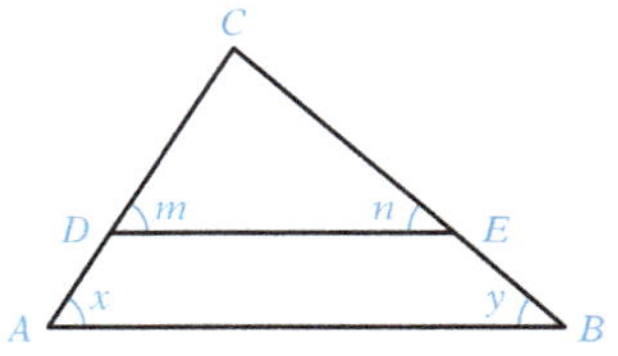

The sum of the three angles of a triangle is 180°. If two angles of one triangle are equal to two angles of another triangle, then the third angles must be equal. Thus we can say that if two angles of one triangle are equal to two angles of another triangle, then the two triangles are similar.

Focus on solving a problem using the properties of similar triangles

The line segments $\overline{AB}$ and $\overline{CD}$ intersect at point O in the figure at the right. Angles C and D are right angles. Find DO.

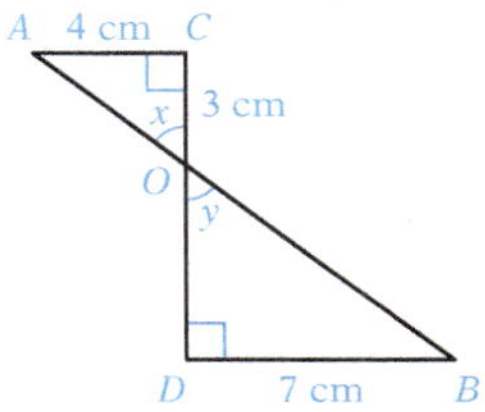

Strategy ▶ Determine whether triangle AOC is similar to triangle BOD.

▶ Use a proportion to find the length of the unknown side.

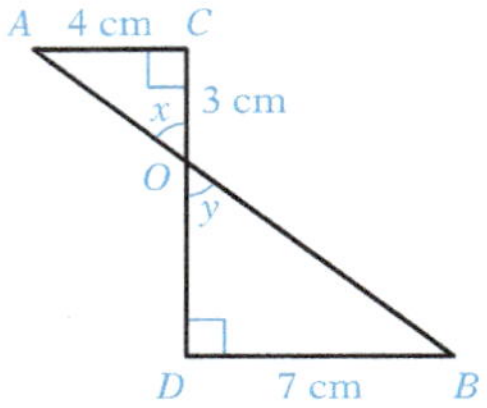

Solution $\angle C = \angle D$ because they are both right angles. $\angle x = \angle y$ because they are vertical angles. Therefore, triangle AOC is similar to triangle BOD because two angles of one triangle are equal to two angles of the other triangle.

Write a proportion using the ratios of corresponding sides. $\dfrac{AC}{DB} = \dfrac{CO}{DO}$

$AC = 4, DB = 7, CO = 3$ $\dfrac{4}{7} = \dfrac{3}{DO}$

The LCM of 7 and DO is $7(DO)$. Multiply each side of the equation by $7(DO)$. $7(DO) \cdot \dfrac{4}{7} = 7(DO) \cdot \dfrac{3}{DO}$

Simplify. $4(DO) = 7(3)$

Solve for DO.

$$4(DO) = 21$$
$$DO = 5.25$$

The length of DO is 5.25 cm.

EXAMPLE 6 In the figure at the right, $\overline{AB}$ is parallel to $\overline{DC}$, and angles B and D are right angles. $AB = 12$ m, $CD = 4$ m, and $AC = 18$ m. Find CO.

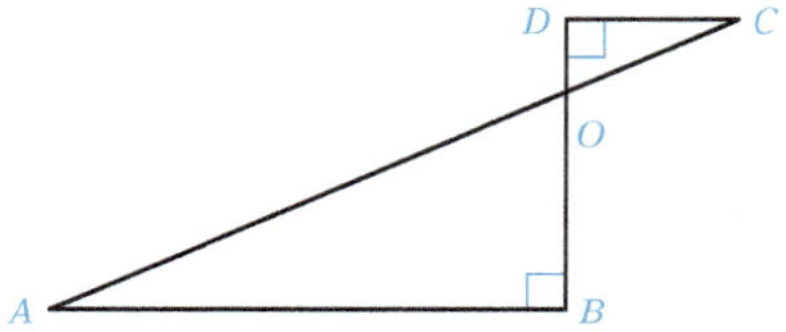

Strategy Triangle AOB is similar to triangle COD. Solve a proportion to find CO. Let x represent CO. Then $18 - x$ represents AO.

Solution

$$\frac{CD}{AB} = \frac{CO}{AO}$$
• **Write a proportion using the ratios of the lengths of corresponding sides.**

$$\frac{4}{12} = \frac{x}{18 - x}$$
• **$CD = 4$, $AB = 12$, $CO = x$, $AO = 18 - x$**

$$12(18 - x) \cdot \frac{4}{12} = 12(18 - x) \cdot \frac{x}{18 - x}$$
• **Multiply each side of the equation by $12(18 - x)$.**

$$(18 - x)4 = 12x$$
• **Simplify.**

$$72 - 4x = 12x$$
• **Use the Distributive Property.**

$$72 = 16x$$
$$4.5 = x$$
• **Solve for x.**

The length of CO is 4.5 m.

Problem 6 In the figure at the right, $\overline{AB}$ is parallel to $\overline{DC}$, and angles A and D are right angles. $AB = 10$ cm, $CD = 4$ cm, and $DO = 3$ cm. Find the area of triangle AOB.

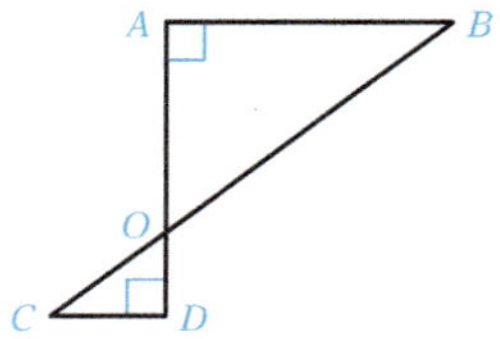

Solution See page S22.

Try Exercise 95, page 436.

9.5 Exercises

CONCEPT CHECK

1. The process of clearing denominators in an equation containing fractions is an application of which property of equations?

2. If the denominator of a fraction is $x + 3$, for what value of x is the fraction undefined?

3. Explain why you can clear denominators in part (a) below but not in part (b).

   a. $\frac{x}{2} + \frac{1}{3} = \frac{5}{2}$ **b.** $\frac{x}{2} + \frac{1}{3} + \frac{5}{2}$

4. Explain the difference between a ratio and a rate.

5. Explain the difference between a ratio and a proportion.

6. Identify each of the following as a ratio or a rate. Then write it in simplest form.

   a. $\frac{50\text{ ft}}{4\text{ s}}$ **b.** $\frac{28\text{ in.}}{21\text{ in.}}$ **c.** $\frac{20\text{ mi}}{2\text{ h}}$ **d.** $\frac{3\text{ gal}}{18\text{ gal}}$

For Exercises 7 and 8, use the pair of similar triangles shown at the right. Triangle PQR is similar to triangle XYZ.

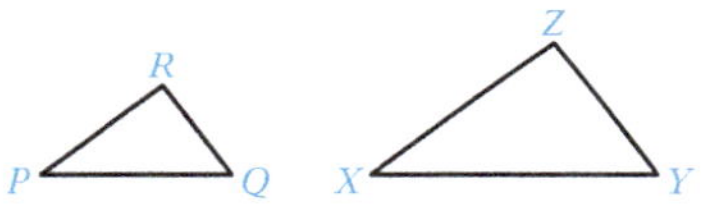

7. **a.** The corresponding part for side $\overline{RP}$ is ___?___.

   b. The corresponding part for side $\overline{YX}$ is ___?___.

   c. The corresponding part for $\angle X$ is ___?___.

8. **a.** Complete this proportion: $\frac{QR}{?} = \frac{PR}{XZ}$.

   b. Complete this equality: $\angle Z =$ ___?___.

1 Solve equations containing fractions (See pages 423–425.)

9. After solving an equation containing fractions, why must we check the solution?

GETTING READY

For Exercises 10 to 13, use the equation $\frac{7}{x} = \frac{3}{x - 4}$.

10. The first step in solving the equation is to clear denominators by multiplying each side of the equation by the LCM of the denominators ___?___ and ___?___. The LCD is ___?___.

11. **a.** When you multiply the left side of the equation by the LCD $x(x - 4)$, the left side simplifies to ___?___.

    b. When you multiply the right side of the equation by the LCD $x(x - 4)$, the right side simplifies to ___?___.

12. Use your answers to Exercise 11 to write the equation that results from clearing the denominators: __?__ = __?__. The solution of this equation is __?__.

13. The solution of a rational equation must be checked in the original equation. To check the solution you found in Exercise 12, substitute __?__ for x in the original equation $\frac{7}{x} = \frac{3}{x-4}$: $\frac{7}{7} = \frac{3}{7-4}$ simplifies to __?__ = __?__. The solution checks.

When a proposed solution of a rational equation does not check in the original equation, it is because it results in an expression involving division by zero. For each equation, state the values of x that would result in division by zero when substituted into the original equation.

14. $\frac{1}{x+5} = \frac{x}{x-3} + \frac{2}{x^2+2x-15}$

15. $\frac{3}{x-9} = \frac{1}{x^2-9x} + 2$

16. $\frac{6x}{x+1} - \frac{x}{x-2} = 4$

Solve.

17. $2 + \frac{5}{x} = 7$

18. $3 + \frac{8}{n} = 5$

19. $1 - \frac{9}{x} = 4$

20. $3 - \frac{12}{x} = 7$

21. $\frac{2}{y} + 5 = 9$

22. $\frac{6}{x} + 3 = 11$

23. $\frac{4x}{x-4} + 5 = \frac{5x}{x-4}$

24. $\frac{2x}{x+2} - 5 = \frac{7x}{x+2}$

25. $2 + \frac{3}{a-3} = \frac{a}{a-3}$

26. $\frac{x}{x+4} = 3 - \frac{4}{x+4}$

27. $\frac{x}{x-1} = \frac{8}{x+2}$

28. $\frac{x}{x+12} = \frac{1}{x+5}$

29. $\frac{2x}{x+4} = \frac{3}{x-1}$

30. $\frac{5}{3n-8} = \frac{n}{n+2}$

31. $x + \frac{6}{x-2} = \frac{3x}{x-2}$

32. $x - \frac{6}{x-3} = \frac{2x}{x-3}$

33. $\frac{x}{x+2} + \frac{2}{x-2} = \frac{x+6}{x^2-4}$

34. $\frac{x}{x+4} = \frac{11}{x^2-16} + 2$

35. $\frac{8}{y} = \frac{2}{y-2} + 1$

36. $\frac{8}{r} + \frac{3}{r-1} = 3$

2 Solve proportions (See pages 425–427.)

GETTING READY

37. The quotient of two quantities that have the same unit is called a __?__. The quotient of two quantities that have different units is called a __?__.

38. An equation that states the equality of two ratios or rates is called a __?__.

Solve.

39. $\frac{x}{12} = \frac{3}{4}$

40. $\frac{6}{x} = \frac{2}{3}$

41. $\frac{4}{9} = \frac{x}{27}$

42. $\frac{16}{9} = \frac{64}{x}$

43. $\frac{x+3}{12} = \frac{5}{6}$

44. $\frac{3}{5} = \frac{x-4}{10}$

45. $\frac{18}{x+4}=\frac{9}{5}$

46. $\frac{2}{11}=\frac{20}{x-3}$

47. $\frac{2}{x}=\frac{4}{x+1}$

48. $\frac{16}{x-2}=\frac{8}{x}$

49. $\frac{x+3}{4}=\frac{x}{8}$

50. $\frac{x-6}{3}=\frac{x}{5}$

51. $\frac{2}{x-1}=\frac{6}{2x+1}$

52. $\frac{9}{x+2}=\frac{3}{x-2}$

53. $\frac{2x}{7}=\frac{x-2}{14}$

54. True or false? If $\frac{a}{b}=\frac{c}{d}$, then $\frac{d}{b}=\frac{c}{a}$.

55. True or false? If $\frac{a}{b}=\frac{c}{d}$, then $\frac{b}{a}=\frac{d}{c}$.

3 Applications of proportions (See pages 427–428.)

GETTING READY

56. The scale on a map shows that a distance of 3 cm on the map represents an actual distance of 10 mi. This rate can be expressed as the quotient $\frac{10\text{ mi}}{?}$ or as the quotient $\frac{3\text{ cm}}{?}$.

57. On the map described in Exercise 56, would a distance of 8 cm represent an actual distance that is greater than 30 mi or less than 30 mi?

58. **Elections** An exit poll showed that 4 out of every 7 voters cast a ballot in favor of an amendment to a city charter. At this rate, how many people voted in favor of the amendment if 35,000 people voted?

59. **Business** A quality control inspector found 3 defective transistors in a shipment of 500 transistors. At this rate, how many transistors would be defective in a shipment of 2000 transistors?

60. **Health Insurance** See the news clipping at the right. How many Americans do not have health insurance? Use a figure of 300 million for the population of the United States.

61. **Poverty** See the news clipping at the right. How many American children live in poverty? Use a figure of 75 million for the number of children living in the United States.

62. **Construction** An air conditioning specialist recommends 2 air vents for every 300 ft^2 of floor space. At this rate, how many air vents are required for an office building of 21,000 ft^2?

63. **Television** In a city of 25,000 homes, a survey was taken to determine the number with Wi-Fi access. Of the 300 homes surveyed, 210 had Wi-Fi access. Estimate the number of homes in the city that have Wi-Fi access.

Fossils For Exercises 64 and 65, use the information in the article at the right. Assume that all scorpions have approximately the same ratio of claw length to body length.

64. Estimate the length, in feet, of the longest previously known prehistoric sea scorpion's claw. Round to the nearest hundredth.

65. Today, scorpions range in length from about 0.5 in. to about 8 in. Estimate the length, in inches, of a claw of a 7-inch scorpion. Round to the nearest hundredth. (*Hint:* Convert 8.2 ft to inches.)

In the News

Room for Improvement

According to a U.N. publication, the United States ranks 13th in the world in the areas of human development. Government data show that in regard to health, 3 in 20 Americans do not have health insurance. With respect to standard of living, 1 in 6 American children lives in poverty.

Sources: www.undp.org, www.census.gov, Human Development Report 2009; Income, Poverty, and Health Insurance Coverage in the United States, 2008

In the News

390-Million-Year-Old Scorpion Fossil Found

Scientists have announced the unearthing of the largest fossil sea scorpion claw ever discovered. Based on the 18-inch claw length, scientists estimate that the scorpion would have measured 8.2 ft in length. The longest previously known prehistoric sea scorpion was estimated to be 6.7 ft long.

Source: news.nationalgeographic.com

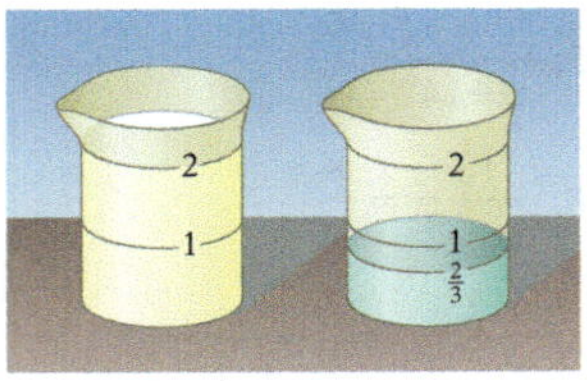

66. Cooking A simple syrup is made by dissolving 2 c of sugar in $\frac{2}{3}$ c of boiling water. At this rate, how many cups of sugar are required for 2 c of boiling water?

67. Energy The lighting for a billboard is provided by solar energy. If 3 energy panels generate 10 watts of power, how many panels are needed to provide 600 watts of power?

68. Conservation As part of a conservation effort for a lake, 40 fish were caught, tagged, and then released. Later, 80 fish were caught from the lake. Four of these 80 fish were found to have tags. Estimate the number of fish in the lake.

69. Business A company will accept a shipment of 10,000 computer disks if there are 2 or fewer defects in a sample of 100 randomly chosen disks. Assume that there are 300 defective disks in the shipment and that the rate of defective disks in the sample is the same as the rate in the shipment. Will the shipment be accepted?

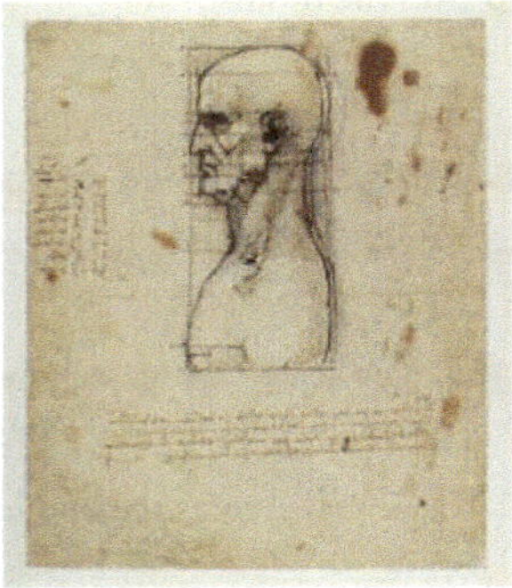
Cameraphoto Arte, Venice/Art Resource, NY

70. Business A company will accept a shipment of 20,000 precision bearings if there are 3 or fewer defects in a sample of 100 randomly chosen bearings. Assume that there are 400 defective bearings in the shipment and that the rate of defective bearings in the sample is the same as the rate in the shipment. Will the shipment be accepted?

71. Art Leonardo da Vinci measured various distances on the human body in order to make accurate drawings. He determined that in general, the ratio of the kneeling height of a person to his or her standing height is $\frac{3}{4}$. Using this ratio, determine the standing height of a person who has a kneeling height of 48 in.

72. Art In one of Leonardo da Vinci's notebooks, he wrote that ". . . from the top to the bottom of the chin is the sixth part of a face, and it is the fifty-fourth part of the man." Suppose the distance from the top to the bottom of a person's chin is 1.25 in. Using da Vinci's measurements, find the height of the person.

73. Cartography On a map, two cities are $2\frac{5}{8}$ in. apart. If $\frac{3}{8}$ in. on the map represents 25 mi, find the number of miles between the two cities.

74. Cartography On a map, two cities are $5\frac{5}{8}$ in. apart. If $\frac{3}{4}$ in. on the map represents 100 mi, find the number of miles between the two cities.

75. Rocketry The engine of a small rocket burns 170,000 lb of fuel in 1 min. At this rate, how many pounds of fuel does the rocket burn in 45 s?

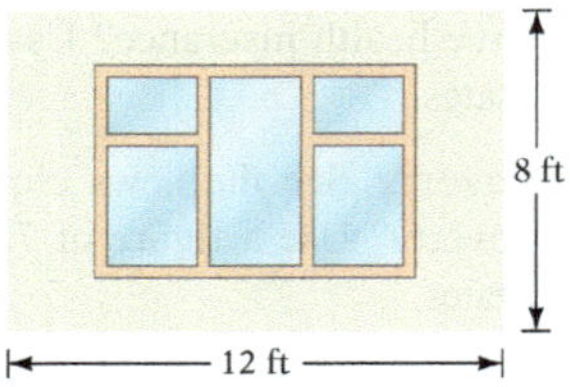

76. Construction To conserve energy and still allow for as much natural lighting as possible, an architect suggests that the ratio of the area of a window to the area of the total wall surface be 5 to 12. Using this ratio, determine the recommended area of a window to be installed in a wall that measures 8 ft by 12 ft.

77. Paint Mixtures A green paint is created by mixing 3 parts of yellow with every 5 parts of blue. How many gallons of yellow paint are needed to make 60 gal of this green paint?

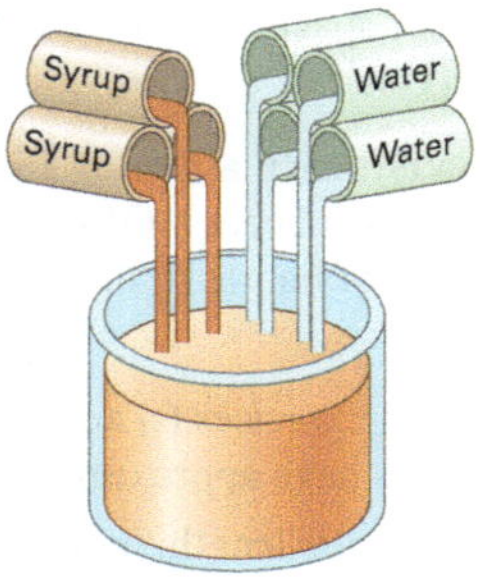

78. Food Industry A soft drink is made by mixing 4 parts of carbonated water with every 3 parts of syrup. How many milliliters of carbonated water are in 280 ml of soft drink?

79. Redecorating A painter estimates that 5 gal of paint will cover 1200 ft^2 of wall surface. How many additional gallons are required to cover 1680 ft^2?

80. Agriculture A 50-acre field yields 1100 bushels of wheat annually. How many additional acres must be planted so that the annual yield will be 1320 bushels?

81. Taxes The sales tax on a car that sold for $18,000 is $1170. At the same rate, how much greater is the sales tax on a car that sells for $19,500?

82. Catering A caterer estimates that 5 gal of coffee will serve 50 people. How much additional coffee is necessary to serve 70 people?

4 Problems involving similar triangles (See pages 428–430.)

Solve. Triangles ABC and DEF in Exercises 83 to 90 are similar. Round answers to the nearest tenth.

83. Find AC.

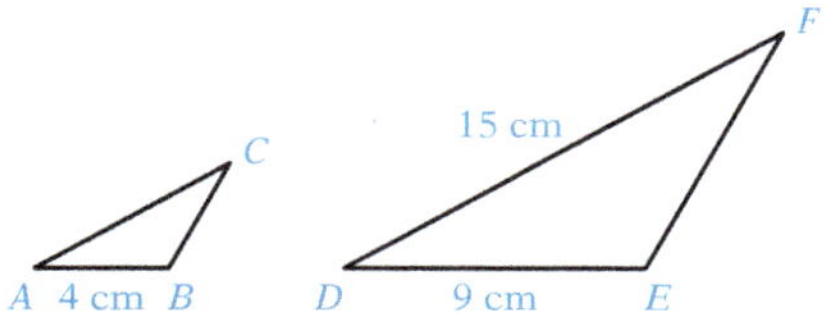

84. Find DE.

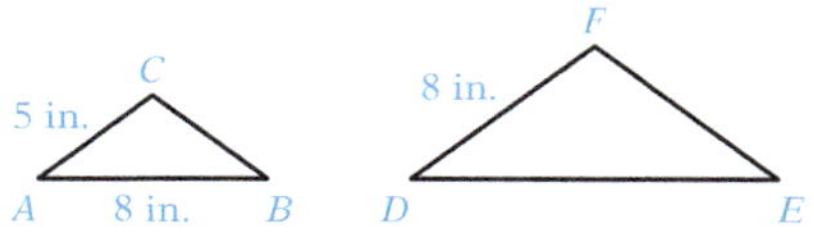

85. Find the height of triangle ABC.

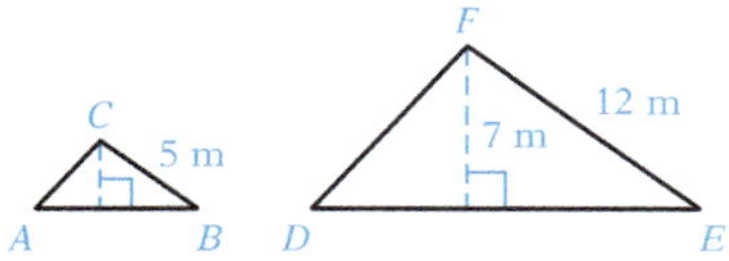

86. Find the height of triangle DEF.

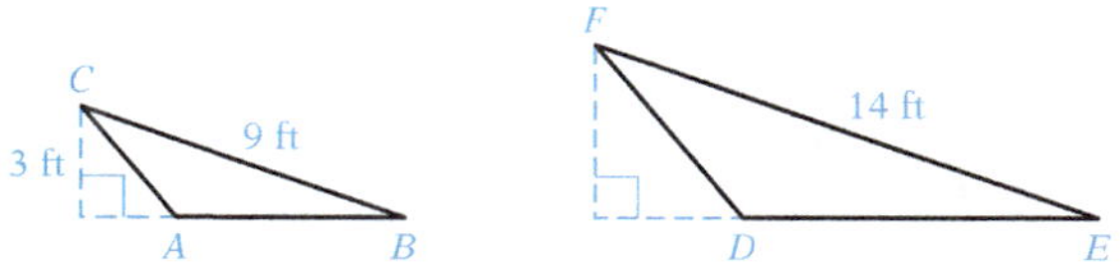

87. Find the perimeter of triangle DEF.

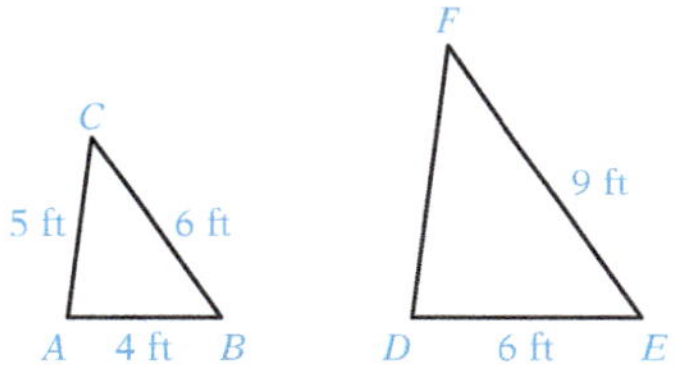

88. Find the perimeter of triangle ABC.

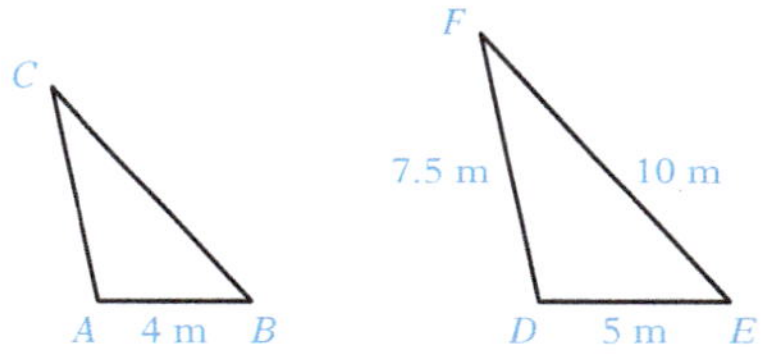

89. Find the area of triangle ABC.

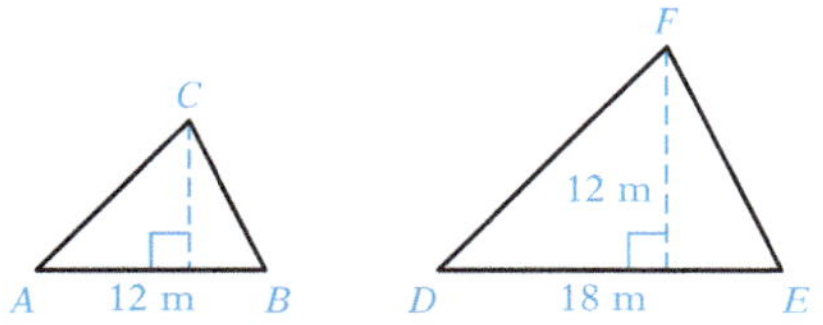

90. Find the area of triangle ABC.

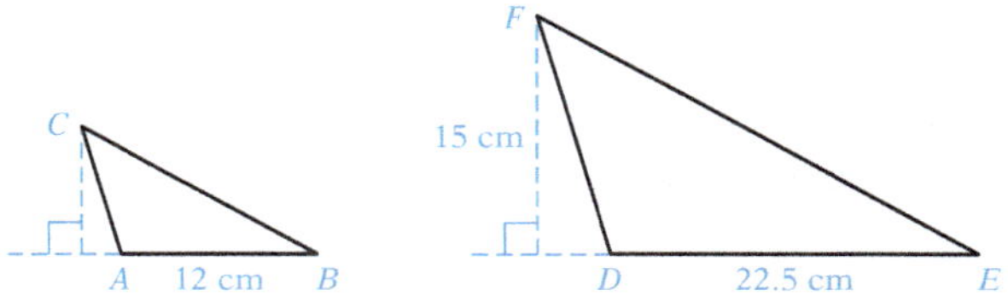

91. True or false? If the ratio of the corresponding sides of two similar triangles is $\frac{2}{3}$, the ratio of the perimeters of the two triangles is also $\frac{2}{3}$.

92. True or false? If the ratio of the corresponding sides of two similar triangles is $\frac{5}{4}$, the ratio of the areas of the two triangles is also $\frac{5}{4}$.

GETTING READY

93. Look at the diagram in Exercise 95, in which $\overline{BD}$ is parallel to $\overline{AE}$.

a. Triangle CBD is similar to triangle __?__.

b. Complete this proposition: $\frac{BD}{AE} = \frac{CD}{?}$.

94. Look at the diagram in Exercise 99.

a. Triangle MNO is similar to triangle __?__.

b. Complete this proportion: $\frac{OP}{?} = \frac{PQ}{MN}$.

95. Given $\overline{BD} \parallel \overline{AE}$, BD measures 5 cm, AE measures 8 cm, and AC measures 10 cm, find BC.

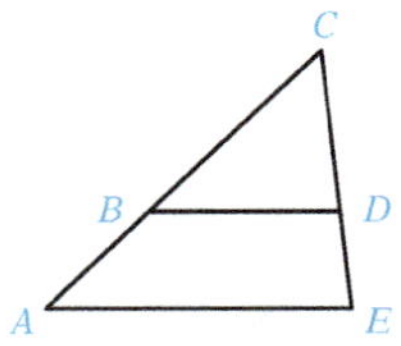

96. Given $\overline{AC} \parallel \overline{DE}$, BD measures 8 m, AD measures 12 m, and BE measures 6 m, find BC.

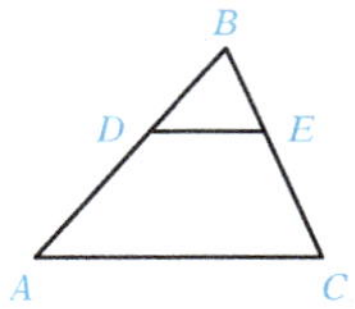

97. Given $\overline{DE} \parallel \overline{AC}$, DE measures 6 in., AC measures 10 in., and AB measures 15 in., find DA.

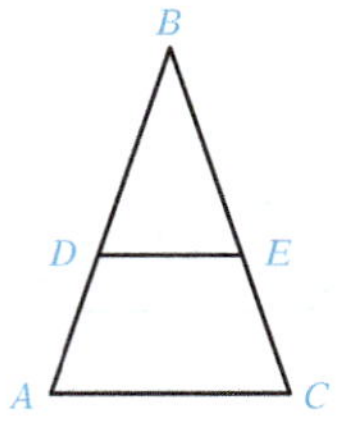

98. Given $\overline{MP}$ and $\overline{NQ}$ intersect at O, NO measures 25 ft, MO measures 20 ft, and PO measures 8 ft, find QO.

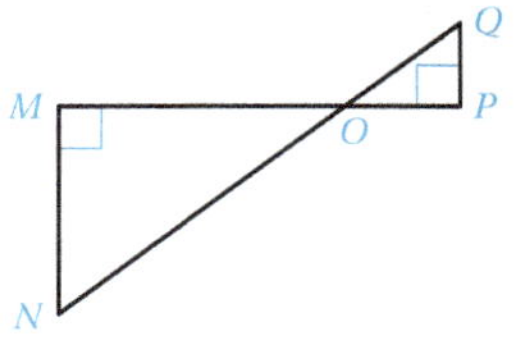

99. Given $\overline{MP}$ and $\overline{NQ}$ intersect at O, NO measures 24 cm, MN measures 10 cm, MP measures 39 cm, and QO measures 12 cm, find OP.

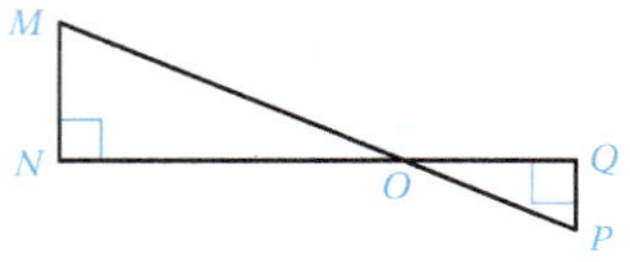

100. Given $\overline{MQ}$ and $\overline{NP}$ intersect at O, NO measures 12 m, MN measures 9 m, PQ measures 3 m, and MQ measures 20 m, find the perimeter of triangle OPQ.

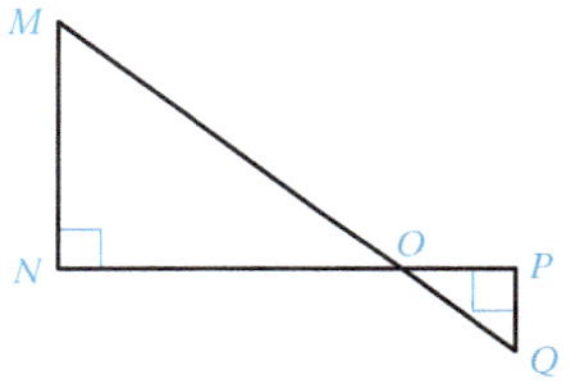

101. Similar triangles can be used as an indirect way of measuring inaccessible distances. The diagram at the right represents a river of width DC. The triangles AOB and DOC are similar. The distances AB, BO, and OC can be measured. Find the width of the river.

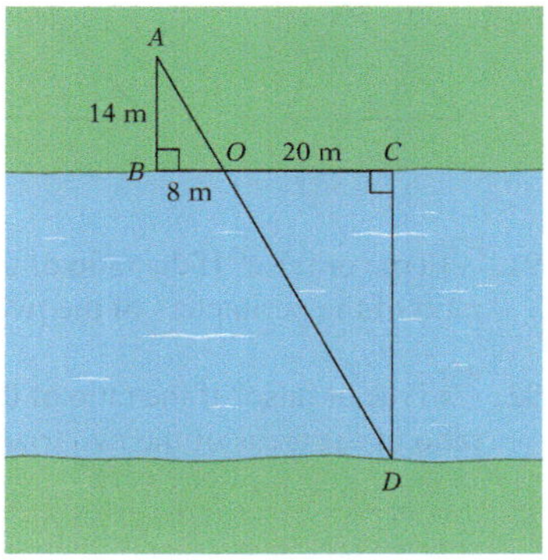

102. The sun's rays cast a shadow as shown in the diagram at the right. Find the height of the flagpole. Write the answer in terms of feet.

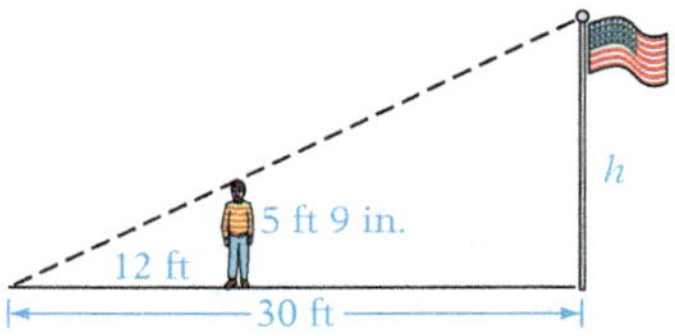

APPLYING CONCEPTS

103. Number Problem The sum of a number and its reciprocal is $\frac{26}{5}$. Find the number.

104. Lotteries Three people put their money together to buy lottery tickets. The first person put in \$25, the second person put in \$30, and the third person put in \$35. One of the tickets was a winning ticket. If they won \$4.5 million, what was the first person's share of the winnings?

105. Sports A basketball player has made 5 out of every 6 foul shots attempted. If 42 foul shots were missed in the player's career, how many foul shots were made in the player's career?

Dennis Ku/Shutterstock.com

106. Fundraising No one belongs to both the Math Club and the Photography Club, but the two clubs join to hold a car wash. Ten members of the Math Club and 6 members of the Photography Club participate. The profits from the car wash are \$120. If each club's profits are proportional to the number of members participating, what share of the profits does the Math Club receive?

107. History Eratosthenes, the fifth librarian of Alexandria (230 B.C.), was familiar with certain astronomical data that enabled him to calculate the circumference of Earth by using a proportion. He knew that on a midsummer day, the sun was directly overhead at Syene, as shown in the diagram. At the same time, at Alexandria, the sun was at a 7.5° angle from the zenith. The distance from Syene to Alexandria was 5000 stadia, or about 520 mi. Eratosthenes reasoned that the ratio of the 7.5° angle to one revolution was equal to the ratio of the arc length of 520 mi to the circumference of Earth. From this, he wrote and solved a proportion.

a. What did Eratosthenes calculate to be the circumference of Earth?

b. Find the difference between his calculation and the accepted value of 24,800 mi.

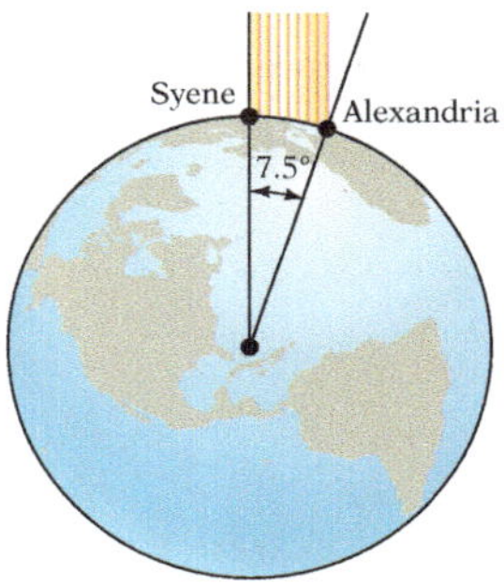

PROJECTS OR GROUP ACTIVITIES

108. Population The table below shows the projected population of people age 65 and older and the projected total population of the United States for each decade from 2020 to 2050.

Year	Population 65+, in thousands	Total Population, in thousands
2020	54,632	335,805
2030	71,453	363,584
2040	80,049	391,948
2050	86,705	419,854

a. What is the percent increase, for each 10-year period, in the population of people age 65 and older? Round to the nearest tenth of a percent.

b. What is the percent increase, for each 10-year period, in the total population of the United States?

c. Are there any 10-year periods during which the total population of the United States is increasing more rapidly (at a faster rate) than the population of people age 65 and older?

d. Many people, as they age, need more medical care. The population of people in the 20-to-44-year age group who provide that care is given below. For each decade, find the ratio of the number of potential caregivers to the number of people in the 65-and-older age group. Round to the nearest whole number.

Year	2020	2030	2040	2050
Population 20–44, in millions	108,632	114,747	121,659	130,897

e. Write a sentence that explains the meaning of the ratios you found in part (d). Are the ratios increasing or decreasing?

f. Write a few sentences that summarize the calculations in parts (a) through (d).

9.6 Variation

OBJECTIVE 1 Direct and inverse variation problems

Many important relationships in business, science, and engineering involve quantities that are directly proportional. That is, an increase in one quantity leads to a proportional increase in another quantity. Often these relationships can be described by an equation of the form $y = kx$, where k is a constant called the **constant of variation.** The equation $y = kx$ is the general form of a **direct variation equation** and is read "y varies directly as x."

Here are some examples of direct variation.

The distance traveled by a car traveling at a constant rate of 55 mph is represented by the direct variation equation $y = 55x$, where x is the number of hours traveled and y is the total distance traveled. The number 55 is the constant of variation.

The cost of filling the tank of your car with gasoline costing \$3.75 per gallon is directly proportional to the number of gallons purchased. The direct variation equation is $y = 3.75x$, where x is the number of gallons of gasoline purchased and y is the cost to fill the tank. The number 3.75 is the constant of variation.

DIRECT VARIATION

The equation $y = kx$ states that y varies directly as x.

In this equation, k is a constant called the *constant of variation.*

EXAMPLE 1 Find the constant of variation if y varies directly as x, and $y = 35$ when $x = 5$. Then write the specific direct variation equation that relates y and x.

Strategy

- ▶ To find the constant of variation, replace y with 35 and x with 5 in the general form of a direct variation equation $y = kx$. Solve for k.
- ▶ To write the specific direct variation equation, replace k by its value in the general direct variation equation.

Solution

$y = kx$ • Use the general form of a direct variation equation.

$35 = k \cdot 5$ • Replace y by 35 and x by 5.

$7 = k$ • Solve for k by dividing both sides by 5.

The constant of variation is 7.

$y = 7x$ • Write the specific direct variation equation by substituting 7 for k in $y = kx$.

The direct variation equation is $y = 7x$.

Problem 1 Find the constant of variation if p varies directly as q, and $p = 120$ when $q = 8$. Then write the specific direct variation equation that relates p and q.

Solution See page S22.

Try Exercise 7, page 442.

EXAMPLE 2 The amount (A) of medication prescribed for a person is directly related to the person's weight (W). For a 50-kilogram person, 2 ml of medication are prescribed. How many milliliters of medication are required for a person who weighs 75 kg?

Strategy

- This is a direct variation. To find the value of k, write the general form of a direct variation equation, replace the variables by the given values, and solve for k.
- Write the specific direct variation equation, replacing k by its value. Substitute 75 for W and solve for A.

Solution

$A = kW$ • Use the general form of a direct variation equation.

$2 = k \cdot 50$ • Replace A by 2 and W by 50.

$0.04 = k$ • Solve for k by dividing both sides by 50.

$A = 0.04W$ • Write the specific direct variation equation by substituting 0.04 for k.

$A = 0.04(75)$ • Replace W with 75 to find A when W is 75.

$A = 3$ • Simplify.

The required amount of medication is 3 ml.

Problem 2 A nurse's total wage (w) is directly proportional to the number of hours (h) worked. If the nurse earns \$264 for working 12 h, what is the nurse's total wage for working 18 h?

Solution See pages S22–S23.

Try Exercise 27, page 443.

Two quantities are **inversely proportional** if an increase in one quantity leads to a proportional decrease in the other quantity, or if a decrease in one leads to an increase in the other. An **inverse variation** is one that can be written in the form $y = \frac{k}{x}$, where k is a constant. The equation $y = \frac{k}{x}$ is the general form of an inverse variation equation and is read "y varies inversely as x."

The number of items N that can be purchased for a given amount of money is inversely proportional to the price P of an item. The inverse variation equation that relates N and P is $N = \frac{k}{P}$. The greater the price of an item, the fewer items you can purchase with a given amount of money. The lower the price of an item, the more items you can buy with that same amount of money.

© Jeff Greenberg/Alamy

It is important to note that in the direct variation equation $y = kx$, as the quantity x increases, y increases. For example, if you are buying power bars, twice as many power bars will cost twice as much money. In the inverse variation equation $y = \frac{k}{x}$, as the quantity x increases, y decreases, or as x decreases, y increases. The greater the price of a power bar, the fewer power bars you can purchase with \$10. The lower the price of a power bar, the more power bars you can buy with \$10.

Another example of inverse variation relates to uniform motion. The time required to travel a given distance is inversely proportional to the speed of travel. The faster you travel, the shorter the time to reach your destination. The more slowly you travel, the longer it takes to reach your destination. For example, suppose you are driving to a destination 200 mi away. If you drive at a rate of 50 mph, you will reach your destination in 4 h. If you drive at 40 mph (a slower speed), you will reach your destination in 5 h (a longer time).

INVERSE VARIATION

The equation $y = \frac{k}{x}$ states that y varies inversely as x.

In this equation, k is a constant called the constant of variation.

EXAMPLE 3 Find the constant of variation if y varies inversely as x, and $y = 30$ when $x = 10$. Then write the specific inverse variation equation that relates y and x.

Strategy

- To find the constant of variation, replace y with 30 and x with 10 in the general form of an inverse variation equation $y = \frac{k}{x}$. Solve for k.
- To write the specific inverse variation equation, replace k by its value in the general inverse variation equation.

Solution

$y = \frac{k}{x}$ • **Use the general form of an inverse variation equation.**

$30 = \frac{k}{10}$ • **Replace y by 30 and x by 10.**

$300 = k$ • **Solve for k by multiplying both sides by 10.**

The constant of variation is 300.

$y = \frac{300}{x}$ • **Write the specific inverse variation equation by substituting 300 for k in $y = \frac{k}{x}$.**

The inverse variation equation is $y = \frac{300}{x}$.

Problem 3 Find the constant of variation if s varies inversely as t, and $s = 12$ when $t = 8$. Then write the specific inverse variation equation that relates s and t.

Solution See page S23.

Try Exercise 9, page 442.

EXAMPLE 4 In an automobile cylinder, the volume (V) of a gas varies inversely with the pressure (P), given that the temperature does not change. If the volume of gas in the cylinder is 300 cm^3 when the pressure is 20 pounds per square inch (psi), what is the volume when the pressure is increased to 80 psi?

Strategy

- This is an inverse variation. To find the value of k, write the general inverse variation equation, replace the variables by the given values, and solve for k.
- Write the specific inverse variation equation, replacing k by its value. Substitute 80 for P and solve for V.

Solution

$V = \frac{k}{P}$ • Use the general form of an inverse variation equation.

$300 = \frac{k}{20}$ • Replace V by 300 and P by 20.

$6000 = k$ • Solve for k by multiplying both sides by 20.

$V = \frac{6000}{P}$ • Write the specific inverse variation equation by substituting 6000 for k.

$V = \frac{6000}{80}$ • Replace P with 80 to find V when P is 80.

$V = 75$ • Simplify.

When the pressure is 80 psi, the volume is 75 cm^3.

Problem 4 At an assembly plant, the number of hours (h) it takes to complete the daily quota is inversely proportional to the number of assembly machines (m) operating. If five assembly machines can complete the daily quota in 9 h, how many hours does it take for four assembly machines to complete the daily quota?

Solution See page S23.

Try Exercise 35, page 443.

9.6 Exercises

CONCEPT CHECK

1. **a.** When are two quantities directly proportional?
   b. When are two quantities inversely proportional?

2. State whether the two quantities vary directly or inversely.
   a. One acre planted in wheat will produce 45 bushels of wheat.
   b. The loudness of an amplifier is 80 decibels when you are standing at a distance of 1 ft from it.
   c. A truck travels 17 mi on 1 gal of fuel.
   d. The intensity of light is 15 foot-candles at a distance of 10 m.

3. Determine whether the statement is true or false.
   a. In the direct variation equation $y = kx$, if x increases, then y increases.
   b. In the inverse variation equation $y = \frac{k}{x}$, if x increases, then y increases.

1 Direct and inverse variation problems (See pages 438-441.)

GETTING READY

4. P varies directly as Q. The direct variation equation is $P = \underline{\quad ? \quad} Q$.

5. W varies inversely as L. The inverse variation equation is $W = \dfrac{?}{L}$.

6. Find the constant of variation when y varies directly as x, and $y = 15$ when $x = 2$. Then write the specific direct variation equation that relates y and x.

7. Find the constant of variation when t varies directly as s, and $t = 24$ when $s = 120$. Then write the specific direct variation equation that relates s and t.

8. Find the constant of variation when y varies inversely as x, and $y = 10$ when $x = 5$. Then write the specific inverse variation equation that relates y and x.

9. Find the constant of proportionality when T varies inversely as S, and $T = 0.2$ when $S = 8$. Then write the specific inverse variation equation that relates T and S.

10. If y varies directly as x, and $x = 10$ when $y = 4$, find y when $x = 15$.

11. Given that L varies directly as P, and $L = 24$ when $P = 21$, find P when $L = 80$.

12. Given that P varies directly as R, and $P = 20$ when $R = 5$, find P when $R = 6$.

13. Given that T varies directly as S, and $T = 36$ when $S = 9$, find T when $S = 2$.

14. Given that M is directly proportional to P, and $M = 15$ when $P = 30$, find M when $P = 20$.

15. Given that A is directly proportional to B, and $A = 6$ when $B = 18$, find A when $B = 21$.

16. If y varies inversely as x, and $y = 500$ when $x = 4$, find y when $x = 10$.

17. If W varies inversely as L, and $W = 20$ when $L = 12$, find L when $W = 90$.

18. If z varies inversely as w, and $z = 32$ when $w = 3$, find z when $w = 12$.

19. If C varies inversely as D, and $C = 4.5$ when $D = 2$, find C when $D = 9$.

Val Thoermer/Shutterstock.com

GETTING READY

20. A bookkeeper earns \$24 per hour. The bookkeeper's total wage (w) is directly proportional to the number of hours worked (h). The variation equation is $w = \underline{\quad ? \quad}$. The constant of variation is $\underline{\quad ? \quad}$.

21. An athlete jogs at a rate of 4 mph. The distance the athlete jogs (d) is directly proportional to the time spent jogging (t). The variation equation is $d = \underline{\quad ? \quad}$. The constant of variation is $\underline{\quad ? \quad}$.

In Exercises 22 to 25, do not do any calculations. Determine the answer by thinking about the relationship between the variables x and y.

22. If $y = 5x$, as x grows larger, does y grow larger or smaller?

23. If $y = \frac{100}{x}$, as x grows larger, does y grow larger or smaller?

24. Given the variation equation $y = \frac{6.7}{x}$, which number is closest to the value of y when $x = 3.2$?

(i) 2 (ii) $\frac{1}{2}$ (iii) 20

25. Given the variation equation $y = \frac{101}{x}$, which number is closest to the value of x when $y = 26$?

(i) 4 (ii) $\frac{1}{4}$ (iii) 40

26. Compensation A worker's wage (w) varies directly as the number of hours (h) worked. If \$82 is earned for working 8 h, how much is earned for working 30 h?

27. Sound The distance (d) sound travels varies directly as the time (t) it travels. If sound travels 8920 ft in 8 s, find the distance sound travels in 3 s.

28. Physics The distance (d) a spring will stretch varies directly as the force (F) applied to the spring. If a force of 12 lb is required to stretch a spring 3 in., what force is required to stretch the spring 5 in.?

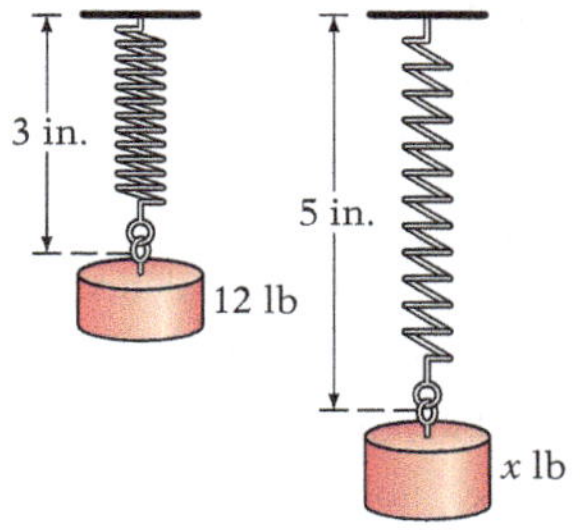

29. Scuba Diving The pressure (P) on a diver in the water varies directly as the depth (d). If the pressure is 2.25 psi when the depth is 5 ft, what is the pressure when the depth is 12 ft?

30. Clerical Work The number of words typed (w) varies directly as the time (t) spent typing. A typist can type 260 words in 4 min. Find the number of words typed in 15 min.

31. Electricity The current (I) varies directly as the voltage (V) in an electric circuit. If the current is 4 amperes when the voltage is 100 volts, find the current when the voltage is 75 volts.

32. Travel The distance traveled (d) varies directly as the time (t) of travel, assuming the speed is constant. If it takes 45 min to travel 50 mi, how many hours would it take to travel 180 mi?

33. Consumerism The number of items (N) that can be purchased for a given amount of money is inversely proportional to the cost (C) of the item. If 390 items can be purchased when the cost per item is \$.50, how many items can be purchased when the cost per item is \$.20?

34. Geometry The length (L) of a rectangle of fixed area varies inversely as the width (W). If the length of the rectangle is 8 ft when the width is 5 ft, find the length of the rectangle when the width is 4 ft.

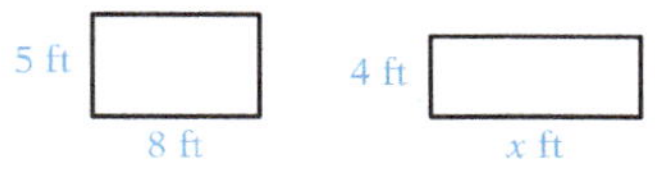

35. Travel The time (t) of travel on an automobile trip varies inversely as the speed (v) of the automobile. Traveling at an average speed of 65 mph, a trip took 4 h. The return trip took 5 h. Find the average speed on the return trip.

36. Electricity The current (I) in an electric circuit is inversely proportional to the resistance (R). If the resistance is 8 ohms when the current is 0.25 ampere, find the resistance when the current is 1.2 amperes.

37. Physics For a constant temperature, the pressure (P) of a gas varies inversely as the volume (V). If the pressure is 25 psi when the volume is 400 ft^3, find the pressure when the volume is 150 ft^3.

38. Physics The volume (V) of a gas varies inversely as the pressure (P) on the gas. If the volume of a gas is 12 ft^3 when the pressure is 15 lb/ft^2, find the volume of the gas when the pressure is 4 lb/ft^2.

39. Business A computer company that produces personal computers has determined that the number of computers it can sell (S) is inversely proportional to the price (P) of a computer. If the price is \$1800, 1800 computers can be sold. How many computers can be sold if the price is \$1500?

40. Gold See the news clipping at the right. Write a specific direct variation equation that relates the price (p) of gold and the number of ounces (n) purchased. Use the variation equation to determine the price of 8 oz of gold.

In the News

Gold Prices Reach Another Record High

Gold prices hit another record high this week, reaching \$1250 an ounce on the New York Mercantile Exchange. According to market analysts, the rise in gold prices is a consequence of concern about the stability of the U.S. economy.

Source: finance.yahoo.com

APPLYING CONCEPTS

41. Explain why the formula for the area of a circle is a direct variation.

42. You order an extra large pizza, cut into 12 slices, to be delivered to your home.

a. If six people are to share the pizza, how many slices does each person get?

b. If two people leave before the pizza arrives, how many slices does each person get?

c. Is this a direct variation or an inverse variation?

d. What is the constant of variation?

e. Write the equation that represents the variation.

43. The monthly interest charged on the unpaid balance on a credit card varies directly as the unpaid balance.

a. If the interest on \$475 is \$6.65, what is the constant of variation?

b. What does the constant of variation represent in this situation?

c. For what unpaid balance would the monthly interest charge be \$4.06?

PROJECTS OR GROUP ACTIVITIES

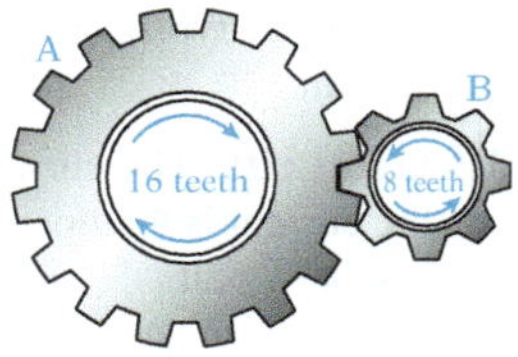

Gears The relationship between the size of a gear or pulley and the speed with which it rotates is an inverse variation.

For the diagram at the right, gear A has twice as many teeth as gear B. Therefore, when A makes 1 turn, B will make 2 turns.

Suppose gear A has three times as many teeth as gear B. Then when A makes 1 turn, B will make 3 turns. If gear A has four times as many teeth as gear B, then A will make 1 turn while B makes 4 turns. The fewer teeth gear B has compared to gear A, the more turns it will make for each 1 turn of gear A. **The speed of the gear is inversely proportional to the number of teeth.** We can express this relationship with the inverse variation equation

$$R = \frac{k}{T}$$

where R is the speed of the gear in revolutions per minute (rpm) and T is the number of teeth.

44. The speed of a gear varies inversely as the number of teeth. If a gear that has 40 teeth makes 15 rpm, how many revolutions per minute will a gear that has 32 teeth make?

Christian Lagerek/Shutterstock.com

45. The speed of a gear varies inversely as the number of teeth. If a gear that has 48 teeth makes 20 rpm, how many revolutions per minute will a gear that has 30 teeth make?

46. Gear A has 20 teeth and turns at a rate of 240 rpm. If gear B has 48 teeth, how many revolutions per minute does gear B make?

47. Gear A has 30 teeth and gear B has 12 teeth. If the speed of gear A is 150 rpm, what is the speed of gear B?

48. A 12-tooth gear mounted on a motor shaft drives a larger gear. The motor shaft rotates at 1450 rpm. The larger gear has 40 teeth. Find the speed of the larger gear.

9.7 Literal Equations

OBJECTIVE 1 Solve a literal equation for one of the variables

A **literal equation** is an equation that contains more than one variable. Examples of literal equations are shown at the right.

$$2x + 3y = 6$$
$$4w - 2x + z = 0$$

Formulas are used to express a relationship among physical quantities. A **formula** is a literal equation that states rules about measurements. Examples of formulas are shown below.

$\frac{1}{R_1} + \frac{1}{R_2} = \frac{1}{R}$ (Physics)

$s = a + (n - 1)d$ (Mathematics)

$A = P + Prt$ (Business)

The Addition and Multiplication Properties can be used to solve a literal equation for one of the variables. The goal is to rewrite the equation so that the letter being solved for is alone on one side of the equation, and all numbers and other variables are on the other side.

Focus on solving a literal equation

Solve $A = P(1 + i)$ for i.

The goal is to rewrite the equation so that i is on one side of the equation, and all other variables are on the other side.

$$A = P(1 + i)$$

Use the Distributive Property to remove parentheses.

$$A = P + Pi$$

Subtract P from each side of the equation.

$$A - P = P - P + Pi$$
$$A - P = Pi$$

Divide each side of the equation by P.

$$\frac{A - P}{P} = \frac{Pi}{P}$$

i is alone on one side of the equation. We have solved the equation for i.

$$\frac{A - P}{P} = i$$

EXAMPLE 1 **A.** Solve $I = \frac{E}{R + r}$ for R. **B.** Solve $L = a(1 + ct)$ for c.
C. Solve $S = C - rC$ for C.

Solution **A.**

$$I = \frac{E}{R + r}$$

$$(R + r)I = (R + r)\frac{E}{R + r}$$ • Multiply each side of the equation by $R + r$.

$$RI + rI = E$$ • Use the Distributive Property on the left side.

$$RI + rI - rI = E - rI$$ • Subtract rI from each side.
$$RI = E - rI$$

$$\frac{RI}{I} = \frac{E - rI}{I}$$ • Divide each side by I.

$$R = \frac{E - rI}{I}$$

B.

$$L = a(1 + ct)$$
$$L = a + act \quad \text{• Use the Distributive Property.}$$
$$L - a = a - a + act \quad \text{• Subtract } a \text{ from each side.}$$
$$L - a = act$$
$$\frac{L - a}{at} = \frac{act}{at} \quad \text{• Divide each side by } at.$$
$$\frac{L - a}{at} = c$$

C.

$$S = C - rC$$
$$S = C(1 - r) \quad \text{• Factor } C \text{ from } C - rC.$$
$$\frac{S}{1 - r} = \frac{C(1 - r)}{1 - r} \quad \text{• Divide each side of the equation by } 1 - r.$$
$$\frac{S}{1 - r} = C$$

Problem 1 **A.** Solve $s = \frac{A + L}{2}$ for L. **B.** Solve $S = a + (n - 1)d$ for n.
C. Solve $S = rS + C$ for S.

Solution See page S23.

Try Exercise 19, page 447.

9.7 Exercises

CONCEPT CHECK

For Exercises 1 to 4, determine whether the statement is true or false.

1. An equation that contains more than one variable is a literal equation.
2. The linear equation $y = 4x - 5$ is not a literal equation.
3. Literal equations are solved using the same properties of equations that are used to solve equations in one variable.
4. In solving a literal equation, the goal is to get the variable being solved for alone on one side of the equation and all numbers and other variables on the other side of the equation.

GETTING READY

5. In solving $I = \frac{E}{R + r}$ for R, the goal is to get ___?___ alone on one side of the equation.
6. In solving $L = a(1 + ct)$ for c, the goal is to get ___?___ alone on one side of the equation.

1 Solve a literal equation for one of the variables (See pages 445–446.)

Solve the formula for the variable given.

7. $A = \frac{1}{2}bh$; h (Geometry)

8. $P = a + b + c$; b (Geometry)

9. $d = rt; t$ (Physics)

10. $E = IR; R$ (Physics)

11. $PV = nRT; T$ (Chemistry)

12. $A = bh; h$ (Geometry)

13. $P = 2L + 2W; L$ (Geometry)

14. $F = \frac{9}{5}C + 32; C$ (Temperature conversion)

15. $A = \frac{1}{2}h(b_1 + b_2); b_1$ (Geometry)

16. $C = \frac{5}{9}(F - 32); F$ (Temperature conversion)

17. $V = \frac{1}{3}Ah; h$ (Geometry)

18. $P = R - C; C$ (Business)

19. $R = \frac{C - S}{t}; S$ (Business)

20. $P = \frac{R - C}{n}; R$ (Business)

21. $A = P + Prt; P$ (Business)

22. $T = fm - gm; m$ (Engineering)

23. $A = Sw + w; w$ (Physics)

24. $a = S - Sr; S$ (Mathematics)

25. When asked to solve $A = P(1 + i)$ for i, one student answered $i = \frac{A}{P} - 1$ and another student answered $i = \frac{A - P}{P}$. Are the two answers equivalent?

26. When asked to solve $A = P(1 + i)$ for i, one student answered $i = -\frac{P - A}{P}$ and another student answered $i = \frac{A - P}{P}$. Are the two answers equivalent?

APPLYING CONCEPTS

27. Solve for x: $cx - y = bx + 5$

28. Solve the physics formula $\frac{1}{R_1} + \frac{1}{R_2} = \frac{1}{R}$ for R_2.

PROJECTS OR GROUP ACTIVITIES

Business When markup is based on selling price, the selling price of a product is given by the formula $S = C + rC$, where C is the cost of the product and r is the markup rate.

29. **a.** Solve the formula $S = C + rC$ for r.

b. Use your answer to part (a) to find the markup rate on an e-reader when the cost is \$108 and the selling price is \$180.

c. Use your answer to part (a) to find the markup rate on a GPS navigation system when the cost is \$110 and the selling price is \$150.

Business Break-even analysis is a method used to determine the sales volume required for a company to "break even," or experience neither a profit nor a loss on the sale of its product. The break-even point represents the number of units that must be made and sold for income from sales to equal the cost of producing the product. The break-even point can be calculated using the formula $B = \frac{F}{S - V}$, where F is the fixed costs, S is the selling price per unit, and V is the variable costs per unit.

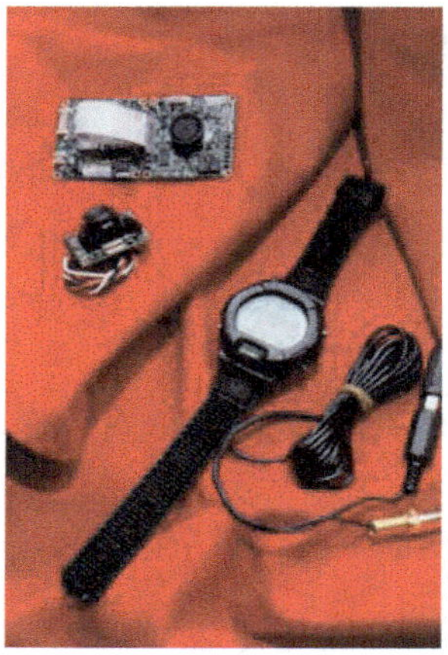

30. a. Solve the formula $B = \frac{F}{S - V}$ for S.

b. Use your answer to part (a) to find the selling price per button pinhole video spycam required for a company to break even. The fixed costs are \$15,000, the variable costs per spycam are \$60, and the company plans to make and sell 200 spycams.

c. Use your answer to part (a) to find the selling price per spy camera video watch required for a company to break even. The fixed costs are \$18,000, the variable costs per watch are \$65, and the company plans to make and sell 600 watches.

Electricity Resistors are used to control the flow of current. The total resistance of two resistors in a circuit can be given by the formula $R = \dfrac{1}{\dfrac{1}{R_1} + \dfrac{1}{R_2}}$, where R_1 and R_2 are the resistances of the two resistors in the circuit. Resistance is measured in ohms.

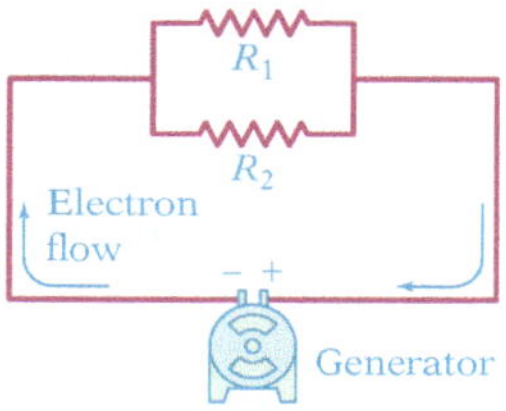

31. a. Solve the formula $R = \dfrac{1}{\dfrac{1}{R_1} + \dfrac{1}{R_2}}$ for R_1.

b. Use your answer to part (a) to find the resistance in R_1 if the resistance in R_2 is 30 ohms and the total resistance is 12 ohms.

c. Use your answer to part (a) to find the resistance in R_1 if the resistance in R_2 is 15 ohms and the total resistance is 6 ohms.

9.8 Application Problems

OBJECTIVE 1 Work problems

If a painter can paint a room in 4 h, then in 1 h the painter can paint $\frac{1}{4}$ of the room. The painter's rate of work is $\frac{1}{4}$ of the room each hour. The **rate of work** is that part of a task that is completed in 1 unit of time.

A pipe can fill a tank in 30 min. This pipe can fill $\frac{1}{30}$ of the tank in 1 min. The rate of work is $\frac{1}{30}$ of the tank each minute. If a second pipe can fill the tank in x min, the rate of work for the second pipe is $\frac{1}{x}$ of the tank each minute.

In solving a work problem, the goal is to determine the time it takes to complete a task. The basic equation that is used to solve work problems is

Rate of work · Time worked = Part of task completed

For example, if a faucet can fill a sink in 6 min, then in 5 min the faucet will fill $\frac{1}{6} \cdot 5 = \frac{5}{6}$ of the sink. In 5 min, the faucet completes $\frac{5}{6}$ of the task.

Solve: A painter can paint a ceiling in 60 min. The painter's apprentice can paint the same ceiling in 90 min. How long will it take them to paint the ceiling if they work together?

STRATEGY FOR SOLVING A WORK PROBLEM

▶ For each person or machine, write a numerical or variable expression for the rate of work, the time worked, and the part of the task completed. The results can be recorded in a table.

Unknown time to paint the ceiling working together: t

Take Note

Use the information given in the problem to fill in the rate and time columns of the table. Fill in the "part of task completed" column by multiplying the two expressions you wrote in each row.

	Rate of work	·	Time worked	=	Part of task completed
Painter	$\frac{1}{60}$	·	t	=	$\frac{t}{60}$
Apprentice	$\frac{1}{90}$	·	t	=	$\frac{t}{90}$

▶ Determine how the parts of the task completed are related. Use the fact that the sum of the parts of the task completed must equal 1, the complete task.

$$\frac{t}{60} + \frac{t}{90} = 1$$
• The part of the task completed by the painter plus the part of the task completed by the apprentice must equal 1.

$$180\left(\frac{t}{60} + \frac{t}{90}\right) = 180 \cdot 1$$
• Multiply each side of the equation by the LCD of $\frac{t}{60}$ and $\frac{t}{90}$.

$$3t + 2t = 180$$
• Use the Distributive Property.

$$5t = 180$$
• Combine like terms.

$$t = 36$$
• Divide by 5.

Working together, they will paint the ceiling in 36 min.

EXAMPLE 1 A small water pipe takes four times longer to fill a tank than does a large water pipe. With both pipes open, it takes 3 h to fill the tank. Find the time it would take the small pipe, working alone, to fill the tank.

Strategy

▶ Time for large pipe to fill the tank: t

▶ Time for small pipe to fill the tank: $4t$

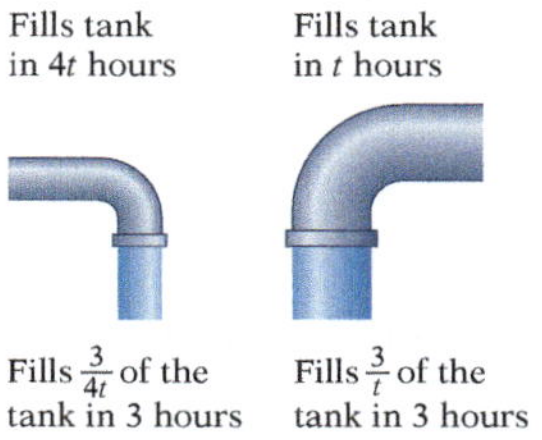

	Rate	·	Time	=	Part
Small pipe	$\frac{1}{4t}$	·	3	=	$\frac{3}{4t}$
Large pipe	$\frac{1}{t}$	·	3	=	$\frac{3}{t}$

▶ The sum of the parts of the task completed must equal 1.

Solution

$$\frac{3}{4t} + \frac{3}{t} = 1$$

- The part of the task completed by the small pipe plus the part of the task completed by the large pipe must equal 1.

$$4t\left(\frac{3}{4t} + \frac{3}{t}\right) = 4t \cdot 1$$

- Multiply each side of the equation by $4t$, the LCD of $\frac{3}{4t}$ and $\frac{3}{t}$.

$$3 + 12 = 4t$$

- Use the Distributive Property.

$$15 = 4t$$

$$\frac{15}{4} = t$$

- t is the time for the large pipe to fill the tank.

$$4t = 4\left(\frac{15}{4}\right) = 15$$

- Substitute the value of t into the variable expression for the time for the small pipe to fill the tank.

The small pipe, working alone, takes 15 h to fill the tank.

Problem 1 Two computer printers that work at the same rate are working together to print the payroll checks for a corporation. After they work together for 3 h, one of the printers fails. The second printer requires 2 more hours to complete the checks. Find the time it would take one printer, working alone, to print the checks.

Solution See page S23.

Try Exercise 19, page 454.

OBJECTIVE 2

Uniform motion problems

A car that travels constantly in a straight line at 30 mph is in *uniform motion.* When an object is in **uniform motion**, its speed does not change.

The basic equation used to solve uniform motion problems is

$$\textbf{Distance} = \textbf{Rate} \cdot \textbf{Time}$$

An alternative form of this equation is written by solving the equation for time:

$$\frac{\textbf{Distance}}{\textbf{Rate}} = \textbf{Time}$$

This form of the equation is useful when the total time of travel for two objects is known or the times of travel for two objects are equal.

Solve: The speed of a boat in still water is 20 mph. The boat traveled 120 mi down a river in the same amount of time it took the boat to travel 80 mi up the river. Find the rate of the river's current.

STRATEGY FOR SOLVING A UNIFORM MOTION PROBLEM

► For each object, write a numerical or variable expression for the distance, rate, and time. The results can be recorded in a table.

The unknown rate of the river's current: r

Take Note
Use the information given in the problem to fill in the distance and rate columns of the table. Fill in the time column by dividing the two expressions you wrote in each row.

	Distance	÷	Rate	=	Time
Down river	120	÷	$20 + r$	=	$\frac{120}{20 + r}$
Up river	80	÷	$20 - r$	=	$\frac{80}{20 - r}$

► Determine how the times traveled by the two objects are related. For example, it may be known that the times are equal, or the total time may be known.

$$\frac{120}{20 + r} = \frac{80}{20 - r}$$

- The time spent traveling down the river is equal to the time spent traveling up the river.

$$(20 + r)(20 - r)\frac{120}{20 + r} = (20 + r)(20 - r)\frac{80}{20 - r}$$

- Multiply each side of the equation by $(20 + r)(20 - r)$.

$$(20 - r)120 = (20 + r)80$$

- Simplify.

$$2400 - 120r = 1600 + 80r$$

- Use the Distributive Property.

$$-200r = -800$$

- Solve for r.

$$r = 4$$

The rate of the river's current is 4 mph.

EXAMPLE 2 A cyclist rode the first 20 mi of a trip at a constant rate. For the next 16 mi, the cyclist reduced the speed by 2 mph. The total time for the 36 mi was 4 h. Find the rate of the cyclist for each leg of the trip.

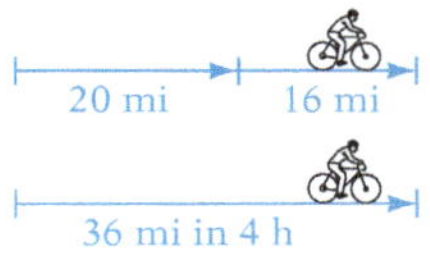

Strategy

► Rate for the first 20 mi: r

► Rate for the next 16 mi: $r - 2$

	Distance	÷	Rate	=	Time
First 20 mi	20	÷	r	=	$\frac{20}{r}$
Next 16 mi	16	÷	$r - 2$	=	$\frac{16}{r - 2}$

► The total time for the trip was 4 h.

Solution

$$\frac{20}{r} + \frac{16}{r - 2} = 4$$

- The time spent riding the first 20 mi plus the time spent riding the next 16 mi is equal to 4 h.

$$r(r - 2)\left[\frac{20}{r} + \frac{16}{r - 2}\right] = r(r - 2) \cdot 4$$

- Multiply each side of the equation by $r(r - 2)$.

$$(r - 2)20 + 16r = (r^2 - 2r)4$$
$$20r - 40 + 16r = 4r^2 - 8r$$
$$36r - 40 = 4r^2 - 8r$$
$$0 = 4r^2 - 44r + 40$$
$$0 = 4(r^2 - 11r + 10)$$

- This is a quadratic equation. Write it in standard form.

$$0 = r^2 - 11r + 10$$ • **Divide each side by 4.**

$$0 = (r - 10)(r - 1)$$ • **Solve for r by factoring.**

$$r - 10 = 0 \qquad r - 1 = 0$$
$$r = 10 \qquad r = 1$$
$$r - 2 = 10 - 2 = 8 \qquad r - 2 = 1 - 2 = -1$$

• **Find the rate for the last 16 mi.**

The solution $r = 1$ is not possible because the rate on the last 16 mi would be -1 mph.

The rate for the first 20 mi was 10 mph.
The rate for the next 16 mi was 8 mph.

Problem 2 The total time for a sailboat to sail back and forth across a lake 6 km wide was 3 h. The rate sailing back was twice the rate sailing across the lake. Find the rate of the sailboat going across the lake.

Solution See page S24.

Try Exercise 39, page 455.

9.8 Exercises

CONCEPT CHECK

1. If it takes a janitorial crew 5 h to clean a company's offices, then in x hours the crew has completed __?__ of the job.
2. If it takes an automotive crew x minutes to service a car, then the rate of work is __?__ of the job each minute.
3. Only two people worked on a job, and together they completed it. If one person completed $\frac{t}{30}$ of the job and the other person completed $\frac{t}{20}$ of the job, then $\frac{t}{30} + \frac{t}{20} =$ __?__.
4. If a plane flies 300 mph in calm air and the rate of the wind is r miles per hour, then the rate of the plane flying with the wind can be represented as __?__, and the rate of the plane flying against the wind can be represented as __?__.
5. If Jen can paint a wall in 30 min and Amelia can paint the same wall in 45 min, who has the greater rate of work?
6. Kim and Les painted a fence together in 8 h. Working alone, it would have taken Kim 12 h to paint the fence. What fraction of the fence did Kim paint? What fraction of the fence did Les paint?
7. Suppose you have a powerboat with the throttle set to move the boat at 8 mph in calm water, and the rate of the current is 4 mph. **a.** What is the speed of the boat when traveling with the current? **b.** What is the speed of the boat when traveling against the current?
8. The speed of a plane is 500 mph. There is a headwind of 50 mph. What is the speed of the plane relative to an observer on the ground?

1 Work problems (See pages 448–450.)

GETTING READY

9. One electrician can do a wiring job in 10 h. It would take the electrician's assistant 12 h to complete the same wiring job. Let t represent the amount of time it would take the electrician and the assistant to complete the job if they worked together. Complete the following table.

	Rate of work	$\cdot$	Time worked	=	Part of task completed
Electrician	?	$\cdot$	?	=	?
Assistant	?	$\cdot$	?	=	?

10. Refer to the situation presented in Exercise 9. When the wiring job is finished, the "part of task completed" is the whole task, so the sum of the parts completed by the electrician and the assistant is __?__. Use this fact and the expressions in the table in Exercise 9 to write an equation that can be solved to find the amount of time it takes for the electrician and the assistant to complete the job together: __?__ + __?__ = __?__.

© Tom Carter/Photo Edit

11. A park has two sprinklers that are used to fill a fountain. One sprinkler can fill the fountain in 3 h, whereas the second sprinkler can fill the fountain in 6 h. How long will it take to fill the fountain with both sprinklers operating?

12. One grocery clerk can stock a shelf in 20 min. A second clerk requires 30 min to stock the same shelf. How long would it take to stock the shelf if the two clerks worked together?

13. One person with a skiploader requires 12 h to transfer a large quantity of earth. With a larger skiploader, the same amount of earth can be transferred in 4 h. How long would it take to transfer the earth if both skiploaders were operated together?

14. It takes Doug 6 days to reroof a house. If Doug's son helps him, the job can be completed in 4 days. How long would it take Doug's son, working alone, to do the job?

Lee Morris/Shutterstock.com

15. One computer can solve a complex prime factorization problem in 75 h. A second computer can solve the same problem in 50 h. How long would it take both computers, working together, to solve the problem?

16. A new machine makes 10,000 aluminum cans three times faster than an older machine. With both machines operating, it takes 9 h to make 10,000 cans. How long would it take the new machine, working alone, to make 10,000 cans?

17. A small air conditioner can cool a room 5°F in 60 min. A larger air conditioner can cool the room 5°F in 40 min. How long would it take to cool the room 5°F with both air conditioners working?

18. One printing press can print the first edition of a book in 55 min. A second printing press requires 66 min to print the same number of copies. How long would it take to print the first edition of the book with both presses operating?

Lisa F. Young/Shutterstock.com

19. Two welders working together can complete a job in 6 h. One of the welders, working alone, can complete the task in 10 h. How long would it take the second welder, working alone, to complete the task?

20. Working together, Pat and Chris can reseal a driveway in 6 h. Working alone, Pat can reseal the driveway in 15 h. How long would it take Chris, working alone, to reseal the driveway?

21. Two oil pipelines can fill a small tank in 30 min. One of the pipelines, working alone, would require 45 min to fill the tank. How long would it take the second pipeline, working alone, to fill the tank?

22. A cement mason can construct a retaining wall in 8 h. A second mason requires 12 h to do the same job. After working alone for 4 h, the first mason quits. How long will it take the second mason to complete the wall?

23. With two reapers operating, a field can be harvested in 1 h. If only the newer reaper is used, the crop can be harvested in 1.5 h. How long would it take to harvest the field using only the older reaper?

24. A manufacturer of prefabricated homes has the company's employees work in teams. Team 1 can erect the Silvercrest model in 15 h. Team 2 can erect the same model in 10 h. How long would it take for Team 1 and Team 2, working together, to erect the Silvercrest model home?

bikeriderlondon/Shutterstock.com

25. One technician can wire a security alarm in 4 h, whereas it takes 6 h for a second technician to do the same job. After working alone for 2 h, the first technician quits. How long will it take the second technician to complete the wiring?

26. A wallpaper hanger requires 2 h to hang the wallpaper on one wall of a room. A second wallpaper hanger requires 4 h to hang the same amount of wallpaper. The first wallpaper hanger works alone for 1 h and then quits. How long will it take the second hanger, working alone, to finish papering the wall?

27. A large heating unit and a small heating unit are being used to heat the water in a pool. The large unit, working alone, requires 8 h to heat the pool. After both units have been operating for 2 h, the large unit is turned off. The small unit requires 9 more hours to heat the pool. How long would it take the small unit, working alone, to heat the pool?

Sudheer Sakthan/Shutterstock.com

28. Two machines fill cereal boxes at the same rate. After the two machines work together for 7 h, one machine breaks down. The second machine requires 14 more hours to finish filling the boxes. How long would it have taken one of the machines, working alone, to fill the boxes?

29. A mechanic requires 2 h to repair a transmission, whereas an apprentice requires 6 h to make the same repairs. The mechanic worked alone for 1 h and then stopped. How long will it take the apprentice, working alone, to complete the repairs?

30. A large drain and a small drain are opened to drain a pool. The large drain can empty the pool in 6 h. After both drains have been open for 1 h, the large drain becomes clogged and is closed. The small drain remains open and requires 9 more hours to empty the pool. How long would it have taken the small drain, working alone, to empty the pool?

31. It takes Sam h hours to rake the yard, and it takes Emma k hours to rake the yard, where $h > k$. Let t be the amount of time it takes Sam and Emma to rake the yard together. Is t less than k, between k and h, or greater than k?

32. Zachary and Eli picked a row of peas together in m minutes. It would have taken Zachary n minutes to pick the row of peas by himself. What fraction of the row of peas did Zachary pick? What fraction of the row of peas did Eli pick?

2 Uniform motion problems (See pages 450–452.)

GETTING READY

33. a. A plane can fly 380 mph in calm air. In the time it takes the plane to fly 1440 mi against a headwind, it could fly 1600 mi if it were flying with the wind. Let r represent the rate of the wind. Complete the following table.

	Distance	÷	Rate	=	Time
Against the wind	___?___	÷	___?___	=	___?___
With the wind	___?___	÷	___?___	=	___?___

b. Use the relationship between the expressions in the last column of the table to write an equation that can be solved to find the rate of the wind: ___?___ = ___?___.

34. Use the equation from part (b) of Exercise 33. The first step in solving this equation is to multiply each side of the equation by ___?___.

35. A camper drove 80 mi to a recreational area and then hiked 4 mi into the woods. The rate of the camper while driving was ten times the rate while hiking. The time spent hiking and driving was 3 h. Find the rate at which the camper hiked.

36. The president of a company traveled 1800 mi by jet and 300 mi on a prop plane. The rate of the jet was four times the rate of the prop plane. The entire trip took 5 h. Find the rate of the jet.

37. To assess the damage done by a fire, a forest ranger traveled 1080 mi by jet and then an additional 180 mi by helicopter. The rate of the jet was four times the rate of the helicopter. The entire trip took 5 h. Find the rate of the jet.

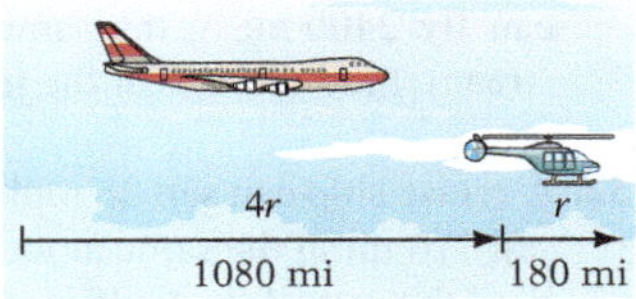

38. An engineer traveled 165 mi by car and then an additional 660 mi by plane. The rate of the plane was four times the rate of the car. The total trip took 6 h. Find the rate of the car.

39. After sailing 15 mi, a sailor changed direction and increased the boat's speed by 2 mph. An additional 19 mi was sailed at the increased speed. The total sailing time was 4 h. Find the rate of the boat for the first 15 mi.

40. On a recent trip, a trucker traveled 330 mi at a constant rate. Because of road conditions, the trucker then reduced the speed by 25 mph. An additional 30 mi was traveled at the reduced rate. The entire trip took 7 h. Find the rate of the trucker for the first 330 mi.

41. In calm water, the rate of a small rental motorboat is 15 mph. The rate of the current on the river is 3 mph. How far down the river can a family travel and still return the boat in 3 h?

42. The rate of a small aircraft in calm air is 125 mph. If the wind is currently blowing south at a rate of 15 mph, how far north can a pilot fly the plane and return it within 2 h?

43. Commuting from work to home, a lab technician traveled 10 mi at a constant rate through congested traffic. Upon reaching the expressway, the technician increased the speed by 20 mph. An additional 20 mi was traveled at the increased speed. The total time for the trip was 1 h. At what rate did the technician travel through the congested traffic?

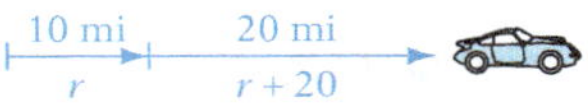

44. As part of a conditioning program, a jogger ran 8 mi in the same amount of time it took a cyclist to ride 20 mi. The rate of the cyclist was 12 mph faster than the rate of the jogger. Find the rate of the jogger and the rate of the cyclist.

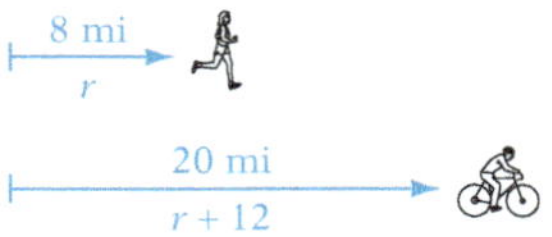

45. The speed of a boat in still water is 20 mph. The Jacksons traveled 75 mi down the Woodset River in this boat in the same amount of time it took them to return 45 mi up the river. Find the rate of the river's current.

46. A backpacker hiking into a wilderness area walked 9 mi at a constant rate and then reduced this rate by 1 mph. Another 4 mi was hiked at this reduced rate. The time required to hike the 4 mi was 1 h less than the time required to walk the 9 mi. Find the rate at which the hiker walked the first 9 mi.

© Lee Cohen/Corbis

47. An express train traveled 600 mi in the same amount of time it took a freight train to travel 360 mi. The rate of the express train was 20 mph faster than the rate of the freight train. Find the rate of each train.

48. A twin-engine plane flies 800 mi in the same amount of time it takes a single-engine plane to fly 600 mi. The rate of the twin-engine plane is 50 mph faster than the rate of the single-engine plane. Find the rate of the twin-engine plane.

49. A small motor on a fishing boat can move the boat at a rate of 6 mph in calm water. Traveling with the current, the boat can travel 24 mi in the same amount of time it takes to travel 12 mi against the current. Find the rate of the current.

50. A car is traveling at a rate that is 36 mph faster than the rate of a cyclist. The car travels 384 mi in the same time it takes the cyclist to travel 96 mi. Find the rate of the car.

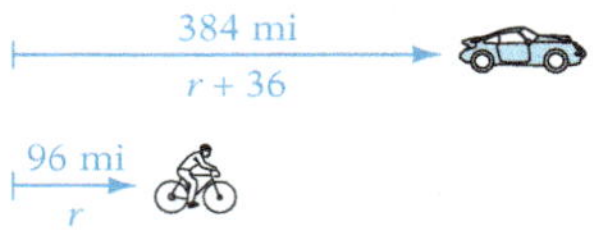

51. A commercial jet can fly 550 mph in calm air. Traveling with the jet stream, the plane can fly 2400 mi in the same amount of time it takes to fly 2000 mi against the jet stream. Find the rate of the jet stream.

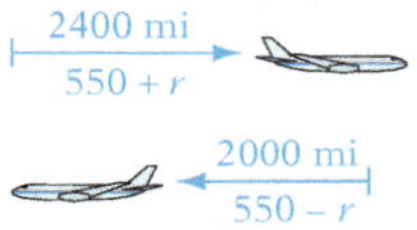

52. A cruise ship can sail 28 mph in calm water. Sailing with the Gulf Stream, the ship can sail 170 mi in the same amount of time it takes to sail 110 mi against the Gulf Stream. Find the rate of the Gulf Stream.

53. Rowing with the current of a river, a rowing team can row 25 mi in the same amount of time it takes to row 15 mi against the current. The rate of the rowing team in calm water is 20 mph. Find the rate of the current.

John Kropewnicki/Shutterstock.com

54. A plane can fly 180 mph in calm air. Flying with the wind, the plane can fly 600 mi in the same amount of time it takes to fly 480 mi against the wind. Find the rate of the wind.

For Exercises 55 and 56, use the following problem situation: A plane can fly 380 mph in calm air. In the time it takes the plane to fly 1440 mi against a headwind, it could fly 1600 mi with the wind. Use the equation $\frac{1440}{380 - r} = \frac{1600}{380 + r}$ to find the rate r of the wind.

55. Explain the meaning of $380 - r$ and $380 + r$ in terms of the problem situation.

56. Explain the meaning of $\frac{1440}{380 - r}$ and $\frac{1600}{380 + r}$ in terms of the problem situation.

APPLYING CONCEPTS

57. Work Problem One pipe can fill a tank in 2 h, a second pipe can fill the tank in 4 h, and a third pipe can fill the tank in 5 h. How long would it take to fill the tank with all three pipes operating?

58. **Work Problem** A mason can construct a retaining wall in 10 h. The mason's more experienced apprentice can do the same job in 15 h. How long would it take the mason's less experienced apprentice to do the job if, working together, all three can complete the wall in 5 h?

59. **Uniform Motion** An Outing Club traveled 18 mi by canoe and then hiked 3 mi. The rate by canoe was three times the rate on foot. The time spent walking was 1 h less than the time spent canoeing. Find the amount of time spent traveling by canoe.

60. **Uniform Motion** A motorist drove 120 mi before running out of gas and walking 4 mi to a gas station. The rate of the motorist in the car was ten times the rate walking. The time spent walking was 2 h less than the time spent driving. How long did it take for the motorist to drive the 120 mi?

PROJECTS OR GROUP ACTIVITES

61. **Uniform Motion** Because of bad weather, a bus driver reduced the usual speed along a 150-mile bus route by 10 mph. The bus arrived only 30 min later than its usual arrival time. How fast does the bus usually travel?

62. A construction project must be completed in 15 days. Twenty-five workers did one-half of the job in 10 days. How many workers are needed to complete the job on schedule?

CHAPTER 9 Summary

Key Words	Objective and Page Reference	Examples
A **rational expression** is a fraction in which the numerator or denominator is a variable expression. A rational expression is in **simplest form** when the numerator and denominator have no common factors other than 1.	[9.1.1, p. 396]	$\frac{x+4}{x-3}$ is a rational expression in simplest form.
The **reciprocal** of a fraction is that fraction with the numerator and denominator interchanged.	[9.1.3, p. 399]	The reciprocal of $\frac{x-7}{y}$ is $\frac{y}{x-7}$.
The **least common multiple** (LCM) of two or more numbers is the smallest number that contains the prime factorization of each number.	[9.2.1, p. 405]	The LCM of 8 and 12 is 24 because 24 is the smallest number that is a multiple of both 8 and 12.
The **least common denominator** (LCD) of two or more fractions is the LCM of the denominators of the fractions.	[9.2.2, p. 406]	The LCD of $\frac{1}{3x}$ and $\frac{3}{8x^2y}$ is $24x^2y$.

A **complex fraction** is a fraction in which the numerator or denominator contains one or more fractions.	[9.4.1, p. 418]	$\dfrac{x + \frac{1}{x}}{2 - \frac{3}{x}}$ is a complex fraction.
A **ratio** is the quotient of two quantities that have the same unit. A **rate** is the quotient of two quantities that have different units. A **proportion** is an equation that states the equality of two ratios or rates.	[9.5.2, pp. 425–426]	$\frac{3}{8} = \frac{15}{40}$ and $\frac{20 \text{ m}}{5 \text{ s}} = \frac{80 \text{ m}}{20 \text{ s}}$ are examples of proportions.
The equation $y = kx$ is the general form of a **direct variation equation.** The equation $y = \dfrac{k}{x}$ is the general form of an **inverse variation equation.** In these equations, k is the **constant of variation.**	[9.6.1, pp. 438, 439]	$y = 50x$ is a direct variation equation. $y = \frac{12}{x}$ is an inverse variation equation.
A **literal equation** is an equation that contains more than one variable. A **formula** is a literal equation that states rules about measurements.	[9.7.1, p. 445]	$2x + 3y = 6$ is an example of a literal equation. $A = \pi r^2$ is a literal equation that is also a formula.

Essential Rules and Procedures	Objective and Page Reference	Examples
Simplifying Rational Expressions Factor the numerator and denominator. Divide by the common factors.	[9.1.1, p. 396]	$\dfrac{x^2 - 3x - 10}{x^2 - 25} = \dfrac{(x + 2)\overset{1}{\cancel{(x - 5)}}}{(x + 5)\underset{1}{\cancel{(x - 5)}}} = \dfrac{x + 2}{x + 5}$
Multiplying Rational Expressions Multiply the numerators. Multiply the denominators. Write the answer in simplest form. $\dfrac{a}{b} \cdot \dfrac{c}{d} = \dfrac{ac}{bd}$	[9.1.2, p. 398]	$\dfrac{x^2 - 3x}{x^2 + x} \cdot \dfrac{x^2 + 5x + 4}{x^2 - 4x + 3} = \dfrac{x(x - 3)}{x(x + 1)} \cdot \dfrac{(x + 4)(x + 1)}{(x - 3)(x - 1)} = \dfrac{\overset{1}{\cancel{x}}\overset{1}{\cancel{(x - 3)}}(x + 4)\overset{1}{\cancel{(x + 1)}}}{\underset{1}{\cancel{x}}\underset{1}{\cancel{(x + 1)}}\underset{1}{\cancel{(x - 3)}}(x - 1)} = \dfrac{x + 4}{x - 1}$
Dividing Rational Expressions To divide two fractions, multiply by the reciprocal of the divisor. $\dfrac{a}{b} \div \dfrac{c}{d} = \dfrac{a}{b} \cdot \dfrac{d}{c} = \dfrac{ad}{bc}$	[9.1.3, p. 400]	$\dfrac{3x^3y^2}{8a^4b^5} \div \dfrac{6x^4y}{4a^3b^2} = \dfrac{3x^3y^2}{8a^4b^5} \cdot \dfrac{4a^3b^2}{6x^4y} = \dfrac{3x^3y^2 \cdot 4a^3b^2}{8a^4b^5 \cdot 6x^4y} = \dfrac{y}{4ab^3x}$
Finding the LCM of Polynomials Factor each polynomial completely. The LCM is the product of each factor the greatest number of times it occurs in any one factorization.	[9.2.1, p. 405]	$3x^2 + 6x = 3x(x + 2)$ $x^2 + 3x + 2 = (x + 2)(x + 1)$ $\text{LCM} = 3x(x + 2)(x + 1)$

Adding and Subtracting Rational Expressions

1. Find the LCD.
2. Write each fraction in terms of the LCD.
3. Add or subtract the numerators. The denominator of the sum or difference is the common denominator.
4. Write the answer in simplest form.

[9.3.2, pp. 411–412]

$$\frac{4x-2}{3x+12}-\frac{x-2}{x+4}=\frac{4x-2}{3(x+4)}-\frac{x-2}{x+4}$$
$$=\frac{4x-2}{3(x+4)}-\frac{x-2}{x+4}\cdot\frac{3}{3}$$
$$=\frac{4x-2}{3(x+4)}-\frac{3x-6}{3(x+4)}$$
$$=\frac{4x-2-(3x-6)}{3(x+4)}$$
$$=\frac{x+4}{3(x+4)}=\frac{1}{3}$$

Simplifying Complex Fractions

Multiply the numerator and denominator of the complex fraction by the LCD of the fractions in the numerator and denominator.

[9.4.1, p. 419]

$$\frac{\frac{1}{x}+\frac{1}{y}}{\frac{1}{x}-\frac{1}{y}}=\frac{\frac{1}{x}+\frac{1}{y}}{\frac{1}{x}-\frac{1}{y}}\cdot\frac{xy}{xy}=\frac{\frac{1}{x}\cdot xy+\frac{1}{y}\cdot xy}{\frac{1}{x}\cdot xy-\frac{1}{y}\cdot xy}$$
$$=\frac{y+x}{y-x}$$

Solving Equations Containing Fractions

Clear denominators by multiplying each side of the equation by the LCD of the fractions. Then solve for the variable.

[9.5.1, pp. 423–424]

$$\frac{1}{2a}=\frac{2}{a}-\frac{3}{8}$$
$$8a\left(\frac{1}{2a}\right)=8a\left(\frac{2}{a}-\frac{3}{8}\right)$$
$$4=16-3a$$
$$-12=-3a$$
$$4=a$$

Similar Triangles

Similar triangles have the same shape but not necessarily the same size. The ratios of corresponding sides of similar triangles are equal. The ratio of corresponding heights is equal to the ratio of corresponding sides. Proportions are used to find the length of an unknown side or the unknown height of one of two similar triangles.

[9.5.4, p. 428]

Triangles ABC and DEF are similar triangles. The ratio of corresponding sides and of corresponding heights is $\frac{1}{3}$.

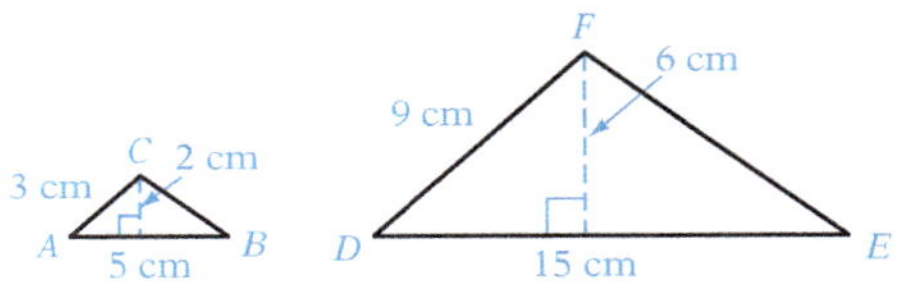

Direct and Inverse Variation Problems

Find the value of k by replacing the variables in the variation equation by the given values and solving for k. Then write the specific variation equation, replacing k by its value. Finally, substitute the given value of one of the variables into the equation and solve for the other variable.

[9.6.1, pp. 438–441]

The distance (d) a spring will stretch varies directly as the force (F) applied to the spring. If a force of 16 lb is required to stretch a spring 4 in., what distance will a force of 20 lb stretch the spring?

$$d=kF$$
$$4=k(16)$$
$$0.25=k$$

$$d=0.25F$$
$$d=0.25(20)$$
$$d=5$$

A force of 20 lb will stretch the spring 5 in.

Solving Literal Equations Rewrite the equation so that the letter being solved for is alone on one side of the equation, and all numbers and other variables are on the other side.	[9.7.1, p. 445]	Solve $A = \frac{1}{2}bh$ for b. $2(A) = 2\left(\frac{1}{2}bh\right)$ $2A = bh$ $\frac{2A}{h} = \frac{bh}{h}$ $\frac{2A}{h} = b$
Work Problems Rate of work · Time worked = Part of task completed	[9.8.1, pp. 448–449]	Pat can do a certain job in 3 h. Chris can do the same job in 5 h. How long would it take them, working together, to get the job done? $\frac{t}{3} + \frac{t}{5} = 1$
Uniform Motion Problems Distance ÷ Rate = Time	[9.8.2, pp. 450–451]	Train A's speed is 15 mph faster than Train B's speed. Train A travels 150 mi in the same amount of time it takes Train B to travel 120 mi. Find the rate of Train B. $\frac{120}{r} = \frac{150}{r+15}$

CHAPTER 9 Review Exercises

1. Multiply: $\frac{8ab^2}{15x^3y} \cdot \frac{5xy^4}{16a^2b}$

2. Add: $\frac{5}{3x-4} + \frac{4}{2x+3}$

3. Solve $4x + 3y = 12$ for x.

4. Simplify: $\frac{16x^5y^3}{24xy^{10}}$

5. Divide: $\frac{20x^2-45x}{6x^3+4x^2} \div \frac{40x^3-90x^2}{12x^2+8x}$

6. Simplify: $\dfrac{x - \dfrac{16}{5x-2}}{3x - 4 - \dfrac{88}{5x-2}}$

7. Find the LCM of $24a^2b^5$ and $36a^3b$.

8. Subtract: $\frac{5x}{3x+7} - \frac{x}{3x+7}$

9. Write each fraction in terms of the LCD.
$\frac{3}{16x}, \frac{5}{8x^2}$

10. Simplify: $\frac{2x^2-13x-45}{2x^2-x-15}$

11. Divide: $\frac{x^2-5x-14}{x^2-3x-10} \div \frac{x^2-4x-21}{x^2-9x+20}$

12. Add: $\frac{2y}{5y-7} + \frac{3}{7-5y}$

13. Multiply: $\frac{3x^3+10x^2}{10x-2} \cdot \frac{20x-4}{6x^4+20x^3}$

14. Subtract: $\frac{5x+3}{2x^2+5x-3} - \frac{3x+4}{2x^2+5x-3}$

15. Find the LCM of $5x^4(x - 7)^2$ and $15x(x - 7)$.

16. Solve: $\dfrac{6}{x - 7} = \dfrac{8}{x - 6}$

17. Solve: $\dfrac{x + 8}{x + 4} = 1 + \dfrac{5}{x + 4}$

18. Simplify: $\dfrac{12a^2b(4x - 7)}{15ab^2(7 - 4x)}$

19. Simplify: $\dfrac{5x - 1}{x^2 - 9} + \dfrac{4x - 3}{x^2 - 9} - \dfrac{8x - 1}{x^2 - 9}$

20. Triangles *ABC* and *DEF* below are similar. Find the perimeter of triangle *ABC*.

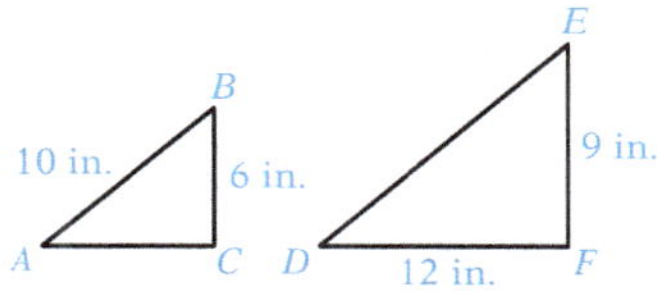

21. Solve: $\dfrac{20}{2x + 3} = \dfrac{17x}{2x + 3} - 5$

22. Simplify: $\dfrac{\dfrac{5}{x - 1} - \dfrac{3}{x + 3}}{\dfrac{6}{x + 3} + \dfrac{2}{x - 1}}$

23. Add: $\dfrac{x - 1}{x + 2} + \dfrac{3x - 2}{5 - x} + \dfrac{5x^2 + 15x - 11}{x^2 - 3x - 10}$

24. Divide: $\dfrac{18x^2 + 25x - 3}{9x^2 - 28x + 3} \div \dfrac{2x^2 + 9x + 9}{x^2 - 6x + 9}$

25. Simplify: $\dfrac{x^2 + x - 30}{15 + 2x - x^2}$

26. Solve: $\dfrac{5}{7} + \dfrac{x}{2} = 2 - \dfrac{x}{7}$

27. Simplify: $\dfrac{x + \dfrac{6}{x - 5}}{1 + \dfrac{2}{x - 5}}$

28. Multiply: $\dfrac{3x^2 + 4x - 15}{x^2 - 11x + 28} \cdot \dfrac{x^2 - 5x - 14}{3x^2 + x - 10}$

29. Solve: $\dfrac{x}{5} = \dfrac{x + 12}{9}$

30. Solve: $\dfrac{3}{20} = \dfrac{x}{80}$

31. Simplify: $\dfrac{1 - \dfrac{1}{x}}{1 - \dfrac{8x - 7}{x^2}}$

32. Solve $x - 2y = 15$ for x.

33. Add: $\dfrac{6}{a} + \dfrac{9}{b}$

34. Find the LCM of $10x^2 - 11x + 3$ and $20x^2 - 17x + 3$.

35. If y varies directly as x, and $x = 20$ when $y = 5$, find y when $x = 12$.

36. If y varies inversely as x, and $y = 400$ when $x = 5$, find y when $x = 20$.

37. Solve $i = \dfrac{100m}{c}$ for c.

38. Solve: $\dfrac{15}{x} = \dfrac{3}{8}$

39. Solve: $\dfrac{22}{2x + 5} = 2$

40. Add: $\dfrac{x + 7}{15x} + \dfrac{x - 2}{20x}$

41. Multiply: $\dfrac{16a^2 - 9}{16a^2 - 24a + 9} \cdot \dfrac{8a^2 - 13a - 6}{4a^2 - 5a - 6}$

42. Write each fraction in terms of the LCD.

$\dfrac{x}{12x^2 + 16x - 3}; \dfrac{4x^2}{6x^2 + 7x - 3}$

43. Add: $\dfrac{3}{4ab} + \dfrac{5}{4ab}$

44. Solve: $\dfrac{20}{x+2} = \dfrac{5}{16}$

45. Solve: $\dfrac{5x}{3} - \dfrac{2}{5} = \dfrac{8x}{5}$

46. Divide: $\dfrac{6a^2b^7}{25x^3y} \div \dfrac{12a^3b^4}{5x^2y^2}$

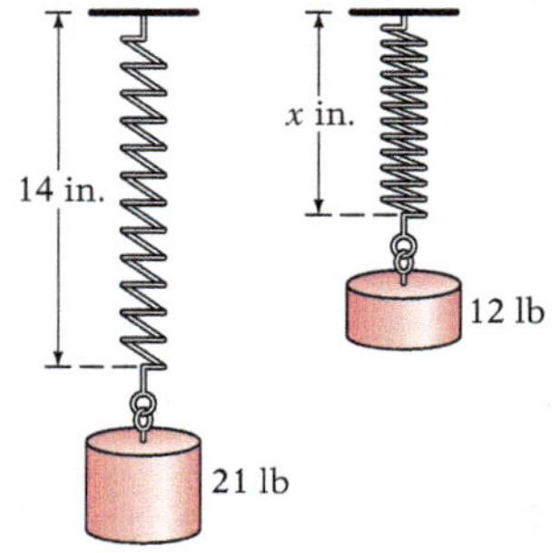

47. Masonry A brick mason can construct a patio in 3 h. If the mason works with an apprentice, they can construct the patio in 2 h. How long would it take the apprentice, working alone, to construct the patio?

48. Physics A weight of 21 lb stretches a spring 14 in. At the same rate, how far would a weight of 12 lb stretch the spring?

49. Uniform Motion The rate of a jet is 400 mph in calm air. Traveling with the wind, the jet can fly 2100 mi in the same amount of time it takes to fly 1900 mi against the wind. Find the rate of the wind.

50. Gardening A gardener uses 4 oz of insecticide to make 2 gal of garden spray. At this rate, how much additional insecticide is necessary to make 10 gal of the garden spray?

51. Work Problem One hose can fill a pool in 15 h. A second hose can fill the pool in 10 h. How long would it take to fill the pool using both hoses?

52. Uniform Motion A car travels 315 mi in the same amount of time it takes a bus to travel 245 mi. The rate of the car is 10 mph faster than that of the bus. Find the rate of the car.

53. Compensation A worker's wage (w) varies directly as the number of hours (h) worked. If \$82 is earned for working 4 h, how much is earned for working 20 h?

54. Travel The time (t) of travel on a car trip varies inversely as the speed (r) of the car. Traveling at an average speed of 55 mph, a trip took 4 h. The return trip took 5 h. Find the average rate of speed on the return trip.

CHAPTER 9 Test

1. Divide: $\dfrac{x^2+3x+2}{x^2+5x+4} \div \dfrac{x^2-x-6}{x^2+2x-15}$

2. Subtract: $\dfrac{2x}{x^2+3x-10} - \dfrac{4}{x^2+3x-10}$

3. Find the LCM of $6x - 3$ and $2x^2 + x - 1$.

4. Solve: $\dfrac{3}{x+4} = \dfrac{5}{x+6}$

5. Multiply: $\dfrac{x^3y^4}{x^2-4x+4} \cdot \dfrac{x^2-x-2}{x^6y^4}$

6. Simplify: $\dfrac{1+\dfrac{1}{x}-\dfrac{12}{x^2}}{1+\dfrac{2}{x}-\dfrac{8}{x^2}}$

7. Write each fraction in terms of the LCD.

$\dfrac{3}{x^2-2x}, \dfrac{x}{x^2-4}$

8. Solve $3x + 5y + 15 = 0$ for x.

9. Solve: $\frac{6}{x} - 2 = 1$

10. Subtract: $\frac{2}{2x - 1} - \frac{3}{3x + 1}$

11. Divide: $\frac{x^2 - x - 56}{x^2 + 8x + 7} \div \frac{x^2 - 13x + 40}{x^2 - 4x - 5}$

12. Subtract: $\frac{3x}{x^2 + 5x - 24} - \frac{9}{x^2 + 5x - 24}$

13. Find the LCM of $3x^2 + 6x$ and $2x^2 + 8x + 8$.

14. Simplify: $\frac{x^2 - 7x + 10}{25 - x^2}$

15. Solve: $\frac{3x}{x - 3} - 2 = \frac{10}{x - 3}$

16. Solve $f = v + at$ for t.

17. Simplify: $\frac{12x^4y^2}{18xy^7}$

18. Subtract: $\frac{2}{2x - 1} - \frac{1}{x + 1}$

19. Solve: $\frac{2}{x - 2} = \frac{12}{x + 3}$

20. Multiply: $\frac{x^5y^3}{x^2 - x - 6} \cdot \frac{x^2 - 9}{x^2y^4}$

21. Write each fraction in terms of the LCD.
$\frac{3y}{x(1 - x)}; \frac{x}{(x + 1)(x - 1)}$

22. Simplify: $\dfrac{1 - \frac{2}{x} - \frac{15}{x^2}}{1 - \frac{25}{x^2}}$

23. **Physics** The distance (d) a spring will stretch varies directly as the weight (w) on the spring. A weight of 5 lb will stretch a spring 2 in. How far will a weight of 28 lb stretch the spring?

24. **Mixture Problem** A salt solution is formed by mixing 4 lb of salt with 10 gal of water. At this rate, how many additional pounds of salt are required for 15 gal of water?

25. **Uniform Motion** A small plane can fly 110 mph in calm air. Flying with the wind, the plane can fly 260 mi in the same amount of time it takes to fly 180 mi against the wind. Find the rate of the wind.

26. **Work Problem** One pipe can fill a tank in 9 min. A second pipe requires 18 min to fill the tank. How long would it take both pipes, working together, to fill the tank?

27. **Physics** The volume (V) of a gas varies inversely as the pressure (P), assuming the temperature remains constant. The pressure of the gas in a balloon is 6 psi when the volume is 2.5 ft^3. Find the volume of the balloon when the pressure is increased to 12 psi.

Cumulative Review Exercises

1. Evaluate: $-|-17|$

2. Evaluate: $-\frac{3}{4} \cdot (2)^3$

3. Simplify: $\left(\frac{2}{3}\right)^2 \div \left(\frac{3}{2} - \frac{2}{3}\right) + \frac{1}{2}$

4. Evaluate $-a^2 + (a - b)^2$ when $a = -2$ and $b = 3$.

5. Simplify: $-2x - (-3y) + 7x - 5y$

6. Simplify: $2[3x - 7(x - 3) - 8]$

7. Solve: $3 - \frac{1}{4}x = 8$

8. Solve: $3[x - 2(x - 3)] = 2(3 - 2x)$

9. Find $16\frac{2}{3}\%$ of 60.

10. Solve: $\frac{5}{9}x < 1$

11. Solve: $x - 2 \geq 4x - 47$

12. Graph: $y = 2x - 1$

13. Graph: $f(x) = 3x + 2$

14. Graph: $x = 3$

15. Graph the solution set of $5x + 2y < 6$.

16. Find the range of the function given by the equation $f(x) = -2x + 11$ if the domain is $\{-8, -4, 0, 3, 7\}$.

17. Evaluate $f(x) = 4x - 3$ at $x = 10$.

18. Solve by substitution: $\begin{aligned} 6x - y &= 1 \\ y &= 3x + 1 \end{aligned}$

19. Solve by the addition method: $\begin{aligned} 2x - 3y &= 4 \\ 4x + y &= 1 \end{aligned}$

20. Multiply: $(3xy^4)(-2x^3y)$

21. Simplify: $(a^4b^3)^5$

22. Simplify: $\frac{a^2b^{-5}}{a^{-1}b^{-3}}$

23. Multiply: $(a - 3b)(a + 4b)$

24. Divide: $\frac{15b^4 - 5b^2 + 10b}{5b}$

25. Divide: $(x^3 - 8) \div (x - 2)$

26. Factor: $12x^2 - x - 1$

27. Factor: $y^2 - 7y + 6$

28. Factor: $2a^3 + 7a^2 - 15a$

29. Factor: $4b^2 - 100$

30. Solve: $(x + 3)(2x - 5) = 0$

31. Simplify: $\frac{x^2 + 3x - 28}{16 - x^2}$

32. Divide: $\frac{x^2 - 3x - 10}{x^2 - 4x - 12} \div \frac{x^2 - x - 20}{x^2 - 2x - 24}$

33. Subtract: $\frac{6}{3x - 1} - \frac{2}{x + 1}$

34. Solve: $\frac{4x}{x - 3} - 2 = \frac{8}{x - 3}$

35. Solve $f = v + at$ for a.

36. Translate "the difference between five times a number and eighteen is the opposite of three" into an equation and solve.

37. **Investments** An investment of $5000 is made at an annual simple interest rate of 7%. How much additional money must be invested at an annual simple interest rate of 11% so that the total interest earned is 9% of the total investment?

38. **Metallurgy** A silversmith mixes 60 g of an alloy that is 40% silver with 120 g of another silver alloy. The resulting alloy is 60% silver. Find the percent of silver in the 120-gram alloy.

39. **Geometry** The length of the base of a triangle is 2 in. less than twice the height. The area of the triangle is 30 in^2. Find the base and height of the triangle.

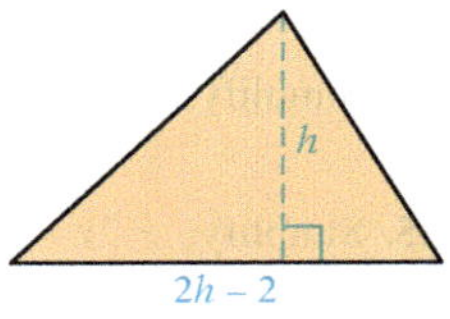

40. **Work Problem** One water pipe can fill a tank in 12 min. A second pipe requires 24 min to fill the tank. How long would it take both pipes, working together, to fill the tank?

CHAPTER 10

Radical Expressions

Digital Vision

Focus on Success

What resources do you use when you need help in this course? You know to read and reread the text when you are having difficulty understanding a concept. Instructors are available to help you during their office hours. Most schools have a math center where students can get help. Some schools have a tutoring program. You might also ask a student who has been successful in this class for assistance. (See Habits of Successful Students, page AIM-6.)

OBJECTIVES

10.1
1. Simplify numerical radical expressions
2. Simplify variable radical expressions

10.2
1. Add and subtract radical expressions

10.3
1. Multiply radical expressions
2. Divide radical expressions

10.4
1. Solve equations containing one or more radical expressions
2. Application problems

PREP TEST

Are you ready to succeed in this chapter?
Take the Prep Test below to find out if you are ready to learn the new material.

1. Evaluate $-|-14|$.
2. Simplify: $3x^2y - 4xy^2 - 5x^2y$
3. Solve: $1.5h = 21$
4. Solve: $3x - 2 = 5 - 2x$
5. Simplify: $x^3 \cdot x^3$
6. Simplify: $(x + y)^2$
7. Simplify: $(2x - 3)^2$
8. Simplify: $(a - 5)(a + 5)$
9. Simplify: $(2 - 3v)(2 + 3v)$
10. Simplify: $\dfrac{2x^4y^3}{18x^2y}$

10.1 Introduction to Radical Expressions

OBJECTIVE 1 Simplify numerical radical expressions

Point of Interest

The radical symbol was first used in 1525, when it was written as $\surd$. Some historians suggest that the radical symbol also developed into the symbols for "less than" and "greater than." Because typesetters of that time did not want to make additional symbols, the radical was rotated to the position $\setminus$ and used as a "greater than" symbol and rotated to $\leftarrow$ and used for the "less than" symbol. Other evidence, however, suggests that the "less than" and "greater than" symbols were developed independently of the radical symbol.

A **square root** of a positive number x is a number whose square is x.

A square root of 16 is 4 because $4^2 = 16$.

A square root of 16 is -4 because $(-4)^2 = 16$.

Every positive number has two square roots, one a positive number and one a negative number. The symbol $\sqrt{}$, called a **radical sign,** is used to indicate the positive or **principal square root** of a number. For example, $\sqrt{16} = 4$ and $\sqrt{25} = 5$. The number or variable expression under the radical sign is called the **radicand.**

When the negative square root of a number is to be found, a negative sign is placed in front of the radical. For example, $-\sqrt{16} = -4$ and $-\sqrt{25} = -5$.

The square of an integer is a **perfect square.** 49, 81, and 144 are examples of perfect squares.

$$7^2 = 49$$
$$9^2 = 81$$
$$12^2 = 144$$

The principal square root of a perfect-square integer is a positive integer.

$$\sqrt{49} = 7$$
$$\sqrt{81} = 9$$
$$\sqrt{144} = 12$$

The chart below shows square roots of some perfect squares.

Square Roots of Perfect Squares

$\sqrt{1} = 1$	$\sqrt{16} = 4$	$\sqrt{49} = 7$	$\sqrt{100} = 10$
$\sqrt{4} = 2$	$\sqrt{25} = 5$	$\sqrt{64} = 8$	$\sqrt{121} = 11$
$\sqrt{9} = 3$	$\sqrt{36} = 6$	$\sqrt{81} = 9$	$\sqrt{144} = 12$

Take Note

Recall that a factor of a number divides the number evenly. For example, 6 is a factor of 18, and 9 is also a factor of 18. Note that 9 is a *perfect-square factor* of 18, whereas 6 is not a perfect-square factor of 18.

If an integer is not a perfect square, its square root can only be approximated. For example, 2 and 7 are not perfect squares. Thus their square roots can only be approximated. These numbers are **irrational numbers.** Their decimal representations never terminate or repeat.

$$\sqrt{2} \approx 1.4142135\ldots$$
$$\sqrt{7} \approx 2.6457513\ldots$$

The approximate square roots of numbers that are not perfect squares can be found using a calculator. The square roots can be rounded to any given place value.

Radical expressions that contain radicands that are not perfect squares are generally written in *simplest form.* A radical expression is in **simplest form** when the radicand contains no factor greater than 1 that is a perfect square. For example, $\sqrt{50}$ is not in simplest form because 25 is a perfect-square factor of 50. The radical expression $\sqrt{15}$ is in simplest form because there is no perfect-square factor of 15 that is greater than 1.

The Product Property of Square Roots is used to simplify radical expressions.

THE PRODUCT PROPERTY OF SQUARE ROOTS

If a and b are positive real numbers, then $\sqrt{ab} = \sqrt{a} \cdot \sqrt{b}$.

Focus on simplifying a numerical radical expression

Simplify. **A.** $\sqrt{72}$ **B.** $\sqrt{360}$ **C.** $\sqrt{-16}$

A. Write the radicand as the product of a perfect square and a factor that does not contain a perfect square. $\sqrt{72} = \sqrt{36 \cdot 2}$

Use the Product Property of Square Roots to write the expression as a product. $= \sqrt{36}\sqrt{2}$

Simplify $\sqrt{36}$. $= 6\sqrt{2}$

Note in this example that 72 must be written as the product of a perfect square and *a factor that does not contain a perfect square.* Therefore, it would not be correct to rewrite $\sqrt{72}$ as $\sqrt{9 \cdot 8}$ and simplify the expression as shown at the right. Although 9 is a perfect-square factor of 72, 8 contains a perfect square $(8 = 4 \cdot 2)$, and $\sqrt{4}$ can be simplified. Therefore, $\sqrt{8}$ is not in simplest form. Remember to find the *largest* perfect square that is a factor of the radicand.

$$\begin{aligned}\sqrt{72} &= \sqrt{9 \cdot 8}\\ &= \sqrt{9}\sqrt{8}\\ &= 3\sqrt{8}\end{aligned}$$

Not in simplest form

B. Write the radicand as the product of a perfect square and a factor that does not contain a perfect square. $\sqrt{360} = \sqrt{36 \cdot 10}$

Use the Product Property of Square Roots to write the expression as a product. $= \sqrt{36}\sqrt{10}$

Simplify $\sqrt{36}$. $= 6\sqrt{10}$

In this example, note that $\sqrt{360} = 6\sqrt{10}$. The two expressions are different representations of the same number. Using a calculator, we find that $\sqrt{360} \approx 18.973666$ and $6\sqrt{10} \approx 6(3.1622777) = 18.973666$.

C. $\sqrt{-16}$

Because the square of any real number is positive, there is no real number whose square is -16.

$\sqrt{-16}$ is not a real number.

EXAMPLE 1 Simplify: $\sqrt{252}$

Solution $\sqrt{252} = \sqrt{36 \cdot 7}$
- **Write the radicand as the product of a perfect square and a factor that does not contain a perfect square.**

$= \sqrt{36}\sqrt{7}$
- **Use the Product Property of Square Roots to write the expression as a product.**

$= 6\sqrt{7}$
- **Simplify $\sqrt{36}$.**

Problem 1 Simplify: $\sqrt{216}$

Solution See page S24.

Try Exercise 19, page 471.

EXAMPLE 2 Simplify: $-3\sqrt{90}$

Solution

$-3\sqrt{90} = -3\sqrt{9 \cdot 10}$ • Write the radicand as the product of a perfect square and a factor that does not contain a perfect square.

$= -3\sqrt{9}\,\sqrt{10}$ • Use the Product Property of Square Roots to write the expression as a product.

$= -3 \cdot 3\sqrt{10}$ • Simplify $\sqrt{9}$.

$= -9\sqrt{10}$ • Multiply $-3 \cdot 3$.

Problem 2 Simplify: $-5\sqrt{32}$

Solution See page S24.

Try Exercise 23, page 471.

OBJECTIVE 2 Simplify variable radical expressions

How It's Used

Statisticians use math to make sense of large quantities of raw data. Many statistics formulas involve a radical expression. One such formula is the formula for *standard deviation,* a measure of how spread out a data set is from its average value.

Variable expressions that contain radicals do not always represent real numbers.

The variable expression at the right does not represent a real number when x is a negative number, such as -4.

$\sqrt{x^3}$

$\sqrt{(-4)^3} = \sqrt{-64} \leftarrow$ Not a real number

Now consider the expression $\sqrt{x^2}$ and evaluate this expression for $x = -2$ and $x = 2$.

$\sqrt{x^2}$

$\sqrt{(-2)^2} = \sqrt{4} = 2 = |-2|$

$\sqrt{2^2} = \sqrt{4} = 2 = |2|$

This result suggests the following.

THE SQUARE ROOT OF a^2

For any real number a, $\sqrt{a^2} = |a|$.

If $a \geq 0$, then $\sqrt{a^2} = a$.

EXAMPLES

1. $\sqrt{5^2} = 5$
2. $\sqrt{9^2} = 9$

In order to avoid variable expressions that do not represent real numbers, and so that absolute value signs are not needed for certain expressions, the variables in this chapter will represent *positive* numbers unless otherwise stated.

A variable or a product of variables written in exponential form is a **perfect square** when each exponent is an even number.

To find the square root of a perfect square, remove the radical sign and divide each exponent by 2.

Focus on finding the square root of a variable expression in which the radicand is a perfect square

Simplify: $\sqrt{a^6}$

Remove the radical sign, and divide the exponent by 2. $\sqrt{a^6} = a^3$

A variable radical expression is not in simplest form when the radicand contains a factor greater than 1 that is a perfect square.

Focus on simplifying the square root of a variable raised to an odd power

Simplify: $\sqrt{x^7}$

Write x^7 as the product of x and a perfect square. $\sqrt{x^7} = \sqrt{x^6 \cdot x}$

Use the Product Property of Square Roots. $= \sqrt{x^6}\sqrt{x}$

Simplify the square root of the perfect square. $= x^3\sqrt{x}$

EXAMPLE 3 Simplify: $\sqrt{b^{15}}$

Solution $\sqrt{b^{15}} = \sqrt{b^{14} \cdot b} = \sqrt{b^{14}} \cdot \sqrt{b} = b^7\sqrt{b}$

Problem 3 Simplify: $\sqrt{y^{19}}$

Solution See page S24.

Try Exercise 63, page 472.

Focus on simplifying the square root of a variable radical expression

Simplify: $3x\sqrt{8x^3y^{13}}$

Write the radicand as the product of a perfect square and factors that do not contain a perfect square. $3x\sqrt{8x^3y^{13}} = 3x\sqrt{4x^2y^{12}(2xy)}$

Use the Product Property of Square Roots. $= 3x\sqrt{4x^2y^{12}}\sqrt{2xy}$

Simplify. $= 3x \cdot 2xy^6\sqrt{2xy}$

$= 6x^2y^6\sqrt{2xy}$

EXAMPLE 4 Simplify. **A.** $\sqrt{24x^5}$ **B.** $2a\sqrt{18a^3b^{10}}$

Solution **A.** $\sqrt{24x^5} = \sqrt{4x^4 \cdot 6x}$
- **Write the radicand as the product of a perfect square and factors that do not contain a perfect square.**

$= \sqrt{4x^4}\sqrt{6x}$
- **Use the Product Property of Square Roots.**

$= 2x^2\sqrt{6x}$
- **Simplify $\sqrt{4x^4}$.**

B. $2a\sqrt{18a^3b^{10}} = 2a\sqrt{9a^2b^{10} \cdot 2a}$
- **Write the radicand as the product of a perfect square and factors that do not contain a perfect square.**

$= 2a\sqrt{9a^2b^{10}}\sqrt{2a}$
- **Use the Product Property of Square Roots.**

$= 2a \cdot 3ab^5\sqrt{2a}$
- **Simplify $\sqrt{9a^2b^{10}}$.**

$= 6a^2b^5\sqrt{2a}$
- **Multiply $2a$ and $3ab^5$.**

Problem 4 Simplify. **A.** $\sqrt{45b^7}$ **B.** $3a\sqrt{28a^9b^{18}}$

Solution See page S24.

Try Exercise 75, page 472.

Focus on simplifying the square root of an expression containing the square of a binomial

Simplify: $\sqrt{25(x+2)^2}$

25 is a perfect square. $(x+2)^2$ is a perfect square.

$$\sqrt{25(x+2)^2} = 5(x+2) = 5x + 10$$

EXAMPLE 5 Simplify: $\sqrt{16(x+5)^2}$

Solution $\sqrt{16(x+5)^2} = 4(x+5)$ • 16 is a perfect square. $(x+5)^2$ is a perfect square.

$= 4x + 20$

Problem 5 Simplify: $\sqrt{49(a+3)^2}$

Solution See page S24.

Try Exercise 95, page 472.

10.1 Exercises

CONCEPT CHECK

1. The symbol $\sqrt{}$ is called a __?__.
2. The number or variable expression under the radical sign is called the __?__.
3. The square of an integer is a(n) __?__.
4. The square root of an integer that is not a perfect square is a(n) __?__ number.
5. The __?__ Property of Square Roots states that, if a and b are positive real numbers, then $\sqrt{ab} = \sqrt{a} \cdot \sqrt{b}$.
6. When a perfect square is written in exponential notation, the exponents are multiples of the number __?__.

1 Simplify numerical radical expressions (See pages 466–468.)

7. Show by example that a positive number has two square roots.
8. Why is the square root of a negative number not a real number?
9. Which of the numbers 2, 9, 20, 25, 50, 81, and 100 are *not* perfect squares?
10. Write down a number that has a perfect-square factor that is greater than 1.
11. Write a sentence or two that you could email a friend to explain the concept of a perfect-square factor.
12. Name the perfect-square factors of 540. What is the largest perfect-square factor of 540?

GETTING READY

13. In the expression $\sqrt{a}$, the symbol $\sqrt{\ }$ is called the __?__, and a is called the __?__.

14. Simplify: $\sqrt{98}$

$\sqrt{98} = \sqrt{\underline{\ ?\ } \cdot 2}$ • Write 98 as the product of a perfect-square factor and a factor that does not contain a perfect square.

$= \sqrt{\underline{\ ?\ }}\sqrt{\underline{\ ?\ }}$ • Use the Product Property of Square Roots to write the expression as the product of two square roots.

$= \underline{\ ?\ }\sqrt{2}$ • Take the square root of the perfect square.

Simplify.

15. $\sqrt{16}$ **16.** $\sqrt{64}$ **17.** $\sqrt{49}$ **18.** $\sqrt{144}$ **19.** $\sqrt{32}$ **20.** $\sqrt{50}$

21. $\sqrt{8}$ **22.** $\sqrt{12}$ **23.** $-6\sqrt{18}$ **24.** $-3\sqrt{48}$ **25.** $5\sqrt{40}$ **26.** $2\sqrt{28}$

27. $\sqrt{15}$ **28.** $\sqrt{21}$ **29.** $\sqrt{29}$ **30.** $\sqrt{13}$ **31.** $-9\sqrt{72}$ **32.** $-11\sqrt{80}$

33. $\sqrt{45}$ **34.** $\sqrt{0}$ **35.** $6\sqrt{128}$ **36.** $9\sqrt{288}$ **37.** $\sqrt{300}$ **38.** $5\sqrt{180}$

39. $7\sqrt{98}$ **40.** $\sqrt{250}$ **41.** $\sqrt{120}$ **42.** $\sqrt{96}$ **43.** $\sqrt{160}$ **44.** $\sqrt{444}$

Complete Exercises 45 and 46 without using a calculator.

45. Is the given number a rational number, an irrational number, or a number that is not a real number?

a. $\sqrt{0.16}$ **b.** $-\sqrt{18}$ **c.** $\sqrt{\sqrt{36}}$ **d.** $\sqrt{-25}$

46. Find consecutive integers m and n such that the given number is between m and n, or state that the given number is not a real number.

a. $-\sqrt{115}$ **b.** $-\sqrt{-90}$ **c.** $\sqrt{\sqrt{64}}$ **d.** $\sqrt{200}$

Find the decimal approximation to the nearest thousandth.

47. $\sqrt{240}$ **48.** $\sqrt{300}$ **49.** $\sqrt{288}$ **50.** $\sqrt{600}$

51. $\sqrt{245}$ **52.** $\sqrt{525}$ **53.** $\sqrt{352}$ **54.** $\sqrt{363}$

2 Simplify variable radical expressions (See pages 468–470.)

55. How can you tell whether a variable exponential expression is a perfect square?

56. When is a radical expression in simplest form?

GETTING READY

57. $16x^6y^8$ is a perfect square because $16 = (\underline{\ ?\ })^2$, $x^6 = (\underline{\ ?\ })^2$, and $y^8 = (\underline{\ ?\ })^2$.

58. To simplify $\sqrt{75x^{15}}$, write $75x^{15}$ as the product of a perfect square and a factor that does not contain a perfect square. Then use the Product Property of Square Roots and simplify:

$\sqrt{75x^{15}} = \sqrt{\underline{\ ?\ } \cdot 3x} = \sqrt{\underline{\ ?\ }}\sqrt{3x} = \underline{\ ?\ }\sqrt{3x}$

Simplify.

59. $\sqrt{x^6}$

60. $\sqrt{x^{12}}$

61. $\sqrt{y^{15}}$

62. $\sqrt{y^{11}}$

63. $\sqrt{a^{20}}$

64. $\sqrt{a^{16}}$

65. $\sqrt{x^4y^4}$

66. $\sqrt{x^{12}y^8}$

67. $\sqrt{4x^4}$

68. $\sqrt{25y^8}$

69. $\sqrt{24x^2}$

70. $\sqrt{18y^4}$

71. $\sqrt{x^3y^7}$

72. $\sqrt{a^{15}b^5}$

73. $\sqrt{a^3b^{11}}$

74. $\sqrt{x^9y^7}$

75. $\sqrt{60x^5}$

76. $\sqrt{72y^7}$

77. $\sqrt{49a^4b^8}$

78. $\sqrt{144x^2y^8}$

79. $\sqrt{18x^5y^7}$

80. $\sqrt{32a^5b^{15}}$

81. $\sqrt{40x^{11}y^7}$

82. $\sqrt{72x^9y^3}$

83. $\sqrt{80a^9b^{10}}$

84. $\sqrt{96a^5b^7}$

85. $-2\sqrt{16a^2b^3}$

86. $-5\sqrt{25a^4b^7}$

87. $x\sqrt{x^4y^2}$

88. $y\sqrt{x^3y^6}$

89. $-4\sqrt{20a^4b^7}$

90. $-5\sqrt{12a^3b^4}$

91. $3x\sqrt{12x^2y^7}$

92. $4y\sqrt{18x^5y^4}$

93. $2x^2\sqrt{8x^2y^3}$

94. $3y^2\sqrt{27x^4y^3}$

95. $\sqrt{25(a+4)^2}$

96. $\sqrt{81(x+y)^4}$

97. $\sqrt{4(x+2)^4}$

98. $\sqrt{9(x+2)^2}$

99. $\sqrt{x^2+4x+4}$

100. $\sqrt{b^2+8b+16}$

101. $\sqrt{y^2+2y+1}$

102. $\sqrt{a^2+6a+9}$

For Exercises 103 to 106, a is a positive integer. State whether the expression represents a rational number or an irrational number.

103. $\sqrt{100a^6}$

104. $\sqrt{9a^9}$

105. $\sqrt{\sqrt{25a^{16}}}$

106. $\sqrt{\sqrt{81a^8}}$

APPLYING CONCEPTS

107. **Credit Cards** See the news clipping at the right. The equation $N = 2.3\sqrt{S}$, where S is a student's year in college, can be used to find the average number of credit cards N that a student has. Use this equation to find the average number of credit cards for **a.** a first-year student, **b.** a sophomore, **c.** a junior, and **d.** a senior. Round to the nearest tenth.

In the News

Student Credit Card Debt Grows

With each advancing year in college, students acquire more credit cards and accumulate more debt. The average credit card balance for a first-year student is $1585, for a sophomore is $1581, for a junior is $2000, and for a senior or fifth-year student is $2864.

Source: Nellie Mae

108. **Traffic Safety** Traffic accident investigators can estimate the speed S, in miles per hour, of a car from the length of its skid mark by using the formula $S = \sqrt{30fl}$, where f is the coefficient of friction (which depends on the type of road surface) and l is the length of the skid mark in feet. Suppose the coefficient of friction is 1.2 and the length of a skid mark is 60 ft. Determine the speed of the car **a.** as a radical expression in simplest form and **b.** rounded to the nearest integer.

109. **Aviation** The distance a pilot in an airplane can see to the horizon can be approximated by the equation $d = 1.2\sqrt{h}$, where d is the distance to the horizon in miles and h is the height of the plane in feet. For a pilot flying at an altitude of 5000 ft, what is the distance to the horizon? Round to the nearest tenth.

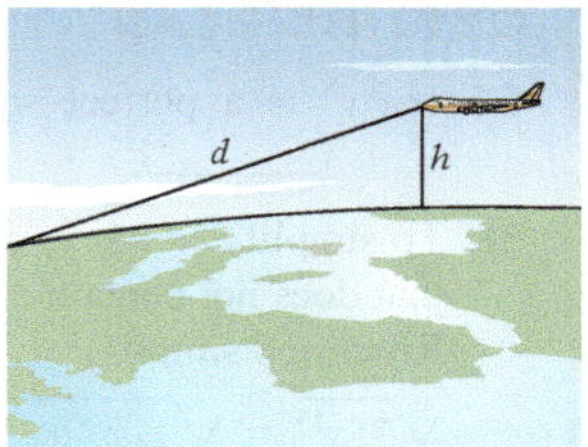

110. Given $f(x) = \sqrt{2x-1}$, find each of the following. Write your answer in simplest form.

a. $f(1)$ **b.** $f(5)$ **c.** $f(14)$

111. **Geometry** The area of a square is 76 cm^2. Find the length of a side of the square. Round to the nearest tenth.

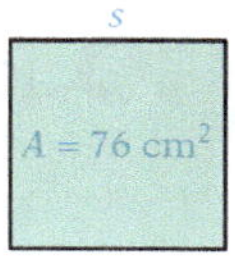

Assuming x can be any real number, for what values of x is the radical expression a real number? Write the answer as an inequality or write "all real numbers."

112. $\sqrt{x}$

113. $\sqrt{4x}$

114. $\sqrt{x-2}$

115. $\sqrt{x+5}$

116. $\sqrt{6-4x}$

117. $\sqrt{5-2x}$

118. $\sqrt{x^2+7}$

119. $\sqrt{x^2+1}$

120. Describe in your own words how to simplify a radical expression.

121. Explain why $2\sqrt{2}$ is in simplest form and $\sqrt{8}$ is not in simplest form.

PROJECTS OR GROUP ACTIVITIES

122. If a and b are positive real numbers, does $\sqrt{a+b} = \sqrt{a} + \sqrt{b}$? If not, give an example in which the expressions are not equal.

123. **Expressway On-Ramp Curves** Highway engineers design expressway on-ramps with both efficiency and safety in mind. An on-ramp cannot be built with too sharp a curve, or the speed at which drivers take the curve will cause them to skid off the road. However, an on-ramp built with too wide a curve requires more building material, takes longer to travel, and uses more land. The following formula states the relationship between the maximum speed at which a car can travel around a curve without skidding: $v = \sqrt{2.5r}$. In this formula, v is speed in miles per hour and r is the radius, in feet, of an unbanked curve.

a. Use a graphing calculator to graph this equation. Use a window of Xmin = 0, Xmax = 2000, Ymin = 0, Ymax = 100. What do the values used for Xmin, Xmax, Ymin, and Ymax represent?

b. As the radius of the curve increases, does the maximum safe speed increase or decrease? Explain your answer on the basis of the graph of the equation.

c. Use the graph to approximate the maximum safe speed when the radius of the curve is 100 ft. Round to the nearest whole number.

d. Use the graph to approximate the radius of the curve for which the maximum safe speed is 40 mph. Round to the nearest whole number.

e. Dario Franchitti won the Indianapolis 500 in 2010 driving at an average speed of 161.623 mph. At this speed, should the radius of an unbanked curve be more or less than 1 mi?

10.2 Addition and Subtraction of Radical Expressions

OBJECTIVE 1 Add and subtract radical expressions

The Distributive Property is used to simplify the sum or difference of radical expressions with the same radicand.

$$5\sqrt{2} + 3\sqrt{2} = (5 + 3)\sqrt{2} = 8\sqrt{2}$$
$$6\sqrt{2x} - 4\sqrt{2x} = (6 - 4)\sqrt{2x} = 2\sqrt{2x}$$

Radical expressions that are in simplest form and have different radicands cannot be simplified by the Distributive Property.

$2\sqrt{3} + 4\sqrt{2}$ cannot be simplified by the Distributive Property.

To simplify the sum or difference of radical expressions, simplify each term. Then use the Distributive Property.

Focus on subtracting numerical radical expressions

Subtract: $4\sqrt{8} - 10\sqrt{2}$

Simplify each term.

$$\begin{aligned} 4\sqrt{8} - 10\sqrt{2} &= 4\sqrt{4\cdot 2} - 10\sqrt{2} \\ &= 4\sqrt{4}\sqrt{2} - 10\sqrt{2} \\ &= 4\cdot 2\sqrt{2} - 10\sqrt{2} \\ &= 8\sqrt{2} - 10\sqrt{2} \\ &= (8 - 10)\sqrt{2} \\ &= -2\sqrt{2} \end{aligned}$$

Subtract by using the Distributive Property.

EXAMPLE 1 Simplify. **A.** $5\sqrt{2} - 3\sqrt{2} + 12\sqrt{2}$ **B.** $3\sqrt{12} - 5\sqrt{27}$

Solution **A.** $5\sqrt{2} - 3\sqrt{2} + 12\sqrt{2}$

$= (5 - 3 + 12)\sqrt{2}$ • **Use the Distributive Property.**

$= 14\sqrt{2}$

B. $3\sqrt{12} - 5\sqrt{27} = 3\sqrt{4\cdot 3} - 5\sqrt{9\cdot 3}$ • **Simplify each term.**

$= 3\sqrt{4}\sqrt{3} - 5\sqrt{9}\sqrt{3}$

$= 3\cdot 2\sqrt{3} - 5\cdot 3\sqrt{3}$

$= 6\sqrt{3} - 15\sqrt{3}$

$= (6 - 15)\sqrt{3}$ • **Subtract by using the Distributive Property.**

$= -9\sqrt{3}$

Problem 1 Simplify. **A.** $9\sqrt{3} + 3\sqrt{3} - 18\sqrt{3}$ **B.** $2\sqrt{50} - 5\sqrt{32}$

Solution See page S24.

Try Exercise 45, page 476.

Focus on subtracting variable radical expressions

Subtract: $8\sqrt{18x} - 2\sqrt{32x}$

Simplify each term.

$$8\sqrt{18x} - 2\sqrt{32x} = 8\sqrt{9}\,\sqrt{2x} - 2\sqrt{16}\,\sqrt{2x}$$
$$= 8 \cdot 3\sqrt{2x} - 2 \cdot 4\sqrt{2x}$$
$$= 24\sqrt{2x} - 8\sqrt{2x}$$

Use the Distributive Property to subtract the radical expressions.

$$= (24 - 8)\sqrt{2x}$$
$$= 16\sqrt{2x}$$

EXAMPLE 2 Simplify.

A. $3\sqrt{12x^3} - 2x\sqrt{3x}$ **B.** $2x\sqrt{8y} - 3\sqrt{2x^2y} + 2\sqrt{32x^2y}$

Solution **A.** $3\sqrt{12x^3} - 2x\sqrt{3x} = 3\sqrt{4x^2}\,\sqrt{3x} - 2x\sqrt{3x}$ • Simplify each term.

$$= 3 \cdot 2x\sqrt{3x} - 2x\sqrt{3x}$$
$$= 6x\sqrt{3x} - 2x\sqrt{3x}$$
$$= (6x - 2x)\sqrt{3x}$$ • Subtract by using the Distributive Property.
$$= 4x\sqrt{3x}$$

B. $2x\sqrt{8y} - 3\sqrt{2x^2y} + 2\sqrt{32x^2y}$

$$= 2x\sqrt{4}\,\sqrt{2y} - 3\sqrt{x^2}\,\sqrt{2y} + 2\sqrt{16x^2}\,\sqrt{2y}$$ • Simplify each term.
$$= 2x \cdot 2\sqrt{2y} - 3 \cdot x\sqrt{2y} + 2 \cdot 4x\sqrt{2y}$$
$$= 4x\sqrt{2y} - 3x\sqrt{2y} + 8x\sqrt{2y}$$
$$= (4x - 3x + 8x)\sqrt{2y}$$ • Subtract by using the Distributive Property.
$$= 9x\sqrt{2y}$$

Problem 2 Simplify.

A. $y\sqrt{28y} + 7\sqrt{63y^3}$ **B.** $2\sqrt{27a^5} - 4a\sqrt{12a^3} + a^2\sqrt{75a}$

Solution See page S24.

Try Exercise 63, page 477.

10.2 Exercises

CONCEPT CHECK

Can the expression be simplified?

1. $5\sqrt{3} + 6\sqrt{3}$
2. $3\sqrt{5} + 3\sqrt{6}$
3. $7\sqrt{11} - \sqrt{11}$
4. $4\sqrt{2x} - 8\sqrt{2x}$
5. $4\sqrt{2y} - 8\sqrt{y}$
6. $3\sqrt{5x} + 5\sqrt{3x}$

1 Add and subtract radical expressions (See pages 474-475.)

GETTING READY

7. Simplify: $5\sqrt{5} - 8\sqrt{5}$

$5\sqrt{5} - 8\sqrt{5} = (\underline{\ ?\ } - \underline{\ ?\ })\sqrt{5}$ • The radicands are the same. Use the Distributive Property.

$= \underline{\ ?\ }\sqrt{5}$ • Subtract.

Simplify.

8. $2\sqrt{2} + \sqrt{2}$

9. $3\sqrt{5} + 8\sqrt{5}$

10. $-3\sqrt{7} + 2\sqrt{7}$

11. $4\sqrt{5} - 10\sqrt{5}$

12. $-3\sqrt{11} - 8\sqrt{11}$

13. $-3\sqrt{3} - 5\sqrt{3}$

14. $2\sqrt{x} + 8\sqrt{x}$

15. $3\sqrt{y} + 2\sqrt{y}$

16. $8\sqrt{y} - 10\sqrt{y}$

17. $-5\sqrt{2a} + 2\sqrt{2a}$

18. $-2\sqrt{3b} - 9\sqrt{3b}$

19. $-7\sqrt{5a} - 5\sqrt{5a}$

20. $3x\sqrt{2} - x\sqrt{2}$

21. $2y\sqrt{3} - 9y\sqrt{3}$

22. $2a\sqrt{3a} - 5a\sqrt{3a}$

23. $-5b\sqrt{3x} - 2b\sqrt{3x}$

24. $3\sqrt{xy} - 8\sqrt{xy}$

25. $-4\sqrt{xy} + 6\sqrt{xy}$

26. $\sqrt{45} + \sqrt{125}$

27. $\sqrt{32} - \sqrt{98}$

28. $2\sqrt{2} + 3\sqrt{8}$

29. $4\sqrt{128} - 3\sqrt{32}$

30. $5\sqrt{18} - 2\sqrt{75}$

31. $5\sqrt{75} - 2\sqrt{18}$

32. $5\sqrt{4x} - 3\sqrt{9x}$

33. $-3\sqrt{25y} + 8\sqrt{49y}$

GETTING READY

34. Simplify: $3\sqrt{x^3} + 2x\sqrt{x}$

$3\sqrt{x^3} + 2x\sqrt{x} = 3\sqrt{\underline{\ ?\ }}\sqrt{x} + 2x\sqrt{x}$ • Write $\sqrt{x^3}$ as the product of the square root of a perfect square and $\sqrt{x}$.

$= 3\underline{\ ?\ }\sqrt{x} + 2x\sqrt{x}$ • Simplify $\sqrt{x^2}$.

$= (\underline{\ ?\ } + \underline{\ ?\ })\sqrt{x}$ • Use the Distributive Property.

$= \underline{\ ?\ }\sqrt{x}$ • Add.

Simplify.

35. $3\sqrt{3x^2} - 5\sqrt{27x^2}$

36. $-2\sqrt{8y^2} + 5\sqrt{32y^2}$

37. $2x\sqrt{xy^2} - 3y\sqrt{x^2y}$

38. $4a\sqrt{b^2a} - 3b\sqrt{a^2b}$

39. $3x\sqrt{12x} - 5\sqrt{27x^3}$

40. $2a\sqrt{50a} + 7\sqrt{32a^3}$

41. $4y\sqrt{8y^3} - 7\sqrt{18y^5}$

42. $2a\sqrt{8ab^2} - 2b\sqrt{2a^3}$

43. $b^2\sqrt{a^5b} + 3a^2\sqrt{ab^5}$

44. $y^2\sqrt{x^5y} + x\sqrt{x^3y^5}$

45. $4\sqrt{2} - 5\sqrt{2} + 8\sqrt{2}$

46. $3\sqrt{3} + 8\sqrt{3} - 16\sqrt{3}$

47. $5\sqrt{x} - 8\sqrt{x} + 9\sqrt{x}$

48. $\sqrt{x} - 7\sqrt{x} + 6\sqrt{x}$

49. $8\sqrt{2} - 3\sqrt{y} - 8\sqrt{2}$

50. $8\sqrt{3} - 5\sqrt{2} - 5\sqrt{3}$

51. $8\sqrt{8} - 4\sqrt{32} - 9\sqrt{50}$

52. $2\sqrt{12} - 4\sqrt{27} + \sqrt{75}$

53. $-2\sqrt{3} + 5\sqrt{27} - 4\sqrt{45}$

54. $-2\sqrt{8} - 3\sqrt{27} + 3\sqrt{50}$

55. $4\sqrt{75} + 3\sqrt{48} - \sqrt{99}$

56. $2\sqrt{75} - 5\sqrt{20} + 2\sqrt{45}$

57. $\sqrt{25x} - \sqrt{9x} + \sqrt{16x}$

58. $\sqrt{4x} - \sqrt{100x} - \sqrt{49x}$

59. $3\sqrt{3x} + \sqrt{27x} - 8\sqrt{75x}$

60. $5\sqrt{5x} + 2\sqrt{45x} - 3\sqrt{80x}$

61. $2a\sqrt{75b} - a\sqrt{20b} + 4a\sqrt{45b}$

62. $2b\sqrt{75a} - 5b\sqrt{27a} + 2b\sqrt{20a}$

63. $x\sqrt{3y^2} - 2y\sqrt{12x^2} + xy\sqrt{3}$

64. $a\sqrt{27b^2} + 3b\sqrt{147a^2} - ab\sqrt{3}$

65. $3\sqrt{ab^3} + 4a\sqrt{a^2b} - 5b\sqrt{4ab}$

66. $5\sqrt{a^3b} + a\sqrt{4ab} - 3\sqrt{49a^3b}$

67. $3a\sqrt{2ab^2} - \sqrt{a^2b^2} + 4b\sqrt{3a^2b}$

68. $2\sqrt{4a^2b^2} - 3a\sqrt{9ab^2} + 4b\sqrt{a^2b}$

69. Determine whether the statement is true or false.
a. $7x\sqrt{x} + x\sqrt{x} = 7x^2\sqrt{x}$
b. $\sqrt{10a} - \sqrt{a} = \sqrt{a}(\sqrt{10} - 1)$
c. $\sqrt{9 + y^2} = 3 + y$
d. $\sqrt{27x^4} + \sqrt{3} = \sqrt{3}\,(3x^2 + 1)$

70. Which expression is equivalent to $\sqrt{2ab} + \sqrt{2ab}$?
(i) $2\sqrt{ab}$ (ii) $\sqrt{4ab}$ (iii) $2ab$ (iv) $\sqrt{8ab}$

APPLYING CONCEPTS

Add or subtract.

71. $5\sqrt{x+2} + 3\sqrt{x+2}$

72. $8\sqrt{a+5} - 4\sqrt{a+5}$

73. $\frac{1}{2}\sqrt{8x^2y} + \frac{1}{3}\sqrt{18x^2y}$

74. $\frac{1}{4}\sqrt{48ab^2} + \frac{1}{5}\sqrt{75ab^2}$

75. $\frac{a}{3}\sqrt{54ab^3} + \frac{b}{4}\sqrt{96a^3b}$

76. $\frac{x}{6}\sqrt{72xy^5} + \frac{y}{7}\sqrt{98x^3y^3}$

77. Geometry The lengths of the sides of a triangle are $4\sqrt{3}$ cm, $2\sqrt{3}$ cm, and $2\sqrt{15}$ cm. Find the perimeter of the triangle.

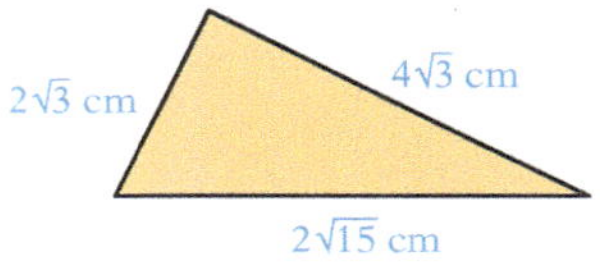

78. Geometry The length of a rectangle is $3\sqrt{2}$ cm. The width is $\sqrt{2}$ cm. Find the perimeter of the rectangle.

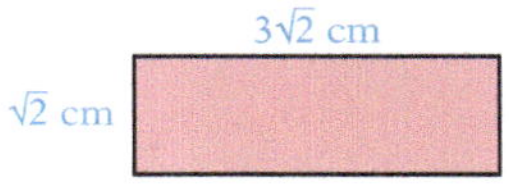

79. Geometry The length of a rectangle is $4\sqrt{5}$ cm. The width is $\sqrt{5}$ cm. Find the decimal approximation of the perimeter. Round to the nearest tenth.

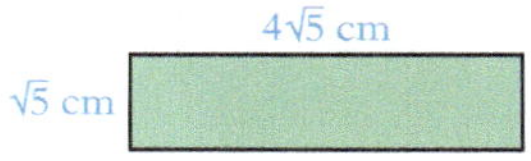

80. Given $G(x) = \sqrt{x+5} + \sqrt{5x+3}$, write $G(3)$ in simplest form.

81. Use complete sentences to explain the steps in simplifying the radical expression $a\sqrt{32ab^2} + b\sqrt{50a^3}$.

PROJECTS OR GROUP ACTIVITIES

82. Write a paragraph that compares adding two monomials to adding two radical expressions. For example, compare the addition of $5x + 3x$ to the addition of $5\sqrt{x} + 3\sqrt{x}$.

10.3 Multiplication and Division of Radical Expressions

OBJECTIVE 1 Multiply radical expressions

The Product Property of Square Roots is used to multiply variable radical expressions.

$$\sqrt{2x}\sqrt{3y} = \sqrt{2x \cdot 3y} = \sqrt{6xy}$$

Focus on multiplying two radical expressions

Multiply: $\sqrt{2x^2}\sqrt{32x^5}$

Use the Product Property of Square Roots to multiply the radicands.

$$\sqrt{2x^2}\sqrt{32x^5} = \sqrt{2x^2 \cdot 32x^5} = \sqrt{64x^7}$$

Simplify.

$$= \sqrt{64x^6}\sqrt{x} = 8x^3\sqrt{x}$$

EXAMPLE 1 Multiply: $\sqrt{3x^4}\sqrt{2x^2y}\sqrt{6xy^2}$

Solution

$$\sqrt{3x^4}\sqrt{2x^2y}\sqrt{6xy^2} = \sqrt{3x^4 \cdot 2x^2y \cdot 6xy^2}$$

- Use the Product Property of Square Roots to multiply the radicands.

$$= \sqrt{36x^7y^3} = \sqrt{36x^6y^2}\sqrt{xy} = 6x^3y\sqrt{xy}$$

- Simplify.

Problem 1 Multiply: $\sqrt{5a}\sqrt{15a^3b^4}\sqrt{3b^5}$

Solution See page S24.

Try Exercise 19, page 483.

When the expression $(\sqrt{x})^2$ is simplified by using the Product Property of Square Roots, the result is x.

$$(\sqrt{x})^2 = \sqrt{x}\sqrt{x} = \sqrt{x \cdot x} = \sqrt{x^2} = x$$

THE SQUARE OF $\sqrt{a}$

For $a > 0$, $(\sqrt{a})^2 = a$.

EXAMPLES

1. $(\sqrt{3})^2 = 3$
2. $(\sqrt{5y})^2 = 5y$

Focus on multiplying two radical expressions by using the Distributive Property

Multiply: $\sqrt{2x}(x + \sqrt{2x})$

Use the Distributive Property to remove parentheses.

$\sqrt{2x}(x + \sqrt{2x}) = \sqrt{2x}(x) + \sqrt{2x}\sqrt{2x}$
$= x\sqrt{2x} + (\sqrt{2x})^2$

Simplify $(\sqrt{2x})^2$.

$= x\sqrt{2x} + 2x$

EXAMPLE 2 Multiply: $\sqrt{3ab}(\sqrt{3a} + \sqrt{9b})$

Solution $\sqrt{3ab}(\sqrt{3a} + \sqrt{9b})$

$= \sqrt{3ab}(\sqrt{3a}) + \sqrt{3ab}(\sqrt{9b})$ • Use the Distributive Property to remove parentheses.

$= \sqrt{9a^2b} + \sqrt{27ab^2}$

$= \sqrt{9a^2}\sqrt{b} + \sqrt{9b^2}\sqrt{3a}$ • Simplify each radical expression.

$= 3a\sqrt{b} + 3b\sqrt{3a}$

Problem 2 Multiply: $\sqrt{5x}(\sqrt{5x} - \sqrt{25y})$

Solution See page S24.

Try Exercise 31, page 483.

Focus on multiplying two radical expressions by using the FOIL method

Multiply: $(\sqrt{2} - 3x)(\sqrt{2} + x)$

Use the FOIL method.

$(\sqrt{2} - 3x)(\sqrt{2} + x) = (\sqrt{2})^2 + x\sqrt{2} - 3x\sqrt{2} - 3x^2$
$= 2 + (x - 3x)\sqrt{2} - 3x^2$

Simplify.

$= 2 - 2x\sqrt{2} - 3x^2$

EXAMPLE 3 Multiply: $(2\sqrt{x} - \sqrt{y})(5\sqrt{x} - 2\sqrt{y})$

Solution $(2\sqrt{x} - \sqrt{y})(5\sqrt{x} - 2\sqrt{y})$

$= 10(\sqrt{x})^2 - 4\sqrt{xy} - 5\sqrt{xy} + 2(\sqrt{y})^2$ • Use the FOIL method.

$= 10x - 9\sqrt{xy} + 2y$ • Simplify.

Problem 3 Multiply: $(3\sqrt{x} - \sqrt{y})(5\sqrt{x} - 2\sqrt{y})$

Solution See page S24.

Try Exercise 41, page 483.

The expressions $a + b$ and $a - b$, which are the sum and difference of two terms, are called **conjugates** of each other.

The product of conjugates is the difference of two squares.

$(a + b)(a - b) = a^2 - b^2$

$(2 + \sqrt{7})(2 - \sqrt{7}) = 2^2 - (\sqrt{7})^2 = 4 - 7 = -3$

$(3 + \sqrt{y})(3 - \sqrt{y}) = 3^2 - (\sqrt{y})^2 = 9 - y$

EXAMPLE 4 Multiply: $(\sqrt{a} - \sqrt{b})(\sqrt{a} + \sqrt{b})$

Solution $(\sqrt{a} - \sqrt{b})(\sqrt{a} + \sqrt{b}) = (\sqrt{a})^2 - (\sqrt{b})^2$
$= a - b$

• The expressions are conjugates of each other.

Problem 4 Multiply: $(2\sqrt{x} + 7)(2\sqrt{x} - 7)$

Solution See page S24.

Try Exercise 43, page 483.

OBJECTIVE 2 Divide radical expressions

The square root of a quotient is equal to the quotient of the square roots.

THE QUOTIENT PROPERTY OF SQUARE ROOTS

If a and b are positive real numbers, then $\sqrt{\dfrac{a}{b}} = \dfrac{\sqrt{a}}{\sqrt{b}}$.

EXAMPLES

1. $\sqrt{\dfrac{a^2}{b^4}} = \dfrac{\sqrt{a^2}}{\sqrt{b^4}} = \dfrac{a}{b^2}$
2. $\sqrt{\dfrac{4x^2}{z^6}} = \dfrac{\sqrt{4x^2}}{\sqrt{z^6}} = \dfrac{2x}{z^3}$

Point of Interest

A radical expression that occurs in Einstein's Theory of Relativity is

$$\frac{1}{\sqrt{1 - \dfrac{v^2}{c^2}}}$$

where v is the velocity of an object and c is the speed of light.

Focus on simplifying a radical expression by using the Quotient Property of Square Roots

Simplify. **A.** $\sqrt{\dfrac{24x^3y^7}{3x^7y^2}}$ **B.** $\dfrac{\sqrt{4x^2y}}{\sqrt{xy}}$

A. Simplify the radicand. $\sqrt{\dfrac{24x^3y^7}{3x^7y^2}} = \sqrt{\dfrac{8y^5}{x^4}}$

Use the Quotient Property of Square Roots. $= \dfrac{\sqrt{8y^5}}{\sqrt{x^4}}$

Simplify. $= \dfrac{\sqrt{4y^4}\sqrt{2y}}{\sqrt{x^4}}$

$= \dfrac{2y^2\sqrt{2y}}{x^2}$

B. Use the Quotient Property of Square Roots. $\dfrac{\sqrt{4x^2y}}{\sqrt{xy}} = \sqrt{\dfrac{4x^2y}{xy}}$

Simplify the radicand. $= \sqrt{4x}$

Simplify the radical expression. $= \sqrt{4}\sqrt{x}$

$= 2\sqrt{x}$

A radical expression is not in simplest form if a radical remains in the denominator. The procedure used to remove a radical from the denominator is called **rationalizing the denominator.**

Focus on rationalizing the denominator of a radical expression when the denominator contains a radical expression with one term

Simplify: $\dfrac{2}{\sqrt{3}}$

The expression $\frac{2}{\sqrt{3}}$ has a radical expression in the denominator. Multiply the expression by $\frac{\sqrt{3}}{\sqrt{3}}$, which equals 1.

$$\frac{2}{\sqrt{3}} = \frac{2}{\sqrt{3}} \cdot \frac{\sqrt{3}}{\sqrt{3}}$$

Simplify.

$$= \frac{2\sqrt{3}}{(\sqrt{3})^2}$$

$$= \frac{2\sqrt{3}}{3}$$

Thus $\frac{2}{\sqrt{3}} = \frac{2\sqrt{3}}{3}$. Note that $\frac{2}{\sqrt{3}}$ is not in simplest form. However, because no radical remains in the denominator and the radical in the numerator contains no perfect-square factors other than 1, $\frac{2\sqrt{3}}{3}$ *is* in simplest form.

When the denominator contains a radical expression with two terms, simplify the radical expression by multiplying the numerator and denominator by the conjugate of the denominator.

Focus on rationalizing the denominator of a radical expression when the denominator contains a radical expression with two terms

Simplify: $\dfrac{\sqrt{2y}}{\sqrt{y} + 3}$

Multiply the numerator and denominator by $\sqrt{y} - 3$, the conjugate of $\sqrt{y} + 3$.

$$\frac{\sqrt{2y}}{\sqrt{y} + 3} = \frac{\sqrt{2y}}{\sqrt{y} + 3} \cdot \frac{\sqrt{y} - 3}{\sqrt{y} - 3}$$

Simplify.

$$= \frac{\sqrt{2y^2} - 3\sqrt{2y}}{(\sqrt{y})^2 - 3^2}$$

$$= \frac{y\sqrt{2} - 3\sqrt{2y}}{y - 9}$$

The following list summarizes our discussion about radical expressions in simplest form.

RADICAL EXPRESSIONS IN SIMPLEST FORM

A radical expression is in simplest form if:

1. The radicand contains no factor greater than 1 that is a perfect square.
2. There is no fraction under the radical sign.
3. There is no radical in the denominator of a fraction.

EXAMPLE 5 Simplify. **A.** $\dfrac{\sqrt{4x^2y^5}}{\sqrt{3x^4y}}$ **B.** $\dfrac{\sqrt{2}}{\sqrt{2}-\sqrt{x}}$ **C.** $\dfrac{3-\sqrt{5}}{2+3\sqrt{5}}$

Solution **A.** $\dfrac{\sqrt{4x^2y^5}}{\sqrt{3x^4y}} = \sqrt{\dfrac{4x^2y^5}{3x^4y}}$ • Use the Quotient Property of Square Roots.

$= \sqrt{\dfrac{4y^4}{3x^2}}$ • Simplify the radicand.

$= \dfrac{\sqrt{4y^4}}{\sqrt{3x^2}}$ • Use the Quotient Property of Square Roots.

$= \dfrac{2y^2}{x\sqrt{3}}$ • Simplify the radical expressions in the numerator and denominator.

$= \dfrac{2y^2}{x\sqrt{3}} \cdot \dfrac{\sqrt{3}}{\sqrt{3}}$ • Rationalize the denominator by multiplying the expression by $\frac{\sqrt{3}}{\sqrt{3}}$, which equals 1.

$= \dfrac{2y^2\sqrt{3}}{3x}$

B. $\dfrac{\sqrt{2}}{\sqrt{2}-\sqrt{x}}$

$= \dfrac{\sqrt{2}}{\sqrt{2}-\sqrt{x}} \cdot \dfrac{\sqrt{2}+\sqrt{x}}{\sqrt{2}+\sqrt{x}}$ • Rationalize the denominator by multiplying the numerator and denominator by the conjugate of the denominator.

$= \dfrac{2+\sqrt{2x}}{2-x}$

C. $\dfrac{3-\sqrt{5}}{2+3\sqrt{5}} = \dfrac{3-\sqrt{5}}{2+3\sqrt{5}} \cdot \dfrac{2-3\sqrt{5}}{2-3\sqrt{5}}$ • Rationalize the denominator by multiplying the numerator and denominator by the conjugate of the denominator.

$= \dfrac{6-9\sqrt{5}-2\sqrt{5}+3\cdot 5}{4-9\cdot 5}$ • Use the FOIL method to multiply the numerators.

$= \dfrac{21-11\sqrt{5}}{-41}$ • Simplify.

$= -\dfrac{21-11\sqrt{5}}{41}$

Problem 5 Simplify. **A.** $\dfrac{\sqrt{15x^6y^7}}{\sqrt{3x^7y^9}}$ **B.** $\dfrac{\sqrt{y}}{\sqrt{y}+3}$ **C.** $\dfrac{5+\sqrt{y}}{1-2\sqrt{y}}$

Solution See pages S24–S25.

Try Exercises 79, 91, and 107, pages 484–485.

10.3 Exercises

CONCEPT CHECK

Determine the conjugate of each of the following.

1. $3+\sqrt{5}$ **2.** $6-\sqrt{x}$ **3.** $\sqrt{2a}-8$

Find the product of the expression and its conjugate.

4. $4 + \sqrt{3}$

5. $5 - \sqrt{y}$

By what form of 1 should the expression be multiplied to rationalize the denominator?

6. $\frac{2}{\sqrt{6}}$

7. $\frac{3}{\sqrt{x}}$

8. $\frac{2 - \sqrt{x}}{\sqrt{y}}$

1 Multiply radical expressions (See pages 478–480.)

GETTING READY

9. $\sqrt{3} \cdot \sqrt{5} = \sqrt{\underline{\quad ? \quad}}$
 - Use the Product Property of Square Roots to multiply the radicands.

10. $\sqrt{3} \cdot \sqrt{15} = \sqrt{\underline{\quad ? \quad}}$
 - Use the Product Property of Square Roots to multiply the radicands.

$= \sqrt{9 \cdot \underline{\quad ? \quad}}$
 - Write the radicand as the product of a perfect square and a factor that does not contain a perfect square.

$= \sqrt{9}\sqrt{\underline{\quad ? \quad}}$
 - Use the Product Property of Square Roots.

$= \underline{\quad ? \quad}\sqrt{\underline{\quad ? \quad}}$
 - Simplify the square root of the perfect square.

Simplify.

11. $\sqrt{5}\sqrt{5}$

12. $\sqrt{11}\sqrt{11}$

13. $\sqrt{3}\sqrt{12}$

14. $\sqrt{2}\sqrt{8}$

15. $(\sqrt{7y})^2$

16. $(\sqrt{11b})^2$

17. $\sqrt{xy^3}\sqrt{x^5y}$

18. $\sqrt{a^3b^5}\sqrt{ab^5}$

19. $\sqrt{3a^2b^5}\sqrt{6ab^7}$

20. $\sqrt{5x^3y}\sqrt{10x^2y}$

21. $\sqrt{6a^3b^2}\sqrt{24a^5b}$

22. $\sqrt{8ab^5}\sqrt{12a^7b}$

23. $\sqrt{2ac}\sqrt{5ab}\sqrt{10cb}$

24. $\sqrt{3xy}\sqrt{6x^3y}\sqrt{2y^2}$

25. $\sqrt{2}(\sqrt{2} - \sqrt{3})$

26. $3(\sqrt{12} - \sqrt{3})$

27. $\sqrt{8}(\sqrt{2} - \sqrt{5})$

28. $\sqrt{10}(\sqrt{20} - \sqrt{a})$

29. $\sqrt{5}(\sqrt{10} - \sqrt{x})$

30. $\sqrt{6}(\sqrt{y} - \sqrt{18})$

31. $\sqrt{x}(\sqrt{x} - \sqrt{y})$

32. $\sqrt{b}(\sqrt{a} - \sqrt{b})$

33. $\sqrt{3a}(\sqrt{3a} - \sqrt{3b})$

34. $\sqrt{5x}(\sqrt{10x} - \sqrt{x})$

35. $(\sqrt{x} - 3)^2$

36. $(2\sqrt{a} - y)^2$

37. $(\sqrt{5} + 3)(\sqrt{5} + 7)$

38. $(4\sqrt{3} + 1)(\sqrt{3} - 1)$

39. $(2\sqrt{x} - 5)(\sqrt{x} - 2)$

40. $(\sqrt{y} - 3)(3\sqrt{y} + 2)$

41. $(3\sqrt{x} - 2y)(5\sqrt{x} - 4y)$

42. $(5\sqrt{x} + 2\sqrt{y})(3\sqrt{x} - \sqrt{y})$

43. $(\sqrt{2} - \sqrt{y})(\sqrt{2} + \sqrt{y})$

44. $(\sqrt{3x} + 4)(\sqrt{3x} - 4)$

45. $(5 + \sqrt{6})(5 - \sqrt{6})$

46. $(7 - \sqrt{11})(7 + \sqrt{11})$

47. $(2\sqrt{x} + \sqrt{y})(5\sqrt{x} + 4\sqrt{y})$

48. $(5\sqrt{x} - 2\sqrt{y})(3\sqrt{x} - 4\sqrt{y})$

49. For $a > 0$, is $(\sqrt{a} - 1)(\sqrt{a} + 1)$ less than, equal to, or greater than a?

50. For $a > 0$, is $\sqrt{a}(\sqrt{2a} - \sqrt{a})$ less than, equal to, or greater than a?

2 Divide radical expressions (See pages 480–482.)

51. Explain why $\frac{\sqrt{5}}{5}$ is in simplest form and $\frac{1}{\sqrt{5}}$ is not in simplest form.

52. Why can we multiply $\frac{1}{\sqrt{3}}$ by $\frac{\sqrt{3}}{\sqrt{3}}$ without changing the value of $\frac{1}{\sqrt{3}}$?

53. State whether the expression is in simplest form.

a. $\frac{a}{\sqrt{b}}$ **b.** $\frac{\sqrt{a}}{b}$ **c.** $\sqrt{\frac{a}{b}}$ **d.** $\frac{\sqrt{a}}{\sqrt{b}}$

GETTING READY

54. The Quotient Property of Square Roots states that $\sqrt{\frac{a}{b}} = \underline{\quad ? \quad}$.

Simplify.

55. $\frac{\sqrt{32}}{\sqrt{2}}$ 56. $\frac{\sqrt{45}}{\sqrt{5}}$ 57. $\frac{\sqrt{98}}{\sqrt{2}}$ 58. $\frac{\sqrt{48}}{\sqrt{3}}$

59. $\frac{\sqrt{27a}}{\sqrt{3a}}$ 60. $\frac{\sqrt{72x^5}}{\sqrt{2x}}$ 61. $\frac{\sqrt{15x^3y}}{\sqrt{3xy}}$ 62. $\frac{\sqrt{40x^5y^2}}{\sqrt{5xy}}$

63. $\frac{\sqrt{2a^5b^4}}{\sqrt{98ab^4}}$ 64. $\frac{\sqrt{48x^5y^2}}{\sqrt{3x^3y}}$ 65. $\frac{1}{\sqrt{3}}$ 66. $\frac{1}{\sqrt{8}}$

67. $\frac{15}{\sqrt{75}}$ 68. $\frac{6}{\sqrt{72}}$ 69. $\frac{3}{\sqrt{x}}$ 70. $\frac{4}{\sqrt{2x}}$

71. $\frac{6}{\sqrt{12x}}$ 72. $\frac{14}{\sqrt{7y}}$ 73. $\frac{8}{\sqrt{32x}}$ 74. $\frac{15}{\sqrt{50x}}$

75. $\frac{\sqrt{4x^2}}{\sqrt{9y}}$ 76. $\frac{\sqrt{16a}}{\sqrt{49ab}}$ 77. $\frac{5\sqrt{8}}{4\sqrt{50}}$ 78. $\frac{5\sqrt{18}}{9\sqrt{27}}$

79. $\frac{\sqrt{12a^3b}}{\sqrt{24a^2b^2}}$ 80. $\frac{\sqrt{3xy}}{\sqrt{27x^3y^2}}$ 81. $\frac{\sqrt{9xy^2}}{\sqrt{27x}}$ 82. $\frac{\sqrt{4x^2y}}{\sqrt{3xy^3}}$

83. $\frac{1}{\sqrt{2} - 3}$ 84. $\frac{5}{\sqrt{7} - 3}$ 85. $\frac{3}{5 + \sqrt{5}}$

86. $\frac{7}{\sqrt{2} - 7}$ 87. $\frac{\sqrt{xy}}{\sqrt{x} - \sqrt{y}}$ 88. $\frac{\sqrt{x}}{\sqrt{x} - \sqrt{y}}$

89. $\frac{\sqrt{16x^3y^2}}{\sqrt{8x^3y}}$ 90. $\frac{\sqrt{2}}{1 - \sqrt{2}}$ 91. $\frac{\sqrt{5}}{\sqrt{2} - \sqrt{5}}$

92. $\frac{\sqrt{6}}{\sqrt{3} - \sqrt{2}}$ 93. $\frac{\sqrt{x}}{\sqrt{x} + 3}$ 94. $\frac{\sqrt{y}}{2 - \sqrt{y}}$

95. $\frac{5\sqrt{3} - 7\sqrt{3}}{4\sqrt{3}}$ 96. $\frac{10\sqrt{7} - 2\sqrt{7}}{2\sqrt{7}}$ 97. $\frac{5\sqrt{8} - 3\sqrt{2}}{\sqrt{2}}$

98. $\frac{5\sqrt{12} - \sqrt{3}}{\sqrt{27}}$ 99. $\frac{3\sqrt{2} - 8\sqrt{2}}{\sqrt{2}}$ 100. $\frac{5\sqrt{3} - 2\sqrt{3}}{2\sqrt{3}}$

101. $\dfrac{2 + \sqrt{3}}{2 - \sqrt{3}}$

102. $\dfrac{5 + \sqrt{2}}{3 - \sqrt{2}}$

103. $\dfrac{3 - \sqrt{6}}{5 - 2\sqrt{6}}$

104. $\dfrac{6 - 2\sqrt{3}}{4 + 3\sqrt{3}}$

105. $\dfrac{\sqrt{2} + 2\sqrt{6}}{2\sqrt{2} - 3\sqrt{6}}$

106. $\dfrac{2\sqrt{3} - \sqrt{6}}{5\sqrt{3} + 2\sqrt{6}}$

107. $\dfrac{3 + \sqrt{x}}{2 - \sqrt{x}}$

108. $\dfrac{\sqrt{a} - 4}{2\sqrt{a} + 2}$

109. $\dfrac{3 + 2\sqrt{y}}{2 - \sqrt{y}}$

110. $\dfrac{2 + \sqrt{y}}{\sqrt{y} - 3}$

111. $\dfrac{\sqrt{x} + \sqrt{y}}{\sqrt{x} - \sqrt{y}}$

APPLYING CONCEPTS

Simplify.

112. $-\sqrt{1.3}\,\sqrt{1.3}$

113. $\sqrt{\dfrac{5}{8}}\sqrt{\dfrac{5}{8}}$

114. $-\sqrt{\dfrac{16}{81}}$

115. $\sqrt{1\dfrac{9}{16}}$

116. $\sqrt{2\dfrac{1}{4}}$

117. $-\sqrt{6\dfrac{1}{4}}$

Geometry Find the area of the geometric figure. All dimensions are given in meters.

118.

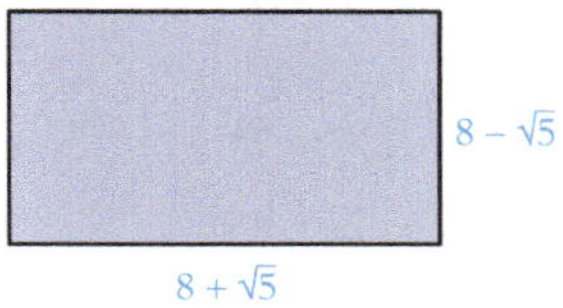

119.

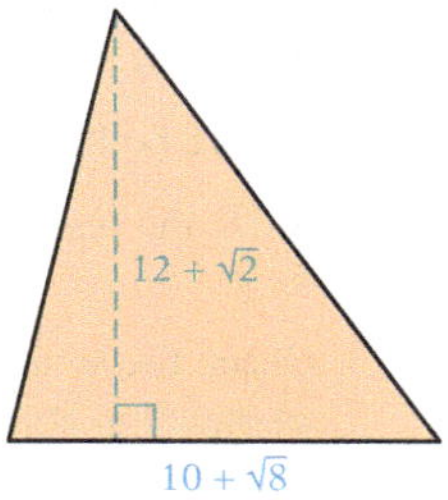

120. Is 16 a solution of the equation $\sqrt{x} - \sqrt{x + 9} = 1$?

121. Show that 2 is a solution of the equation $\sqrt{x + 2} + \sqrt{x - 1} = 3$.

122. Show that $1 + \sqrt{6}$ and $1 - \sqrt{6}$ are solutions of the equation $x^2 - 2x - 5 = 0$.

123. In your own words, describe the process of rationalizing the denominator.

PROJECTS OR GROUP ACTIVITIES

124. The number $\dfrac{\sqrt{5} + 1}{2}$ is called the golden ratio. Research the golden ratio and write a few paragraphs about this number and its applications.

10.4 Solving Equations Containing Radical Expressions

OBJECTIVE 1 Solve equations containing one or more radical expressions

An equation that contains a variable expression in a radicand is a **radical equation.**

$$\left.\begin{aligned}\sqrt{x} &= 4\\ \sqrt{x+2} &= \sqrt{x-7}\end{aligned}\right\}\text{ Radical equations}$$

The following property of equality states that if two numbers are equal, then the squares of the numbers are equal. This property is used to solve radical equations.

> **PROPERTY OF SQUARING BOTH SIDES OF AN EQUATION**
>
> If a and b are real numbers and $\boldsymbol{a = b}$, then $\boldsymbol{a^2 = b^2}$.

Focus on solving an equation containing one radical expression

Solve: $\sqrt{x-2} - 7 = 0$

Rewrite the equation with the radical on one side of the equation and the constant on the other side.

$$\sqrt{x-2} - 7 = 0$$
$$\sqrt{x-2} = 7$$

Square both sides of the equation.

$$(\sqrt{x-2})^2 = 7^2$$

Solve the resulting equation.

$$x - 2 = 49$$
$$x = 51$$

Check the solution. When both sides of an equation are squared, the resulting equation may have a solution that is not a solution of the original equation.

Check: $\sqrt{x-2} - 7 = 0$

$\sqrt{x-2} - 7$	0
$\sqrt{51-2} - 7$	0
$\sqrt{49} - 7$	0
$7 - 7$	0

$0 = 0$

The solution is 51.

> **Take Note**
>
> Any time each side of an equation is squared, you must check the proposed solution of the equation.

EXAMPLE 1 Solve: $\sqrt{3x} + 2 = 5$

Solution

$$\sqrt{3x} + 2 = 5$$
$$\sqrt{3x} = 3$$

- Rewrite the equation so that the radical is alone on one side of the equation.

$$(\sqrt{3x})^2 = 3^2$$

- Square both sides of the equation.

$$3x = 9$$
$$x = 3$$

- Solve for x.

Check $\sqrt{3x} + 2 = 5$

- Both sides of the equation were squared. The solution must be checked.

$\sqrt{3x} + 2$	5
$\sqrt{3 \cdot 3} + 2$	5
$\sqrt{9} + 2$	5
$3 + 2$	5

$5 = 5$

- This is a true equation. The solution checks.

The solution is 3.

Problem 1 Solve: $\sqrt{4x} + 3 = 7$

Solution See page S25.

Try Exercise 17, page 491.

EXAMPLE 2 Solve. **A.** $0 = 3 - \sqrt{2x - 3}$ **B.** $\sqrt{2x - 5} + 3 = 0$

Solution **A.**

$$0 = 3 - \sqrt{2x - 3}$$
$$\sqrt{2x - 3} = 3$$

- **Rewrite the equation so that the radical is alone on one side of the equation.**

$$(\sqrt{2x - 3})^2 = 3^2$$
$$2x - 3 = 9$$
$$2x = 12$$
$$x = 6$$

- **Square both sides of the equation.**
- **Solve for x.**

Check

$0 = 3 - \sqrt{2x - 3}$	
0	$3 - \sqrt{2 \cdot 6 - 3}$
0	$3 - \sqrt{12 - 3}$
0	$3 - \sqrt{9}$
0	$3 - 3$

$0 = 0$

- **This is a true equation. The solution checks.**

The solution is 6.

Solution **B.**

$$\sqrt{2x - 5} + 3 = 0$$
$$\sqrt{2x - 5} = -3$$

- **Rewrite the equation so that the radical is alone on one side of the equation.**

$$(\sqrt{2x - 5})^2 = (-3)^2$$
$$2x - 5 = 9$$
$$2x = 14$$
$$x = 7$$

- **Square each side of the equation.**
- **Solve for x.**

Check

$\sqrt{2x - 5} + 3 = 0$	
$\sqrt{2 \cdot 7 - 5} + 3$	0
$\sqrt{14 - 5} + 3$	0
$\sqrt{9} + 3$	0
$3 + 3$	0

$6 \neq 0$

- **This is not a true equation. The solution does not check.**

There is no solution.

Problem 2 Solve. **A.** $\sqrt{3x - 2} - 5 = 0$ **B.** $\sqrt{4x - 7} + 5 = 0$

Solution See page S25.

Try Exercise 31, page 491.

The following example illustrates the procedure for solving a radical equation containing two radical expressions. Note that the process of squaring both sides of the equation is performed twice.

Focus on solving an equation containing two radical expressions

Solve: $\sqrt{5 + x} + \sqrt{x} = 5$

$$\sqrt{5+x} + \sqrt{x} = 5$$

Solve for one of the radical expressions. Square each side.

$$\sqrt{5+x} = 5 - \sqrt{x}$$
$$(\sqrt{5+x})^2 = (5 - \sqrt{x})^2$$

Recall that $(a-b)^2 = a^2 - 2ab + b^2$.

$$5 + x = 25 - 10\sqrt{x} + x$$

Simplify.

$$-20 = -10\sqrt{x}$$

This is still a radical equation.

$$2 = \sqrt{x}$$

Square each side.

$$2^2 = (\sqrt{x})^2$$
$$4 = x$$

4 checks as the solution. The solution is 4.

EXAMPLE 3 Solve: $\sqrt{x} - \sqrt{x-5} = 1$

Solution

$$\sqrt{x} - \sqrt{x-5} = 1$$
$$\sqrt{x} = 1 + \sqrt{x-5}$$
$$(\sqrt{x})^2 = (1 + \sqrt{x-5})^2$$
$$x = 1 + 2\sqrt{x-5} + (x-5)$$
$$4 = 2\sqrt{x-5}$$
$$2 = \sqrt{x-5}$$

- **Solve for one of the radical expressions.**
- **Square each side.**
- **Simplify.**

$$2^2 = (\sqrt{x-5})^2$$
$$4 = x - 5$$
$$9 = x$$

- **This is still a radical equation.**
- **Square each side.**
- **Simplify.**

Check

$$\sqrt{x} - \sqrt{x-5} = 1$$

$\sqrt{9} - \sqrt{9-5}$	1
$3 - \sqrt{4}$	1
$3 - 2$	1
$1 = 1$	

The solution is 9.

Problem 3 Solve: $\sqrt{x} + \sqrt{x+9} = 9$

Solution See page S25.

Try Exercise 39, page 492.

OBJECTIVE 2 Application problems

Pythagoras (c. 580 B.C.–520 B.C.)

A **right triangle** contains one 90° angle. The side opposite the 90° angle is called the **hypotenuse.** The other two sides are called **legs.**

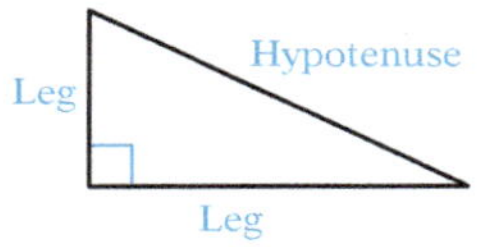

The angles in a right triangle are usually labeled with the capital letters A, B, and C, with C reserved for the right angle. The side opposite angle A is side a, the side opposite angle B is side b, and c is the hypotenuse.

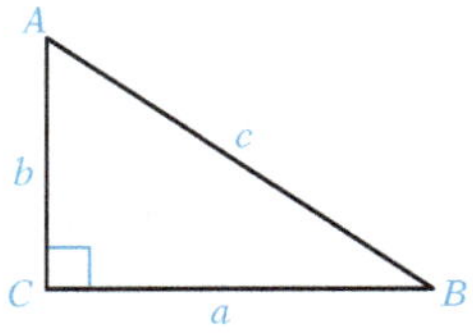

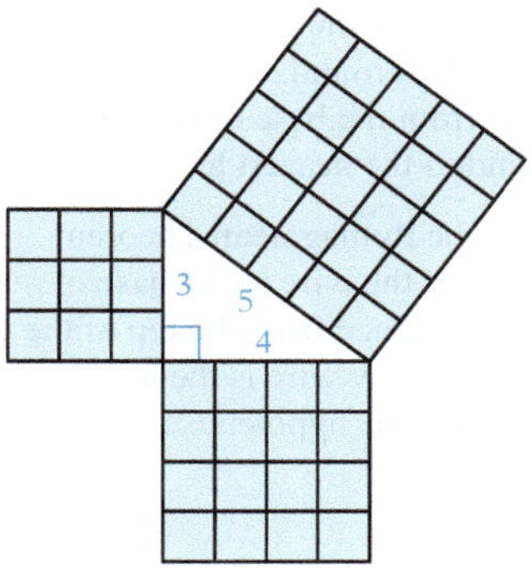

The Greek mathematician Pythagoras is generally credited with the discovery that the square of the hypotenuse of a right triangle is equal to the sum of the squares of the two legs. This is called the **Pythagorean Theorem.**

The figure at the left is a right triangle with legs measuring 3 units and 4 units and a hypotenuse measuring 5 units. Each side of the triangle is also the side of a square. The number of square units in the area of the largest square is equal to the sum of the numbers of square units in the areas of the smaller squares.

Square of the hypotenuse	=	Sum of the squares of the two legs

$$5^2 = 3^2 + 4^2$$
$$25 = 9 + 16$$
$$25 = 25$$

Point of Interest

The first known proof of the Pythagorean Theorem occurs in a Chinese text, *Arithmetic Classic*, which was written around 600 B.C. (but there are no existing copies) and revised over a period of 500 years. The earliest known copy of this text dates from approximately 100 B.C.

PYTHAGOREAN THEOREM

If a and b are the lengths of the legs of a right triangle and c is the length of the hypotenuse, then $c^2 = a^2 + b^2$.

If the lengths of two sides of a right triangle are known, the Pythagorean Theorem can be used to find the length of the third side.

The Pythagorean Theorem is used to find the hypotenuse when the two legs are known.

$$\text{Hypotenuse} = \sqrt{(\text{leg})^2 + (\text{leg})^2}$$
$$c = \sqrt{a^2 + b^2}$$
$$c = \sqrt{(5)^2 + (12)^2}$$
$$c = \sqrt{25 + 144}$$
$$c = \sqrt{169}$$
$$c = 13$$

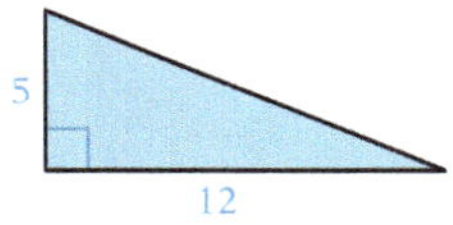

Take Note

If we let $a = 12$ and $b = 5$, the result is the same.

The Pythagorean Theorem is used to find the length of a leg when one leg and the hypotenuse are known.

$$\text{Leg} = \sqrt{(\text{hypotenuse})^2 - (\text{leg})^2}$$
$$a = \sqrt{c^2 - b^2}$$
$$a = \sqrt{(25)^2 - (20)^2}$$
$$a = \sqrt{625 - 400}$$
$$a = \sqrt{225}$$
$$a = 15$$

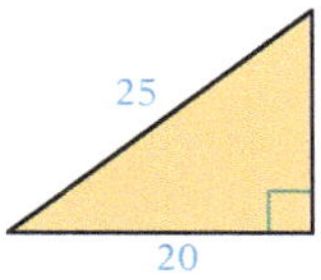

Example 4 and Problem 4 illustrate the use of the Pythagorean Theorem. Example 5 and Problem 5 illustrate other applications of radical equations.

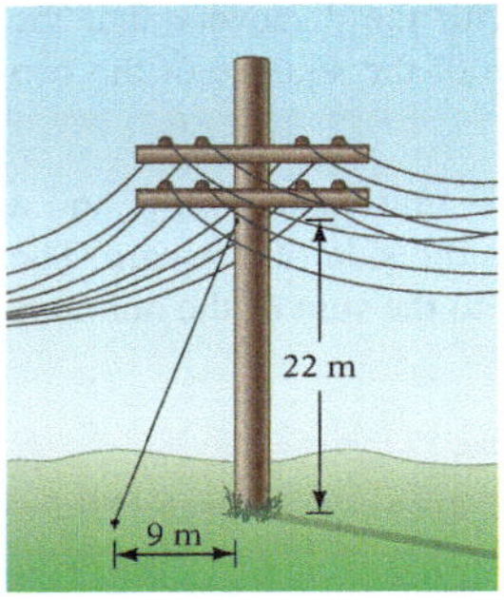

EXAMPLE 4 A guy wire is attached to a point 22 m above the ground on a telephone pole that is perpendicular to the ground. The wire is anchored to the ground at a point 9 m from the base of the pole. Find the length of the guy wire. Round to the nearest hundredth.

Strategy To find the length of the guy wire, use the Pythagorean Theorem. One leg is the distance from the bottom of the wire to the base of the telephone pole. The other leg is the distance from the top of the wire to the base of the telephone pole. The guy wire is the hypotenuse. Solve the Pythagorean Theorem for the hypotenuse.

Solution

$c = \sqrt{a^2 + b^2}$

$c = \sqrt{(22)^2 + (9)^2}$ • $a = 22, b = 9$

$c = \sqrt{484 + 81}$ • Simplify the radicand.

$c = \sqrt{565}$

$c \approx 23.77$ • Use a calculator to approximate $\sqrt{565}$.

The guy wire has a length of 23.77 m.

Problem 4 A ladder 12 ft long is resting against a building. How high on the building will the ladder reach when the bottom of the ladder is 5 ft from the building? Round to the nearest hundredth.

Solution See page S25.

Try Exercise 67, page 493.

EXAMPLE 5 How far above the water would a submarine periscope have to be for the lookout to locate a ship 5 mi away? The equation for the distance in miles that the lookout can see is $d = \sqrt{1.5h}$, where h is the height in feet above the surface of the water. Round to the nearest hundredth.

Strategy To find the height above the water, replace d in the equation with the given value. Then solve for h.

Solution

$d = \sqrt{1.5h}$

$5 = \sqrt{1.5h}$ • $d = 5$

$5^2 = (\sqrt{1.5h})^2$ • Square both sides of the equation.

$25 = 1.5h$

$\frac{25}{1.5} = h$ • Solve for h.

$16.67 \approx h$

The periscope must be 16.67 ft above the water.

Problem 5 Find the length of a pendulum that makes one swing in 1.5 s. The equation for the time of one swing is $T = 2\pi\sqrt{\frac{L}{32}}$, where T is the time in seconds and L is the length in feet. Round to the nearest hundredth.

Solution See page S25.

Try Exercise 69, page 494.

10.4 Exercises

CONCEPT CHECK

1. Which of the following equations are radical equations?
(i) $8 = \sqrt{5} + x$ (ii) $\sqrt{x - 7} = 9$ (iii) $\sqrt{x} + 4 = 6$ (iv) $12 = \sqrt{3}x$

Determine whether the statement is always true, sometimes true, or never true.

2. A radical equation is an equation that contains a radical.
3. We can square both sides of an equation without changing the solutions of the equation.
4. We use the Property of Squaring Both Sides of an Equation in order to eliminate a radical expression from an equation.
5. The first step in solving a radical equation is to square both sides of the equation.

1 Solve equations containing one or more radical expressions (See pages 486–488.)

6. What does the Property of Squaring Both Sides of an Equation state?

GETTING READY

7. Solve: $\sqrt{x - 2} = 5$. You will check the solution in Exercise 8.

$\sqrt{x - 2} = 5$ • **The radical is alone on one side.**
$(\sqrt{x - 2})^{?} = 5^{?}$ • **Square each side of the equation.**
$\underline{\ ?\ } = \underline{\ ?\ }$ • **Simplify.**
$x = \underline{\ ?\ }$ • **Add ___?___ to each side of the equation.**

8. Check the solution that you found in Exercise 7.

$\sqrt{x - 2} = 5$		
$\sqrt{27 - 2}$	5	• **Replace *x* with ___?___.**
$\sqrt{\underline{\ ?\ }}$	5	• **Subtract.**
$\underline{\ ?\ }$	= 5	• **Simplify the radical expression. This is a true equation. The solution checks.**

Solve and check.

9. $\sqrt{x} = 5$
10. $\sqrt{y} = 7$
11. $\sqrt{a} = 12$
12. $\sqrt{a} = 9$
13. $\sqrt{5x} = 5$
14. $\sqrt{3x} = 4$
15. $\sqrt{4x} = 8$
16. $\sqrt{6x} = 3$
17. $\sqrt{2x} - 4 = 0$
18. $3 - \sqrt{5x} = 0$
19. $\sqrt{4x} + 5 = 2$
20. $\sqrt{3x} + 9 = 4$
21. $\sqrt{3x - 2} = 4$
22. $\sqrt{5x + 6} = 1$
23. $\sqrt{2x + 1} = 7$
24. $\sqrt{5x + 4} = 3$
25. $\sqrt{5x + 2} = 0$
26. $\sqrt{3x - 7} = 0$
27. $\sqrt{3x} - 6 = -4$
28. $\sqrt{5x} + 8 = 23$
29. $0 = 2 - \sqrt{3 - x}$
30. $0 = 5 - \sqrt{10 + x}$
31. $0 = \sqrt{3x - 9} - 6$
32. $0 = \sqrt{2x + 7} - 3$

33. $\sqrt{5x - 1} = \sqrt{3x + 9}$

34. $\sqrt{3x + 4} = \sqrt{12x - 14}$

35. $\sqrt{5x - 3} = \sqrt{4x - 2}$

36. $\sqrt{5x - 9} = \sqrt{2x - 3}$

37. $\sqrt{x^2 - 5x + 6} = \sqrt{x^2 - 8x + 9}$

38. $\sqrt{x^2 - 2x + 4} = \sqrt{x^2 + 5x - 12}$

39. $\sqrt{x} = \sqrt{x + 3} - 1$

40. $\sqrt{x + 5} = \sqrt{x} + 1$

41. $\sqrt{2x + 5} = 5 - \sqrt{2x}$

42. $\sqrt{2x} + \sqrt{2x + 9} = 9$

43. $\sqrt{3x} - \sqrt{3x + 7} = 1$

44. $\sqrt{x} - \sqrt{x + 9} = 1$

45. Without solving the equations, identify which equation has no solution.
(i) $-\sqrt{2x - 5} = -3$ (ii) $\sqrt{2x} - 5 = -3$ (iii) $\sqrt{2x - 5} = -3$

46. How many times will the Property of Squaring Both Sides of an Equation be used in solving each equation? Do not solve.
(i) $\sqrt{x - 3} = \sqrt{4x - 7}$ (ii) $\sqrt{x - 6} = \sqrt{x} - 3$

2 Application problems (See pages 488–490.)

47. **Integer Problem** Five added to the square root of the product of four and a number is equal to seven. Find the number.

48. **Integer Problem** Two added to the square root of the sum of a number and five is equal to six. Find the number.

GETTING READY

49. In a right triangle, the hypotenuse is the side opposite the ___?___ angle. The other two sides are called ___?___.

50. Label the right triangle shown at the right. Include the right angle symbol, the three angles, the two legs, and the hypotenuse.

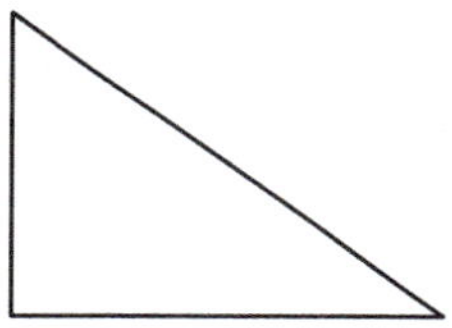

51. What does the Pythagorean Theorem state?

52. A 20-foot ladder leans against the side of a building with its bottom *d* feet from the building. The ladder reaches a height of *h* feet. Which of the following distances is not possible as a value for *h*?
(i) 4 ft (ii) 10 ft (iii) 16 ft (iv) 22 ft

Solve. Round to the nearest hundredth.

53. **Geometry** The two legs of a right triangle measure 5 cm and 9 cm. Find the length of the hypotenuse.

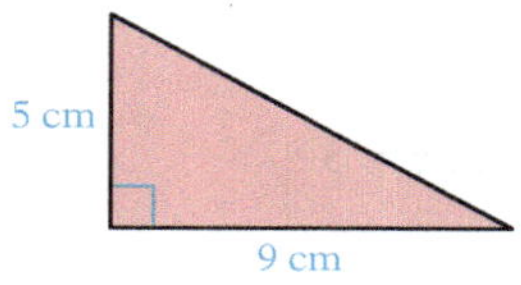

54. **Geometry** The two legs of a right triangle measure 8 in. and 4 in. Find the length of the hypotenuse.

55. **Geometry** The hypotenuse of a right triangle measures 12 ft. One leg of the triangle measures 7 ft. Find the length of the other leg of the triangle.

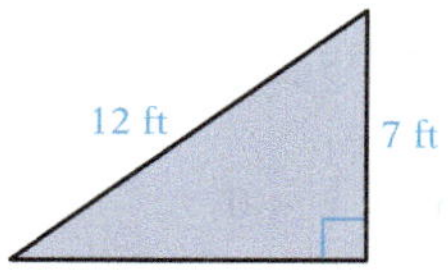

56. **Geometry** The hypotenuse of a right triangle measures 20 cm. One leg of the triangle measures 16 cm. Find the length of the other leg of the triangle.

57. **Geometry** The diagonal of a rectangle is a line drawn from one vertex to the opposite vertex. Find the length of the diagonal in the rectangle shown at the right. Round to the nearest tenth.

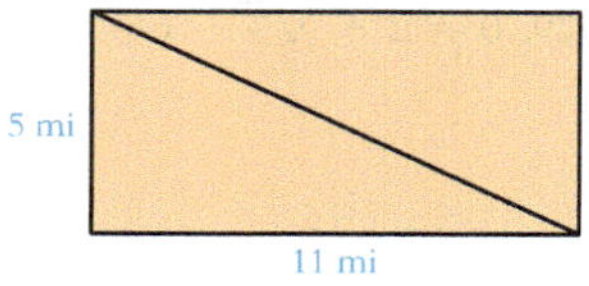

GETTING READY

58. The equation $v = 4\sqrt{d}$ gives the speed v, in feet per second, of a dropped object after the object has fallen d feet. A stone is dropped from a bridge. To find the distance the stone has fallen when it reaches a speed of 48 ft/s, replace __?__ with 48 and solve for __?__.

59. Is the equation equivalent to the equation $C = \sqrt{32H}$ given in Exercise 61? Assume C and H are positive.

a. $\frac{C^2}{H} = 32$ **b.** $C = 4\sqrt{2H}$ **c.** $\frac{C^2}{32} = H$ **d.** $\left(\frac{C}{4}\right)^2 = 2H$

60. Education One method used to "curve" the grades on an exam is to use the formula $R = 10\sqrt{O}$, where R is the revised score and O is the original score. Use this formula to find the original score on an exam that has a revised score of 75. Round to the nearest whole number.

61. Physics A formula used in the study of shallow-water wave motion is $C = \sqrt{32H}$, where C is the wave velocity in feet per second and H is the depth in feet. Use this formula to find the depth of the water when the wave velocity is 20 ft/s.

62. Physics See the news clipping at the right. The time it takes an object to fall a certain distance is given by the equation $t = \sqrt{\frac{d}{16}}$, where t is the time in seconds and d is the distance in feet. Use this equation to find the height from which the hay was dropped.

In the News

Hay Drop for Stranded Cattle

The Wyoming and Colorado National Guards have come to the aid of thousands of cattle stranded by the blizzard that has paralyzed southeastern Colorado. Flying low over the cattle, the guardsmen drop bales of hay that 6 s later smash into the ground, break apart, and provide food for the animals, which would otherwise starve.

Sources: The Denver Post; www.af.mil

63. Sports The infield of a baseball diamond is a square. The distance between successive bases is 90 ft. The pitcher's mound is on the diagonal between home plate and second base at a distance of 60.5 ft from home plate. Is the pitcher's mound more or less than halfway between home plate and second base?

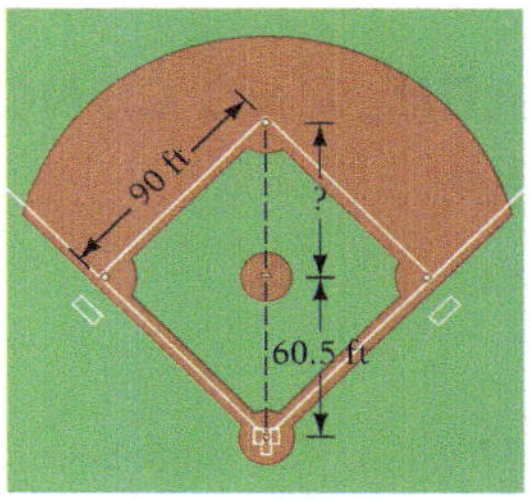

64. Sports The infield of a softball diamond is a square. The distance between successive bases is 60 ft. The pitcher's mound is on the diagonal between home plate and second base at a distance of 46 ft from home plate. Is the pitcher's mound more or less than halfway between home plate and second base?

65. Communications Marta Lightfoot leaves a dock in her sailboat and sails 2.5 mi due east. She then tacks and sails 4 mi due north. The walkie-talkie Marta has on board has a range of 5 mi. Will she be able to call a friend on the dock from her location using the walkie-talkie?

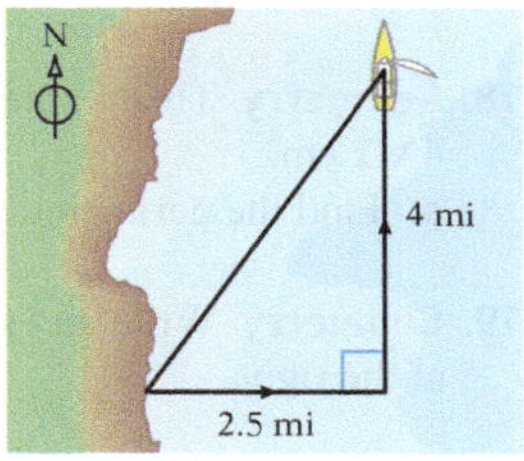

66. Navigation How far above the water would a submarine periscope have to be for the lookout to locate a ship 4 mi away? The equation for the distance in miles that the lookout can see is $d = \sqrt{1.5h}$, where h is the height in feet above the surface of the water. Round to the nearest hundredth.

67. Home Maintenance Rick Wyman needs to clean the gutters of his home. The gutters are 24 ft above the ground. For safety, the distance a ladder reaches up a wall should be four times the distance from the bottom of the ladder to the base of the side of the house. Therefore, the bottom of the ladder must be 6 ft from the base of the house. Will a 25-foot ladder be long enough to reach the gutters?

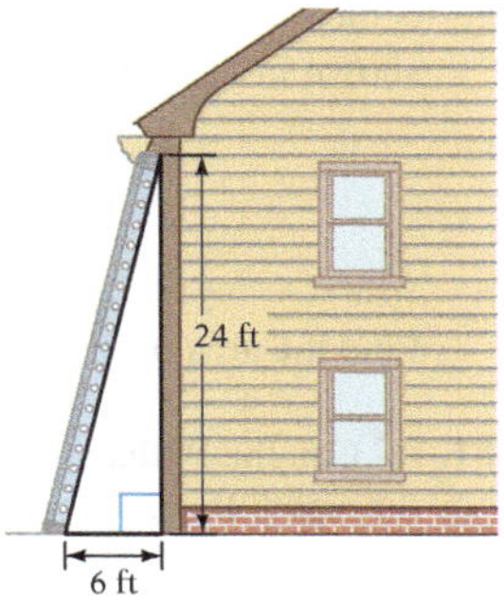

68. Physics The speed of a child riding a merry-go-round at a carnival is given by the equation $v = \sqrt{12r}$, where v is the speed in feet per second and r is the distance in feet from the center of the merry-go-round to the rider. If a child is moving at 15 ft/s, how far is the child from the center of the merry-go-round?

69. **Physics** Find the length of a pendulum that makes one swing in 3 s. The equation for the time of one swing is $T = 2\pi\sqrt{\frac{L}{32}}$, where T is the time in seconds and L is the length in feet.

70. **Aviation** A commuter plane leaves an airport traveling due south at 400 mph. Another plane leaving at the same time travels due east at 300 mph. Find the distance between the two planes after 2 h.

71. **Physics** A stone is dropped from a bridge and hits the water 2 s later. How high is the bridge? The equation for the distance an object falls in T seconds is $T = \sqrt{\frac{d}{16}}$, where d is the distance in feet.

72. **Physics** A stone is dropped into a mine shaft and hits the bottom 3.5 s later. How deep is the mine shaft? The equation for the distance an object falls in T seconds is $T = \sqrt{\frac{d}{16}}$, where d is the distance in feet.

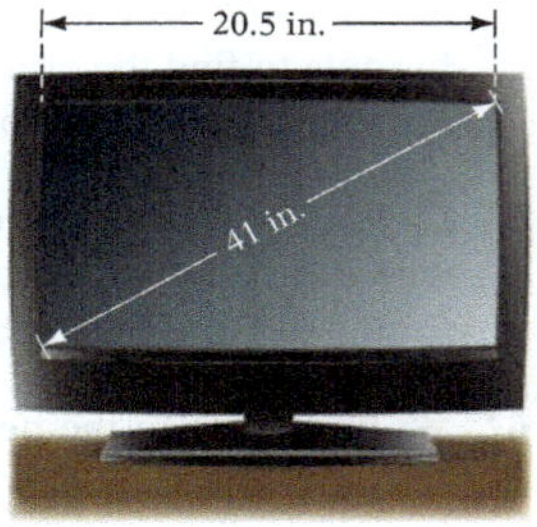

73. **Home Entertainment** The measure of a television screen is given by the length of a diagonal across the screen. A 41-inch television has a width of 20.5 in. Find the height of the screen to the nearest tenth of an inch.

APPLYING CONCEPTS

Solve.

74. $\sqrt{\frac{5y + 2}{3}} = 3$

75. $\sqrt{\frac{3y}{5}} - 1 = 2$

76. $\sqrt{9x^2 + 49} + 1 = 3x + 2$

77. **Geometry** In the coordinate plane, a triangle is formed by drawing lines between the points (0, 0) and (5, 0), (5, 0) and (5, 12), and (5, 12) and (0, 0). Find the number of units in the perimeter of the triangle.

78. **Geometry** The hypotenuse of a right triangle is $5\sqrt{2}$ cm, and the length of one leg is $4\sqrt{2}$ cm.
a. Find the perimeter of the triangle. **b.** Find the area of the triangle.

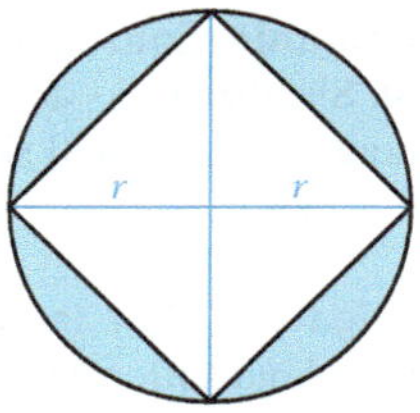

79. **Geometry** Write an expression in factored form for the shaded region in the diagram at the right.

80. **Geometry** A circular fountain is being designed for a triangular plaza in a cultural center. The fountain is placed so that each side of the triangle touches the fountain as shown in the diagram at the right. The formula for the radius of the circle is given by

$$r = \sqrt{\frac{(s - a)(s - b)(s - c)}{s}}$$

where $s = \frac{1}{2}(a + b + c)$, and a, b, and c are the lengths of the sides of the triangle. Find the area of the fountain. Round to the nearest hundredth.

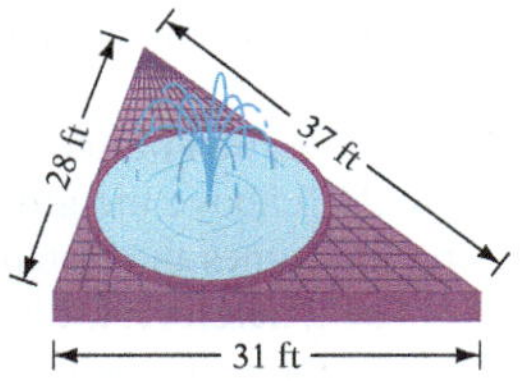

81. Can the Pythagorean Theorem be used to find the length of side c of the triangle at the right? If so, determine c. If not, explain why the theorem cannot be used.

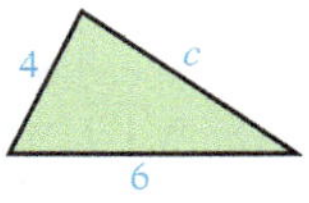

PROJECTS OR GROUP ACTIVITIES

82. Complete the statement using the symbol <, =, or >. Explain how you determined which symbol to use.

a. For an acute triangle with side c the longest side, $a^2 + b^2$________ c^2.

b. For a right triangle with side c the longest side, $a^2 + b^2$________ c^2.

c. For an obtuse triangle with side c the longest side, $a^2 + b^2$________ c^2.

83. The length of a side of the outer square in the diagram at the right is $2x$ inches. The corners of the inner square are the midpoints of the sides of the outer square.

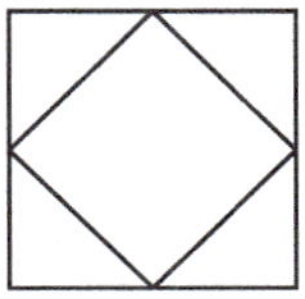

a. What is the length of a side of the inner square?

b. What is the area of the inner square?

84. Three squares are lined up along the x-axis as shown at the right. Find AB. Round to the nearest tenth.

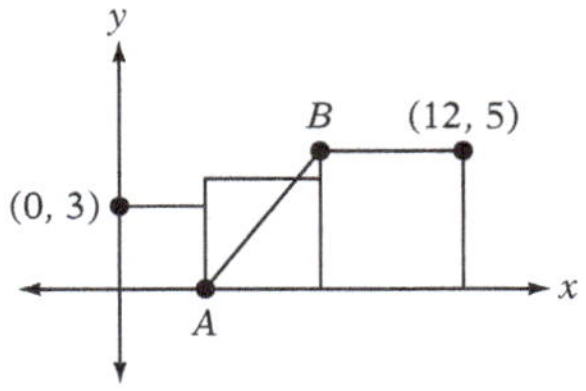

85. The sum of the squares of the lengths of all four sides of the rectangle shown at the right is 100. The rectangle is not a square. Find the length of the diagonal of the rectangle.

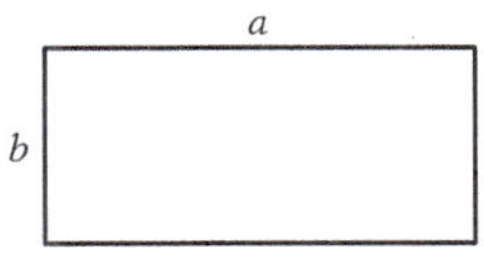

CHAPTER 10 Summary

Key Words	Objective and Page Reference	Examples
A **square root** of a positive number x is a number whose square is x. The square root of a negative number is not a real number.	[10.1.1, p. 466]	A square root of 25 is 5 because $5^2 = 25$. A square root of 25 is -5 because $(-5)^2 = 25$. $\sqrt{-9}$ is not a real number.
The **principal square root** of a number is the positive square root. The symbol $\sqrt{\ }$ is called a **radical sign** and is used to indicate the principal square root of a number. The negative square root of a number is indicated by placing a negative sign in front of the radical. The **radicand** is the expression under the radical symbol.	[10.1.1, p. 466]	$\sqrt{25} = 5$ $-\sqrt{25} = -5$ In the expression $\sqrt{25}$, 25 is the radicand.

The square of an integer is a **perfect square.** If a number is not a perfect square, its square root can only be approximated. Such numbers are **irrational numbers.** Their decimal representations never terminate or repeat.	[10.1.1, p. 466]	$2^2 = 4, 3^2 = 9, 4^2 = 16, 5^2 = 25, 6^2 = 36, \ldots$ 4, 9, 16, 25, 36, ... are perfect squares. 5 is not a perfect square. $\sqrt{5}$ is an irrational number.
Conjugates are expressions with two terms that differ only in the sign of the second term. The expressions $a + b$ and $a - b$ are conjugates.	[10.3.1, p. 479]	$3 + \sqrt{7}$ and $3 - \sqrt{7}$ are conjugates. $\sqrt{x} + 2$ and $\sqrt{x} - 2$ are conjugates.
A **radical equation** is an equation that contains a variable expression in a radicand.	[10.4.1, p. 486]	$\sqrt{2x} + 8 = 12$ is a radical equation. $2x + \sqrt{8} = 12$ is not a radical equation.

Essential Rules and Procedures	Objective and Page Reference	Examples
The Product Property of Square Roots If a and b are positive real numbers, then $\sqrt{ab} = \sqrt{a} \cdot \sqrt{b}$.	[10.1.1–10.1.2, pp. 466, 469]	$\sqrt{12} = \sqrt{4 \cdot 3} = \sqrt{4}\,\sqrt{3} = 2\sqrt{3}$ $\sqrt{b^{11}} = \sqrt{b^{10} \cdot b} = \sqrt{b^{10}}\sqrt{b} = b^5\sqrt{b}$
Adding and Subtracting Radical Expressions The Distributive Property is used to simplify the sum or difference of like radical expressions.	[10.2.1, p. 474]	$16\sqrt{3x} - 4\sqrt{3x} = (16 - 4)\sqrt{3x} = 12\sqrt{3x}$
Multiplying Radical Expressions The Product Property of Square Roots is used to multiply radical expressions. Use the Distributive Property to remove parentheses. Use the FOIL method to multiply radical expressions with two terms.	[10.3.1, pp. 478–479]	$\sqrt{5x}\,\sqrt{7y} = \sqrt{5x \cdot 7y} = \sqrt{35xy}$ $\sqrt{y}(2 + \sqrt{3x}) = 2\sqrt{y} + \sqrt{3xy}$ $(5 - \sqrt{x})(11 + \sqrt{x}) = 55 + 5\sqrt{x} - 11\sqrt{x} - (\sqrt{x})^2 = 55 - 6\sqrt{x} - x$
The Quotient Property of Square Roots If a and b are positive real numbers, then $\sqrt{\frac{a}{b}} = \frac{\sqrt{a}}{\sqrt{b}}$.	[10.3.2, p. 480]	$\sqrt{\frac{9x^2}{y^8}} = \frac{\sqrt{9x^2}}{\sqrt{y^8}} = \frac{3x}{y^4}$
Dividing Radical Expressions The Quotient Property of Square Roots is used to divide radical expressions.	[10.3.2, p. 480]	$\frac{\sqrt{3x^5y}}{\sqrt{75xy^3}} = \sqrt{\frac{3x^5y}{75xy^3}} = \sqrt{\frac{x^4}{25y^2}} = \frac{\sqrt{x^4}}{\sqrt{25y^2}} = \frac{x^2}{5y}$
Rationalizing the Denominator A radical expression is not in simplest form if there is a radical in the denominator. The procedure used to remove a radical from the denominator is called rationalizing the denominator.	[10.3.2, p. 481]	$\frac{5}{\sqrt{7}} = \frac{5}{\sqrt{7}} \cdot \frac{\sqrt{7}}{\sqrt{7}} = \frac{5\sqrt{7}}{7}$

Radical Expressions in Simplest Form
A radical expression is in simplest form if:
1. The radicand contains no factor greater than 1 that is a perfect square.
2. There is no fraction under the radical sign.
3. There is no radical in the denominator of a fraction.

[10.3.2, p. 481]

$\sqrt{12}$, $\sqrt{\frac{3}{4}}$, and $\frac{1}{\sqrt{2}}$ are not in simplest form.

$2\sqrt{3}$, $\frac{\sqrt{3}}{2}$, and $\frac{\sqrt{2}}{2}$ are in simplest form.

Property of Squaring Both Sides of an Equation
If a and b are real numbers and $a = b$, then $a^2 = b^2$.

[10.4.1, p. 486]

$$\begin{aligned} \sqrt{x+3} &= 4 \\ (\sqrt{x+3})^2 &= 4^2 \\ x + 3 &= 16 \\ x &= 13 \end{aligned}$$

Solving Radical Equations
When an equation contains one radical expression, write the equation with the radical alone on one side of the equation. Square both sides of the equation. Solve for the variable. Whenever both sides of an equation are squared, the solutions must be checked.

[10.4.1, p. 486]

$$\begin{aligned} \sqrt{2x} - 1 &= 5 \\ \sqrt{2x} &= 6 \\ (\sqrt{2x})^2 &= 6^2 \\ 2x &= 36 \\ x &= 18 \quad \text{The solution checks.} \end{aligned}$$

Pythagorean Theorem
If a and b are the lengths of the legs of a right triangle and c is the length of the hypotenuse, then $c^2 = a^2 + b^2$.

[10.4.2, p. 489]

Two legs of a right triangle measure 4 cm and 7 cm. To find the length of the hypotenuse, use the equation

$c = \sqrt{a^2 + b^2}$

$c = \sqrt{4^2 + 7^2} = \sqrt{16 + 49} = \sqrt{65}$

The hypotenuse measures $\sqrt{65}$ cm.

CHAPTER 10 Review Exercises

1. Subtract: $5\sqrt{3} - 16\sqrt{3}$
2. Simplify: $\dfrac{2x}{\sqrt{3} - \sqrt{5}}$
3. Simplify: $\sqrt{x^2 + 16x + 64}$
4. Solve: $3 - \sqrt{7x} = 5$
5. Simplify: $\dfrac{5\sqrt{y} - 2\sqrt{y}}{3\sqrt{y}}$
6. Add: $6\sqrt{7} + \sqrt{7}$
7. Multiply: $(6\sqrt{a} + 5\sqrt{b})(2\sqrt{a} + 3\sqrt{b})$
8. Simplify: $\sqrt{49(x + 3)^4}$
9. Simplify: $2\sqrt{36}$
10. Solve: $\sqrt{b} = 4$
11. Subtract: $9x\sqrt{5} - 5x\sqrt{5}$
12. Multiply: $(\sqrt{5ab} - \sqrt{7})(\sqrt{5ab} + \sqrt{7})$
13. Solve: $\sqrt{2x + 9} = \sqrt{8x - 9}$
14. Simplify: $\sqrt{35}$
15. Add: $2x\sqrt{60x^3y^3} + 3x^2y\sqrt{15xy}$
16. Simplify: $\dfrac{\sqrt{3x^3y}}{\sqrt{27xy^5}}$
17. Simplify: $(3\sqrt{x} - \sqrt{y})^2$
18. Simplify: $\sqrt{(a + 4)^2}$

19. Simplify: $5\sqrt{48}$

20. Add: $3\sqrt{12x} + 5\sqrt{48x}$

21. Simplify: $\dfrac{8}{\sqrt{x} - 3}$

22. Multiply: $\sqrt{6a}(\sqrt{3a} + \sqrt{2a})$

23. Simplify: $3\sqrt{18a^5b}$

24. Simplify: $-3\sqrt{120}$

25. Subtract: $\sqrt{20a^5b^9} - 2ab^2\sqrt{45a^3b^5}$

26. Simplify: $\dfrac{\sqrt{98x^7y^9}}{\sqrt{2x^3y}}$

27. Solve: $\sqrt{5x + 1} = \sqrt{20x - 8}$

28. Simplify: $\sqrt{c^{18}}$

29. Simplify: $\sqrt{450}$

30. Simplify: $6a\sqrt{80b} - \sqrt{180a^2b} + 5a\sqrt{b}$

31. Simplify: $\dfrac{16}{\sqrt{a}}$

32. Solve: $6 - \sqrt{2y} = 2$

33. Multiply: $\sqrt{a^3b^4c}\sqrt{a^7b^2c^3}$

34. Simplify: $7\sqrt{630}$

35. Simplify: $y\sqrt{24y^6}$

36. Simplify: $\dfrac{\sqrt{250}}{\sqrt{10}}$

37. Solve: $\sqrt{x^2 + 5x + 4} = \sqrt{x^2 + 7x - 6}$

38. Multiply: $(4\sqrt{y} - \sqrt{5})(2\sqrt{y} + 3\sqrt{5})$

39. Find the decimal approximation of $\sqrt{9900}$ to the nearest thousandth.

40. Multiply: $\sqrt{7}\sqrt{7}$

41. Simplify: $5x\sqrt{150x^7}$

42. Simplify: $2x^2\sqrt{18x^2y^5} + 6y\sqrt{2x^6y^3} - 9xy^2\sqrt{8x^4y}$

43. Simplify: $\dfrac{\sqrt{54a^3}}{\sqrt{6a}}$

44. Simplify: $4\sqrt{250}$

45. Solve: $\sqrt{5x} = 10$

46. Simplify: $\dfrac{3a\sqrt{3} + 2\sqrt{12a^2}}{\sqrt{27}}$

47. Multiply: $\sqrt{2}\sqrt{50}$

48. Simplify: $\sqrt{36x^4y^5}$

49. Simplify: $4y\sqrt{243x^{17}y^9}$

50. Solve: $\sqrt{x + 1} - \sqrt{x - 2} = 1$

51. Simplify: $\sqrt{400}$

52. Simplify: $-4\sqrt{8x} + 7\sqrt{18x} - 3\sqrt{50x}$

53. Solve: $0 = \sqrt{10x + 4} - 8$

54. Multiply: $\sqrt{3}(\sqrt{12} - \sqrt{3})$

55. **Integer Problem** The square root of the sum of two consecutive odd integers is equal to 10. Find the larger integer.

56. **Space Travel** The weight of an object is related to the object's distance from the surface of Earth. An equation for this relationship is $d = 4000\sqrt{\dfrac{W_0}{W_a}}$, where W_0 is the object's weight on the surface of Earth and W_a is the object's weight at a distance of d miles above Earth's surface. A space explorer weighs 36 lb when 8000 mi above Earth's surface. Find the explorer's weight on the surface of Earth.

57. **Geometry** The hypotenuse of a right triangle measures 18 cm. One leg of the triangle measures 11 cm. Find the length of the other leg of the triangle. Round to the nearest hundredth.

58. Bicycling A bicycle will overturn if it rounds a corner too sharply or too quickly. The equation for the maximum velocity at which a cyclist can turn a corner without tipping over is given by the equation $v = 4\sqrt{r}$, where v is the velocity of the bicycle in miles per hour and r is the radius of the corner in feet. Find the radius of the sharpest corner that a cyclist can safely turn when riding at a speed of 20 mph.

59. Tsunamis A tsunami is a great sea wave produced by underwater earthquakes or volcanic eruption. The velocity of a tsunami as it approaches land is approximated by the equation $v = 3\sqrt{d}$, where v is the velocity in feet per second and d is the depth of the water in feet. Find the depth of the water when the velocity of a tsunami is 30 ft/s.

60. Guy Wires A guy wire is attached to a point 25 ft above the ground on a telephone pole that is perpendicular to the ground. The wire is anchored to the ground at a point 8 ft from the base of the pole. Find the length of the guy wire. Round to the nearest hundredth.

CHAPTER 10 Test

1. Simplify: $\sqrt{121x^8y^2}$

2. Subtract: $5\sqrt{8} - 3\sqrt{50}$

3. Multiply: $\sqrt{3x^2y}\sqrt{6x^2}\sqrt{2x}$

4. Simplify: $\sqrt{45}$

5. Simplify: $\sqrt{72x^7y^2}$

6. Simplify: $3\sqrt{8y} - 2\sqrt{72x} + 5\sqrt{18x}$

7. Multiply: $(\sqrt{y} + 3)(\sqrt{y} + 5)$

8. Simplify: $\dfrac{4}{\sqrt{8}}$

9. Simplify: $\dfrac{\sqrt{162}}{\sqrt{2}}$

10. Solve: $\sqrt{5x - 6} = 7$

11. Find the decimal approximation of $\sqrt{500}$ to the nearest thousandth.

12. Simplify: $\sqrt{32a^5b^{11}}$

13. Multiply: $\sqrt{a}(\sqrt{a} - \sqrt{b})$

14. Multiply: $\sqrt{8x^3y}\sqrt{10xy^4}$

15. Simplify: $\dfrac{\sqrt{98a^6b^4}}{\sqrt{2a^3b^2}}$

16. Solve: $\sqrt{9x} + 3 = 18$

17. Simplify: $\sqrt{192x^{13}y^5}$

18. Simplify: $2a\sqrt{2ab^3} + b\sqrt{8a^3b} - 5ab\sqrt{ab}$

19. Multiply: $(\sqrt{a} - 2)(\sqrt{a} + 2)$

20. Simplify: $\dfrac{3}{2 - \sqrt{5}}$

21. Solve: $3 = 8 - \sqrt{5x}$

22. Multiply: $\sqrt{3}(\sqrt{6} - \sqrt{x^2})$

23. Subtract: $3\sqrt{a} - 9\sqrt{a}$

24. Simplify: $\sqrt{108}$

25. Find the decimal approximation of $\sqrt{63}$ to the nearest thousandth.

26. Simplify: $\dfrac{\sqrt{108a^7b^3}}{\sqrt{3a^4b}}$

27. Solve: $\sqrt{x} - \sqrt{x + 3} = 1$

28. Integer Problem The square root of the sum of two consecutive integers is equal to 9. Find the smaller integer.

29. **Physics** Find the length of a pendulum that makes one swing in 2.5 s. The equation for the time of one swing of a pendulum is $T = 2\pi\sqrt{\frac{L}{32}}$, where T is the time in seconds and L is the length in feet. Round to the nearest hundredth.

30. **Home Maintenance** A ladder 16 ft long is resting against a building. How high on the building will the ladder reach when the bottom of the ladder is 5 ft from the building? Round to the nearest tenth.

Cumulative Review Exercises

1. Simplify: $\left(\frac{2}{3}\right)^2\left(\frac{3}{4} - \frac{3}{2}\right) + \left(\frac{1}{2}\right)^2$

2. Simplify: $-3[x - 2(3 - 2x) - 5x] + 2x$

3. Solve: $2x - 4[3x - 2(1 - 3x)] = 2(3 - 4x)$

4. Solve: $3(x - 7) \geq 5x - 12$

5. Find the slope of the line that contains the points whose coordinates are $(2, -5)$ and $(-4, 3)$.

6. Find the equation of the line that contains the point whose coordinates are $(-2, -3)$ and has slope $\frac{1}{2}$.

7. Evaluate $f(x) = \frac{5}{2}x - 8$ at $x = -4$.

8. Graph: $f(x) = -4x + 2$

9. Graph the solution set of $2x + y < -2$.

10. Solve by graphing: $\begin{aligned} 3x - 2y &= 8 \\ 4x + 5y &= 3 \end{aligned}$

11. Solve by substitution: $\begin{aligned} 4x - 3y &= 1 \\ 2x + y &= 3 \end{aligned}$

12. Solve by the addition method: $\begin{aligned} 5x + 4y &= 7 \\ 3x - 2y &= 13 \end{aligned}$

13. Simplify: $(-3x^2y)(-2x^3y^{-4})$

14. Simplify: $\dfrac{12b^4 - 6b^2 + 2}{-6b^2}$

15. Factor: $12x^3y^2 - 9x^2y^3$

16. Factor: $9b^2 + 3b - 20$

17. Factor: $2a^3 - 16a^2 + 30a$

18. Multiply: $\dfrac{3x^3 - 6x^2}{4x^2 + 4x} \cdot \dfrac{3x - 9}{9x^3 - 45x^2 + 54x}$

19. Simplify: $\dfrac{1 - \dfrac{2}{x} - \dfrac{15}{x^2}}{1 - \dfrac{9}{x^2}}$

20. Subtract: $\dfrac{x + 2}{x - 4} - \dfrac{6}{(x - 4)(x - 3)}$

21. Solve: $\dfrac{x}{2x - 5} - 2 = \dfrac{3x}{2x - 5}$

22. Simplify: $2\sqrt{27a} - 5\sqrt{49a} + 8\sqrt{48a}$

23. Simplify: $\dfrac{\sqrt{320}}{\sqrt{5}}$

24. Solve: $\sqrt{2x - 3} - 5 = 0$

25. Integer Problem Three-eighths of a number is less than negative twelve. Find the largest integer that satisfies the inequality.

26. Markup The selling price of a book is \$29.40. The markup rate used by the bookstore is 20% of the cost. Find the cost of the book.

27. Mixture Problem How many ounces of pure water must be added to 40 oz of a 12% salt solution to make a salt solution that is 5% salt?

28. Integer Problem The sum of two numbers is twenty-one. Three more than twice the smaller number is equal to the larger. Find the two numbers.

29. Work Problem A small water pipe takes twice as long to fill a tank as does a larger water pipe. With both pipes open, it takes 16 h to fill the tank. Find the time it would take the small pipe, working alone, to fill the tank.

30. Integer Problem The square root of the sum of two consecutive integers is equal to 7. Find the smaller integer.

CHAPTER

11

Quadratic Equations

Focus on Success

The end of the semester is generally a very busy and stressful time. You may have papers or projects due and assignments to complete. You may be dealing with the anxiety of taking final exams. You have covered a great deal of material in this course, and reviewing all of it may be daunting. You might begin by reviewing the Chapter Summary for each chapter that you were assigned during the term. Then take the final exam on page 558. The answer to each exercise is given at the back of the book, along with the objective the question relates to. If you have trouble with any of the questions, restudy the objectives the questions are taken from and retry some of the exercises in those objectives. (See Ace the Test, page AIM-11).

OBJECTIVES

11.1
1. Solve quadratic equations by factoring
2. Solve quadratic equations by taking square roots

11.2
1. Solve quadratic equations by completing the square

11.3
1. Solve quadratic equations by using the quadratic formula

11.4
1. Simplify complex numbers
2. Add and subtract complex numbers
3. Multiply complex numbers
4. Divide complex numbers
5. Solve quadratic equations with complex number solutions

11.5
1. Graph a quadratic equation of the form $y = ax^2 + bx + c$

11.6
1. Application problems

PREP TEST

Are you ready to succeed in this chapter?
Take the Prep Test below to find out if you are ready to learn the new material.

1. Evaluate $b^2 - 4ac$ when $a = 2$, $b = -3$, and $c = -4$.
2. Solve: $5x + 4 = 3$
3. Factor: $x^2 + x - 12$
4. Factor: $4x^2 - 12x + 9$
5. Is $x^2 - 10x + 25$ a perfect square trinomial?
6. Solve: $\dfrac{5}{x-2} = \dfrac{15}{x}$
7. Graph: $y = -2x + 3$
8. Simplify: $\sqrt{28}$
9. If a is *any* real number, simplify $\sqrt{a^2}$.
10. Walking at a constant speed of 4.5 mph, Lucy and Sam traveled from the beginning to the end of a hiking trail. When they reached the end, they immediately started back along the same path at a constant speed of 3 mph. If the round trip took 2 h, what is the length of the hiking trail?

11.1 Solving Quadratic Equations by Factoring or by Taking Square Roots

OBJECTIVE 1 Solve quadratic equations by factoring

In Section 5 of the chapter on factoring, we solved quadratic equations by factoring. In this section, we will review that material and then solve quadratic equations by taking square roots.

An equation that can be written in the form $ax^2 + bx + c = 0$, $a \neq 0$, is a **quadratic equation.**

$$4x^2 - 3x + 1 = 0, \qquad a = 4, b = -3, c = 1$$
$$3x^2 - 4 = 0, \qquad a = 3, b = 0, c = -4$$

A quadratic equation is also called a **second-degree equation.**

A quadratic equation is in **standard form** when the polynomial is in descending order and equal to zero. The two quadratic equations above are in standard form.

Recall that the Principle of Zero Products states that if the product of two factors is zero, then at least one of the factors must be zero.

If $a \cdot b = 0$, then $a = 0$ or $b = 0$.

The Principle of Zero Products can be used in solving quadratic equations.

Focus on solving a quadratic equation by factoring

Solve by factoring: $2x^2 - x = 1$

This is a quadratic equation. Write it in standard form.

$$2x^2 - x = 1$$
$$2x^2 - x - 1 = 0$$

Factor the left side of the equation.

$$(2x + 1)(x - 1) = 0$$

If $(2x + 1)(x - 1) = 0$, then either $2x + 1 = 0$ or $x - 1 = 0$.

$$2x + 1 = 0 \qquad x - 1 = 0$$

Solve each equation for x.

$$2x = -1 \qquad x = 1$$
$$x = -\frac{1}{2}$$

The solutions are $-\frac{1}{2}$ and 1.

```
-1/2→X
                -.5
2X²-X
                  1
```

A graphing calculator can be used to check the solutions of a quadratic equation. Consider the preceding example. The solutions appear to be $-\frac{1}{2}$ and 1. To check these solutions, store one value of x, $-\frac{1}{2}$, in the calculator. Note that after you store $-\frac{1}{2}$ in the calculator's memory, your calculator rewrites the fraction as the decimal -0.5. Evaluate the expression on the left side of the original equation: $2x^2 - x$. The result is 1, which is the number on the right side of the original equation. The solution -0.5 checks. Now store the other value of x, 1, in the calculator. Evaluate the expression on the left side of the original equation: $2x^2 - x$. The result is 1, which is the number on the right side of the equation. The solution 1 checks.

Focus on solving a quadratic equation with a double root

Solve by factoring: $x^2 = 10x - 25$

$$x^2 = 10x - 25$$

This is a quadratic equation. Write it in standard form. $x^2 - 10x + 25 = 0$

Factor the left side of the equation. $(x - 5)(x - 5) = 0$

Let each factor equal 0. $x - 5 = 0 \quad x - 5 = 0$

Solve each equation for x. $x = 5 \quad x = 5$

The factorization in this example produced two identical factors. Because $x - 5$ occurs twice in the factored form of the equation, 5 is a **double root** of the equation.

EXAMPLE 1 Solve by factoring: $\frac{z^2}{2} + \frac{z}{3} - \frac{1}{6} = 0$

Solution

$$\frac{z^2}{2} + \frac{z}{3} - \frac{1}{6} = 0$$

$$6\left(\frac{z^2}{2} + \frac{z}{3} - \frac{1}{6}\right) = 6(0)$$

- To eliminate the fractions, multiply each side of the equation by 6, the LCD of the fractions.

$$3z^2 + 2z - 1 = 0$$

- The quadratic equation is in standard form.

$$(3z - 1)(z + 1) = 0$$

- Factor the left side of the equation.

$$3z - 1 = 0 \qquad z + 1 = 0$$

- Let each factor equal zero.

$$3z = 1 \qquad z = -1$$

$$z = \frac{1}{3}$$

- Solve each equation for z.

The solutions are $\frac{1}{3}$ and -1.

- $\frac{1}{3}$ and -1 check as solutions.

Problem 1 Solve by factoring: $\frac{3y^2}{2} + y - \frac{1}{2} = 0$

Solution See page S26.

Try Exercise 55, page 509.

OBJECTIVE 2 Solve quadratic equations by taking square roots

> **Take Note**
>
> Recall that the solution of the equation $|x| = 5$ is ± 5. This principle is used when solving an equation by taking square roots. Remember that $\sqrt{x^2} = |x|$. Therefore,
>
> $$x^2 = 25$$
> $$\sqrt{x^2} = \sqrt{25}$$
> $$|x| = 5$$
> $$x = \pm 5$$

Consider a quadratic equation of the form $x^2 = a$. This equation can be solved by factoring.

$$x^2 = 25$$
$$x^2 - 25 = 0$$
$$(x + 5)(x - 5) = 0$$
$$x + 5 = 0 \qquad x - 5 = 0$$
$$x = -5 \qquad x = 5$$

The solutions are -5 and 5.

Solutions that are plus or minus the same number are frequently written using $\pm$. For the example above, this would be written: "The solutions are ± 5." Because the solutions ± 5 can be written as $\pm\sqrt{25}$, an alternative method of solving this equation is suggested.

PRINCIPLE OF TAKING THE SQUARE ROOT OF EACH SIDE OF AN EQUATION

If $x^2 = a$, then $x = \pm\sqrt{a}$.

EXAMPLES

1. If $x^2 = 3$, then $x = \pm\sqrt{3}$.
2. If $x^2 = 16$, then $x = \pm\sqrt{16} = \pm 4$.

We can solve the equation $x^2 = 25$ by using this principle.

$$x^2 = 25$$

Take the square root of each side of the equation.

$$\sqrt{x^2} = \sqrt{25}$$

Simplify.

$$x = \pm\sqrt{25} = \pm 5$$

Write the solutions.

The solutions are 5 and -5.

Focus on solving a quadratic equation by taking square roots

Solve by taking square roots: $3x^2 = 36$

$$3x^2 = 36$$

Solve for x^2.

$$x^2 = 12$$

Take the square root of each side of the equation.

$$\sqrt{x^2} = \sqrt{12}$$

Simplify.

$$x = \pm\sqrt{12} = \pm 2\sqrt{3}$$

Check:

$3x^2 = 36$	
$3(2\sqrt{3})^2$	36
$3(12)$	36
$36 = 36$	

$3x^2 = 36$	
$3(-2\sqrt{3})^2$	36
$3(12)$	36
$36 = 36$	

These are true equations. The solutions check.

Write the solutions.

The solutions are $2\sqrt{3}$ and $-2\sqrt{3}$.

Take Note

You should always check your solutions by substituting them back into the *original* equation.

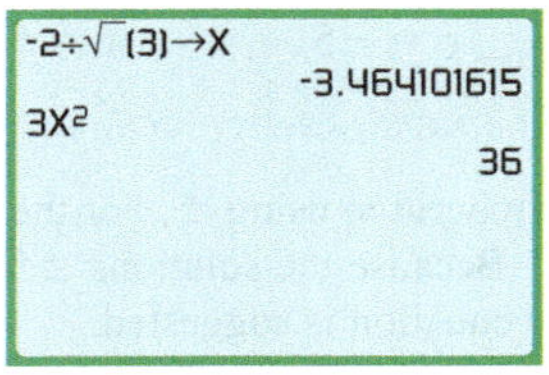

A graphing calculator can be used to check irrational solutions of a quadratic equation. Consider the preceding example; the solutions $2\sqrt{3}$ and $-2\sqrt{3}$ checked. To check these solutions on a calculator, store one value of x, $2\sqrt{3}$, in the calculator. Note that after you store this number in the calculator's memory, the calculator rewrites the expression as the decimal 3.464101615, which is an approximation of $2\sqrt{3}$. Evaluate the expression on the left side of the original equation. The result is 36, which is the number on the right side of the original equation. The solution $2\sqrt{3}$ checks. Now store the other value of x, $-2\sqrt{3}$, in the calculator. Note that after you store this number in the calculator's memory, the calculator rewrites the expression as the decimal −3.464101615, which is an approximation of $-2\sqrt{3}$. Evaluate the expression on the left side of the original equation. The result is 36, which is the number on the right side of the equation. The solution $-2\sqrt{3}$ checks.

EXAMPLE 2 Solve by taking square roots: $2x^2 - 72 = 0$

Solution

$$2x^2 - 72 = 0$$
$$2x^2 = 72 \quad \text{• Solve for } x^2.$$
$$x^2 = 36$$
$$\sqrt{x^2} = \sqrt{36} \quad \text{• Take the square root of each side of the equation. Then simplify.}$$
$$x = \pm\sqrt{36}$$
$$x = \pm 6$$

The solutions are 6 and -6.

Problem 2 Solve by taking square roots: $4x^2 - 96 = 0$

Solution See page S26.

Try Exercise 73, page 510.

An equation containing the square of a binomial can be solved by taking square roots.

Focus on solving an equation that contains the square of a binomial

Solve by taking square roots: $2(x - 1)^2 - 36 = 0$

$$2(x - 1)^2 - 36 = 0$$

Solve for $(x - 1)^2$.
$$2(x - 1)^2 = 36$$

Divide each side of the equation by 2.
$$(x - 1)^2 = 18$$

Take the square root of each side of the equation.
$$\sqrt{(x - 1)^2} = \sqrt{18}$$

Simplify.
$$x - 1 = \pm\sqrt{18}$$
$$x - 1 = \pm 3\sqrt{2}$$

Solve for x.
$$x - 1 = 3\sqrt{2} \qquad x - 1 = -3\sqrt{2}$$
$$x = 1 + 3\sqrt{2} \qquad x = 1 - 3\sqrt{2}$$

Check the solutions.

Check:

$2(x - 1)^2 - 36$	$= 0$
$2(1 + 3\sqrt{2} - 1)^2 - 36$	0
$2(3\sqrt{2})^2 - 36$	0
$2(18) - 36$	0
$36 - 36$	0
$0 = 0$	

$2(x - 1)^2 - 36$	$= 0$
$2(1 - 3\sqrt{2} - 1)^2 - 36$	0
$2(-3\sqrt{2})^2 - 36$	0
$2(18) - 36$	0
$36 - 36$	0
$0 = 0$	

Write the solutions.

The solutions are $1 + 3\sqrt{2}$ and $1 - 3\sqrt{2}$.

EXAMPLE 3 Solve by taking square roots: $(x - 6)^2 = 12$

Solution

$$(x - 6)^2 = 12$$
$$\sqrt{(x - 6)^2} = \sqrt{12} \quad \text{• Take the square root of each side of the equation.}$$
$$x - 6 = \pm\sqrt{12} \quad \text{• Then simplify.}$$
$$x - 6 = \pm 2\sqrt{3}$$

$$x - 6 = 2\sqrt{3} \qquad x - 6 = -2\sqrt{3} \quad \text{• Solve for } x.$$
$$x = 6 + 2\sqrt{3} \qquad x = 6 - 2\sqrt{3}$$

The solutions are $6 + 2\sqrt{3}$ and $6 - 2\sqrt{3}$.

Problem 3 Solve by taking square roots: $(x + 5)^2 = 20$

Solution See page S26.

Try Exercise 95, page 510.

11.1 Exercises

CONCEPT CHECK

1. $2x^2 - 3x + 9 = 0$ is a _______ equation.
2. By the Principle of Zero Products, if $(3x + 4)(x - 7) = 0$, then _______ $= 0$ or _______ $= 0$.
3. To solve the equation $3x^2 - 26x = 9$ by factoring, first write the equation in _______ form.
4. The solutions of an equation are $x = \pm 6$. This means that $x =$ _____ or $x =$ _____.

1 Solve quadratic equations by factoring (See pages 504–505.)

For the given quadratic equation, find the values of *a, b,* and *c.*

5. $3x^2 - 4x + 1 = 0$
6. $x^2 + 2x - 5 = 0$
7. $2x^2 - 5 = 0$
8. $4x^2 + 1 = 0$
9. $6x^2 - 3x = 0$
10. $-x^2 + 7x = 0$

Write the quadratic equation in standard form.

11. $x^2 - 8 = 3x$
12. $2x^2 = 4x - 1$
13. $x^2 = 16$
14. $x + 5 = x(x - 3)$
15. $2(x + 3)^2 = 5$
16. $4(x - 1)^2 = 3$
17. How does a quadratic equation differ from a linear equation?
18. What does the Principle of Zero Products state?
19. Explain why the equation $x(x + 5) = 12$ cannot be solved by writing $x = 12$ or $x + 5 = 12$.
20. What is a double root of a quadratic equation?

GETTING READY

21. Solve: $x^2 - 36 = 0$

$x^2 - 36 = 0$		• The equation is in standard form.
(___?___)(___?___) = 0		• Factor the difference of two squares.
$x + 6 =$ ___?___	$x - 6 =$ ___?___	• Use the Principle of Zero Products to set each factor equal to 0.
$x =$ ___?___	$x =$ ___?___	• Solve each equation for *x*.

22. Solve: $x^2 + 4x - 12 = 0$

$x^2 + 4x - 12 = 0$	• **The equation is in standard form.**
$(x + \underline{\quad ? \quad})(x - \underline{\quad ? \quad}) = 0$	• **Factor the trinomial.**
$\underline{\quad ? \quad} = 0 \qquad \underline{\quad ? \quad} = 0$	• **Use the Principle of Zero Products to set each factor equal to 0.**
$x = \underline{\quad ? \quad} \qquad x = \underline{\quad ? \quad}$	• **Solve each equation for *x*.**

Solve for x.

23. $(x + 3)(x - 5) = 0$ **24.** $(x - 9)(x + 11) = 0$ **25.** $x(x - 7) = 0$

26. $x(x + 10) = 0$ **27.** $(2x + 5)(3x - 1) = 0$ **28.** $(2x - 7)(3x + 4) = 0$

Solve by factoring.

29. $x^2 + 2x - 15 = 0$ **30.** $t^2 + 3t - 10 = 0$ **31.** $z^2 - 4z + 3 = 0$

32. $s^2 - 5s + 4 = 0$ **33.** $p^2 + 3p + 2 = 0$ **34.** $v^2 + 6v + 5 = 0$

35. $x^2 - 6x + 9 = 0$ **36.** $y^2 - 8y + 16 = 0$ **37.** $6x^2 - 9x = 0$

38. $12y^2 + 8y = 0$ **39.** $r^2 - 10 = 3r$ **40.** $t^2 - 12 = 4t$

41. $3v^2 - 5v + 2 = 0$ **42.** $2p^2 - 3p - 2 = 0$ **43.** $3s^2 + 8s = 3$

44. $3x^2 + 5x = 12$ **45.** $6r^2 = 12 - r$ **46.** $4t^2 = 4t + 3$

47. $5y^2 + 11y = 12$ **48.** $4v^2 - 4v + 1 = 0$ **49.** $9s^2 - 6s + 1 = 0$

50. $x^2 - 9 = 0$ **51.** $t^2 - 16 = 0$ **52.** $4y^2 - 1 = 0$

53. $9z^2 - 4 = 0$ **54.** $x + 15 = x(x - 1)$ **55.** $\dfrac{3x^2}{2} = 4x - 2$

56. $\dfrac{2x^2}{5} = 3x - 5$ **57.** $\dfrac{2x^2}{9} + x = 2$ **58.** $\dfrac{3x^2}{8} - x = 2$

59. $\dfrac{3}{4}z^2 - z = -\dfrac{1}{3}$ **60.** $\dfrac{r^2}{2} = 1 - \dfrac{r}{12}$

61. $p + 18 = p(p - 2)$ **62.** $r^2 - r - 2 = (2r - 1)(r - 3)$

63. $s^2 + 5s - 4 = (2s + 1)(s - 4)$ **64.** $x^2 + x + 5 = (3x + 2)(x - 4)$

65. $ax^2 + bx + c = 0$ is a quadratic equation that can be solved by factoring. For each case, state whether the equation has two positive solutions, two negative solutions, or one positive and one negative solution.

a. $a > 0, b < 0, c > 0$ **b.** $a > 0, b < 0, c < 0$ **c.** $a > 0, b > 0, c < 0$

66. Let a be a positive integer. Which equation has a positive double root?

(i) $x^2 - a^2 = 0$ (ii) $x^2 + 2ax - a^2 = 0$

(iii) $x^2 + 2ax + a^2 = 0$ (iv) $x^2 - 2ax + a^2 = 0$

2 Solve quadratic equations by taking square roots (See pages 505–508.)

GETTING READY

67. To solve the equation $4x^2 - 9 = 0$ by taking square roots, the first step is to add $\underline{\quad ? \quad}$ to each side of the equation.

68. To solve the equation $4(x-3)^2 = 100$ by taking square roots, the first step is to divide both sides of the equation by __?__.

Solve by taking square roots.

69. $x^2 = 36$

70. $y^2 = 49$

71. $v^2 - 1 = 0$

72. $z^2 - 64 = 0$

73. $4x^2 - 49 = 0$

74. $9w^2 - 64 = 0$

75. $9y^2 = 4$

76. $4z^2 = 25$

77. $16v^2 - 9 = 0$

78. $25x^2 - 64 = 0$

79. $y^2 - 12 = 0$

80. $z^2 - 32 = 0$

81. $w^2 - 24 = 0$

82. $v^2 - 48 = 0$

83. $(x-1)^2 = 36$

84. $(y+2)^2 = 49$

85. $2(x+5)^2 = 8$

86. $4(z-3)^2 = 100$

87. $2(x+1)^2 = 50$

88. $3(x-4)^2 = 27$

89. $4(x+5)^2 = 64$

90. $9(x-3)^2 = 81$

91. $12(x+3)^2 = 27$

92. $8(x-4)^2 = 50$

93. $9(x-1)^2 - 16 = 0$

94. $4(y+3)^2 - 81 = 0$

95. $(x-4)^2 - 20 = 0$

96. $(y+5)^2 - 50 = 0$

97. $(x+1)^2 - 18 = 0$

98. $(y-7)^2 - 98 = 0$

99. $2\left(z - \frac{1}{2}\right)^2 = 12$

100. $3\left(v + \frac{3}{4}\right)^2 = 36$

For Exercises 101 to 103, a and b are both positive numbers. In each case, state how many solutions the equation has.

101. $(x+a)^2 = 0$

102. $(x+a)^2 = b$

103. $ax^2 - b = 0$

APPLYING CONCEPTS

104. For what value of x^2 is $x = \pm 8$?

105. Evaluate $2n^2 - 7n - 4$ given $n(n-2) = 15$.

106. Evaluate $3y^2 + 5y - 2$ given $y(y+3) = 28$.

107. One solution of the equation $2x^2 - 5x + c = 0$ is 3. Find the other solution.

PROJECTS OR GROUP ACTIVITIES

108. **Number Problem** Use the formula $S = \frac{n^2 + n}{2}$, where S is the sum of the first n natural numbers. How many consecutive natural numbers beginning with 1 will give a sum of 55?

109. Sports Use the formula $N = \frac{t^2 - t}{2}$, where N is the number of football games that must be scheduled in a league with t teams if each team is to play every other team once. A league has 21 games scheduled. How many teams are in the league if each team plays every other team once?

110. Investments The value P of an initial investment of A dollars after 2 years is given by $P = A(1 + r)^2$, where r is the annual percentage rate earned by the investment. If an initial investment of \$1500 grew to a value of \$1782.15 in 2 years, what was the annual percentage rate?

111. Energy The kinetic energy of a moving body is given by $E = \frac{1}{2}mv^2$, where E is the kinetic energy in newton-meters, m is the mass in kilograms, and v is the speed in meters per second. What is the speed of a moving body whose mass is 5 kg and whose kinetic energy is 250 newton-meters?

112. Automobiles On a certain type of street surface, the equation $d = 0.055v^2$ can be used to approximate the distance d, in feet, a car traveling v miles per hour will slide when its brakes are applied. After applying the brakes, the owner of a car involved in an accident skidded 40 ft. Did the traffic officer investigating the accident issue the car owner a ticket for speeding if the speed limit was 30 mph?

11.2 Solving Quadratic Equations by Completing the Square

OBJECTIVE 1 Solve quadratic equations by completing the square

Recall that a perfect-square trinomial is the square of a binomial.

Perfect-square Trinomial		Square of a Binomial
$x^2 + 6x + 9$	=	$(x + 3)^2$
$x^2 - 10x + 25$	=	$(x - 5)^2$
$x^2 + 8x + 16$	=	$(x + 4)^2$

For each perfect-square trinomial above, the square of $\frac{1}{2}$ of the coefficient of x equals the constant term.

$$x^2 + 6x + 9, \qquad \left(\frac{1}{2} \cdot 6\right)^2 = 9$$

$$x^2 - 10x + 25, \qquad \left[\frac{1}{2}(-10)\right]^2 = 25 \qquad \left(\frac{1}{2} \text{ coefficient of } x\right)^2 = \text{Constant term}$$

$$x^2 + 8x + 16, \qquad \left(\frac{1}{2} \cdot 8\right)^2 = 16$$

This relationship can be used to write the constant term for a perfect-square trinomial. Adding to a binomial the constant term that makes it a perfect-square trinomial is called **completing the square.**

Focus on completing the square

Complete the square of the binomial. Write the resulting perfect-square trinomial as the square of a binomial.

A. $x^2 - 8x$ **B.** $y^2 + 5y$

A.

$x^2 - 8x$

Find the constant term.

$\left[\frac{1}{2}(-8)\right]^2 = 16$

Complete the square of $x^2 - 8x$ by adding the constant term.

$x^2 - 8x + 16$

Write the resulting perfect-square trinomial as the square of a binomial.

$x^2 - 8x + 16 = (x - 4)^2$

B.

$y^2 + 5y$

Find the constant term.

$\left(\frac{1}{2} \cdot 5\right)^2 = \left(\frac{5}{2}\right)^2 = \frac{25}{4}$

Complete the square of $y^2 + 5y$ by adding the constant term.

$y + 5y + \frac{25}{4}$

Write the resulting perfect-square trinomial as the square of a binomial.

$y^2 + 5y + \frac{25}{4} = \left(y + \frac{5}{2}\right)^2$

A quadratic equation of the form $x^2 + bx + c = 0$, $x \neq 0$, that cannot be solved by factoring can be solved by completing the square. The procedure is:

1. Write the equation in the form $x^2 + bx = c$.
2. Add to each side of the equation the term that completes the square of $x^2 + bx$.
3. Factor the perfect-square trinomial. Write it as the square of a binomial.
4. Take the square root of each side of the equation.
5. Solve for x.

Focus on solving a quadratic equation of the form $x^2 + bx + c = 0$ by completing the square

Solve by completing the square: $x^2 - 6x - 3 = 0$

Add the opposite of the constant term to each side of the equation.

$$x^2 - 6x - 3 = 0$$
$$x^2 - 6x = 3$$

Find the constant term that completes the square of $x^2 - 6x$.

$$\left[\frac{1}{2}(-6)\right]^2 = 9$$

Add this term to each side of the equation.

$$x^2 - 6x + 9 = 3 + 9$$

Factor the perfect-square trinomial.

$$(x - 3)^2 = 12$$

Take the square root of each side of the equation.

$$\sqrt{(x - 3)^2} = \sqrt{12}$$

Simplify.

$$x - 3 = \pm\sqrt{12}$$
$$x - 3 = \pm 2\sqrt{3}$$

Solve for x.

$$x - 3 = 2\sqrt{3} \qquad x - 3 = -2\sqrt{3}$$
$$x = 3 + 2\sqrt{3} \qquad x = 3 - 2\sqrt{3}$$

Check:

$x^2 - 6x - 3 = 0$	
$(3 + 2\sqrt{3})^2 - 6(3 + 2\sqrt{3}) - 3$	0
$9 + 12\sqrt{3} + 12 - 18 - 12\sqrt{3} - 3$	0
$0 = 0$	

$x^2 - 6x - 3 = 0$	
$(3 - 2\sqrt{3})^2 - 6(3 - 2\sqrt{3}) - 3$	0
$9 - 12\sqrt{3} + 12 - 18 + 12\sqrt{3} - 3$	0
$0 = 0$	

Write the solutions. The solutions are $3 + 2\sqrt{3}$ and $3 - 2\sqrt{3}$.

For the method of completing the square to be used, the coefficient of the x^2 term must be 1. If it is not 1, we must multiply each side of the equation by the reciprocal of the coefficient of x^2. This is illustrated in the example below.

Focus on solving a quadratic equation of the form $ax^2 + bx + c = 0$ by completing the square

Solve by completing the square: $2x^2 - x - 1 = 0$

Add the opposite of the constant term to each side of the equation.

$$2x^2 - x - 1 = 0$$
$$2x^2 - x = 1$$

The coefficient of the x^2 term must be 1. Multiply each side by the reciprocal of the coefficient of x^2.

$$\frac{1}{2}(2x^2 - x) = \frac{1}{2} \cdot 1$$
$$x^2 - \frac{1}{2}x = \frac{1}{2}$$

Find the constant term that completes the square of $x^2 - \frac{1}{2}x$.

$$\left[\frac{1}{2}\left(-\frac{1}{2}\right)\right]^2 = \left(-\frac{1}{4}\right)^2 = \frac{1}{16}$$

Add this term to each side of the equation.

$$x^2 - \frac{1}{2}x + \frac{1}{16} = \frac{1}{2} + \frac{1}{16}$$

Factor the perfect-square trinomial.

$$\left(x - \frac{1}{4}\right)^2 = \frac{9}{16}$$

Take the square root of each side of the equation.

$$\sqrt{\left(x - \frac{1}{4}\right)^2} = \sqrt{\frac{9}{16}}$$

Simplify.

$$x - \frac{1}{4} = \pm\sqrt{\frac{9}{16}}$$
$$x - \frac{1}{4} = \pm\frac{3}{4}$$

Solve for x.

$$x - \frac{1}{4} = \frac{3}{4} \qquad x - \frac{1}{4} = -\frac{3}{4}$$
$$x = 1 \qquad x = -\frac{1}{2}$$

Check:

$2x^2 - x - 1 = 0$	
$2(1)^2 - 1 - 1$	0
$2(1) - 1 - 1$	0
$2 - 1 - 1$	0
$0 = 0$	

$$2x^2 - x - 1 = 0$$

$$\begin{array}{c|c} 2\left(-\frac{1}{2}\right)^2 - \left(-\frac{1}{2}\right) - 1 & 0 \\ 2\left(\frac{1}{4}\right) - \left(-\frac{1}{2}\right) - 1 & 0 \\ \frac{1}{2} + \frac{1}{2} - 1 & 0 \end{array}$$

$$0 = 0$$

Write the solutions. The solutions are 1 and $-\frac{1}{2}$.

EXAMPLE 1 Solve by completing the square: $2x^2 - 4x - 1 = 0$

Solution

$2x^2 - 4x - 1 = 0$

$2x^2 - 4x = 1$
- **Add the opposite of −1 to each side of the equation.**

$\frac{1}{2}(2x^2 - 4x) = \frac{1}{2} \cdot 1$
- **The coefficient of the x^2 term must be 1. Multiply each side of the equation by $\frac{1}{2}$.**

$x^2 - 2x = \frac{1}{2}$

$x^2 - 2x + 1 = \frac{1}{2} + 1$
- **Complete the square of $x^2 - 2x$. Add 1 to each side of the equation.**

$(x - 1)^2 = \frac{3}{2}$
- **Factor the perfect-square trinomial.**

$\sqrt{(x - 1)^2} = \sqrt{\frac{3}{2}}$
- **Take the square root of each side of the equation.**

$x - 1 = \pm\sqrt{\frac{3}{2}}$
- **Simplify.**

$x - 1 = \pm\frac{\sqrt{3}}{\sqrt{2}}$
- **Quotient Property of Square Roots**

$x - 1 = \pm\frac{\sqrt{6}}{2}$
- **Rationalize the denominator.** $\frac{\sqrt{3}}{\sqrt{2}} = \frac{\sqrt{3}}{\sqrt{2}} \cdot \frac{\sqrt{2}}{\sqrt{2}} = \frac{\sqrt{6}}{2}$

- **Solve for x.**

$$x - 1 = \frac{\sqrt{6}}{2} \qquad\qquad x - 1 = -\frac{\sqrt{6}}{2}$$

$$x = 1 + \frac{\sqrt{6}}{2} \qquad\qquad x = 1 - \frac{\sqrt{6}}{2}$$

$$= \frac{2}{2} + \frac{\sqrt{6}}{2} \qquad\qquad = \frac{2}{2} - \frac{\sqrt{6}}{2}$$

$$= \frac{2 + \sqrt{6}}{2} \qquad\qquad = \frac{2 - \sqrt{6}}{2}$$

The solutions are $\frac{2 + \sqrt{6}}{2}$ and $\frac{2 - \sqrt{6}}{2}$.

Problem 1 Solve by completing the square: $3x^2 - 6x - 2 = 0$

Solution See page S26.

Try Exercise 45, page 516.

EXAMPLE 2 Solve by completing the square: $x^2 + 6x + 4 = 0$
Approximate the solutions to the nearest thousandth.

Solution

$x^2 + 6x + 4 = 0$

$x^2 + 6x = -4$ • Subtract 4 from each side of the equation.

$x^2 + 6x + 9 = -4 + 9$ • Complete the square of $x^2 + 6x$. Add 9 to each side of the equation.

$(x + 3)^2 = 5$ • Factor the perfect-square trinomial.

$\sqrt{(x + 3)^2} = \sqrt{5}$ • Take the square root of each side of the equation.

$x + 3 = \pm\sqrt{5}$ • Solve for x.

$x + 3 \approx \pm 2.2361$ • Use a calculator to approximate the solutions.

$x + 3 \approx 2.236$	$x + 3 \approx -2.236$
$x \approx -3 + 2.236$	$x \approx -3 - 2.236$
≈ -0.764	≈ -5.236

The solutions are approximately -0.764 and -5.236.

Problem 2 Solve by completing the square: $x^2 + 8x + 8 = 0$
Approximate the solutions to the nearest thousandth.

Solution See page S26.

Try Exercise 63, page 516.

11.2 Exercises

CONCEPT CHECK

1. When we square a binomial, the result is a __?__.
2. The constant term in a perfect-square trinomial is equal to the square of __?__.
3. To "complete the square" means to add to $x^2 + bx$ the constant term that will make it a __?__.
4. When solving the equation $x^2 - 8x + 16 = 18$ by completing the square, the next step after writing the equation in the form $(x - 4)^2 = 18$ is to __?__.

1 Solve quadratic equations by completing the square (See pages 511–515.)

GETTING READY

5. To complete the square of $x^2 + 18x$, find the constant term that completes the square: $\left[\frac{1}{2}(\underline{\ ?\ })\right]^2 = \underline{\ ?\ }$.
6. To complete the square of $x^2 + 7x$, find the constant term that completes the square: $\left[\frac{1}{2}(\underline{\ ?\ })\right]^2 = \underline{\ ?\ }$.

Complete the square. Write the resulting perfect-square trinomial as the square of a binomial.

7. $x^2 + 12x$

8. $x^2 - 4x$

9. $x^2 + 10x$

10. $x^2 + 3x$

11. $x^2 - x$

12. $x^2 + 5x$

Solve by completing the square.

13. $x^2 + 2x - 3 = 0$

14. $y^2 + 4y - 5 = 0$

15. $v^2 + 4v + 1 = 0$

16. $y^2 - 2y - 5 = 0$

17. $x^2 = 4x - 4$

18. $z^2 = 8z - 16$

19. $z^2 = 2z + 1$

20. $y^2 = 10y - 20$

Solve. First try to solve the equation by factoring. If you are unable to solve the equation by factoring, solve the equation by completing the square.

21. $p^2 + 3p = 1$

22. $r^2 + 5r = 2$

23. $w^2 + 7w = 8$

24. $y^2 + 5y = -4$

25. $x^2 + 6x + 4 = 0$

26. $y^2 - 8y - 1 = 0$

27. $r^2 - 8r = -2$

28. $s^2 + 6s = 5$

29. $t^2 - 3t = -2$

30. $y^2 = 4y + 12$

31. $w^2 = 3w + 5$

32. $x^2 = 1 - 3x$

33. $x^2 - x - 1 = 0$

34. $y^2 - 5y + 3 = 0$

35. $y^2 = 7 - 10y$

36. $v^2 = 14 + 16v$

37. $s^2 + 3s = -1$

38. $r^2 - 3r = 5$

39. $t^2 - t = 4$

40. $y^2 + y - 4 = 0$

41. $2t^2 - 3t + 1 = 0$

42. $2x^2 - 7x + 3 = 0$

43. $2r^2 + 5r = 3$

44. $2y^2 - 3y = 9$

45. $4v^2 - 4v - 1 = 0$

46. $2s^2 - 4s - 1 = 0$

47. $4z^2 - 8z = 1$

48. $3r^2 - 2r = 2$

49. $3y - 5 = (y - 1)(y - 2)$

50. $4p + 2 = (p - 1)(p + 3)$

51. $\frac{x^2}{4} - \frac{x}{2} = 3$

52. $\frac{x^2}{6} - \frac{x}{3} = 1$

53. $\frac{2x^2}{3} = 2x + 3$

54. $\frac{3x^2}{2} = 3x + 2$

55. $\frac{x}{3} + \frac{3}{x} = \frac{8}{3}$

56. $\frac{x}{4} - \frac{2}{x} = \frac{3}{4}$

Complete Exercises 57 and 58 without using a calculator.

57. A quadratic equation has solutions $-3 \pm \sqrt{5}$. Are both of these solutions negative, are both positive, or is one negative and one positive?

58. A quadratic equation has solutions $2 \pm \sqrt{7}$. Are both of these solutions negative, are both positive, or is one negative and one positive?

Solve by completing the square. Approximate the solutions to the nearest thousandth.

59. $y^2 + 3y = 5$

60. $w^2 + 5w = 2$

61. $2z^2 - 3z = 7$

62. $2x^2 + 3x = 11$

63. $4x^2 + 6x - 1 = 0$

64. $4x^2 + 2x - 3 = 0$

APPLYING CONCEPTS

65. If $(x + 6)^2 = 9$, then $x + 6$ is equal to what number?

66. Find the solutions of the quadratic equation in which $a = 1$, $b = 8$, and $c = -14$.

67. Evaluate $2b^2$ given $b^2 - 6b + 7 = 0$.

68. Evaluate $2y^2$ given $y^2 - 2y - 7 = 0$.

Solve.

69. $\sqrt{2x + 7} - 4 = x$

70. $\dfrac{x + 1}{2} + \dfrac{3}{x - 1} = 4$

71. $\dfrac{x - 2}{3} + \dfrac{2}{x + 2} = 4$

PROJECTS OR GROUP ACTIVITIES

72. What number is equal to three less than its square?

73. The equation $x^2 - 2x - 11 = 0$ has solutions a and b. Find the value of $a^2 + b^2$.

74. Explain why the equation $(x - 2)^2 = -4$ does not have a real number solution.

11.3 Solving Quadratic Equations by Using the Quadratic Formula

OBJECTIVE 1 Solve quadratic equations by using the quadratic formula

Any quadratic equation can be solved by completing the square. Applying this method to the standard form of a quadratic equation produces a formula that can be used to solve any quadratic equation.

To solve $ax^2 + bx + c = 0$, $a \neq 0$, by completing the square, subtract the constant term from each side of the equation.

$$ax^2 + bx + c = 0$$
$$ax^2 + bx + c - c = 0 - c$$
$$ax^2 + bx = -c$$

Multiply each side of the equation by the reciprocal of a, the coefficient of x^2.

$$\frac{1}{a}(ax^2 + bx) = \frac{1}{a}(-c)$$
$$x^2 + \frac{b}{a}x = -\frac{c}{a}$$

Complete the square by adding $\left(\frac{1}{2}\cdot\frac{b}{a}\right)^2$ to each side of the equation.

$$x^2 + \frac{b}{a}x + \left(\frac{1}{2}\cdot\frac{b}{a}\right)^2 = \left(\frac{1}{2}\cdot\frac{b}{a}\right)^2 - \frac{c}{a}$$
$$x^2 + \frac{b}{a}x + \frac{b^2}{4a^2} = \frac{b^2}{4a^2} - \frac{c}{a}$$

Simplify the right side of the equation.

$$x^2 + \frac{b}{a}x + \frac{b^2}{4a^2} = \frac{b^2}{4a^2} - \left(\frac{c}{a}\cdot\frac{4a}{4a}\right)$$
$$x^2 + \frac{b}{a}x + \frac{b^2}{4a^2} = \frac{b^2}{4a^2} - \frac{4ac}{4a^2}$$
$$x^2 + \frac{b}{a}x + \frac{b^2}{4a^2} = \frac{b^2 - 4ac}{4a^2}$$

Factor the perfect-square trinomial on the left side of the equation.

$$\left(x + \frac{b}{2a}\right)^2 = \frac{b^2 - 4ac}{4a^2}$$

Take the square root of each side of the equation.

$$\sqrt{\left(x + \frac{b}{2a}\right)^2} = \sqrt{\frac{b^2 - 4ac}{4a^2}}$$

$$x + \frac{b}{2a} = \pm\frac{\sqrt{b^2 - 4ac}}{2a}$$

Solve for x.

$$x + \frac{b}{2a} = \frac{\sqrt{b^2 - 4ac}}{2a} \qquad x + \frac{b}{2a} = -\frac{\sqrt{b^2 - 4ac}}{2a}$$

$$x = -\frac{b}{2a} + \frac{\sqrt{b^2 - 4ac}}{2a} \qquad x = -\frac{b}{2a} - \frac{\sqrt{b^2 - 4ac}}{2a}$$

$$x = \frac{-b + \sqrt{b^2 - 4ac}}{2a} \qquad x = \frac{-b - \sqrt{b^2 - 4ac}}{2a}$$

THE QUADRATIC FORMULA

If $ax^2 + bx + c = 0$, $a \neq 0$, then

$$\boldsymbol{x = \frac{-b + \sqrt{b^2 - 4ac}}{2a}} \quad \text{or} \quad \boldsymbol{x = \frac{-b - \sqrt{b^2 - 4ac}}{2a}}$$

The quadratic formula is frequently written in the form

$$\boldsymbol{x = \frac{-b \pm \sqrt{b^2 - 4ac}}{2a}}$$

Focus on solving a quadratic equation by using the quadratic formula

Solve by using the quadratic formula: $2x^2 = 4x - 1$

Write the equation in standard form. $a = 2$, $b = -4$, and $c = 1$.

$$2x^2 = 4x - 1$$
$$2x^2 - 4x + 1 = 0$$

Replace a, b, and c in the quadratic formula by their values.

$$x = \frac{-b \pm \sqrt{b^2 - 4ac}}{2a}$$
$$= \frac{-(-4) \pm \sqrt{(-4)^2 - 4 \cdot 2 \cdot 1}}{2 \cdot 2}$$

Simplify.

$$= \frac{4 \pm \sqrt{16 - 8}}{4}$$
$$= \frac{4 \pm \sqrt{8}}{4}$$
$$= \frac{4 \pm 2\sqrt{2}}{4} = \frac{2(2 \pm \sqrt{2})}{2 \cdot 2} = \frac{2 \pm \sqrt{2}}{2}$$

Check:

$2x^2$	$= 4x - 1$
$2\left(\frac{2+\sqrt{2}}{2}\right)^2$	$4\left(\frac{2+\sqrt{2}}{2}\right) - 1$
$2\left(\frac{4+4\sqrt{2}+2}{4}\right)$	$2(2+\sqrt{2}) - 1$
$2\left(\frac{6+4\sqrt{2}}{4}\right)$	$4 + 2\sqrt{2} - 1$
$\frac{6+4\sqrt{2}}{2}$	$3 + 2\sqrt{2}$

$3 + 2\sqrt{2} = 3 + 2\sqrt{2}$

$2x^2$	$= 4x - 1$
$2\left(\frac{2-\sqrt{2}}{2}\right)^2$	$4\left(\frac{2-\sqrt{2}}{2}\right) - 1$
$2\left(\frac{4-4\sqrt{2}+2}{4}\right)$	$2(2-\sqrt{2}) - 1$
$2\left(\frac{6-4\sqrt{2}}{4}\right)$	$4 - 2\sqrt{2} - 1$
$\frac{6-4\sqrt{2}}{2}$	$3 - 2\sqrt{2}$

$3 - 2\sqrt{2} = 3 - 2\sqrt{2}$

Write the solutions. The solutions are $\frac{2+\sqrt{2}}{2}$ and $\frac{2-\sqrt{2}}{2}$.

EXAMPLE 1 Solve by using the quadratic formula.

A. $2x^2 - 3x + 1 = 0$ **B.** $2x^2 = 8x - 5$

Solution **A.** $2x^2 - 3x + 1 = 0$

- **This is a quadratic equation in standard form. $a = 2, b = -3, c = 1$**

$$x = \frac{-(-3) \pm \sqrt{(-3)^2 - 4(2)(1)}}{2 \cdot 2}$$

- **Replace a, b, and c in the quadratic formula by their values.**

$$= \frac{3 \pm \sqrt{9 - 8}}{4} = \frac{3 \pm \sqrt{1}}{4} = \frac{3 \pm 1}{4}$$

- **Simplify.**

$$x = \frac{3+1}{4} \qquad x = \frac{3-1}{4}$$

$$= \frac{4}{4} = 1 \qquad = \frac{2}{4} = \frac{1}{2}$$

The solutions are 1 and $\frac{1}{2}$.

B.

$$2x^2 = 8x - 5$$

$$2x^2 - 8x + 5 = 0$$

- **This is a quadratic equation.**

$$x = \frac{-(-8) \pm \sqrt{(-8)^2 - 4(2)(5)}}{2 \cdot 2}$$

- **Replace a, b, and c in the quadratic formula by their values.**

$$= \frac{8 \pm \sqrt{64 - 40}}{4}$$

- **Simplify.**

$$= \frac{8 \pm \sqrt{24}}{4}$$

$$= \frac{8 \pm 2\sqrt{6}}{4}$$

$$= \frac{2(4 \pm \sqrt{6})}{2 \cdot 2} = \frac{4 \pm \sqrt{6}}{2}$$

The solutions are $\frac{4+\sqrt{6}}{2}$ and $\frac{4-\sqrt{6}}{2}$.

Problem 1 Solve by using the quadratic formula.

A. $3x^2 + 4x - 4 = 0$ **B.** $x^2 + 2x = 1$

Solution See page S26.

Try Exercise 33, page 521.

11.3 Exercises

CONCEPT CHECK

1. If a quadratic equation is solved by using the quadratic formula and the result is $x = \dfrac{1 \pm \sqrt{13}}{2}$, what are the solutions of the equation?

2. If a quadratic equation is solved by using the quadratic formula and the result is $x = \dfrac{2 \pm 6}{4}$, what are the solutions of the equation?

Determine whether the statement is true or false.

3. Any quadratic equation can be solved by using the quadratic formula.

4. The solutions of a quadratic equation can be written as $x = \dfrac{-b + \sqrt{b^2 - 4ac}}{2a}$ and $x = \dfrac{-b - \sqrt{b^2 - 4ac}}{2a}$.

5. The equation $4x^2 - 3x = 9$ is a quadratic equation in standard form.

6. In the quadratic formula $x = \dfrac{-b \pm \sqrt{b^2 - 4ac}}{2a}$, b is the coefficient of x^2.

1 Solve quadratic equations by using the quadratic formula (See pages 517–520.)

7. Write the quadratic formula. Explain what each variable in the formula represents.

8. Explain what the quadratic formula is used for.

GETTING READY

9. To write the equation $x^2 = 4x + 2$ in standard form, subtract __?__ and __?__ from each side of the equation. The resulting equation is __?__ = 0. Then a = __?__, b = __?__, and c = __?__.

10. Substitute the values of a, b, and c from Exercise 9 into the quadratic formula.

$$x = \frac{-b \pm \sqrt{b^2 - 4ac}}{2a}$$

$$= \frac{-(\underline{\ ?\ }) \pm \sqrt{(\underline{\ ?\ })^2 - 4(\underline{\ ?\ })(\underline{\ ?\ })}}{2(\underline{\ ?\ })}$$

Solve by using the quadratic formula.

11. $z^2 + 6z - 7 = 0$
12. $s^2 + 3s - 10 = 0$
13. $w^2 = 3w + 18$
14. $r^2 = 5 - 4r$

15. $t^2 - 2t = 5$
16. $y^2 - 4y = 6$
17. $t^2 + 6t - 1 = 0$
18. $z^2 + 4z + 1 = 0$

19. $w^2 + 3w - 5 = 0$
20. $x^2 - 3x - 6 = 0$
21. $w^2 = 4w + 9$
22. $y^2 = 8y + 3$

Solve. First try to solve the equation by factoring. If you are unable to solve the equation by factoring, solve the equation by using the quadratic formula.

23. $p^2 - p = 0$
24. $2v^2 + v = 0$
25. $4t^2 - 4t - 1 = 0$

26. $4x^2 - 8x - 1 = 0$
27. $4t^2 - 9 = 0$
28. $4s^2 - 25 = 0$

29. $3x^2 - 6x + 2 = 0$
30. $5x^2 - 6x = 3$
31. $3t^2 = 2t + 3$

32. $4n^2 = 7n - 2$
33. $2y^2 + 3 = 8y$
34. $5x^2 - 1 = x$

35. $3t^2 = 7t + 6$
36. $3x^2 = 10x + 8$
37. $3y^2 - 4 = 5y$

38. $6x^2 - 5 = 3x$
39. $3x^2 = x + 3$
40. $2n^2 = 7 - 3n$

41. $5d^2 - 2d - 8 = 0$
42. $x^2 - 7x - 10 = 0$
43. $5z^2 + 11z = 12$

44. $4v^2 = v + 3$
45. $v^2 + 6v + 1 = 0$
46. $s^2 + 4s - 8 = 0$

47. $4t^2 - 12t - 15 = 0$
48. $4w^2 - 20w + 5 = 0$
49. $9y^2 + 6y - 1 = 0$

50. $9s^2 - 6s - 2 = 0$
51. $6s^2 - s - 2 = 0$
52. $6y^2 + 5y - 4 = 0$

53. $4p^2 + 16p = -11$
54. $4y^2 - 12y = -1$
55. $4x^2 = 4x + 11$

56. $4s^2 + 12s = 3$
57. $9v^2 = -30v - 23$
58. $9t^2 = 30t + 17$

59. $\frac{x^2}{2} - \frac{x}{3} = 1$
60. $\frac{x^2}{4} - \frac{x}{2} = 5$
61. $\frac{2x^2}{5} = x + 1$

62. $\frac{3x^2}{2} + 2x = 1$
63. $\frac{x}{5} + \frac{5}{x} = \frac{12}{5}$
64. $\frac{x}{4} + \frac{3}{x} = \frac{5}{2}$

65. True or false? If you solve $ax^2 + bx + c = 0$ by using the quadratic formula and the solutions are rational numbers, then the equation could have been solved by factoring.

66. True or false? If the value of $b^2 - 4ac$ in the quadratic formula is 0, then $ax^2 + bx + c = 0$ has one solution, a double root.

Solve by using the quadratic formula. Approximate the solutions to the nearest thousandth.

67. $x^2 - 2x - 21 = 0$

68. $y^2 + 4y - 11 = 0$

69. $s^2 - 6s - 13 = 0$

70. $w^2 + 8w - 15 = 0$

71. $2p^2 - 7p - 10 = 0$

72. $3t^2 - 8t - 1 = 0$

73. $4z^2 + 8z - 1 = 0$

74. $4x^2 + 7x + 1 = 0$

75. $5v^2 - v - 5 = 0$

APPLYING CONCEPTS

76. Find the solutions of the quadratic equation in which $a = 4$, $b = -8$, and $c = 1$.

77. Find the difference between the larger root and the smaller root of $x^2 - 6x = 14$.

Solve.

78. $\sqrt{x^2 + 2x + 1} = x - 1$

79. $\dfrac{x+2}{3} - \dfrac{4}{x-2} = 2$

80. $\dfrac{x+1}{5} - \dfrac{3}{x-1} = 2$

81. Explain why the equation $0x^2 + 3x + 4 = 0$ cannot be solved by the quadratic formula.

PROJECTS OR GROUP ACTIVITIES

For a quadratic equation of the form $x^2 + bx + c = 0$, the sum of the solutions is equal to the opposite of b, and the product of the solutions is equal to c. For example, the solutions of the equation $x^2 + 5x + 6 = 0$ are -2 and -3. The sum of the solutions is -5, the opposite of the coefficient of x. The product of the solutions is 6, the constant term. This is one way to check the solutions of a quadratic equation. Use this method to determine whether the given numbers are solutions of the equation. If they are not solutions of the equation, find the solutions.

82. $x^2 - 4x - 21 = 0$; -3 and 7

83. $x^2 - 4x - 3 = 0$; $2 + \sqrt{7}$ and $2 - \sqrt{7}$

84. $x^2 - 4x + 1 = 0$; $2 + \sqrt{3}$ and $2 - \sqrt{3}$

85. $x^2 - 8x - 14 = 0$; $-4 + \sqrt{15}$ and $-4 - \sqrt{15}$

86. Factoring, completing the square, and using the quadratic formula are three methods of solving quadratic equations. Describe each method, and cite the advantages and disadvantages of each.

11.4 Complex Numbers

OBJECTIVE 1 Simplify complex numbers

Point of Interest

It may seem strange to just invent new numbers, but that is how mathematics evolves. For instance, negative numbers were not an accepted part of mathematics until well into the 13th century. In fact, these numbers were often referred to as "fictitious numbers."

In the 17th century, René Descartes called square roots of negative numbers "imaginary numbers," an unfortunate choice of words, and started using the letter *i* to denote these numbers. These numbers were subjected to the same skepticism as negative numbers.

Recall that $\sqrt{16} = 4$ because $4^2 = 16$. Now consider the expression $\sqrt{-16}$. To find $\sqrt{-16}$, we need to find a number n such that $n^2 = -16$. However, the square of any real number (except zero) is a *positive* number. To find a number n such that $n^2 = -16$, we introduce a new number.

DEFINITION OF *i*

The **imaginary unit,** designated by the letter i, is the number whose square is -1. Symbolically, this is written as follows:

$$i^2 = -1$$

Using this definition, we can now write $\sqrt{-16} = 4i$ because

$$\begin{aligned}(4i)^2 &= 4^2 i^2 && \text{• Use the Rule for Simplifying Powers of Products.}\\ &= 16(-1) && \text{• } i^2 = -1\\ &= -16\end{aligned}$$

The principal square root of a negative number is defined in terms of i.

PRINCIPAL SQUARE ROOT OF A NEGATIVE NUMBER

If a is a positive real number, then $\sqrt{-a} = \sqrt{a}i = i\sqrt{a}$. The number $i\sqrt{a}$ is called an **imaginary number.** When $a = 1$, we have $\sqrt{-1} = i$.

EXAMPLES of Imaginary Numbers

1. $-12i$ **2.** $\frac{3}{2}i$ **3.** $i\sqrt{15}$

In example (3) above, note that we wrote $i\sqrt{15}$ rather than $\sqrt{15}i$. It is customary, when a radical occurs in an imaginary number, to write the i in front of the radical. This is done to avoid confusion of $\sqrt{15}i$ with $\sqrt{15i}$.

EXAMPLE 1 Write each expression in terms of i.

A. $\sqrt{-23}$ **B.** $\sqrt{-36}$ **C.** $\sqrt{-45}$ **D.** $5\sqrt{-18}$

Solution

A. $\sqrt{-23} = \sqrt{-1 \cdot 23} = \sqrt{-1}\sqrt{23} = i\sqrt{23}$

B. $\sqrt{-36} = \sqrt{-1 \cdot 36} = \sqrt{-1}\sqrt{36} = i\sqrt{36} = i \cdot 6 = 6i$

C. $\sqrt{-45} = \sqrt{-1 \cdot 45} = \sqrt{-1}\sqrt{45} = i\sqrt{45} = i\sqrt{9 \cdot 5} = i\sqrt{9}\sqrt{5} = i \cdot 3\sqrt{5} = 3i\sqrt{5}$

D. $5\sqrt{-18} = 5\sqrt{-1 \cdot 18} = 5\sqrt{-1}\sqrt{18} = 5i\sqrt{18} = 5i\sqrt{9 \cdot 2} = 5i\sqrt{9}\sqrt{2} = 5i \cdot 3\sqrt{2} = 15i\sqrt{2}$

Problem 1 Write each expression in terms of i.

A. $\sqrt{-17}$ **B.** $\sqrt{-81}$ **C.** $\sqrt{-63}$ **D.** $4\sqrt{-50}$

Solution See page S27.

Try Exercise 15, page 531.

The real numbers and the imaginary numbers make up the *complex numbers*.

How It's Used

Complex numbers have applications in physics and engineering. In the field of *fluid dynamics*, complex numbers are useful in the analysis of the flow of gases (such as the flow of air over airplane wings) and liquids (such as the flow of oil through a pipeline).

COMPLEX NUMBERS

A **complex number** is a number that can be written in the form $a + bi$, where a and b are real numbers and $i = \sqrt{-1}$, the imaginary unit. The number a is the **real part of a complex number,** and b is the **imaginary part of a complex number.**

EXAMPLES of Complex Numbers and Their Real and Imaginary Parts

1. $-3 + 5i$ — The real part is -3. The imaginary part is 5.
2. $2 - 6i$ — The real part is 2. The imaginary part is -6. Note that $2 - 6i = 2 + (-6i)$.
3. 4 — The real part is 4. The imaginary part is 0. Note that $4 = 4 + 0i$.
4. $7i$ — The real part is 0. The imaginary part is 7. Note that $7i = 0 + 7i$.

From the examples above, note that a real number, such as 5, is a complex number whose imaginary part is zero and that an imaginary number, such as $7i$, is a complex number whose real part is zero.

Although $3 + \sqrt{-4}$ is a complex number, the preferred form is $3 + 2i$. This is called the **standard form of a complex number.** If the imaginary part of a complex number is negative, such as $2 + (-5i)$, we normally write $2 - 5i$, for example, as the standard form of the complex number.

Focus on writing a complex number in standard form

Write $7 - 2\sqrt{-20}$ in standard form.

Rewrite $\sqrt{-20}$ as $i\sqrt{20}$. Simplify $\sqrt{20}$. Multiply $2i$ times 2.

$$\begin{aligned} 7 - 2\sqrt{-20} &= 7 - 2i\sqrt{20} \\ &= 7 - 2i\sqrt{4}\sqrt{5} \\ &= 7 - 2i \cdot 2\sqrt{5} \\ &= 7 - 4i\sqrt{5} \end{aligned}$$

EXAMPLE 2 Write $\dfrac{4 + 8\sqrt{-45}}{6}$ in standard form.

Solution

$$\frac{4 + 8\sqrt{-45}}{6} = \frac{4 + 8i\sqrt{45}}{6}$$ • $\sqrt{-45} = i\sqrt{45}$

$$= \frac{4 + 8i\sqrt{9}\sqrt{5}}{6} = \frac{4 + 8i \cdot 3\sqrt{5}}{6}$$ • Simplify $\sqrt{45}$.

$$= \frac{4 + 24i\sqrt{5}}{6}$$

$$= \frac{4}{6} + \frac{24i\sqrt{5}}{6} = \frac{2}{3} + 4i\sqrt{5}$$ • Write the answer in the form $a + bi$.

Problem 2 Write $\dfrac{5 - 3\sqrt{-50}}{10}$ in standard form.

Solution See page S27.

Try Exercise 33, page 532.

OBJECTIVE 2 Add and subtract complex numbers

To add two complex numbers, add the real parts and add the imaginary parts. To subtract two complex numbers, subtract the real parts and subtract the imaginary parts.

EXAMPLE 3 Simplify.

A. $(-3 + 2i) + (5 - 6i)$ **B.** $(-9 + 4i) - (2 - 7i)$

Solution **A.** Add the real parts.

$$(-3 + 2i) + (5 - 6i) = (-3 + 5) + (2 + (-6))i$$
$$= 2 + (-4)i$$
$$= 2 - 4i$$

Add the imaginary parts.

B. Subtract the real parts.

$$(-9 + 4i) - (2 - 7i) = (-9 - 2) + (4 - (-7))i$$
$$= -11 + 11i$$

Subtract the imaginary parts.

Problem 3 Simplify.

A. $(7 - 3i) + (-8 + i)$ **B.** $(-5 + 4i) - (-3 - 7i)$

Solution See page S27.

Try Exercise 51, page 532.

OBJECTIVE 3 Multiply complex numbers

When multiplying complex numbers, recall that $i^2 = -1$. Here are two examples.

$$3i(5i) = 15i^2 = 15(-1) = -15$$
$$-2i(6i) = -12i^2 = -12(-1) = 12$$

EXAMPLE 4 Multiply: $3i(2 - 4i)$

Solution

$3i(2 - 4i) = 6i - 12i^2$ • **Use the Distributive Property to remove parentheses.**

$= 6i - 12(-1)$ • **Replace i^2 with -1.**

$= 6i + 12$

$= 12 + 6i$ • **Write the answer in the form $a + bi$.**

Problem 4 Multiply: $5i(3 + 6i)$

Solution See page S27.

Try Exercise 69, page 533.

The product of two complex numbers of the form $a + bi$, $a \neq 0$, $b \neq 0$, can be found by using the FOIL method and the fact that $i^2 = -1$.

EXAMPLE 5 Multiply: $(5 - 4i)(2 + 3i)$

Solution

$$\begin{aligned}(5 - 4i)(2 + 3i) &= 5(2) + 5(3i) + (-4i)2 + (-4i)3i && \text{• Use the FOIL method.}\\ &= 10 + 15i - 8i - 12i^2\\ &= 10 + 7i - 12i^2 && \text{• Combine like terms.}\\ &= 10 + 7i - 12(-1) && \text{• Replace } i^2 \text{ with } -1.\\ &= 10 + 7i + 12\\ &= 22 + 7i && \text{• Write the answer in the form } a + bi.\end{aligned}$$

Problem 5 Multiply: $(-1 + 5i)(2 - 3i)$

Solution See page S27.

Try Exercise 73, page 533.

The square of a complex number is found by using the FOIL method.

EXAMPLE 6 Write $(2 - 3i)^2$ in standard form.

Solution

$$\begin{aligned}(2 - 3i)^2 &= (2 - 3i)(2 - 3i) && \text{• Multiply } 2 - 3i \text{ times itself.}\\ &= 4 - 6i - 6i + 9i^2 && \text{• Use the FOIL method.}\\ &= 4 - 12i + 9i^2 && \text{• Combine like terms.}\\ &= 4 - 12i + 9(-1) && \text{• Replace } i^2 \text{ with } -1.\\ &= 4 - 12i - 9\\ &= -5 - 12i && \text{• Write the answer in the form } a + bi.\end{aligned}$$

Problem 6 Write $(5 - 4i)^2$ in standard form.

Solution See page S27.

Try Exercise 79, page 533.

The complex numbers $a + bi$ and $a - bi$ are called **complex conjugates** or **conjugates** of each other. Here are some examples.

Number	Conjugate
$2 + 5i$	$2 - 5i$
$-3 + 6i$	$-3 - 6i$
$5 - 4i$	$5 + 4i$

Consider the product of a complex number and its conjugate. For instance,

$$\begin{aligned}(2 + 5i)(2 - 5i) &= 4 - 10i + 10i - 25i^2\\ &= 4 - 25i^2\\ &= 4 - 25(-1)\\ &= 4 + 25 = 29\end{aligned}$$

Note that the product of complex conjugates is a *real* number. This is always true.

PRODUCT OF COMPLEX CONJUGATES

The product of a complex number and its conjugate is a real number. That is, $(a + bi)(a - bi) = a^2 + b^2$.

EXAMPLES

1. $(5 + 3i)(5 - 3i) = 5^2 + 3^2 = 25 + 9 = 34$
2. $(6 + 4i)(6 - 4i) = 6^2 + 4^2 = 36 + 16 = 52$

EXAMPLE 7 Multiply: $(8 + 3i)(8 - 3i)$

Solution $(8 + 3i)(8 - 3i) = 8^2 + 3^2$ • $(a + bi)(a - bi) = a^2 + b^2$, $a = 8, b = 3$

$= 64 + 9$

$= 73$

Problem 7 Multiply: $(-2 + 5i)(-2 - 5i)$

Solution See page S27.

Try Exercise 83, page 533.

OBJECTIVE 4

Divide complex numbers

Recall that the number $\frac{3}{\sqrt{2}}$ is not in simplest form because there is a radical expression in the denominator. Similarly, $\frac{3}{i}$ is not in simplest form because $i = \sqrt{-1}$. To write this expression in simplest form, multiply the numerator and denominator by i.

$$\frac{3}{i} \cdot \frac{i}{i} = \frac{3i}{i^2} = \frac{3i}{-1} = -3i$$

This process is used to divide a complex number by an imaginary number.

EXAMPLE 8 Divide: $\frac{3 - 6i}{2i}$

Solution

$$\frac{3 - 6i}{2i} = \frac{3 - 6i}{2i} \cdot \frac{i}{i}$$ • Multiply the numerator and denominator by i.

$$= \frac{3i - 6i^2}{2i^2} = \frac{3i - 6(-1)}{2(-1)}$$ • $i^2 = -1$

$$= \frac{3i + 6}{-2} = \frac{3}{-2}i + \frac{6}{-2} = -3 - \frac{3}{2}i$$ • Write the answer in the form $a + bi$.

Problem 8 Divide: $\frac{-12 + 8i}{-4i}$

Solution See page S27.

Try Exercise 99, page 534.

Recall that to simplify the quotient $\frac{2 + \sqrt{3}}{5 + 2\sqrt{3}}$, multiply the numerator and denominator by the conjugate of $5 + 2\sqrt{3}$, which is $5 - 2\sqrt{3}$. In a similar manner, to simplify the quotient of the two complex numbers, multiply the numerator and denominator by the conjugate of the complex number in the denominator.

EXAMPLE 9 Divide: $\frac{3 - 6i}{4 - 3i}$

Solution

$$\frac{3 - 6i}{4 - 3i} = \frac{3 - 6i}{4 - 3i} \cdot \frac{4 + 3i}{4 + 3i}$$
- **Multiply the numerator and denominator by the conjugate of the denominator.**

$$= \frac{12 + 9i - 24i - 18i^2}{4^2 + 3^2}$$
- **$(a + bi)(a - bi) = a^2 + b^2$**

$$= \frac{12 - 15i - 18i^2}{16 + 9}$$
- **Combine like terms in the numerator. Simplify the denominator.**

$$= \frac{12 - 15i - 18(-1)}{25}$$
- **Replace i^2 with -1.**

$$= \frac{12 - 15i + 18}{25}$$
- **Simplify.**

$$= \frac{30 - 15i}{25}$$

$$= \frac{30}{25} - \frac{15}{25}i = \frac{6}{5} - \frac{3}{5}i$$
- **Write the answer in the form $a + bi$.**

Problem 9 Divide: $\frac{16 - 11i}{5 + 2i}$

Solution See page S27.

Try Exercise 109, page 534.

OBJECTIVE 5 Solve quadratic equations with complex number solutions

Allowing complex numbers to be solutions of quadratic equations enables us to solve quadratic equations that have no real number solutions.

Focus on solving a quadratic equation that has complex number solutions by taking the square root of each side of the equation

Solve: $2x^2 + 128 = 0$	$2x^2 + 128 = 0$
Subtract 128 from each side of the equation.	$2x^2 = -128$
Divide each side of the equation by 2.	$x^2 = -64$
Take the square root of each side of the equation.	$\sqrt{x^2} = \sqrt{-64}$
Rewrite $\sqrt{-64}$ as $i\sqrt{64}$.	$x = \pm\sqrt{-64} = \pm i\sqrt{64}$
Simplify $\sqrt{64}$.	$x = \pm 8i$

The solutions are $-8i$ and $8i$.

Take Note

You should check these solutions. For instance, here is the check for $-8i$.

$2x^2 + 128 = 0$	
$2(-8i)^2 + 128$	0
$2(-8)^2 i^2 + 128$	0
$2(64)(-1) + 128$	0
$-128 + 128$	0

$0 = 0$

The solution $-8i$ checks. You should check that $8i$ is a solution.

EXAMPLE 10 Solve $(x-1)^2+9=1$ and check.

Solution

$$(x-1)^2+9=1$$

$$(x-1)^2=-8$$
• Subtract 9 from each side of the equation.

$$\sqrt{(x-1)^2}=\sqrt{-8}$$
• Take the square root of each side of the equation.

$$x-1=\pm\sqrt{-8}=\pm i\sqrt{8}$$
• Rewrite $\sqrt{-8}$ as $i\sqrt{8}$.

$$x-1=\pm 2i\sqrt{2}$$
• Simplify $\sqrt{8}$.

$$x=1\pm 2i\sqrt{2}$$
• Add 1 to each side of the equation.

Check:

$(x-1)^2+9$	1
$(1+2i\sqrt{2}-1)^2+9$	1
$(2i\sqrt{2})^2+9$	1
$2^2(i^2)(\sqrt{2})^2+9$	1
$4(-1)(2)+9$	1
$-8+9$	1
$1=1$	

$(x-1)^2+9$	1
$(1-2i\sqrt{2}-1)^2+9$	1
$(-2i\sqrt{2})^2+9$	1
$(-2)^2(i^2)(\sqrt{2})^2+9$	1
$4(-1)(2)+9$	1
$-8+9$	1
$1=1$	

The solutions are $1+2i\sqrt{2}$ and $1-2i\sqrt{2}$.

Problem 10 Solve: $8z^2+17=-7$ and check.

Solution See page S27.

Try Exercise 125, page 534.

In Example 10, we were able to solve the equation by taking the square root of each side of the equation. In many instances, however, the quadratic formula will be the best choice for solving a quadratic equation with complex number solutions. We restate the quadratic formula here for your reference.

The solutions of the quadratic equation $ax^2+bx+c=0$, $a\neq 0$, are given by

$$x=\frac{-b\pm\sqrt{b^2-4ac}}{2a}$$

Focus on solving a quadratic equation that has complex number solutions by using the quadratic formula

Solve: $4x^2-4x+2=0$

$$x=\frac{-b\pm\sqrt{b^2-4ac}}{2a}$$

Substitute into the quadratic formula: $a=4$, $b=-4$, $c=2$.

$$x=\frac{-(-4)\pm\sqrt{(-4)^2-4(4)(2)}}{2(4)}$$

$$=\frac{4\pm\sqrt{16-32}}{8}=\frac{4\pm\sqrt{-16}}{8}$$

Rewrite $\sqrt{-16}$ as $i\sqrt{16}$. $\sqrt{16}=4$

$$=\frac{4\pm i\sqrt{16}}{8}=\frac{4\pm 4i}{8}$$

Write the solution in the form $a+bi$.

$$=\frac{4}{8}\pm\frac{4i}{8}=\frac{1}{2}\pm\frac{1}{2}i$$

Check:

$$\begin{array}{r|l} 4x^2 - 4x + 2 = 0 & \\ \hline 4\left(\frac{1}{2} + \frac{1}{2}i\right)^2 - 4\left(\frac{1}{2} + \frac{1}{2}i\right) + 2 & 0 \\ 4\left(\frac{1}{4} + \frac{1}{4}i + \frac{1}{4}i + \frac{1}{4}i^2\right) - 2 - 2i + 2 & 0 \\ 1 + i + i + i^2 - 2 - 2i + 2 & 0 \\ 1 + 2i + (-1) - 2i & 0 \\ 0 = 0 & \end{array}$$

$$\begin{array}{r|l} 4x^2 - 4x + 2 = 0 & \\ \hline 4\left(\frac{1}{2} - \frac{1}{2}i\right)^2 - 4\left(\frac{1}{2} - \frac{1}{2}i\right) + 2 & 0 \\ 4\left(\frac{1}{4} - \frac{1}{4}i - \frac{1}{4}i + \frac{1}{4}i^2\right) - 2 + 2i + 2 & 0 \\ 1 - i - i + i^2 - 2 + 2i + 2 & 0 \\ 1 - 2i + (-1) + 2i & 0 \\ 0 = 0 & \end{array}$$

The solutions are $\frac{1}{2} + \frac{1}{2}i$ and $\frac{1}{2} - \frac{1}{2}i$.

EXAMPLE 11 Solve $x^2 + 4x = -6$ and check.

Solution

$$x^2 + 4x = -6$$

$$x^2 + 4x + 6 = 0$$

- Write the quadratic equation in standard form.

$$x = \frac{-b \pm \sqrt{b^2 - 4ac}}{2a}$$

$$= \frac{-4 \pm \sqrt{4^2 - 4(1)(6)}}{2(1)}$$

- $a = 1, b = 4, c = 6$

$$= \frac{-4 \pm \sqrt{16 - 24}}{2} = \frac{-4 \pm \sqrt{-8}}{2}$$

$$= \frac{-4 \pm i\sqrt{8}}{2} = \frac{-4 \pm 2i\sqrt{2}}{2}$$

- Rewrite $\sqrt{-8}$ as $i\sqrt{8}$ and then rewrite $i\sqrt{8}$ as $2i\sqrt{2}$.

$$= -\frac{4}{2} \pm \frac{2i\sqrt{2}}{2} = -2 \pm i\sqrt{2}$$

- Write the solution in the form $a + bi$.

Check

$$\begin{array}{r|l} x^2 + 4x = -6 & \\ \hline (-2 + i\sqrt{2})^2 + 4(-2 + i\sqrt{2}) & -6 \\ 4 - 2i\sqrt{2} - 2i\sqrt{2} + i^2(\sqrt{2})^2 - 8 + 4i\sqrt{2} & -6 \\ 4 - 4i\sqrt{2} + (-1)2 - 8 + 4i\sqrt{2} & -6 \\ 4 - 4i\sqrt{2} + (-2) - 8 + 4i\sqrt{2} & -6 \\ -6 = -6 & \end{array}$$

$$\begin{array}{r|l} x^2 + 4x = -6 & \\ \hline (-2 - i\sqrt{2})^2 + 4(-2 - i\sqrt{2}) & -6 \\ 4 + 2i\sqrt{2} + 2i\sqrt{2} + i^2(\sqrt{2})^2 - 8 - 4i\sqrt{2} & -6 \\ 4 + 4i\sqrt{2} + (-1)2 - 8 - 4i\sqrt{2} & -6 \\ 4 + 4i\sqrt{2} + (-2) - 8 - 4i\sqrt{2} & -6 \\ -6 = -6 & \end{array}$$

The solutions are $-2 + i\sqrt{2}$ and $-2 - i\sqrt{2}$.

Problem 11 Solve $x^2 - 6x + 4 = -6$ and check.

Solution See pages S27–S28.

Try Exercise 137, page 534.

11.4 Exercises

CONCEPT CHECK

Complete the equation.

1. $i^2 =$ ___?___

2. $i = \sqrt{\underline{\quad ? \quad}}$

3. $\sqrt{-7} = i\sqrt{\underline{\quad ? \quad}}$

4. $\sqrt{-11} =$ ___?___ $\sqrt{11}$

What is the conjugate of the number?

5. $6 + 5i$

6. $7 - 9i$

7. $-8i + 2$

What must you multiply the expression by in order to simplify it?

8. $\dfrac{7}{2i}$

9. $\dfrac{5i}{1 - 6i}$

10. $\dfrac{3 - 4i}{8 + 5i}$

1 Simplify complex numbers (See pages 523–524.)

GETTING READY

11. a. The real part of the complex number $5 - 8i$ is ___?___.
b. The imaginary part of the complex number $5 - 8i$ is ___?___.

12. a. If the complex number $a + bi$ is a real number, then ___?___ $= 0$.
b. If the complex number $a + bi$ is an imaginary number, then ___?___ $= 0$.

Write each expression in terms of i.

13. $\sqrt{-81}$

14. $\sqrt{-64}$

15. $5\sqrt{-49}$

16. $-3\sqrt{-16}$

17. $\sqrt{-13}$

18. $\sqrt{-26}$

19. $\sqrt{-75}$

20. $\sqrt{-90}$

21. $-4\sqrt{-54}$

22. $3\sqrt{-80}$

23. Let a and b be positive numbers. Is $a - bi$ equivalent to $a + \sqrt{-b^2}$ or to $a - \sqrt{-b^2}$?

24. Let a be a positive number. Which expression is equivalent to $\sqrt{a^2} + \sqrt{-a^2}$?
(i) 0 (ii) $a + ai$ (iii) $a - ai$ (iv) $2ai$

Write each number in standard form.

25. $5 + \sqrt{-4}$

26. $-3 + \sqrt{-25}$

27. $-7 - 2\sqrt{-36}$

28. $5 - 3\sqrt{-100}$

29. $5 + 3\sqrt{-48}$

30. $2 - 2\sqrt{-50}$

31. $7 - 6\sqrt{-28}$

32. $-2 + 5\sqrt{-72}$

33. $\dfrac{6 + 5\sqrt{-4}}{2}$

34. $\dfrac{5 - 3\sqrt{-25}}{5}$

35. $\dfrac{-6 + 2\sqrt{-54}}{6}$

36. $\dfrac{-4 + 2\sqrt{-72}}{4}$

37. $\dfrac{-2 - 4\sqrt{-90}}{5}$

38. $\dfrac{4 - 5\sqrt{-63}}{6}$

2 Add and subtract complex numbers (See page 525.)

GETTING READY

39. Add: $(6 - 8i) + (-2 + 3i)$

$(6 - 8i) + (-2 + 3i)$

$= (6 + \underline{\quad ? \quad}) + (-8 + \underline{\quad ? \quad})i$ • **Add the real parts. Add the imaginary parts.**

$= \underline{\quad ? \quad}$ • **Simplify.**

40. Subtract: $10 - (8 - 7i)$

$10 - (8 - 7i) = (10 - \underline{\quad ? \quad}) - \underline{\quad ? \quad}i$ • **Subtract the real parts.**

$= \underline{\quad ? \quad}$ • **Simplify.**

Simplify.

41. $3i + 7i$

42. $-9i + 6i$

43. $-2i - 7i$

44. $5i - (-4i)$

45. $6i + (7 + 3i)$

46. $3 + (4 + 2i)$

47. $-2i - (8 - 9i)$

48. $5 - (7 + 6i)$

49. $(7 + 4i) + (6 - 2i)$

50. $(-2 + 12i) + (7 - 15i)$

51. $(5 - 8i) - (3 + 2i)$

52. $(-7 - i) - (6 - 2i)$

53. $(-10 + 4i) - (-2 + 6i)$

54. $(6 - 11i) + (-7 - i)$

55. $(8 + 4i) + (-2 - 4i)$

56. $(7 - 8i) + (-7 + 3i)$

57. $(-2 + 11i) - (-2 + 11i)$

58. $(3 - 13i) - (3 + 13i)$

For Exercises 59 to 62, use the complex numbers $m = a - bi$ and $n = c + di$.

59. True or false? If $a = c$ and $b = d$, then $m - n = 0$.

60. True or false? If $m - n$ is a real number, then b and d are opposites.

61. Suppose $b > 0$, $d > 0$, and $b > d$. Is the imaginary part of $m + n$ positive or negative?

62. Suppose $a > 0$, $c > 0$, and $a > c$. Is the real part of $m - n$ positive or negative?

3 Multiply complex numbers (See pages 525–527.)

GETTING READY

63. $4i(7i) = 28\underline{\quad ? \quad} = 28(\underline{\quad ? \quad}) = \underline{\quad ? \quad}$

64. Multiply: $-8i(1 + 7i)$

$-8i(1 + 7i) = (\underline{\quad ? \quad})(1) + (\underline{\quad ? \quad})(7i)$ • Use the Distributive Property.

$= \underline{\quad ? \quad} - (\underline{\quad ? \quad})i^2$ • Simplify.

$= -8i - 56(\underline{\quad ? \quad})$ • $i^2 = \underline{\quad ? \quad}$.

$= \underline{\quad ? \quad}$ • Write in the form $a + bi$.

Multiply.

65. $4i \cdot 5i$

66. $6i \cdot 8i$

67. $(-2i)(9i)$

68. $6i(-9i)$

69. $5i(2 + 4i)$

70. $3i(2 - 7i)$

71. $(3 + 2i)(4 + 5i)$

72. $(4 - 3i)(2 + i)$

73. $(-3 + 2i)(1 + 3i)$

74. $(5 - 3i)(-2 + 4i)$

75. $(3 - 7i)(-2 - 5i)$

76. $(-5 - 3i)(4 - 6i)$

77. $(2 + 5i)^2$

78. $(-3 + 2i)^2$

79. $(1 - 2i)^2$

80. $(-4 - 3i)^2$

81. $(2 - 5i)(-5 + 2i)$

82. $(-1 + 3i)(3 - i)$

83. $(2 + 4i)(2 - 4i)$

84. $(-3 + 4i)(-3 - 4i)$

85. $(4 - i)(4 + i)$

86. True or false? For all nonzero real numbers a and b, the product $(a + bi)(a - bi)$ is a positive real number.

87. True or false? The product of two imaginary numbers is never a positive real number.

4 Divide complex numbers (See pages 527–528.)

GETTING READY

88. Simplify: $\frac{4}{i}$

$\frac{4}{i} = \frac{4}{i} \cdot \frac{\underline{\quad ? \quad}}{\underline{\quad ? \quad}} = \frac{4i}{\underline{\quad ? \quad}} = \frac{4i}{-1} = \underline{\quad ? \quad}$

89. Simplify: $\frac{2}{3i}$

$\frac{2}{3i} = \frac{2}{3i} \cdot \frac{\underline{\quad ? \quad}}{\underline{\quad ? \quad}} = \frac{2i}{\underline{\quad ? \quad}} = \frac{2i}{3(-1)} = \frac{2i}{\underline{\quad ? \quad}} = -\frac{2i}{3}$

Simplify.

90. $\frac{6}{5i}$

91. $\frac{9}{2i}$

92. $\frac{3}{-8i}$

93. $\frac{4}{-7i}$

Divide.

94. $\frac{2 + 3i}{i}$

95. $\frac{6 + i}{i}$

96. $\frac{-2 + 7i}{-i}$

97. $\frac{-5 - 9i}{-i}$

98. $\frac{-9 + 15i}{3i}$

99. $\frac{6 - 4i}{2i}$

100. $\frac{6 + 12i}{8i}$

101. $\frac{4 - 10i}{6i}$

102. $\frac{-15 + 6i}{-9i}$

103. $\frac{4 - 15i}{-6i}$

104. $\frac{3}{1 + i}$

105. $\frac{-4}{1 - 2i}$

106. $\frac{-10i}{3 + i}$

107. $\frac{12i}{2 - 2i}$

108. $\frac{13 - i}{2 + i}$

109. $\frac{12 + i}{1 - 2i}$

110. $\frac{18 + 4i}{1 + 3i}$

111. $\frac{-18 + 13i}{1 + 4i}$

112. $\frac{8 - 14i}{-2 - 3i}$

113. $\frac{5 + 15i}{3 + 4i}$

114. True or false? The quotient of two imaginary numbers is a real number.

115. True or false? The reciprocal of an imaginary number is a real number.

5 Solve quadratic equations with complex number solutions (See pages 528–531.)

GETTING READY

116. To write the equation $x^2 = 3x + 8$ in standard form, subtract ___?___ and subtract ___?___ from each side of the equation. The resulting equation is ___?___ = 0. Then a = ___?___, b = ___?___, and c = ___?___.

117. Substitute the values of a, b, and c from Exercise 116 into the quadratic formula.

$$x = \frac{-b \pm \sqrt{b^2 - 4ac}}{2a} = \frac{-(\underline{\ ?\ }) \pm \sqrt{(\underline{\ ?\ })^2 - 4(\underline{\ ?\ })(\underline{\ ?\ })}}{2(\underline{\ ?\ })}$$

Solve.

118. $x^2 + 16 = 0$

119. $z^2 + 25 = 0$

120. $2z^2 + 49 = -49$

121. $3x^2 + 54 = -54$

122. $x^2 + 11 = -16$

123. $z^2 + 12 = -28$

124. $3z^2 + 16 = -20$

125. $4z^2 - 13 = -21$

126. $(x + 1)^2 + 36 = 0$

127. $(x - 3)^2 + 81 = 0$

128. $(x - 2)^2 + 12 = 0$

129. $(z + 4)^2 + 18 = 0$

130. $(z + 2)^2 + 27 = -23$

131. $(x - 5)^2 + 17 = 9$

132. $(x + 3)^2 + 9 = 5$

133. $(z - 1)^2 + 25 = 9$

134. $z^2 - 2z + 2 = 0$

135. $z^2 - 2z + 5 = 0$

136. $x^2 + 4x = -5$

137. $x^2 + 4x = -8$

138. $x^2 + 13 = 6x$

139. $x^2 + 10 = 2x$

140. $x^2 - 2x + 3 = 0$

141. $x^2 + 2x + 13 = 0$

142. $4x^2 + 8x = -5$

143. $4x^2 + 12x + 25 = 0$

144. $4x^2 - 4x + 13 = 0$

For Exercises 145 and 146, a is a positive real number. State how many solutions the equation has and whether the solutions are real, imaginary, or complex numbers.

145. $x^2 + a = 0$

146. $(x + a)^2 = 0$

APPLYING CONCEPTS

Complete the equation.

147. $2i +$ ___?___ $= 7i$

148. $4i -$ ___?___ $= -3i$

Write each of the following in the form $a + bi$.

149. $(1 - i)(1 + i)(2 - 3i)$

150. $(2 - i)(2 + i)(1 + 5i)$

151. $(1 + i)^3$

152. $(2 - 3i)^3$

153. $(4 - 3i)(1 + 2i) + (3 - 2i)(2 - i)$

154. $(2 + 5i)(1 - 3i) - (3 + i)(4 - 2i)$

Solve.

155. $(x - 1)(x + 1) + 5 = 0$

156. $(x + 2)(x - 2) + 12 = 0$

157. $(x + 4)(x + 6) + 4 = 0$

158. $(x - 5)(x + 1) = -14$

159. $(x - 2)(x - 4) + 4 = 0$

160. $(x - 4)(x - 2) = -6$

161. Show that $i\sqrt{5}$ and $-i\sqrt{5}$ are solutions of the equation $4x^2 + 20 = 0$.

162. Show that $1 + 4i$ and $1 - 4i$ are solutions of the equation $x^2 = 2x - 17$.

PROJECTS OR GROUP ACTIVITIES

163. We can evaluate i^3 as $i^3 = (i \cdot i) \cdot i = i^2 \cdot i = (-1) \cdot i = -i$. Using a similar method, evaluate each of the following.

a. i^4 **b.** i^5 **c.** i^6 **d.** i^7

Using the fact that $i^9 = i$, find each of the following.

e. i^{18} **f.** i^{27} **g.** i^{36}

Classify each equation as linear or quadratic. Then solve the equation.

164. $2x^2 + 3x - 2 = 0$

165. $2x = 3x - 2$

166. $4x(x - 2) = 1$

167. $4(x - 2) = 12$

168. $9x^2 = 6x - 2$

11.5 Graphing Quadratic Equations in Two Variables

OBJECTIVE 1 Graph a quadratic equation of the form $y = ax^2 + bx + c$

An equation of the form $y = ax^2 + bx + c$, $a \neq 0$, is a **quadratic equation in two variables.** Examples of quadratic equations in two variables are shown at the right.

$$y = 3x^2 - x + 1$$
$$y = -x^2 - 3$$
$$y = 2x^2 - 5x$$

For these equations, y is a function of x, and we can write $f(x) = ax^2 + bx + c$. This equation represents a **quadratic function.**

EXAMPLE 1 Evaluate $f(x) = 2x^2 - 3x + 4$ when $x = -2$.

Solution

$$\begin{aligned} f(x) &= 2x^2 - 3x + 4 \\ f(-2) &= 2(-2)^2 - 3(-2) + 4 && \bullet \text{ Replace } x \text{ with } -2. \\ &= 2(4) - 3(-2) + 4 && \bullet \text{ Simplify.} \\ &= 8 + 6 + 4 \\ &= 18 \end{aligned}$$

The value of the function when $x = -2$ is 18.

Problem 1 Evaluate $f(x) = -x^2 + 5x - 2$ when $x = -1$.

Solution See page S28.

Try Exercise 15, page 540.

The graph of $y = ax^2 + bx + c$ or $f(x) = ax^2 + bx + c$ is a **parabola.** The graph is U-shaped and opens up when a is positive and down when a is negative. The graphs of two parabolas are shown below.

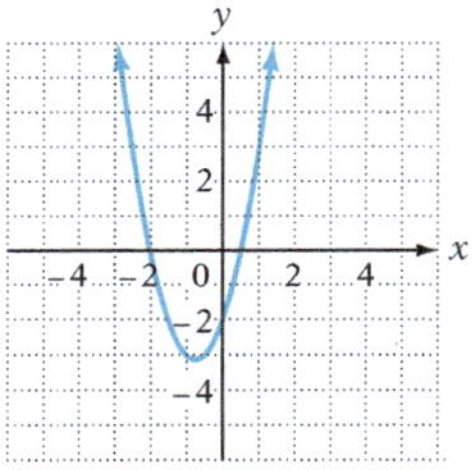

$y = 2x^2 + 3x - 2$
$a = 2$, a positive number
Parabola opens up

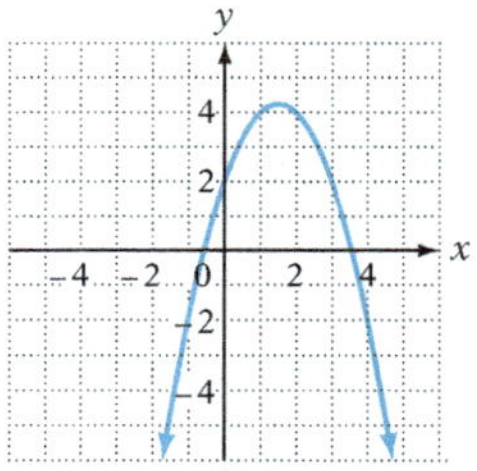

$f(x) = -x^2 + 3x + 2$
$a = -1$, a negative number
Parabola opens down

Focus on graphing a quadratic equation in two variables

Graph $y = x^2 - x - 6$.

Find several solutions of the equation. Because the graph is not a straight line, several solutions must be found in order to determine the U-shape. Record the ordered pairs in a table.

x	$y = x^2 - x - 6$
-3	6
-2	0
-1	-4
0	-6
1	-6
2	-4
3	0
4	6

Graph the ordered-pair solutions on a rectangular coordinate system. Draw a parabola through the points.

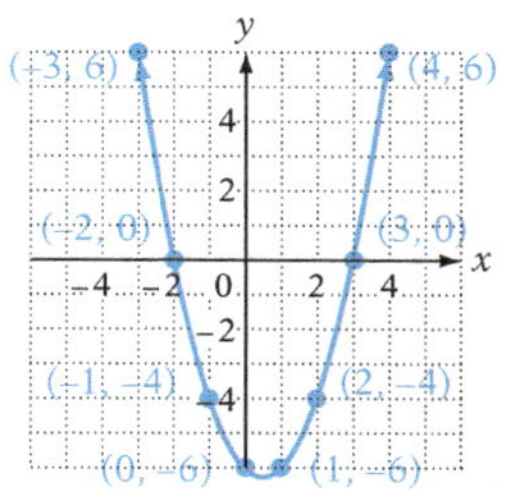

Note that the graph of $y = x^2 - x - 6$ crosses the x-axis at $(-2, 0)$ and at $(3, 0)$. These points are confirmed in the table of values. The x-intercepts of the graph are $(-2, 0)$ and $(3, 0)$.

We can find the x-intercepts algebraically by letting $y = 0$ and solving for x.

$$y = x^2 - x - 6$$

$$0 = x^2 - x - 6$$

• **Replace y with 0 and solve for x.**

$$0 = (x + 2)(x - 3)$$

• **This equation can be solved by factoring. However, it will be necessary to use the quadratic formula to solve some quadratic equations.**

$$x + 2 = 0 \qquad x - 3 = 0$$

$$x = -2 \qquad x = 3$$

When $y = 0$, $x = -2$ or $x = 3$. The x-intercepts are $(-2, 0)$ and $(3, 0)$.

The y-intercept of the graph of $y = x^2 - x - 6$ is the point at which the graph crosses the y-axis. At this point, $x = 0$. From the graph or the table, we can see that the y-intercept is $(0, -6)$.

We can find the y-intercept algebraically by letting $x = 0$ and solving for y.

$$y = x^2 - x - 6$$

$$y = 0^2 - 0 - 6$$

• **Replace x with 0 and simplify.**

$$y = 0 - 0 - 6$$

$$y = -6$$

When $x = 0$, $y = -6$. The y-intercept is $(0, -6)$.

Using a graphing utility, enter the equation $y = x^2 - x - 6$ and verify the graph shown above. (Refer to the Appendix for instructions on using a graphing calculator to graph a quadratic equation.) Verify that $(-2, 0)$, $(3, 0)$, and $(0, -6)$ are coordinates of points on the graph.

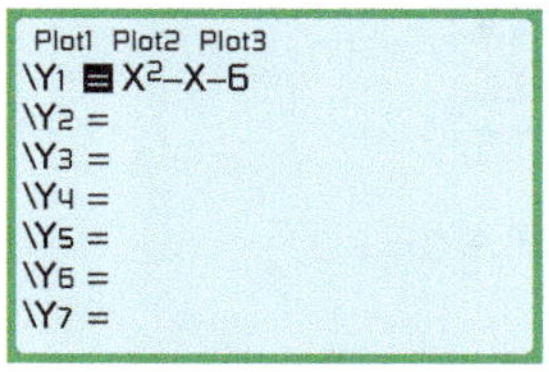

The graph of $y = -2x^2 + 1$ is shown below.

x	$y = -2x^2 + 1$
0	1
1	−1
−1	−1
2	−7
−2	−7

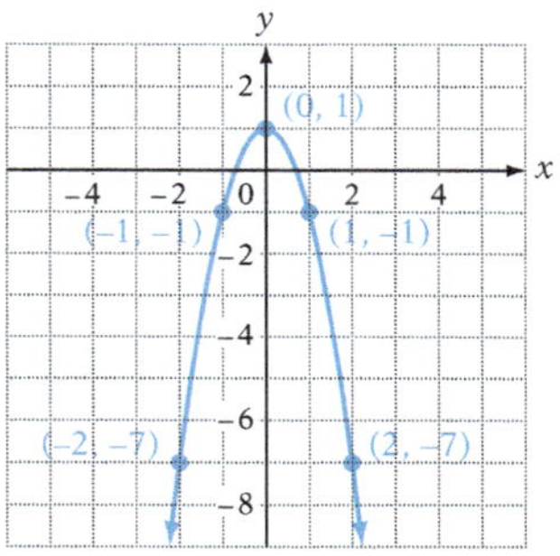

Note that the graph of $y = x^2 - x - 6$, shown on the previous page, opens up and that the coefficient of x^2 is positive. The graph of $y = -2x^2 + 1$ opens down, and the coefficient of x^2 is negative. As stated earlier, for any quadratic equation in two variables, the coefficient of x^2 determines whether the parabola opens up or down. When a is positive, the parabola opens up. When a is negative, the parabola opens down.

Point of Interest

Mirrors in some telescopes are ground into the shape of a parabola. The mirror at the Mount Palomar Observatory is 2 ft thick at the ends and weighs 14.75 tons. The mirror has been ground to a true paraboloid (the three-dimensional version of a parabola) to within 0.0000015 in.

Every parabola has an **axis of symmetry** and a **vertex** that is on the axis of symmetry. If the parabola opens up, the vertex is the lowest point on the graph. If the parabola opens down, the vertex is the highest point on the graph.

To understand the axis of symmetry, think of folding the paper along that axis. The two halves of the graph will match up.

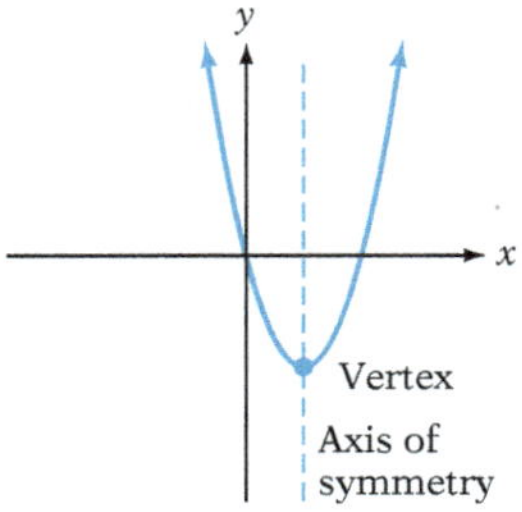

When graphing a quadratic equation in two variables, use the value of a to determine whether the parabola opens up or down. After graphing ordered-pair solutions of the equation, use symmetry to help you draw the parabola.

EXAMPLE 2 Graph. **A.** $y = x^2 - 2x$ **B.** $y = -x^2 + 4x - 4$

Solution **A.** $a = 1$. a is positive.
The parabola opens up.

x	y
0	0
1	-1
-1	3
2	0
3	3

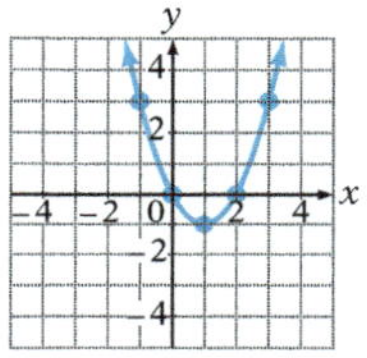

B. $a = -1$. a is negative.
The parabola opens down.

x	y
0	-4
1	-1
2	0
3	-1
4	-4

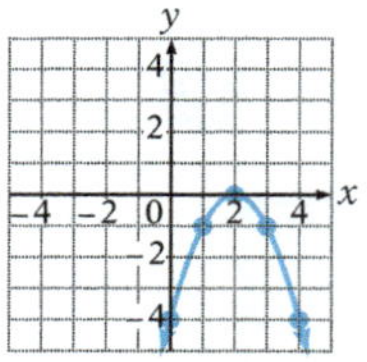

Problem 2 Graph. **A.** $y = x^2 + 2$ **B.** $y = -x^2 - 2x + 3$

Solution See page S28.

Try Exercise 33, page 540.

EXAMPLE 3 Find the x- and y-intercepts of the graph of $y = x^2 - 2x - 5$.

Solution
$y = x^2 - 2x - 5$
$0 = x^2 - 2x - 5$ • **To find the x-intercepts, let $y = 0$ and solve for x.**

$$x = \frac{-b \pm \sqrt{b^2 - 4ac}}{2a}$$

$$= \frac{-(-2) \pm \sqrt{(-2)^2 - 4(1)(-5)}}{2(1)}$$

$$= \frac{2 \pm \sqrt{24}}{2} = \frac{2 \pm 2\sqrt{6}}{2} = 1 \pm \sqrt{6}$$

- **$x^2 - 2x - 5$ is nonfactorable over the integers. Use the quadratic formula.**
- **$a = 1, b = -2, c = -5$**

The x-intercepts are $(1 + \sqrt{6}, 0)$ and $(1 - \sqrt{6}, 0)$.

$y = x^2 - 2x - 5$
$y = 0^2 - 2(0) - 5 = -5$

- **To find the y-intercept, let $x = 0$ and solve for y.**

The y-intercept is $(0, -5)$.

Problem 3 Find the x- and y-intercepts of the graph of $y = x^2 - 6x + 9$.

Solution See page S28.

Try Exercise 41, page 541.

11.5 Exercises

CONCEPT CHECK

1. What is the name of the graph of a quadratic equation in two variables?
2. What is the clue in the equation $y = x^2 - 4x + 3$ that the graph will be a parabola and not a straight line?
3. What is the clue in the equation $y = x^2 - 5x + 4$ that the graph will open up?
4. Explain how to find the x-intercepts of the graph of $y = x^2 - 4x + 3$.
5. Explain how to find the y-intercept of the graph of $y = x^2 - 5x + 4$.

Determine whether the graph of the equation opens up or down.

6. $y = -x^2 + 4$
7. $y = \frac{1}{2}x^2 - 2$
8. $y = x^2 - 3$
9. $y = -\frac{1}{3}x^2 + 5$
10. $y = -x^2 + 4x - 1$
11. $y = x^2 - 2x + 3$

1 Graph a quadratic equation of the form $y = ax^2 + bx + c$ (See pages 536–539.)

12. What is a parabola?

GETTING READY

13. Use the quadratic equation $y = -x^2 + 9$.
   a. The equation is in the standard form $y = ax^2 + bx + c$, where $a =$ ___?___, $b =$ ___?___, and $c =$ ___?___.
   b. Because $a < 0$, the graph of the equation opens ___?___.

c. The equation in part (b) can be written in function notation as $f(x) = -x^2 + 9$. To evaluate the function for $x = 2$, find $f(2)$. $f(2) = -(\underline{\ ?\ })^2 + 9$
$= \underline{\ ?\ } + 9 = \underline{\ ?\ }$.

14. a. To find the x-intercepts of the graph of $y = x^2 - 3x - 4$, let $\underline{\ ?\ } = 0$ and solve for $\underline{\ ?\ }$.

b. The equation $0 = x^2 - 3x - 4$ can be solved by factoring: $0 = (x - \underline{\ ?\ })(x + \underline{\ ?\ })$. The solutions are $\underline{\ ?\ }$ and $\underline{\ ?\ }$.

c. The x-intercepts of the graph of $y = x^2 - 3x - 4$ are $(\underline{\ ?\ }, 0)$ and $(\underline{\ ?\ }, 0)$.

Evaluate the function for the given value of x.

15. $f(x) = x^2 - 2x + 1; x = 3$

16. $f(x) = 2x^2 + x - 1; x = -2$

17. $f(x) = 4 - x^2; x = -3$

18. $f(x) = x^2 + 6x + 9; x = -3$

19. $f(x) = -x^2 + 5x - 6; x = -4$

20. $f(x) = -2x^2 + 2x - 1; x = -1$

Graph.

21. $y = x^2$

22. $y = -x^2$

23. $y = -x^2 + 1$

24. $y = x^2 - 1$

25. $y = 2x^2$

26. $y = \frac{1}{2}x^2$

27. $y = -\frac{1}{2}x^2 + 1$

28. $y = 2x^2 - 1$

29. $y = x^2 - 4x$

30. $y = x^2 + 4x$

31. $y = x^2 - 2x + 3$

32. $y = x^2 - 4x + 2$

33. $y = -x^2 + 2x + 3$

34. $y = x^2 + 2x + 1$

35. $y = -x^2 + 3x - 4$

36. $y = -x^2 + 6x - 9$

37. $y = 2x^2 + x - 3$

38. $y = -2x^2 - 3x + 3$

For Exercises 39 and 40, $y = ax^2 + bx + c$ is a quadratic equation whose graph is a parabola.

39. True or false? If $a < 0$, then no point on the graph has a negative y-coordinate.

40. True or false? The parabola has only one x-intercept. If the coordinates of the x-intercept are $(n, 0)$, then the point on the parabola with x-coordinate $n - m$ has the same y-coordinate as the point on the parabola with x-coordinate $n + m$.

Determine the x- and y-intercepts of the graph of the equation.

41. $y = x^2 - 5x + 6$

42. $y = x^2 + 5x - 6$

43. $f(x) = 9 - x^2$

44. $f(x) = x^2 + 12x + 36$

45. $y = x^2 + 2x - 6$

46. $y = x^2 + 4x - 2$

47. $f(x) = 2x^2 - x - 3$

48. $f(x) = 2x^2 - 13x + 15$

Graph using a graphing utility. Verify that the graph is a graph of a parabola opening up if a is positive or opening down if a is negative.

49. $y = x^2 - 2$

50. $y = -x^2 + 3$

51. $y = x^2 + 2x$

52. $y = -2x^2 + 4x$

53. $y = \frac{1}{2}x^2 - x$

54. $y = -\frac{1}{2}x^2 + 2$

55. $y = x^2 - x - 2$

56. $y = x^2 - 3x + 2$

57. $y = -x^2 - 2x - 1$

APPLYING CONCEPTS

State whether the graph is the graph of a linear function, a quadratic function, or neither.

58.

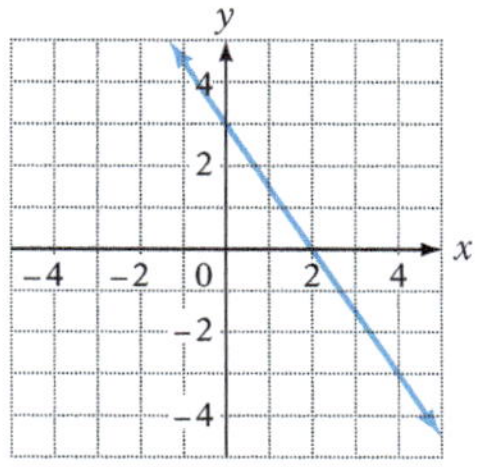

59.

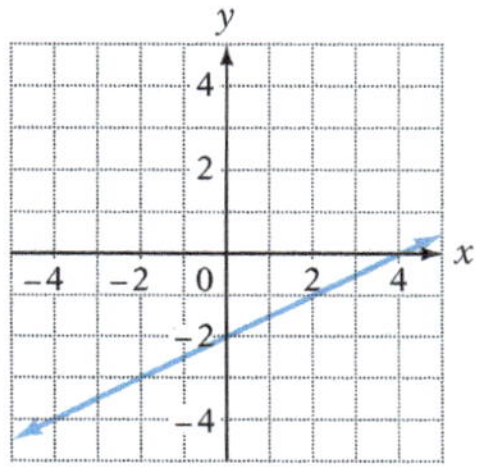

60.

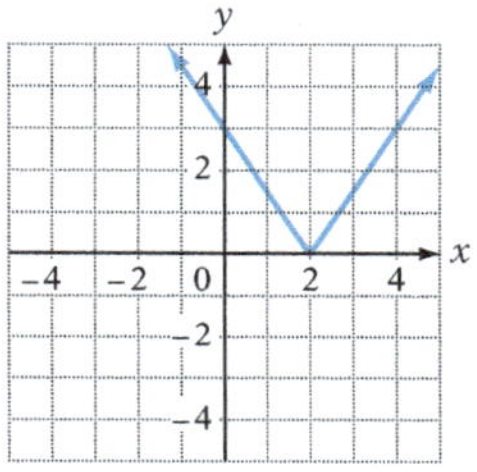

61. 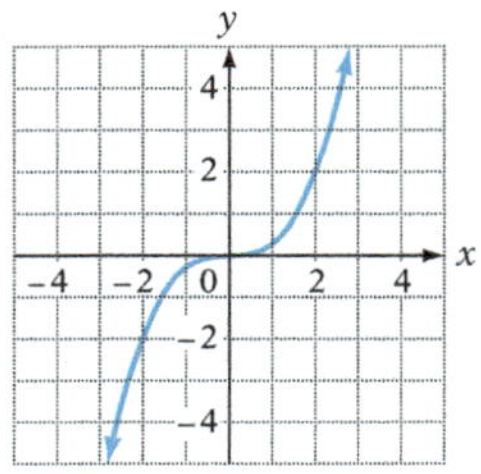

62. 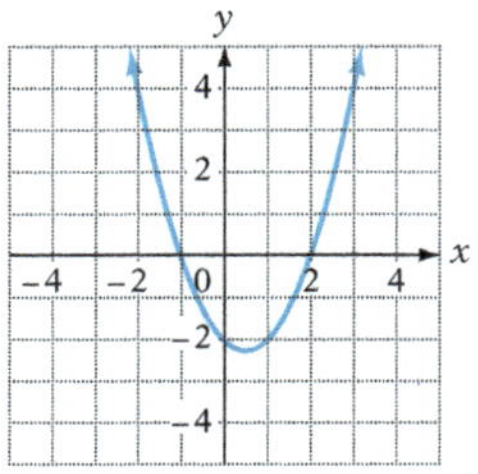

63. 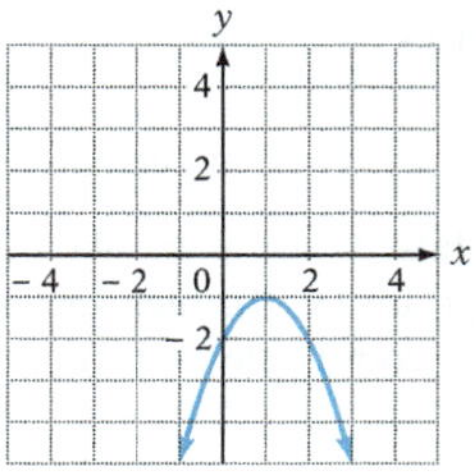

64. Consider the graph shown at the right.
   a. What are the x-intercepts?
   b. What is the y-intercept?
   c. What do you know about the value of a?
   d. What are the coordinates of the vertex?
   e. Where is the axis of symmetry?
   f. What is the value of y when $x = 1$?

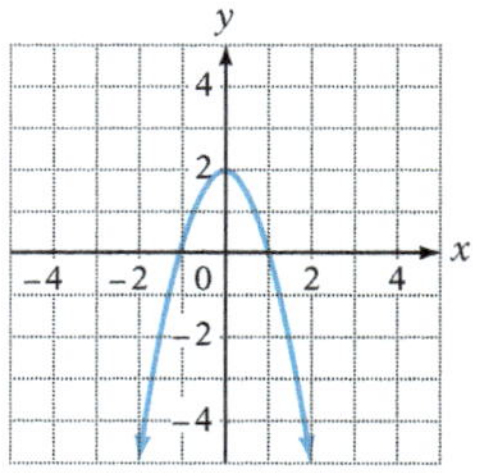

65. Consider the graph shown at the right.
   a. What are the x-intercepts?
   b. What is the y-intercept?
   c. What do you know about the value of a?
   d. What are the coordinates of the vertex?
   e. Where is the axis of symmetry?
   f. What is the value of y when $x = -1$?

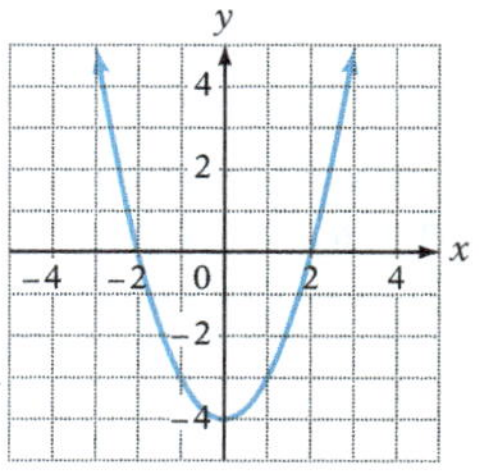

PROJECTS OR GROUP ACTIVITIES

66. Draw a parabola that opens up and has $(-2, -4)$ as its vertex.

67. Draw a parabola that opens down and has $(3, 1)$ as its vertex.

68. The point whose coordinates are (x_1, y_1) lies in quadrant I and is a point on the graph of the equation $y = 2x^2 - 2x + 1$. Given $y_1 = 13$, find x_1.

69. The point whose coordinates are (x_1, y_1) lies in quadrant II and is a point on the graph of the equation $y = 2x^2 - 3x - 2$. Given $y_1 = 12$, find x_1.

11.6 Application Problems

OBJECTIVE 1

Grafton Marshall Smith/Corbis

Application problems

The application problems in this section are varieties of those problems solved earlier in the text. Each of the strategies for the problems in this section results in a quadratic equation.

Solve: It took a motorboat a total of 7 h to travel 48 mi down a river and then 48 mi back again. The rate of the current was 2 mph. Find the rate of the motorboat in calm water.

STRATEGY FOR SOLVING AN APPLICATION PROBLEM

▶ Determine the type of problem. For example, is it a uniform motion problem, a geometry problem, or a work problem?

The problem is a uniform motion problem.

▶ Choose a variable to represent the unknown quantity. Write numerical or variable expressions for all the remaining quantities. These results can be recorded in a table.

The unknown rate of the motorboat: r

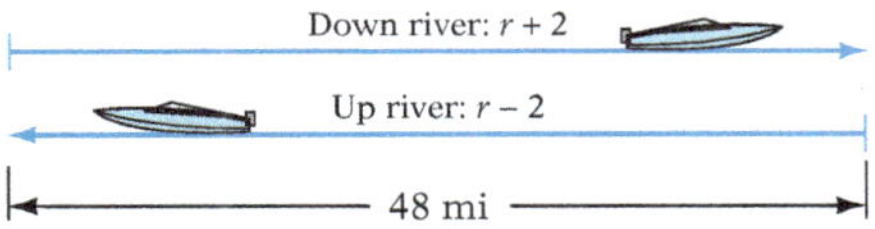

	Distance	÷	Rate	=	Time
Down river	48	÷	$r+2$	=	$\frac{48}{r+2}$
Up river	48	÷	$r-2$	=	$\frac{48}{r-2}$

▶ Determine how the quantities are related. If necessary, review the strategies presented in the chapter "Solving Equations and Inequalities: Applications."

The time spent traveling down the river plus the time spent traveling up the river equals 7 h.

$$\frac{48}{r+2} + \frac{48}{r-2} = 7$$

$$(r+2)(r-2)\left(\frac{48}{r+2} + \frac{48}{r-2}\right) = (r+2)(r-2)7$$

• Clear denominators by multiplying each side of the equation by the LCD of the fractions.

$$(r-2)48 + (r+2)48 = (r^2-4)7$$

• Simplify.

$$48r - 96 + 48r + 96 = 7r^2 - 28$$

$$96r = 7r^2 - 28$$

• This is a quadratic equation.

$$0 = 7r^2 - 96r - 28$$

• Write the quadratic equation in standard form.

$$0 = (7r+2)(r-14)$$

• Solve by factoring.

$$7r + 2 = 0 \qquad r - 14 = 0$$

$$7r = -2 \qquad r = 14$$

$$r = -\frac{2}{7}$$

The solution $-\frac{2}{7}$ is not possible, because the rate cannot be negative.

The rate of the motorboat in calm water is 14 mph.

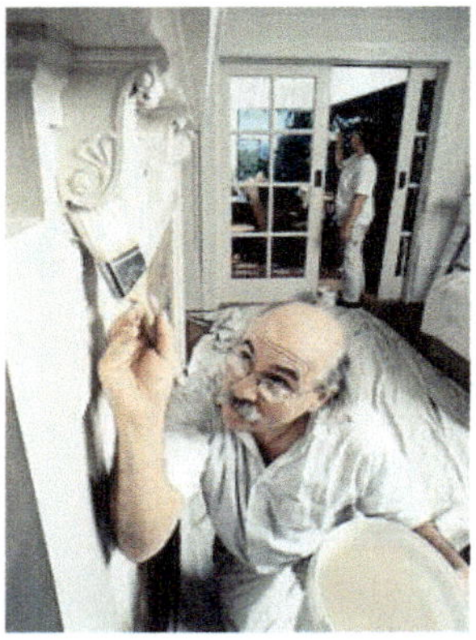
Billy Hustace/Getty Images

EXAMPLE 1 Working together, a painter and the painter's apprentice can paint a room in 4 h. Working alone, the apprentice requires 6 more hours to paint the room than the painter requires working alone. How long does it take the painter, working alone, to paint the room?

Strategy

► This is a work problem.

► Time for the painter to paint the room: t
Time for the apprentice to paint the room: $t + 6$

	Rate	·	Time	=	Part
Painter	$\frac{1}{t}$	·	4	=	$\frac{4}{t}$
Apprentice	$\frac{1}{t+6}$	·	4	=	$\frac{4}{t+6}$

► The sum of the parts of the task completed must equal 1.

Solution

$$\frac{4}{t} + \frac{4}{t+6} = 1$$

$$t(t+6)\left(\frac{4}{t} + \frac{4}{t+6}\right) = t(t+6) \cdot 1$$

• Clear denominators by multiplying each side of the equation by the LCD.

$$(t+6)4 + t(4) = t(t+6)$$

• Simplify.

$$4t + 24 + 4t = t^2 + 6t$$

• This is a quadratic equation.

$$0 = t^2 - 2t - 24$$

• Write the quadratic equation in standard form.

$$0 = (t-6)(t+4)$$

• Solve by factoring.

$$t - 6 = 0 \qquad t + 4 = 0$$

$$t = 6 \qquad t = -4$$

The solution $t = -4$ is not possible because the time cannot be negative.

Working alone, the painter requires 6 h to paint the room.

Problem 1 The length of a rectangle is 3 m more than the width. The area is 40 m^2. Find the width.

Solution See page S28.

Try Exercise 11, page 546.

11.6 Exercises

CONCEPT CHECK

1. If the length of a rectangle is three more than twice the width and the width is represented by W, then the length is represented by ______.

2. If it takes one pipe 15 min longer to fill a tank than it does a second pipe, then the rate of work for the second pipe can be represented by $\frac{1}{t}$, and the rate of work for the first pipe can be represented by ______.

3. If a plane's rate of speed is r and the rate of the wind is 30 mph, then the plane's rate of speed flying with the wind is $r + 30$, and the plane's rate of speed flying against the wind is ______.

4. When using the quadratic formula to solve the equation $2 = -16t^2 + 24t + 4$ for t, substitute ______ for a in the quadratic formula, ______ for b, and ______ for c.

1 Application problems (See pages 543–545.)

GETTING READY

5. The length of a rectangle is 2 ft less than twice the width. Let w be the width of the rectangle.
   a. In terms of w, the length of the rectangle is __?__.
   b. In terms of w, the area of the rectangle is __?__.
   c. The area of the rectangle is 84 ft^2. Use your answer to part (b) and the fact that the area of the rectangle is 84 ft^2 to write an equation that can be solved to find the width of the rectangle: __?__ = __?__.

6. a. Solve the equation that you wrote in part (c) of Exercise 5.

$(2w - 2)w = 84$

__?__ − __?__ $= 84$ • **Use the Distributive Property.**

$2w^2 - 2w - 84 = 0$ • **Subtract __?__ from each side of the equation.**

$w^2 - w - 42 = 0$ • **Divide each side of the equation by __?__.**

$(w -$ __?__$)(w +$ __?__$) = 0$ • **Factor the left side.**

__?__ $= 0$ __?__ $= 0$ • **Use the Principle of Zero Products.**

$w =$ __?__ $w =$ __?__ • **Solve for w.**

7. **Geometry** The height of a triangle is 2 m more than twice the length of the base. The area of the triangle is 20 m^2. Find the height of the triangle and the length of the base.

8. **Geometry** The length of a rectangle is 4 ft more than twice the width. The area of the rectangle is 160 ft^2. Find the length and width of the rectangle.

9. **Sports** The area of the batter's box on a major-league baseball field is 24 ft^2. The length of the batter's box is 2 ft more than the width. Find the length and width of the rectangular batter's box.

Copyright © Robert Brenner/Photo Edit

10. **Sports** The length of the batter's box on a softball field is 1 ft more than twice the width. The area of the batter's box is 21 ft^2. Find the length and width of the rectangular batter's box.

11. **Work Problem** A tank has two drains. One drain takes 16 min longer to empty the tank than does the second drain. With both drains open, the tank is emptied in 6 min. How long would it take each drain, working alone, to empty the tank?

12. **Work Problem** One computer takes 21 min longer than a second computer to calculate the value of a complex expression. Working together, these computers can complete the calculation in 10 min. How long would it take each computer, working alone, to calculate the value?

13. **Sports** The length of a swimming pool is twice the width. The area of the pool is 5000 ft^2. Find the length and width of the pool.

14. **Sports** Read the article at the right. The Longhorns' old scoreboard was a rectangle with a length 30 ft greater than its width. Find the length and width of the old scoreboard.

In the News

Long Board for the Longhorns

The University of Texas Longhorns have replaced their stadium's old 2800-square-foot scoreboard with a new, state-of-the-art, 7370-square-foot scoreboard designed and built by local business Daktronics, Inc.

Sources: Business Wire, www.engadget.com

15. **Transportation** Using one engine of a ferryboat, it takes 6 h longer to cross a channel than it does using a second engine alone. With both engines operating, the ferryboat can make the crossing in 4 h. How long would it take each engine, working alone, to power the ferryboat across the channel?

16. **Work Problem** An apprentice mason takes 8 h longer than an experienced mason to build a small fireplace. Working together, they can build the fireplace in 3 h. How long would it take the experienced mason, working alone, to build the fireplace?

17. **Uniform Motion** It took a small plane 2 h longer to fly 375 mi against the wind than it took the plane to fly the same distance with the wind. The rate of the wind was 25 mph. Find the rate of the plane in calm air.

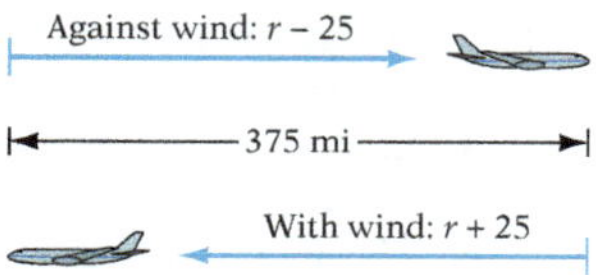

18. **Uniform Motion** It took a motorboat 1 h longer to travel 36 mi against the current than it took the boat to travel 36 mi with the current. The rate of the current was 3 mph. Find the rate of the boat in calm water.

19. **Uniform Motion** A motorcycle traveled 150 mi at a constant rate before its speed was decreased by 15 mph. Another 35 mi was driven at the decreased speed. The total time for the 185-mile trip was 4 h. Find the cyclist's rate during the first 150 mi.

Ljupco Smokovski/Shutterstock.com

20. **Uniform Motion** A cruise ship sailed through a 20-mile inland passageway at a constant rate before its speed was increased by 15 mph. Another 75 mi was traveled at the increased rate. The total time for the 95-mile trip was 5 h. Find the rate of the ship during the last 75 mi.

21. **Physics** An arrow is projected into the air with an initial velocity of 48 ft/s. At what times will the arrow be 32 ft above the ground? Use the equation $h = 48t - 16t^2$, where h is the height, in feet, above the ground after t seconds.

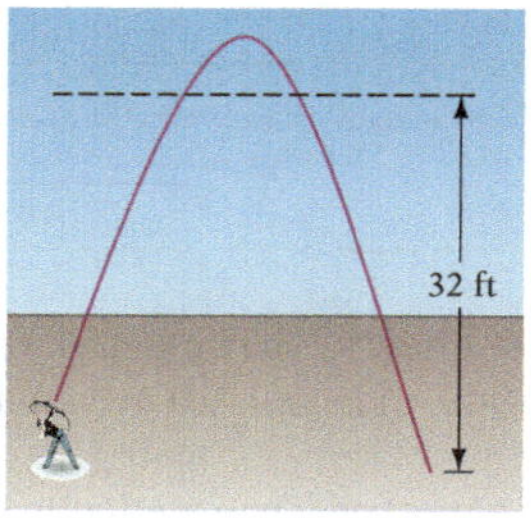

22. **Physics** A model rocket is launched with an initial velocity of 200 ft/s. The height h of the rocket t seconds after launch is given by $h = -16t^2 + 200t$. How many seconds after the launch will the rocket be 300 ft above the ground? Round to the nearest hundredth.

23. Botany Botanists have determined that some species of weeds grow in a circular pattern. For one such weed, the area of growth A, in square meters, can be approximated by $A(t) = 0.005\pi t^2$, where t is the time in days after the growth of the weed first can be observed. How many days after the growth is first observed will this weed cover an area of 10 m^2? Round to the nearest whole number.

Malgorzata Kistryn/Shutterstock.com

24. Physics The kinetic energy of a moving body is given by $E = \frac{1}{2}mv^2$, where E is the kinetic energy, m is the mass, and v is the velocity in meters per second. What is the velocity of a moving body whose mass is 5 kg and whose kinetic energy is 250 newton-meters?

25. Demography See the news clipping at the right. Approximate the year in which there will be 50 million people aged 65 and older in the United States. Use the equation $y = 0.03x^2 + 0.36x + 34.6$, where y is the population, in millions, in year x, where $x = 0$ corresponds to the year 2000.

In the News

Boomers Turn 65

By the time the last baby boomer turns 65, the population of people aged 65 and older will have more than doubled, from 35 million to 71 million.

Source: Census Bureau

26. Sports A basketball player shoots at a basket 25 ft away. The height h, in feet, of the ball above the ground at time t, in seconds, is given by $h = -16t^2 + 32t + 6.5$. How many seconds after the ball is released does it hit the basket? Round to the nearest hundredth. (*Hint:* When it hits the basket, $h = 10$ ft.)

27. Sports In a slow-pitch softball game, the height of the ball thrown by a pitcher can be modeled by the equation $h = -16t^2 + 24t + 4$, where h is the height of the ball in feet and t is the time, in seconds, since it was released by the pitcher. If the batter hits the ball when it is 2 ft off the ground, how many seconds has the ball been in the air? Round to the nearest hundredth.

In the News

Alzheimer's Diagnoses Rising

As the population of senior citizens grows, so will the number of people diagnosed with Alzheimer's, the disease that afflicted former president Ronald Reagan for the last 10 years of his life.

Source: The Alzheimer's Association

28. Alzheimer's See the news clipping at the right. Find the year in which 15 million Americans are expected to have Alzheimer's. Use the equation $y = 0.002x^2 + 0.05x + 2$, where y is the population with Alzheimer's, in millions, in year x, where $x = 0$ corresponds to the year 1980.

29. Sports The hang time of a football that is kicked on the opening kickoff is given by $s = -16t^2 + 88t + 1$, where s is the height, in feet, of the football t seconds after leaving the kicker's foot. What is the hang time of a kickoff that hits the ground without being caught? Round to the nearest tenth.

30. The Internet See the news clipping at the right. Find the year in which consumer Internet traffic will reach 55 million terabytes per month. Use the equation $y = 0.932x^2 - 12.6x + 49.4$, where y is consumer Internet traffic in millions of terabytes per month and x is the year, where $x = 10$ corresponds to the year 2010.

In the News

72 Million Years of Video

Transmission of video through the Internet is increasing so quickly that before long you will need 72 million years to watch the video content that will be transmitted in 1 year. Total consumer Internet traffic is projected to reach 55 million terabytes per month before 2015, with over 90% of that traffic being video content.

Source: businessweek.com

Complete Exercises 31 and 32 *without* writing and solving an equation. Use this situation: A small pipe takes 12 min longer to fill a tank than does a larger pipe. Working together, the pipes can fill the tank in 4 min.

31. True or false? The amount of time it takes for the larger pipe to fill the tank is less than 4 min.

32. True or false? The amount of time it takes for the small pipe to fill the tank is greater than 16 min.

APPLYING CONCEPTS

33. Geometry The hypotenuse of a right triangle measures $\sqrt{13}$ cm. One leg is 1 cm shorter than twice the length of the other leg. Find the lengths of the legs of the right triangle.

34. Integer Problem The sum of the squares of four consecutive integers is 86. Find the four integers.

35. Geometry Find the radius of a right circular cone that has a volume of 800 cm^3 and a height of 12 cm. Round to the nearest hundredth.

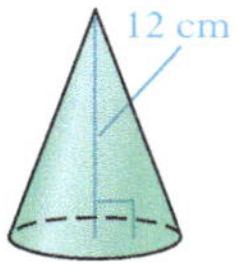

36. Food Industry The radius of a large pizza is 1 in. less than twice the radius of a small pizza. The difference between the areas of the two pizzas is 33π in^2. Find the radius of the large pizza.

37. Food Industry A square piece of cardboard is to be formed into a box to transport pizzas. The box is formed by cutting 2-inch-square corners from the cardboard and folding them up as shown in the figure at the right. If the volume of the box is 512 in^3, what are the dimensions of the cardboard?

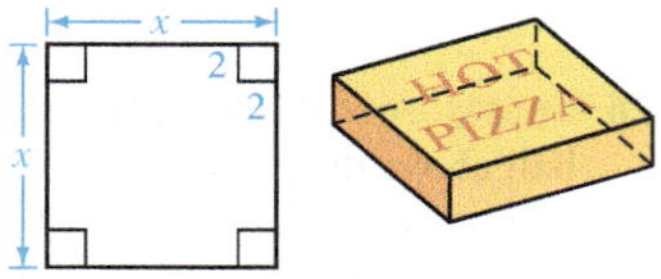

PROJECTS OR GROUP ACTIVITIES

38. Metalwork A wire 8 ft long is cut into two pieces. A circle is formed from one piece and a square is formed from the other. The total area of both figures is given by $A = \frac{1}{16}(8 - x)^2 + \frac{x^2}{4\pi}$. What is the length of each piece of wire if the total area is 4.5 ft^2? Round to the nearest thousandth.

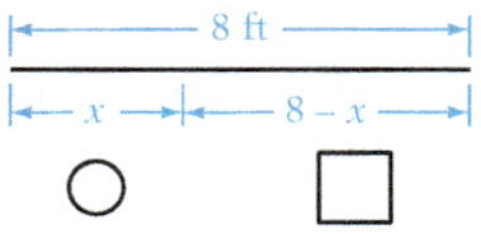

39. Geometry Consider the two rectangles shown below. Both rectangles have the same perimeter, but the areas of the two rectangles are different.

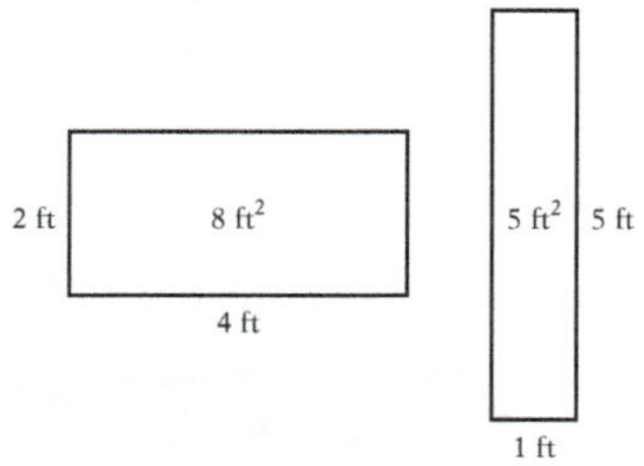

a. Using L for the length and W for the width, write the perimeter formula for a rectangle whose perimeter is 12 ft.

b. Using A for the area, L for the length, and W for the width, write the formula for the area of a rectangle.

c. Solve the formula in part (a) for L. Substitute your expression for L into the formula in part (b). Then simplify.

d. The formula you wrote in part (c) gives the area of a rectangle in terms of the width. Experiment with this formula until you find the dimensions of the rectangle of perimeter 12 ft that has the largest area.

CHAPTER 11 Summary

Key Words	Objective and Page Reference	Examples
A **quadratic equation** is an equation that can be written in the form $ax^2 + bx + c = 0$, $a \neq 0$. A quadratic equation is also called a **second-degree equation.**	[11.1.1, p. 504]	The equation $4x^2 - 3x + 6 = 0$ is a quadratic equation. In this equation, $a = 4$, $b = -3$, and $c = 6$.
A quadratic equation is in **standard form** when the polynomial is in descending order and equal to zero.	[11.1.1, p. 504]	The quadratic equation $5x^2 + 3x - 1 = 0$ is in standard form.
Adding to a binomial the constant term that makes it a perfect-square trinomial is called **completing the square.**	[11.2.1, p. 511]	Adding to $x^2 - 6x$ the constant term 9 results in a perfect-square trinomial: $x^2 - 6x + 9 = (x - 3)(x - 3) = (x - 3)^2$
A **complex number** is a number that can be written in the form $a + bi$, where a and b are real numbers and $i = \sqrt{-1}$. The form $a + bi$ is the **standard form of a complex number**. For the complex number $a + bi$, a is the **real part** of the complex number and b is the **imaginary part** of the complex number.	[11.4.1, pp. 523–524]	$5 + 2i$ is a complex number. 5 is the real part and 2 is the imaginary part of the complex number.
The complex numbers $a + bi$ and $a - bi$ are called **conjugates** of each other.	[11.4.3, p. 526]	$4 + 3i$ and $4 - 3i$ are conjugates. $6 - 8i$ and $6 + 8i$ are conjugates.
An equation of the form $y = ax^2 + bx + c$, $a \neq 0$, is a **quadratic equation in two variables.** For such equations, y is a function of x, and we can write $f(x) = ax^2 + bx + c$. This equation represents a **quadratic function.** The graph of $y = ax^2 + bx + c$ or $f(x) = ax^2 + bx + c$ is a **parabola.** Every parabola has an **axis of symmetry** and a **vertex** that is on the axis of symmetry. If $a > 0$, the parabola opens up and the vertex is the lowest point on the graph. If $a < 0$, the parabola opens down and the vertex is the highest point on the graph.	[11.5.1, pp. 536–538]	$y = 2x^2 - 4x + 3$ is a quadratic equation in two variables. $f(x) = 2x^2 - 4x + 3$ is the same equation written in function notation.

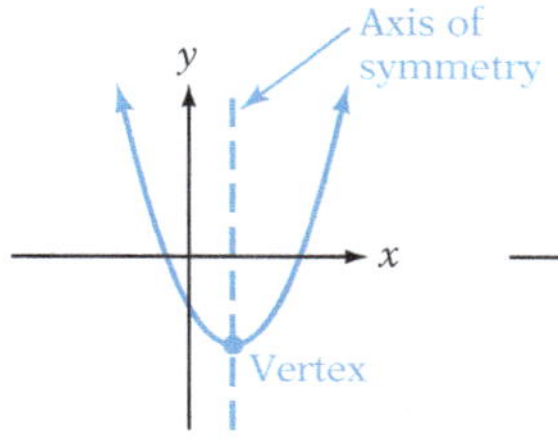

Parabola that opens up

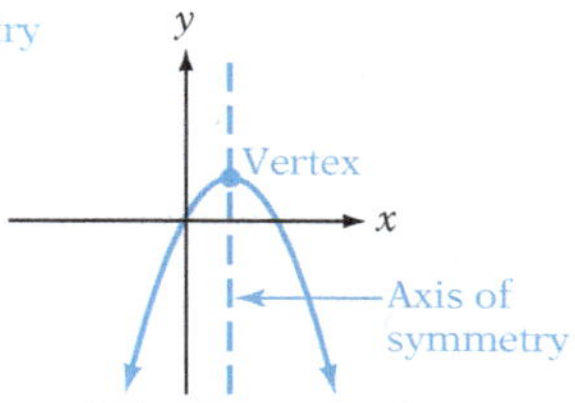

Parabola that opens down

Essential Rules and Procedures	Objective and Page Reference	Examples
Solving a Quadratic Equation by Factoring Write the equation in standard form, factor the left side of the equation, apply the Principle of Zero Products, and solve for the variable.	[11.1.1, p. 504]	$x^2 + 2x = 15$ $x^2 + 2x - 15 = 0$ $(x - 3)(x + 5) = 0$ $x - 3 = 0$ $\quad$ $x + 5 = 0$ $x = 3$ $\quad$ $x = -5$

Principle of Taking the Square Root of Each Side of an Equation
If $x^2 = a$, then $x = \pm\sqrt{a}$.

This principle is used to solve quadratic equations by taking square roots.

[11.1.2, p. 506]

$$\begin{aligned} 3x^2 - 48 &= 0 \\ 3x^2 &= 48 \\ x^2 &= 16 \\ \sqrt{x^2} &= \sqrt{16} \\ x &= \pm\sqrt{16} \\ x &= \pm 4 \end{aligned}$$

Completing the Square
Add to a binomial the constant term that makes it a perfect-square trinomial.

$$\left(\frac{1}{2}\text{ coefficient of } x\right)^2 = \text{constant term}$$

[11.2.1, p. 511]

Complete the square of $x^2 + 12x$.

$$\left(\frac{1}{2}\cdot 12\right)^2 = 36$$

$$x^2 + 12x + 36$$

Solving a Quadratic Equation by Completing the Square

1. Write the equation in the form $x^2 + bx = c$.
2. Add to each side of the equation the term that completes the square of $x^2 + bx$.
3. Factor the perfect-square trinomial. Write it as the square of a binomial.
4. Take the square root of each side of the equation.
5. Solve for x.

[11.2.1, p. 512]

$$\begin{aligned} x^2 + 6x &= 5 \\ x^2 + 6x + 9 &= 5 + 9 \\ (x+3)^2 &= 14 \\ \sqrt{(x+3)^2} &= \sqrt{14} \\ x + 3 &= \pm\sqrt{14} \\ x &= -3 \pm \sqrt{14} \end{aligned}$$

The Quadratic Formula
The solutions of $ax^2 + bx + c = 0, a \neq 0$, are $x = \dfrac{-b \pm \sqrt{b^2 - 4ac}}{2a}$.

[11.3.1, p. 518]

$$\begin{aligned} 3x^2 &= x + 5 \\ 3x^2 - x - 5 &= 0 \end{aligned}$$

$$\begin{aligned} x &= \frac{-b \pm \sqrt{b^2 - 4ac}}{2a} \\ &= \frac{-(-1) \pm \sqrt{(-1)^2 - 4(3)(-5)}}{2(3)} \\ &= \frac{1 \pm \sqrt{1 + 60}}{6} = \frac{1 \pm \sqrt{61}}{6} \end{aligned}$$

Principal Square Root of a Negative Number
If a is a positive real number, then the **principal square root of negative a** is the imaginary number $i\sqrt{a}$: $\sqrt{-a} = i\sqrt{a}$. The number $i\sqrt{a}$ is called an **imaginary number.** When $a = 1$, we have $\sqrt{-1} = i$.

[11.4.1, p. 523]

$$\sqrt{-8} = i\sqrt{8} = 2i\sqrt{2}$$

Addition and Subtraction of Complex Numbers
To add two complex numbers, add the real parts and add the imaginary parts. To subtract two complex numbers, subtract the real parts and subtract the imaginary parts.

[11.4.2, p. 525]

$$\begin{aligned} (3 + 5i) + (2 + 4i) &= (3 + 2) + (5 + 4)i \\ &= 5 + 9i \end{aligned}$$

$$\begin{aligned} (5 + 2i) - (8 + 4i) &= (5 - 8) + (2 - 4)i \\ &= -3 - 2i \end{aligned}$$

Multiplication of Complex Numbers
The product of two complex numbers of the form $a + bi, a \neq 0, b \neq 0$, can be found by using the FOIL method and the fact that $i^2 = -1$.

[11.4.3, p. 526]

$$\begin{aligned} (2 - 3i)(5 + 4i) &= 10 + 8i - 15i - 12i^2 \\ &= 10 - 7i - 12(-1) \\ &= 22 - 7i \end{aligned}$$

Product of Complex Conjugates
The product of a complex number and its conjugate is a real number.

$(a + bi)(a - bi) = a^2 + b^2$

[11.4.3, p. 527]

$$(3 + 5i)(3 - 5i) = 3^2 + 5^2 = 9 + 25 = 34$$

Division of Complex Numbers
To divide a complex number by an imaginary number, multiply the numerator and denominator by i.

To simplify the quotient of two complex numbers, multiply the numerator and denominator by the conjugate of the complex number in the denominator.

[11.4.4, pp. 527–528]

$$\frac{5}{i} = \frac{5}{i} \cdot \frac{i}{i} = \frac{5i}{i^2} = \frac{5i}{-1} = -5i$$

$$\frac{1 + 2i}{1 + i} = \frac{1 + 2i}{1 + i} \cdot \frac{1 - i}{1 - i} = \frac{1 + i - 2i^2}{1^2 + 1^2} = \frac{1 + i - 2(-1)}{1 + 1} = \frac{3 + i}{2} = \frac{3}{2} + \frac{1}{2}i$$

Graphing a Quadratic Equation in Two Variables
Find several solutions of the equation. Graph the ordered-pair solutions on a rectangular coordinate system. Draw a parabola through the points.

[11.5.1, pp. 536–537]

$y = x^2 - x - 2$

x	y
0	−2
1	−2
−1	0
2	0
−2	4
3	4

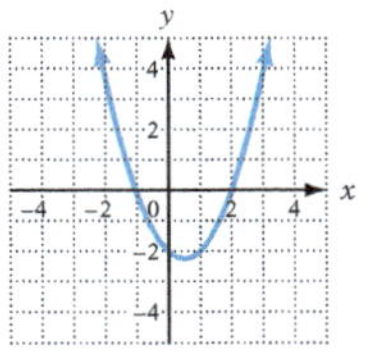

To find the x-intercepts of the graph of a parabola, let $y = 0$ and solve for x. The x-intercepts of the graph at the right are $(-1, 0)$ and $(2, 0)$. To find the y-intercept, let $x = 0$ and solve for y. The y-intercept of the graph at the right is $(0, -2)$.

[11.5.1, p. 537]

CHAPTER 11 Review Exercises

1. Solve: $b^2 - 16 = 0$

2. Solve: $x^2 - x - 3 = 0$

3. Solve: $x^2 - 3x - 5 = 0$

4. Solve: $49x^2 = 25$

5. Graph: $y = -\frac{1}{4}x^2$

6. Graph: $y = -3x^2$

7. Simplify: $3\sqrt{-28}$

8. Solve: $6x(x + 1) = x - 1$

9. Solve: $4y^2 + 9 = 0$

10. Solve: $5x^2 + 20x + 12 = 0$

11. Add: $(-4 + i) + (6 - 5i)$

12. Solve: $x^2 - x = 30$

13. Solve: $6x^2 + 13x - 28 = 0$

14. Solve: $x^2 = 40$

15. Subtract: $(-1 + 2i) - (2 - 3i)$

16. Solve: $x^2 - 2x - 10 = 0$

17. Solve: $x^2 - 12x + 27 = 0$

18. Solve: $(x - 7)^2 = 81$

19. Graph: $y = 2x^2 + 1$

20. Graph: $y = \frac{1}{2}x^2 - 1$

21. Solve: $(y + 4)^2 - 25 = 0$

22. Solve: $4x^2 + 16x = 7$

23. Solve: $24x^2 + 34x + 5 = 0$

24. Solve: $x^2 = 4x - 8$

25. Multiply: $4i(3 - 5i)$

26. Solve: $25(2x^2 - 2x + 1) = (x + 3)^2$

27. Solve: $\left(x - \frac{1}{2}\right)^2 = \frac{9}{4}$

28. Multiply: $(-2 + 4i)(3 - i)$

29. Solve: $x^2 + 7x = 3$

30. Solve: $12x^2 + 10 = 29x$

31. Solve: $4(x - 3)^2 = 20$

32. Solve: $x^2 + 8x - 3 = 0$

33. Graph: $y = x^2 - 3x$

34. Graph: $y = x^2 - 4x + 3$

35. Solve: $(x + 9)^2 = x + 11$

36. Simplify: $\dfrac{3 - i}{2 - 4i}$

37. Solve: $x^2 + 6x + 12 = 0$

38. Solve: $x^2 + 6x - 2 = 0$

39. Solve: $18x^2 - 52x = 6$

40. Solve: $2x^2 + 5x = 1$

41. Graph: $y = -x^2 + 4x - 5$

42. Simplify: $\dfrac{2 - 6i}{2i}$

43. Solve: $2x^2 + 5 = 7x$

44. Graph: $y = 4 - x^2$

45. Uniform Motion It took an air balloon 1 h longer to fly 60 mi against the wind than it took to fly 60 mi with the wind. The rate of the wind was 5 mph. Find the rate of the air balloon in calm air.

46. Geometry The height of a triangle is 2 m more than twice the length of the base. The area of the triangle is 20 m². Find the height of the triangle and the length of the base.

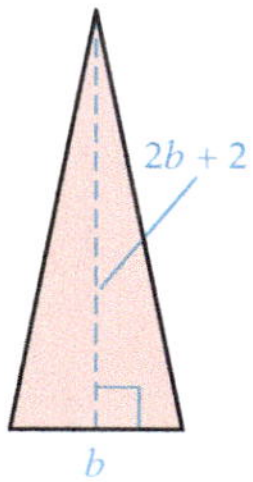

47. Travel In Germany there are no speed limits on some portions of the autobahn (highway). Other portions have a speed limit of 180 km/h (approximately 112 mph). The distance d (in meters) required to stop a car traveling at v kilometers per hour is $d = 0.0056v^2 + 0.14v$. What is the maximum speed at which a driver can be traveling and still be able to stop within 150 m? Round to the nearest tenth.

German Autobahn System

48. **Uniform Motion** In 5 h, two campers rowed 12 mi down a stream and then rowed back to their campsite. The rate of the stream's current was 1 mph. Find the rate of the boat in still water.

49. **Physics** An object is thrown into the air with an initial velocity of 32 ft/s. At what times will the object be 12 ft above the ground? Use the equation $h = 32t - 16t^2$, where h is the height, in feet, above the ground after t seconds.

50. **Work Problem** A smaller drain takes 8 h longer to empty a tank than does a larger drain. Working together, the drains can empty the tank in 3 h. How long would it take each drain, working alone, to empty the tank?

CHAPTER 11 Test

1. Solve: $3(x + 4)^2 - 60 = 0$

2. Solve: $2x^2 + 8x = 3$

3. Solve: $3x^2 + 7x = 20$

4. Simplify: $4\sqrt{-45}$

5. Solve: $x^2 + 4x - 16 = 0$

6. Graph: $y = x^2 + 2x - 4$

7. Solve: $x^2 + 4x + 2 = 0$

8. Subtract: $(-5 + 6i) - (3 - 7i)$

9. Solve: $2x^2 - 5x - 3 = 0$

10. Solve: $2x^2 - 6x + 1 = 0$

11. Multiply: $6i(5 - 4i)$

12. Solve: $2(x - 5)^2 = 36$

13. Solve: $x^2 - 6x - 5 = 0$

14. Multiply: $(-5 + 2i)(4 - i)$

15. Solve: $x^2 - 5x = 2$

16. Solve: $6x^2 - 17x = -5$

17. Solve: $x^2 + 3x - 7 = 0$

18. Simplify: $\frac{3 - i}{2i}$

19. Solve: $x^2 - 8x + 17 = 0$

20. Graph: $y = x^2 - 2x - 3$

21. Simplify: $\frac{4 - 2i}{1 - i}$

22. Geometry The length of a rectangle is 2 ft less than twice the width. The area of the rectangle is 40 ft^2. Find the length and width of the rectangle.

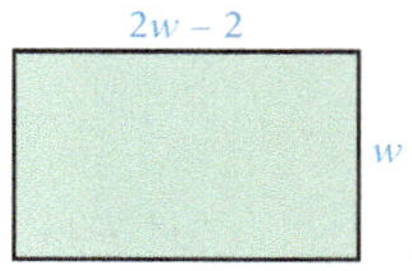

23. Uniform Motion It took a motorboat 1 h more to travel 60 mi against the current than it took to go 60 mi with the current. The rate of the current was 1 mph. Find the rate of the boat in calm water.

24. Integer Problem The sum of the squares of three consecutive odd integers is 83. Find the middle odd integer.

25. Uniform Motion A jogger ran 7 mi at a constant rate and then reduced the rate by 3 mph. An additional 8 mi was run at the reduced rate. The total time spent jogging the 15 mi was 3 h. Find the rate for the last 8 mi.

Cumulative Review Exercises

1. Simplify: $2x - 3[2x - 4(3 - 2x) + 2] - 3$

2. Solve: $-\frac{3}{5}x = -\frac{9}{10}$

3. Solve: $2x - 3(4x - 5) = -3x - 6$

4. Solve: $2x - 3(2 - 3x) > 2x - 5$

5. Find the x- and y-intercepts of the line $4x - 3y = 12$.

6. Find the equation of the line that contains the point whose coordinates are $(-3, 2)$ and has slope $-\frac{4}{3}$.

7. Find the domain and range of the relation $\{(-2, -8), (-1, -1), (0, 0), (1, 1), (2, 8)\}$. Is the relation a function?

8. Evaluate $f(x) = -3x + 10$ at $x = -9$.

9. Graph: $y = \frac{1}{4}x - 2$

10. Graph the solution set of $2x - 3y > 6$.

11. Solve by substitution: $\begin{aligned} 3x - y &= 5 \\ y &= 2x - 3 \end{aligned}$

12. Solve by the addition method: $\begin{aligned} 3x + 2y &= 2 \\ 5x - 2y &= 14 \end{aligned}$

13. Simplify: $\dfrac{(2a^{-2}b)^2}{-3a^{-5}b^4}$

14. Divide: $(x^2 - 8) \div (x - 2)$

15. Factor: $4y(x - 4) - 3(x - 4)$

16. Factor: $3x^3 + 2x^2 - 8x$

17. Divide: $\dfrac{3x^2 - 6x}{4x - 6} \div \dfrac{2x^2 + x - 6}{6x^2 - 24x}$

18. Subtract: $\dfrac{x}{2(x - 1)} - \dfrac{1}{(x - 1)(x + 1)}$

19. Simplify: $\dfrac{1 - \frac{7}{x} + \frac{12}{x^2}}{2 - \frac{1}{x} - \frac{15}{x^2}}$

20. Solve: $\dfrac{x}{x + 6} = \dfrac{3}{x}$

21. Multiply: $(\sqrt{a} - \sqrt{2})(\sqrt{a} + \sqrt{2})$

22. Solve: $3 = 8 - \sqrt{5x}$

23. Solve: $2x^2 - 7x = -3$

24. Solve: $3(x - 2)^2 = 36$

25. Solve: $3x^2 - 4x - 5 = 0$

26. Graph: $y = x^2 - 3x - 2$

27. Taxes In a certain state, the sales tax is $7\frac{1}{4}$%. The sales tax on a chemistry textbook is \$5.22. Find the cost of the textbook before the tax is added.

28. Mixture Problem Find the cost per pound of a mixture made from 20 lb of cashews that cost \$7 per pound and 50 lb of peanuts that cost \$3.50 per pound.

29. Investments A stock investment of 100 shares paid a dividend of $215. At this rate, how many additional shares are required to earn a dividend of $752.50?

30. Uniform Motion A 720-mile trip from one city to another takes 3 h when a plane is flying with the wind. The return trip, against the wind, takes 4.5 h. Find the rate of the plane in calm air and the rate of the wind.

31. Test Scores A student received a 70, a 91, an 85, and a 77 on four tests in a math class. What scores on the fifth test will enable the student to receive a minimum of 400 points?

32. Guy Wires A guy wire is attached to a point 30 m above the ground on a telephone pole that is perpendicular to the ground. The wire is anchored to the ground at a point 10 m from the base of the pole. Find the length of the guy wire. Round to the nearest hundredth.

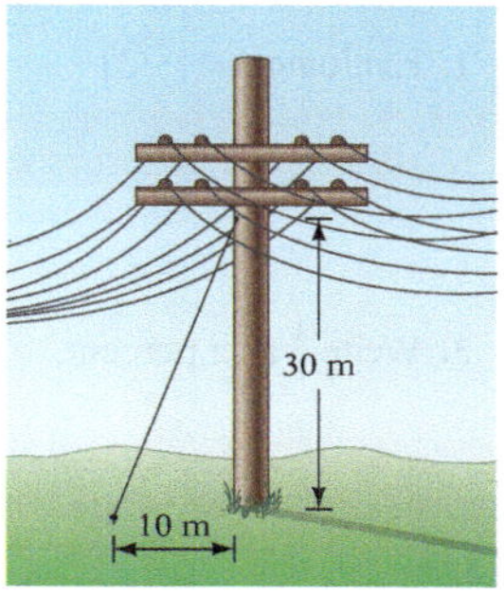

33. Sports The length of a singles tennis court is 24 ft more than twice the width. The area is 2106 ft^2. Find the length and width of the singles tennis court.

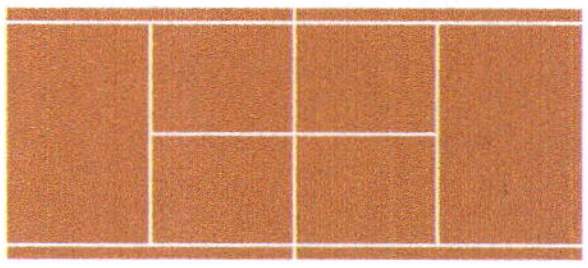

Final Exam

1. Evaluate: $-|-3|$

2. Subtract: $-15 - (-12) - 3$

3. Write $\frac{1}{8}$ as a percent.

4. Simplify: $-2^4 \cdot (-2)^4$

5. Simplify: $-7 - \dfrac{12 - 15}{2 - (-1)} \cdot (-4)$

6. Evaluate $\dfrac{a^2 - 3b}{2a - 2b^2}$ when $a = 3$ and $b = -2$.

7. Simplify: $6x - (-4y) - (-3x) + 2y$

8. Multiply: $(-15z)\left(-\dfrac{2}{5}\right)$

9. Simplify: $-2[5 - 3(2x - 7) - 2x]$

10. Solve: $20 = -\dfrac{2}{5}x$

11. Solve: $4 - 2(3x + 1) = 3(2 - x) + 5$

12. Find 19% of 80.

13. Solve: $4 - x \geq 7$

14. Solve: $2 - 2(y - 1) \leq 2y - 6$

15. Find the slope of the line that contains the points whose coordinates are $(-1, -3)$ and $(2, -1)$.

16. Find the equation of the line that contains the point whose coordinates are $(3, -4)$ and has slope $-\frac{2}{3}$.

17. Graph the line with slope $-\frac{1}{2}$ and y-intercept $(0, -3)$.

18. Graph: $f(x) = \dfrac{2}{3}x - 4$

19. Find the range of the function given by the equation $f(x) = -x + 5$ if the domain is $\{-6, -3, 0, 3, 6\}$.

20. Graph the solution set of $3x - 2y \geq 6$.

21. Solve by substitution: $y = 4x - 7$
$y = 2x + 5$

22. Solve by the addition method: $4x - 3y = 11$
$2x + 5y = -1$

23. Subtract: $(2x^2 - 5x + 1) - (5x^2 - 2x - 7)$

24. Simplify: $(-3xy^3)^4$

25. Multiply: $(3x^2 - x - 2)(2x + 3)$

26. Simplify: $\dfrac{(-2x^2y^3)^3}{(-4x^{-1}y^4)^2}$

27. Simplify: $(4x^{-2}y)^3(2xy^{-2})^{-2}$

28. Divide: $\dfrac{12x^3y^2 - 16x^2y^2 - 20y^2}{4xy^2}$

29. Divide: $(5x^2 - 2x - 1) \div (x + 2)$

30. Write 0.000000039 in scientific notation.

31. Factor: $2a(4 - x) - 6(x - 4)$

32. Factor: $x^2 - 5x - 6$

33. Factor: $2x^2 - x - 3$

34. Factor: $6x^2 - 5x - 6$

35. Factor: $8x^3 - 28x^2 + 12x$

36. Factor: $25x^2 - 16$

37. Factor: $75y - 12x^2y$

38. Solve: $2x^2 = 7x - 3$

39. Multiply: $\dfrac{2x^2 - 3x + 1}{4x^2 - 2x} \cdot \dfrac{4x^2 + 4x}{x^2 - 2x + 1}$

40. Subtract: $\dfrac{5}{x + 3} - \dfrac{3x}{2x - 5}$

41. Simplify: $\dfrac{x - \dfrac{3}{2x - 1}}{1 - \dfrac{2}{2x - 1}}$

42. Solve: $\dfrac{5x}{3x - 5} - 3 = \dfrac{7}{3x - 5}$

43. Solve $a = 3a - 2b$ for a.

44. Simplify: $\sqrt{49x^6}$

45. Add: $2\sqrt{27a} + 8\sqrt{48a}$

46. Simplify: $\dfrac{\sqrt{3}}{\sqrt{5} - 2}$

47. Solve: $\sqrt{x + 4} - \sqrt{x - 1} = 1$

48. Solve: $(x - 3)^2 = 7$

49. Solve: $4x^2 - 2x - 1 = 0$

50. Graph: $y = x^2 - 4x + 3$

51. Translate and simplify "the sum of twice a number and three times the difference between the number and two."

52. **Depreciation** Because of depreciation, the value of an office machine is now \$2400. This is 80% of its original value. Find the original value of the machine.

53. **Medicine** The recommended dose (d) of a children's medication varies directly with the weight of the child in pounds (w). The proper dosage for a child weighing 36 lb is 16 mg. Find the correct dosage for a child weighing 45 lb.

54. **Geometry** The measure of one angle of a triangle is 10° more than the measure of the second angle. The measure of the third angle is 10° more than the measure of the first angle. Find the measure of each angle of the triangle.

55. **Markup** The manufacturer's cost for a laser printer is \$900. The manufacturer sells the printer for \$1485. Find the markup rate.

56. **Investments** An investment of \$3000 is made at an annual simple interest rate of 8%. How much additional money must be invested at 11% so that the total interest earned is 10% of the total investment?

57. **Food Mixtures** A grocer mixes 4 lb of peanuts that cost \$2 per pound with 2 lb of walnuts that cost \$5 per pound. What is the cost per pound of the resulting mixture?

58. **Mixture Problem** A pharmacist mixes 20 L of a solution that is 60% acid with 30 L of a solution that is 20% acid. What is the percent concentration of acid in the resulting mixture?

59. **Uniform Motion** A small plane flew at a constant rate for 1 h. The pilot then doubled the plane's speed. An additional 1.5 h were flown at the increased speed. If the entire flight was 860 km, how far did the plane travel during the first hour?

60. **Uniform Motion** With the current, a motorboat travels 50 mi in 2.5 h. Against the current, it takes twice as long to travel 50 mi. Find the rate of the boat in calm water and the rate of the current.

61. **Geometry** The length of a rectangle is 5 m more than the width. The area of the rectangle is 50 m^2. Find the dimensions of the rectangle.

62. Paint A paint formula requires 2 oz of dye for every 15 oz of base paint. How many ounces of dye are required for 120 oz of base paint?

63. Work Problem It takes a chef 1 h to prepare a dinner. The chef's apprentice can prepare the same dinner in 1.5 h. How long would it take the chef and the apprentice, working together, to prepare the dinner?

64. Geometry The hypotenuse of a right triangle measures 14 cm. One leg of the triangle measures 8 cm. Find the length of the other leg of the triangle. Round to the nearest tenth.

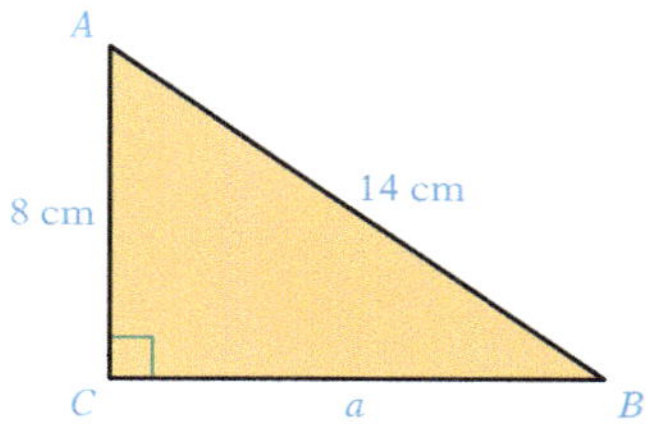

65. Uniform Motion It took a plane $\frac{1}{2}$ h more to fly 500 mi against the wind than it took to fly the same distance with the wind. The rate of the plane in calm air is 225 mph. Find the rate of the wind.

Digital Vision

Table of Properties

APPENDIX

Properties of Real Numbers

The Associative Property of Addition
If a, b, and c are real numbers, then $(a + b) + c = a + (b + c)$.

The Commutative Property of Addition
If a and b are real numbers, then $a + b = b + a$.

The Addition Property of Zero
If a is a real number, then $a + 0 = 0 + a = a$.

The Multiplication Property of Zero
If a is a real number, then $a \cdot 0 = 0 \cdot a = 0$.

The Inverse Property of Addition
If a is a real number, then $a + (-a) = (-a) + a = 0$.

The Associative Property of Multiplication
If a, b, and c are real numbers, then $(a \cdot b) \cdot c = a \cdot (b \cdot c)$.

The Commutative Property of Multiplication
If a and b are real numbers, then $a \cdot b = b \cdot a$.

The Multiplication Property of One
If a is a real number, then $a \cdot 1 = 1 \cdot a = a$.

The Inverse Property of Multiplication
If a is a real number and $a \neq 0$, then $a \cdot \frac{1}{a} = \frac{1}{a} \cdot a = 1$.

Distributive Property
If a, b, and c are real numbers, then $a(b + c) = ab + ac$ or $(b + c)a = ba + ca$.

Properties of Equations

Addition Property of Equations
The same number or variable term can be added to each side of an equation without changing the solution of the equation.

Multiplication Property of Equations
Each side of an equation can be multiplied by the same nonzero number without changing the solution of the equation.

Properties of Inequalities

Addition Property of Inequalities
If $a > b$, then $a + c > b + c$.

If $a < b$, then $a + c < b + c$.

Multiplication Property of Inequalities
If $a > b$ and $c > 0$, then $ac > bc$.

If $a < b$ and $c > 0$, then $ac < bc$.

If $a > b$ and $c < 0$, then $ac < bc$.

If $a < b$ and $c < 0$, then $ac > bc$.

Properties of Exponents

If m and n are integers, then $x^m \cdot x^n = x^{m+n}$.

If m and n are integers, then $(x^m)^n = x^{mn}$.

If $x \neq 0$, then $x^0 = 1$.

If m and n are integers and $x \neq 0$, then
$\frac{x^m}{x^n} = x^{m-n}$.

If m, n, and p are integers, then $(x^m \cdot y^n)^p = x^{mp}y^{np}$.

If n is a positive integer and $x \neq 0$, then
$x^{-n} = \frac{1}{x^n}$ and $\frac{1}{x^{-n}} = x^n$.

If m, n, and p are integers and $y \neq 0$, then
$\left(\frac{x^m}{y^n}\right)^p = \frac{x^{mp}}{y^{np}}$.

Principle of Zero Products

If $a \cdot b = 0$, then $a = 0$ or $b = 0$.

Properties of Radical Expressions

If a and b are positive real numbers, then
$\sqrt[n]{ab} = \sqrt[n]{a}\sqrt[n]{b}$.

If a and b are positive real numbers, then
$\sqrt[n]{\frac{a}{b}} = \frac{\sqrt[n]{a}}{\sqrt[n]{b}}$.

Property of Squaring Both Sides of an Equation

If a and b are real numbers and $a = b$, then $a^2 = b^2$.

Keystroke Guide for the TI-83 Plus and TI-84 Plus

APPENDIX

Digital Vision

Basic Operations

Numerical calculations are performed on the **home screen.** You can always return to the home screen by pressing [2ND] QUIT. Pressing [CLEAR] erases the home screen.

To evaluate the expression $-2(3 + 5) - 8 \div 4$, use the following keystrokes.

[(-)] 2 [(] 3 [+] 5 [)] [−] 8 [÷] 4 [ENTER]

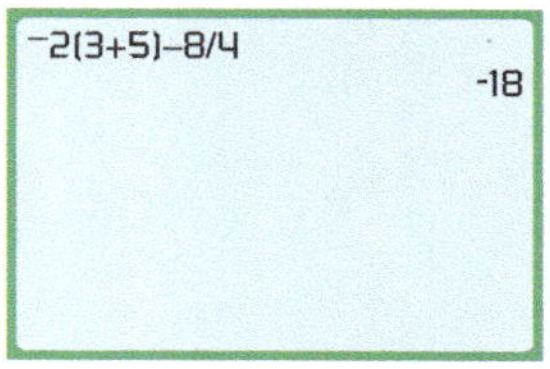

Note: There is a difference between the key to enter a negative number, [(-)], and the key for subtraction, [−]. You cannot use these keys interchangeably.

Take Note

The descriptions in the margins (for example, Basic Operations and Evaluating Functions) are the same as those used in the text and are arranged alphabetically.

The [2ND] key is used to access the commands in blue writing above a key. For instance, to evaluate the $\sqrt{49}$, press [2ND] √ 49 [)] [ENTER].

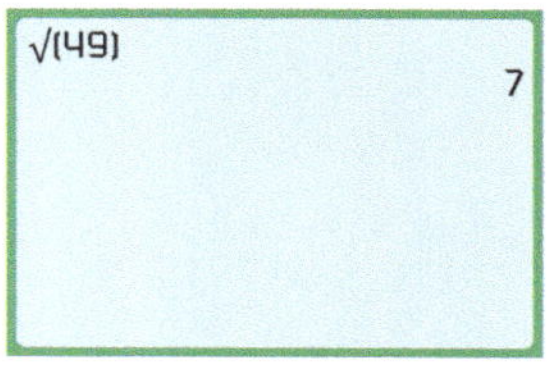

The [ALPHA] key is used to place a letter on the screen. One reason to do this is to store a value of a variable. The following keystrokes give A the value of 5.

5 [STO▸] [ALPHA] A [ENTER]

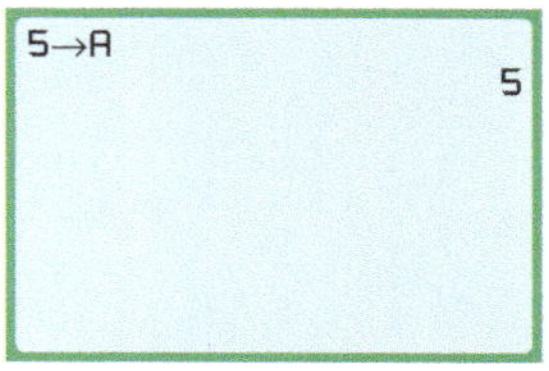

This value is now available in calculations. For instance, we can find the value of $3a^2$ by using the following keystrokes: 3 [ALPHA] A [x^2]. To display the value of the variable on the screen, press [2ND] RCL [ALPHA] A.

Note: When you use the [ALPHA] key, only capital letters are available on the TI calculators.

Evaluating Functions

There are various methods of evaluating a function, but all methods require that the expression be entered as one of the ten functions Y1 to Y0. To evaluate $f(x) = \frac{x^2}{x - 1}$ when $x = -3$, enter the expression into, for instance, Y1, and then press [VARS] [▸] 11 [(] [(-)] 3 [)] [ENTER].

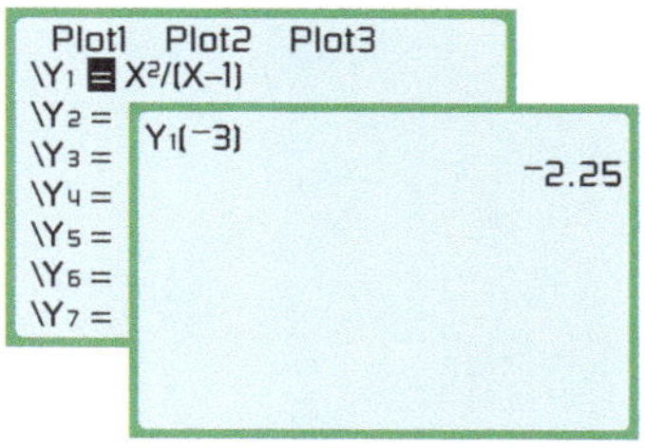

Take Note

Use the down arrow key to scroll past Y_7 to see Y_8, Y_9, and Y_0.

Note: If you try to evaluate a function at a number that is not in the domain of the function, you will get an error message. For instance, 1 is not in the domain of $f(x) = \frac{x^2}{x-1}$. If we try to evaluate the function at 1, the error screen at the right appears.

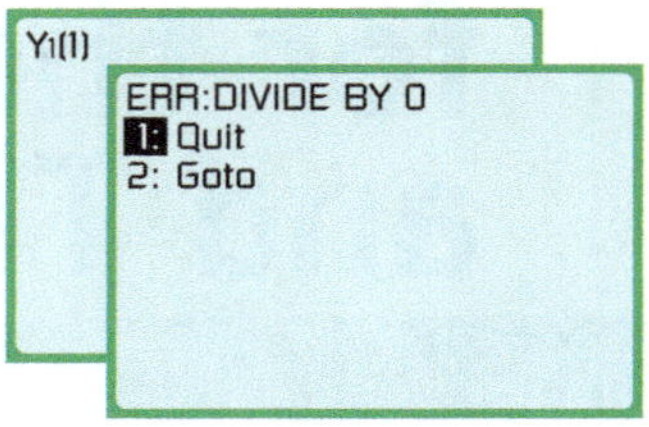

Evaluating Variable Expressions

To evaluate a variable expression, first store the values of each variable. Then enter the variable expression. For instance, to evaluate $s^2 + 2sl$ when $s = 4$ and $l = 5$, use the following keystrokes.

4 STO→ ALPHA S ENTER 5 STO→ ALPHA L ENTER ALPHA S x^2 +
2 ALPHA S ALPHA L ENTER

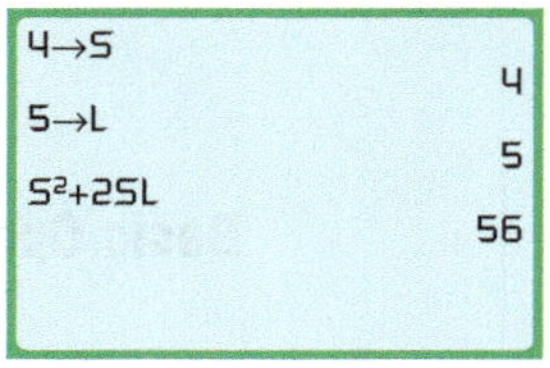

Graph

To graph a function, use the Y= key to enter the expression for the function, select a suitable viewing window, and then press GRAPH. For instance, to graph $f(x) = 0.1x^3 - 2x - 1$ in the standard viewing window, use the following keystrokes.

Y= 0.1 X,T,θ,n ^ 3 − 2 X,T,θ,n − 1 ZOOM (scroll to 6) ENTER

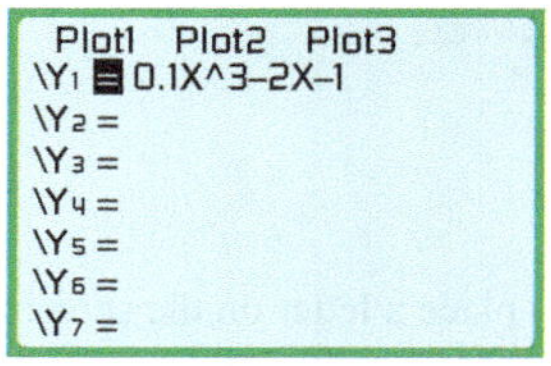

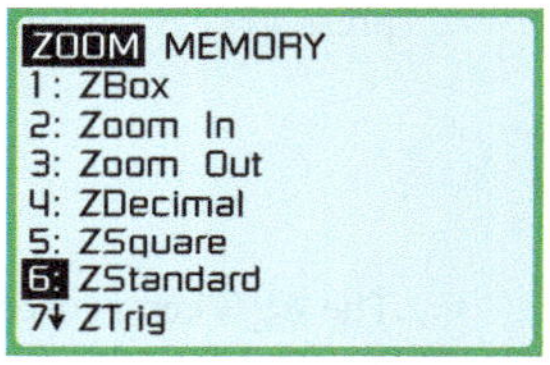

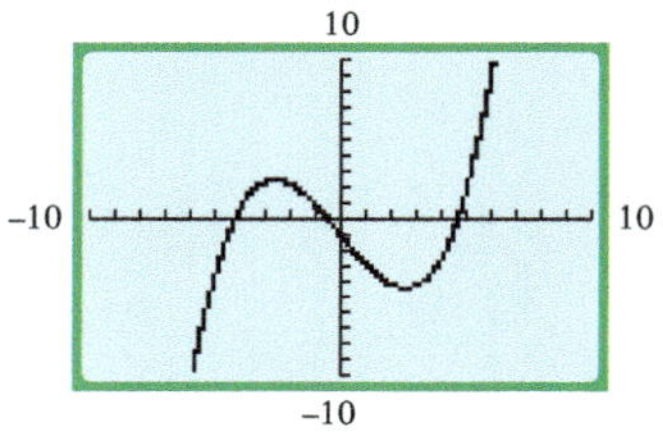

Note: For the keystrokes above, you do not have to scroll to 6. Alternatively, use ZOOM 6. This will select the standard viewing window and automatically start the graph. Use the WINDOW key to create a custom window for a graph.

Graphing Inequalities

To illustrate this feature, we will graph $y \le 2x - 1$. Enter $2x - 1$ into Y1. Because $y \le 2x - 1$, we want to shade below the graph. Move the cursor to the left of Y1 and press ENTER three times. Press GRAPH.

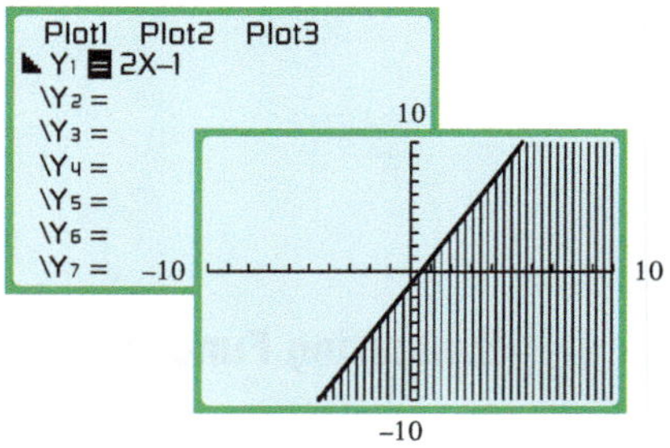

Note: To shade above the graph, move the cursor to the left of Y1 and press ENTER two times. An inequality with the symbol $\le$ or $\ge$ should be graphed with a solid line, and an inequality with the symbol $<$ or $>$ should be graphed with a dashed line. However, the graph of a linear inequality on a graphing calculator does not distinguish between a solid line and a dashed line.

To graph the solution set of a system of inequalities, solve each inequality for y and graph each inequality. The solution set is the intersection of the two inequalities. The solution set of $\begin{matrix} 3x + 2y > 10 \\ 4x - 3y \le 5 \end{matrix}$ is shown at the right.

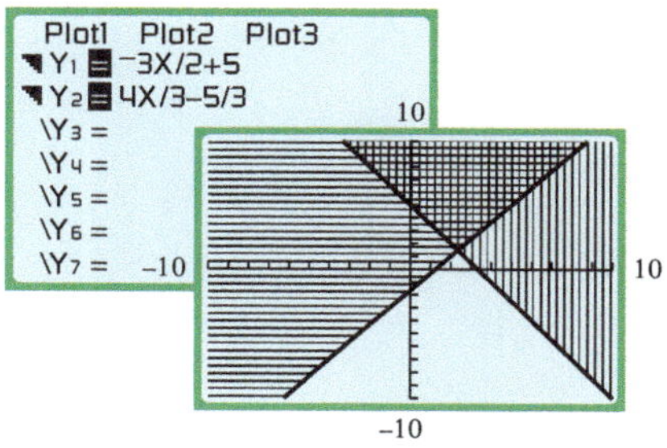

Intersect

The INTERSECT feature is used to solve a system of equations. To illustrate this feature, we will use the system of equations $\begin{matrix} 2x - 3y = 13 \\ 3x + 4y = -6 \end{matrix}$.

Note: Some equations can be solved by this method. See the section "Solve an equation" on the next page. Also, this method is used to find a number in the domain of a function for a given number in the range. See the section "Find a domain element."

Solve each of the equations in the system of equations for y. In this case, we have $y = \frac{2}{3}x - \frac{13}{3}$ and $y = -\frac{3}{4}x - \frac{3}{2}$.

Use the Y= editor to enter $\frac{2}{3}x - \frac{13}{3}$ into Y1 and $-\frac{3}{4}x - \frac{3}{2}$ into Y2. Graph the two functions in the standard viewing window. (If the window does not show the point of intersection of the two graphs, adjust the window until you can see the point of intersection.)

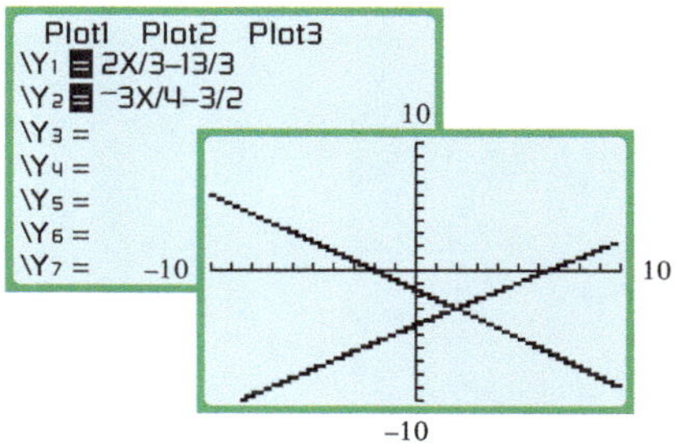

Press 2ND CALC (scroll to 5, intersect) ENTER.

Alternatively, you can just press 2ND CALC 5.

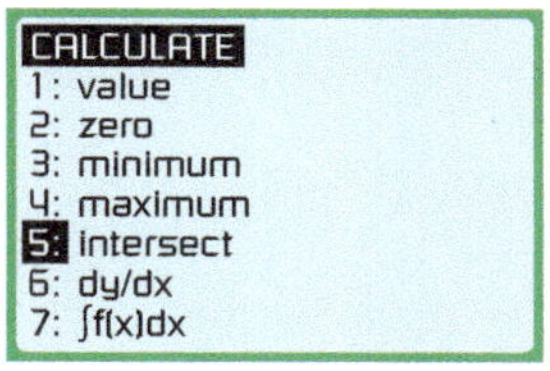

First curve? is shown at the bottom of the screen and identifies one of the two graphs on the screen. Press ENTER.

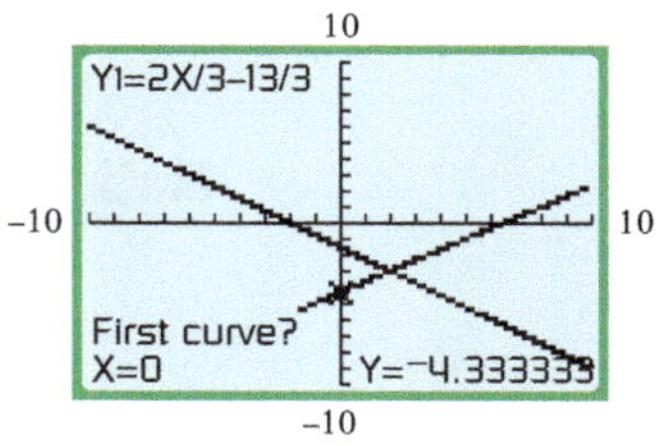

Second curve? is shown at the bottom of the screen and identifies the second of the two graphs on the screen. Press ENTER.

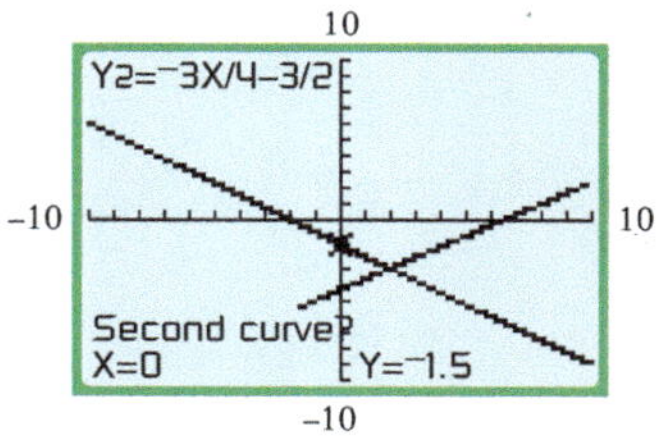

Guess? shown at the bottom of the screen asks you to use the left or right arrow key to move the cursor to the *approximate* location of the point of intersection. (If there are two or more points of intersection, it does not matter which one you choose first.) Press ENTER.

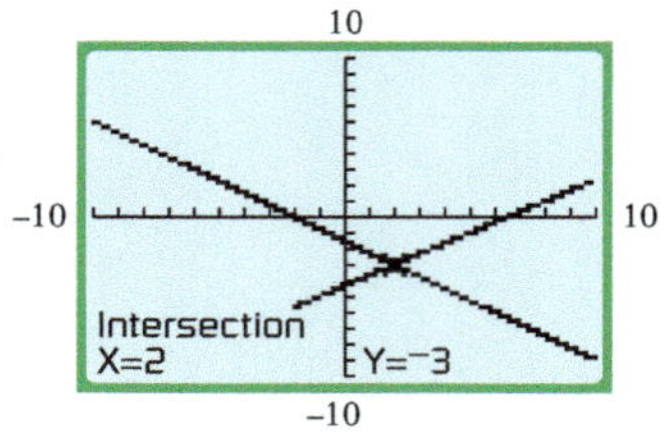

The solution of the system of equations is $(2, -3)$.

Solve an equation To illustrate the steps involved, we will solve the equation $2x + 4 = -3x - 1$. The idea is to write the equation as the system of equations $\begin{array}{l} y = 2x + 4 \\ y = -3x - 1 \end{array}$ and then use the steps for solving a system of equations.

Use the Y= editor to enter the left and right sides of the equation into Y1 and Y2. Graph the two functions and then follow the steps for Intersect.

The solution is -1, the x-coordinate of the point of intersection.

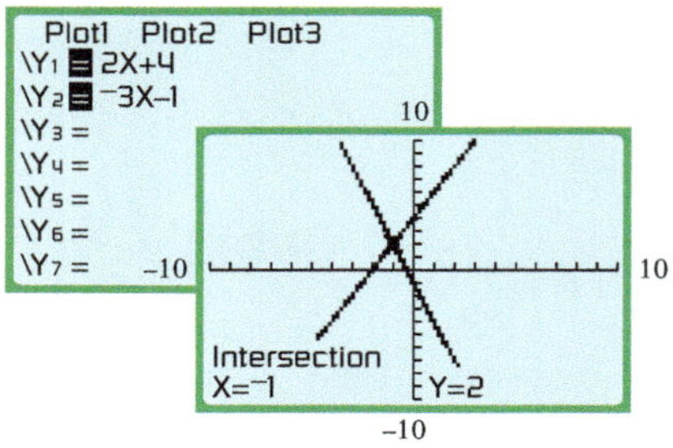

Find a domain element For this example, we will find a number in the domain of $f(x) = -\frac{2}{3}x + 2$ that corresponds to 4 in the range of the function. This is like solving the system of equations $y = -\frac{2}{3}x + 2$ and $y = 4$.

Use the Y= editor to enter the expression for the function in Y1 and the desired output, 4, in Y2. Graph the two functions and then follow the steps for Intersect.

The point of intersection is $(-3, 4)$. The number -3 in the domain of f produces an output of 4 in the range of f.

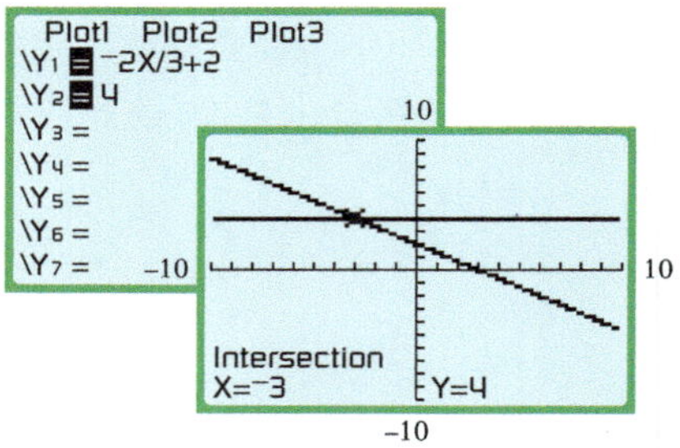

Math

Pressing MATH gives you access to many built-in functions. The following keystrokes will convert 0.125 to a fraction: .125 MATH 1 ENTER.

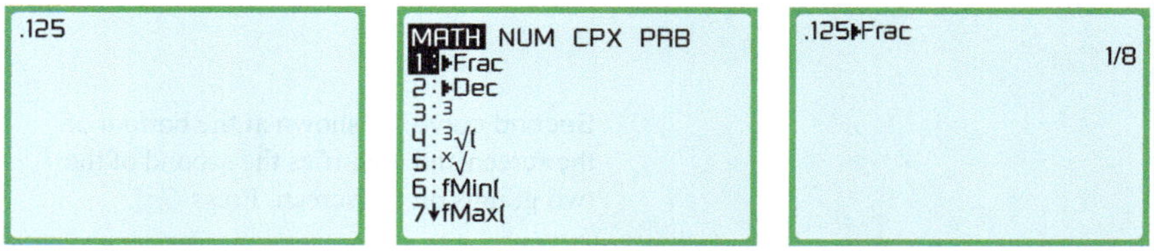

Additional built-in functions under MATH can be found by pressing MATH ▸. For instance, to evaluate $-|-25|$, press (−) MATH ▸ 1 (−) 25) ENTER.

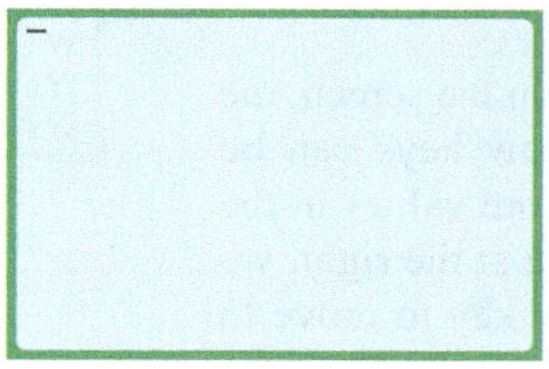

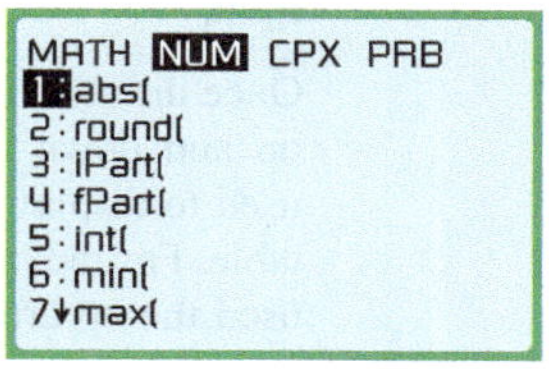

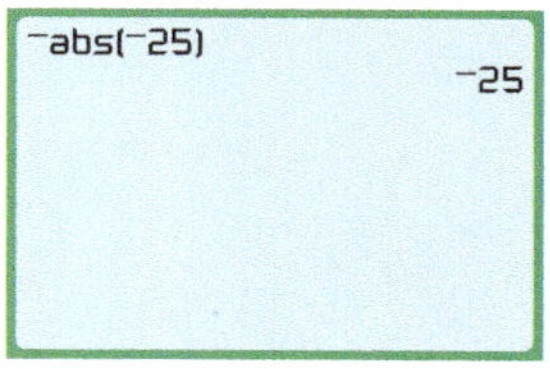

See your owner's manual for assistance with other functions under the MATH key.

Radical Expressions

To evaluate a square-root expression, press 2ND √‾.

For instance, to evaluate $0.15\sqrt{p^2 + 4p + 10}$ when $p = 100{,}000$, first store 100,000 in P. Then press 0.15 2ND √‾ ALPHA P x^2 + 4 ALPHA P + 10) ENTER.

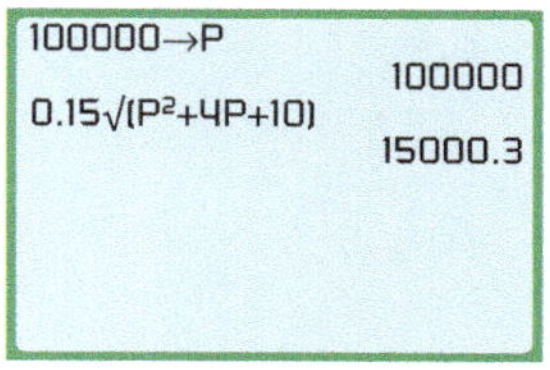

To evaluate a radical expression other than a square root, access $\sqrt[x]{}$ by pressing MATH. For instance, to evaluate $\sqrt[4]{67}$, press 4 (the index of the radical) MATH (scroll to 5) ENTER 67 ENTER.

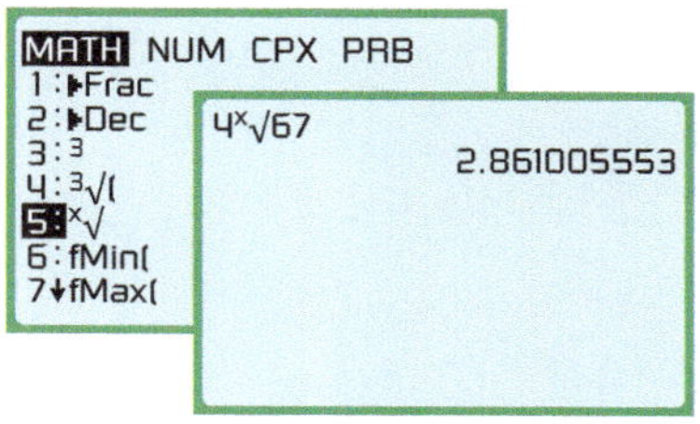

Scientific Notation

To enter a number in scientific notation, use 2ND EE. For instance, to find $\frac{3.45 \times 10^{-12}}{1.5 \times 10^{25}}$, press 3.45 2ND EE (−) 12 ÷ 1.5 2ND EE 25 ENTER. The answer is 2.3×10^{-37}.

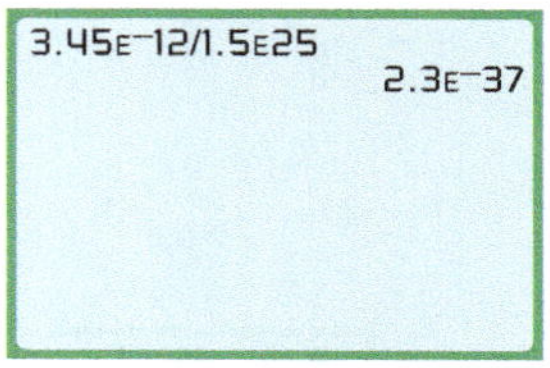

Table

There are three steps in creating an input/output table for a function. First use the Y= editor to input the function. The second step is setting up the table, and the third step is displaying the table.

To set up the table, press 2ND TBLSET. TblStart is the first value of the independent variable in the input/output table. ΔTbl is the difference between successive values. Setting this to 1 means that, for this table, the input values are $-2, -1, 0, 1, 2, \ldots$. If ΔTbl $= 0.5$, then the input values are $-2, -1.5, -1, -0.5, 0, 0.5, \ldots$.

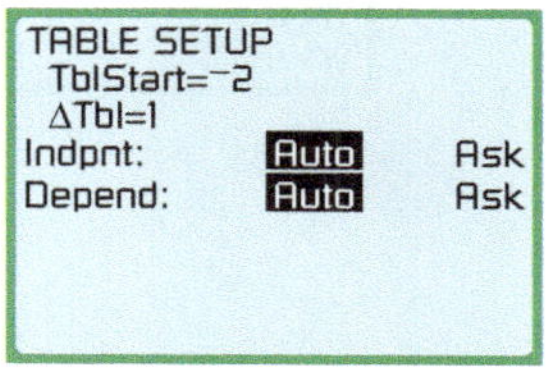

Indpnt is the independent variable. When this is set to Auto, values of the independent variable are automatically entered into the table. Depend is the dependent variable. When this is set to Auto, values of the dependent variable are automatically entered into the table.

To display the table, press 2ND TABLE. An input/output table for $f(x) = x^2 - 1$ is shown at the right.

Once the table is on the screen, the up and down arrow keys can be used to display more values in the table. For the table at the right, we used the up arrow key to move to $x = -7$.

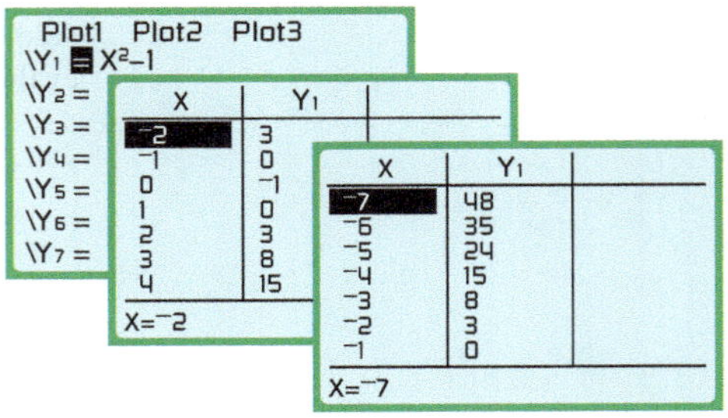

An input/output table for any given input can be created by selecting Ask for the independent variable. The table at the right shows an input/output table for $f(x) = \frac{4x}{x-2}$ for selected values of x. Note that the word ERROR or ERR: appeared when 2 was entered. This occurred because f is not defined when $x = 2$.

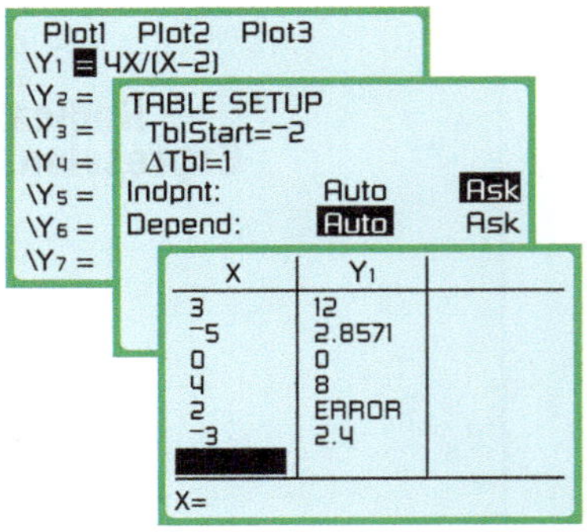

Note: Using the table feature in Ask mode is the same as evaluating a function for given values of the independent variable. For instance, from the table above, we have $f(4) = 8$.

Test

The TEST feature has many uses, one of which is to graph the solution set of a linear inequality in one variable. To illustrate this feature, we will graph the solution set of $x - 1 < 4$. Press Y= X,T,θ,n − 1 2ND TEST (scroll to 5) ENTER 4 GRAPH.

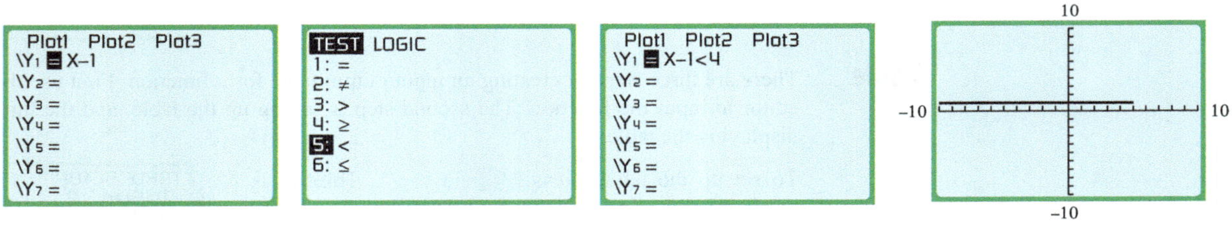

Trace

Once a graph is drawn, pressing TRACE will place a cursor on the screen, and the coordinates of the point below the cursor are shown at the bottom of the screen. Use the left and right arrow keys to move the cursor along the graph. For the graph at the right, we have $f(4.8) = 3.4592$, where $f(x) = 0.1x^3 - 2x + 2$ is shown at the top left of the screen.

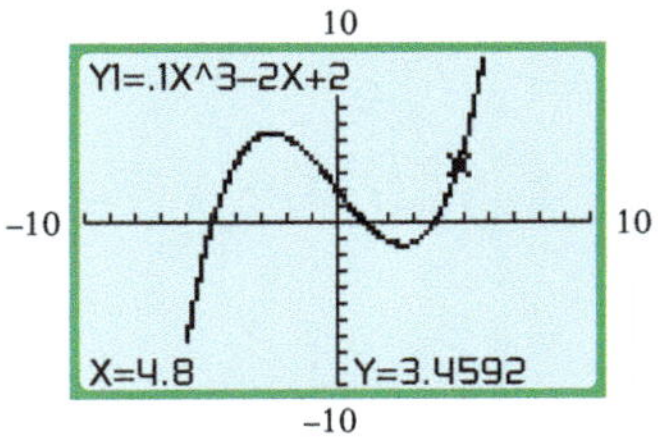

In TRACE mode, you can evaluate a function at any value of the independent variable that is within Xmin and Xmax. To do this, first graph the function. Now press TRACE (the value of x) ENTER. For the graph at the left below, we used $x = -3.5$. If a value of x is chosen outside the window, an error message is displayed.

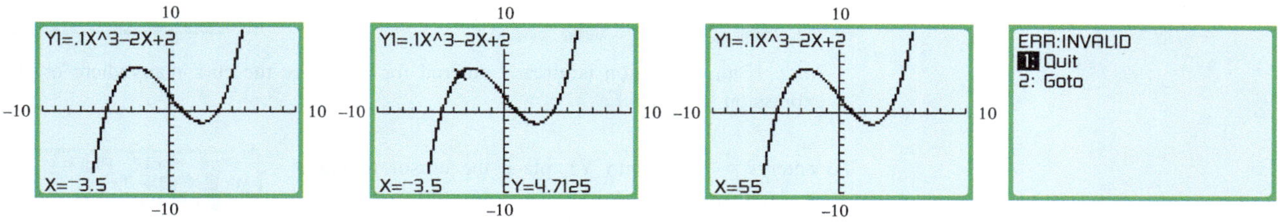

In the example above where we entered -3.5 for x, the value of the function was calculated as 4.7125. This means that $f(-3.5) = 4.7125$. The keystrokes 2ND QUIT VARS ▸ 11 MATH 1 ENTER will convert the decimal value to a fraction.

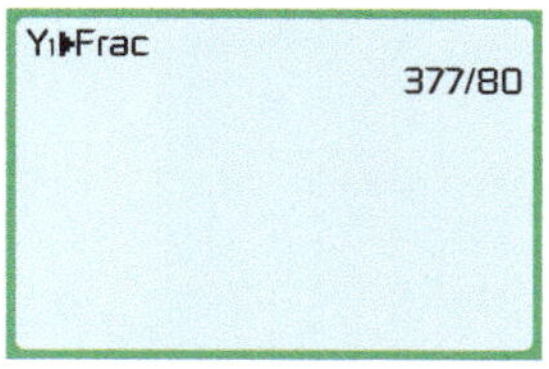

When the TRACE feature is used with two or more graphs, the up and down arrow keys are used to move between the graphs. The following two graphs are for the functions $f(x) = 0.1x^3 - 2x + 2$ and $g(x) = 2x - 3$. By using the up and down arrows, we can place the cursor on either graph. The right and left arrows are used to move along the graph.

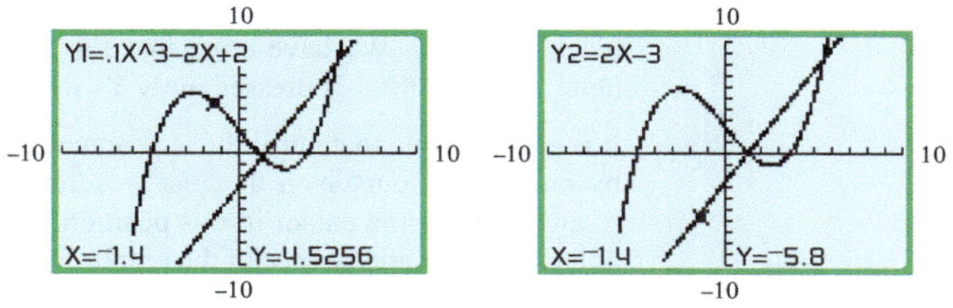

Window

The viewing window for a graph is controlled by pressing WINDOW. Xmin and Xmax are the minimum value and maximum value, respectively, of the independent variable shown on the graph. Xscl is the distance between tick marks on the x-axis. Ymin and Ymax are the minimum value and maximum value, respectively, of the dependent variable shown on the graph. Yscl is the distance between tick marks on the y-axis. Leave Xres as 1.

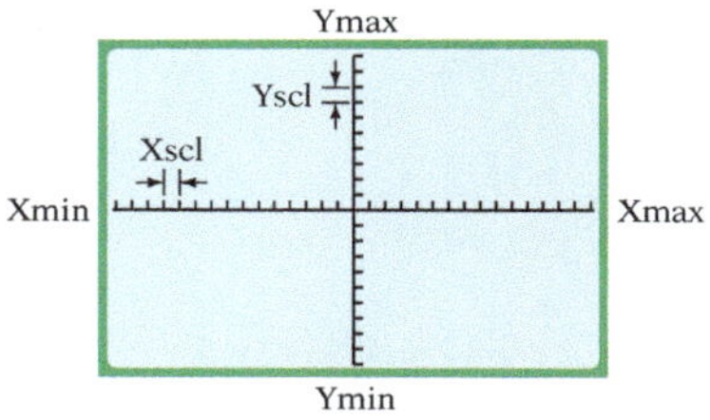

Note: In the standard viewing window, the distance between tick marks on the x-axis is different from the distance between tick marks on the y-axis. This will distort a graph. A more accurate picture of a graph can be created by using a square viewing window. See ZOOM.

Y= The Y= editor is used to enter the expression for a function. There are ten possible functions, labeled Y_1 to Y_0, that can be active at any one time. For instance, to enter $f(x) = x^2 + 3x - 2$ as Y_1, use the following keystrokes.

Y= X,T,θ,n x² + 3 X,T,θ,n − 2

```
Plot1 Plot2 Plot3
\Y1 ■ X²+3X−2
\Y2 =
\Y3 =
\Y4 =
\Y5 =
\Y6 =
\Y7 =
```

Note: If an expression is already entered for Y_1, place the cursor anywhere on that expression and press CLEAR.

To enter $s = \frac{2v - 1}{v^3 - 3}$ into Y_2, place the cursor to the right of the equals sign for Y_2. Then press (2 X,T,θ,n − 1) ÷ (X,T,θ,n ^ 3 − 3).

```
Plot1 Plot2 Plot3
\Y1 ■ X²+3X−2
\Y2 ■ (2X−1)/(X^3−3)
\Y3 =
\Y4 =
\Y5 =
\Y6 =
\Y7 =
```

Note: When we enter an equation, the independent variable (v in the expression above) is entered using X,T,θ,n. The dependent variable (s in the expression above) is one of Y_1 to Y_0. Also note the use of parentheses to ensure the correct order of operations.

Observe the black rectangle that covers the equals sign for the two examples we have shown. This rectangle means that the function is "active." If we were to press GRAPH, then the graph of both functions would appear. You can make a function inactive by using the arrow keys to move the cursor over the equals sign of that function and then pressing ENTER. This will remove the black rectangle. We have done that for Y_2, as shown here. Now if GRAPH is pressed, only Y_1 will be graphed.

```
Plot1 Plot2 Plot3
\Y1 ■ X²+3X−2
\Y2 = (2X−1)/(X^3−3)
\Y3 =
\Y4 =
\Y5 =
\Y6 =
\Y7 =
```

It is also possible to control the appearance of the graph by moving the cursor on the Y= screen to the left of any Y. With the cursor in this position, pressing ENTER will change the appearance of the graph. The options are shown at the right.

```
Plot1 Plot2 Plot3
\Y1 = Default graph line
\Y2 = Bold graph line
◥Y3 = Shade above graph
◣Y4 = Shade below graph
-oY5 = Draw path of graph
oY6 = Travel path of graph
:Y7 = Dashed graph line
```

Zero

The ZERO feature of a graphing calculator is used for various calculations: to find the x-intercepts of a function, to solve some equations, and to find the zero of a function.

x-intercepts To illustrate the procedure for finding x-intercepts, we will use $f(x) = x^2 + x - 2$.

First, use the Y= editor to enter the expression for the function and then graph the function in the standard viewing window. (It may be necessary to adjust this window so that the intercepts are visible.) Once the graph is displayed, use the keystrokes below to find the x-intercepts of the graph of the function.

Press 2ND CALC (scroll to 2 for zero of the function) ENTER.

Alternatively, you can just press 2ND CALC 2.

```
CALCULATE
1: value
2: zero
3: minimum
4: maximum
5: intersect
6: dy/dx
7: ∫f(x)dx
```

Left Bound? shown at the bottom of the screen asks you to use the left or right arrow key to move the cursor to the *left* of the desired x-intercept. Press ENTER.

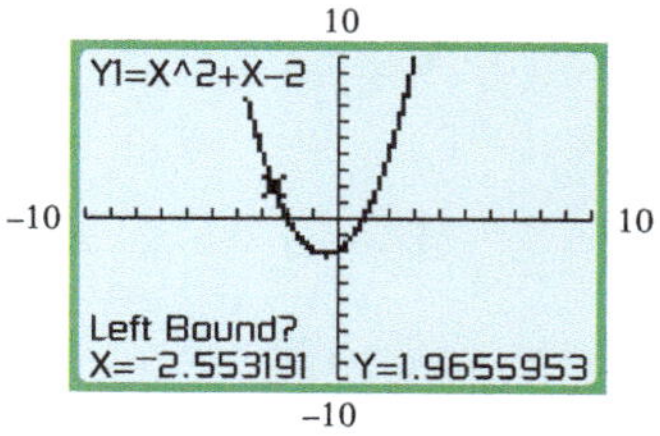

Right Bound? shown at the bottom of the screen asks you to use the left or right arrow key to move the cursor to the *right* of the desired x-intercept. Press ENTER.

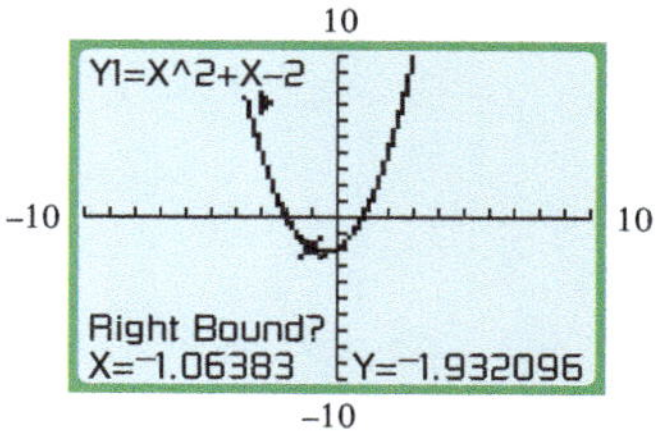

Guess? shown at the bottom of the screen asks you to use the left or right arrow key to move the cursor to the *approximate* location of the desired x-intercept. Press ENTER.

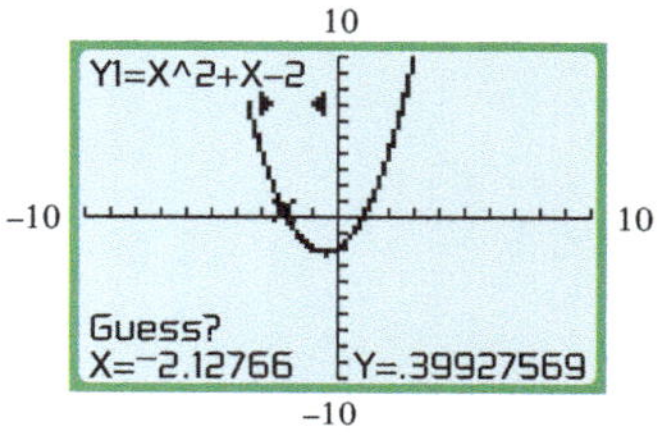

The x-coordinate of an x-intercept is -2. Therefore, an x-intercept is $(-2, 0)$.

To find the other x-intercept, follow the same steps as above. The screens for this calculation are shown below.

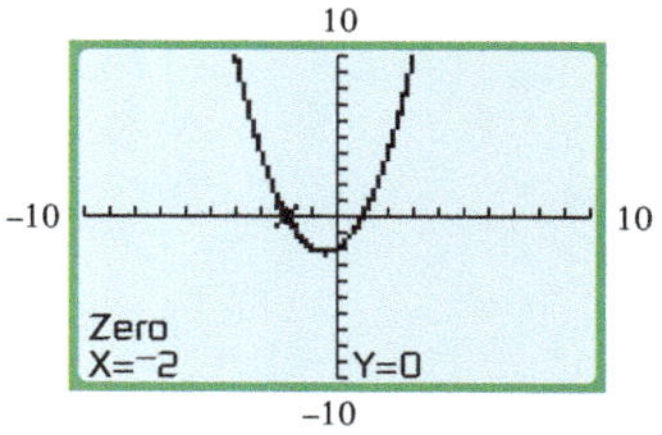

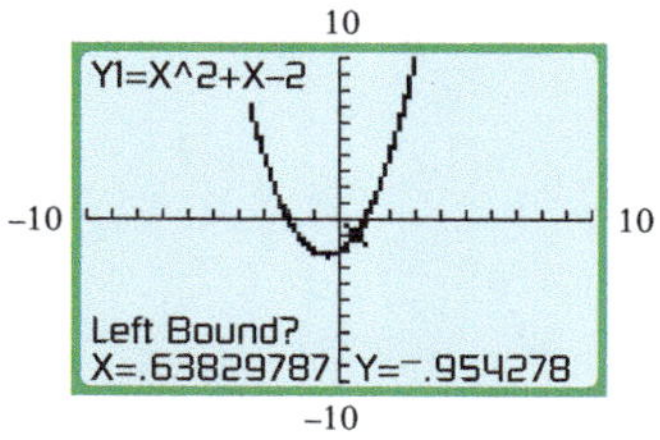

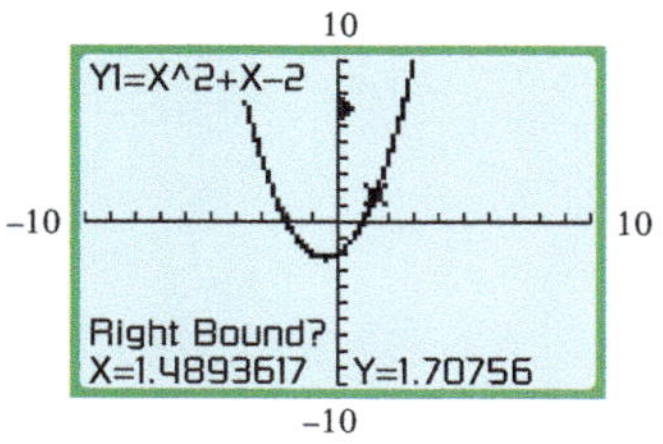

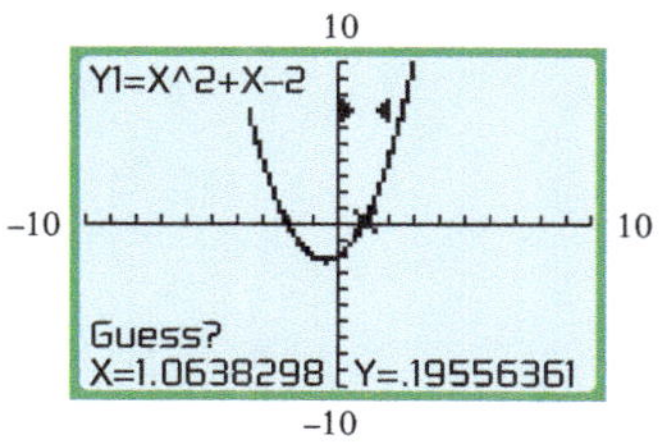

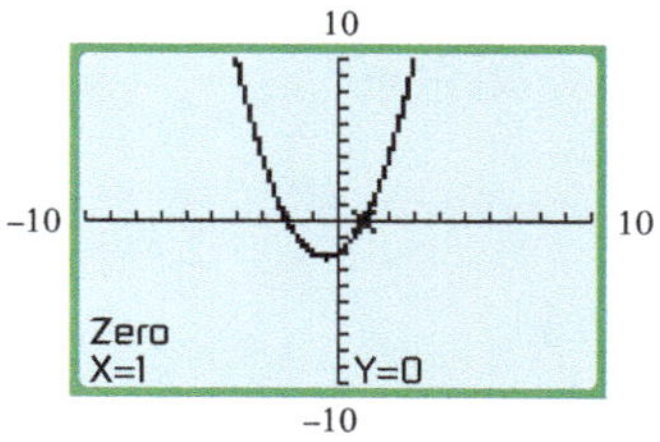

A second x-intercept is $(1, 0)$.

Solve an equation To use the ZERO feature to solve an equation, first rewrite the equation with all terms on one side. For instance, one way to solve the equation $x^3 - x + 1 = -2x + 3$ is first to rewrite it as $x^3 + x - 2 = 0$. Enter $x^3 + x - 2$ into Y1 and then follow the steps for finding x-intercepts.

Find the real zeros of a function To find the real zeros of a function, follow the steps for finding x-intercepts.

Zoom

Pressing ZOOM allows you to select some preset viewing windows. This key also gives you access to ZBox, Zoom In, and Zoom Out. These functions enable you to redraw a selected portion of a graph in a new window. Some windows that are used frequently in this text are shown below.

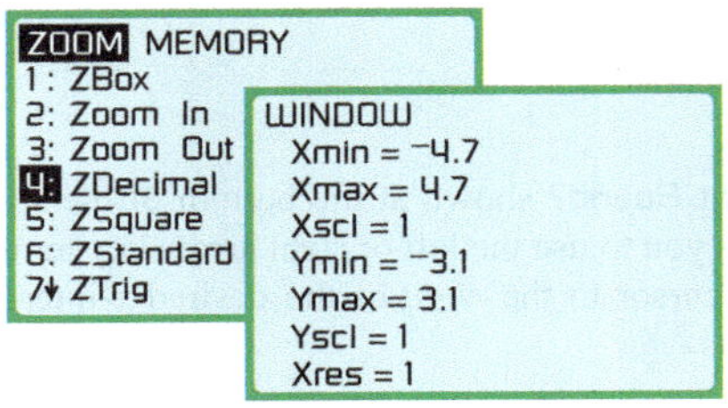

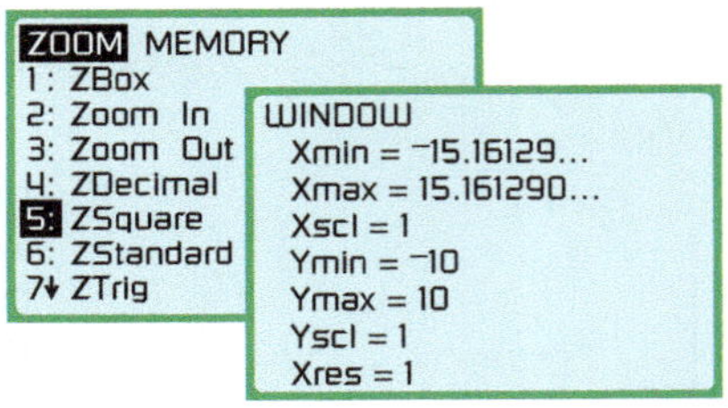

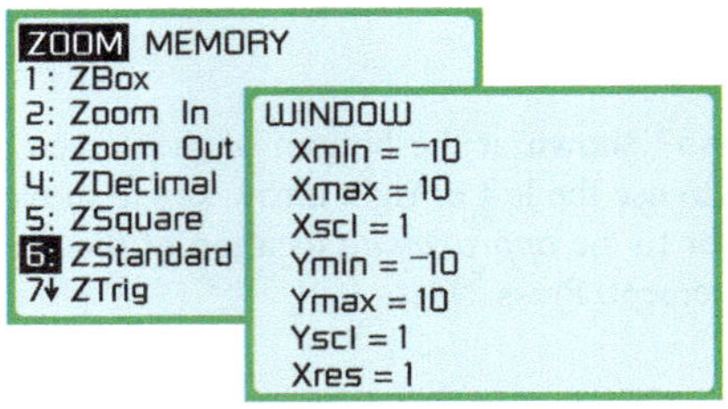

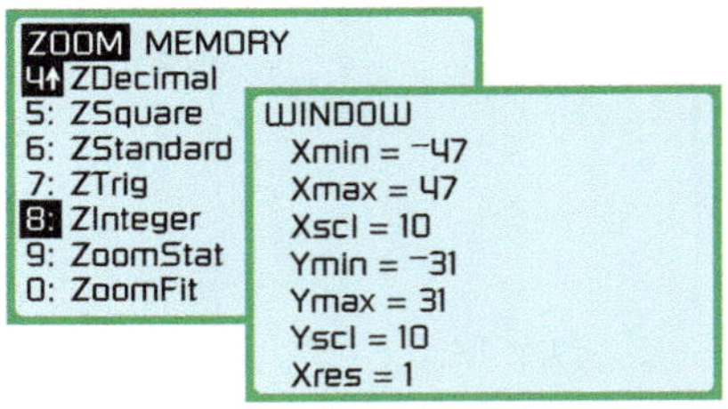

Digital Vision

Solutions to Chapter Problems

Solutions to Chapter 1 Problems

SECTION 1.1

Problem 1

$A = \{1, 2, 3, 4\}$ • **The roster method encloses a list of elements in braces.**

Problem 2

$$\begin{aligned} -5 &< -1 \\ -1 &= -1 \\ 5 &> -1 \end{aligned}$$
• **Find the order relation between each element of set B and -1.**

The element 5 is greater than -1.

Problem 3

A. The opposite of -9 is 9.
B. The opposite of 62 is -62.

Problem 4

A. $|-5| = 5$ **B.** $-|-9| = -9$

SECTION 1.2

Problem 1

A. $-162 + 98 = -64$
B. $-154 + (-37) = -191$
C. $\begin{aligned}[t] -36 + 17 + (-21) &= -19 + (-21) \\ &= -40 \end{aligned}$

Problem 2

$$\begin{aligned} -8 - 14 &= -8 + (-14) \\ &= -22 \end{aligned}$$
• **Rewrite subtraction as addition of the opposite.**

Problem 3

$$\begin{aligned} &4 - (-3) - 12 - (-7) - 20 \\ &= 4 + 3 + (-12) + 7 + (-20) \\ &= 7 + (-12) + 7 + (-20) \\ &= -5 + 7 + (-20) \\ &= 2 + (-20) \\ &= -18 \end{aligned}$$
• **Rewrite subtraction as addition of the opposite.**

Problem 4

A. $-38 \cdot 51 = -1938$
B. $\begin{aligned}[t] -7(-8)(9)(-2) &= 56(9)(-2) \\ &= 504(-2) \\ &= -1008 \end{aligned}$

Problem 5

A. $(-135) \div (-9) = 15$
B. $\dfrac{84}{-6} = -14$
C. $-\dfrac{36}{-12} = -(-3) = 3$

Problem 6

Strategy To find the average daily high temperature:
▶ Add the seven temperature readings.
▶ Divide the sum by 7.

Solution
$$\begin{aligned} &-5 + (-6) + 3 + 0 + (-4) + (-7) + (-2) \\ &= -11 + 3 + 0 + (-4) + (-7) + (-2) \\ &= -8 + 0 + (-4) + (-7) + (-2) \\ &= -8 + (-4) + (-7) + (-2) \\ &= -12 + (-7) + (-2) \\ &= -19 + (-2) \\ &= -21 \end{aligned}$$
$-21 \div 7 = -3$

The average daily high temperature was -3°C.

SECTION 1.3

Problem 1

$$\begin{array}{r} 0.16 \\ 25\overline{)4.00} \\ \underline{-2\,5} \\ 1\,50 \\ \underline{-1\,50} \\ 0 \end{array}$$
• **This is a terminating decimal.**
• **The remainder is zero.**

$\dfrac{4}{25} = 0.16$

Problem 2

$$\begin{array}{r} 0.444... \\ 9\overline{)4.000} \\ \underline{-3\,6} \\ 40 \\ \underline{-36} \\ 40 \\ \underline{-36} \\ 4 \end{array}$$
• **This is a repeating decimal.**

$\dfrac{4}{9} = 0.\overline{4}$ • **The bar shows that the digit repeats.**

Problem 3

$$-\frac{7}{12}\cdot\frac{9}{14} = -\frac{7\cdot 9}{12\cdot 14} = -\frac{\overset{1}{\cancel{7}}\cdot\overset{1}{\cancel{3}}\cdot 3}{2\cdot 2\cdot\underset{1}{\cancel{3}}\cdot 2\cdot\underset{1}{\cancel{7}}} = -\frac{3}{8}$$

• Divide by the common factors.

Problem 4

$$-\frac{3}{8}\div\left(-\frac{5}{12}\right) = \frac{3}{8}\cdot\frac{12}{5}$$

• Change division to multiplication and invert the divisor.

$$= \frac{3\cdot 12}{8\cdot 5} = \frac{3\cdot\overset{1}{\cancel{2}}\cdot\overset{1}{\cancel{2}}\cdot 3}{\underset{1}{\cancel{2}}\cdot\underset{1}{\cancel{2}}\cdot 2\cdot 5} = \frac{9}{10}$$

Problem 5

$$\begin{array}{r} 5.44 \\ \times\ \ 3.8 \\ \hline 4352 \\ 1632\ \\ \hline 20.672 \end{array}$$

• Multiply the absolute values.

$(-5.44)(3.8) = -20.672$ • The product is negative.

Problem 6

$$\begin{array}{r} 4.88 \\ 0.27.\overline{)1.32.00} \\ -108\ \ \ \\ \hline 24\ 0\ \\ -21\ 6\ \\ \hline 2\ 40 \\ -2\ 16 \\ \hline 24 \end{array}$$

• Move the decimal point two places to the right in the divisor and dividend. Place the decimal point in the quotient.

$1.32 \div 0.27 \approx 4.9$

Problem 7

Prime factorization of 9 and 12:

$9 = 3\cdot 3 \qquad 12 = 2\cdot 2\cdot 3$

$\text{LCM} = 2\cdot 2\cdot 3\cdot 3 = 36$ • The LCM is the common denominator.

$$\frac{5}{9}-\frac{11}{12} = \frac{20}{36}-\frac{33}{36} = \frac{20}{36}+\frac{-33}{36}$$

$$= \frac{20+(-33)}{36} = \frac{-13}{36} = -\frac{13}{36}$$

Problem 8

$$-\frac{7}{8}-\frac{5}{6}+\frac{1}{2} = -\frac{21}{24}-\frac{20}{24}+\frac{12}{24}$$

$$= \frac{-21}{24}+\frac{-20}{24}+\frac{12}{24}$$

$$= \frac{-21+(-20)+12}{24} = \frac{-29}{24} = -\frac{29}{24}$$

Problem 9

$$\begin{array}{r} 3.097 \\ 4.9\ \ \ \\ +3.09\ \\ \hline 11.087 \end{array}$$

• Align the decimal points vertically.

Problem 10

$$\begin{array}{r} 67.910 \\ -16.127 \\ \hline 51.783 \end{array}$$

• Subtract the absolute values.

$16.127 - 67.91 = -51.783$ • Attach the sign of the number with the larger absolute value.

Problem 11

$$125\% = 125\left(\frac{1}{100}\right) = \frac{125}{100} = 1\frac{1}{4}$$

$$125\% = 125(0.01) = 1.25$$

Problem 12

$$16\frac{2}{3}\% = 16\frac{2}{3}\left(\frac{1}{100}\right) = \frac{50}{3}\left(\frac{1}{100}\right) = \frac{1}{6}$$

• Write $16\frac{2}{3}$ as $\frac{50}{3}$.

Problem 13

$6.08\% = 6.08(0.01) = 0.0608$

Problem 14

A. $0.043 = 0.043(100\%) = 4.3\%$

B. $2.57 = 2.57(100\%) = 257\%$

Problem 15

$$\frac{5}{9} = \frac{5}{9}(100\%) = \frac{500}{9}\% \approx 55.6\%$$

Problem 16

$$\frac{9}{16} = \frac{9}{16}(100\%) = \frac{900}{16}\% = 56\frac{1}{4}\%$$

SECTION 1.4

Problem 1

$(-5)^3 = (-5)(-5)(-5) = 25(-5) = -125$ • Cube −5.

$-5^3 = -(5\cdot 5\cdot 5) = -(25\cdot 5) = -125$ • The opposite of 5^3.

Problem 2

$(-3)^3 = (-3)(-3)(-3) = 9(-3) = -27$

$(-3)^4 = (-3)(-3)(-3)(-3)$

$\quad = 9(-3)(-3) = -27(-3) = 81$

Problem 3

$(3^3)(-2)^3 = (3)(3)(3)\cdot(-2)(-2)(-2)$

$\quad = 27\cdot(-8) = -216$

$$\left(-\frac{2}{5}\right)^2 = \left(-\frac{2}{5}\right)\left(-\frac{2}{5}\right) = \frac{2\cdot 2}{5\cdot 5} = \frac{4}{25}$$

Problem 4

$36 \div (8-5)^2 - (-3)^2\cdot 2$

$= 36 \div (3)^2 - (-3)^2\cdot 2$ • Simplify within grouping symbols.

$= 36 \div 9 - 9\cdot 2$ • Simplify exponential expressions.

$= 4 - 9\cdot 2$ • Multiply and divide, left to right.

$= 4 - 18$

$= -14$ • Add and subtract, left to right.

Problem 5

$27 \div 3^2 + (-3)^2\cdot 4$

$= 27 \div 9 + 9\cdot 4$ • Simplify exponential expressions.

$= 3 + 9\cdot 4$ • Multiply and divide.

$= 3 + 36$

$= 39$ • Add and subtract.

Problem 6

$4 - 3[4 - 2(6 - 3)] \div 2$
$= 4 - 3[4 - 2(3)] \div 2$ • Perform operations within inner grouping symbols.
$= 4 - 3[4 - 6] \div 2$
$= 4 - 3(-2) \div 2$ • Perform operations within outer grouping symbols.
$= 4 - (-6) \div 2$
$= 4 - (-3)$
$= 7$

SECTION 1.5

Problem 1

To find the complement of 87°, subtract 87° from 90°.

$$90° - 87° = 3°$$

3° is the complement of 87°.

Problem 2

To find the supplement of 87°, subtract 87° from 180°.

$$180° - 87° = 93°$$

93° is the supplement of 87°.

Problem 3

$m\angle x$ is the sum of the measures of two angles.

$$m\angle x = 34° + 95° = 129°$$

Problem 4

Perimeter $= 4 \cdot \text{side}$
$= 4 \cdot 4.2 \text{ m} = 16.8 \text{ m}$

The perimeter is 16.8 m.

Problem 5

Circumference $= \pi \cdot \text{diameter}$
$\approx 3.14 \cdot 5 \text{ in.}$
$= 15.7 \text{ in.}$

The circumference is 15.7 in.

Problem 6

Strategy To find the cost of the metal strip:
- ▶ Find the circumference of the table in inches.
- ▶ Convert inches to feet.
- ▶ Multiply the circumference by the per-foot cost of the metal strip.

Solution Circumference $= \pi \cdot \text{diameter}$
$\approx 3.14 \cdot 36 \text{ in.}$
$= 113.04 \text{ in.}$

$\frac{113.04}{12} = 9.42 \text{ ft}$

Cost: $9.42(3.21) = 30.2382$

The cost is $30.24.

Problem 7

Area $= \frac{1}{2} \cdot \text{base} \cdot \text{height}$

$= \frac{1}{2} \cdot 5 \text{ ft} \cdot 3 \text{ ft} = 7.5 \text{ ft}^2$

The area is 7.5 ft^2.

Problem 8

Area $= \pi \cdot (\text{radius})^2$
$\approx 3.14 \cdot (6 \text{ in.})^2$
$= 113.04 \text{ in}^2$

The area is 113.04 in^2.

Problem 9

Area $= \text{base} \cdot \text{height}$
$= 28 \text{ in.} \cdot 15 \text{ in.} = 420 \text{ in}^2$

The area is 420 in^2.

Problem 10

Strategy To find how much more expensive the wool rug is than the nylon rug:
- ▶ Find the area of the rug.
- ▶ Multiply the area of the rug by the per-square-foot cost of the wool rug.
- ▶ Multiply the area of the rug by the per-square-foot cost of the nylon rug.
- ▶ Find the difference between the two costs.

Solution Area $= \text{length} \cdot \text{width}$
$= 15 \text{ ft} \cdot 4 \text{ ft} = 60 \text{ ft}^2$

Cost of wool rug = $5.93 · 60 = $355.80

Cost of nylon rug = $3.25 · 60 = $195

Difference in cost = $355.80 − $195 = $160.80

The difference in cost is $160.80.

Solutions to Chapter 2 Problems

SECTION 2.1

Problem 1

-4

Problem 2

$2xy + y^2$

$2(-4)(2) + (2)^2 = 2(-4)(2) + 4$ • Replace each variable with the number it represents. Then simplify.
$= -8(2) + 4$
$= -16 + 4$
$= -12$

Problem 3

$\frac{a^2 + b^2}{a + b}$

$\frac{(5)^2 + (-3)^2}{5 + (-3)} = \frac{25 + 9}{5 + (-3)}$ • Replace each variable with the number it represents. Then simplify.
$= \frac{34}{2}$
$= 17$

Problem 4

$x^3 - 2(x + y) + z^2$

$(2)^3 - 2[2 + (-4)] + (-3)^2$ • Replace each variable with the number it represents.
$= (2)^3 - 2(-2) + (-3)^2$ • Use the Order of Operations Agreement to simplify.
$= 8 - 2(-2) + 9$
$= 8 + 4 + 9$
$= 12 + 9$
$= 21$

Problem 5

$V = \frac{1}{3}\pi r^2 h$ • **Formula for the volume of a right circular cone**

$V = \frac{1}{3}\pi(4.5)^2(9.5)$ • $r = \frac{1}{2}d = \frac{1}{2}(9) = 4.5$

$V = \frac{1}{3}\pi(20.25)(9.5)$

$V \approx 201.5$ • **Use the π key on your calculator.**

The volume is approximately 201.5 cm^3.

SECTION 2.2

Problem 1

$7 + (-7) = 0$

Problem 2

The Associative Property of Addition

Problem 3

A. $9x + 6x = (9 + 6)x = 15x$ • **Distributive Property**

B. $-4y - 7y = [-4 + (-7)]y = -11y$ • **Distributive Property**

Problem 4

A. $3a - 2b + 5a = 3a + 5a - 2b$
$= (3a + 5a) - 2b$
$= 8a - 2b$

B. $x^2 - 7 + 9x^2 - 14$
$= x^2 + 9x^2 - 7 - 14$
$= (x^2 + 9x^2) + (-7 - 14)$
$= 10x^2 - 21$

Problem 5

A. $-7(-2a) = [-7(-2)]a$ • **Associative Property of Multiplication**
$= 14a$

B. $-\frac{5}{6}(-30y^2) = \left[-\frac{5}{6}(-30)\right]y^2$ • **Associative Property of Multiplication**
$= 25y^2$

C. $(-5x)(-2) = (-2)(-5x)$ • **Commutative Property of Multiplication**
$= [-2(-5)]x$ • **Associative Property of Multiplication**
$= 10x$

Problem 6

A. $7(4 + 2y) = 7(4) + 7(2y)$ • **Distributive Property**
$= 28 + 14y$

B. $-(5x - 12) = -1(5x - 12)$
$= -1(5x) - (-1)(12)$ • **Distributive Property**
$= -5x + 12$

C. $(3a - 1)5 = (3a)(5) - (1)(5)$ • **Distributive Property**
$= 15a - 5$

D. $-3(6a^2 - 8a + 9)$
$= -3(6a^2) - (-3)(8a) + (-3)(9)$ • **Distributive Property**
$= -18a^2 + 24a - 27$

Problem 7

$7(x - 2y) - 3(-x - 2y)$
$= 7x - 14y + 3x + 6y$ • **Distributive Property**
$= 10x - 8y$ • **Combine like terms.**

Problem 8

$3y - 2[x - 4(2 - 3y)]$
$= 3y - 2[x - 8 + 12y]$ • **Distributive Property**
$= 3y - 2x + 16 - 24y$ • **Distributive Property**
$= -2x - 21y + 16$ • **Combine like terms.**

SECTION 2.3

Problem 1

A. eighteen less than the cube of x

$x^3 - 18$

B. y decreased by the sum of z and nine

$y - (z + 9)$

C. the difference between q and the sum of r and t

$q - (r + t)$

Problem 2

the unknown number: n
the square of the number: n^2
the product of five and the square of the number: $5n^2$
$5n^2 + n$

Problem 3

the unknown number: n
twice the number: $2n$
the sum of seven and twice the number: $7 + 2n$
$3(7 + 2n)$

Problem 4

the unknown number: n
twice the number: $2n$
the difference between twice the number and 17:
$2n - 17$
$n - (2n - 17) = n - 2n + 17$
$= -n + 17$

Problem 5

the unknown number: n
three-fourths of the number: $\frac{3}{4}n$
one-fifth of the number: $\frac{1}{5}n$

$\frac{3}{4}n + \frac{1}{5}n = \frac{15}{20}n + \frac{4}{20}n$
$= \frac{19}{20}n$

Problem 6

the time required by the newer model: t
the time required by the older model is twice the time required by the newer model: $2t$

Problem 7

the length of one piece: L
the length of the second piece: $6 - L$

Solutions to Chapter 3 Problems

SECTION 3.1

Problem 1

$5 - 4x$	$= 8x + 2$	
$5 - 4\left(\frac{1}{4}\right)$	$8\left(\frac{1}{4}\right) + 2$	• Replace x by $\frac{1}{4}$.
$5 - 1$	$2 + 2$	• Evaluate.
$4 = 4$		• Compare the results.

Yes, $\frac{1}{4}$ is a solution.

Problem 2

$10x - x^2$	$= 3x - 10$	
$10(5) - (5)^2$	$3(5) - 10$	• Replace x by 5.
$50 - 25$	$15 - 10$	• Evaluate.
$25 \neq 5$		• Compare the results.

No, 5 is not a solution.

Problem 3

$$x - \frac{1}{3} = -\frac{3}{4}$$

$$x - \frac{1}{3} + \frac{1}{3} = -\frac{3}{4} + \frac{1}{3}$$ • Add $\frac{1}{3}$ to each side.

$$x + 0 = -\frac{9}{12} + \frac{4}{12}$$ • Simplify.

$$x = -\frac{5}{12}$$ • The form *variable* = *constant*

Check

$x - \frac{1}{3}$	$= -\frac{3}{4}$	
$-\frac{5}{12} - \frac{1}{3}$	$-\frac{3}{4}$	
$-\frac{3}{4} = -\frac{3}{4}$		• A true equation. The solution checks.

The solution is $-\frac{5}{12}$.

Problem 4

$$-8 = 5 + x$$

$$-8 - 5 = 5 - 5 + x$$ • Subtract 5 from each side.

$$-13 = x$$ • Simplify.

The solution is -13.

Problem 5

$$-\frac{2x}{5} = 6$$

$$\left(-\frac{5}{2}\right)\left(-\frac{2}{5}x\right) = \left(-\frac{5}{2}\right)(6)$$ • Multiply each side by $-\frac{5}{2}$.

$$1x = -15$$ • Simplify.

$$x = -15$$ • The form *variable* = *constant*

The solution is -15.

Problem 6

$$6x = 10$$

$$\frac{6x}{6} = \frac{10}{6}$$

$$x = \frac{5}{3}$$

Check

$6x$	$= 10$
$6\left(\frac{5}{3}\right)$	10
$10 = 10$	

The solution is $\frac{5}{3}$.

Problem 7

$$4x - 8x = 16$$

$$-4x = 16$$ • Combine like terms.

$$\frac{-4x}{-4} = \frac{16}{-4}$$ • Divide each side by -4.

$$x = -4$$

The solution is -4.

SECTION 3.2

Problem 1

$$PB = A$$

$$P(60) = 27$$ • $B = 60, A = 27$

$$60P = 27$$

$$\frac{60P}{60} = \frac{27}{60}$$ • Divide each side by 60.

$$P = 0.45$$ • $0.45 = 45\%$

27 is 45% of 60.

Problem 2

Strategy To find the percent of the questions answered correctly, solve the basic percent equation using $B = 80$ and $A = 72$. The percent is unknown.

Solution

$$PB = A$$

$$P(80) = 72$$ • $B = 80, A = 72$

$$80P = 72$$

$$\frac{80P}{80} = \frac{72}{80}$$ • Divide each side by 80.

$$P = 0.9$$ • $0.9 = 90\%$

90% of the questions were answered correctly.

Problem 3

Strategy To determine the sales tax, solve the basic percent equation using $B = 895$ and $P = 6\% = 0.06$. The amount is unknown.

Solution

$$PB = A$$

$$0.06(895) = A$$ • $P = 0.06, B = 895$

$$53.7 = A$$ • Multiply.

The sales tax is $53.70.

Problem 4

Strategy To determine how much she must deposit into the bank account:

▶ Find the amount of interest earned on the municipal bond by solving the equation $I = Prt$ for I using $P = 1000$, $r = 6.4\% = 0.064$, and $t = 1$.

► Solve $I = Prt$ for P using the amount of interest earned on the municipal bond for I, $r = 8\% = 0.08$, and $t = 1$.

Solution $I = Prt$

$I = 1000(0.064)(1)$ • $P = 1000, r = 0.064, t = 1$

$I = 64$ • Multiply.

The interest earned on the municipal bond is \$64.

$I = Prt$

$64 = P(0.08)(1)$ • $I = 64, r = 0.08, t = 1$

$64 = 0.08P$ • Multiply.

$\frac{64}{0.08} = \frac{0.08P}{0.08}$ • Divide each side by 0.08.

$800 = P$

Clarissa must deposit \$800 into the bank account.

Problem 5

Strategy To find the number of ounces of cereal in the bowl, solve the equation $Q = Ar$ for A using $Q = 2$ and $r = 25\% = 0.25$.

Solution $Q = Ar$

$2 = A(0.25)$ • $Q = 2, r = 0.25$

$\frac{2}{0.25} = \frac{A(0.25)}{0.25}$ • Divide each side by 0.25.

$8 = A$

The cereal bowl contains 8 oz of cereal.

Problem 6

Strategy The distance is 80 mi. Therefore, $d = 80$. The cyclists are moving in opposite directions, so the rate at which the distance between them is changing is the sum of the rates of the two cyclists. The rate is 18 mph + 14 mph = 32 mph. Therefore, $r = 32$. To find the time, solve the equation $d = rt$ for t.

Solution $d = rt$

$80 = 32t$ • $d = 80, r = 32$

$\frac{80}{32} = \frac{32t}{32}$ • Divide each side by 32.

$2.5 = t$

They will meet 2.5 h after they begin.

Problem 7

Strategy Because the plane is flying against the wind, the rate of the plane is its speed in calm air (250 mph) minus the speed of the headwind (25 mph): 250 mph − 25 mph = 225 mph. Therefore, $r = 225$. The time is 3 h, so $t = 3$. To find the distance, solve the equation $d = rt$ for d.

Solution $d = rt$

$d = 225(3)$ • $r = 225, t = 3$

$d = 675$ • Multiply.

The plane can fly 675 mi in 3 h.

SECTION 3.3

Problem 1

$5x + 7 = 10$

$5x + 7 - 7 = 10 - 7$ • Subtract 7 from each side.

$5x = 3$ • Simplify.

$\frac{5x}{5} = \frac{3}{5}$ • Divide each side by 5.

$x = \frac{3}{5}$ • The form *variable* = *constant*

The solution is $\frac{3}{5}$.

Problem 2

$11 = 11 + 3x$

$11 - 11 = 11 - 11 + 3x$ • Subtract 11 from each side.

$0 = 3x$ • Simplify.

$\frac{0}{3} = \frac{3x}{3}$ • Divide each side by 3.

$0 = x$ • Simplify.

The solution is 0.

Problem 3

$\frac{5}{2} - \frac{2}{3}x = \frac{1}{2}$ • Find the LCM of the denominators.

$6\left(\frac{5}{2} - \frac{2}{3}x\right) = 6\left(\frac{1}{2}\right)$ • The LCM is 6. Multiply each side by 6.

$6\left(\frac{5}{2}\right) - 6\left(\frac{2}{3}x\right) = 3$ • Distributive Property

$15 - 4x = 3$ • The equation now has no fractions.

$15 - 15 - 4x = 3 - 15$

$-4x = -12$

$\frac{-4x}{-4} = \frac{-12}{-4}$

$x = 3$

The solution is 3.

Problem 4

$\frac{x}{4} + \frac{3}{2} = \frac{3x}{8}$ • Find the LCM of the denominators.

$8\left(\frac{x}{4} + \frac{3}{2}\right) = 8\left(\frac{3x}{8}\right)$ • The LCM is 8. Multiply each side by 8.

$8\left(\frac{x}{4}\right) + 8\left(\frac{3}{2}\right) = 3x$ • Distributive Property

$2x + 12 = 3x$ • The equation now has no fractions.

$2x - 2x + 12 = 3x - 2x$

$12 = x$

The solution is 12.

Problem 5

$5x + 4 = 6 + 10x$

$5x - 10x + 4 = 6 + 10x - 10x$ • Subtract $10x$ from each side.

$-5x + 4 = 6$ • Simplify.

$-5x + 4 - 4 = 6 - 4$ • Subtract 4 from each side.

$-5x = 2$ • Simplify.

$\frac{-5x}{-5} = \frac{2}{-5}$ • **Divide each side by −5.**

$x = -\frac{2}{5}$ • **The form *variable = constant***

The solution is $-\frac{2}{5}$.

Problem 6

$5x - 4(3 - 2x) = 2(3x - 2) + 6$

$5x - 12 + 8x = 6x - 4 + 6$ • **Distributive Property**

$13x - 12 = 6x + 2$ • **Simplify.**

$13x - 6x - 12 = 6x - 6x + 2$ • **Subtract 6*x* from each side.**

$7x - 12 = 2$

$7x - 12 + 12 = 2 + 12$ • **Add 12 to each side.**

$7x = 14$

$\frac{7x}{7} = \frac{14}{7}$ • **Divide each side by 7.**

$x = 2$

The solution is 2.

Problem 7

$-2[3x - 5(2x - 3)] = 3x - 8$

$-2[3x - 10x + 15] = 3x - 8$ • **Distributive Property**

$-2[-7x + 15] = 3x - 8$ • **Simplify in brackets.**

$14x - 30 = 3x - 8$ • **Distributive Property**

$14x - 3x - 30 = 3x - 3x - 8$ • **Subtract 3*x* from each side.**

$11x - 30 = -8$

$11x - 30 + 30 = -8 + 30$ • **Add 30 to each side.**

$11x = 22$

$\frac{11x}{11} = \frac{22}{11}$ • **Divide each side by 11.**

$x = 2$

The solution is 2.

Problem 8

Strategy To find the number of years, replace V with 10,200 in the given equation and solve for t.

Solution

$V = 450t + 7500$

$10{,}200 = 450t + 7500$

$10{,}200 - 7500 = 450t + 7500 - 7500$

$2700 = 450t$

$\frac{2700}{450} = \frac{450t}{450}$

$6 = t$

The value of the investment will be \$10,200 in 6 years.

Problem 9

Strategy The lever is 14 ft long, so $d = 14$. One force is 6 ft from the fulcrum, so $x = 6$. This force is 40 lb, so $F_1 = 40$. To find the other force when the system balances, replace the variables F_1, x, and d in the lever system equation with the given values, and solve for F_2.

Solution

$F_1x = F_2(d - x)$

$40(6) = F_2(14 - 6)$ • **$F_1 = 40$, $x = 6$, $d = 14$**

$240 = F_2(8)$ • **Simplify.**

$240 = 8F_2$

$\frac{240}{8} = \frac{8F_2}{8}$ • **Divide each side by 8.**

$30 = F_2$

A 30-pound force must be applied to the other end.

SECTION 3.4

Problem 1

The solution set is the numbers greater than −2.

Problem 2

$x + 2 < -2$

$x + 2 - 2 < -2 - 2$ • **Subtract 2 from each side.**

$x < -4$

The solution set is $x < -4$.

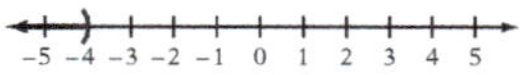

Problem 3

$5x + 3 > 4x + 5$

$5x - 4x + 3 > 4x - 4x + 5$ • **Subtract 4*x* from each side.**

$x + 3 > 5$

$x + 3 - 3 > 5 - 3$ • **Subtract 3 from each side.**

$x > 2$

The solution set is $x > 2$.

Problem 4

$-\frac{3}{4}x \ge 18$

$-\frac{4}{3}\left(-\frac{3}{4}x\right) \le -\frac{4}{3}(18)$ • **Multiply each side by $-\frac{4}{3}$. Reverse the inequality symbol.**

$x \le -24$

The solution set is $x \le -24$.

Problem 5

$3x < 9$

$\frac{3x}{3} < \frac{9}{3}$ • **Divide each side by 3.**

$x < 3$

The solution set is $x < 3$.

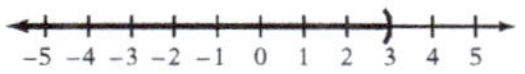

Problem 6

$5 - 4x > 9 - 8x$

$5 - 4x + 8x > 9 - 8x + 8x$ • **Add 8*x* to each side.**

$5 + 4x > 9$

$5 - 5 + 4x > 9 - 5$ • **Subtract 5 from each side.**

$4x > 4$

$\frac{4x}{4} > \frac{4}{4}$ • **Divide each side by 4.**

$x > 1$

The solution set is $x > 1$.

Problem 7

$8 - 4(3x + 5) \le 6(x - 8)$

$8 - 12x - 20 \le 6x - 48$ • **Distributive Property**

$-12 - 12x \le 6x - 48$

$-12 - 12x - 6x \le 6x - 6x - 48$ • **Subtract 6x from each side.**

$-12 - 18x \le -48$

$-12 + 12 - 18x \le -48 + 12$ • **Add 12 to each side.**

$-18x \le -36$

$\frac{-18x}{-18} \ge \frac{-36}{-18}$ • **Divide each side by −18. Reverse the inequality symbol.**

$x \ge 2$

The solution set is $x \ge 2$.

Solutions to Chapter 4 Problems

SECTION 4.1

Problem 1

the unknown number: n

nine less than twice a number	is	five times the sum of the number and twelve

$$2n - 9 = 5(n + 12)$$
$$2n - 9 = 5n + 60$$
$$2n - 5n - 9 = 5n - 5n + 60$$
$$-3n - 9 = 60$$
$$-3n - 9 + 9 = 60 + 9$$
$$-3n = 69$$
$$\frac{-3n}{-3} = \frac{69}{-3}$$
$$n = -23$$

The number is −23.

Problem 2

Strategy ► First consecutive integer: n
Second consecutive integer: $n + 1$
Third consecutive integer: $n + 2$
► The sum of the three integers is −12.

Solution $n + (n + 1) + (n + 2) = -12$
$$3n + 3 = -12$$
$$3n = -15$$
$$n = -5$$

$n + 1 = -5 + 1 = -4$
$n + 2 = -5 + 2 = -3$

The three consecutive integers are −5, −4, and −3.

Problem 3

Strategy To find the number of carbon atoms, write and solve an equation using n to represent the number of carbon atoms.

Solution

eight	represents	twice the number of carbon atoms

$$8 = 2n$$
$$\frac{8}{2} = \frac{2n}{2}$$
$$4 = n$$

There are 4 carbon atoms in a butane gas molecule.

Problem 4

Strategy To find the number of 10-speed bicycles made each day, write and solve an equation using n to represent the number of 10-speed bicycles and $160 - n$ to represent the number of 3-speed bicycles.

Solution

four times the number of 3-speed bicycles made	equals	30 less than the number of 10-speed bicycles made

$$4(160 - n) = n - 30$$
$$640 - 4n = n - 30$$
$$640 - 4n - n = n - n - 30$$
$$640 - 5n = -30$$
$$640 - 640 - 5n = -30 - 640$$
$$-5n = -670$$
$$\frac{-5n}{-5} = \frac{-670}{-5}$$
$$n = 134$$

There are 134 10-speed bicycles made each day.

SECTION 4.2

Problem 1

Strategy ► Each equal side: s
► Use the equation for the perimeter of a square.

Solution $P = 4s$
$52 = 4s$
$13 = s$

The length of each side is 13 ft.

Problem 2

Strategy The angles labeled are adjacent angles of intersecting lines and are therefore supplementary angles. To find x, write an equation and solve for x.

Solution $x + (3x + 20°) = 180°$
$$4x + 20° = 180°$$
$$4x = 160°$$
$$x = 40°$$

Problem 3

Strategy $2x = y$ because alternate exterior angles have the same measure.

$y + (x + 15°) = 180°$ because adjacent angles of intersecting lines are supplementary angles.

Substitute $2x$ for y and solve for x.

Solution $y + (x + 15°) = 180°$
$$2x + (x + 15°) = 180°$$
$$3x + 15° = 180°$$
$$3x = 165°$$
$$x = 55°$$

Problem 4

Strategy
- ► To find the measure of $\angle a$, use the fact that $\angle a$ and $\angle y$ are vertical angles.
- ► To find the measure of $\angle b$, use the fact that the sum of the interior angles of a triangle is 180°.
- ► To find the measure of $\angle d$, use the fact that $\angle b$ and $\angle d$ are supplementary angles.

Solution

$\angle a = \angle y = 55°$ • **$\angle a$ and $\angle y$ are vertical angles.**

$\angle a + \angle b + 90° = 180°$
$55° + \angle b + 90° = 180°$
$145° + \angle b = 180°$
$\angle b = 35°$
• **The sum of the measures of the interior angles of a triangle is 180°.**

$\angle b + \angle d = 180°$
$35° + \angle d = 180°$
$\angle d = 145°$
• **$\angle b$ and $\angle d$ are supplementary angles.**

Problem 5

Strategy To find the measure of the third angle, use the fact that the sum of the measures of the interior angles of a triangle is 180°. Write an equation using x to represent the measure of the third angle. Solve the equation for x.

Solution

$x + 90° + 27° = 180°$
$x + 117° = 180°$
$x = 63°$

The measure of the third angle is 63°.

SECTION 4.3

Problem 1

Strategy Given: $C = \$120$
$S = \$180$
Unknown markup rate: r
Use the equation $S = C + rC$.

Solution

$S = C + rC$
$180 = 120 + 120r$ • **Substitute the values of C and S.**
$60 = 120r$ • **Subtract 120 from each side.**
$0.5 = r$ • **Divide each side by 120.**

The markup rate is 50%. • **Change 0.5 to a percent.**

Problem 2

Strategy Given: $r = 40\% = 0.40$
$S = \$266$
Unknown cost: C
Use the equation $S = C + rC$.

Solution

$S = C + rC$
$266 = C + 0.4C$ • **Substitute the values of S and r.**
$266 = 1.40C$ • **$C + 0.40C = 1C + 0.40C = (1 + 0.40)C$**
$190 = C$

The cost is \$190.

Problem 3

Strategy Given: $R = \$39.80$
$S = \$29.85$
Unknown discount rate: r
Use the equation $S = R - rR$.

Solution

$S = R - rR$
$29.85 = 39.80 - 39.80r$ • **Substitute for S and R.**
$-9.95 = -39.80r$ • **Subtract 39.80 from each side.**
$0.25 = r$ • **Divide each side by −39.80.**

The discount rate is 25%.

Problem 4

Strategy Given: $S = \$43.50$
$r = 25\% = 0.25$
Unknown regular price: R
Use the equation $S = R - rR$.

Solution

$S = R - rR$
$43.50 = R - 0.25R$ • **Substitute for S and r.**
$43.50 = 0.75R$ • **$R - 0.25R = 1R - 0.25R = (1 - 0.25)R$**
$58 = R$

The regular price is \$58.

SECTION 4.4

Problem 1

Strategy ► Amount invested at 4%: x
Amount invested at 6%: $18{,}000 - x$

	Principal	Rate	Interest
Amount at 4%	x	0.04	$0.04x$
Amount at 6%	$18{,}000 - x$	0.06	$0.06(18{,}000 - x)$

► The interest earned on one account is equal to the interest earned on the other account.

Solution

$0.04x = 0.06(18{,}000 - x)$ • **The two accounts earn the same amount of money.**
$0.04x = 1080 - 0.06x$
$0.1x = 1080$
$x = 10{,}800$ • **The amount invested at 4% is \$10,800.**

Substitute the value of x into the variable expression for the amount invested at 6%.

$18{,}000 - x = 18{,}000 - 10{,}800 = 7200$

The amount invested at 4% is \$10,800.
The amount invested at 6% is \$7200.

SECTION 4.5

Problem 1

Strategy ► Pounds of \$.75 fertilizer: x

	Amount	Cost	Value
\$.90 fertilizer	20	0.90	0.90(20)
\$.75 fertilizer	x	0.75	$0.75x$
\$.85 fertilizer	$20 + x$	0.85	$0.85(20 + x)$

► The sum of the values before mixing equals the value after mixing.

Solution

$$\begin{aligned} 0.90(20) + 0.75x &= 0.85(20 + x) \\ 18 + 0.75x &= 17 + 0.85x \\ 18 - 0.10x &= 17 \\ -0.10x &= -1 \\ x &= 10 \end{aligned}$$

• **The value of the \$.90 fertilizer plus the value of the \$.75 fertilizer equals the value of the mixture.**

10 lb of the \$.75 fertilizer must be added.

Problem 2

Strategy ► Pure orange juice is 100% orange juice. 100% = 1.00

Amount of pure orange juice: x

	Amount	Percent	Quantity
Pure orange juice	x	1.00	$1.00x$
Fruit drink	5	0.10	0.10(5)
Orange drink	$x + 5$	0.25	$0.25(x + 5)$

► The sum of the quantities before mixing is equal to the quantity after mixing.

Solution

$$\begin{aligned} 1.00x + 0.10(5) &= 0.25(x + 5) \\ 1.00x + 0.5 &= 0.25x + 1.25 \\ 0.75x + 0.5 &= 1.25 \\ 0.75x &= 0.75 \\ x &= 1 \end{aligned}$$

• **The amount of pure orange juice plus the amount of fruit drink equals the amount of orange drink.**

To make the orange drink, 1 qt of pure orange juice is added to the fruit drink.

SECTION 4.6

Problem 1

Strategy ► Rate of the first train: r

Rate of the second train: $2r$

	Rate	Time	Distance
First train	r	3	$3r$
Second train	$2r$	3	$3(2r)$

► The sum of the distances traveled by each train equals 306 mi.

Solution

$$\begin{aligned} 3r + 3(2r) &= 306 \\ 3r + 6r &= 306 \\ 9r &= 306 \\ r &= 34 \end{aligned}$$

• **The distance traveled by the first train plus the distance traveled by the second train is 306 mi.**

$2r = 2(34) = 68$

The first train is traveling at 34 mph.
The second train is traveling at 68 mph.

Problem 2

Strategy ► Time spent flying out: t

Time spent flying back: $7 - t$

	Rate	Time	Distance
Out	120	t	$120t$
Back	90	$7 - t$	$90(7 - t)$

► The distance out equals the distance back.

Solution

$$\begin{aligned} 120t &= 90(7 - t) \\ 120t &= 630 - 90t \\ 210t &= 630 \\ t &= 3 \end{aligned}$$

• **The time flying out was 3 h. Find the distance.**

Distance $= 120t = 120(3) = 360$ mi

The parcel of land was 360 mi away.

SECTION 4.7

Problem 1

Strategy To find the minimum selling price, write and solve an inequality using S to represent the selling price.

Solution

$$340 < 0.70S$$

$$\frac{340}{0.70} < \frac{0.70S}{0.70}$$

• **Divide each side by 0.70.**

$$485.71429 < S$$

The minimum selling price is \$485.72.

Problem 2

Strategy To find the maximum number of miles:

► Write an expression for the cost of each car, using x to represent the number of miles driven outside the radius during the week.

► Write and solve an inequality.

Solution

cost of a Company A car	is less than	cost of a Company B car

$$\begin{aligned} 40(7) + 0.20x &< 45(7) + 0.15x \\ 280 + 0.20x &< 315 + 0.15x \\ 280 + 0.20x - 0.15x &< 315 + 0.15x - 0.15x \\ 280 + 0.05x &< 315 \\ 280 - 280 + 0.05x &< 315 - 280 \\ 0.05x &< 35 \\ \frac{0.05x}{0.05} &< \frac{35}{0.05} \\ x &< 700 \end{aligned}$$

The maximum number of miles is 699.

Solutions to Chapter 5 Problems

SECTION 5.1

Problem 1

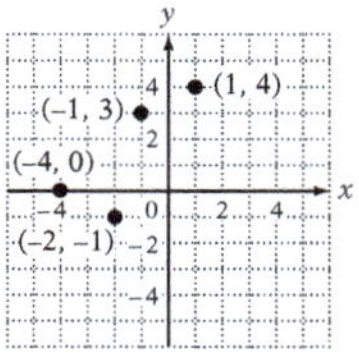

Problem 2

$A(4, 2), B(-3, 4), C(-3, 0), D(0, 0)$

Problem 3

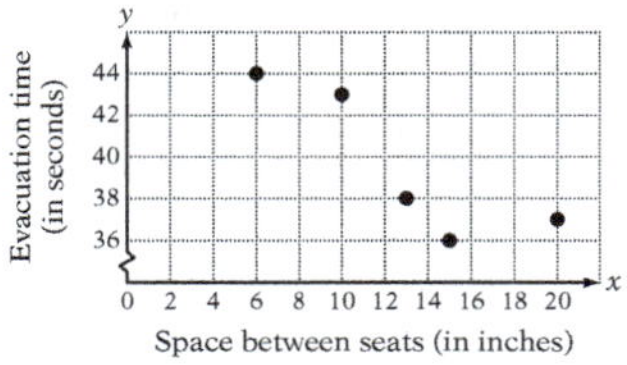

Problem 4

In 1900, the population was 76 million: (1900, 76)
In 2000, the population was 281 million: (2000, 281)
Average rate of change

$$= \frac{\text{change in } y}{\text{change in } x}$$
$$= \frac{\text{population in 2000} - \text{population in 1900}}{2000 - 1900}$$
$$= \frac{281 - 76}{2000 - 1900}$$
$$= \frac{205}{100} = 2.05$$

The average rate of change in the population was 2.05 million people per year.

SECTION 5.2

Problem 1

$$y = -\frac{1}{2}x - 3$$

-4	$-\frac{1}{2}(2) - 3$
	$-1 - 3$
	-4

$-4 = -4$

• **Replace x with 2 and y with −4.**

Yes, $(2, -4)$ is a solution of $y = -\frac{1}{2}x - 3$.

Problem 2

$$y = -\frac{1}{4}x + 1$$
$$= -\frac{1}{4}(4) + 1$$ • **Substitute 4 for x.**
$$= -1 + 1$$ • **Solve for y.**
$$= 0$$ • **When $x = 4$, $y = 0$.**

The ordered-pair solution is $(4, 0)$.

Problem 3

x	$y = 3x + 1$
0	1
1	4
−1	−2

• **Choose three values for x. Find the corresponding values of y.**

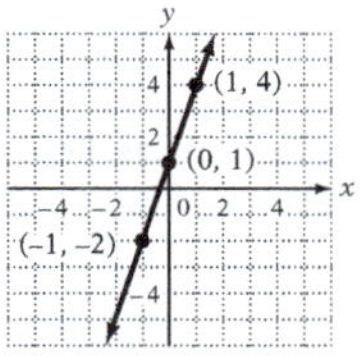

• **Graph the ordered-pair solutions. Draw a straight line through the points.**

Problem 4

x	$y = \frac{1}{3}x - 3$
0	−3
3	−2
−3	−4

• **The value of m is $\frac{1}{3}$. Choose three values for x that are multiples of the denominator 3. Find the corresponding values of y.**

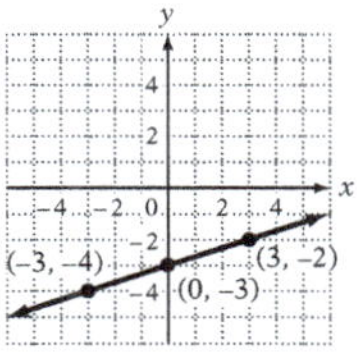

• **Graph the ordered-pair solutions. Draw a straight line through the points.**

Problem 5

$$5x - 2y = 10$$
$$5x - 5x - 2y = -5x + 10$$ • **Subtract $5x$ from each side.**
$$-2y = -5x + 10$$
$$\frac{-2y}{-2} = \frac{-5x + 10}{-2}$$ • **Divide each side by −2.**
$$y = \frac{-5x}{-2} + \frac{10}{-2}$$
$$y = \frac{5}{2}x - 5$$

Problem 6

A. $5x - 2y = 10$ • **Solve the equation for y.**
$$-2y = -5x + 10$$
$$y = \frac{5}{2}x - 5$$

x	$y = \frac{5}{2}x - 5$
0	−5
2	0
4	5

• **The value of m is $\frac{5}{2}$. Choose three values for x that are multiples of the denominator 2. Find the corresponding values of y.**

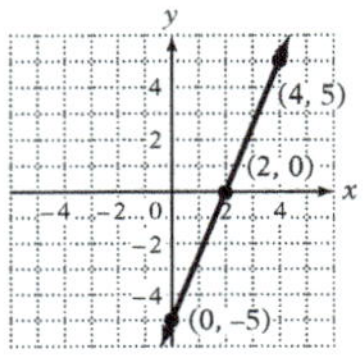

• Graph the ordered-pair solutions. Draw a straight line through the points.

B. $x - 3y = 9$ • Solve the equation for y.

$$-3y = -x + 9$$
$$y = \frac{1}{3}x - 3$$

x	$y = \frac{1}{3}x - 3$
0	−3
3	−2
−3	−4

• The value of m is $\frac{1}{3}$. Choose three values for x that are multiples of the denominator 3. Find the corresponding values of y.

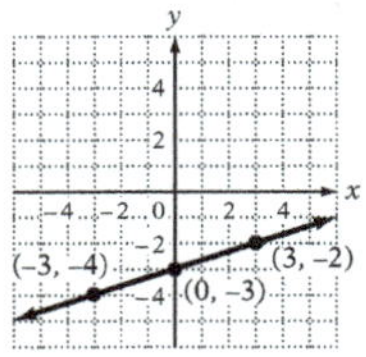

• Graph the ordered-pair solutions. Draw a straight line through the points.

Problem 7

x-intercept: $4x - y = 4$

$$4x - 0 = 4$$ • To find the x-intercept, let $y = 0$.
$$4x = 4$$
$$x = 1$$

The x-intercept is (1, 0).

y-intercept: $4x - y = 4$

$$4(0) - y = 4$$ • To find the y-intercept, let $x = 0$.
$$-y = 4$$
$$y = -4$$

The y-intercept is (0, −4).

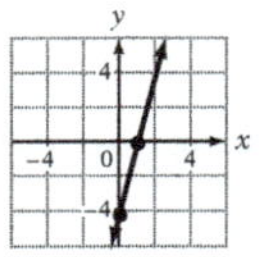

• Graph the ordered pairs (1, 0) and (0, −4). Draw a line through the points.

Problem 8

A. The graph of the equation $y = 3$ is a horizontal line through (0, 3).

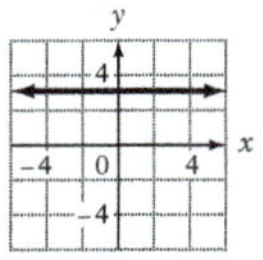

B. The graph of the equation $x = -4$ is a vertical line through (−4, 0).

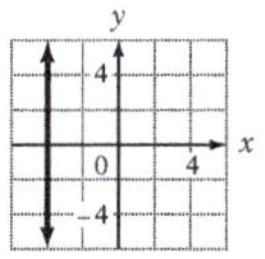

SECTION 5.3

Problem 1

A. $m = \dfrac{y_2 - y_1}{x_2 - x_1} = \dfrac{3 - 2}{1 - (-1)} = \dfrac{1}{2}$

The slope is $\frac{1}{2}$.

B. $m = \dfrac{y_2 - y_1}{x_2 - x_1} = \dfrac{-5 - 2}{4 - 1} = \dfrac{-7}{3}$

The slope is $-\frac{7}{3}$.

C. $m = \dfrac{y_2 - y_1}{x_2 - x_1} = \dfrac{7 - 3}{2 - 2} = \dfrac{4}{0}$

The slope is undefined.

D. $m = \dfrac{y_2 - y_1}{x_2 - x_1} = \dfrac{-3 - (-3)}{-5 - 1} = \dfrac{0}{-6} = 0$

The slope is 0.

Problem 2

Find the slope of the line through (−2, −3) and (7, 1).

$$m = \frac{y_2 - y_1}{x_2 - x_1} = \frac{1 - (-3)}{7 - (-2)} = \frac{4}{9}$$

Find the slope of the line through (1, 4) and (−5, 6).

$$m = \frac{y_2 - y_1}{x_2 - x_1} = \frac{6 - 4}{-5 - 1} = \frac{2}{-6} = -\frac{1}{3}$$

$$\frac{4}{9} \neq -\frac{1}{3}$$

The slopes are not equal.
The lines are not parallel.

Problem 3

Find the slope of the line through (4, −5) and (−2, 3).

$$m = \frac{y_2 - y_1}{x_2 - x_1} = \frac{3 - (-5)}{-2 - 4} = \frac{8}{-6} = -\frac{4}{3}$$

Find the slope of the line through (−1, 6) and (−5, 3).

$$m = \frac{y_2 - y_1}{x_2 - x_1} = \frac{3 - 6}{-5 - (-1)} = \frac{-3}{-4} = \frac{3}{4}$$

The product of the slopes is $\left(-\frac{4}{3}\right)\left(\frac{3}{4}\right) = -1$.

The lines are perpendicular.

Problem 4

$$m = \frac{y_2 - y_1}{x_2 - x_1} = \frac{8650 - 6100}{1 - 4} = \frac{2550}{-3} = -850$$

A slope of −850 means that the value of the car is decreasing at the rate of $850 per year.

Problem 5

y-intercept $= (0, b) = (0, -1)$

$$m = -\frac{1}{4} = \frac{-1}{4}$$

A slope of $\frac{-1}{4}$ means to move down 1 unit (change in y) and then right 4 units (change in x).

Beginning at the y-intercept (0, −1), move down 1 unit and then right 4 units.

The point (4, −2) is a second point on the graph. Draw a line through (0, −1) and (4, −2).

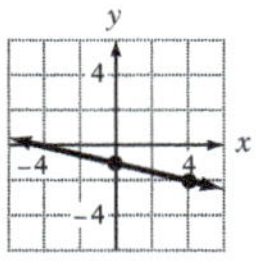

Problem 6

Solve the equation for y.

$$x - 2y = 4$$
$$-2y = -x + 4$$
$$y = \frac{1}{2}x - 2$$

y-intercept $= (0, b) = (0, -2)$; $m = \frac{1}{2}$

Beginning at the y-intercept $(0, -2)$, move up 1 unit (change in y) and then right 2 units (change in x).

The point $(2, -1)$ is a second point on the graph. Draw a line through $(0, -2)$ and $(2, -1)$.

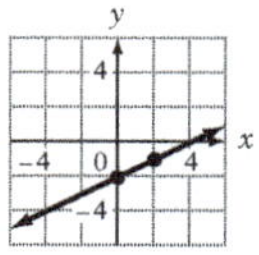

SECTION 5.4

Problem 1

$y = mx + b$

The given slope, $\frac{4}{3}$, is m. Replace m with $\frac{4}{3}$.

$$y = \frac{4}{3}x + b$$

The given point, $(0, -1)$, is the y-intercept. Replace b with -1.

$$y = \frac{4}{3}x - 1$$

The equation of the line is $y = \frac{4}{3}x - 1$.

Problem 2

$y = \frac{3}{2}x + b$ • **Replace *m* with the given slope.**

$-2 = \frac{3}{2}(4) + b$ • **Replace *x* and *y* with the coordinates of the given point.**

$-2 = 6 + b$

$-8 = b$ • **Solve for *b*.**

$y = \frac{3}{2}x - 8$ • **Replace *b* by its value.**

Problem 3

$m = \frac{2}{5} \qquad (x_1, y_1) = (5, 4)$

$y - y_1 = m(x - x_1)$ • **Point-slope formula**

$y - 4 = \frac{2}{5}(x - 5)$ • **Substitute 5 for x_1, 4 for y_1, and $\frac{2}{5}$ for *m*.**

$y - 4 = \frac{2}{5}x - 2$

$y = \frac{2}{5}x + 2$ • **Rewrite in the form $y = mx + b$.**

Problem 4

Let $(x_1, y_1) = (-3, -9)$ and $(x_2, y_2) = (1, -1)$.

$$m = \frac{y_2 - y_1}{x_2 - x_1} = \frac{-1 - (-9)}{1 - (-3)} = \frac{8}{4} = 2$$

$y - y_1 = m(x - x_1)$ • **Point-slope formula**

$y - (-9) = 2(x - (-3))$ • **$m = 2$; $(x_1, y_1) = (-3, -9)$**

$y + 9 = 2(x + 3)$

$y + 9 = 2x + 6$ • **Distributive Property**

$y = 2x - 3$ • **Slope-intercept form**

The equation of the line is $y = 2x - 3$.

SECTION 5.5

Problem 1

The domain is $\{1\}$.
The range is $\{0, 1, 2, 3, 4\}$.
There are ordered pairs with the same first coordinate and different second coordinates. The relation is not a function.

Problem 2

$f(x) = -5x + 1$ • **Write the function.**

$f(2) = -5(2) + 1$ • **Replace *x* by 2 and evaluate.**

$f(2) = -10 + 1$

$f(2) = -9$ • **$f(2)$ is the number that is paired with 2.**

The ordered pair $(2, -9)$ is an element of the function.

Problem 3

$f(x) = 4x - 3$ • **Write the function.**

$f(-5) = 4(-5) - 3 = -20 - 3 = -23$ • **Replace *x* by each member of the domain.**

$f(-3) = 4(-3) - 3 = -12 - 3 = -15$

$f(-1) = 4(-1) - 3 = -4 - 3 = -7$

$f(1) = 4(1) - 3 = 4 - 3 = 1$

The range is $\{-23, -15, -7, 1\}$.
The ordered pairs $(-5, -23)$, $(-3, -15)$, $(-1, -7)$, and $(1, 1)$ belong to the function.

Problem 4

$$f(x) = -\frac{1}{2}x - 3$$

$y = -\frac{1}{2}x - 3$ • **Think of the function as the equation $y = -\frac{1}{2}x - 3$.**

The graph is a straight line with y-intercept $(0, -3)$ and slope $-\frac{1}{2}$.

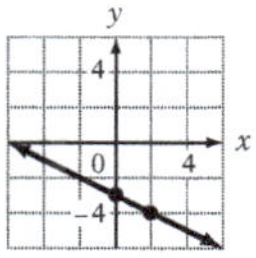

Problem 5

A. The distance d the car travels depends on the amount of time t it has been traveling. Distance d is a function of time t.

$$d = 40t$$
$$f(t) = 40t$$

B.

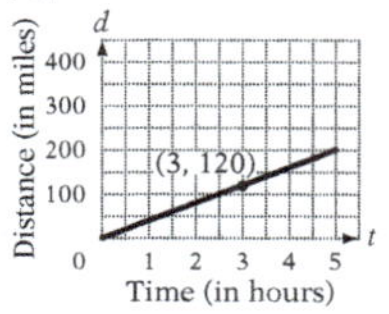

• **Some ordered pairs of the function are (1, 40), (2, 80), and (3, 120).**

C. The ordered pair $(3, 120)$ means that in 3 h, the car travels a distance of 120 mi.

SECTION 5.6

Problem 1

$$
\begin{aligned}
x - 3y &< 2 \\
x - x - 3y &< -x + 2 \\
-3y &< -x + 2 \\
\frac{-3y}{-3} &> \frac{-x + 2}{-3} \\
y &> \frac{1}{3}x - \frac{2}{3}
\end{aligned}
$$

• Solve the inequality for y.

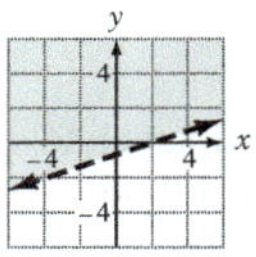

• Graph $y = \frac{1}{3}x - \frac{2}{3}$ as a dashed line. Shade the upper half-plane.

Problem 2

$x < 3$

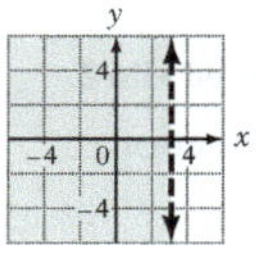

• Graph $x = 3$ as a dashed line. Shade to the left of the line.

Solutions to Chapter 6 Problems

SECTION 6.1

Problem 1

Substitute -1 for x and -2 for y in each equation.

$2x - 5y = 8$		$-x + 3y = -5$	
$2(-1) - 5(-2)$	8	$-(-1) + 3(-2)$	-5
$-2 + 10$	8	$1 + (-6)$	-5
$8 = 8$		$-5 = -5$	

Yes, $(-1, -2)$ is a solution of the system of equations.

Problem 2

A. Graph each line. Find the point of intersection.

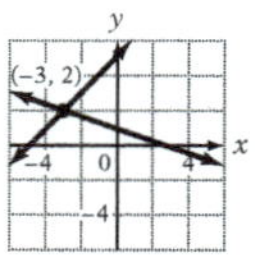

The solution is $(-3, 2)$.

B. Graph each line.

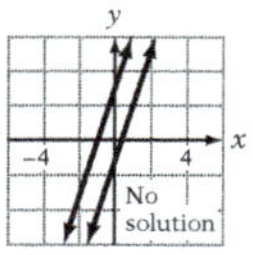

The lines are parallel and therefore do not intersect. The system of equations is inconsistent and has no solution.

SECTION 6.2

Problem 1

(1) $7x - y = 4$

(2) $3x + 2y = 9$

$$
\begin{aligned}
7x - y &= 4 \\
-y &= -7x + 4 \\
y &= 7x - 4
\end{aligned}
$$

• Solve equation (1) for y.

$$
\begin{aligned}
3x + 2y &= 9 \\
3x + 2(7x - 4) &= 9 \\
3x + 14x - 8 &= 9 \\
17x - 8 &= 9 \\
17x &= 17 \\
x &= 1
\end{aligned}
$$

• Substitute $7x - 4$ for y in equation (2).
• Solve for x.

$$
\begin{aligned}
7x - y &= 4 \\
7(1) - y &= 4 \\
7 - y &= 4 \\
-y &= -3 \\
y &= 3
\end{aligned}
$$

• Substitute the value of x in equation (1).
• Solve for y.

The solution is $(1, 3)$.

Problem 2

(1) $3x - y = 4$

(2) $y = 3x + 2$

$$
\begin{aligned}
3x - y &= 4 \\
3x - (3x + 2) &= 4 \\
3x - 3x - 2 &= 4 \\
-2 &= 4
\end{aligned}
$$

• Substitute $3x + 2$ for y in equation (1).
• A false equation

The system of equations is inconsistent and has no solution.

Problem 3

(1) $y = -2x + 1$

(2) $6x + 3y = 3$

$$
\begin{aligned}
6x + 3y &= 3 \\
6x + 3(-2x + 1) &= 3 \\
6x - 6x + 3 &= 3 \\
3 &= 3
\end{aligned}
$$

• Substitute $-2x + 1$ for y in equation (2).
• A true equation

The system of equations is dependent. The solutions are the ordered pairs that satisfy the equation $y = -2x + 1$.

SECTION 6.3

Problem 1

(1) $x - 2y = 1$

(2) $2x + 4y = 0$

$$
\begin{aligned}
2(x - 2y) &= 2 \cdot 1 \\
2x + 4y &= 0
\end{aligned}
$$

• Eliminate y. Multiply each side of equation (1) by 2.

$$
\begin{aligned}
2x - 4y &= 2 \\
2x + 4y &= 0 \\
\hline
4x &= 2 \\
x &= \frac{2}{4} = \frac{1}{2}
\end{aligned}
$$

• Add the equations.
• Solve for x.

$$
\begin{aligned}
2x + 4y &= 0 \\
2\left(\frac{1}{2}\right) + 4y &= 0 \\
1 + 4y &= 0 \\
4y &= -1 \\
y &= -\frac{1}{4}
\end{aligned}
$$

• Replace x in equation (2).
• Solve for y.

The solution is $\left(\frac{1}{2}, -\frac{1}{4}\right)$.

Problem 2

(1) $4x = y - 6$
(2) $2x + 5y = 8$

$4x - y = -6$
$2x + 5y = 8$
• Write equation (1) in the form $Ax + By = C$.

$4x - y = -6$
$-2(2x + 5y) = -2(8)$
• Eliminate x. Multiply each side of equation (2) by -2.

$4x - y = -6$
$-4x - 10y = -16$

$-11y = -22$ • Add the equations.
$y = 2$ • Solve for y.

$4x = 2 - 6$ • Replace y with 2 in equation (1).
$4x = -4$
$x = -1$

The solution is $(-1, 2)$.

Problem 3

(1) $2x - 3y = 4$
(2) $-4x + 6y = -8$

$2(2x - 3y) = 2 \cdot 4$
$-4x + 6y = -8$
• Eliminate y. Multiply each side of equation (1) by 2.

$4x - 6y = 8$
$-4x + 6y = -8$

$0 + 0 = 0$ • Add the equations.
$0 = 0$ • A true equation

The system of equations is dependent. The solutions are the ordered pairs that satisfy the equation $2x - 3y = 4$.

SECTION 6.4

Problem 1

Strategy ► Rate of the current: c
Rate of the canoeist in calm water: r

	Rate	Time	Distance
With current	$r + c$	3	$3(r + c)$
Against current	$r - c$	4	$4(r - c)$

► The distance traveled with the current is 24 mi.
The distance traveled against the current is 24 mi.

Solution

$3(r + c) = 24 \rightarrow \frac{3(r + c)}{3} = \frac{24}{3} \rightarrow r + c = 8$

$4(r - c) = 24 \rightarrow \frac{4(r - c)}{4} = \frac{24}{4} \rightarrow r - c = 6$

$2r = 14$
$r = 7$

$r + c = 8$
$7 + c = 8$
$c = 1$

The rate of the current is 1 mph.
The rate of the canoeist in calm water is 7 mph.

Problem 2

Strategy ► Cost for a black-and-white copy: x
Cost for a color copy: y

Tuesday:

	Amount	Unit cost	Value
Black-and-white	85	x	$85x$
Color	25	y	$25y$

Wednesday:

	Amount	Unit cost	Value
Black-and-white	75	x	$75x$
Color	5	y	$5y$

► The total of your trip on Tuesday was \$14.20. The total of your colleague's trip on Wednesday was \$6.90.

Solution

(1) $85x + 25y = 14.20$
(2) $75x + 5y = 6.90$

$85x + 25y = 14.20$
$-5(75x + 5y) = -5(6.90)$
• Eliminate y. Multiply each side of equation (2) by -5.

$85x + 25y = 14.20$
$-375x - 25y = -34.50$

$-290x = -20.30$ • Add the equations.
$x = 0.07$ • Solve for x.

The cost of a black-and-white copy is \$.07.

Solutions to Chapter 7 Problems

SECTION 7.1

Problem 1

$2x^2 + 4x - 3$ • Arrange like terms in the same column.
$5x^2 - 6x$

$7x^2 - 2x - 3$ • Combine the terms in each column.

Problem 2

$(-4x^2 - 3xy + 2y^2) + (3x^2 - 4y^2)$
$= (-4x^2 + 3x^2) - 3xy + (2y^2 - 4y^2)$ • Group like terms.
$= -x^2 - 3xy - 2y^2$ • Combine like terms.

Problem 3

The opposite of $2y^2 - xy + 5x^2$ is $-2y^2 + xy - 5x^2$.

$8y^2 - 4xy + x^2$ • Write like terms in the same column.
$-2y^2 + xy - 5x^2$

$6y^2 - 3xy - 4x^2$ • Combine like terms.

Problem 4

$(-3a^2 - 4a + 2) - (5a^3 + 2a - 6)$
$= (-3a^2 - 4a + 2) + (-5a^3 - 2a + 6)$ • Rewrite subtraction as addition of the opposite.
$= -5a^3 - 3a^2 + (-4a - 2a) + (2 + 6)$
$= -5a^3 - 3a^2 - 6a + 8$

SECTION 7.2

Problem 1

$(3x^2)(6x^3) = (3 \cdot 6)(x^2 \cdot x^3) = 18x^5$

Problem 2

$(-3xy^2)(-4x^2y^3)$
$= [(-3)(-4)](x \cdot x^2)(y^2 \cdot y^3)$ • Rearrange and group factors.
$= 12x^3y^5$

Problem 3

$(3x)(2x^2y)^3 = (3x)(2^3x^6y^3) = (3x)(8x^6y^3)$
$= (3 \cdot 8)(x \cdot x^6)y^3 = 24x^7y^3$

SECTION 7.3

Problem 1

A. $5x(3x^2 - 2x + 4)$
$= 5x(3x^2) - 5x(2x) + 5x(4)$ • Distributive Property
$= 15x^3 - 10x^2 + 20x$ • Rule for Multiplying Exponential Expressions

B. $(-2y + 3)(-4y)$
$= (-2y)(-4y) + 3(-4y)$ • Distributive Property
$= 8y^2 - 12y$ • Rule for Multiplying Exponential Expressions

C. $-a^2(3a^2 + 2a - 7)$
$= -3a^4 - 2a^3 + 7a^2$

Problem 2

$$\begin{array}{r} 2y^3 + 2y^2 - 3 \\ 3y \; - 1 \\ \hline -2y^3 - 2y^2 + 3 \\ 6y^4 + 6y^3 - 9y \\ \hline 6y^4 + 4y^3 - 2y^2 - 9y + 3 \end{array}$$

Problem 3

$$\begin{array}{r} 3x^3 - 2x^2 + x - 3 \\ 2x + 5 \\ \hline 15x^3 - 10x^2 + 5x - 15 \\ 6x^4 - 4x^3 + 2x^2 - 6x \\ \hline 6x^4 + 11x^3 - 8x^2 - x - 15 \end{array}$$

Problem 4

$(4y - 5)(3y - 3)$
$= 12y^2 - 12y - 15y + 15$ • Use FOIL.
$= 12y^2 - 27y + 15$ • Combine like terms.

Problem 5

$(3a + 2b)(3a - 5b)$
$= 9a^2 - 15ab + 6ab - 10b^2$ • Use FOIL.
$= 9a^2 - 9ab - 10b^2$ • Combine like terms.

Problem 6

$(2a + 5c)(2a - 5c)$ • Product of the sum and difference of two terms
$= (2a)^2 - (5c)^2$
$= 4a^2 - 25c^2$

Problem 7

$(3x + 2y)^2$ • Square of a binomial
$= (3x)^2 + 2(3x)(2y) + (2y)^2$
$= 9x^2 + 12xy + 4y^2$

Problem 8

$(6x - y)^2$ • Square of a binomial
$= (6x)^2 - 2(6x)(y) + (y)^2$
$= 36x^2 - 12xy + y^2$

Problem 9

Strategy To find the area in terms of x, replace the variables L and W in the equation $A = LW$ with the given values. Simplify the expression on the right side of the equation.

Solution
$A = LW$
$A = (x + 7)(x - 4)$
$A = x^2 - 4x + 7x - 28$ • Use FOIL.
$A = x^2 + 3x - 28$ • Combine like terms.

The area is $(x^2 + 3x - 28)$ m^2.

Problem 10

Strategy To find the area of the triangle in terms of x, replace the variables b and h in the equation $A = \frac{1}{2}bh$ with the given values, and simplify.

Solution
$A = \frac{1}{2}bh$
$A = \frac{1}{2}(x + 3)(4x - 6)$
$A = \frac{1}{2}(4x^2 + 6x - 18)$ • Use FOIL. Combine like terms.
$A = 2x^2 + 3x - 9$ • Distributive Property

The area is $(2x^2 + 3x - 9)$ cm^2.

SECTION 7.4

Problem 1

$\frac{2^{-2}}{2^3} = 2^{-2-3} = 2^{-5} = \frac{1}{2^5} = \frac{1}{32}$

Problem 2

A. $x^5y^{-7} = \frac{x^5}{y^7}$ • Rewrite y^{-7} with a positive exponent.

B. $\frac{b^8}{a^{-5}b^6} = a^5b^{8-6}$ • Divide variables with the same base by subtracting exponents.
$= a^5b^2$

C. $4c^{-3} = 4 \cdot \frac{1}{c^3}$ • Definition of Negative Exponents
$= \frac{4}{c^3}$

Problem 3

A. $\dfrac{12x^{-8}y^4}{-16xy^{-3}} = -\dfrac{3x^{-9}y^7}{4} = -\dfrac{3y^7}{4x^9}$

B. $(-3ab)(2a^3b^{-2})^{-3} = (-3ab)(2^{-3}a^{-9}b^6)$

$= -\dfrac{3a^{-8}b^7}{2^3} = -\dfrac{3b^7}{8a^8}$

Problem 4

A. $57{,}000{,}000{,}000 = 5.7 \times 10^{10}$

B. $0.000000017 = 1.7 \times 10^{-8}$

Problem 5

A. $5 \times 10^{12} = 5{,}000{,}000{,}000{,}000$

B. $4.0162 \times 10^{-9} = 0.0000000040162$

Problem 6

A. Multiply 2.4 and 1.6. Add the exponents on 10.

$(2.4 \times 10^{-9})(1.6 \times 10^3) = 3.84 \times 10^{-6}$

B. Divide 5.4 by 1.8. Subtract the exponents on 10.

$\dfrac{5.4 \times 10^{-2}}{1.8 \times 10^{-4}} = 3 \times 10^2$

SECTION 7.5

Problem 1

$\dfrac{4x^3y + 8x^2y^2 - 4xy^3}{2xy} = \dfrac{4x^3y}{2xy} + \dfrac{8x^2y^2}{2xy} - \dfrac{4xy^3}{2xy}$

$= 2x^2 + 4xy - 2y^2$

Problem 2

$\dfrac{24x^2y^2 - 18xy + 6y}{6xy} = \dfrac{24x^2y^2}{6xy} - \dfrac{18xy}{6xy} + \dfrac{6y}{6xy}$

$= 4xy - 3 + \dfrac{1}{x}$

Problem 3

$$\begin{array}{r} x^2 + 2x + 2 \\ x - 2\overline{)x^3 + 0x^2 - 2x - 4} \\ \underline{x^3 - 2x^2} \\ 2x^2 - 2x \\ \underline{2x^2 - 4x} \\ 2x - 4 \\ \underline{2x - 4} \\ 0 \end{array}$$

- **There is no x^2 term. Insert $0x^2$ for the missing term.**

$(x^3 - 2x - 4) \div (x - 2) = x^2 + 2x + 2$

Solutions to Chapter 8 Problems

SECTION 8.1

Problem 1

$4x^6y = 2 \cdot 2 \cdot x^6 \cdot y$ • **Factor each monomial.**

$18x^2y^6 = 2 \cdot 3 \cdot 3 \cdot x^2 \cdot y^6$

$\text{GCF} = 2 \cdot x^2 \cdot y = 2x^2y$ • **The common variable factors are x^2 and y.**

The GCF of $4x^6y$ and $18x^2y^6$ is $2x^2y$.

Problem 2

A. $14a^2 = 2 \cdot 7 \cdot a^2$ • **Find the GCF of the terms.**

$21a^4b = 3 \cdot 7 \cdot a^4 \cdot b$

The GCF is $7a^2$.

$\dfrac{14a^2}{7a^2} = 2, \dfrac{-21a^4b}{7a^2} = -3a^2b$ • **Divide each term by the GCF.**

$14a^2 - 21a^4b = 7a^2(2) + 7a^2(-3a^2b)$ • **Write each term as a product.**

$= 7a^2(2 - 3a^2b)$

B. $6x^4y^2 = 2 \cdot 3 \cdot x^4 \cdot y^2$ • **Find the GCF of the terms.**

$9x^3y^2 = 3 \cdot 3 \cdot x^3 \cdot y^2$

$12x^2y^4 = 2 \cdot 2 \cdot 3 \cdot x^2 \cdot y^4$

The GCF is $3x^2y^2$.

$\dfrac{6x^4y^2}{3x^2y^2} = 2x^2, \dfrac{-9x^3y^2}{3x^2y^2} = -3x, \dfrac{12x^2y^4}{3x^2y^2} = 4y^2$ • **Divide each term by the GCF.**

$6x^4y^2 - 9x^3y^2 + 12x^2y^4$

$= 3x^2y^2(2x^2) + 3x^2y^2(-3x) + 3x^2y^2(4y^2)$ • **Write each term as a product.**

$= 3x^2y^2(2x^2 - 3x + 4y^2)$

Problem 3

$a(b - 7) + b(b - 7) = (b - 7)(a + b)$ • **The common binomial factor is $b - 7$.**

Problem 4

$3y(5x - 2) - 4(2 - 5x)$

$= 3y(5x - 2) + 4(5x - 2)$ • **Rewrite $2 - 5x$ as $-(5x - 2)$.**

$= (5x - 2)(3y + 4)$

Problem 5

A. $y^5 - 5y^3 + 4y^2 - 20$

$= (y^5 - 5y^3) + (4y^2 - 20)$ • **Group the terms.**

$= y^3(y^2 - 5) + 4(y^2 - 5)$ • **Factor the GCF from each group.**

$= (y^2 - 5)(y^3 + 4)$ • **Factor out the common binomial factor.**

B. $2y^3 - 2y^2 - 3y + 3$

$= (2y^3 - 2y^2) - (3y - 3)$ • **Group the terms.**

$= 2y^2(y - 1) - 3(y - 1)$ • **Factor the GCF from each group.**

$= (y - 1)(2y^2 - 3)$ • **Factor out the common binomial factor.**

SECTION 8.2

Problem 1

Factors of 15	Sum
−1, −15	−16
−3, −5	**−8**

• **The factors must both be negative.**

$x^2 - 8x + 15 = (x - 3)(x - 5)$ • **Write the factors.**

Problem 2

Factors of −18	Sum
1, −18	−17
−1, 18	17
2, −9	−7
−2, 9	7
3, −6	−3
−3, 6	**3**

• **The factors must be of opposite signs.**

$x^2 + 3x - 18 = (x + 6)(x - 3)$ • **Write the factors.**

Problem 3

The GCF is $3b$. • **Find the GCF.**

$3a^2b - 18ab - 81b = 3b(a^2 - 6a - 27)$ • **Factor out the GCF.**

Factor the trinomial $a^2 - 6a - 27$.

Factors of -27	Sum
$1, -27$	-26
$-1, 27$	26
$\mathbf{3, -9}$	$\mathbf{-6}$
$-3, 9$	6

• **The factors must have opposite signs.**

$3a^2b - 18ab - 81b = 3b(a + 3)(a - 9)$

Problem 4

The GCF is 4. • **Find the GCF.**

$4x^2 - 40xy + 84y^2$

$= 4(x^2 - 10xy + 21y^2)$ • **Factor out the GCF.**

Factor the trinomial $x^2 - 10xy + 21y^2$.

Factors of 21	Sum
$-1, -21$	-22
$\mathbf{-3, -7}$	$\mathbf{-10}$

• **Both factors must be negative.**

$4x^2 - 40xy + 84y^2 = 4(x - 3y)(x - 7y)$

SECTION 8.3

Problem 1

Factors of 6	Factors of 5
1, 6 2, 3	$-1, -5$

Trial Factors	Middle Term
$(x - 1)(6x - 5)$	$-5x - 6x = \mathbf{-11x}$

$6x^2 - 11x + 5 = (x - 1)(6x - 5)$

Problem 2

Factors of 8	Factors of -15
1, 8 2, 4	$1, -15$ $-1, 15$ $3, -5$ $-3, 5$

Trial Factors	Middle Term
$(x + 1)(8x - 15)$	$-15x + 8x = -7x$
$(x - 1)(8x + 15)$	$15x - 8x = 7x$
$(x + 3)(8x - 5)$	$-5x + 24x = 19x$
$(x - 3)(8x + 5)$	$5x - 24x = -19x$
$(2x + 1)(4x - 15)$	$-30x + 4x = -26x$
$(2x - 1)(4x + 15)$	$30x - 4x = 26x$
$(2x + 3)(4x - 5)$	$-10x + 12x = 2x$
$(2x - 3)(4x + 5)$	$10x - 12x = -2x$
$(8x + 1)(x - 15)$	$-120x + x = -119x$
$(8x - 1)(x + 15)$	$120x - x = 119x$
$(8x + 3)(x - 5)$	$-40x + 3x = -37x$
$(8x - 3)(x + 5)$	$40x - 3x = 37x$
$(4x + 1)(2x - 15)$	$-60x + 2x = -58x$
$(4x - 1)(2x + 15)$	$60x - 2x = 58x$
$(4x + 3)(2x - 5)$	$-20x + 6x = -14x$
$(4x - 3)(2x + 5)$	$20x - 6x = \mathbf{14x}$

$8x^2 + 14x - 15 = (4x - 3)(2x + 5)$

Problem 3

Factors of 24	Factors of -1
1, 24 2, 12 3, 8 4, 6	$1, -1$

Trial Factors	Middle Term
$(1 - y)(24 + y)$	$y - 24y = -23y$
$(2 - y)(12 + y)$	$2y - 12y = -10y$
$(3 - y)(8 + y)$	$3y - 8y = -5y$
$(4 - y)(6 + y)$	$4y - 6y = \mathbf{-2y}$

$24 - 2y - y^2 = (4 - y)(6 + y)$

Problem 4

The GCF is $2a^2$. • **Find the GCF.**

$4a^2b^2 - 30a^2b + 14a^2$

$= 2a^2(2b^2 - 15b + 7)$ • **Factor out the GCF.**

Factor the trinomial $2b^2 - 15b + 7$.

Factors of 2	Factors of 7
1, 2	$-1, -7$

Trial Factors	Middle Term
$(b - 1)(2b - 7)$	$-7b - 2b = -9b$
$(b - 7)(2b - 1)$	$-b - 14b = \mathbf{-15b}$

$4a^2b^2 - 30a^2b + 14a^2$

$= 2a^2(b - 7)(2b - 1)$

Problem 5

$a \cdot c = -14$ • **Find $a \cdot c$.**

$-1(14) = -14, -1 + 14 = 13$ • **Find two numbers whose product is -14 and whose sum is 13.**

$2a^2 + 13a - 7$

$= 2a^2 - a + 14a - 7$ • **Write $13a$ as $-a + 14a$.**

$= (2a^2 - a) + (14a - 7)$ • **Factor by grouping.**

$= a(2a - 1) + 7(2a - 1)$

$= (2a - 1)(a + 7)$

Problem 6

$a \cdot c = -12$ • **Find $a \cdot c$.**

$1(-12) = -12, 1 - 12 = -11$ • **Find two numbers whose product is -12 and whose sum is -11.**

$4a^2 - 11a - 3$

$= 4a^2 + a - 12a - 3$ • **Write $-11a$ as $a - 12a$.**

$= (4a^2 + a) - (12a + 3)$ • **Factor by grouping.**

$= a(4a + 1) - 3(4a + 1)$

$= (4a + 1)(a - 3)$

Problem 7

The GCF is $5x$.

$15x^3 + 40x^2 - 80x = 5x(3x^2 + 8x - 16)$

$3(-16) = -48$ • **Find $a \cdot c$.**

$-4(12) = -48, -4 + 12 = 8$ • **Find two numbers whose product is −48 and whose sum is 8.**

$3x^2 + 8x - 16$
$= 3x^2 - 4x + 12x - 16$ • **Write 8x as −4x + 12x.**
$= (3x^2 - 4x) + (12x - 16)$ • **Factor by grouping.**
$= x(3x - 4) + 4(3x - 4)$
$= (3x - 4)(x + 4)$

$15x^3 + 40x^2 - 80x = 5x(3x^2 + 8x - 16)$
$= 5x(3x - 4)(x + 4)$

SECTION 8.4

Problem 1

A. $25a^2 - b^2 = (5a)^2 - b^2$ • **Difference of squares**
$= (5a + b)(5a - b)$

B. $6x^2 - 1$ is nonfactorable over the integers.

C. $n^8 - 36 = (n^4)^2 - 6^2$ • **Difference of squares**
$= (n^4 + 6)(n^4 - 6)$

Problem 2

$n^4 - 81 = (n^2 + 9)(n^2 - 9)$ • **Difference of squares**
$= (n^2 + 9)(n + 3)(n - 3)$ • **Factor $n^2 - 9$.**

Problem 3

A. $16y^2 = (4y)^2, 1 = 1^2$
$(4y + 1)^2 = (4y)^2 + 2(4y)(1) + (1)^2$
$= 16y^2 + 8y + 1$ • **The factorization checks.**

$16y^2 + 8y + 1 = (4y + 1)^2$

B. $x^2 = (x)^2, 36 = 6^2$
$(x + 6)^2 = x^2 + 2(x)(6) + 6^2$
$= x^2 + 12x + 36$ • **The factorization does not check.**

The polynomial is not a perfect square.
$x^2 + 14x + 36$ is nonfactorable over the integers.

SECTION 8.5

Problem 1

A. $12x^3 - 75x = 3x(4x^2 - 25)$
$= 3x(2x + 5)(2x - 5)$

B. $a^2b - 7a^2 - b + 7$
$= (a^2b - 7a^2) - (b - 7)$
$= a^2(b - 7) - (b - 7)$
$= (b - 7)(a^2 - 1)$
$= (b - 7)(a + 1)(a - 1)$

C. $4x^3 + 28x^2 - 120x = 4x(x^2 + 7x - 30)$
$= 4x(x + 10)(x - 3)$

SECTION 8.6

Problem 1

$2x^2 - 50 = 0$ • **A quadratic equation**
$2(x^2 - 25) = 0$ • **Factor out 2.**
$x^2 - 25 = 0$ • **Divide each side by 2.**
$(x + 5)(x - 5) = 0$ • **Factor.**

$x + 5 = 0$ $x - 5 = 0$ • **Set each factor equal to zero.**
$x = -5$ $x = 5$

The solutions are −5 and 5.

Problem 2

$(x + 2)(x - 7) = 52$
$x^2 - 5x - 14 = 52$ • **Multiply.**
$x^2 - 5x - 66 = 0$ • **Write in standard form.**
$(x - 11)(x + 6) = 0$ • **Factor.**

$x - 11 = 0$ $x + 6 = 0$ • **Set each factor equal to zero.**
$x = 11$ $x = -6$

The solutions are −6 and 11.

Problem 3

Strategy First positive integer: n
Second positive integer: $n + 1$
Square of the first positive integer: n^2
Square of the second positive integer: $(n + 1)^2$

The sum of the squares of the two integers is 85.

Solution
$n^2 + (n + 1)^2 = 85$
$n^2 + n^2 + 2n + 1 = 85$ • **Square $n + 1$.**
$2n^2 + 2n - 84 = 0$ • **Standard form**
$2(n^2 + n - 42) = 0$ • **Factor out 2.**
$n^2 + n - 42 = 0$ • **Divide each side by 2.**
$(n + 7)(n - 6) = 0$ • **Factor.**

$n + 7 = 0$ $n - 6 = 0$ • **Set each factor equal to zero.**
$n = -7$ $n = 6$

−7 is not a positive integer.

$n + 1 = 6 + 1 = 7$

The two integers are 6 and 7.

Problem 4

Strategy Width $= W$
Length $= 2W + 3$
The area of the rectangle is 90 m².
Use the equation $A = LW$.

Solution
$A = LW$
$90 = (2W + 3)W$ • **Replace A with 90.**
$90 = 2W^2 + 3W$ • **Multiply.**
$0 = 2W^2 + 3W - 90$ • **Standard form**
$0 = (2W + 15)(W - 6)$ • **Factor.**

$2W + 15 = 0$ $W - 6 = 0$ • **Set each factor equal to zero.**
$2W = -15$ $W = 6$
$W = -\frac{15}{2}$

The width cannot be a negative number.

$2W + 3 = 2(6) + 3 = 12 + 3 = 15$

The width is 6 m. The length is 15 m.

Solutions to Chapter 9 Problems

SECTION 9.1

Problem 1

$$\frac{6x^5y}{12x^2y^3} = \frac{\overset{1}{\cancel{2}} \cdot \overset{1}{\cancel{3}} \cdot x^5y}{\underset{1}{\cancel{2}} \cdot 2 \cdot \underset{1}{\cancel{3}} \cdot x^2y^3}$$ • Simplify the coefficients.

$$= \frac{x^3}{2y^2}$$ • Simplify the variable parts.

Problem 2

$$\frac{x^2 + 2x - 24}{16 - x^2} = \frac{(x - 4)(x + 6)}{(4 - x)(4 + x)}$$ • Factor the numerator and denominator.

$$= \frac{\overset{-1}{\cancel{(x - 4)}}(x + 6)}{\underset{1}{\cancel{(4 - x)}}(4 + x)}$$ • Divide by the common factors.

$$= -\frac{x + 6}{x + 4}$$

Problem 3

A. $$\frac{12x^2 + 3x}{10x - 15} \cdot \frac{8x - 12}{9x + 18}$$

$$= \frac{3x(4x + 1)}{5(2x - 3)} \cdot \frac{4(2x - 3)}{9(x + 2)}$$ • Factor the numerators and denominators.

$$= \frac{\overset{1}{\cancel{3}}x(4x + 1) \cdot 2 \cdot 2\overset{1}{\cancel{(2x - 3)}}}{5\underset{1}{\cancel{(2x - 3)}} \cdot \underset{1}{\cancel{3}} \cdot 3(x + 2)}$$ • Multiply. Divide by the common factors.

$$= \frac{4x(4x + 1)}{15(x + 2)}$$ • Simplify.

B. $$\frac{x^2 + 2x - 15}{9 - x^2} \cdot \frac{x^2 - 3x - 18}{x^2 - 7x + 6}$$

$$= \frac{(x - 3)(x + 5)}{(3 - x)(3 + x)} \cdot \frac{(x + 3)(x - 6)}{(x - 1)(x - 6)}$$

$$= \frac{\overset{-1}{\cancel{(x - 3)}}(x + 5) \cdot \overset{1}{\cancel{(x + 3)}}\overset{1}{\cancel{(x - 6)}}}{\underset{1}{\cancel{(3 - x)}}\underset{1}{\cancel{(3 + x)}} \cdot (x - 1)\underset{1}{\cancel{(x - 6)}}}$$

$$= -\frac{x + 5}{x - 1}$$

Problem 4

A. $$\frac{a^2}{4bc^2 - 2b^2c} \div \frac{a}{6bc - 3b^2}$$

$$= \frac{a^2}{4bc^2 - 2b^2c} \cdot \frac{6bc - 3b^2}{a}$$ • Write division as multiplication by the reciprocal.

$$= \frac{a^2 \cdot 3b\overset{1}{\cancel{(2c - b)}}}{2bc\underset{1}{\cancel{(2c - b)}} \cdot a} = \frac{3a}{2c}$$ • Multiply. Divide by the common factors.

B. $$\frac{3x^2 + 26x + 16}{3x^2 - 7x - 6} \div \frac{2x^2 + 9x - 5}{x^2 + 2x - 15}$$

$$= \frac{3x^2 + 26x + 16}{3x^2 - 7x - 6} \cdot \frac{x^2 + 2x - 15}{2x^2 + 9x - 5}$$ • Write division as multiplication by the reciprocal.

$$= \frac{\overset{1}{\cancel{(3x + 2)}}(x + 8)}{\underset{1}{\cancel{(3x + 2)}}\underset{1}{\cancel{(x - 3)}}} \cdot \frac{\overset{1}{\cancel{(x + 5)}}\overset{1}{\cancel{(x - 3)}}}{(2x - 1)\underset{1}{\cancel{(x + 5)}}}$$ • Factor. Divide by the common factors.

$$= \frac{x + 8}{2x - 1}$$

SECTION 9.2

Problem 1

$8uv^2 = 2 \cdot 2 \cdot 2 \cdot u \cdot v \cdot v$ • Factor each monomial.

$12uw = 2 \cdot 2 \cdot 3 \cdot u \cdot w$

$\text{LCM} = 2 \cdot 2 \cdot 2 \cdot 3 \cdot u \cdot v \cdot v \cdot w$ • The LCM is the product of each factor the greatest number of times it occurs in any one factorization.

$= 24uv^2w$

Problem 2

$m^2 - 6m + 9 = (m - 3)(m - 3)$ • Factor each polynomial.

$m^2 - 2m - 3 = (m + 1)(m - 3)$

$\text{LCM} = (m - 3)(m - 3)(m + 1)$ • The LCM is the product of each factor the greatest number of times it occurs in any one factorization.

Problem 3

The LCD is $36xy^2z$.

$$\frac{x - 3}{4xy^2} = \frac{x - 3}{4xy^2} \cdot \frac{9z}{9z} = \frac{9xz - 27z}{36xy^2z}$$ • The product of $4xy^2$ and $9z$ is the LCD.

$$\frac{2x + 1}{9y^2z} = \frac{2x + 1}{9y^2z} \cdot \frac{4x}{4x} = \frac{8x^2 + 4x}{36xy^2z}$$ • The product of $9y^2z$ and $4x$ is the LCD.

Problem 4

The LCD is $(x + 2)(x - 5)(x + 5)$.

$$\frac{x + 4}{x^2 - 3x - 10} = \frac{x + 4}{(x + 2)(x - 5)} \cdot \frac{x + 5}{x + 5}$$

$$= \frac{x^2 + 9x + 20}{(x + 2)(x - 5)(x + 5)}$$

$$\frac{2x}{x^2 - 25} = \frac{2x}{(x - 5)(x + 5)} \cdot \frac{x + 2}{x + 2}$$

$$= \frac{2x^2 + 4x}{(x + 2)(x - 5)(x + 5)}$$

SECTION 9.3

Problem 1

A. $$\frac{3}{xy} + \frac{12}{xy} = \frac{3 + 12}{xy} = \frac{15}{xy}$$ • The denominators are the same. Add the numerators.

B. $$\frac{2x^2}{x^2 - x - 12} - \frac{7x + 4}{x^2 - x - 12} = \frac{2x^2 - (7x + 4)}{x^2 - x - 12}$$

$$= \frac{2x^2 - 7x - 4}{x^2 - x - 12} = \frac{(2x + 1)\overset{1}{\cancel{(x - 4)}}}{(x + 3)\underset{1}{\cancel{(x - 4)}}} = \frac{2x + 1}{x + 3}$$

Problem 2

A. The LCD is $24y$.

$$\frac{z}{8y} - \frac{4z}{3y} + \frac{5z}{4y} = \frac{z}{8y}\cdot\frac{3}{3} - \frac{4z}{3y}\cdot\frac{8}{8} + \frac{5z}{4y}\cdot\frac{6}{6}$$

$$= \frac{3z}{24y} - \frac{32z}{24y} + \frac{30z}{24y} = \frac{3z - 32z + 30z}{24y} = \frac{z}{24y}$$

B. The LCD is $x - 2$.

$$\frac{5x}{x-2} - \frac{3}{2-x} = \frac{5x}{x-2} - \frac{3}{-(x-2)}\cdot\frac{-1}{-1}$$

$$= \frac{5x}{x-2} - \frac{-3}{x-2} = \frac{5x-(-3)}{x-2} = \frac{5x+3}{x-2}$$

C. The LCD is $y - 7$.

$$y + \frac{5}{y-7} = \frac{y}{1} + \frac{5}{y-7} = \frac{y}{1}\cdot\frac{y-7}{y-7} + \frac{5}{y-7}$$

$$= \frac{y^2-7y}{y-7} + \frac{5}{y-7} = \frac{y^2-7y+5}{y-7}$$

Problem 3

A. The LCD is $(3x-1)(x+4)$.

$$\frac{4x}{3x-1} - \frac{9}{x+4}$$

$$= \frac{4x}{3x-1}\cdot\frac{x+4}{x+4} - \frac{9}{x+4}\cdot\frac{3x-1}{3x-1}$$

$$= \frac{4x^2+16x}{(3x-1)(x+4)} - \frac{27x-9}{(3x-1)(x+4)}$$

$$= \frac{(4x^2+16x)-(27x-9)}{(3x-1)(x+4)}$$ • Subtract the fractions.

$$= \frac{4x^2+16x-27x+9}{(3x-1)(x+4)}$$

$$= \frac{4x^2-11x+9}{(3x-1)(x+4)}$$ • Simplify the numerator.

B. The LCD is $(x+5)(x-5)$.

$$\frac{2x-1}{x^2-25} + \frac{2}{5-x}$$

$$= \frac{2x-1}{(x+5)(x-5)} + \frac{2}{-(x-5)}\cdot\frac{-1(x+5)}{-1(x+5)}$$

$$= \frac{2x-1}{(x+5)(x-5)} + \frac{-2(x+5)}{(x+5)(x-5)}$$

$$= \frac{(2x-1)+(-2)(x+5)}{(x+5)(x-5)}$$ • Add the fractions.

$$= \frac{2x-1-2x-10}{(x+5)(x-5)}$$ • Simplify the numerator.

$$= \frac{-11}{(x+5)(x-5)} = -\frac{11}{(x+5)(x-5)}$$

SECTION 9.4

Problem 1

A. The LCM of the denominators, 3, x, 9, and x^2, is $9x^2$.

$$\frac{\frac{1}{3}-\frac{1}{x}}{\frac{1}{9}-\frac{1}{x^2}} = \frac{\frac{1}{3}-\frac{1}{x}}{\frac{1}{9}-\frac{1}{x^2}}\cdot\frac{9x^2}{9x^2}$$ • Multiply the numerator and denominator of the complex fraction by $9x^2$.

$$= \frac{\frac{1}{3}\cdot 9x^2 - \frac{1}{x}\cdot 9x^2}{\frac{1}{9}\cdot 9x^2 - \frac{1}{x^2}\cdot 9x^2}$$ • Distributive Property

$$= \frac{3x^2-9x}{x^2-9} = \frac{3x\cancel{(x-3)}^{1}}{\cancel{(x-3)}_{1}(x+3)} = \frac{3x}{x+3}$$

B. The LCM of the denominators, x and x^2, is x^2.

$$\frac{1+\frac{4}{x}+\frac{3}{x^2}}{1+\frac{10}{x}+\frac{21}{x^2}} = \frac{1+\frac{4}{x}+\frac{3}{x^2}}{1+\frac{10}{x}+\frac{21}{x^2}}\cdot\frac{x^2}{x^2}$$ • Multiply the numerator and denominator by x^2.

$$= \frac{1\cdot x^2+\frac{4}{x}\cdot x^2+\frac{3}{x^2}\cdot x^2}{1\cdot x^2+\frac{10}{x}\cdot x^2+\frac{21}{x^2}\cdot x^2}$$ • Distributive Property

$$= \frac{x^2+4x+3}{x^2+10x+21}$$

$$= \frac{(x+1)\cancel{(x+3)}^{1}}{\cancel{(x+3)}_{1}(x+7)} = \frac{x+1}{x+7}$$

SECTION 9.5

Problem 1

The LCD is $3x$.

$$x + \frac{1}{3} = \frac{4}{3x}$$

$$3x\left(x+\frac{1}{3}\right) = 3x\left(\frac{4}{3x}\right)$$ • Multiply each side by $3x$.

$$3x\cdot x + \frac{3x}{1}\cdot\frac{1}{3} = \frac{3x}{1}\cdot\frac{4}{3x}$$

$$3x^2 + x = 4$$ • Quadratic equation

$$3x^2 + x - 4 = 0$$ • Write in standard form.

$$(3x+4)(x-1) = 0$$ • Factor.

$3x + 4 = 0$ $\quad$ $x - 1 = 0$

$3x = -4$ $\quad$ $x = 1$

$x = -\frac{4}{3}$

Both $-\frac{4}{3}$ and 1 check as solutions. The solutions are $-\frac{4}{3}$ and 1.

Problem 2

The LCD is $x + 2$.

$$\frac{5x}{x+2} = 3 - \frac{10}{x+2}$$

$$\frac{(x+2)}{1} \cdot \frac{5x}{x+2} = \frac{(x+2)}{1}\left(3 - \frac{10}{x+2}\right)$$

$$5x = (x+2)3 - 10$$

$$5x = 3x + 6 - 10 \quad \text{• Distributive Property}$$

$$5x = 3x - 4 \quad \text{• Solve for } x.$$

$$2x = -4$$

$$x = -2$$

-2 does not check as a solution. The equation has no solution.

Problem 3

A.

$$\frac{2}{5} = \frac{6}{5x+5} \quad \text{• The LCD is } 5(5x+5).$$

$$5(5x+5)\frac{2}{5} = 5(5x+5)\frac{6}{5x+5}$$

$$(5x+5)2 = (5)6$$

$$10x + 10 = 30 \quad \text{• Distributive Property}$$

$$10x = 20 \quad \text{• Solve for } x.$$

$$x = 2$$

The solution is 2.

B.

$$\frac{5}{2x-3} = \frac{10}{x+3} \quad \text{• The LCD is } (2x-3)(x+3).$$

$$(x+3)(2x-3)\frac{5}{2x-3} = (x+3)(2x-3)\frac{10}{x+3}$$

$$(x+3)5 = (2x-3)10$$

$$5x + 15 = 20x - 30 \quad \text{• Distributive Property}$$

$$-15x + 15 = -30 \quad \text{• Solve for } x.$$

$$-15x = -45$$

$$x = 3$$

The solution is 3.

Problem 4

Strategy To find the total area that 270 ceramic tiles will cover, write and solve a proportion using x to represent the number of square feet that 270 tiles will cover.

Solution

$$\frac{4}{9} = \frac{x}{270}$$

• The numerators represent square feet covered. The denominators represent numbers of tiles.

$$270\left(\frac{4}{9}\right) = 270\left(\frac{x}{270}\right) \quad \text{• Multiply by the LCD.}$$

$$120 = x$$

A 120-square-foot area can be tiled using 270 ceramic tiles.

Problem 5

Strategy To find the additional amount of medication required for a 180-pound adult, write and solve a proportion using x to represent the additional medication. Then $3 + x$ is the total amount required for a 180-pound adult.

Solution

$$\frac{120}{3} = \frac{180}{3+x}$$

• The numerators represent the weights. The denominators represent the amounts of medication.

$$\frac{40}{1} = \frac{180}{3+x}$$

$$(3+x)40 = (3+x)\frac{180}{3+x} \quad \text{• Multiply by the LCD.}$$

$$120 + 40x = 180$$

$$40x = 60$$

$$x = 1.5$$

1.5 additional ounces are required for a 180-pound adult.

Problem 6

Strategy Triangle AOB is similar to triangle DOC. Solve a proportion to find AO. Then use the formula for the area of a triangle to find the area of triangle AOB.

Solution

$$\frac{CD}{AB} = \frac{DO}{AO} \quad \text{• Write a proportion.}$$

$$\frac{4}{10} = \frac{3}{x} \quad \text{• Let } x \text{ represent } AO.$$

$$10x\left(\frac{4}{10}\right) = 10x\left(\frac{3}{x}\right) \quad \text{• Multiply by } 10x.$$

$$4x = 30 \quad \text{• Solve for } x.$$

$$x = 7.5$$

$$A = \frac{1}{2}bh \quad \text{• Area formula}$$

$$A = \frac{1}{2}(10)(7.5) \quad \text{• } b = 10, h = 7.5$$

$$A = 37.5$$

The area of triangle AOB is 37.5 cm^2.

SECTION 9.6

Problem 1

Strategy

- ► To find the constant of variation, replace p with 120 and q with 8 in the general form of a direct variation equation, $p = kq$. Solve for k.
- ► To write the specific direct variation equation, replace k by its value in the general direct variation equation.

Solution

$$p = kq \quad \text{• Use the general form of a direct variation equation.}$$

$$120 = k \cdot 8 \quad \text{• Replace } p \text{ by 120 and } q \text{ by 8.}$$

$$15 = k \quad \text{• Solve for } k \text{ by dividing both sides by 8.}$$

The constant of variation is 15.

$$p = 15q \quad \text{• Write the specific direct variation equation by substituting 15 for } k \text{ in } p = kq.$$

The direct variation equation is $p = 15q$.

Problem 2

Strategy

- ► This is a direct variation. To find the value of k, write the general form of a direct variation equation, replace the variables by the given values, and solve for k.

► Write the specific direct variation equation, replacing k by its value. Substitute 18 for h and solve for w.

Solution

$w = kh$ • Use the general form of a direct variation equation.

$264 = k \cdot 12$ • Replace w by 264 and h by 12.

$22 = k$ • Solve for k by dividing both sides by 12.

$w = 22h$ • Write the specific direct variation equation by substituting 22 for k.

$w = 22(18)$ • Replace h with 18 to find w when h is 18.

$w = 396$

The nurse's total wage for working 18 h is $396.

Problem 3

Strategy ► To find the constant of variation, replace s with 12 and t with 8 in the general form of an inverse variation equation, $s = \frac{k}{t}$. Solve for k.

► To write the specific inverse variation equation, replace k by its value in the general inverse variation equation.

Solution

$s = \dfrac{k}{t}$ • Use the general form of an inverse variation equation.

$12 = \dfrac{k}{8}$ • Replace s by 12 and t by 8.

$96 = k$ • Solve for k by multiplying both sides by 8.

The constant of variation is 96.

$s = \dfrac{96}{t}$ • Write the specific inverse variation equation by substituting 96 for k in $s = \frac{k}{t}$.

The inverse variation equation is $s = \frac{96}{t}$.

Problem 4

Strategy ► This is an inverse variation. To find the value of k, write the general inverse variation equation, replace the variables by the given values, and solve for k.

► Write the specific direct variation equation, replacing k by its value. Substitute 4 for m and solve for h.

Solution

$h = \dfrac{k}{m}$ • Use the general form of an inverse variation equation.

$9 = \dfrac{k}{5}$ • Replace h by 9 and m by 5.

$45 = k$ • Solve for k by multiplying both sides by 5.

$h = \dfrac{45}{m}$ • Write the specific inverse variation equation by substituting 45 for k.

$h = \dfrac{45}{4}$ • Replace m with 4 to find h when m is 4.

$h = 11.25$

It takes 11.25 h for four assembly machines to complete the daily quota.

SECTION 9.7

Problem 1

A.

$s = \dfrac{A + L}{2}$

$2 \cdot s = 2\left(\dfrac{A + L}{2}\right)$ • Multiply each side by 2.

$2s = A + L$

$2s - A = A - A + L$ • Subtract A from each side.

$2s - A = L$

B.

$S = a + (n - 1)d$

$S = a + nd - d$ • Distributive Property

$S - a = a - a + nd - d$ • Subtract a from each side.

$S - a = nd - d$

$S - a + d = nd - d + d$ • Add d to each side.

$S - a + d = nd$

$\dfrac{S - a + d}{d} = \dfrac{nd}{d}$ • Divide each side by d.

$\dfrac{S - a + d}{d} = n$

C.

$S = rS + C$

$S - rS = C$ • Subtract rS from each side.

$S(1 - r) = C$ • Factor S from the left side.

$S = \dfrac{C}{1 - r}$ • Divide each side by $1 - r$.

SECTION 9.8

Problem 1

Strategy ► Time for one printer to complete the job: t

	Rate	·	Time	=	Part
1st printer	$\frac{1}{t}$	·	3	=	$\frac{3}{t}$
2nd printer	$\frac{1}{t}$	·	5	=	$\frac{5}{t}$

► The sum of the parts of the task completed must equal 1.

Solution

$\dfrac{3}{t} + \dfrac{5}{t} = 1$

$t\left(\dfrac{3}{t} + \dfrac{5}{t}\right) = t \cdot 1$ • Multiply each side by t, the LCD.

$3 + 5 = t$ • Distributive Property

$8 = t$

Working alone, one printer takes 8 h to print the payroll.

Problem 2

Strategy ▶ Rate sailing across the lake: r
Rate sailing back: $2r$

	Distance	÷	Rate	=	Time
Across	6	÷	r	=	$\frac{6}{r}$
Back	6	÷	$2r$	=	$\frac{6}{2r}$

▶ The total time for the trip was 3 h.

Solution

$$\frac{6}{r} + \frac{6}{2r} = 3$$ • The time spent sailing across plus the time spent sailing back equals 3 h.

$$2r\left(\frac{6}{r} + \frac{6}{2r}\right) = 2r(3)$$ • Multiply each side by $2r$.

$$2r \cdot \frac{6}{r} + 2r \cdot \frac{6}{2r} = 6r$$ • Distributive Property

$$12 + 6 = 6r$$ • Solve for r.

$$18 = 6r$$

$$3 = r$$

The rate across the lake was 3 km/h.

Solutions to Chapter 10 Problems

SECTION 10.1

Problem 1

$$\sqrt{216} = \sqrt{36 \cdot 6}$$
$$= \sqrt{36}\,\sqrt{6}$$ • Product Property of Square Roots
$$= 6\sqrt{6}$$ • Simplify $\sqrt{36}$.

Problem 2

$$-5\sqrt{32} = -5\sqrt{16 \cdot 2}$$
$$= -5\sqrt{16}\,\sqrt{2}$$ • Product Property of Square Roots
$$= -5 \cdot 4\sqrt{2}$$ • Simplify $\sqrt{16}$.
$$= -20\sqrt{2}$$ • Multiply $-5 \cdot 4$.

Problem 3

$$\sqrt{y^{19}} = \sqrt{y^{18} \cdot y} = \sqrt{y^{18}}\,\sqrt{y} = y^9\sqrt{y}$$

Problem 4

A. $$\sqrt{45b^7} = \sqrt{9b^6 \cdot 5b}$$ • Write the radicand as the product of a perfect square and factors that do not contain a perfect square.
$$= \sqrt{9b^6}\,\sqrt{5b}$$ • Product Property of Square Roots
$$= 3b^3\sqrt{5b}$$ • Simplify $\sqrt{9b^6}$.

B. $$3a\sqrt{28a^9b^{18}} = 3a\sqrt{4a^8b^{18} \cdot 7a}$$
$$= 3a\sqrt{4a^8b^{18}}\,\sqrt{7a}$$
$$= 3a \cdot 2a^4b^9\sqrt{7a}$$
$$= 6a^5b^9\sqrt{7a}$$

Problem 5

49 is a perfect square.
$(a + 3)^2$ is a perfect square.
$$\sqrt{49(a + 3)^2} = 7(a + 3) = 7a + 21$$

SECTION 10.2

Problem 1

A. $$9\sqrt{3} + 3\sqrt{3} - 18\sqrt{3}$$
$$= (9 + 3 - 18)\sqrt{3}$$ • Distributive Property
$$= -6\sqrt{3}$$

B. $$2\sqrt{50} - 5\sqrt{32}$$
$$= 2\sqrt{25 \cdot 2} - 5\sqrt{16 \cdot 2}$$ • Simplify each term.
$$= 2\sqrt{25}\,\sqrt{2} - 5\sqrt{16}\,\sqrt{2}$$
$$= 2 \cdot 5\sqrt{2} - 5 \cdot 4\sqrt{2}$$
$$= 10\sqrt{2} - 20\sqrt{2}$$
$$= (10 - 20)\sqrt{2}$$ • Distributive Property
$$= -10\sqrt{2}$$

Problem 2

A. $$y\sqrt{28y} + 7\sqrt{63y^3}$$
$$= y\sqrt{4}\,\sqrt{7y} + 7\sqrt{9y^2}\,\sqrt{7y}$$
$$= y \cdot 2\sqrt{7y} + 7 \cdot 3y\sqrt{7y}$$
$$= 2y\sqrt{7y} + 21y\sqrt{7y} = 23y\sqrt{7y}$$

B. $$2\sqrt{27a^5} - 4a\sqrt{12a^3} + a^2\sqrt{75a}$$
$$= 2\sqrt{9a^4}\,\sqrt{3a} - 4a\sqrt{4a^2}\,\sqrt{3a} + a^2\sqrt{25}\,\sqrt{3a}$$
$$= 2 \cdot 3a^2\sqrt{3a} - 4a \cdot 2a\sqrt{3a} + a^2 \cdot 5\sqrt{3a}$$
$$= 6a^2\sqrt{3a} - 8a^2\sqrt{3a} + 5a^2\sqrt{3a} = 3a^2\sqrt{3a}$$

SECTION 10.3

Problem 1

$$\sqrt{5a}\,\sqrt{15a^3b^4}\,\sqrt{3b^5}$$
$$= \sqrt{225a^4b^9}$$ • Product Property of Square Roots
$$= \sqrt{225a^4b^8}\,\sqrt{b}$$ • Simplify.
$$= 15a^2b^4\sqrt{b}$$

Problem 2

$$\sqrt{5x}(\sqrt{5x} - \sqrt{25y})$$
$$= (\sqrt{5x})^2 - \sqrt{125xy}$$ • Distributive Property
$$= 5x - \sqrt{25}\,\sqrt{5xy}$$ • Product Property of Square Roots
$$= 5x - 5\sqrt{5xy}$$ • Simplify.

Problem 3

$$(3\sqrt{x} - \sqrt{y})(5\sqrt{x} - 2\sqrt{y})$$
$$= 15(\sqrt{x})^2 - 6\sqrt{xy} - 5\sqrt{xy} + 2(\sqrt{y})^2$$ • Use FOIL.
$$= 15x - 11\sqrt{xy} + 2y$$ • Simplify.

Problem 4

$$(2\sqrt{x} + 7)(2\sqrt{x} - 7)$$ • Conjugates
$$= (2\sqrt{x})^2 - 7^2 = 4x - 49$$

Problem 5

A. $$\frac{\sqrt{15x^6y^7}}{\sqrt{3x^7y^9}} = \sqrt{\frac{15x^6y^7}{3x^7y^9}} = \sqrt{\frac{5}{xy^2}} = \frac{\sqrt{5}}{\sqrt{xy^2}}$$
$$= \frac{\sqrt{5}}{y\sqrt{x}} = \frac{\sqrt{5}}{y\sqrt{x}} \cdot \frac{\sqrt{x}}{\sqrt{x}} = \frac{\sqrt{5x}}{xy}$$

B. $$\frac{\sqrt{y}}{\sqrt{y} + 3} = \frac{\sqrt{y}}{\sqrt{y} + 3} \cdot \frac{\sqrt{y} - 3}{\sqrt{y} - 3} = \frac{y - 3\sqrt{y}}{y - 9}$$

C. $\dfrac{5+\sqrt{y}}{1-2\sqrt{y}} = \dfrac{5+\sqrt{y}}{1-2\sqrt{y}} \cdot \dfrac{1+2\sqrt{y}}{1+2\sqrt{y}}$

$= \dfrac{5+10\sqrt{y}+\sqrt{y}+2y}{1-4y}$

$= \dfrac{5+11\sqrt{y}+2y}{1-4y}$

SECTION 10.4

Problem 1

$\sqrt{4x}+3=7$ • **Subtract 3 from each side.**

$\sqrt{4x}=4$

$(\sqrt{4x})^2=4^2$ • **Square both sides.**

$4x=16$ • **Solve for x.**

$x=4$

Check $\sqrt{4x}+3=7$

$\sqrt{4\cdot4}+3$	7
$\sqrt{16}+3$	7
$4+3$	7

$7=7$ • **A true equation**

The solution is 4.

Problem 2

A. $\sqrt{3x-2}-5=0$

$\sqrt{3x-2}=5$ • **Add 5 to each side.**

$(\sqrt{3x-2})^2=5^2$ • **Square each side.**

$3x-2=25$

$3x=27$ • **Solve for x.**

$x=9$

Check $\sqrt{3x-2}-5=0$

$\sqrt{3\cdot9-2}-5$	0
$\sqrt{27-2}-5$	0
$\sqrt{25}-5$	0
$5-5$	0

$0=0$ • **A true equation**

The solution is 9.

B. $\sqrt{4x-7}+5=0$

$\sqrt{4x-7}=-5$ • **Subtract 5 from each side.**

$(\sqrt{4x-7})^2=(-5)^2$ • **Square each side.**

$4x-7=25$

$4x=32$ • **Solve for x.**

$x=8$

Check $\sqrt{4x-7}+5=0$

$\sqrt{4\cdot8-7}+5$	0
$\sqrt{32-7}+5$	0
$\sqrt{25}+5$	0
$5+5$	0

$10\neq0$ • **A false equation**

There is no solution.

Problem 3

$\sqrt{x}+\sqrt{x+9}=9$

$\sqrt{x}=9-\sqrt{x+9}$ • **Solve for one radical.**

$(\sqrt{x})^2=(9-\sqrt{x+9})^2$ • **Square each side.**

$x=81-18\sqrt{x+9}+(x+9)$

$18\sqrt{x+9}=90$ • **Divide each side by 18.**

$\sqrt{x+9}=5$ • **Still a radical**

$(\sqrt{x+9})^2=5^2$ • **Square each side.**

$x+9=25$ • **Simplify.**

$x=16$

Check $\sqrt{x}+\sqrt{x+9}=9$

$\sqrt{16}+\sqrt{16+9}$	9
$4+\sqrt{25}$	9
$4+5$	9

$9=9$ • **A true equation**

The solution is 16.

Problem 4

Strategy To find the distance, use the Pythagorean Theorem. The hypotenuse is the length of the ladder. One leg is the distance from the bottom of the ladder to the base of the building. The distance along the building from the ground to the top of the ladder is the unknown leg.

Solution $a=\sqrt{c^2-b^2}$

$a=\sqrt{(12)^2-(5)^2}$ • **$c = 12, b = 5$**

$a=\sqrt{144-25}$ • **Simplify.**

$a=\sqrt{119}$ • **Use a calculator.**

$a\approx10.91$

The distance is 10.91 ft.

Problem 5

Strategy To find the length of the pendulum, replace T in the equation with the given value and solve for L.

Solution $T=2\pi\sqrt{\dfrac{L}{32}}$

$1.5=2\pi\sqrt{\dfrac{L}{32}}$ • **$T = 1.5$**

$\dfrac{1.5}{2\pi}=\sqrt{\dfrac{L}{32}}$ • **Divide by 2π.**

$\left(\dfrac{1.5}{2\pi}\right)^2=\left(\sqrt{\dfrac{L}{32}}\right)^2$ • **Square each side.**

$\left(\dfrac{1.5}{2\pi}\right)^2=\dfrac{L}{32}$ • **Solve for L.**

$32\left(\dfrac{1.5}{2\pi}\right)^2=L$ • **Use the π key on your calculator.**

$1.82\approx L$

The length of the pendulum is 1.82 ft.

Solutions to Chapter 11 Problems

SECTION 11.1

Problem 1

$$\frac{3y^2}{2} + y - \frac{1}{2} = 0$$

$$2\left(\frac{3y^2}{2} + y - \frac{1}{2}\right) = 2(0)$$ • **Multiply each side by 2, the LCD.**

$$3y^2 + 2y - 1 = 0$$ • **Standard form of a quadratic equation**

$$(3y - 1)(y + 1) = 0$$ • **Factor the left side.**

$3y - 1 = 0 \qquad y + 1 = 0$ • **Let each factor equal zero.**

$3y = 1 \qquad y = -1$

$y = \frac{1}{3}$

The solutions are $\frac{1}{3}$ and -1.

Problem 2

$$4x^2 - 96 = 0$$

$$4x^2 = 96$$ • **Solve for x^2.**

$$x^2 = 24$$

$$\sqrt{x^2} = \sqrt{24}$$ • **Take the square root of each side.**

$$x = \pm\sqrt{24}$$

$$x = \pm 2\sqrt{6}$$ • **Simplify. Check the solutions.**

The solutions are $2\sqrt{6}$ and $-2\sqrt{6}$.

Problem 3

$$(x + 5)^2 = 20$$

$$\sqrt{(x + 5)^2} = \sqrt{20}$$ • **Take the square root of each side.**

$$x + 5 = \pm\sqrt{20}$$

$$x + 5 = \pm 2\sqrt{5}$$ • **Simplify.**

$x + 5 = 2\sqrt{5} \qquad x + 5 = -2\sqrt{5}$ • **Solve for x.**

$x = -5 + 2\sqrt{5} \qquad x = -5 - 2\sqrt{5}$

The solutions are $-5 + 2\sqrt{5}$ and $-5 - 2\sqrt{5}$.

SECTION 11.2

Problem 1

$$3x^2 - 6x - 2 = 0$$ • **Add 2 to each side.**

$$3x^2 - 6x = 2$$

$$\frac{1}{3}(3x^2 - 6x) = \frac{1}{3} \cdot 2$$ • **The coefficient of x^2 must be 1. Multiply each side by $\frac{1}{3}$.**

$$x^2 - 2x = \frac{2}{3}$$

Complete the square.

$$x^2 - 2x + 1 = \frac{2}{3} + 1$$ • **Add 1 to each side.**

$$(x - 1)^2 = \frac{5}{3}$$ • **Factor the perfect-square trinomial.**

$$\sqrt{(x - 1)^2} = \sqrt{\frac{5}{3}}$$ • **Take square roots.**

$$x - 1 = \pm\sqrt{\frac{5}{3}}$$ • **Simplify.**

$$x - 1 = \pm\frac{\sqrt{15}}{3}$$ • **Rationalize the denominator.**

$x - 1 = \frac{\sqrt{15}}{3} \qquad x - 1 = -\frac{\sqrt{15}}{3}$ • **Solve for x.**

$x = 1 + \frac{\sqrt{15}}{3} \qquad x = 1 - \frac{\sqrt{15}}{3}$

$= \frac{3 + \sqrt{15}}{3} \qquad = \frac{3 - \sqrt{15}}{3}$

The solutions are $\frac{3 + \sqrt{15}}{3}$ and $\frac{3 - \sqrt{15}}{3}$.

Problem 2

$$x^2 + 8x + 8 = 0$$

$$x^2 + 8x = -8$$ • **Subtract 8 from each side.**

$$x^2 + 8x + 16 = -8 + 16$$ • **Complete the square of $x^2 + 8x$.**

$$(x + 4)^2 = 8$$ • **Factor the perfect-square trinomial.**

$$\sqrt{(x + 4)^2} = \sqrt{8}$$ • **Take square roots.**

$$x + 4 = \pm\sqrt{8}$$ • **Solve for x.**

$$x + 4 \approx \pm 2.828$$ • **Use a calculator.**

$x + 4 \approx 2.828 \qquad x + 4 \approx -2.828$

$x \approx -4 + 2.828 \qquad x \approx -4 - 2.828$

$\approx -1.172 \qquad \approx -6.828$

The solutions are approximately -1.172 and -6.828.

SECTION 11.3

Problem 1

A. $3x^2 + 4x - 4 = 0$ • **Quadratic equation in standard form**

$a = 3, b = 4, c = -4$

$$x = \frac{-(4) \pm \sqrt{(4)^2 - 4(3)(-4)}}{2 \cdot 3}$$ • **Replace a, b, and c in the quadratic formula. Simplify.**

$$= \frac{-4 \pm \sqrt{16 + 48}}{6}$$

$$= \frac{-4 \pm \sqrt{64}}{6} = \frac{-4 \pm 8}{6}$$

$x = \frac{-4 + 8}{6} \qquad x = \frac{-4 - 8}{6}$

$= \frac{4}{6} = \frac{2}{3} \qquad = \frac{-12}{6} = -2$

The solutions are $\frac{2}{3}$ and -2.

B. $x^2 + 2x = 1$ • **Quadratic equation**

$x^2 + 2x - 1 = 0$ • **Standard form**

$a = 1, b = 2, c = -1$

$$x = \frac{-(2) \pm \sqrt{(2)^2 - 4(1)(-1)}}{2 \cdot 1}$$ • **Replace a, b, and c in the quadratic formula. Simplify.**

$$= \frac{-2 \pm \sqrt{4 + 4}}{2}$$

$$= \frac{-2 \pm \sqrt{8}}{2}$$

$$= \frac{-2 \pm 2\sqrt{2}}{2}$$

$$= \frac{2(-1 \pm \sqrt{2})}{2} = -1 \pm \sqrt{2}$$

The solutions are $-1 + \sqrt{2}$ and $-1 - \sqrt{2}$.

SECTION 11.4

Problem 1

A. $\sqrt{-17} = \sqrt{-1 \cdot 17} = \sqrt{-1}\sqrt{17} = i\sqrt{17}$

B. $\sqrt{-81} = \sqrt{-1 \cdot 81} = \sqrt{-1}\sqrt{81} = i\sqrt{81}$
$= i \cdot 9 = 9i$

C. $\sqrt{-63} = \sqrt{-1 \cdot 63} = \sqrt{-1}\sqrt{63} = i\sqrt{63} = i\sqrt{9 \cdot 7}$
$= i\sqrt{9}\sqrt{7} = i \cdot 3\sqrt{7} = 3i\sqrt{7}$

D. $4\sqrt{-50} = 4\sqrt{-1 \cdot 50} = 4\sqrt{-1}\sqrt{50} = 4i\sqrt{50}$
$= 4i\sqrt{25 \cdot 2} = 4i\sqrt{25}\sqrt{2} = 4i \cdot 5\sqrt{2} = 20i\sqrt{2}$

Problem 2

$\dfrac{5 - 3\sqrt{-50}}{10} = \dfrac{5 - 3i\sqrt{50}}{10}$ • $\sqrt{-50} = i\sqrt{50}$

$= \dfrac{5 - 3i\sqrt{25}\sqrt{2}}{10} = \dfrac{5 - 3i \cdot 5\sqrt{2}}{10}$ • Simplify $\sqrt{50}$.

$= \dfrac{5 - 15i\sqrt{2}}{10}$

$= \dfrac{5}{10} - \dfrac{15i\sqrt{2}}{10} = \dfrac{1}{2} - \dfrac{3i\sqrt{2}}{2}$

$= \dfrac{1}{2} - \dfrac{3\sqrt{2}}{2}i$ • Write the answer in $a + bi$ form.

Problem 3

A. $(7 - 3i) + (-8 + i) = (7 + (-8)) + (-3 + 1)i$
$= -1 - 2i$

B. $(-5 + 4i) - (-3 - 7i) = (-5 - (-3)) + (4 - (-7))i$
$= -2 + 11i$

Problem 4

$5i(3 + 6i) = 15i + 30i^2$ • Use the Distributive Property to remove parentheses.
$= 15i + 30(-1)$ • Replace i^2 with -1.
$= 15i + (-30)$
$= -30 + 15i$ • Write the answer in the form $a + bi$.

Problem 5

$(-1 + 5i)(2 - 3i)$
$= (-1)(2) + (-1)(-3i) + 5i(2) + 5i(-3i)$ • Use the FOIL method.
$= -2 + 3i + 10i - 15i^2$
$= -2 + 13i - 15i^2$ • Combine like terms.
$= -2 + 13i - 15(-1)$ • Replace i^2 with -1.
$= -2 + 13i + 15$
$= 13 + 13i$ • Write the answer in the form $a + bi$.

Problem 6

$(5 - 4i)^2 = (5 - 4i)(5 - 4i)$ • Multiply $5 - 4i$ by itself.
$= 25 - 20i - 20i + 16i^2$ • Use the FOIL method.
$= 25 - 40i + 16i^2$ • Combine like terms.
$= 25 - 40i + (-1)16$ • Replace i^2 with -1.
$= 25 - 40i - 16$
$= 9 - 40i$ • Write the answer in $a + bi$ form.

Problem 7

$(-2 + 5i)(-2 - 5i)$ • $(a + bi)(a - bi) = a^2 + b^2$; $a = -2, b = 5$
$= (-2)^2 + 5^2$
$= 4 + 25 = 29$

Problem 8

$\dfrac{-12 + 8i}{-4i} = \dfrac{-12 + 8i}{-4i} \cdot \dfrac{i}{i}$ • Multiply the numerator and denominator by i.

$= \dfrac{-12i + 8i^2}{-4i^2}$

$= \dfrac{-12i + 8(-1)}{(-4)(-1)}$ • $i^2 = -1$

$= \dfrac{-12i - 8}{4} = \dfrac{-12i}{4} - \dfrac{8}{4}$

$= -2 - 3i$ • Write in the form $a + bi$.

Problem 9

$\dfrac{16 - 11i}{5 + 2i} = \dfrac{16 - 11i}{5 + 2i} \cdot \dfrac{5 - 2i}{5 - 2i}$ • Multiply the numerator and denominator by the conjugate of the denominator.

$= \dfrac{80 - 32i - 55i + 22i^2}{5^2 + 2^2}$

$= \dfrac{80 - 87i + 22i^2}{25 + 4}$ • Combine like terms. Simplify the denominator.

$= \dfrac{80 - 87i + 22(-1)}{29}$ • $i^2 = -1$

$= \dfrac{80 - 87i - 22}{29}$

$= \dfrac{58 - 87i}{29}$

$= \dfrac{58}{29} - \dfrac{87i}{29} = 2 - 3i$ • Write in the form $a + bi$.

Problem 10

$8z^2 + 17 = -7$
$8z^2 = -24$ • Subtract 17 from each side of the equation.
$z^2 = -3$ • Divide each side of the equation by 8.
$\sqrt{z^2} = \sqrt{-3}$ • Take the square root of each side.
$z = \pm\sqrt{-3}$
$= \pm i\sqrt{3}$ • Rewrite $\sqrt{-3}$ as $i\sqrt{3}$.

Check

$8z^2 + 17$	-7
$8(i\sqrt{3})^2 + 17$	-7
$8(i)^2(\sqrt{3})^2 + 17$	-7
$8(-1)3 + 17$	-7
$-24 + 17$	-7
$-7 =$	-7

$8z^2 + 17$	-7
$8(-i\sqrt{3})^2 + 17$	-7
$8(-i)^2(\sqrt{3})^2 + 17$	-7
$8(-1)3 + 17$	-7
$-24 + 17$	-7
$-7 =$	-7

The solutions are $i\sqrt{3}$ and $-i\sqrt{3}$.

Problem 11

$x^2 - 6x + 4 = -6$
$x^2 - 6x + 10 = 0$ • Add 6 to each side of the equation.

$x = \dfrac{-b \pm \sqrt{b^2 - 4ac}}{2a}$

$x = \dfrac{-(-6) \pm \sqrt{(-6)^2 - 4(1)(10)}}{2(1)}$ • $a = 1, b = -6, c = 10$

$= \dfrac{6 \pm \sqrt{36 - 40}}{2} = \dfrac{6 \pm \sqrt{-4}}{2}$

$= \dfrac{6 \pm i\sqrt{4}}{2} = \dfrac{6 \pm 2i}{2}$ • Rewrite $\sqrt{-4}$ as $i\sqrt{4}$ and then rewrite $i\sqrt{4}$ as $2i$.

$= \dfrac{6}{2} \pm \dfrac{2i}{2} = 3 \pm i$ • Write in the form $a + bi$.

Check

$$\begin{array}{r|l} x^2 - 6x + 4 = & -6 \\ \hline (3+i)^2 - 6(3+i) + 4 & -6 \\ 9 + 6i + i^2 - 18 - 6i + 4 & -6 \\ 9 + 6i + (-1) - 18 - 6i + 4 & -6 \\ -6 = & -6 \end{array}$$

$$\begin{array}{r|l} x^2 - 6x + 4 = & -6 \\ \hline (3-i)^2 - 6(3-i) + 4 & -6 \\ 9 - 6i + i^2 - 18 + 6i + 4 & -6 \\ 9 - 6i + (-1) - 18 + 6i + 4 & -6 \\ -6 = & -6 \end{array}$$

The solutions are $3 + i$ and $3 - i$.

SECTION 11.5

Problem 1

$f(x) = -x^2 + 5x - 2$

$f(-1) = -(-1)^2 + 5(-1) - 2$ • **Replace x with −1.**

$= -(1) + 5(-1) - 2$ • **Simplify.**

$= -1 - 5 - 2$

$= -8$

Problem 2

A. $y = x^2 + 2$

$a = 1$. a is positive.

The parabola opens up.

x	y
0	2
1	3
−1	3
2	6
−2	6

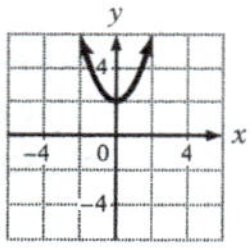

B. $y = -x^2 - 2x + 3$

$a = -1$. a is negative.

The parabola opens down.

x	y
0	3
1	0
−1	4
2	−5
−2	3
−3	0
−4	−5

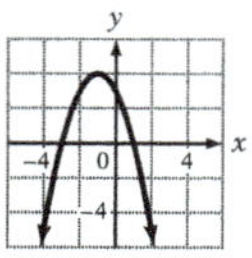

Problem 3

To find the x-intercepts, let $y = 0$ and solve for x.

$y = x^2 - 6x + 9$

$0 = x^2 - 6x + 9$

$0 = (x - 3)(x - 3)$

$x - 3 = 0 \qquad x - 3 = 0$

$x = 3 \qquad x = 3$

The x-intercept is $(3, 0)$.

To find the y-intercept, let $x = 0$ and solve for y.

$y = x^2 - 6x + 9$

$y = 0^2 - 6(0) + 9 = 9$

The y-intercept is $(0, 9)$.

SECTION 11.6

Problem 1

Strategy

- This is a geometry problem.
- Width of the rectangle: W
  Length of the rectangle: $W + 3$
- Use the equation $A = LW$.

Solution

$A = LW$

$40 = (W + 3)W$

$40 = W^2 + 3W$ • **A quadratic equation**

$0 = W^2 + 3W - 40$ • **Write in standard form.**

$0 = (W + 8)(W - 5)$ • **Solve by factoring.**

$W + 8 = 0 \qquad W - 5 = 0$

$W = -8 \qquad W = 5$

The solution -8 is not possible because the width cannot be negative. The width is 5 m.

Answers to Selected Exercises

Answers to Chapter 1 Selected Exercises

PREP TEST

1. 127.16 **2.** 49,743 **3.** 4517 **4.** 11,396 **5.** 24 **6.** 24 **7.** 4 **8.** $3 \cdot 7$ **9.** $\frac{2}{5}$

SECTION 1.1

1. Sometimes true **3.** Always true **5. a.** Negative integer **b.** Positive integer **c.** Negative integer **d.** Neither **e.** Neither **f.** Positive integer **9.** is less than **11.** $>$ **13.** $<$ **15.** $>$ **17.** $>$ **19.** $<$ **21.** $>$ **23.** $>$ **25.** $<$ **27.** i **29.** Yes **31.** $\{1, 2, 3, 4, 5, 6, 7, 8\}$ **33.** $\{1, 2, 3, 4, 5, 6, 7, 8\}$ **35.** $\{-6, -5, -4, -3, -2, -1\}$ **37.** 5 **39.** $-23, -18$ **41.** 21, 37 **43.** $-52, -46, 0$ **45.** $-17, 0, 4, 29$ **47.** 5, 6, 7, 8, 9 **49.** $-10, -9, -8, -7, -6, -5$ **51.** absolute value **53.** -22 **55.** 31 **57.** 168 **59.** -630 **61.** 18 **63.** -49 **65.** 16 **67.** 12 **69.** -29 **71.** -14 **73.** 0 **75.** -34 **77. a.** $8, 5, 2, -1, -3$ **b.** 8, 5, 2, 1, 3 **79.** True **81.** $>$ **83.** $<$ **85.** $>$ **87.** $<$ **89.** $-19, -|-8|, |-5|, 6$ **91.** $-22, -(-3), |-14|, |-25|$ **93. a.** 5°F with a 20 mph wind feels colder. **b.** -25°F with a 10 mph wind feels colder. **95.** $-4, 4$ **97.** $-3, 11$ **99.** Negative **101.** $-5 < 3$ because -5 is to the left of 3 on the number line. $3 > -5$ because 3 is to the right of -5 on the number line. **103.** 7 **105.** -8

SECTION 1.2

1. Sometimes true **3.** Always true **5.** Always true **7.** Negative; minus **11.** 8; -3; 5 **13.** -11 **15.** -9 **17.** -3 **19.** 1 **21.** -5 **23.** -30 **25.** 9 **27.** 1 **29.** -10 **31.** Positive **35.** (-4); -14 **37.** 8 **39.** -7 **41.** 7 **43.** -2 **45.** -28 **47.** -13 **49.** 6 **51.** -9 **53.** 2 **55.** Negative **59.** -10; 7; -70 **61.** positive; 48 **63.** 42 **65.** -20 **67.** -16 **69.** 25 **71.** 0 **73.** -72 **75.** -102 **77.** 140 **79.** -70 **81.** 162 **83.** 120 **85.** 36 **87.** Negative **89.** -15; 3 **91.** $3(-12) = -36$ **93.** $-5(11) = -55$ **95.** -2 **97.** 8 **99.** 0 **101.** -9 **103.** -9 **105.** 9 **107.** -24 **109.** -12 **111.** -13 **113.** -18 **115.** 19 **117.** 26 **119.** Positive **121.** ii **123.** The temperature is 3°C. **125.** The difference is 14°C. **127.** The difference is 399°C. **129.** The difference in elevation is 5670 m. **131.** The difference in elevation is 6051 m. **133.** The difference in elevation is 9261 m. **135.** The average daily low temperature is -3°C. **137.** The temperature rose 49°C. **139.** The difference is 86°F. **141.** The totals for the other players are: Lee Westwood, -13; Anthony Kim, -12; K. J. Choi, -11. **143.** 17 **145.** 3 **147.** $-4, -9, -14$ **149.** $-16, 4, -1$ **151.** 5436 **153.** b **155.** a

159. 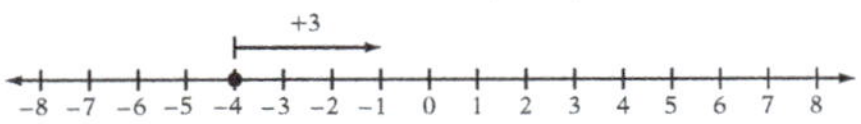

161. 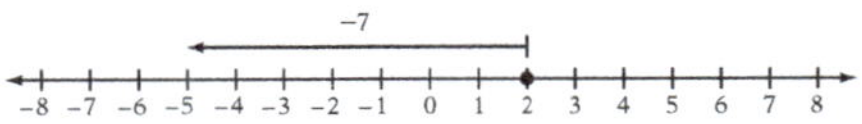

163. 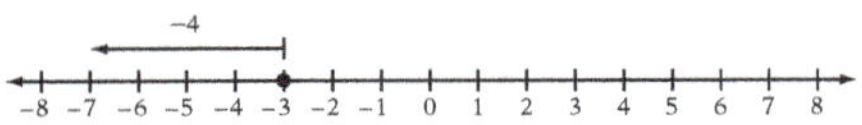

165. To model $-7 + 4$, place 7 red chips and 4 blue chips in a circle. Pair as many red and blue chips as possible. There are 3 red chips remaining, or -3. For $-2 + 6$, use 2 red chips and 6 blue chips. After pairing, there are 4 blue chips remaining, or $+4$. For $-5 + (-3)$, use 5 red chips and then 3 more red chips. There are no red/blue pairs, so there are 8 red chips. The solution is -8.

SECTION 1.3

1. Never true **3.** Always true **5.** Never true **7.** 2; 3; repeating **9.** $0.\overline{3}$ **11.** 0.25 **13.** 0.4 **15.** $0.1\overline{6}$ **17.** 0.125 **19.** $0.\overline{2}$ **21.** $0.\overline{45}$ **23.** $0.58\overline{3}$ **25.** $0.2\overline{6}$ **27.** 0.4375 **29.** 0.24 **31.** 0.225 **33.** $0.6\overline{81}$ **35.** four **37.** $-\frac{3}{8}$ **39.** $\frac{1}{10}$ **41.** $\frac{15}{64}$ **43.** $\frac{3}{2}$ **45.** $-\frac{8}{9}$ **47.** $\frac{2}{3}$ **49.** 4.164 **51.** 4.347 **53.** -4.028 **55. a.** Negative **b.** Positive **57.** 0.75 **59.** -2060.55 **61.** 6 **63.** $\frac{1}{2}$ **65.** -1 **67.** $-\frac{25}{18}$ **69.** $\frac{1}{24}$ **71.** $\frac{17}{18}$ **73.** $-\frac{47}{48}$ **75.** $\frac{3}{8}$ **77.** $-\frac{7}{60}$ **79.** $-\frac{1}{16}$ **81.** $-\frac{7}{24}$ **83.** 7.29 **85.** -3.049 **87.** -1.06

89. −23.845 **91.** −10.7893 **93.** 11.56 **95.** −60.03 **97.** −34.99 **99.** **a.** 2 **b.** 0 **c.** 1 **d.** −1
103. $\frac{1}{100}$; $\frac{1}{100}$; $\frac{4}{5}$ **105.** 100%; 100%; 30% **107.** $\frac{3}{4}$, 0.75 **109.** $\frac{1}{2}$, 0.5 **111.** $\frac{16}{25}$, 0.64 **113.** $1\frac{3}{4}$, 1.75
115. $\frac{19}{100}$, 0.19 **117.** $\frac{1}{20}$, 0.05 **119.** $4\frac{1}{2}$, 4.5 **121.** $\frac{2}{25}$, 0.08 **123.** $\frac{1}{9}$ **125.** $\frac{5}{16}$ **127.** $\frac{1}{200}$ **129.** $\frac{1}{16}$
131. 0.073 **133.** 0.158 **135.** 0.0915 **137.** 0.1823 **139.** 15% **141.** 5% **143.** 17.5% **145.** 115%
147. 0.8% **149.** 6.5% **151.** 54% **153.** 33.3% **155.** 44.4% **157.** 250% **159.** $37\frac{1}{2}$% **161.** $35\frac{5}{7}$%
163. 125% **165.** $155\frac{5}{9}$% **167.** Greater than 100% **169.** $\frac{2}{5}$ of the respondents found their most recent jobs on the Internet.
171. Less than one-quarter of the respondents found their most recent jobs through a newspaper ad. **173.** Natural number, integer, positive integer, rational number, real number **175.** Rational number, real number **177.** Irrational number, real number
179. **a.** The difference is 200.0°F. **b.** The difference is 111.1°C. **181.** The difference is \$138.478 billion. **183.** The deficit in 1985 was 4 times greater than in 1975. **185.** The average normal temperature in the Northeast in February is −3.6°C. **187.** 0.70x

191.

$\frac{2}{3}$	$-\frac{1}{6}$	0
$-\frac{1}{2}$	$\frac{1}{6}$	$\frac{5}{6}$
$\frac{1}{3}$	$\frac{1}{2}$	$-\frac{1}{3}$

193. Answers will vary. For example: $\frac{1}{2} + \left(-\frac{1}{4}\right) = -\frac{3}{4}$; $\frac{1}{2} + \frac{1}{4} = \frac{3}{4}$; $\frac{3}{4} + \left(-\frac{1}{4}\right) = \frac{1}{2}$

SECTION 1.4

1. 9^5 **3.** 7^n **5.** False **7.** True **9.** base; exponent; −5; −5; 25 **11.** 36 **13.** −49 **15.** 9 **17.** 81
19. $\frac{1}{4}$ **21.** 0.09 **23.** −12 **25.** −864 **27.** 12 **29.** 0.216 **31.** 3 **33.** Negative **37.** 27; 54 **39.** 0
41. −11 **43.** 20 **45.** 11 **47.** −11 **49.** 6 **51.** 741 **53.** −17 **55.** 1 **57.** −1 **59.** 0.51
61. $\frac{1}{4}$ **63.** iii **67.** $>$ **69.** It would take the computer 13 s. **71.** Answers will vary. For example: **a.** $r = \frac{1}{2}$
b. $r = 0$ or $r = 1$ **c.** $r = 2$ **73.** 6 **75.** 9

SECTION 1.5

1. True **3.** True **5.** Less than **7.** 90° **9.** 90° **11.** Less than **13.** 28° **15.** 132° **17.** 83° **19.** 91°
21. 132° **23.** 51° **25.** 77° **27.** 79° **29.** 292° **31.** triangle **33.** 9.71 cm **35.** 14 ft 2 in. **37.** 52 in.
39. 131.88 cm **41.** 3.768 m **43.** The wood framing would cost \$76.96. **45.** The cost of the binding is \$19.78.
47. 22 units **49.** width **51.** 32 ft^2 **53.** 378 cm^2 **55.** 50.24 in^2 **57.** 16.81 m^2 **59.** 52.5 cm^2
61. 226.865 in^2 **63.** The cost of carpeting the entire living space would be \$1,600,000. **65.** 36 yd^2 of carpet are necessary.
67. The cost to build the window is \$603. **69.** The area of the reserve is approximately 10,500 mi^2. **71.** Dimensions of 5 units by 5 units will give a maximum area. **73.** Perimeter: 176 ft; area: 1008 ft^2 **75.** 8 ft^2

CHAPTER 1 REVIEW EXERCISES*

1. {1, 2, 3, 4, 5, 6} [1.1.1] **2.** 62.5% [1.3.4] **3.** −4 [1.1.2] **4.** 4 [1.2.2] **5.** 18 cm^2 [1.5.3] **6.** $0.\overline{7}$ [1.3.1]
7. −5.3578 [1.3.2] **8.** 8 [1.4.2] **9.** 4 [1.1.2] **10.** −14 [1.2.2] **11.** 67.2% [1.3.4] **12.** $\frac{159}{200}$ [1.3.4]
13. −9 [1.2.4] **14.** 0.85 [1.3.1] **15.** $-\frac{1}{2}$ [1.3.2] **16.** 9 [1.4.2] **17.** 37° [1.5.1] **18.** −16 [1.2.1]
19. 90 [1.2.3] **20.** −6.881 [1.3.3] **21.** −5, −3 [1.1.1] **22.** 0.07 [1.3.4] **23.** 12 [1.4.1] **24.** $>$ [1.1.1]
25. −3 [1.2.1] **26.** 34° [1.5.1] **27.** $\frac{1}{15}$ [1.3.3] **28.** −108 [1.4.1] **29.** 277.8% [1.3.4] **30.** 2.4 [1.3.4]
31. 152° [1.5.1] **32.** −8 [1.2.4] **33.** 28.26 m^2 [1.5.3] **34.** 12 [1.4.2] **35.** 3 [1.1.2] **36.** $166\frac{7}{10}$% [1.3.4]
37. **a.** 12, 8, 1, −7 **b.** 12, 8, 1, 7 [1.1.2] **38.** $0.\overline{63}$ [1.3.1] **39.** −11.5 [1.3.2] **40.** 8 [1.4.2] **41.** −8 [1.2.1]
42. $-\frac{11}{24}$ [1.3.3] **43.** −17, −9, 0, 4 [1.1.1] **44.** 0.2% [1.3.4] **45.** −4 [1.4.1] **46.** 3.561 [1.3.3]
47. −17 [1.1.2] **48.** −27 [1.2.2] **49.** $-\frac{1}{10}$ [1.3.2] **50.** $<$ [1.1.1] **51.** 44 in. [1.5.2] **52.** −128 [1.4.1]
53. 7.5% [1.3.4] **54.** $54\frac{2}{7}$% [1.3.4] **55.** −18 [1.2.1] **56.** 0 [1.2.3] **57.** 9 [1.4.2] **58.** $\frac{7}{8}$ [1.3.3]
59. 16 [1.2.4] **60.** $<$ [1.1.1] **61.** −3 [1.4.1] **62.** −6 [1.2.1] **63.** 300 [1.2.3] **64.** {−3, −2, −1} [1.1.1]

*The numbers in brackets following the answers in the Chapter Review Exercises refer to the objective that corresponds to that problem. For example, the reference [1.2.1] stands for Section 1.2, Objective 1. This notation will be used for all Prep Tests, Chapter Review Exercises, Chapter Tests, and Cumulative Review Exercises throughout the text.

65. The cost is \$240.96. [1.5.3] **66.** The temperature is 8°C. [1.2.5] **67.** The average low temperature was −2°C. [1.2.5] **68.** The difference is 108°F. [1.2.5] **69.** The temperature is −6°C. [1.2.5] **70.** The difference is 714°C. [1.2.5]

CHAPTER 1 TEST

1. $\frac{11}{20}$ [1.3.4, Example 11] **2.** −8, −6 [1.1.1, Example 2] **3.** 47° [1.5.1, Example 3] **4.** 0.15 [1.3.1, Example 1] **5.** $-\frac{1}{14}$ [1.3.2, Problem 3] **6.** −15 [1.2.4, Example 5B] **7.** $-\frac{8}{3}$ [1.4.1, Example 3] **8.** 2 [1.2.1, Problem 1C] **9.** {1, 2, 3, 4, 5, 6} [1.1.1, Example 1] **10.** 159% [1.3.4, Example 14B] **11.** 29 [1.1.2, Example 4A] **12.** > [1.1.1, Example 2] **13.** $-\frac{23}{18}$ [1.3.3, Problem 7] **14.** 90 cm^2 [1.5.3, Example 9] **15.** 3 [1.4.2, Example 5] **16.** $\frac{5}{6}$ [1.3.2, Problem 4] **17.** 23.1% [1.3.4, Example 15] **18.** 0.062 [1.3.4, Example 13] **19.** 14 [1.2.2, Example 3] **20.** $0.4\overline{3}$ [1.3.1, Example 2] **21.** −2.43 [1.3.2, Problem 5] **22.** 62° [1.5.1, Example 1] **23.** 640 [1.4.1, Example 3] **24.** −34 [1.1.2, Example 4B] **25.** 84.78 in. [1.5.2, Problem 5] **26. a.** 17, 6, −5, −9 [1.1.2, Example 3] **b.** 17, 6, 5, 9 [1.1.2, Example 4] **27.** $69\frac{13}{23}$% [1.3.4, Example 16] **28.** −11.384 [1.3.3, Example 10] **29.** 160 [1.2.3, Example 4B] **30.** −2 [1.4.2, Example 6] **31.** The temperature is 4°C. [1.2.5, Example 6] **32.** The average is −48°F. [1.2.5, Example 6] **33.** The cost is \$4564. [1.5.2, Example 6]

Answers to Chapter 2 Selected Exercises

PREP TEST

1. 3 [1.2.2] **2.** 4 [1.2.4] **3.** $\frac{1}{12}$ [1.3.3] **4.** $-\frac{4}{9}$ [1.3.2] **5.** $\frac{3}{10}$ [1.3.2] **6.** −16 [1.4.1] **7.** $\frac{8}{27}$ [1.4.1] **8.** 48 [1.4.1] **9.** 1 [1.4.2] **10.** 12 [1.4.2]

SECTION 2.1

1. Always true **3.** Sometimes true **5.** Always true **7.** −2; 5; 25; −50; 2; −48 **9.** $-3n^2$; $-4n$; $\underline{7}$ **11.** $-9\underline{b^2}$, $-4\underline{ab}$, $\underline{a^2}$ **13.** $-8\underline{n}$, $-3\underline{n^2}$ **15.** 12, −8, −1 **19.** 12 **21.** −4 **23.** 6 **25.** 6 **27.** −2 **29.** −3 **31.** −5 **33.** 25 **35.** 0 **37.** 10 **39.** −11 **41.** 1 **43.** $\frac{5}{2}$ **45.** $-\frac{5}{6}$ **47.** 28 **49.** 5 **51.** 8 **53.** 1 **55.** 22 **57.** −2 **59.** 9 **61.** 8 **63.** 9 **65.** 4.96 **67.** −5.68 **69. a.** No, V cannot be a whole number. **b.** No, volume is not measured in square units. **71.** The volume is 25.8 in^3. **73.** The area is 93.7 cm^2. **75.** The volume is 484.9 m^3. **77.** 41 **79.** 21 **81.** 24 **83.** 1 **85.** −8 **87.** −23 **89. a.** 4 **b.** 5 **c.** 6 **d.** 7; $n^x > x^n$ if $x \geq n + 1$ **91.** $G + 8$ **93.** $A - 16$

SECTION 2.2

1. iv **3.** True. The Multiplication Property of One **5.** False **7.** False **9.** Commutative **11.** reciprocal (or multiplicative inverse) **13.** 2 **15.** 5 **17.** 6 **19.** −8 **21.** −4 **23.** The Inverse Property of Addition **25.** The Commutative Property of Addition **27.** The Associative Property of Addition **29.** The Commutative Property of Multiplication **31.** The Associative Property of Multiplication **35.** $3x$, $5x$ **37.** $-8a$; Distributive; −3 **39.** $14x$ **41.** $5a$ **43.** $-6y$ **45.** $-3b - 7$ **47.** $5a$ **49.** $-2ab$ **51.** $5xy$ **53.** 0 **55.** $-\frac{5}{6}x$ **57.** $-\frac{1}{24}x^2$ **59.** $11x$ **61.** $7a$ **63.** $-14x^2$ **65.** $-x + 3y$ **67.** $17x - 3y$ **69.** $-2a - 6b$ **71.** $-3x - 8y$ **73.** $-4x^2 - 2x$ **75. a.** Negative **b.** Positive **c.** 0 **77.** Associative; $-48x$ **79.** $12x$ **81.** $-21a$ **83.** $6y$ **85.** $8x$ **87.** $-6a$ **89.** $12b$ **91.** $-15x^2$ **93.** x^2 **95.** x **97.** n **99.** x **101.** n **103.** $2x$ **105.** $-2x$ **107.** $-15a^2$ **109.** $6y$ **111.** $3y$ **113.** $-2x$ **115.** Less than 1 **117.** $8x - 6$ **119.** $-2a - 14$ **121.** $-6y + 24$ **123.** $-x - 2$ **125.** $35 - 21b$ **127.** $-9 + 15x$ **129.** $15x^2 + 6x$ **131.** $2y - 18$ **133.** $-15x - 30$ **135.** $-6x^2 - 28$ **137.** $-6y^2 + 21$ **139.** $3x^2 + 6x - 18$ **141.** $-2y^2 + 4y - 8$ **143.** $-2a^2 - 4a + 6$ **145.** $10x^2 + 15x - 35$ **147.** $-3a^2 - 5a + 4$ **149.** Positive **151.** 12; 28; 2; 6 **153.** $-2x - 16$ **155.** $-12y - 9$ **157.** $7n - 7$ **159.** $-2x + 41$ **161.** $-a - 7b$ **163.** $-4x + 24$ **165.** $-3x + 21$ **167.** $-7x + 24$ **169.** $20x - 41y$ **171.** iii **173.** 0 **175.** $-a + b$ **177. a.** False; $8 \div 4 \neq 4 \div 8$ **b.** False; $(8 \div 4) \div 2 \neq 8 \div (4 \div 2)$ **c.** False; $(7 - 5) - 1 \neq 7 - (5 - 1)$ **d.** False; $6 - 3 \neq 3 - 6$ **e.** True **179.** ◇ **181.** ◇ **183.** ‡ **185.** Δ

SECTION 2.3

1. False **3.** False **5.** True **7.** less than, quotient **9.** subtracted from, product, cube **11.** $6 + c$ **13.** $w + 55$ **15.** $16 + y$ **17.** $b^2 - 30$ **19.** $\frac{4}{5}m + 18$ **21.** $9 + \frac{t}{5}$ **23.** $7(r + 8)$ **25.** $a(a + 13)$ **27.** $\frac{1}{2}z^2 + 14$

29. $9m^3 + m^2$ **31.** $s - \frac{s}{2}$ **33.** $c^2 - (c + 14)$ **35.** $8(b + 5)$ **37.** $13 - n$ **39.** $\frac{3}{7}n$ **41.** $\frac{2n}{5}$ **43.** $n(n + 10)$ **45.** $\frac{3}{4 + n}$ **47.** $n^2 + 3n$ **49.** $7n^2 - 4$ **51.** $n^3 - 12n$ **53.** i, ii, iii **55.** 12; 15; $-3n$ **57.** $5n + n$; $6n$ **59.** $(n + 11) + 8$; $n + 19$ **61.** $(n + 9) + 4$; $n + 13$ **63.** $7(5n)$; $35n$ **65.** $17n + 2n$; $19n$ **67.** $n + 12n$; $13n$ **69.** $3(n^2 + 4)$; $3n^2 + 12$ **71.** $\frac{3}{4}(16n + 4)$; $12n + 3$ **73.** $16 - (n + 9)$; $-n + 7$ **75.** $6(n + 8)$; $6n + 48$ **77.** $7 - (n + 2)$; $-n + 5$ **79.** $\frac{1}{3}(n + 6n)$; $\frac{7n}{3}$ **81.** $(n - 6) + (n + 12)$; $2n + 6$ **83.** $(n - 20) + (n + 9)$; $2n - 11$ **85.** $25 - x$ **87.** $\frac{1}{2}L$ **89.** Let x be one number; x and $18 - x$ **91.** Let d be the diameter of Dione; $d + 253$ **93.** Let G be the number of genes in the roundworm genome; $G + 11{,}000$ **95.** Let N be the total number of Americans; $\frac{3}{4}N$ **97.** Let s be the number of points awarded for a safety; $3s$ **99.** Attendance at major league basketball games: B; attendance at major league baseball games: $B + 50{,}000{,}000$ **101.** Let L be the measure of the largest angle; $\frac{1}{2}L - 10$ **103.** Let h be the number of hours of labor; $238 + 89h$ **105.** Let n be either the number of nickels or the number of dimes; n and $35 - n$ **107.** $2x$ **109.** $\frac{3}{5}x$ **113.** Answers will vary. For example: The sum of p and 8; the total of p and 8; 8 more than p; 8 added to p; p increased by 8 **115.** Answers will vary. For example: 4 times c; the product of 4 and c; 4 multiplied by c **117. a.** Column 1: 1, 2, 3, 4, 5, 6, 7; Column 2: 5, 7, 9, 11, 13, 15, 17 **b.** $n + n + 3$, or $2n + 3$

CHAPTER 2 REVIEW EXERCISES

1. y^2 [2.2.2] **2.** $3x$ [2.2.3] **3.** $-10a$ [2.2.3] **4.** $-4x + 8$ [2.2.4] **5.** $7x + 38$ [2.2.5] **6.** 16 [2.1.1] **7.** 9 [2.2.1] **8.** $36y$ [2.2.3] **9.** $6y - 18$ [2.2.4] **10.** $-3x + 21y$ [2.2.5] **11.** $-8x^2 + 12y^2$ [2.2.4] **12.** $5x$ [2.2.2] **13.** 22 [2.1.1] **14.** $2x$ [2.2.3] **15.** $15 - 35b$ [2.2.4] **16.** $-7x + 33$ [2.2.5] **17.** The Commutative Property of Multiplication [2.2.1] **18.** $24 - 6x$ [2.2.4] **19.** $5x^2$ [2.2.2] **20.** $-7x + 14$ [2.2.5] **21.** $-9y^2 + 9y + 21$ [2.2.4] **22.** $2x + y$ [2.2.5] **23.** 3 [2.1.1] **24.** $36y$ [2.2.3] **25.** $5x - 43$ [2.2.5] **26.** $2x$ [2.2.3] **27.** $-6x^2 + 21y^2$ [2.2.4] **28.** 6 [2.1.1] **29.** $-x + 6$ [2.2.5] **30.** $-5a - 2b$ [2.2.2] **31.** $-10x^2 + 15x - 30$ [2.2.4] **32.** $-9x - 7y$ [2.2.2] **33.** $6a$ [2.2.3] **34.** $17x - 24$ [2.2.5] **35.** $-2x - 5y$ [2.2.2] **36.** $30b$ [2.2.3] **37.** 21 [2.2.1] **38.** $-2x^2 + 4x$ [2.2.2] **39.** $-6x^2$ [2.2.3] **40.** $15x - 27$ [2.2.5] **41.** $-8a^2 + 3b^2$ [2.2.4] **42.** The Multiplication Property of Zero [2.2.1] **43.** $b - 7b$ [2.3.1] **44.** $n + 2n^2$ [2.3.1] **45.** $\frac{6}{n} - 3$ [2.3.1] **46.** $\frac{10}{y - 2}$ [2.3.1] **47.** $8\left(\frac{2n}{16}\right)$; n [2.3.2] **48.** $4(2 + 5n)$; $8 + 20n$ [2.3.2] **49.** Let h be the height of the triangle; $h + 15$ [2.3.3] **50.** Let b be the amount of either espresso beans or mocha java beans; b and $20 - b$ [2.3.3]

CHAPTER 2 TEST

1. $36y$ [2.2.3, Example 5C] **2.** $4x - 3y$ [2.2.2, Example 4B] **3.** $10n - 6$ [2.2.5, Example 7] **4.** 2 [2.1.1, Example 2] **5.** The Multiplication Property of One [2.2.1, Example 2] **6.** $4x - 40$ [2.2.4, Example 6A] **7.** $\frac{1}{12}x^2$ [2.2.2, Example 3B] **8.** $4x$ [2.2.3, Problem 5C] **9.** $-24y^2 + 48$ [2.2.4, Example 6C] **10.** 19 [2.2.1, Problem 1] **11.** 6 [2.1.1, Example 3] **12.** $-3x + 13y$ [2.2.5, Example 7] **13.** b [2.2.2, Problem 4A] **14.** $78a$ [2.2.3, Example 5A] **15.** $3x^2 - 15x + 12$ [2.2.4, Problem 6D] **16.** -32 [2.1.1, Example 4] **17.** $37x - 5y$ [2.2.5, Example 7] **18.** $\frac{n + 8}{17}$ [2.3.1, Example 1B] **19.** $(a + b) - b^2$ [2.3.1, Problem 1C] **20.** $n^2 + 11n$ [2.3.1, Problem 2] **21.** $20(n + 9)$; $20n + 180$ [2.3.2, Example 4] **22.** $(n - 3) + (n + 2)$; $2n - 1$ [2.3.2, Problem 4] **23.** $n - \frac{1}{4}(2n)$; $\frac{1}{2}n$ [2.3.2, Example 5] **24.** Let d be the distance from Earth to the sun; $30d$ [2.3.3, Example 6] **25.** Let L be the length of one piece; L and $9 - L$ [2.3.3, Problem 7]

CUMULATIVE REVIEW EXERCISES

1. -7 [1.2.1] **2.** 5 [1.2.2] **3.** 24 [1.2.3] **4.** -5 [1.2.4] **5.** 1.25 [1.3.1] **6.** $\frac{3}{5}$, 0.60 [1.3.4] **7.** $\{-4, -3, -2, -1\}$ [1.1.1] **8.** 8% [1.3.4] **9.** $\frac{11}{48}$ [1.3.3] **10.** $\frac{5}{18}$ [1.3.2] **11.** $\frac{1}{4}$ [1.3.2] **12.** $\frac{8}{3}$ [1.4.1] **13.** -5 [1.4.2] **14.** $\frac{53}{48}$ [1.4.2] **15.** -8 [2.1.1] **16.** $5x^2$ [2.2.2] **17.** $-a - 12b$ [2.2.2] **18.** $3a$ [2.2.3] **19.** $20b$ [2.2.3] **20.** $20 - 10x$ [2.2.4] **21.** $6y - 21$ [2.2.4] **22.** $-6x^2 + 8y^2$ [2.2.4] **23.** $-8y^2 + 20y + 32$ [2.2.4] **24.** $-10x + 15$ [2.2.5] **25.** $5x - 17$ [2.2.5] **26.** $13x - 16$ [2.2.5] **27.** $6x + 29y$ [2.2.5] **28.** $6 - 12n$ [2.3.1] **29.** $5 + (n - 7)$; $n - 2$ [2.3.2] **30.** Let w be the speed of the wildebeest; $4w$ [2.3.3]

Answers to Chapter 3 Selected Exercises

PREP TEST

1. −4 [1.2.2] **2.** 1 [1.3.2] **3.** −10 [1.3.2] **4.** 0.9 [1.3.4] **5.** 75% [1.3.4] **6.** 63 [2.1.1] **7.** $10x - 5$ [2.2.2] **8.** −9 [2.2.2] **9.** $9x - 18$ [2.2.5]

SECTION 3.1

1. a and d are equations; b, c, and e are expressions. **3.** a, b, and d are equations of the form $x + a = b$. **5.** Sometimes true **7.** Never true **9.** −3; 15, 27; 24, 24; equal **11.** No **13.** No **15.** Yes **17.** Yes **19.** Yes **21.** Yes **23.** Yes **25.** No **27.** Yes **29.** subtract; 7 **33.** 6 **35.** 16 **37.** 7 **39.** −2 **41.** 1 **43.** 0 **45.** 3 **47.** −10 **49.** −3 **51.** −14 **53.** 2 **55.** −9 **57.** −1 **59.** −14 **61.** −5 **63.** −1 **65.** 1 **67.** $-\frac{2}{3}$ **69.** $\frac{5}{12}$ **71.** $\frac{8}{9}$ **73.** 0.6529 **75.** −0.283 **79.** $\frac{3}{2}$; −27 **83.** 3 **85.** −4 **87.** −2 **89.** 9 **91.** 5 **93.** −4 **95.** 0 **97.** −8 **99.** 6 **101.** −10 **103.** −28 **105.** 30 **107.** −24 **109.** 3 **111.** −24 **113.** 9 **115.** 4 **117.** 3 **119.** 4.48 **121.** 2.06 **123.** No **125.** Yes **127.** −21 **129.** −27 **131.** 21 **133.** One possible answer is $x + 7 = 9$. **135.** $\frac{7}{11}$ **137. a.** iv **b.** ii **c.** i **d.** v **e.** iii

SECTION 3.2

1. Amount: 30; base: 40 **3.** Keith **7.** 24% **9.** 7.2 **11.** 400 **13.** 9 **15.** 25% **17.** 5 **19.** 200% **21.** 400 **23.** 7.7 **25.** 200 **27.** 400 **29.** 20 **31.** 80% **33.** 40% of 80 is equal to 80% of 40. **35.** unknown; 30; 24 **37.** 250 seats are wheelchair accessible. **39.** In the average single-family home, 74 gal of water are used per person per day. **41.** There is insufficient information. **43.** 12% of the deaths were not attributed to motor vehicle accidents. **45.** In this country, 96.1 billion kilowatt-hours of electricity are used for home lighting per year. **47.** 34.4% of the U.S. population watched Super Bowl XLIV. **49.** The price of the less expensive model is $1423.41. **51.** The annual simple interest rate is 9%. **53.** Sal earned $240 in interest from the two accounts. **55.** Makana earned $63 in one year. **57.** The interest rate on the combined investment is between 6% and 9%. **59.** The percent concentration of the hydrogen peroxide is 2%. **61.** Apple Dan's has the greater concentration of apple juice. **63.** 12.5 g of the cream are not glycerine. **65.** The percent concentration of salt in the remaining solution is 12.5%. **67.** 20 **69.** 355; 295 **71. a.** Equal to **b.** Less than **73.** The train travels 245 mi in the 5-hour period. **75.** The dietician's average rate of speed is 30 mph. **77.** It will take 4 h to complete the trip. **79.** It would take 31.25 s to walk from one end of the moving sidewalk to the other. **81.** The two joggers will meet 40 min after they start. **83.** The cyclists are 8.5 mi apart. **85.** The trains are 30 mi apart. **87.** The cost of the dinner was $80. **89.** The new value is two times the original value. **91. a.** Northeast: 18.1%; Midwest: 21.8%; South: 36.8%; West: 23.4% **b.** The South has the largest population. The largest percent of the population lives in the South. **c.** 12.3% of the U.S. population lives in California. **d.** 520,000 residents live in Wyoming. **e.** Answers will vary.

SECTION 3.3

1. a and i; b and iii; c and ii; d and iv **3.** True **5.** True **7.** True **9.** 18 **11.** Positive **13.** 2 **15.** 3 **17.** −2 **19.** −2 **21.** 2 **23.** 2 **25.** 2 **27.** 3 **29.** 1 **31.** 6 **33.** −7 **35.** $\frac{1}{2}$ **37.** $\frac{7}{3}$ **39.** $\frac{7}{8}$ **41.** $\frac{1}{3}$ **43.** 1 **45.** 0 **47.** $\frac{2}{5}$ **49.** 18 **51.** 8 **53.** −16 **55.** 25 **57.** 21 **59.** 15 **61.** $\frac{7}{6}$ **63.** $\frac{1}{6}$ **65.** −25 **67.** 1 **69.** $-\frac{21}{4}$ **71.** 2 **73.** 1 **75.** 2 **77.** 3.95 **79.** −0.8 **81.** −11 **83.** 0 **85.** 12; 12; Distributive; 2; 84; $2x$; $2x$; 7; Divide; 7; −12 **87.** −2 **89.** 3 **91.** 8 **93.** 2 **95.** −2 **97.** −3 **99.** 2 **101.** −2 **103.** −2 **105.** −7 **107.** 0 **109.** −2 **111.** −2 **113.** 4 **115.** 10 **117.** 3 **119.** $\frac{3}{4}$ **121.** $\frac{2}{7}$ **123.** $-\frac{3}{4}$ **125.** 3 **127.** 10 **129.** $\frac{4}{3}$ **131.** 3 **133.** −14 **135.** 7 **137.** $3x$; 18 **139.** i and iii **141.** 3 **143.** 4 **145.** 1 **147.** 2 **149.** 2 **151.** −7 **153.** $\frac{4}{7}$ **155.** $\frac{1}{2}$ **157.** $-\frac{1}{3}$ **159.** $\frac{10}{3}$ **161.** $-\frac{1}{4}$ **163.** 0.5 **165.** 0 **167.** −1 **169.** The car will slide 168 ft. **171.** The depth of the diver is 40 ft. **173.** The height of the adult is approximately 182.5 cm. **175.** The initial velocity is 8 ft/s. **177.** The passenger was driven 6 mi. **179.** The average crown spread of the baldcypress is 57 ft. **181.** He made 8 errors. **183.** The break-even point is 350 television sets. **187.** The fulcrum is 5 ft from the other person. **189.** No **191.** No, the seesaw is not balanced. **193.** A force of 25 lb must be applied to the other end of the lever. **195.** The fulcrum must be placed 10 ft from the 128-pound acrobat. **197.** No solution **199.** $-\frac{11}{4}$ **201.** 0 **205.** 7 **207.** Yes; medical care cost more than three times as much in 2010 as during the base years. **209.** A comparable new car would have cost $15,617 during the base years.

SECTION 3.4

1. $x \le 4$ includes the element 4; $x < 4$ does not. **3. a.** No **b.** No **c.** Yes **d.** No **e.** No **f.** Yes
5. Always true **7.** includes; $\le$ **9.** 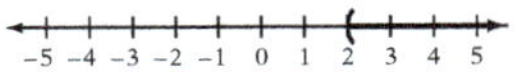 **11.** 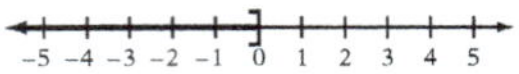 **13.** i, iii
15. $x < 2$  **17.** $x > 3$ 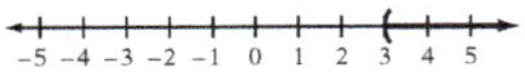 **19.** $n \ge 3$ 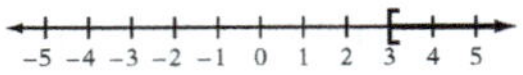
21. $x \le -4$ 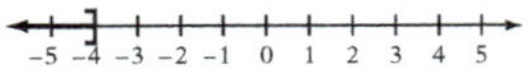 **23.** $x \ge -1$ 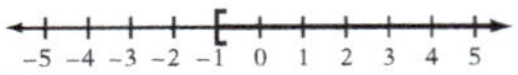 **25.** Only negative numbers
27. Only positive numbers **29.** $y \ge -9$ **31.** $x < 12$ **33.** $x \ge 5$ **35.** $x < -11$ **37.** $x \le 10$ **39.** $x \ge -6$
41. $x > 2$ **43.** $d < -\frac{1}{6}$ **45.** $x \ge -\frac{31}{24}$ **47.** $x < \frac{5}{4}$ **49.** $x > \frac{5}{24}$ **51.** $x \le -1.2$ **53.** $x \le 0.70$ **55.** $x < -7.3$
57. i, ii, iii **59.** is reversed **61.** $x \le -3$ 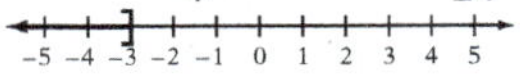 **63.** $x > -2$ 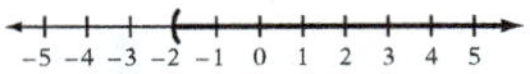
65. $x > 0$ 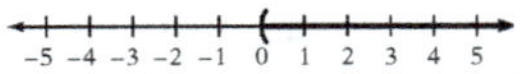 **67.** $n \ge 4$ 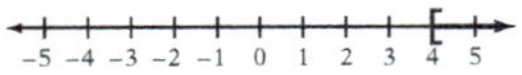 **69.** $x > -2$ 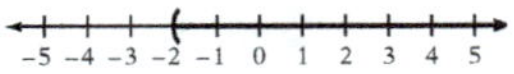
71. $x < \frac{5}{3}$ **73.** $x \ge 5$ **75.** $x > -\frac{5}{2}$ **77.** $x \le -\frac{2}{3}$ **79.** $x < -18$ **81.** $x > -16$ **83.** $b \le 33$ **85.** $n < \frac{3}{4}$
87. $x \le -\frac{6}{7}$ **89.** $y \le \frac{5}{6}$ **91.** $x > -0.5$ **93.** $y \ge -0.8$ **95.** $x < -5.4$ **97.** iv **101.** $2x$; $2x$; $3x$; 0; 9; $-3x$; -9; -3; $\le$; -3; 3 **103.** $x \le 5$ **105.** $x < 0$ **107.** $x < 4$ **109.** $x < -4$ **111.** $x \ge 1$ **113.** $x < 20$
115. $x > 500$ **117.** $x > 2$ **119.** $x \le -5$ **121.** $y \le \frac{5}{2}$ **123.** $x < \frac{25}{11}$ **125.** $x > 11$ **127.** $n \le \frac{11}{18}$
129. $x \ge 6$ **131.** $x \le \frac{2}{5}$ **133.** $t < 1$ **135.** 3 **137.** iii, iv **139.** $\{3, 4, 5\}$ **141.** $\{10, 11, 12, 13\}$ **143.** $x \le -2$
145. The length of the third side is between 8 in. and 28 in. **147.** 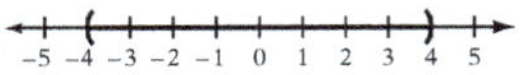
149. 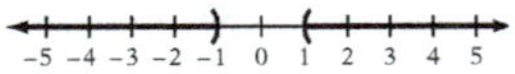 **151.** Always true **153.** Never true **155.** Sometimes true

CHAPTER 3 REVIEW EXERCISES

1. No [3.1.1] **2.** 20 [3.1.2] **3.** -7 [3.1.3] **4.** 7 [3.3.1] **5.** 4 [3.3.2] **6.** $-\frac{1}{5}$ [3.3.3] **7.** 405 [3.2.1]
8. 25 [3.2.1] **9.** 67.5% [3.2.1] **10.** 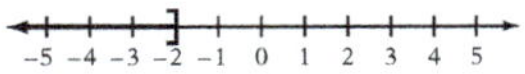 [3.4.1] **11.** $x > 2$ 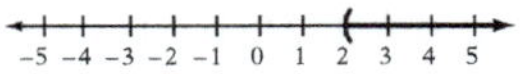 [3.4.1] **12.** $x > -4$ 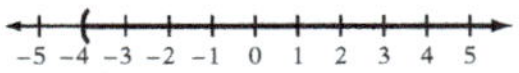 [3.4.2] **13.** $x \ge -4$ [3.4.3] **14.** $x \ge 4$ [3.4.3] **15.** Yes [3.1.1]
16. 2.5 [3.1.2] **17.** -49 [3.1.3] **18.** $\frac{1}{2}$ [3.3.1] **19.** $\frac{1}{3}$ [3.3.2] **20.** 10 [3.3.3] **21.** 16 [3.2.1]
22. 125 [3.2.1] **23.** $16\frac{2}{3}\%$ [3.2.1] **24.** $x < -4$ 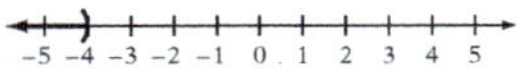 [3.4.1] **25.** $x \le -2$
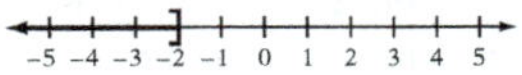 [3.4.2] **26.** -2 [3.3.3] **27.** $\frac{5}{6}$ [3.1.2] **28.** 20 [3.1.3] **29.** 6 [3.3.1]
30. 0 [3.3.3] **31.** $x < 12$ [3.4.3] **32.** 15 [3.2.1] **33.** 5 [3.3.1] **34.** $x > 5$ [3.4.3] **35.** 4% [3.2.1]
36. $x > -18$ [3.4.3] **37.** $x < \frac{1}{2}$ [3.4.3] **38.** The measure of the third angle is 110°. [3.3.4] **39.** A force of 24 lb must be applied to the other end of the lever. [3.3.4] **40.** The width is 6 ft. [3.3.4] **41.** The discount is \$41.99. [3.3.4]
42. Approximately 11,065 plants and animals are at risk of extinction on Earth. [3.2.1] **43.** The depth is 80 ft. [3.3.4]
44. The fulcrum is 3 ft from the 25-pound force. [3.3.4] **45.** The length of the rectangle is 24 ft. [3.3.4] **46.** It will take the motorboat 1.5 h to travel 30 mi. [3.2.2] **47.** She must invest \$625 in an account that earns an annual simple interest rate of 8%. [3.2.1] **48.** The percent concentration of hydrochloric acid is 6%. [3.2.1]

CHAPTER 3 TEST

1. -12 [3.1.3, Example 5] **2.** $-\frac{1}{2}$ [3.3.2, Example 5] **3.** -3 [3.3.1, Example 1] **4.** No [3.1.1, Example 1]
5. $\frac{1}{8}$ [3.1.2, Problem 3] **6.** $-\frac{1}{3}$ [3.3.3, Example 6] **7.** 5 [3.3.1, Example 1] **8.** $\frac{1}{2}$ [3.3.2, Example 5] **9.** -5 [3.1.2, Example 3] **10.** -5 [3.3.2, Example 5] **11.** $-\frac{40}{3}$ [3.1.3, Problem 5] **12.** $-\frac{22}{7}$ [3.3.3, Example 6]
13. 2 [3.3.3, Example 6] **14.** $\frac{12}{11}$ [3.3.3, Problem 6] **15.** -3 [3.3.3, Example 6] **16.** 125% [3.2.1, Problem 1]
17. 40 [3.2.1, 1st Focus On, p. 98] **18.** 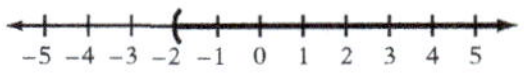 [3.4.1, Example 1] **19.** $x \le -1$
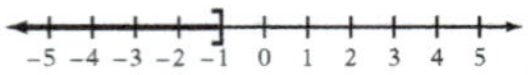 [3.4.1, Example 2] **20.** $x > -2$ 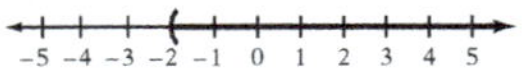 [3.4.2, Example 4]
21. $x > \frac{1}{2}$ [3.4.1, Example 2] **22.** $x \le -\frac{9}{2}$ [3.4.3, Example 7] **23.** $x \ge -16$ [3.4.2, Example 4] **24.** $x \le 2$ [3.4.3, Problem 7] **25.** $x \le -3$ [3.4.3, Example 7] **26.** $x > 2$ [3.4.3, Problem 7] **27.** 4 [3.3.2, Example 5]
28. 24 [3.2.1, Example 1] **29.** $x \ge 3$ 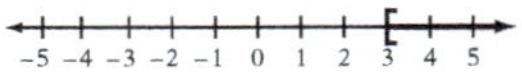 [3.4.2, Example 5] **30.** Yes [3.1.1, Example 1]
31. The astronaut would weigh 30 lb on the moon. [3.2.1, Example 3] **32.** The final temperature of the water after mixing is 60°C. [3.3.4, Example 8] **33.** The number of calculators produced was 200. [3.3.4, Example 8] **34.** They will meet 2 h

after they begin. [3.2.2, Problem 6] **35.** He must invest \$930 in the second account. [3.2.1, Problem 4] **36.** The percent concentration of chocolate syrup in the chocolate milk is 25%. [3.2.1, Example 5]

CUMULATIVE REVIEW EXERCISES

1. 6 [1.2.2] **2.** -48 [1.2.3] **3.** $-\frac{19}{48}$ [1.3.3] **4.** -2 [1.3.2] **5.** 54 [1.4.1] **6.** 24 [1.4.2] **7.** 6 [2.1.1] **8.** $-17x$ [2.2.2] **9.** $-5a - 2b$ [2.2.2] **10.** $2x$ [2.2.3] **11.** $36y$ [2.2.3] **12.** $2x^2 + 6x - 4$ [2.2.4] **13.** $-4x + 14$ [2.2.5] **14.** $6x - 34$ [2.2.5] **15.** $\{-7, -6, -5, -4, -3, -2, -1\}$ [1.1.1] **16.** $87\frac{1}{2}\%$ [1.3.4] **17.** 3.42 [1.3.4] **18.** $\frac{5}{8}$ [1.3.4] **19.** Yes [3.1.1] **20.** -5 [3.1.2] **21.** -25 [3.1.3] **22.** 3 [3.3.1] **23.** -3 [3.3.2] **24.** 13 [3.3.3] **25.** $x < -\frac{8}{9}$ [3.4.2] **26.** $x \geq 12$ [3.4.3] **27.** $x > 9$ [3.4.3] **28.** $8 - \frac{n}{12}$ [2.3.1] **29.** $n + (n + 2)$; $2n + 2$ [2.3.2] **30.** b and $35 - b$ [2.3.3] **31.** Let L be the length of the longer piece; $3 - L$ [2.3.3] **32.** 17% of the computer programmer's salary is deducted for income tax. [3.2.1] **33.** The equation predicts that the first 4-minute mile was run in 1952. [3.3.4] **34.** The final temperature of the water after mixing is 60°C. [3.3.4] **35.** A force of 24 lb must be applied to the other end of the lever. [3.3.4]

Answers to Chapter 4 Selected Exercises

PREP TEST

1. $0.65R$ [2.2.2] **2.** $0.03x + 20$ [2.2.5] **3.** $3n + 6$ [2.2.2] **4.** $5 - 2x$ [2.3.1] **5.** 40% [1.3.4] **6.** 2 [3.3.3] **7.** 0.25 [3.3.1] **8.** $x < 4$ [3.4.3] **9.** $20 - n$ [2.3.3]

SECTION 4.1

1. True **3.** True **5.** True **7. a.** $12 - x$ **b.** $12 - x$ **9.** $n - 15 = 7$; $n = 22$ **11.** $7n = -21$; $n = -3$ **13.** $3n - 4 = 5$; $n = 3$ **15.** $4(2n + 3) = 12$; $n = 0$ **17.** $12 = 6(n - 3)$; $n = 5$ **19.** $22 = 6n - 2$; $n = 4$ **21.** $4n + 7 = 2n + 3$; $n = -2$ **23.** $5n - 8 = 8n + 4$; $n = -4$ **25.** $2(n - 25) = 3n$; $n = -50$ **27.** $3n = 2(20 - n)$; 8 and 12 **29.** $3n + 2(18 - n) = 44$; 8 and 10 **31.** Answers may vary. For example: $n + n + 10 = 14$; $n + n - 10 = 14$ **33.** consecutive **37.** The integers are 17, 18, and 19. **39.** The integers are 26, 28, and 30. **41.** The integers are 17, 19, and 21. **43.** The integers are 8 and 10. **45.** The integers are 7 and 9. **47.** The integers are -9, -8, and -7. **49.** The integers are 10, 12, and 14. **51.** No solution **53.** ii **55.** whole; low-fat **57.** The original value was \$32,000. **59.** There are 58 calories in a medium-size orange. **61.** The amount of mulch is 15 lb. **63.** The intensity of the sound is 140 decibels. **65.** The length is 80 ft. The width is 50 ft. **67.** To replace the water pump, 5 h of labor were required. **69.** You are purchasing 9 tickets. **71.** The length of the shorter piece is 8 ft. **73.** The executive used the phone for 951 min. **75.** The customer pays \$.15 per text message over 300 messages. **77.** The perimeter is 8 ft. **79.** The cyclist will complete the trip in 1 additional hour. **81.** The integers are -12, -10, -8, and -6. **83.** Any three consecutive odd integers satisfy the conditions. **87.** odd **89.** even **91.** even **93.** odd **95.** odd

SECTION 4.2

1. Acute, right, obtuse, straight **3.** i, ii, iii, vi **5.** No **7.** True **9.** True **13. a.** $0.25L$ **b.** L; $0.25L$ **c.** $2.5L$ **15.** The sides measure 50 ft, 50 ft, and 25 ft. **17.** The length is 13 m. The width is 8 m. **19.** The length is 40 ft. The width is 20 ft. **21.** The sides measure 40 cm, 20 cm, and 50 cm. **23.** The length is 130 ft. The width is 39 ft. **25.** The width is 12 ft. **27.** Each side measures 12 in. **29. a.** 90° **b.** 180° **c.** 360° **d.** Between 0° and 90° **e.** Between 90° and 180° **31.** 35° **33.** 20° **35.** 53° **37.** 121° **39.** 15° **41.** 18° **43.** 45° **45.** 49° **47.** 12° **49.** False **51.** True **53.** $\angle a = 122°$, $\angle b = 58°$ **55.** $\angle a = 44°$, $\angle b = 136°$ **57.** Yes **59.** 20° **61.** 40° **63.** 128° **65.** $\angle x = 155°$; $\angle y = 70°$ **67.** $\angle a = 45°$; $\angle b = 135°$ **69.** $90° - x$ **71.** 60° **73.** 35° **75.** 102° **77.** 36°, 36°, and 108° **79.** 41°, 41°, and 98° **81.** False **83.** False **85.** The length is 9 cm. The width is 3 cm. **87.** The length is $16x$. **89.** The angles measure 59°, 60°, and 61°. **91. a.** The sum is 180°.

SECTION 4.3

1. Subtract the cost from the selling price. **3.** Multiply the cost times the markup rate. **5.** Subtract the discount from the regular price. **9.** 40; unknown; 0.25 **11.** The selling price is \$56. **13.** The selling price is \$565.64. **15.** The markup rate is 75%. **17.** The markup rate is 25%. **19.** The cost of the compact disc player is \$120. **21.** The cost of the basketball is \$59. **23.** True **25.** The markup rate is 54%. **29.** 318.75; 375; unknown **31.** The sale price is \$146.25. **33.** The discount price is \$218.50. **35.** The discount rate is 25%. **37.** The markdown rate is 38%. **39.** The regular price is \$710. **41.** The regular price is \$300. **43.** True **45.** The markup is \$18. **47.** The markup rate is 25%. **49.** The cost of the camera is \$230. **51.** The regular price is \$160. **53.** No. The single discount that would give the same sale price is 28%.

SECTION 4.4

1. P is the principal (the amount invested), r is the simple interest rate, and I is the simple interest earned. **3. a.** \$1250 **b.** 5% **c.** \$62.50 **5.** True **7.** False **9.** Row 1: 0.052; $0.052x$; Row 2: $x + 1000$; 0.072; $0.072(x + 1000)$ **11.** \$9000 was invested at 7%, and \$6000 was invested at 6.5%. **13.** \$1500 was invested in the mutual fund. **15.** \$200,000 was deposited at 10%, and \$100,000 was deposited at 8.5%. **17.** Teresa has \$3000 invested in bonds that earn 8% annual simple interest. **19.** She has \$2500 invested at 11%. **21.** The total amount invested was \$650,000. **23.** The total amount invested was \$500,000. **25. a.** 3% and 5% **b.** \$5000 **27.** The amount of the research consultant's investment is \$45,000. **29.** The total annual interest received was \$3040. **31.** The value of the investment in 4 years is \$3859.40. **33. a.** The consultant should have \$298,000 saved for retirement. **b.** The executive's retirement savings should fall between \$524,000 and \$1,144,000. **c.** The manager's retirement savings should fall between \$28,000 and \$75,000.

SECTION 4.5

1. True **3.** True **5.** False **7.** Increase **9.** \$600 **11. a.** Row 1: 1.70; $1.70x$; Row 2: $10 - x$; 0.50; $0.50(10 - x)$; Row 3: 10; 1.30; 13 **b.** $1.70x$; $0.50(10 - x)$; 13 **15.** 56 oz of the \$4.30 alloy and 144 oz of the \$1.80 alloy were used. **17.** 8 lb of chamomile tea are needed. **19.** The cost per ounce is \$5.00. **21.** To make the mixture, 3 lb of caramel are needed. **23.** 20 lb of oak chips and 60 lb of pine chips were used. **25.** 75 gal of fruit juice and 25 gal of ice cream were used. **27.** The cost per pound of the house blend is \$5.50. **29.** The parks department bought 8 bundles of seedlings and 6 bundles of container-grown plants. **31.** The cost per ounce is \$1046.92. **33.** iii, v, and vi **35.** 0.90; 225 **37.** 1.5; 8.5 **41.** The percent concentration is 24%. **43.** 20 gal of the 15% solution must be mixed with 5 gal of the 20% solution. **45.** 30 lb of 25% wool yarn is used. **47.** 6.25 gal of the plant food that is 9% nitrogen are combined with the plant food that is 25% nitrogen. **49.** The percent concentration is 19%. **51.** 30 oz of the potpourri that is 60% lavender are used. **53.** To make the solution, 100 ml of the 7% solution and 200 ml of the 4% solution are used. **55.** The percent concentration is 80%. **57.** 12.5 gal of ethanol must be added. **59.** The percent concentration is 52%. **61.** False **63.** 10 lb of walnuts and 20 lb of cashews were used. **65.** The chemist used 3 L of pure acid and 7 L of water. **67.** 85 adults and 35 children attended the performance. **69.** The percent concentration of acid was 50%.

SECTION 4.6

1. d = distance, r = rate, t = time **3.** The Boeing 767 has been in the air for $(t - 1)$ hours. **5.** The total distance covered is 50 ft. **7.** The speed of the first plane is 105 mph. The speed of the second plane is 130 mph. **9.** The second skater overtakes the first skater after 40 s. **11.** Michael's boat will be alongside the tour boat in 2 h. **13.** The airport is 120 mi from the corporate offices. **15.** The sailboat traveled 36 mi in the first 3 h. **17.** The passenger train is traveling at 50 mph. The freight train is traveling at 30 mph. **19.** It takes the second ship 1 h to catch up to the first ship. **21.** The two trains will pass each other in 4 h. **23.** No, the second car will not overtake the first car. **25.** The bus overtakes the car 180 mi from the starting point. **27.** The plane flew 2 h at 115 mph and 3 h at 125 mph. **29.** The rate of the cyclist is 14 mph. **31.** The campers turned around downstream at 10:15 A.M. **33.** The van overtakes the truck at 2:15 P.M. **37.** It is impossible to average 60 mph. **39.** Chris can skate 128 m.

SECTION 4.7

1. False **3.** True **5.** False **7.** $n \geq 102$ **9.** $n + 0.45n \leq 200$ **11.** 11 **13.** The couple's monthly household income is \$5395 or more. **15.** The organization must collect more than 440 lb of cans. **17.** The student must score 78 or better. **19.** The dollar amount in sales must be more than \$5714. **21.** The height is 45 ft. **23.** The dollar amount the agent expects to sell is \$20,000 or less. **25.** 8 oz or less must be added. **27.** More than 80 oz of the alloy must be used. **29.** The ski area is more than 38 mi away. **31.** The maximum number of miles you can drive is 166. **33.** The integers are 1, 3, and 5 or 3, 5, and 7. **35.** The crew can prepare 8 to 12 aircraft in this period of time. **37.** More than 50 oz of pretzels should be used. **39.** Between $3.\overline{3}$ and 30 lb of the black tea should be used.

CHAPTER 4 REVIEW EXERCISES

1. $2x - 7 = x$; 7 [4.1.1] **2.** The length of the shorter piece is 14 in. [4.1.2] **3.** The two numbers are 8 and 13. [4.1.2] **4.** $2x + 6 = 4(x - 2)$; 7 [4.1.1] **5.** The discount rate is $33\frac{1}{3}\%$. [4.3.2] **6.** \$8000 is invested at 6%, and \$7000 is invested at 7%. [4.4.1] **7.** The angles measure 65°, 65°, and 50°. [4.2.3] **8.** The rate of the motorcyclist is 45 mph. [4.6.1] **9.** The measures of the three angles are 16°, 82°, and 82°. [4.2.3] **10.** The markup rate is 80%. [4.3.1] **11.** The concentration of butterfat is 14%. [4.5.2] **12.** The length is 80 ft. The width is 20 ft. [4.2.1] **13.** A minimum of 1151 copies can be ordered. [4.7.1] **14.** The regular price is \$62. [4.3.2] **15.** 7 qt of cranberry juice and 3 qt of apple juice were used. [4.5.1] **16.** The two numbers are 6 and 30. [4.1.2] **17.** \$5600 is deposited in the 12% account. [4.4.1] **18.** One liter of pure water should be added. [4.5.2] **19.** The lowest score the student can receive is 72. [4.7.1] **20.** $-7 = \frac{1}{2}x - 10$; 6 [4.1.1] **21.** The height is 1063 ft. [4.1.2] **22.** The measures of the three angles are 75°, 60°, and 45°. [4.2.3] **23.** The length of the longer piece is 7 ft. [4.1.2] **24.** The number of hours of consultation was 8. [4.1.2] **25.** The cost is \$671.25. [4.3.1] **26.** They will meet after 4 min. [4.6.1] **27.** The measures of the three sides are 8 in., 12 in., and 15 in. [4.2.1] **28.** The integers are −17, −15, and −13. [4.1.1] **29.** The maximum width is 10 ft. [4.7.1] **30.** $\angle a = 138°$, $\angle b = 42°$ [4.2.2]

CHAPTER 4 TEST

1. $6n + 13 = 3n - 5; n = -6$ [4.1.1, Example 1] **2.** $3n - 15 = 27; n = 14$ [4.1.1, Example 1] **3.** The numbers are 8 and 10. [4.1.1, Example 2] **4.** The lengths are 6 ft and 12 ft. [4.1.2, Example 4] **5.** The cost is $200. [4.3.1, Example 2] **6.** The discount rate is 20%. [4.3.2, Example 3] **7.** 20 gal of the 15% solution must be added. [4.5.2, Problem 2] **8.** The length is 14 m. The width is 5 m. [4.2.1, Focus On, page 151] **9.** The integers are 5, 7, and 9. [4.1.1, Example 2] **10.** Five or more residents are in the nursing home. [4.7.1, Problem 2] **11.** $5000 is invested at 10%, and $2000 is invested at 15%. [4.4.1, Problem 1] **12.** 8 lb of the $7 coffee and 4 lb of the $4 coffee should be used. [4.5.1, Problem 1] **13.** The rate of the first plane is 225 mph. The rate of the second plane is 125 mph. [4.6.1, Example 1] **14.** The measures of the three angles are 48°, 33°, and 99°. [4.2.3, Example 5] **15.** The amounts to be deposited are $1400 at 6.75% and $1000 at 9.45%. [4.4.1, Problem 1] **16.** The minimum length is 24 ft. [4.7.1, Example 2] **17.** The sale price is $79.20. [4.3.2, Example 3]

CUMULATIVE REVIEW EXERCISES

1. $-12, -6$ [1.1.1] **2.** 6 [1.2.2] **3.** $-\frac{1}{6}$ [1.4.1] **4.** $-\frac{11}{6}$ [1.4.2] **5.** -18 [1.1.2] **6.** -24 [2.1.1] **7.** $9x + 4y$ [2.2.2] **8.** $-12 + 8x + 20x^3$ [2.2.4] **9.** $4x + 4$ [2.2.5] **10.** $6x^2$ [2.2.2] **11.** No [3.1.1] **12.** -3 [3.3.1] **13.** -15 [3.1.3] **14.** 3 [3.3.1] **15.** -10 [3.3.3] **16.** $\frac{2}{5}$ [1.3.4] **17.** $x \geq 4$

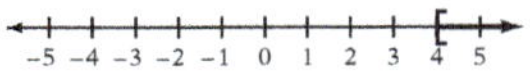

[3.4.2] **18.** $x \geq -3$ [3.4.2] **19.** $x > 18$ [3.4.3] **20.** $x \geq 4$ [3.4.3] **21.** 2.5% [1.3.4] **22.** 12% [1.3.4] **23.** 3 [3.2.1] **24.** 45 [3.2.1] **25.** $8n + 12 = 4n; n = -3$ [4.1.1] **26.** The area of the garage is 600 ft^2. [4.1.2] **27.** The number of hours of labor is 5. [4.1.2] **28.** 20% of the libraries had the reference book. [3.2.1] **29.** The amount of money deposited is $2000. [4.4.1] **30.** The markup rate is 75%. [4.3.1] **31.** 60 g of the gold alloy must be used. [4.5.1] **32.** 30 oz of pure water must be added. [4.5.2] **33.** The measure of one of the equal angles is 47°. [4.2.3] **34.** The middle even integer is 14. [4.1.1]

Answers to Chapter 5 Selected Exercises

PREP TEST

1. -3 [1.4.2] **2.** -1 [2.1.1] **3.** $-3x + 12$ [2.2.4] **4.** -2 [3.3.1] **5.** 5 [3.3.1] **6.** -2 [3.3.1] **7.** 13 [2.1.1] **8.** $\frac{3}{4}x - 4$ [2.2.4] **9.** i, ii, iii [3.4.1]

SECTION 5.1

1. Quadrant II **3.** y-axis **5.** 0 **7.** Answers will vary. For example, $(-3, 2)$ and $(5, 2)$. **9.** right; down

11.

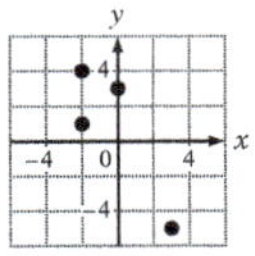

13.

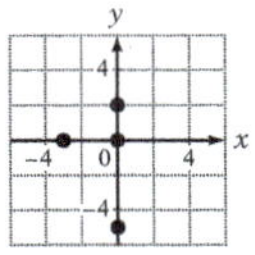

15.

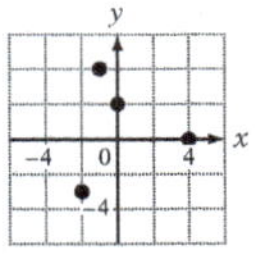

17. $A(2, 3), B(4, 0), C(-4, 1), D(-2, -2)$

19. $A(-2, 5), B(3, 4), C(0, 0), D(-3, -2)$ **21. a.** I **b.** II **c.** IV **d.** III **23.** 18; 51; 22; 48; 18; 51; 22; 48

25.

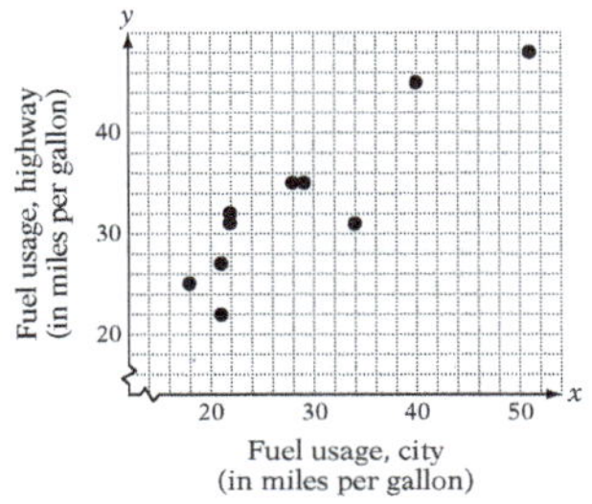

27.

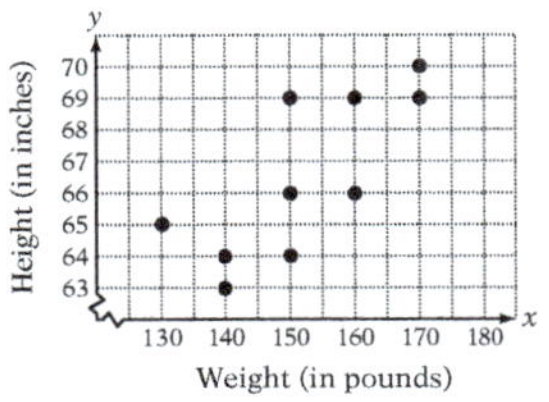

29.

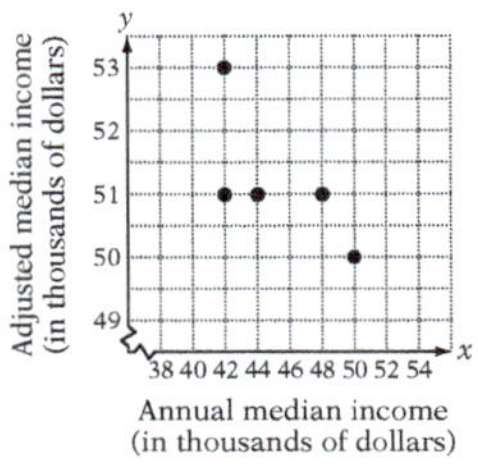

31. Denominator **33.** The average rate of change is 4° per hour. **35. a.** The average rate of change is $0.\overline{3}$ lb per week. **b.** The average rate of change is 0.44 lb per week. **37. a.** The average annual rate of change was 26 students per year. **b.** The average annual rate of change was 566 students per year. **c.** The average annual rate of change in summer enrollment between 2005 and 2009 was greater. **39. a.** The average annual rate of change was -8.95 cents. **b.** The average annual rate of change increased from 1975 to 1980. During that period, the average annual rate of change was 2 cents. **41.** The average rate of change was closest to -1 degree per minute from 30 to 40 min. **43.** 1 unit **45.** 0 units **47.** 1 unit **49.** $(0, 0)$

SECTION 5.2

1. i and ii **3.** b and d are graphs of straight lines. **5. a.** $Ax + By = C$ **b.** $y = mx + b$ **c.** Neither **d.** $Ax + By = C$ **7.** x-coordinate **9.** The line is horizontal. **11.** x; y **13.** Yes **15.** No **17.** No **19.** Yes **21.** No **23.** (3, 7) **25.** (6, 3) **27.** (0, 1) **29.** (−5, 0) **31. a.** 2 **b.** Negative **33. a.** −7; −1; −7 **b.** −4; 0; −4 **c.** −1; 1; −1 **35.** 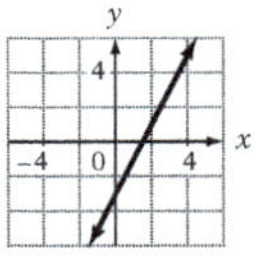 **37.** 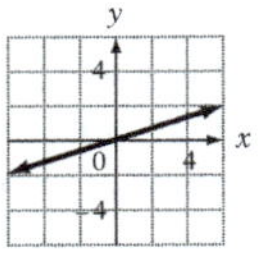 **39.** 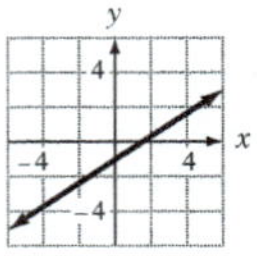 **41.** 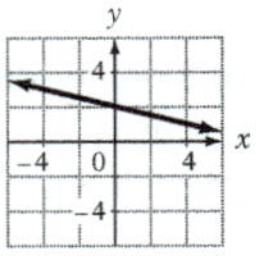

43. 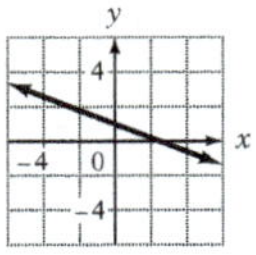 **45.** 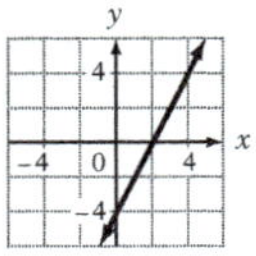 **47.** 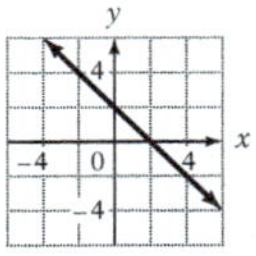 **49.** 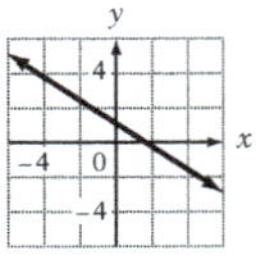 **51.** 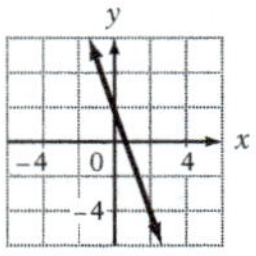

53. 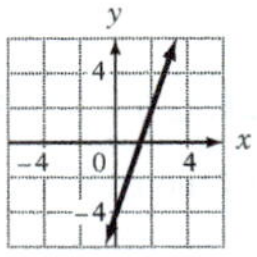 **55.** 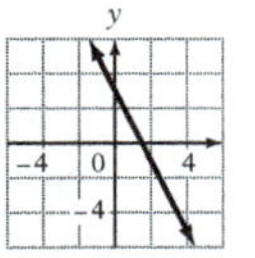 **57.** 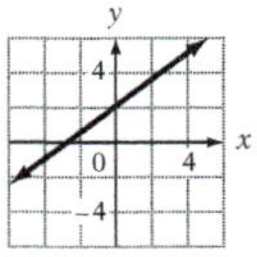 **59.** 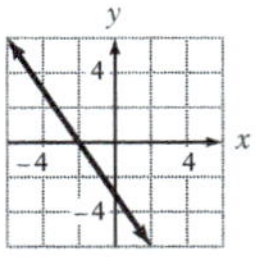 **61.** x; $-2y$; -2; $\frac{1}{2}$; 5 **63.** horizontal

65. $y = -3x + 10$ **67.** $y = 4x - 3$ **69.** $y = -\frac{3}{2}x + 3$ **71.** $y = \frac{2}{5}x - 2$ **73.** $y = -\frac{2}{7}x + 2$ **75.** $y = -\frac{1}{3}x + 2$

77. 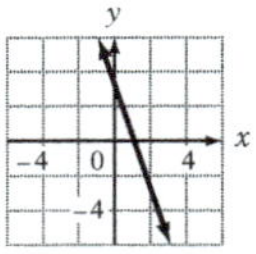 **79.** 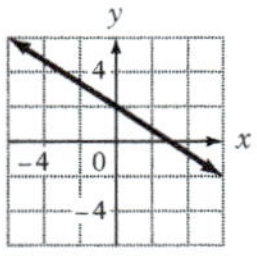 **81.** 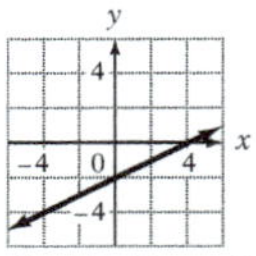 **83.** 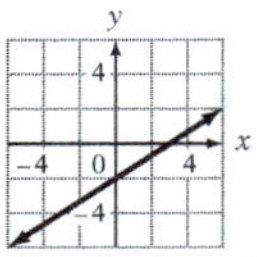 **85.** 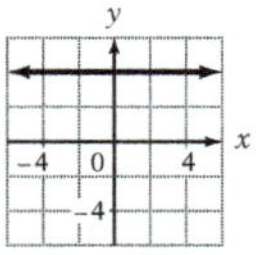

87. 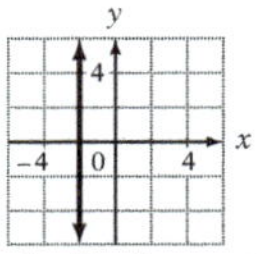 **89.** (3, 0), (0, −3) **91.** (3, 0), (0, −6) **93.** (10, 0), (0, −2) **95.** (−4, 0), (0, 12)

97. (0, 0), (0, 0) **99.** (−6, 0), (0, 3) **101.** 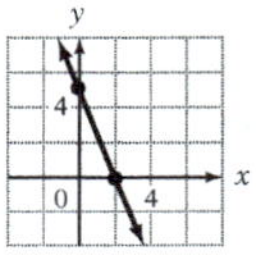 **103.** 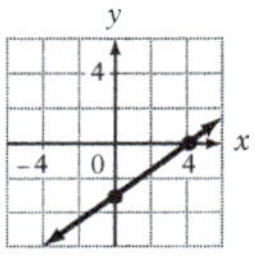 **105.** 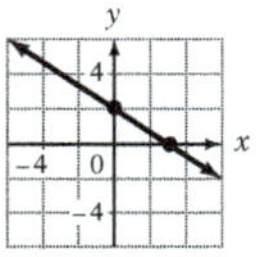

107. 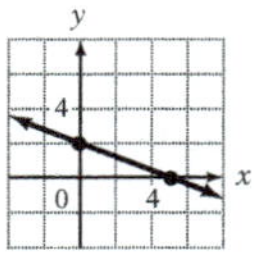 **109.** 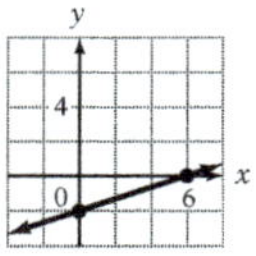 **111.** 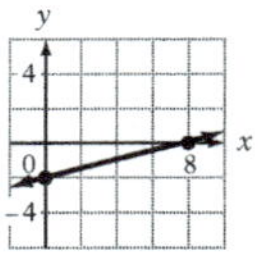 **113.** 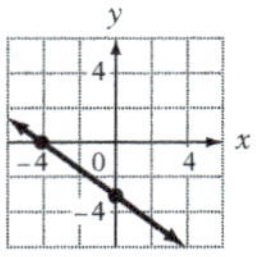 **115.** 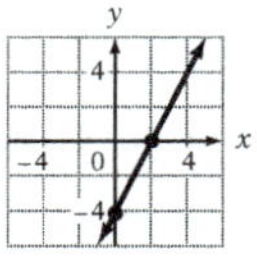

117. To the right **119. a.** $y = -\frac{1}{2}x - 5$ **b.** (−2, −4) **121.** 1 unit **125.** 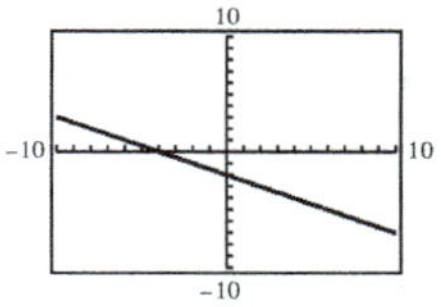

127. 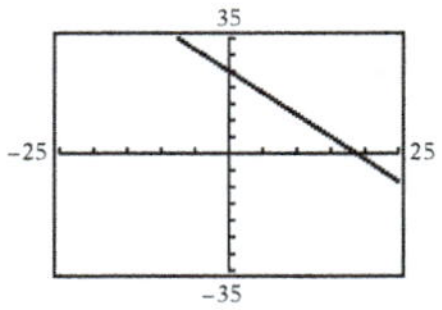

SECTION 5.3

1. m **3.** $\frac{y_2 - y_1}{x_2 - x_1}$ **5.** −4 **7.** $-\frac{2}{3}$ **11.** 2; −4; 3; 1 **13.** −2 **15.** $\frac{1}{3}$ **17.** $-\frac{5}{2}$ **19.** $-\frac{1}{2}$ **21.** −1 **23.** Undefined **25.** 0 **27.** $-\frac{1}{3}$ **29.** 0 **31.** −5 **33.** Undefined **35.** $-\frac{2}{3}$ **37.** 4 **39.** 3 **41.** Yes **43.** Yes **45.** Yes **47.** No **49.** No **51.** Yes **53.** $m = 50$. The electronic mail sorter sorts 50 pieces of mail per minute. **55.** $m = -0.02$. The car uses 0.02 gal of gasoline per mile driven. **57.** $m = 0.8$. The percent of people using seat belts has increased by 0.8% per year. **59. a.** $a = c, b \neq d$ **b.** $b = d, a \neq c$ **c.** Downward to the right

61. 5; (0, −3) **63.**
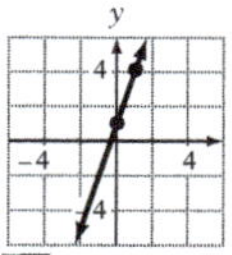
65.
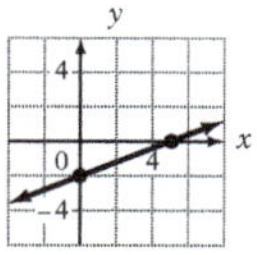
67.
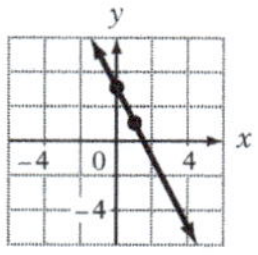
69.
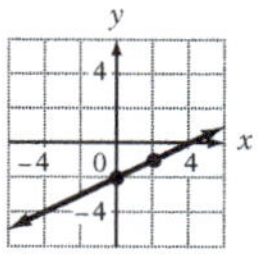
71.
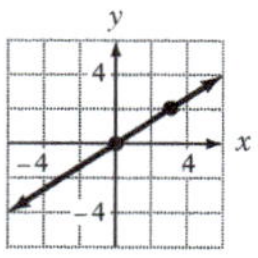
73.
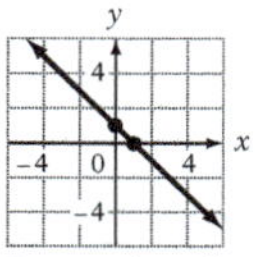
75.
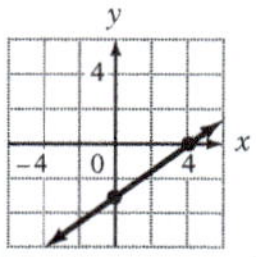
77.
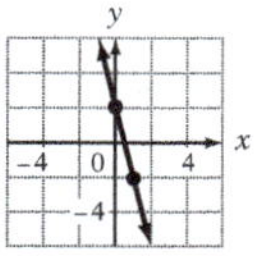
79.
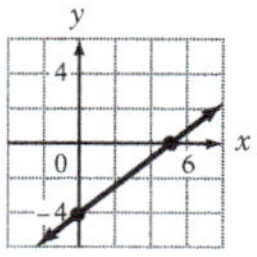
81. Above; downward to the right

83. Increasing the coefficient of x increases the slope. **85.** Increasing the constant term increases the y-intercept. **87.** i and D; ii and C; iii and B; iv and F; v and E; vi and A **89.** No, not all graphs of straight lines have a y-intercept. In general, vertical lines do not have a y-intercept. For example, $x = 2$ does not have a y-intercept. **91.** Line a slants upward to the right. Line b slants downward to the right. **93.** Line a has a y-intercept of (0, 5). Line b has a y-intercept of (0, −5). **95.** Line a is a horizontal line. Line b is a vertical line.

SECTION 5.4

1. 5; 7 **3.** 0; 2 **5.** $y - y_1 = m(x - x_1)$ **7.** 3; 1; $3x + 1$ **9.** Check that the coefficient of x is the given slope. Check that the coordinates of the given point are a solution of your equation. **11.** $y = 2x + 2$ **13.** $y = -3x - 1$ **15.** $y = \frac{1}{3}x$ **17.** $y = \frac{3}{4}x - 5$ **19.** $y = -\frac{3}{5}x$ **21.** $y = \frac{1}{4}x + \frac{5}{2}$ **23.** $y = -\frac{2}{3}x - 7$ **27.** $-\frac{4}{5}$; 3; $-\frac{4}{5}x + 3$ **29.** $y = 2x - 3$ **31.** $y = -2x - 3$ **33.** $y = \frac{2}{3}x$ **35.** $y = \frac{1}{2}x + 2$ **37.** $y = -\frac{3}{4}x - 2$ **39.** $y = \frac{3}{4}x + \frac{5}{2}$ **41.** $y = -\frac{4}{3}x - 9$ **43. a.** $y - b = m(x - 0)$ **b.** Yes **45.** slope **47.** $y = 3x + 4$ **49.** $y = -x - 4$ **51.** $y = -\frac{6}{5}x + 5$ **53.** $y = -\frac{2}{5}x - 7$ **55.** $y = 5$ **57.** $x = 5$ **59.** $y = \frac{4}{5}x + 8$ **61.** $y = -\frac{1}{2}x - 2$ **63.** $y = -\frac{4}{3}x - 3$ **65.** $y = 3x - 3$ **67.** $y = 3x$ **69.** $x = 3$ **71.** Yes; however, all of the points must be on the same line. **73.** Yes; $y = -x + 6$ **75.** No **77.** 7 **79.** −1 **81.** 1 **83.** (3, 0); (0, 2); $-\frac{2}{3}$; $y = -\frac{2}{3}x + 2$ **85.** (−1, 0); (0, −2); −2; $y = -2x - 2$

SECTION 5.5

1. a. Yes; $f(x) = -\frac{3}{5}x - 2$ **b.** Yes; $f(x) = x + 1$ **c.** No **d.** No **e.** Yes; $f(x) = 2x - 4$ **f.** Yes; $f(x) = 6$ **3.** −2, −1, 2, 3 **5.** ordered pairs; domain; range **7.** {(35, 7.50), (45, 7.58), (38, 7.63), (24, 7.78), (47, 7.80), (51, 7.86), (35, 7.89), (48, 7.92)}; No **9.** No **11.** {(4, 411), (5, 514), (6, 618), (7, 720), (8, 823)}; Yes **13.** domain; range **15.** Domain: {0, 2, 4, 6}; range: {0}; yes **17.** Domain: {2}; range: {2, 4, 6, 8}; no **19.** Domain: {0, 1, 2, 3}; range {0, 1, 2, 3}; yes **21.** Domain: {−3, −2, −1, 1, 2, 3}; range: {−3, 2, 3, 4}; yes **23.** 40; (10, 40) **25.** −11; (−6, −11) **27.** 12; (−2, 12) **29.** $\frac{7}{2}$; $\left(\frac{1}{2}, \frac{7}{2}\right)$ **31.** 2; (−5, 2) **33.** 32; (−4, 32) **35.** Range: {−19, −13, −7, −1, 5}; the ordered pairs (−5, −19), (−3, −13), (−1, −7), (1, −1), and (3, 5) belong to the function. **37.** Range: {1, 2, 3, 4, 5}; the ordered pairs (−4, 1), (−2, 2), (0, 3), (2, 4), and (4, 5) belong to the function. **39.** Range: {6, 7, 15}; the ordered pairs (−3, 15), (−1, 7), (0, 6), (1, 7), and (3, 15) belong to the function. **41.** x; y; $f(x)$ **43.**
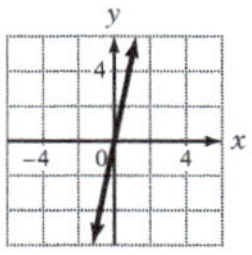
45.
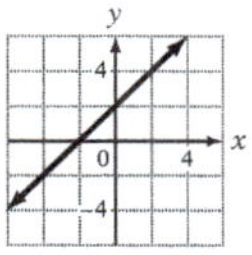
47.
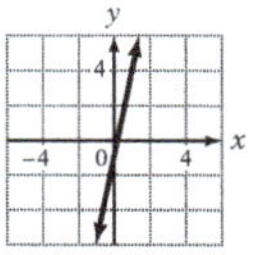
49.
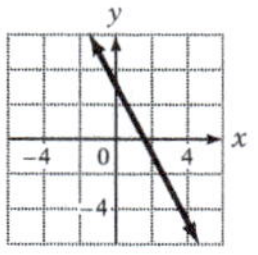
51.
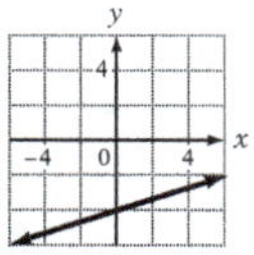
53.
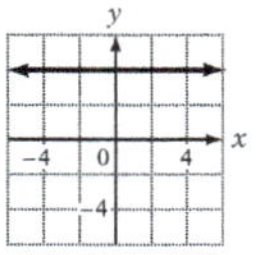
55.
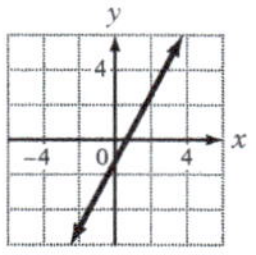
57.
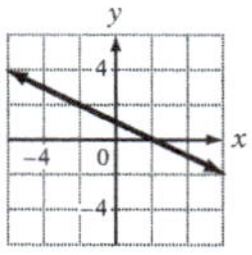
59.
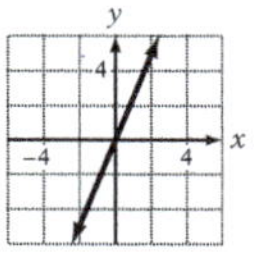
61. a. $f(x) = 2.25m + 2.65$ **b.**
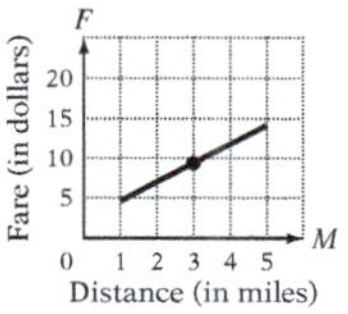

c. A 3-mile taxi ride costs $9.40.

63. **a.** $f(x) = 30{,}000 - 5000x$ **b.** 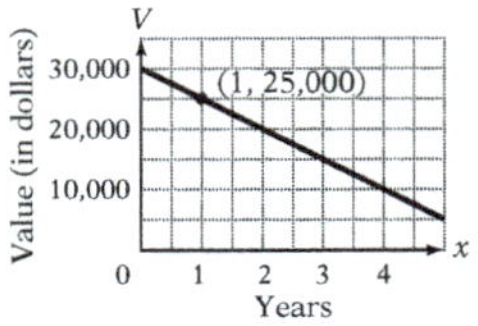

c. The value of the computer after 1 year is \$25,000.

65. **a.** $f(2) = 6$ **b.** $f(1) = 8$ **67.** a and III; b and IV; c and II; d and I **71.** $f(-5) = 11$ and $f(1) = 11$. No

SECTION 5.6

1. **a.** Yes **b.** No **c.** No **d.** Yes **3.** $>$, $<$ **5.** Lower **7.** are; are not **9.** 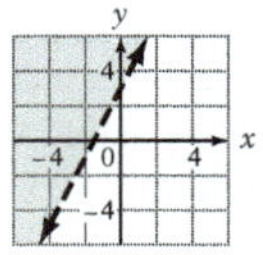 **11.** 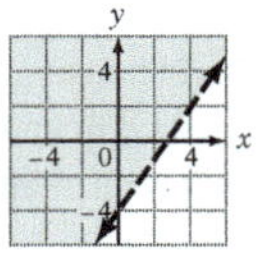

13. 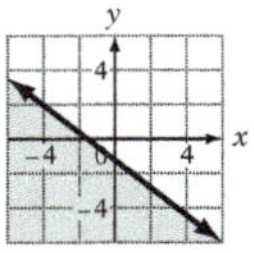 **15.** 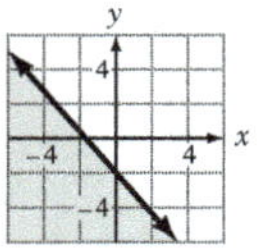 **17.** 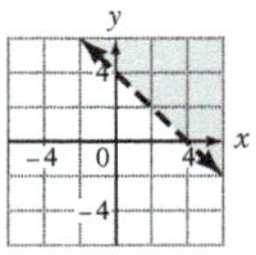 **19.** 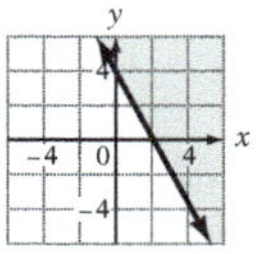 **21.** 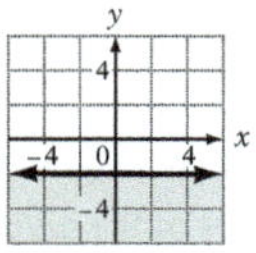 **23.** 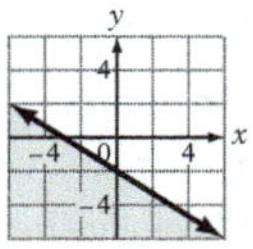

25. 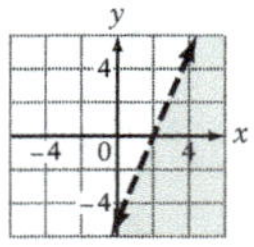 **27.** Negative **29.** $y \geq 2x + 2$ **31.** $y > 2$ **33.** 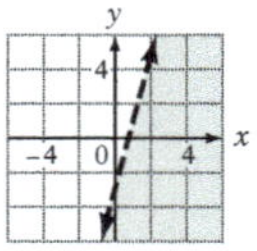 **35.** 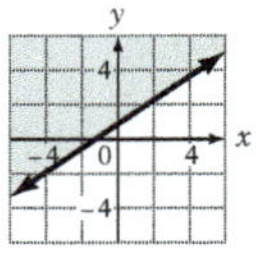

CHAPTER 5 REVIEW EXERCISES

1. (3, 0) [5.2.1] **2.** $y = 3x - 1$ [5.4.1/5.4.2] **3.** 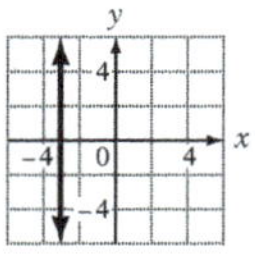 [5.2.3] **4.** 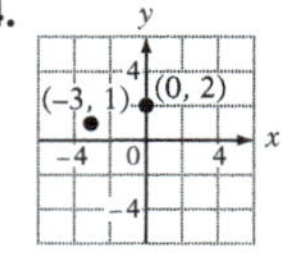

[5.1.1] **5.** 79 [5.5.1]

6. 0 [5.3.1] **7.** 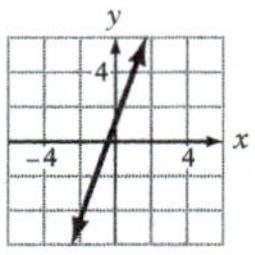 [5.2.2] **8.** 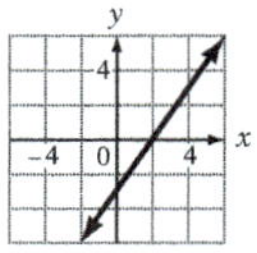 [5.2.3] **9.** $y = -\frac{2}{3}x + \frac{4}{3}$ [5.4.1/5.4.2]

10. (2, 0), (0, −3) [5.2.3] **11.** 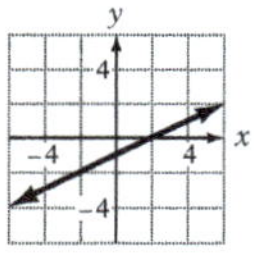 [5.3.2] **12.** 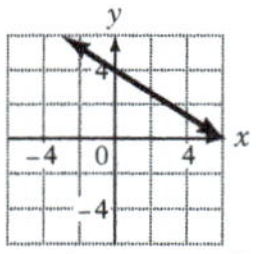 [5.5.2] **13.** $y = \frac{2}{3}x + 3$ [5.4.1/5.4.2]

14. 2 [5.3.1] **15.** −4 [5.5.1] **16.** Domain: $\{-20, -10, 0, 10\}$; range: $\{-10, -5, 0, 5\}$; yes [5.5.1]

17. 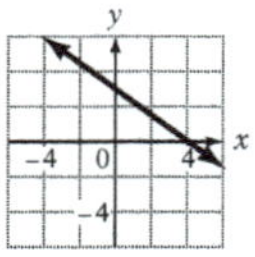 [5.2.2] **18.** 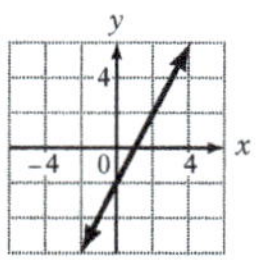 [5.3.2] **19.** 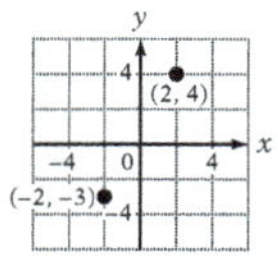

[5.1.1] **20.** 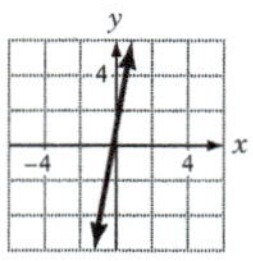 [5.5.2]

21. (6, 0), (0, −4) [5.2.3] **22.** $y = 2x + 2$ [5.4.1/5.4.2] **23.** 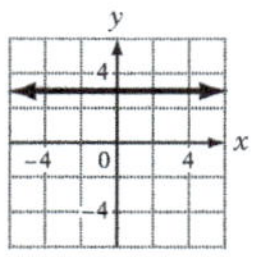 [5.2.3] **24.** 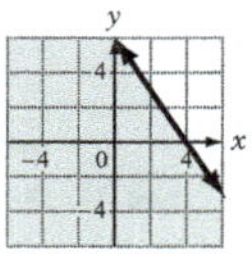 [5.6.1]

25. Yes [5.3.1] **26.** $y = \frac{5}{4}x + 9$ [5.4.3] **27.**
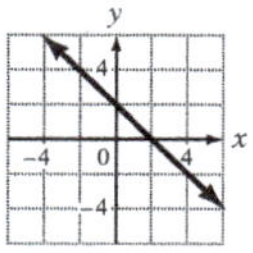
[5.3.2] **28.**
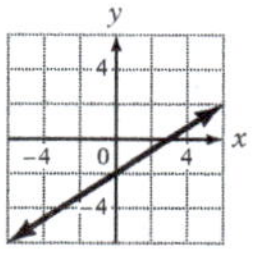
[5.2.3]

29. $(-2, -5)$ [5.2.1] **30.** $y = -3x + 2$ [5.4.1/5.4.2] **31.**
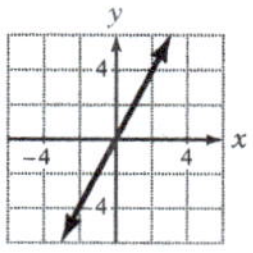
[5.5.2] **32.** 0 [5.5.1]

33. $y = -\frac{1}{3}x - 2$ [5.4.3] **34.** $y = \frac{1}{2}x - 2$ [5.4.1/5.4.2] **35.** Yes [5.2.1] **36.** $(2, -1)$ [5.2.1]

37. $(0, 0)$; $(0, 0)$ [5.2.3] **38.** $y = 3$ [5.4.1/5.4.2] **39.**
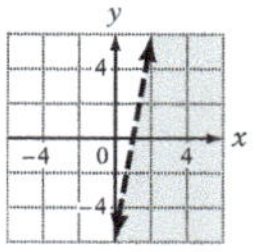
[5.6.1] **40.**
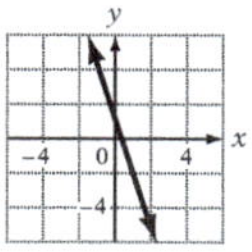
[5.3.2]

41. $y = 3x - 4$ [5.4.1/5.4.2] **42.**
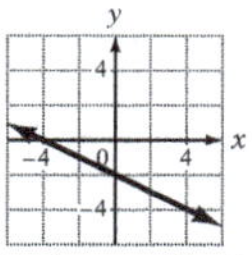
[5.2.3] **43.** Domain: $\{-10, -5, 5\}$; range $\{-5, 0\}$; no [5.5.1]

44. Range: $\{-53, -23, 7, 37, 67\}$ [5.5.1] **45.** Range: $\{2, 3, 4, 5, 6\}$ [5.5.1] **46.** Yes [5.3.1]

47.
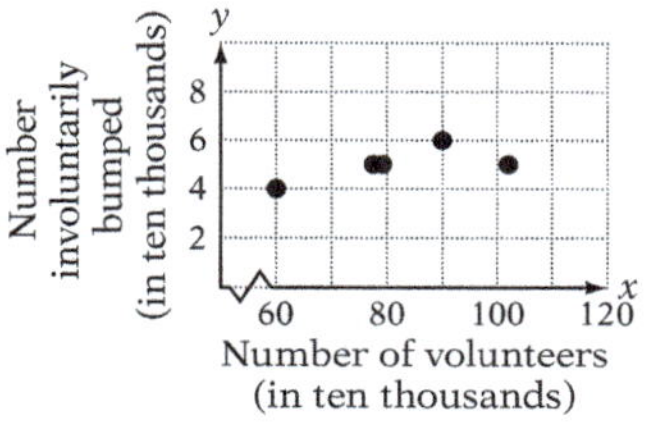

[5.1.2] **48.** $\{(25, 4.8), (35, 3.5), (40, 2.1), (20, 5.5), (45, 1.0)\}$; yes [5.5.1]

49. a. $f(s) = 70s + 40{,}000$ **b.**
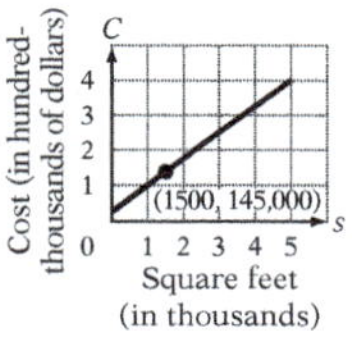

c. The contractor's cost to build a house of 1500 ft^2 is \$145,000. [5.5.2]

50. The average annual rate of change was 40.4 cents per year. [5.1.3]

CHAPTER 5 TEST

1. $y = -\frac{1}{3}x$ [5.4.1, Example 2/5.4.2, Example 3] **2.** $\frac{7}{11}$ [5.3.1, Example 1B] **3.** $(8, 0)$; $(0, -12)$ [5.2.3, Example 7]

4. $(9, -13)$ [5.2.1, Example 2] **5.**
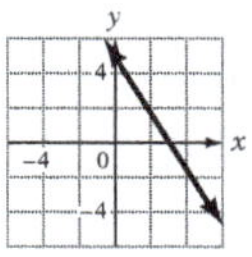
[5.2.3, Example 6B] **6.**
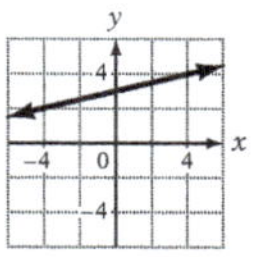
[5.2.2, Example 4]

7. 10 [5.5.1, Example 2] **8.** No [5.2.1, Example 1] **9.**
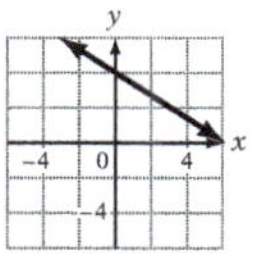
[5.3.2, Example 5]

10.
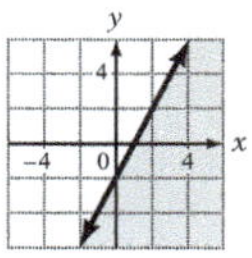
[5.6.1, Example 1] **11.**
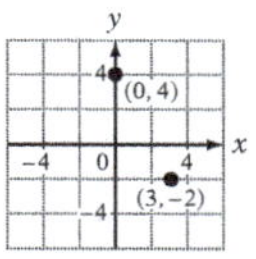

[5.1.1, Example 1] **12.**
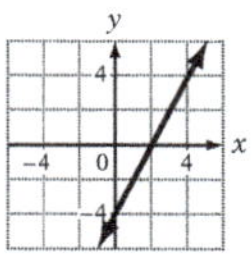
[5.3.2, Example 5]

13. $y = \frac{3}{5}x + 8$ [5.4.3, Example 4] **14.** $y = -\frac{2}{5}x + 7$ [5.4.1, Example 2/5.4.2, Example 3]
15. $y = 4x - 7$ [5.4.1, Example 2/5.4.2, Example 3] **16.** 13 [5.5.1, Problem 2] **17.** 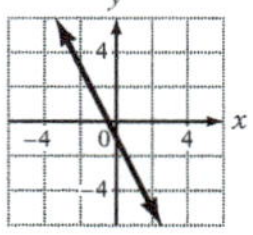 [5.2.2, Example 3]

18. 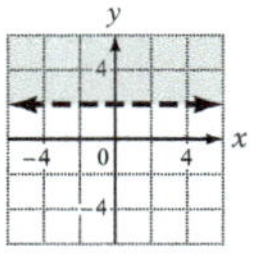 [5.6.1, Example 2] **19.** 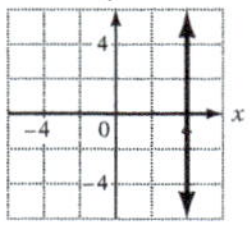 [5.2.3, Example 8B] **20.** 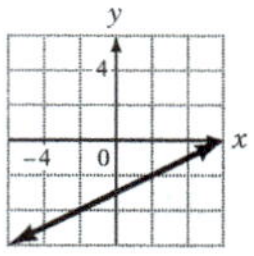 [5.3.2, Problem 5]

21. 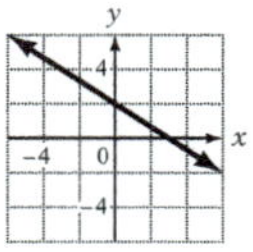 [5.5.2, Example 4] **22.** 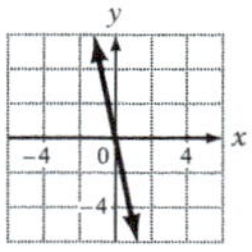 [5.5.2, Problem 4] **23.** Yes [5.3.1, Example 3]

24. $\{-22, -7, 5, 11, 26\}$ [5.5.1, Example 3] **25.** No [5.3.1, Example 2] **26.** The average annual rate of change was 0.5 million drivers per year. [5.1.3, Example 4] **27.** 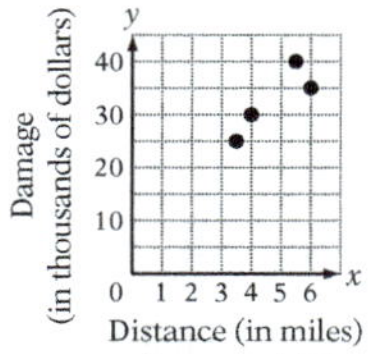

[5.1.2, Example 3] **28. a.** $f(t) = 8t + 1000$

b. 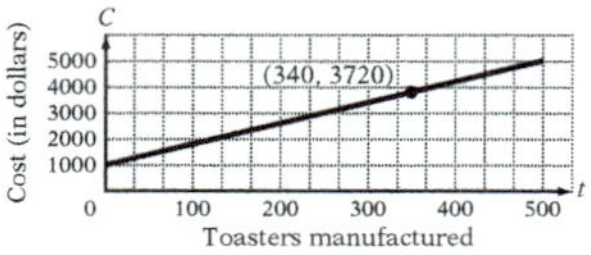

c. The cost to manufacture 340 toasters is \$3720. [5.5.2, Example 5]

29. $\{(8.5, 64), (9.4, 68), (10.1, 76), (11.4, 87), (12.0, 92)\}$; yes [5.5.1, page 250] **30.** The slope is 36. The cost of 1000 board-feet of lumber has been increasing by \$36 per month. [5.3.1, Example 4]

CUMULATIVE REVIEW EXERCISES

1. -12 [1.4.2] **2.** $-\frac{5}{8}$ [2.1.1] **3.** $-17x + 28$ [2.2.5] **4.** $\frac{3}{2}$ [3.3.1] **5.** 1 [3.3.3] **6.** $\frac{1}{15}$ [1.3.4]
7. $\{1, 2, 3, 4, 5, 6, 7, 8\}$ [1.1.1] **8.** $-4, 0, 5$ [1.1.1] **9.** $a \geq -1$ [3.4.3] **10.** $y = \frac{4}{5}x - 3$ [5.2.3]
11. $(-2, -7)$ [5.2.1] **12.** 0 [5.3.1] **13.** $(4, 0)$; $(0, 10)$ [5.2.3] **14.** $y = -x + 5$ [5.4.1/5.4.2]
15. 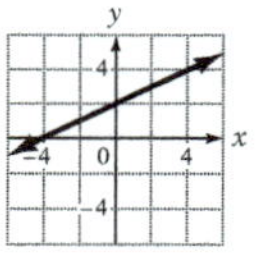 [5.2.2] **16.** 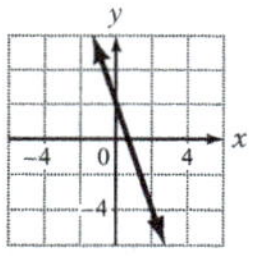 [5.2.3] **17.** 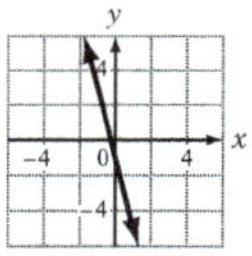 [5.5.2] **18.** 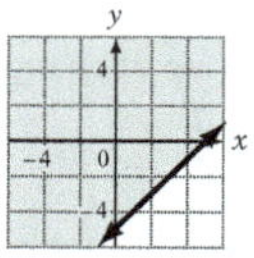 [5.6.1]

19. Domain: $\{0, 1, 2, 3, 4\}$; range: $\{0, 1, 2, 3, 4\}$; yes [5.5.1] **20.** -11 [5.5.1] **21.** $\{-7, -2, 3, 8, 13, 18\}$ [5.5.1]
22. \$8000 was invested in the first account. [3.2.1] **23.** The fulcrum is 7 ft from the 80-pound force. [3.3.4]
24. The length of the first side is 22 ft. [4.2.1] **25.** The sale price is \$62.30. [4.3.2]

Answers to Chapter 6 Selected Exercises

PREP TEST

1. $y = \frac{3}{4}x - 6$ [5.2.3] **2.** 1000 [3.3.3] **3.** $33y$ [2.2.5] **4.** $10x - 10$ [2.2.5] **5.** Yes [5.2.1]
6. $(4, 0)$ and $(0, -3)$ [5.2.3] **7.** Yes [5.3.1] **8.** 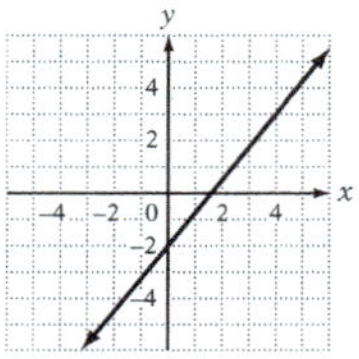 [5.2.2] **9.** The two hikers will be side by side after 1.5 h. [4.6.1]

SECTION 6.1

1. Always true **3.** Sometimes true **5.** Never true **7.** No **9.** No **11.** No **13.** Yes **15.** Yes **19.** Independent **21.** Dependent **23.** Inconsistent **25.** $(2, -1)$ **27.** No solution **29.** $(-2, 4)$ **31.** $(4, 1)$ **33.** $(4, 1)$ **35.** $(4, 3)$ **37.** $(3, -2)$ **39.** $(2, -2)$ **41.** The system of equations is inconsistent and has no solution. **43.** The system of equations is dependent. The solutions are the ordered pairs that satisfy the equation $y = 2x - 2$. **45.** $(1, -4)$ **47.** $(0, 0)$ **49.** The system of equations is inconsistent and has no solution. **51.** $(0, -2)$ **53.** $(1, -1)$ **55.** $(2, 0)$ **57.** $(-3, -2)$ **59.** The system of equations is inconsistent and has no solution. **61.** Inconsistent **63.** $y = 2$; $y = x + 4$ **65.** $y = x$; $y = -x + 2$ **67. a.** Always true **b.** Never true **c.** Always true **69.** a and iii; b and iv; c and i; d and ii

SECTION 6.2

1. Always true **3.** Never true **5.** Never true **7.** 2; 1; 2; 1 **9.** $(2, 1)$ **11.** $(4, 1)$ **13.** $(-1, 1)$ **15.** $(3, 1)$ **17.** $(1, 1)$ **19.** $(-1, 1)$ **21.** The system of equations is inconsistent and has no solution. **23.** The system of equations is inconsistent and has no solution. **25.** $\left(-\frac{3}{4}, -\frac{3}{4}\right)$ **27.** $\left(\frac{9}{5}, \frac{6}{5}\right)$ **29.** $(-7, -23)$ **31.** $(2, 0)$ **33.** $(2, 1)$ **35.** $\left(\frac{9}{19}, -\frac{13}{19}\right)$ **37.** $(1, 7)$ **39.** $\left(\frac{17}{5}, -\frac{7}{5}\right)$ **41.** $(2, 0)$ **43.** $(0, 0)$ **45.** The system of equations is dependent. The solutions are the ordered pairs that satisfy the equation $2x - y = 2$. **47.** $(4, 19)$ **49.** $(3, -10)$ **51.** $(-22, -5)$ **53.** $(-6, -19)$ **55.** Inconsistent **57.** $(1, 5)$ **59.** $(-5, 2)$ **61.** $(2, 100)$ **63.** 2 **65.** 2

SECTION 6.3

1. If the two equations are the same, the system is dependent. If the slopes are the same but the y-intercepts are different, the system is inconsistent. If neither of these is true, the system is independent. **a.** Inconsistent **b.** Independent **c.** Dependent **d.** Independent **3.** True **5. a.** 3 **b.** -4 **7.** $(5, -1)$ **9.** $(1, 3)$ **11.** $(1, 1)$ **13.** $(3, -2)$ **15.** The system of equations is dependent. The solutions are the ordered pairs that satisfy the equation $2x - y = 1$. **17.** $(3, 1)$ **19.** The system of equations is inconsistent and has no solution. **21.** $(1, 3)$ **23.** $(2, 0)$ **25.** $(0, 0)$ **27.** $(5, -2)$ **29.** $(-17, 7)$ **31.** $(3, 4)$ **33.** $(1, -1)$ **35.** The system of equations is dependent. The solutions are the ordered pairs that satisfy the equation $x + 3y = 4$. **37.** $(3, 1)$ **39.** $\left(1, \frac{1}{3}\right)$ **41.** $(1, 1)$ **43.** $\left(\frac{1}{2}, -\frac{1}{2}\right)$ **45.** $\left(\frac{2}{3}, \frac{1}{9}\right)$ **47.** $(1, 4)$ **49.** $(1, 2)$ **51.** $A = 3, B = -1$ **53.** $A = 2$ **55. a.** 14 **b.** $\frac{2}{3}$ **c.** $\frac{1}{2}$ **57.** The graph of the sum of the equations is a line whose graph contains the point of intersection of the graph of the system of equations.

SECTION 6.4

1. True **3.** False **5.** True **7.** Row 1: $b + c$; 2; $2(b + c)$; Row 2: $b - c$; 3; $3(b - c)$ **9.** Greater than **11.** The speed of the whale in calm water is 35 mph. The rate of the current is 5 mph. **13.** The team's rowing rate in calm water was 14 km/h. The rate of the current was 6 km/h. **15.** The rate of the boat in calm water was 7 mph. The rate of the current was 3 mph. **17.** The speed of the jet in calm air was 525 mph. The rate of the wind was 35 mph. **19.** The rate of the plane in calm air was 100 mph. The rate of the wind was 20 mph. **21.** The speed of the helicopter in calm air was 225 mph. The rate of the wind was 45 mph. **23.** Table 1: Row 1: 4; A; $4A$; Row 2: 2; C; $2C$; Table 2: Row 1: 2; A; $2A$; Row 2: 3; C; $3C$ **25.** Less than **27.** The cost is \$65. **29.** The cost of the wheat is \$.96 per pound. The cost of the rye is \$1.89 per pound. **31.** The pastry chef used 15 oz of the 20% solution and 35 oz of the 40% solution. **33.** The team scored 21 two-point baskets and 15 three-point baskets. **35.** Both formulas give the same ideal body weight at 69 in. **37.** You drove 108 city miles and 100 highway miles. **39.** \$3000 is invested in the 5% account. **41.** $x + y = 90, y = 4x$; 18° and 72° **43.** $2L + 2W = 68, L = 3W - 2$; 25 in.

CHAPTER 6 REVIEW EXERCISES

1. $(-1, 1)$ [6.2.1] **2.** $(2, -3)$ [6.1.1] **3.** $(-3, 1)$ [6.3.1] **4.** $(1, 6)$ [6.2.1] **5.** $(1, 1)$ [6.1.1] **6.** $(1, -5)$ [6.3.1] **7.** $(3, 2)$ [6.2.1] **8.** The system of equations is dependent. The solutions are the ordered pairs that satisfy the equation $8x - y = 25$. [6.3.1] **9.** $(3, -1)$ [6.3.1] **10.** $(1, -3)$ [6.1.1] **11.** $(-2, -7)$ [6.3.1] **12.** $(4, 0)$ [6.2.1] **13.** Yes [6.1.1] **14.** $\left(-\frac{5}{6}, \frac{1}{2}\right)$ [6.3.1] **15.** $(-1, -2)$ [6.3.1] **16.** $(3, -3)$ [6.1.1] **17.** The system of equations is inconsistent and has no solution. [6.3.1] **18.** $\left(-\frac{1}{3}, \frac{1}{6}\right)$ [6.2.1] **19.** The system of equations is dependent. The solutions are the ordered pairs that satisfy the equation $y = 2x - 4$. [6.1.1] **20.** The system of equations is dependent. The solutions are the ordered pairs that satisfy the equation $3x + y = -2$. [6.3.1] **21.** $(-2, -13)$ [6.3.1] **22.** The system of equations is dependent. The solutions are the ordered pairs that satisfy the equation $4x + 3y = 12$. [6.2.1] **23.** $(4, 2)$ [6.1.1] **24.** The system of equations is inconsistent and has no solution. [6.3.1] **25.** $\left(\frac{1}{2}, -1\right)$ [6.2.1] **26.** No [6.1.1] **27.** $\left(\frac{2}{3}, -\frac{1}{6}\right)$ [6.3.1] **28.** The system of equations is inconsistent and has no solution. [6.2.1]

29. $\left(\frac{1}{3}, 6\right)$ [6.2.1] **30.** The system of equations is inconsistent and has no solution. [6.1.1] **31.** $(0, -1)$ [6.3.1] **32.** $(-1, -3)$ [6.2.1] **33.** $(0, -2)$ [6.2.1] **34.** $(2, 0)$ [6.3.1] **35.** $(-4, 2)$ [6.3.1] **36.** $(1, 6)$ [6.2.1] **37.** The rate of the plane in calm air is 180 mph. The rate of the wind is 20 mph. [6.4.1] **38.** Sixty adult tickets and 140 children's tickets were sold. [6.4.2] **39.** The rate of the canoeist in calm water was 8 mph. The rate of the current was 2 mph. [6.4.1] **40.** There were 130 mailings requiring 46¢ in postage. [6.4.2] **41.** The rate of the boat in calm water was 14 km/h. The rate of the current was 2 km/h. [6.4.1] **42.** The customer purchased 4 compact discs at $15 each and 6 compact discs at $10 each. [6.4.2] **43.** The rate of the plane in calm air was 125 km/h. The rate of the wind was 15 km/h. [6.4.1] **44.** The rate of the paddle boat in calm water was 3 mph. The rate of the current was 1 mph. [6.4.1] **45.** There are 350 bushels of lentils and 200 bushels of corn in the silo. [6.4.2] **46.** The rate of the plane in calm air was 105 mph. The rate of the wind was 15 mph. [6.4.1] **47.** The investor bought 1300 of the $6 shares and 200 of the $25 shares. [6.4.2] **48.** The rate of the sculling team in calm water was 9 mph. The rate of the current was 3 mph. [6.4.1]

CHAPTER 6 TEST

1. $(3, 1)$ [6.2.1, Example 1] **2.** $(2, 1)$ [6.3.1, Example 1] **3.** Yes [6.1.1, Problem 1] **4.** $(1, -1)$ [6.2.1, Example 1] **5.** $\left(\frac{1}{2}, -1\right)$ [6.3.1, Example 1] **6.** $(-2, 6)$ [6.1.1, Example 2A] **7.** The system of equations is inconsistent and has no solution. [6.2.1, Example 2] **8.** $(2, -1)$ [6.2.1, Example 1] **9.** $(2, -1)$ [6.3.1, Example 3] **10.** $\left(\frac{22}{7}, -\frac{5}{7}\right)$ [6.2.1, Problem 1] **11.** $(1, -2)$ [6.3.1, Example 1] **12.** No [6.1.1, Example 1] **13.** $(2, 1)$ [6.2.1, Problem 1] **14.** $(2, -2)$ [6.3.1, Example 1] **15.** $(2, 0)$ [6.1.1, Problem 2A] **16.** $(4, 1)$ [6.2.1, Example 1] **17.** $(3, -2)$ [6.3.1, Example 1] **18.** The system of equations is dependent. The solutions are the ordered pairs that satisfy the equation $3x + 6y = 2$. [6.1.1, Example 2B] **19.** $(1, 1)$ [6.2.1, Example 1] **20.** $(4, -3)$ [6.3.1, Problem 1] **21.** $(-6, 1)$ [6.2.1, Problem 1] **22.** $(1, -4)$ [6.3.1, Example 1] **23.** The rate of the plane in calm air is 100 mph. The rate of the wind is 20 mph. [6.4.1, Example 1] **24.** The rate of the boat in calm water is 14 mph. The rate of the current is 2 mph. [6.4.1, Problem 1]

CUMULATIVE REVIEW EXERCISES

1. $-8, -4$ [1.1.1] **2.** $\{1, 2, 3, 4, 5, 6, 7, 8, 9, 10\}$ [1.1.1] **3.** 10 [1.4.2] **4.** $40a - 28$ [2.2.5] **5.** $\frac{3}{2}$ [2.1.1] **6.** $-\frac{3}{2}$ [3.1.3] **7.** $-\frac{7}{2}$ [3.3.3] **8.** $-\frac{2}{9}$ [3.3.3] **9.** $x < -5$ [3.4.3] **10.** $x \leq 3$ [3.4.3] **11.** 24% [3.2.1] **12.** $(4, 0), (0, -2)$ [5.2.3] **13.** $-\frac{7}{5}$ [5.3.1] **14.** $y = -\frac{3}{2}x$ [5.4.1/5.4.2] **15.**

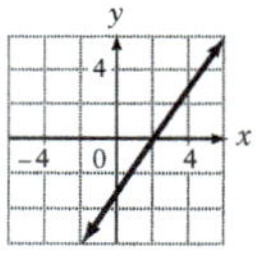

[5.2.3]

16.

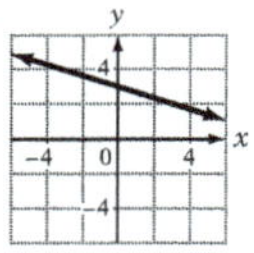

[5.2.2] **17.**

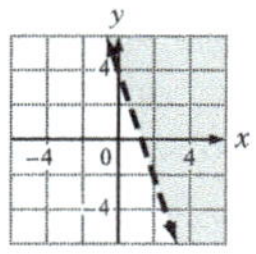

[5.6.1] **18.**

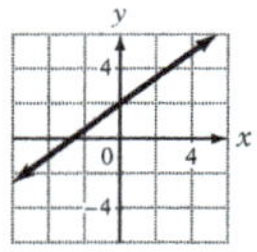

[5.5.2]

19. Domain: $\{-5, 0, 1, 5\}$; range: $\{5\}$; yes [5.5.1] **20.** 3 [5.5.1] **21.** Yes [6.1.1] **22.** $(-2, 1)$ [6.2.1] **23.** $(0, 2)$ [6.1.1] **24.** $(2, 1)$ [6.3.1] **25.** $\{-22, -12, -2, 8, 18\}$ [5.5.1] **26.** 200 cameras were produced during the month. [3.3.4] **27.** $3750 should be invested at 9.6%, and $5000 should be invested at 7.2%. [4.4.1] **28.** The rate of the plane in calm air is 160 mph. The rate of the wind is 30 mph. [6.4.1] **29.** The rate of the boat in calm water is 10 mph. [6.4.1] **30.** The cost of the milk chocolate is $5/lb. The cost of the dark chocolate is $6/lb. [6.4.2]

Answers to Chapter 7 Selected Exercises

PREP TEST

1. 1 [1.2.2] **2.** -18 [1.2.3] **3.** $\frac{2}{3}$ [1.2.4] **4.** 48 [2.1.1] **5.** 0 [1.2.4] **6.** No [2.2.2] **7.** $5x^2 - 9x - 6$ [2.2.2] **8.** 0 [2.2.2] **9.** $-6x + 24$ [2.2.4] **10.** $-7xy + 10y$ [2.2.5]

SECTION 7.1

1. Binomial **3.** Monomial **5.** No **7.** Yes **9.** Yes **11.** descending; like **13.** $-2x^2 + 3x$ **15.** $y^2 - 8$ **17.** $5x^2 + 7x + 20$ **19.** $x^3 + 2x^2 - 6x - 6$ **21.** $2a^3 - 3a^2 - 11a + 2$ **23.** $5x^2 + 8x$ **25.** $7x^2 + xy - 4y^2$ **27.** $3a^2 - 3a + 17$ **29.** $5x^3 + 10x^2 - x - 4$ **31.** $3r^3 + 2r^2 - 11r + 7$ **33.** $-2x^3 + 3x^2 + 10x + 11$ **35.** $Q + R$

37. $-7x^2 - 5x + 3$ **39.** $4x$ **41.** $3y^2 - 4y - 2$ **43.** $-7x - 7$ **45.** $4x^3 + 3x^2 + 3x + 1$ **47.** $y^3 - y^2 + 6y - 6$ **49.** $-y^2 - 13xy$ **51.** $2x^2 - 3x - 1$ **53.** $-2x^3 + x^2 + 2$ **55.** $3a^3 - 2$ **57.** $4y^3 - 2y^2 + 2y - 4$ **59. a.** Positive **b.** Negative **c.** Positive **61.** $-a^2 - 1$ **63.** $-4x^2 + 6x + 3$ **65.** $4x^2 - 6x + 3$ **67.** Yes. For example, $(3x^3 - 2x^2 + 3x - 4) - (3x^3 + 4x^2 - 6x + 5) = -6x^2 + 9x - 9$ **69.** Answers will vary. For example, $3x^2 + 5x - 1$ and $4x^2 + x + 6$ **71.** Answers will vary. For example, $3x^2 - 8$ and $-3x^2 + 9$

SECTION 7.2

1. a. No **b.** Yes **c.** No **d.** No **3. a.** Product **b.** Power **c.** Power **d.** Product **5. a.** Yes **b.** No **c.** No **d.** Yes **7.** 7; 2; x^9 **9.** $2x^2$ **11.** $12x^2$ **13.** $6a^7$ **15.** x^3y^5 **17.** $-10x^9y$ **19.** x^7y^8 **21.** $-6x^3y^5$ **23.** x^4y^5z **25.** $a^3b^5c^4$ **27.** $-a^5b^8$ **29.** $-6a^5b$ **31.** $40y^{10}z^6$ **33.** $-20a^2b^3$ **35.** $x^3y^5z^3$ **37.** $-12a^{10}b^7$ **41.** 4; 6; 2; 6; $x^{24}y^{12}$ **43.** x^9 **45.** x^{14} **47.** 64 **49.** 4 **51.** -64 **53.** x^4 **55.** $4x^2$ **57.** $-8x^6$ **59.** x^4y^6 **61.** $9x^4y^2$ **63.** $27a^8$ **65.** $-8x^7$ **67.** x^8y^4 **69.** a^4b^6 **71.** $16x^{10}y^3$ **73.** $-18x^3y^4$ **75.** $-8a^7b^5$ **77.** $-54a^9b^3$ **79.** $-72a^5b^5$ **81.** $32x^3$ **83.** $-3a^3b^3$ **85.** $13x^5y^5$ **87.** $27a^5b^3$ **89.** $-5x^7y^4$ **91.** a^{2n} **93.** a^{2n} **95.** $12ab$ **97. a.** No; $2^{(3^2)}$ is larger. $(2^3)^2 = 8^2 = 64$, whereas $2^{(3^2)} = 2^9 = 512$. **b.** The order of operations is $x^{(m^n)}$.

SECTION 7.3

1. Always true **3.** Sometimes true **5.** Always true **7.** Sometimes true **9.** Sometimes true **11.** $-3y$; $-3y$; Distributive; $-3y$; $-3y^2$; $21y$ **13.** $x^2 - 2x$ **15.** $-x^2 - 7x$ **17.** $3a^3 - 6a^2$ **19.** $-5x^4 + 5x^3$ **21.** $-3x^5 + 7x^3$ **23.** $12x^3 - 6x^2$ **25.** $6x^2 - 12x$ **27.** $-x^3y + xy^3$ **29.** $2x^4 - 3x^2 + 2x$ **31.** $2a^3 + 3a^2 + 2a$ **33.** $3x^6 - 3x^4 - 2x^2$ **35.** $-6y^4 - 12y^3 + 14y^2$ **37.** $-2a^3 - 6a^2 + 8a$ **39.** $6y^4 - 3y^3 + 6y^2$ **41.** $x^3y - 3x^2y^2 + xy^3$ **43.** $x - 3$; $x - 3$; $x - 3$; $5x$; 15; $17x$; 15 **45.** $x^3 + 4x^2 + 5x + 2$ **47.** $a^3 - 6a^2 + 13a - 12$ **49.** $-2b^3 + 7b^2 + 19b - 20$ **51.** $-6x^3 + 31x^2 - 41x + 10$ **53.** $x^4 - 4x^3 - 3x^2 + 14x - 8$ **55.** $15y^3 - 16y^2 - 70y + 16$ **57.** $5a^4 - 20a^3 - 15a^2 + 62a - 8$ **59.** $y^4 + 4y^3 + y^2 - 5y + 2$ **61. a.** 4 **b.** 5 **63.** $4x$; x; $4x$; 5; -3; x; -3; 5 **65. a.** $3d$; d **b.** $3d$; -1 **c.** 4; d **d.** 4; -1 **67.** $x^2 + 4x + 3$ **69.** $a^2 + a - 12$ **71.** $y^2 - 5y - 24$ **73.** $y^2 - 10y + 21$ **75.** $2x^2 + 15x + 7$ **77.** $3x^2 + 11x - 4$ **79.** $4x^2 - 31x + 21$ **81.** $3y^2 - 2y - 16$ **83.** $9x^2 + 54x + 77$ **85.** $21a^2 - 83a + 80$ **87.** $15b^2 + 47b - 78$ **89.** $2a^2 + 7ab + 3b^2$ **91.** $6a^2 + ab - 2b^2$ **93.** $2x^2 - 3xy - 2y^2$ **95.** $10x^2 + 29xy + 21y^2$ **97.** $6a^2 - 25ab + 14b^2$ **99.** $2a^2 - 11ab - 63b^2$ **101.** $10x^2 - 21xy - 10y^2$ **103.** Positive **105.** $5x$; 1; $25x^2$; 1 **107.** $y^2 - 25$ **109.** $4x^2 - 9$ **111.** $x^2 + 2x + 1$ **113.** $9a^2 - 30a + 25$ **115.** $9x^2 - 49$ **117.** $4a^2 + 4ab + b^2$ **119.** $x^2 - 4xy + 4y^2$ **121.** $16 - 9y^2$ **123.** $25x^2 + 20xy + 4y^2$ **125.** Zero **127.** Positive **129.** $2x + 5$; $2x - 5$; $4x^2 - 25$ **131.** $4x$; $4x$; 1; 1; $16x^2$; $8x$; 1 **133.** The area is $(2x^2 - 9x - 18)$ m². **135.** The area is $(32x^2 - 48x + 18)$ cm². **137.** The area is $(4x^2 - 12x + 9)$ yd². **139.** The area is $(x^2 - 5x - 24)$ in². **141.** The area is $(\pi x^2 - 6\pi x + 9\pi)$ ft². **143.** The area is $(90x + 2025)$ ft². **145.** The one with a length of $(ax + b)$ ft and a width of $(cx + d)$ ft has the larger area. **147.** $4ab$ **149.** $9a^4 - 24a^3 + 28a^2 - 16a + 4$ **151.** $24x^3 - 3x^2$ **153.** 1024 **155.** $12x^2 - x - 20$ **157.** $7x^2 - 11x - 8$ **161.** $-x^3 - 1$ **163.** $-x^5 - 1$ **165.** $-x^7 - 1$

SECTION 7.4

1. False **3.** True **5.** False **9.** 4; 10; 3; $4x^7$ **11.** y^4 **13.** a^3 **15.** p^4 **17.** $2x^3$ **19.** $2k$ **21.** m^5n^2 **23.** $\frac{3r^2}{2}$ **25.** $-\frac{2a}{3}$ **27.** $\frac{1}{x^2}$ **29.** a^6 **31.** $\frac{1}{25}$ **33.** 64 **35.** $\frac{1}{27}$ **37.** 1 **39.** $\frac{4}{x^7}$ **41.** $5b^8$ **43.** $\frac{x^2}{3}$ **45.** 1 **47.** $\frac{1}{y^5}$ **49.** $\frac{1}{a^6}$ **51.** $\frac{1}{3x^3}$ **53.** $\frac{2}{3x^5}$ **55.** $\frac{y^4}{x^2}$ **57.** $\frac{2}{5m^3n^8}$ **59.** $\frac{1}{p^3q}$ **61.** $\frac{1}{2y^3}$ **63.** $\frac{7xz}{8y^3}$ **65.** $-\frac{8x^3}{y^6}$ **67.** $\frac{9}{x^2y^4}$ **69.** $\frac{2}{x^4}$ **71.** $-\frac{5}{a^8}$ **73.** $-\frac{a^5}{8b^4}$ **75.** $\frac{1}{a^5b^6}$ **77.** $\frac{1}{4x^3}$ **79.** $\frac{2y}{x^3}$ **81.** $-\frac{4b}{9}$ **83.** $-\frac{2x^2y^2}{11z^5}$ **85. a.** True **b.** False **87.** 1; 10; 10 **89.** No. 0.8 is not between 1 and 10. **91.** Yes **93.** 9; right; -9 **95.** 7.5×10^4 **97.** 7.6×10^{-5} **99.** 8.19×10^8 **101.** 9.6×10^{-10} **103.** 23,000,000 **105.** 0.000000921 **107.** 5,750,000,000 **109.** 0.0000000354 **111.** 1×10^{21} **113.** 4×10^{-10} **115.** 1.2×10^{10} **117.** 5.88×10^{12} **119.** 5,115,600,000,000,000 **121.** 7.56×10^2 **123.** 7.2×10^{-11} **125.** 2.4×10^{-6} **127.** 2.1×10^{11} **129.** Greater than **131.** $\frac{4}{81}$ **133.** $\frac{1}{9}, \frac{1}{3}$, 1, 3, 9; 9, 3, 1, $\frac{1}{3}, \frac{1}{9}$ **135.** 0.0016 **137.** a **139.** 0 **141.** True **143.** True **147. a.** Each person would have 1047.6 ft² of land. **b.** Each person would have 90 ft² of land. **c.** (i) 57,900 people would fit in a square mile. (ii) 120,900 mi² would be required. **d.** Each person would be allocated 5.2 acres. **e.** The carrying capacity of Earth would be 3300 billion people.

SECTION 7.5

1. $15x^2 + 12x = 3x(5x + 4)$ **3.** True **5.** True **7.** $3y$; $3y$; $6y^4$; 1 **9.** $2a - 5$ **11.** $3a + 2$ **13.** $x - 2$ **15.** $-x + 2$ **17.** $x^2 + 3x - 5$ **19.** $x^4 - 3x^2 - 1$ **21.** $xy + 2$ **23.** $-3y^3 + 5$ **25.** $3x - 2 + \frac{1}{x}$

27. $-3x + 7 - \frac{6}{x}$ **29.** $4a - 5 + 6b$ **31.** $9x + 6 - \frac{3}{y}$ **33.** Multiply $4x$ and $2x^2 - 3x - 1$. $4x(2x^2 - 3x - 1) = 8x^3 - 12x^2 - 4x$ **35.** $x + 2$ **37.** $2x + 1$ **39.** $x + 1 + \frac{2}{x - 1}$ **41.** $2x - 1 - \frac{2}{3x - 2}$ **43.** $b - 5 - \frac{24}{b - 3}$ **45.** $3x + 17 + \frac{64}{x - 4}$ **47.** $5y + 3 + \frac{1}{2y + 3}$ **49.** $3x - 5$ **51.** $x^2 + 2x + 3 + \frac{5}{x + 1}$ **53.** $x^2 - 3$ **55.** False **57.** $x + 4; x - 2; 5$ **59.** $4y^2$ **61.** $x^3 - 4x^2 + 11x - 2$ **63.** $x^2 + 2x - 3$

CHAPTER 7 REVIEW EXERCISES

1. $21y^2 + 4y - 1$ [7.1.1] **2.** $-20x^3y^5$ [7.2.1] **3.** $-8x^3 - 14x^2 + 18x$ [7.3.1] **4.** $10a^2 + 31a - 63$ [7.3.3] **5.** $6x^2 - 7x + 10$ [7.5.1] **6.** $2x^2 + 3x - 8$ [7.1.2] **7.** -729 [7.2.2] **8.** $x^3 - 6x^2 + 7x - 2$ [7.3.2] **9.** $a^2 - 49$ [7.3.4] **10.** 1296 [7.4.1] **11.** $x - 6$ [7.5.2] **12.** $2x^3 + 9x^2 - 3x - 12$ [7.1.1] **13.** $18a^8b^6$ [7.2.1] **14.** $3x^4y - 2x^3y + 12x^2y$ [7.3.1] **15.** $8b^2 - 2b - 15$ [7.3.3] **16.** $-4y + 8$ [7.5.1] **17.** $13y^3 - 12y^2 - 5y - 1$ [7.1.2] **18.** 64 [7.2.2] **19.** $6y^3 + 17y^2 - 2y - 21$ [7.3.2] **20.** $4b^2 - 81$ [7.3.4] **21.** $\frac{b^6c^2}{a^4}$ [7.4.1] **22.** $2y - 9$ [7.5.2] **23.** 3.97×10^{-6} [7.4.2] **24.** $x^4y^8z^4$ [7.2.1] **25.** $-12y^5 + 4y^4 - 18y^3$ [7.3.1] **26.** $18x^2 - 48x + 24$ [7.3.3] **27.** 0.0000623 [7.4.2] **28.** $-7a^2 - a + 4$ [7.1.2] **29.** $9x^4y^6$ [7.2.2] **30.** $12a^3 - 8a^2 - 9a + 6$ [7.3.3] **31.** $25y^2 - 70y + 49$ [7.3.4] **32.** $\frac{x^4y^6}{9}$ [7.4.1] **33.** $x + 5 + \frac{4}{x + 12}$ [7.5.2] **34.** 240,000 [7.4.2] **35.** $a^6b^{11}c^9$ [7.2.1] **36.** $8a^3b^3 - 4a^2b^4 + 6ab^5$ [7.3.1] **37.** $6x^2 - 7xy - 20y^2$ [7.3.3] **38.** $4b^4 + 12b^2 - 1$ [7.5.1] **39.** $-4b^2 - 13b + 28$ [7.1.2] **40.** $100a^{15}b^{13}$ [7.2.2] **41.** $12b^5 - 4b^4 - 6b^3 - 8b^2 + 5$ [7.3.2] **42.** $36 - 25x^2$ [7.3.4] **43.** $\frac{2y^3}{x^3}$ [7.4.1] **44.** $a^2 - 2a + 6$ [7.5.2] **45.** $4b^3 - 5b^2 - 9b + 7$ [7.1.1] **46.** $-54a^{13}b^5c^7$ [7.2.1] **47.** $-18x^4 - 27x^3 + 63x^2$ [7.3.1] **48.** $30y^2 - 109y + 30$ [7.3.3] **49.** 9.176×10^{12} [7.4.2] **50.** $-\frac{3y^7}{4x^6}$ [7.4.1] **51.** $144x^{14}y^{18}z^{16}$ [7.2.2] **52.** $-12x^4 - 17x^3 - 2x^2 - 33x - 27$ [7.3.2] **53.** $64a^2 + 16a + 1$ [7.3.4] **54.** $\frac{2}{ab^6}$ [7.4.1] **55.** $b^2 + 5b + 2 + \frac{7}{b - 7}$ [7.5.2] **56.** The area is $(20x^2 - 35x)$ m². [7.3.5] **57.** The area is $(25x^2 + 40x + 16)$ in². [7.3.5] **58.** The area is $(9x^2 - 4)$ ft². [7.3.5] **59.** The area is $(\pi x^2 - 12\pi x + 36\pi)$ cm². [7.3.5] **60.** The area is $(15x^2 - 28x - 32)$ mi². [7.3.5]

CHAPTER 7 TEST

1. $3x^3 + 6x^2 - 8x + 3$ [7.1.1, Example 2] **2.** $-4x^4 + 8x^3 - 3x^2 - 14x + 21$ [7.3.2, Example 3] **3.** $4x^3 - 6x^2$ [7.3.1, Example 1A] **4.** $-8a^6b^3$ [7.2.2, Example 3] **5.** $-4x^6$ [7.4.1, Example 3A] **6.** $\frac{6b}{a}$ [7.4.1, Example 3B] **7.** $-5a^3 + 3a^2 - 4a + 3$ [7.1.2, Problem 4] **8.** $a^2 + 3ab - 10b^2$ [7.3.3, Example 5] **9.** $4x^4 - 2x^2 + 5$ [7.5.1, Example 1] **10.** $2x + 3 + \frac{2}{2x - 3}$ [7.5.2, Problem 3] **11.** $-6x^3y^6$ [7.2.1, Example 2] **12.** $6y^4 - 9y^3 + 18y^2$ [7.3.1, Example 1A] **13.** $\frac{9}{x^3}$ [7.4.1, Example 3A] **14.** $4x^2 - 20x + 25$ [7.3.4, Example 8] **15.** $10x^2 - 43xy + 28y^2$ [7.3.3, Example 5] **16.** $x^3 - 7x^2 + 17x - 15$ [7.3.2, Example 2] **17.** $\frac{a^4}{b^6}$ [7.4.1, Example 3B] **18.** 2.9×10^{-5} [7.4.2, Example 4B] **19.** $16y^2 - 9$ [7.3.4, Example 6] **20.** $3y^3 + 2y^2 - 10y$ [7.1.2, Example 4] **21.** $9a^6b^4$ [7.2.2, Example 3] **22.** $10a^3 - 39a^2 + 20a - 21$ [7.3.2, Example 3] **23.** $9b^2 + 12b + 4$ [7.3.4, Example 7] **24.** $-\frac{1}{4a^2b^5}$ [7.4.1, Example 3A] **25.** $4x + 8 + \frac{21}{2x - 3}$ [7.5.2, Problem 3] **26.** a^3b^7 [7.2.1, Example 1] **27.** $a^2 + ab - 12b^2$ [7.3.3, Problem 5] **28.** 0.000000035 [7.4.2, Example 5B] **29.** The area is $(4x^2 + 12x + 9)$ m². [7.3.5, Example 10] **30.** The area is $(\pi x^2 - 10\pi x + 25\pi)$ in². [7.3.5, Example 9]

CUMULATIVE REVIEW EXERCISES

1. $\frac{1}{144}$ [1.3.3] **2.** $\frac{25}{9}$ [1.4.1] **3.** $\frac{4}{5}$ [1.4.2] **4.** 87 [1.1.2] **5.** 0.775 [1.3.1] **6.** $-\frac{27}{4}$ [2.1.1] **7.** $-x - 4xy$ [2.2.2] **8.** $-12x$ [2.2.3] **9.** $-22x + 20$ [2.2.5] **10.** 8 [2.2.1] **11.** -18 [3.1.3] **12.** 16 [3.3.2] **13.** 12 [3.3.3] **14.** 80 [3.3.2] **15.** 24% [3.2.1] **16.** $x \geq -3$ [3.4.3] **17.** $-\frac{9}{5}$ [5.3.1] **18.** $y = -\frac{3}{2}x - \frac{3}{2}$ [5.4.1/5.4.2] **19.**

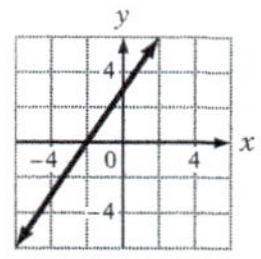

[5.2.3] **20.**

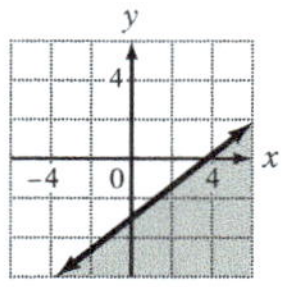

[5.6.1]

21. Domain: $\{-8, -6, -4, -2\}$; range: $\{-7, -5, -2, 0\}$; yes [5.5.1] **22.** -2 [5.5.1] **23.** $(4, 1)$ [6.2.1]

24. (1, −4) [6.3.1] **25.** $5b^3 - 7b^2 + 8b - 10$ [7.1.2] **26.** $15x^3 - 26x^2 + 11x - 4$ [7.3.2]
27. $20b^2 - 47b + 24$ [7.3.3] **28.** $25b^2 + 30b + 9$ [7.3.4] **29.** $-\frac{b^4}{4a}$ [7.4.1] **30.** $5y - 4 + \frac{1}{y}$ [7.5.1]
31. $a - 7$ [7.5.2] **32.** $\frac{9y^2}{x^6}$ [7.4.1] **33.** {1, 9, 17, 21, 25} [5.5.1] **34.** $5(n - 12)$; $5n - 60$ [2.3.2]
35. $8n - 2n = 18$; $n = 3$ [4.1.1] **36.** The length is 15 m. The width is 6 m. [4.2.1] **37.** The selling price is $43.20. [4.3.1]
38. The concentration of orange juice is 28%. [4.5.2] **39.** The car overtakes the cyclist 25 mi from the starting point. [4.6.1]
40. The area is $(9x^2 + 12x + 4)$ ft². [7.3.5]

Answers to Chapter 8 Selected Exercises

PREP TEST

1. $2 \cdot 3 \cdot 5$ [1.3.3] **2.** $-12y + 15$ [2.2.4] **3.** $-a + b$ [2.2.4] **4.** $-3a + 3b$ [2.2.5] **5.** 0 [3.1.3]
6. $-\frac{1}{2}$ [3.3.1] **7.** $x^2 - 2x - 24$ [7.3.3] **8.** $6x^2 - 11x - 10$ [7.3.3] **9.** x^3 [7.4.1] **10.** $3x^3y$ [7.4.1]

SECTION 8.1

1. 4 **3.** **a** is a product. **b** is a sum. **5.** $c - 6$ **9. a.** z **b.** 3 **c.** x **d.** y^2 **e.** $3xy^2$ **11.** x^3
13. xy^4 **15.** xy^4z^2 **17.** $7a^3$ **19.** 1 **21.** $3a^2b^2$ **23.** ab **25.** $2x$ **27.** $3x$ **29.** $5(a + 1)$ **31.** $8(2 - a^2)$
33. $4(2x + 3)$ **35.** $6(5a - 1)$ **37.** $x(7x - 3)$ **39.** $a^2(3 + 5a^3)$ **41.** $2x(x^3 - 2)$ **43.** $2x^2(5x^2 - 6)$ **45.** $xy(x - y^2)$
47. $2a^5b + 3xy^3$ **49.** $6b^2(a^2b - 2)$ **51.** $2abc(3a + 2b)$ **53.** $6x^3y^3(1 - 2x^3y^3)$ **55.** $x(x^2 - 3x - 1)$
57. $2(x^2 + 4x - 6)$ **59.** $b(b^2 - 5b - 7)$ **61.** $3x(x^2 + 2x + 3)$ **63.** $2x^2(x^2 - 2x + 3)$ **65.** $a^2(6a^3 - 3a - 2)$
67. $x^2(8y^2 - 4y + 1)$ **69.** $x^3y^3(4x^2y^2 - 8xy + 1)$ **71.** x^a **73. a.** $4(3x - 2)$; $3x - 2$ **b.** $5x$; 4 **75.** $(a + b)(x + 2)$
77. $(b + 2)(x - y)$ **79.** $(y - 4)(a - b)$ **81.** $a(x - 2) + b(x - 2)$ **83.** $b(a - 7) - 3(a - 7)$
85. $x(a - 2b) + y(a - 2b)$ **87.** $(x - 2)(a - 5)$ **89.** $(y - 3)(b - 3)$ **91.** $(x - y)(a + 2)$ **93.** $(c + 5)(z - 8)$
95. $(3x - 4)(w + 1)$ **97.** $(x + 4)(x^2 + 3)$ **99.** $(y - 4)(3y^2 + 1)$ **101.** $(4 + c)(2 + a^2)$ **103.** $(y - 5)(2y + 7x)$
105. $(a + 3)(b - 2)$ **107.** $(a - 2)(x^2 - 3)$ **109.** $(t + 4)(t - s)$ **111.** $(7x + 2y)(3x - 7)$ **113.** $(2r + a)(a - 1)$
115. ii **117.** P doubles. **119. a.** −1 **b.** $y - x$ **c.** $b - 3a$ **121. a.** $(x + 3)(2x + 5)$ **b.** $(2x + 5)(x + 3)$
123. a. $(a - b)(2a - 3b)$ **b.** $(2a - 3b)(a - b)$

SECTION 8.2

1. −8 **3.** −2 and 6 **5.** Different **7.**

Factors of 18	Sum
1, 18	19
2, 9	11
3, 6	9

9.

Factors of −21	Sum
1, −21	−20
−1, 21	20
3, −7	−4
−3, 7	4

11.

Factors of −28	Sum
1, −28	−27
−1, 28	27
2, −14	−12
−2, 14	12
4, −7	−3
−4, 7	3

13. $(x + 1)(x + 2)$ **15.** $(x + 1)(x - 2)$ **17.** $(a + 4)(a - 3)$

19. $(a - 1)(a - 2)$ **21.** $(a + 2)(a - 1)$ **23.** $(b + 8)(b - 1)$ **25.** $(y + 11)(y - 5)$ **27.** $(y - 2)(y - 3)$
29. $(z - 5)(z - 9)$ **31.** $(p + 7)(p - 5)$ **33.** $(p - 2)(p - 4)$ **35.** $(b + 5)(b + 8)$ **37.** $(x + 14)(x - 5)$
39. $(b + 8)(b - 5)$ **41.** $(y + 8)(y - 9)$ **43.** $(p + 3)(p + 13)$ **45.** Nonfactorable over the integers **47.** $(x + 2)(x + 19)$
49. $(x + 9)(x - 4)$ **51.** $(a - 3)(a - 12)$ **53. a.** Positive **b.** Positive **55. a.** common **b.** 5 **c.** $x^2 - 2x - 8$
d. $x + 2$; $x - 4$ **57.** $2(x + 1)(x + 2)$ **59.** $3(a + 3)(a - 2)$ **61.** $a(b + 5)(b - 3)$ **63.** $x(y - 2)(y - 3)$
65. $2a(a + 1)(a + 2)$ **67.** $4y(y + 6)(y - 3)$ **69.** $5(z + 4)(z - 7)$ **71.** $2a(a + 8)(a - 4)$ **73.** $(x - 2y)(x - 3y)$
75. $(a - 4b)(a - 5b)$ **77.** $(x + 4y)(x - 7y)$ **79.** Nonfactorable over the integers **81.** $z^2(z - 5)(z - 7)$
83. $b^2(b - 10)(b - 12)$ **85.** $2y^2(y + 3)(y - 16)$ **87.** $x^2(x + 8)(x - 1)$ **89.** $3x(x - 3)(x - 9)$ **91.** $(x - 3y)(x - 5y)$
93. $(a - 6b)(a - 7b)$ **95.** $3y(x + 21)(x - 1)$ **97.** $3x(x + 4)(x - 3)$ **99. a.** Yes **b.** Yes **101.** $x - 1$
103. $(c + 4)(c + 5)$ **105.** $a^2(b - 5)(b - 9)$ **107.** 36, 12, −12, −36 **109.** 22, 10, −10, −22 **111.** 3, 4
113. 5, 8, 9 **115.** 2

SECTION 8.3

1. $2x + 5$ **3.** $4x - 3$ **5.** $4, -5$ **7.** $-6x, -2x$ **11.** $(x + 1)(2x + 1)$ **13.** $(y + 3)(2y + 1)$ **15.** $(a - 1)(2a - 1)$ **17.** $(b - 5)(2b - 1)$ **19.** $(x + 1)(2x - 1)$ **21.** $(x - 3)(2x + 1)$ **23.** Nonfactorable over the integers **25.** $(2t - 1)(3t - 4)$ **27.** $(x + 4)(8x + 1)$ **29.** $(2b - 3)(3b - 5)$ **31.** $(p + 8)(3p - 2)$ **33.** $(2x - 3)(3x - 4)$ **35.** $(b + 7)(5b - 2)$ **37.** $(2a - 3)(3a + 8)$ **39.** $(3t + 1)(6t - 5)$ **41.** $(3a + 7)(5a - 3)$ **43.** $(2y - 5)(4y - 3)$ **45.** Nonfactorable over the integers **47.** $(4 + z)(7 - z)$ **49.** $(1 - x)(8 + x)$ **51.** $3(x + 5)(3x - 4)$ **53.** $4(2x - 3)(3x - 2)$ **55.** $a^2(5a + 2)(7a - 1)$ **57.** $5(b - 7)(3b - 2)$ **59.** $2x(x + 1)(5x + 1)$ **61.** $yz(z + 2)(4z - 3)$ **63.** $xy(3x + 2)(3x + 2)$ **65.** m and p **67.** $-18; -17; 18; 17; -9; -7; 9; 7; -6; -3; 6; 3$ **69.** $(t + 2)(2t - 5)$ **71.** $(p - 5)(3p - 1)$ **73.** $(3y - 1)(4y - 1)$ **75.** Nonfactorable over the integers **77.** $(3y + 1)(4y + 5)$ **79.** $(a + 7)(7a - 2)$ **81.** $(z + 2)(4z + 3)$ **83.** $(y + 1)(8y + 9)$ **85.** $(2b - 3)(3b - 2)$ **87.** Nonfactorable over the integers **89.** $(2z - 5)(5z - 2)$ **91.** $2(x + 1)(2x + 1)$ **93.** $5(y - 1)(3y - 7)$ **95.** $x(x - 5)(2x - 1)$ **97.** $(a + 2b)(3a - b)$ **99.** $(y - 2z)(4y - 3z)$ **101.** $(3 - x)(4 + x)$ **103.** $4(9y + 1)(10y - 1)$ **105.** $8(t + 4)(2t - 3)$ **107.** $p(2p + 1)(3p + 1)$ **109.** $3(2z - 5)(5z - 2)$ **111.** $3a(2a + 3)(7a - 3)$ **113.** $y(3x - 5y)(3x - 5y)$ **115.** $xy(3x - 4y)(3x - 4y)$ **117.** Two positive **119.** Two negative **121.** $2y(y - 3)(4y - 1)$ **123.** $ab(a + 4)(a - 6)$ **125.** $5t(3 + 2t)(4 - t)$ **127.** $(y + 3)(2y + 1)$ **129.** $(2x - 1)(5x + 7)$ **131.** $(x + 2)(x - 3)(x - 1)$ **133.** $7, -7, 5, -5$ **135.** $7, -7, 5, -5$ **137.** $11, -11, 7, -7$

SECTION 8.4

1. $4; 25x^6; 100x^4y^4$ **3.** i and iv **5.** Sometimes true **7.** Always true **9. a.** $3x; 2$ **b.** $3x + 2; 3x - 2$ **11.** Answers will vary. For example: **a.** $x^2 - 16$ **b.** $(x + 7)(x - 7)$ **c.** $x^2 + 10x + 25$ **d.** $(x - 3)^2$ **e.** $x^2 + 36$ **13.** $(x + 2)(x - 2)$ **15.** $(a + 9)(a - 9)$ **17.** $(2x + 3)(2x - 3)$ **19.** $(y + 3)^2$ **21.** $(a - 1)^2$ **23.** Nonfactorable over the integers **25.** $(x^3 + 3)(x^3 - 3)$ **27.** $(5x + 2)(5x - 2)$ **29.** $(1 + 7x)(1 - 7x)$ **31.** $(x + y)^2$ **33.** $(2a + 1)^2$ **35.** Nonfactorable over the integers **37.** $(x^2 + y)(x^2 - y)$ **39.** $(3x + 4y)(3x - 4y)$ **41.** $(4b + 3)^2$ **43.** $(2b + 7)^2$ **45.** $(5a + 3b)^2$ **47.** $(xy + 2)(xy - 2)$ **49.** $(x + 1)(9x + 4)$ **51.** i and iii **53.** $(4x + 3)$ ft by $(4x + 3)$ ft; yes; $x > -\frac{3}{4}$ **55.** $(y + 3)(y^2 - 3y + 9)$ **57.** $(x - 1)(x^2 + x + 1)$ **59.** $(x - 5)(x^2 + 5x + 25)$ **61.** $(3y + 1)(9y^2 - 3y + 1)$ **63.** $10, -10$ **65.** $16, -16$ **67.** 4 **69.** 25

SECTION 8.5

1. Determine whether the polynomial is the difference of two squares. If so, factor. **3.** Try to factor the polynomial by grouping. **5.** nonfactorable **7.** $2(x + 3)(x - 3)$ **9.** $x^2(x + 7)(x - 5)$ **11.** $5(b + 3)(b + 12)$ **13.** Nonfactorable over the integers **15.** $2y(x + 11)(x - 3)$ **17.** $x(x^2 - 6x - 5)$ **19.** $3(y^2 - 12)$ **21.** $(2a + 1)(10a + 1)$ **23.** $y^2(x + 1)(x - 8)$ **25.** $5(a + b)(2a - 3b)$ **27.** $2(5 + x)(5 - x)$ **29.** $ab(3a - b)(4a + b)$ **31.** $2(x - 1)(a + b)$ **33.** $3a(2a - 1)^2$ **35.** $3(81 + a^2)$ **37.** $2a(2a - 5)(3a - 4)$ **39.** $(x - 2)(x + 1)(x - 1)$ **41.** $a(2a + 5)^2$ **43.** $3b(3a - 1)^2$ **45.** $6(4 + x)(2 - x)$ **47.** $(x + 2)(x - 2)(a + b)$ **49.** $x^2(x + y)(x - y)$ **51.** $2a(3a + 2)^2$ **53.** $b(2 - 3a)(1 + 2a)$ **55.** $(x - 5)(2 + x)(2 - x)$ **57.** $8x(3y + 1)^2$ **59.** $y^2(5 + x)(3 - x)$ **61.** $y(y + 3)(y - 3)$ **63.** $2x^2y^2(x + 1)(x - 1)$ **65.** $x^5(x^2 + 1)(x + 1)(x - 1)$ **67.** $2xy(3x - 2)(4x + 5)$ **69.** $x^2y^2(2x - 5)^2$ **71.** $(m^2 + 16)(m + 4)(m - 4)$ **73.** $(y^4 + 9)(y^2 + 3)(y^2 - 3)$ **75.** Less than **77.** $(2a + 3 + 5b)(2a + 3 - 5b)$ **79.** $(2x + 3 + 2y)(2x + 3 - 2y)$ **81.** $(40 + 2)(40 - 2) = 1600 - 4 = 1596; (80 + 4)(80 - 4) = 6400 - 16 = 6384$ **83.** 128

SECTION 8.6

1. a. Yes **b.** No **c.** Yes **3. a.** Yes **b.** Yes **c.** No **d.** Yes **e.** No **f.** Yes **5.** False **9.** $0 = x^2 + x - 2$ **11.** $-3, -2$ **13.** $7, 3$ **15.** $0, 5$ **17.** $0, 9$ **19.** $0, -\frac{3}{2}$ **21.** $0, \frac{2}{3}$ **23.** $\frac{1}{3}, -\frac{1}{3}$ **25.** $-4, -2$ **27.** $-7, 2$ **29.** $3, 2$ **31.** $-\frac{1}{2}, 5$ **33.** $-2, 5$ **35.** $0, 7$ **37.** $-1, -4$ **39.** $2, 3$ **41.** $\frac{1}{2}, -4$ **43.** $\frac{1}{3}, 4$ **45.** $3, 9$ **47.** $9, -2$ **49.** $-1, -2$ **51.** $5, -9$ **53.** $-2, -3$ **55.** $-8, -5$ **57.** $-\frac{3}{2}, -2$ **59.** Less than **61.** One **63. a.** $x + 1; x^2 + (x + 1)^2$ **b.** $x^2; (x + 1)^2; 113$ **65.** The number is 6. **67.** The numbers are 2 and 4. **69.** The integers are 4 and 5. **71.** The integers are 15 and 16. **73.** The base is 18 ft. The height is 6 ft. **75.** The length is 18 ft. The width is 8 ft. **77.** The length of a side of the original square is 4 m. **79.** The radius of the original circle is 3.81 in. **81.** The dimensions are 4 in. by 7 in. **83.** The width of the lane is 16 ft. **85.** The object will hit the ground 4 s later. **87.** 15 consecutive natural numbers beginning with 1 will give a sum of 120. **89.** There are 10 teams in the league. **91.** The ball will return to the ground 3 s later. **93.** $\frac{3}{2}, -4$ **95.** $-1, -9$ **97.** $0, 7$ **99.** $18, 1$ **101.** 2 or -128 **105.** The length is 20 in. The width is 10 in.

CHAPTER 8 REVIEW EXERCISES

1. $7y^3(2y^6 - 7y^3 + 1)$ [8.1.1] **2.** $(a - 4)(3a + b)$ [8.1.2] **3.** $(c + 2)(c + 6)$ [8.2.1] **4.** $a(a - 2)(a - 3)$ [8.2.2] **5.** $(2x - 7)(3x - 4)$ [8.3.1/8.3.2] **6.** $(y + 6)(3y - 2)$ [8.3.1/8.3.2] **7.** $(3a + 2)(6a - 5)$ [8.3.1/8.3.2]

8. $(ab + 1)(ab - 1)$ [8.4.1] **9.** $4(y - 2)^2$ [8.5.1] **10.** $0, -\frac{1}{5}$ [8.6.1] **11.** $3ab(4a + b)$ [8.1.1] **12.** $(b - 3)(b - 10)$ [8.2.1] **13.** $(2x + 5)(5x + 2y)$ [8.1.2] **14.** $3(a + 2)(a - 7)$ [8.2.2] **15.** $n^2(n + 1)(n - 3)$ [8.2.2] **16.** Nonfactorable over the integers [8.3.1/8.3.2] **17.** $(2x - 1)(3x - 2)$ [8.3.1/8.3.2] **18.** Nonfactorable over the integers [8.4.1] **19.** $2, \frac{3}{2}$ [8.6.1] **20.** $7(x + 1)(x - 1)$ [8.5.1] **21.** $x^3(3x^2 - 9x - 4)$ [8.1.1] **22.** $(x - 3)(4x + 5)$ [8.1.2] **23.** $(a + 7)(a - 2)$ [8.2.1] **24.** $(y + 9)(y - 4)$ [8.2.1] **25.** $5(x - 12)(x + 2)$ [8.2.2] **26.** $-3, 7$ [8.6.1] **27.** $(a + 2)(7a + 3)$ [8.3.1/8.3.2] **28.** $(x + 20)(4x + 3)$ [8.3.1/8.3.2] **29.** $(3y^2 + 5z)(3y^2 - 5z)$ [8.4.1] **30.** $5(x + 2)(x - 3)$ [8.2.2] **31.** $-\frac{3}{2}, \frac{2}{3}$ [8.6.1] **32.** $2b(2b - 7)(3b - 4)$ [8.3.1/8.3.2] **33.** $5x(x^2 + 2x + 7)$ [8.1.1] **34.** $(x - 2)(x - 21)$ [8.2.1] **35.** $(3a + 2)(a - 7)$ [8.1.2] **36.** $(2x - 5)(4x - 9)$ [8.3.1/8.3.2] **37.** $10x(a - 4)(a - 9)$ [8.2.2] **38.** $(a - 12)(2a + 5)$ [8.3.1/8.3.2] **39.** $(3a - 5b)(7x + 2y)$ [8.1.2] **40.** $(a^3 + 10)(a^3 - 10)$ [8.4.1] **41.** $(4a + 1)^2$ [8.4.1] **42.** $\frac{1}{4}, -7$ [8.6.1] **43.** $10(a + 4)(2a - 7)$ [8.3.1/8.3.2] **44.** $6(x - 3)$ [8.1.1] **45.** $x^2y(3x^2 + 2x + 6)$ [8.1.1] **46.** $(d - 5)(d + 8)$ [8.2.1] **47.** $2(2x - y)(6x - 5)$ [8.5.1] **48.** $4x(x + 1)(x - 6)$ [8.2.2] **49.** $-2, 10$ [8.6.1] **50.** $(x - 5)(3x - 2)$ [8.3.1/8.3.2] **51.** $(2x - 11)(8x - 3)$ [8.3.1/8.3.2] **52.** $(3x - 5)^2$ [8.4.1] **53.** $(2y + 3)(6y - 1)$ [8.3.1/8.3.2] **54.** $3(x + 6)^2$ [8.5.1] **55.** The length is 100 yd. The width is 50 yd. [8.6.2] **56.** The length is 100 yd. The width is 60 yd. [8.6.2] **57.** The two integers are 4 and 5. [8.6.2] **58.** The distance is 20 ft. [8.6.2] **59.** The width of the larger rectangle is 15 ft. [8.6.2] **60.** The length of a side of the original square is 20 ft. [8.6.2]

CHAPTER 8 TEST

1. $3y^2(2x^2 + 3x + 4)$ [8.1.1, Example 2B] **2.** $2x(3x^2 - 4x + 5)$ [8.1.1, Example 2A] **3.** $(p + 2)(p + 3)$ [8.2.1, Example 1] **4.** $(x - 2)(a - b)$ [8.1.2, Example 4] **5.** $\frac{3}{2}, -7$ [8.6.1, Focus On, p. 381] **6.** $(a - 3)(a - 16)$ [8.2.1, Example 2] **7.** $x(x + 5)(x - 3)$ [8.2.2, Example 3] **8.** $4(x + 4)(2x - 3)$ [8.3.1, Example 4] **9.** $(b + 6)(a - 3)$ [8.1.2, Example 5B] **10.** $\frac{1}{2}, -\frac{1}{2}$ [8.6.1, Problem 1] **11.** $(2x + 1)(3x + 8)$ [8.3.2, Example 5] **12.** $(x + 3)(x - 12)$ [8.2.1, Example 2] **13.** $2(b + 4)(b - 4)$ [8.5.1, Example 1A] **14.** $(2a - 3b)^2$ [8.4.1, Example 3A] **15.** $(p + 1)(x - 1)$ [8.1.2, Example 5A] **16.** $5(x^2 - 9x - 3)$ [8.1.1, Example 2A] **17.** Nonfactorable over the integers [8.3.2, Focus On, part C, page 368] **18.** $(2x + 7y)(2x - 7y)$ [8.4.1, Problem 1A] **19.** 3, 5 [8.6.1, Example 2] **20.** $(p + 6)^2$ [8.4.1, Problem 3A] **21.** $2(3x - 4y)^2$ [8.5.1, Example 1C] **22.** $2y^2(y + 1)(y - 8)$ [8.2.2, Problem 4] **23.** The length is 15 cm. The width is 6 cm. [8.6.2, Problem 4] **24.** The length of the base is 12 in. [8.6.2, Problem 4] **25.** The two integers are −13 and −12. [8.6.2, Problem 3]

CUMULATIVE REVIEW EXERCISES

1. −8 [1.2.2] **2.** −8.1 [1.3.2] **3.** 4 [1.4.2] **4.** −7 [2.1.1] **5.** The Associative Property of Addition [2.2.1] **6.** $18x^2$ [2.2.3] **7.** $-6x + 24$ [2.2.5] **8.** $\frac{2}{3}$ [3.1.3] **9.** 5 [3.3.3] **10.** $\frac{7}{4}$ [3.3.2] **11.** 3 [3.3.3] **12.** 35 [3.2.1] **13.** $x \leq -3$ [3.4.3] **14.** $x < \frac{1}{5}$ [3.4.3] **15.**

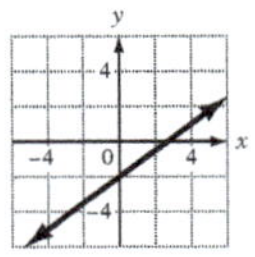

[5.2.2] **16.**

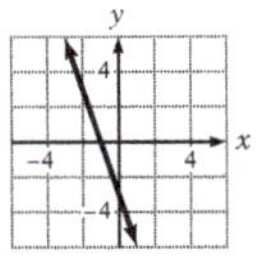

[5.5.2]

17. Domain: {−5, −3, −1, 1, 3}; range: {−4, −2, 0, 2, 4}; yes [5.5.1] **18.** 61 [5.5.1] **19.**

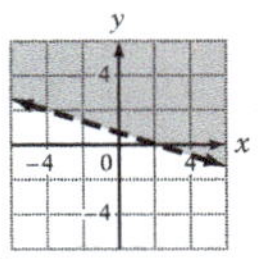

[5.6.1]

20. (2, −5) [6.2.1] **21.** (1, 2) [6.3.1] **22.** $3y^3 - 3y^2 - 8y - 5$ [7.1.1] **23.** $-27a^{12}b^6$ [7.2.2] **24.** $x^3 - 3x^2 - 6x + 8$ [7.3.2] **25.** $4x + 8 + \frac{21}{2x - 3}$ [7.5.2] **26.** $\frac{y^6}{x^{12}}$ [7.4.1] **27.** $(a - b)(3 - x)$ [8.1.2] **28.** $(x + 5y)(x - 2y)$ [8.2.2] **29.** $2a^2(3a + 2)(a + 3)$ [8.3.1/8.3.2] **30.** $(5a + 6b)(5a - 6b)$ [8.4.1] **31.** $3(2x - 3y)^2$ [8.5.1] **32.** $\frac{4}{3}, -5$ [8.6.1] **33.** {−11, −7, −3, 1, 5} [5.5.1] **34.** The average daily high temperature was −3°C. [1.2.5] **35.** The length is 15 cm. The width is 6 cm. [4.2.1] **36.** The pieces measure 4 ft and 6 ft. [4.1.2] **37.** The maximum distance you can drive a Company A car is 559 mi. [4.7.1] **38.** $6500 must be invested at an annual simple interest rate of 11%. [4.4.1] **39.** The discount rate is 40%. [4.3.2] **40.** The three integers are 10, 12, and 14. [4.1.1]

Answers to Chapter 9 Selected Exercises

PREP TEST

1. 36 [1.3.3] **2.** $\frac{3x}{y^3}$ [7.4.1] **3.** $-\frac{5}{36}$ [1.3.3] **4.** $-\frac{10}{11}$ [1.3.2] **5.** $\frac{19}{8}$ [3.3.1] **6.** 130° [4.2.2]
7. $(x-6)(x+2)$ [8.2.1] **8.** $(2x-3)(x+1)$ [8.3.1/8.3.2] **9.** Jean will catch up to Anthony at 9:40 A.M. [4.6.1]

SECTION 9.1

1. True **3.** False **5.** False **11.** 2; 2; 2*x*; $x-2$; 2*x*; $x+2$ **13.** $\frac{2x}{3y^2}$ **15.** $2x-1$ **17.** -1 **19.** $\frac{2x}{3}$
21. $-\frac{2x^2}{3}$ **23.** $\frac{x}{2}$ **25.** $-\frac{y}{3}$ **27.** $\frac{x+2}{x+5}$ **29.** $\frac{a+8}{a+7}$ **31.** $\frac{x+4}{x-6}$ **33.** $\frac{2-y}{y-5}$ **35.** $\frac{x+2}{2(x-2)}$
37. $\frac{2n-1}{2n+3}$ **39.** 2; *d*; 12*a*; *bd* **41.** $\frac{5ab^2}{12x^2y}$ **43.** $\frac{4x^3y^3}{3a^2}$ **45.** $\frac{3}{4}$ **47.** ab^2 **49.** $\frac{x^2(x-1)}{y(x+3)}$ **51.** $\frac{y(x-1)}{x^2(x+10)}$
53. $\frac{x(x+2)}{2(x-1)}$ **55.** $\frac{x+5}{x+4}$ **57.** 1 **59.** $-\frac{n-10}{n-7}$ **61.** $-\frac{x+2}{x-6}$ **63.** $\frac{x+5}{x-12}$ **65.** $\frac{3y+2}{3y+1}$ **67.** $\frac{7a^3y^2}{40bx}$
69. $\frac{4}{3}$ **71.** $\frac{3a}{2}$ **73.** $\frac{x^2(x+4)}{y^2(x+2)}$ **75.** $\frac{x(x-2)}{y(x-6)}$ **77.** $\frac{(x+6)(x-3)}{(x+7)(x-6)}$ **79.** 1 **81.** $-\frac{x+8}{x-4}$ **83.** $\frac{2n+1}{2n-3}$
85. $\frac{(5-4x)(2+3x)(5x-4)}{(4x+5)(2x-3)(3x-5)}$ **87.** Yes **89.** No **91.** 2, -5 **93.** 5, -1 **95.** $\frac{4}{3}, -\frac{1}{2}$ **97.** $\frac{x}{x+8}$
101. a. The recommended percent of oxygen is 8%. **b.** The recommended percent of oxygen decreases. **c.** The oxygen content is more than the recommended percent.

SECTION 9.2

1. True **3.** False **5.** True **7.** $24x^3y^2$ **9.** $30x^4y^2$ **11.** $8x^2(x+2)$ **13.** $6x^2y(x+4)$ **15.** $6(x+1)^2$
17. $(x-1)(x+2)(x+3)$ **19.** $(2x+3)^2(x-5)$ **21.** $(x-1)(x-2)$ **23.** $(x+2)(x-3)(x+4)$
25. $(x+1)(x+4)(x-7)$ **27.** $(x+6)(x-6)(x+4)$ **29.** $(x-10)(x+3)(x-8)$ **31.** $(x-3)(3x-2)(x+2)$
33. $(x+3)(x-5)$ **35. a.** One **b.** Zero **c.** Two **37.** $y(y-3)^2$ **39.** y **41.** $\frac{4x}{x^2}; \frac{3}{x^2}$
43. $\frac{4x}{12y^2}; \frac{3yz}{12y^2}$ **45.** $\frac{xy}{x^2(x-3)}; \frac{6x-18}{x^2(x-3)}$ **47.** $\frac{9x}{x(x-1)^2}; \frac{6x-6}{x(x-1)^2}$ **49.** $\frac{3x}{x(x-3)}; \frac{5}{x(x-3)}$
51. $\frac{3}{(x-5)^2}; -\frac{2x-10}{(x-5)^2}$ **53.** $\frac{3x}{x^2(x+2)}; \frac{4x+8}{x^2(x+2)}$ **55.** $\frac{x^2-6x+8}{(x+3)(x-4)}; \frac{x^2+3x}{(x+3)(x-4)}$
57. $\frac{3}{(x+2)(x-1)}; \frac{x^2-x}{(x+2)(x-1)}$ **59.** $\frac{5}{(2x-5)(x-2)}; \frac{x^2-3x+2}{(2x-5)(x-2)}$
61. $\frac{x^2-3x}{(x+3)(x-3)(x-2)}; \frac{2x^2-4x}{(x+3)(x-3)(x-2)}$ **63.** $-\frac{x^2-3x}{(x-3)^2(x+3)}; \frac{x^2+2x-3}{(x-3)^2(x+3)}$ **65.** $\frac{300}{10^4}; \frac{5}{10^4}$
67. $\frac{b^2}{b}; \frac{5}{b}$ **69.** $\frac{y-1}{y-1}; \frac{y}{y-1}$ **71.** $\frac{x^2+1}{(x-1)^3}; \frac{x^2-1}{(x-1)^3}; \frac{x^2-2x+1}{(x-1)^3}$ **73.** $\frac{x-2}{(x+y)(x+2)(x-2)}; \frac{x+2}{(x+y)(x+2)(x-2)}$

SECTION 9.3

1. False **3.** False **5.** numerators; 8*a* **7.** $\frac{11}{y^2}$ **9.** $-\frac{7}{x+4}$ **11.** $\frac{8x}{2x+3}$ **13.** $\frac{5x+7}{x-3}$ **15.** $\frac{2x-5}{x+9}$
17. $\frac{-3x-4}{2x+7}$ **19.** $\frac{1}{x+5}$ **21.** $\frac{1}{x-6}$ **23.** $\frac{3}{2y-1}$ **25.** $\frac{1}{x-5}$ **27.** Added **29.** Subtracted
31. common denominator; LCM **33.** False **35.** $\frac{4y+5x}{xy}$ **37.** $\frac{19}{2x}$ **39.** $\frac{5}{12x}$ **41.** $\frac{19x-12}{6x^2}$ **43.** $\frac{52y-35x}{20xy}$
45. $\frac{13x+2}{15x}$ **47.** $\frac{7}{24}$ **49.** $\frac{x+90}{45x}$ **51.** $\frac{x^2+2x+2}{2x^2}$ **53.** $\frac{2x^2+3x-10}{4x^2}$ **55.** $\frac{3y^2+8}{3y}$
57. $\frac{x^2+4x+4}{x+4}$ **59.** $\frac{4x+7}{x+1}$ **61.** $\frac{-3x^2+16x+2}{12x^2}$ **63.** $\frac{x^2-x+2}{x^2y}$ **65.** $\frac{16xy-12y+6x^2+3x}{12x^2y^2}$
67. $\frac{3xy-6y-2x^2-14x}{24x^2y}$ **69.** $\frac{9x+2}{(x-2)(x+3)}$ **71.** $\frac{2(x+23)}{(x-7)(x+3)}$ **73.** $\frac{2x^2-7x-1}{(x+1)(x-3)}$ **75.** $\frac{4x^2-34x+5}{(2x-1)(x-6)}$
77. $\frac{2a-5}{a-7}$ **79.** $\frac{2(x+1)}{x-6}$ **81.** $\frac{2(b^2+1)}{(b-1)(b+1)}$ **83.** $\frac{4x+9}{(x+3)(x-3)}$ **85.** $\frac{-x+9}{(x+2)(x-3)}$ **87.** $\frac{14}{(x-5)^2}$

89. $\frac{-2(x+7)}{(x+6)(x-7)}$ **91.** $\frac{x-2}{2x}$ **93.** $\frac{x^2+x+4}{(x-2)(x+1)(x-1)}$ **95.** $\frac{x-4}{x-6}$ **97.** 2 **99.** $\frac{b^2+b-7}{b+4}$ **101.** $\frac{4n}{(n-1)^2}$

103. 1 **105.** $\frac{x^2-x+2}{(x+5)(x+1)}$ **107.** $\frac{5}{a}+\frac{4}{b}$ **109.** $\frac{3}{y^2}+\frac{4}{xy}$ **111.** $\frac{2n+1}{n(n+1)}$ **113. a.** $\frac{x^2-2x-1}{x^2+x}$ **b.** $\frac{6x+5}{4x^2+2x}$

SECTION 9.4

1. $\frac{2}{2}$ **3.** $\frac{(x-1)(x+4)}{(x-1)(x+4)}$ **5.** True **7.** $\frac{5}{x}; \frac{6}{x^2}, \frac{4}{x^2}; x^2$ **9.** x^2-4; $x+2$; $x-2$ **11.** $\frac{3x}{x-9}$ **13.** $\frac{2}{3}$ **15.** $\frac{y+3}{y-4}$

17. $\frac{3}{4}$ **19.** $\frac{x-2}{x+2}$ **21.** $\frac{x+2}{x+3}$ **23.** $\frac{x-6}{x+5}$ **25.** $-\frac{x-2}{x+1}$ **27.** $x-1$ **29.** $\frac{1}{2x-1}$ **31.** $\frac{x-3}{x+5}$

33. $\frac{x-7}{x-8}$ **35.** $\frac{x-2}{x+1}$ **37.** $\frac{x-1}{x+4}$ **39.** $\frac{-x-1}{4x-3}$ **41.** $\frac{b+11}{4b-21}$ **43.** True **45.** $\frac{5}{3}$ **47.** $-\frac{1}{x-1}$

49. $\frac{ab}{b+a}$ **53.** $\frac{6}{5}$ ohms

SECTION 9.5

1. Multiplication Property of Equations **3.** We can clear denominators in an *equation,* as in part (a), but not in an *expression,* as in part (b). **5.** A ratio is a *quotient;* a proportion is an *equation* that states the equality of two ratios (or rates).

7. a. $\overline{ZX}$ **b.** $\overline{QP}$ **c.** $\angle P$ **11. a.** $7(x-4)$ **b.** $3x$ **13.** 7; 1; 1 **15.** 0, 9 **17.** 1 **19.** -3

21. $\frac{1}{2}$ **23.** 5 **25.** No solution **27.** 2, 4 **29.** $-\frac{3}{2}, 4$ **31.** 3 **33.** -1 **35.** 4 **37.** ratio; rate

39. 9 **41.** 12 **43.** 7 **45.** 6 **47.** 1 **49.** -6 **51.** 4 **53.** $-\frac{2}{3}$ **55.** True **57.** Less than

59. Twelve transistors would be defective. **61.** 12.5 million American children live in poverty. **63.** Approximately 17,500 homes have Wi-Fi access. **65.** The length of a claw is approximately 1.28 in. **67.** 180 panels are needed. **69.** No, the shipment will not be accepted. **71.** The person's standing height is 64 in. **73.** The distance between the two cities is 175 mi. **75.** The rocket burns 127,500 lb of fuel in 45 s. **77.** 22.5 gal of yellow point are needed. **79.** Two additional gallons are needed. **81.** The sales tax is \$97.50 greater. **83.** The length of side *AC* is 6.7 cm. **85.** The height of triangle *ABC* is 2.9 m. **87.** The perimeter of triangle *DEF* is 22.5 ft. **89.** The area of triangle *ABC* is 48 m^2. **91.** True

93. a. *CAE* **b.** *CE* **95.** The length of *BC* is 6.25 cm. **97.** The length of *DA* is 6 in. **99.** The length of *OP* is 13 cm.

101. The width of the river is 35 m. **103.** 5 or $\frac{1}{5}$ **105.** The player made 210 foul shots. **107. a.** Eratosthenes calculated the circumference of Earth to be 24,960 mi. **b.** The difference is 160 mi.

SECTION 9.6

3. a. True **b.** False **5.** k **7.** The constant of variation is 0.2. The direct variation equation is $t = 0.2s$.

9. The constant of proportionality is 1.6. The inverse variation equation is $T = \frac{1.6}{S}$. **11.** $P = 70$ when $L = 80$.

13. $T = 8$ when $S = 2$. **15.** $A = 7$ when $B = 21$. **17.** $L = 2.67$ when $W = 90$. **19.** $C = 1$ when $D = 9$. **21.** $4t$; 4

23. Smaller **25.** i **27.** Sound travels 3345 ft in 3 s. **29.** The pressure is 5.4 psi when the depth is 12 ft.

31. The current is 3 amperes when the voltage is 75 volts. **33.** 975 items can be purchased when the cost per item is \$.20.

35. The average speed on the return trip was 52 mph. **37.** The pressure is $66.\overline{6}$ psi when the volume is 150 ft^3.

39. 2160 computers can be sold if the price is \$1500. **43. a.** The constant of variation is 0.014. **b.** The constant of variation represents the monthly interest rate, 1.4%. **c.** The monthly interest charge would be \$4.06 for an unpaid balance of \$290.

45. A gear that has 30 teeth makes 32 rpm. **47.** The speed of gear B is 375 rpm.

SECTION 9.7

1. True **3.** True **5.** R **7.** $h = \frac{2A}{b}$ **9.** $t = \frac{d}{r}$ **11.** $T = \frac{PV}{nR}$ **13.** $L = \frac{P-2W}{2}$ **15.** $b_1 = \frac{2A-hb_2}{h}$

17. $h = \frac{3V}{A}$ **19.** $S = C - Rt$ **21.** $P = \frac{A}{1+rt}$ **23.** $w = \frac{A}{S+1}$ **25.** Yes **27.** $x = \frac{y+5}{c-b}$

29. a. $r = \frac{S-C}{C}$ **b.** The markup rate is $66.\overline{6}$%. **c.** The markup rate is $36.\overline{36}$%. **31. a.** $R_1 = \frac{RR_2}{R_2-R}$

b. The resistance is 20 ohms. **c.** The resistance is 10 ohms.

SECTION 9.8

1. $\frac{x}{5}$ **3.** 1 **5.** Jen has the greater rate of work. **7. a.** The speed traveling with the current is 12 mph.

b. The speed traveling against the current is 4 mph. **9.** Row 1: $\frac{1}{10}$; t; $\frac{t}{10}$; Row 2: $\frac{1}{12}$; t; $\frac{t}{12}$ **11.** It will take 2 h to fill the fountain with both sprinklers operating. **13.** It would take 3 h to move the earth. **15.** It would take the computers 30 h.

17. It would take 24 min to cool the room 5°F. **19.** It would take the second welder 15 h. **21.** It would take the second pipeline 90 min to fill the tank. **23.** It would take 3 h to harvest the field using only the older reaper. **25.** It will take the second technician 3 h. **27.** It would take the small unit $14\frac{2}{3}$ h. **29.** It will take the apprentice 3 h to complete the repairs. **31.** Less than k **33. a.** Row 1: 1440; $380 - r$; $\frac{1440}{380 - r}$; Row 2: 1600; $380 + r$; $\frac{1600}{380 + r}$ **b.** $\frac{1440}{380 - r} = \frac{1600}{380 + r}$ **35.** The camper hiked at a rate of 4 mph. **37.** The rate of the jet was 360 mph. **39.** The rate of the boat for the first 15 mi was 7.5 mph. **41.** The family can travel 21.6 mi down the river. **43.** The technician traveled at a rate of 20 mph through the congested traffic. **45.** The rate of the river's current was 5 mph. **47.** The rate of the freight train was 30 mph. The rate of the express train was 50 mph. **49.** The rate of the current is 2 mph. **51.** The rate of the jet stream is 50 mph. **53.** The rate of the current is 5 mph. **57.** It would take $1\frac{1}{19}$ h to fill the tank. **59.** The amount of time spent traveling by canoe was 2 h. **61.** The bus usually travels at a rate of 60 mph.

CHAPTER 9 REVIEW EXERCISES

1. $\frac{by^3}{6ax^2}$ [9.1.2] **2.** $\frac{22x - 1}{(3x - 4)(2x + 3)}$ [9.3.2] **3.** $x = -\frac{3}{4}y + 3$ [9.7.1] **4.** $\frac{2x^4}{3y^7}$ [9.1.1] **5.** $\frac{1}{x^2}$ [9.1.3] **6.** $\frac{x - 2}{3x - 10}$ [9.4.1] **7.** $72a^3b^5$ [9.2.1] **8.** $\frac{4x}{3x + 7}$ [9.3.1] **9.** $\frac{3x}{16x^2}, \frac{10}{16x^2}$ [9.2.2] **10.** $\frac{x - 9}{x - 3}$ [9.1.1] **11.** $\frac{x - 4}{x + 3}$ [9.1.3] **12.** $\frac{2y - 3}{5y - 7}$ [9.3.2] **13.** $\frac{1}{x}$ [9.1.2] **14.** $\frac{1}{x + 3}$ [9.3.1] **15.** $15x^4(x - 7)^2$ [9.2.1] **16.** 10 [9.5.2] **17.** No solution [9.5.1] **18.** $-\frac{4a}{5b}$ [9.1.1] **19.** $\frac{1}{x + 3}$ [9.3.1] **20.** The perimeter is 24 in. [9.5.4] **21.** 5 [9.5.1] **22.** $\frac{x + 9}{4x}$ [9.4.1] **23.** $\frac{3x - 1}{x - 5}$ [9.3.2] **24.** $\frac{x - 3}{x + 3}$ [9.1.3] **25.** $-\frac{x + 6}{x + 3}$ [9.1.1] **26.** 2 [9.5.1] **27.** $x - 2$ [9.4.1] **28.** $\frac{x + 3}{x - 4}$ [9.1.2] **29.** 15 [9.5.2] **30.** 12 [9.5.2] **31.** $\frac{x}{x - 7}$ [9.4.1] **32.** $x = 2y + 15$ [9.7.1] **33.** $\frac{3(3a + 2b)}{ab}$ [9.3.2] **34.** $(5x - 3)(2x - 1)(4x - 1)$ [9.2.1] **35.** 3 [9.6.1] **36.** 100 [9.6.1] **37.** $c = \frac{100m}{i}$ [9.7.1] **38.** 40 [9.5.2] **39.** 3 [9.5.2] **40.** $\frac{7x + 22}{60x}$ [9.3.2] **41.** $\frac{8a + 3}{4a - 3}$ [9.1.2] **42.** $\frac{3x^2 - x}{(6x - 1)(2x + 3)(3x - 1)}; \frac{24x^3 - 4x^2}{(6x - 1)(2x + 3)(3x - 1)}$ [9.2.2] **43.** $\frac{2}{ab}$ [9.3.1] **44.** 62 [9.5.2] **45.** 6 [9.5.1] **46.** $\frac{b^3y}{10ax}$ [9.1.3] **47.** It would take the apprentice 6 h. [9.8.1] **48.** The spring would stretch 8 in. [9.5.3] **49.** The rate of the wind is 20 mph. [9.8.2] **50.** 10 gal of the garden spray will require 16 additional ounces of insecticide. [9.5.3] **51.** Using both hoses, it would take 6 h to fill the pool. [9.8.1] **52.** The rate of the car is 45 mph. [9.8.2] **53.** \$410 is earned for working 20 h. [9.6.1] **54.** The average rate of speed on the return trip was 44 mph. [9.6.1]

CHAPTER 9 TEST

1. $\frac{x + 5}{x + 4}$ [9.1.3, Example 4B] **2.** $\frac{2}{x + 5}$ [9.3.1, Example 1B] **3.** $3(2x - 1)(x + 1)$ [9.2.1, Example 2] **4.** -1 [9.5.2, Example 2] **5.** $\frac{x + 1}{x^3(x - 2)}$ [9.1.2, Example 3B] **6.** $\frac{x - 3}{x - 2}$ [9.4.1, Example 1B] **7.** $\frac{3x + 6}{x(x - 2)(x + 2)}; \frac{x^2}{x(x - 2)(x + 2)}$ [9.2.2, Example 4] **8.** $x = -\frac{5}{3}y - 5$ [9.7.1, Focus On, page 445] **9.** 2 [9.5.1, Problem 1] **10.** $\frac{5}{(2x - 1)(3x + 1)}$ [9.3.2, Example 3A] **11.** 1 [9.1.3, Example 4B] **12.** $\frac{3}{x + 8}$ [9.3.1, Problem 1B] **13.** $6x(x + 2)^2$ [9.2.1, Example 2] **14.** $-\frac{x - 2}{x + 5}$ [9.1.1, Problem 2] **15.** 4 [9.5.1, Problem 2] **16.** $t = \frac{f - v}{a}$ [9.7.1, Example 1B] **17.** $\frac{2x^3}{3y^5}$ [9.1.1, Example 1] **18.** $\frac{3}{(2x - 1)(x + 1)}$ [9.3.2, Example 3A] **19.** 3 [9.5.2, Problem 3B] **20.** $\frac{x^3(x + 3)}{y(x + 2)}$ [9.1.2, Example 3B] **21.** $-\frac{3xy + 3y}{x(x + 1)(x - 1)}; \frac{x^2}{x(x + 1)(x - 1)}$ [9.2.2, Example 4] **22.** $\frac{x + 3}{x + 5}$ [9.4.1, Example 1B] **23.** A weight of 28 lb will stretch the spring 11.2 in. [9.6.1, Example 2] **24.** For 15 gal of water, 2 additional pounds of salt will be required. [9.5.3, Problem 5] **25.** The rate of the wind is 20 mph. [9.8.2, Example 2] **26.** It would take both pipes 6 min to fill the tank. [9.8.1, Example 1] **27.** The volume of the balloon is 1.25 ft³. [9.6.1, Example 4]

CUMULATIVE REVIEW EXERCISES

1. -17 [1.1.2] **2.** -6 [1.4.1] **3.** $\frac{31}{30}$ [1.4.2] **4.** 21 [2.1.1] **5.** $5x - 2y$ [2.2.2] **6.** $-8x + 26$ [2.2.5] **7.** -20 [3.3.1] **8.** -12 [3.3.3] **9.** 10 [3.2.1] **10.** $x < \frac{9}{5}$ [3.4.2] **11.** $x \le 15$ [3.4.3]

12. 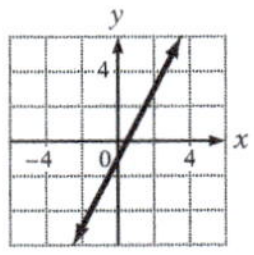 [5.2.2] **13.** 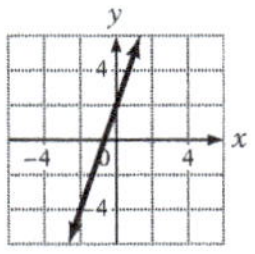 [5.5.2] **14.** 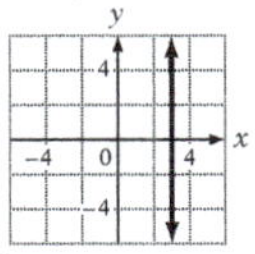 [5.2.3] **15.** 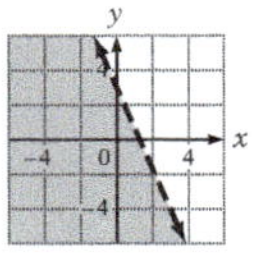 [5.6.1]

16. $\{-3, 5, 11, 19, 27\}$ [5.5.1] **17.** 37 [5.5.1] **18.** $\left(\frac{2}{3}, 3\right)$ [6.2.1] **19.** $\left(\frac{1}{2}, -1\right)$ [6.3.1] **20.** $-6x^4y^5$ [7.2.1] **21.** $a^{20}b^{15}$ [7.2.2] **22.** $\frac{a^3}{b^2}$ [7.4.1] **23.** $a^2 + ab - 12b^2$ [7.3.3] **24.** $3b^3 - b + 2$ [7.5.1] **25.** $x^2 + 2x + 4$ [7.5.2] **26.** $(4x + 1)(3x - 1)$ [8.3.1/8.3.2] **27.** $(y - 6)(y - 1)$ [8.2.1] **28.** $a(2a - 3)(a + 5)$ [8.3.1/8.3.2] **29.** $4(b + 5)(b - 5)$ [8.5.1] **30.** -3 and $\frac{5}{2}$ [8.6.1] **31.** $-\frac{x + 7}{x + 4}$ [9.1.1] **32.** 1 [9.1.3] **33.** $\frac{8}{(3x - 1)(x + 1)}$ [9.3.2] **34.** 1 [9.5.1] **35.** $a = \frac{f - v}{t}$ [9.7.1] **36.** $5x - 18 = -3$; $x = 3$ [4.1.1] **37.** $5000 must be invested at an annual simple interest rate of 11%. [4.4.1] **38.** There is 70% silver in the 120-gram alloy. [4.5.2] **39.** The base is 10 in. The height is 6 in. [8.6.2] **40.** It would take both pipes 8 min to fill the tank. [9.8.1]

Answers to Chapter 10 Selected Exercises

PREP TEST

1. -14 [1.1.2] **2.** $-2x^2y - 4xy^2$ [2.2.2] **3.** 14 [3.1.3] **4.** $\frac{7}{5}$ [3.3.2] **5.** x^6 [7.2.1] **6.** $x^2 + 2xy + y^2$ [7.3.4] **7.** $4x^2 - 12x + 9$ [7.3.4] **8.** $a^2 - 25$ [7.3.4] **9.** $4 - 9v^2$ [7.3.4] **10.** $\frac{x^2y^2}{9}$ [7.4.1]

SECTION 10.1

1. radical sign **3.** perfect square **5.** Product **9.** 2, 20, 50 **13.** radical sign; radicand **15.** 4 **17.** 7 **19.** $4\sqrt{2}$ **21.** $2\sqrt{2}$ **23.** $-18\sqrt{2}$ **25.** $10\sqrt{10}$ **27.** $\sqrt{15}$ **29.** $\sqrt{29}$ **31.** $-54\sqrt{2}$ **33.** $3\sqrt{5}$ **35.** $48\sqrt{2}$ **37.** $10\sqrt{3}$ **39.** $49\sqrt{2}$ **41.** $2\sqrt{30}$ **43.** $4\sqrt{10}$ **45. a.** Rational **b.** Irrational **c.** Irrational **d.** Not a real number **47.** 15.492 **49.** 16.971 **51.** 15.652 **53.** 18.762 **57.** 4; x^3; y^4 **59.** x^3 **61.** $y^7\sqrt{y}$ **63.** a^{10} **65.** x^2y^2 **67.** $2x^2$ **69.** $2x\sqrt{6}$ **71.** $xy^3\sqrt{xy}$ **73.** $ab^5\sqrt{ab}$ **75.** $2x^2\sqrt{15x}$ **77.** $7a^2b^4$ **79.** $3x^2y^3\sqrt{2xy}$ **81.** $2x^5y^3\sqrt{10xy}$ **83.** $4a^4b^5\sqrt{5a}$ **85.** $-8ab\sqrt{b}$ **87.** x^3y **89.** $-8a^2b^3\sqrt{5b}$ **91.** $6x^2y^3\sqrt{3y}$ **93.** $4x^3y\sqrt{2y}$ **95.** $5a + 20$ **97.** $2x^2 + 8x + 8$ **99.** $x + 2$ **101.** $y + 1$ **103.** Rational **105.** Irrational **107. a.** A first-year student has an average of 2.3 credit cards. **b.** A sophomore has an average of 3.3 credit cards. **c.** A junior has an average of 4.0 credit cards. **d.** A senior has an average of 4.6 credit cards. **109.** The distance to the horizon is 84.9 mi. **111.** The length of a side of the square is 8.7 cm. **113.** $x \ge 0$ **115.** $x \ge -5$ **117.** $x \le \frac{5}{2}$ **119.** All real numbers

123. a. 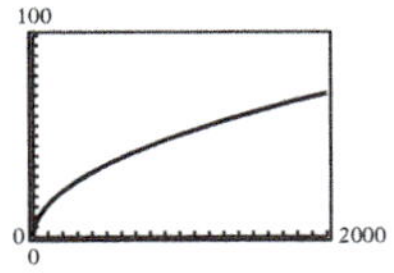

The Xmin and Xmax values represent the minimum and maximum radii of an unbanked curve. The Ymin and Ymax values represent the minimum and maximum speeds of a car. **b.** As the radius increases, the maximum safe speed increases. On the graph, as r increases, v increases. **c.** The maximum safe speed is 16 mph. **d.** The radius is 640 ft. **e.** More than 1 mi

SECTION 10.2

1. Yes **3.** Yes **5.** No **7.** 5; 8; -3 **9.** $11\sqrt{5}$ **11.** $-6\sqrt{5}$ **13.** $-8\sqrt{3}$ **15.** $5\sqrt{y}$ **17.** $-3\sqrt{2a}$ **19.** $-12\sqrt{5a}$ **21.** $-7y\sqrt{3}$ **23.** $-7b\sqrt{3x}$ **25.** $2\sqrt{xy}$ **27.** $-3\sqrt{2}$ **29.** $20\sqrt{2}$ **31.** $25\sqrt{3} - 6\sqrt{2}$ **33.** $41\sqrt{y}$ **35.** $-12x\sqrt{3}$ **37.** $2xy\sqrt{x} - 3xy\sqrt{y}$ **39.** $-9x\sqrt{3x}$ **41.** $-13y^2\sqrt{2y}$ **43.** $4a^2b^2\sqrt{ab}$ **45.** $7\sqrt{2}$ **47.** $6\sqrt{x}$ **49.** $-3\sqrt{y}$ **51.** $-45\sqrt{2}$ **53.** $13\sqrt{3} - 12\sqrt{5}$ **55.** $32\sqrt{3} - 3\sqrt{11}$ **57.** $6\sqrt{x}$ **59.** $-34\sqrt{3x}$ **61.** $10a\sqrt{3b} + 10a\sqrt{5b}$ **63.** $-2xy\sqrt{3}$ **65.** $-7b\sqrt{ab} + 4a^2\sqrt{b}$ **67.** $3ab\sqrt{2a} - ab + 4ab\sqrt{3b}$ **69. a.** False **b.** True **c.** False **d.** True **71.** $8\sqrt{x + 2}$ **73.** $2x\sqrt{2y}$ **75.** $2ab\sqrt{6ab}$ **77.** The perimeter is $(6\sqrt{3} + 2\sqrt{15})$ cm. **79.** The perimeter is 22.4 cm.

SECTION 10.3

1. $3 - \sqrt{5}$ **3.** $\sqrt{2a} + 8$ **5.** $25 - y$ **7.** $\frac{\sqrt{x}}{\sqrt{x}}$ **9.** 15 **11.** 5 **13.** 6 **15.** $7y$ **17.** x^3y^2 **19.** $3ab^6\sqrt{2a}$ **21.** $12a^4b\sqrt{b}$ **23.** $10abc$ **25.** $2 - \sqrt{6}$ **27.** $4 - 2\sqrt{10}$ **29.** $5\sqrt{2} - \sqrt{5x}$ **31.** $x - \sqrt{xy}$ **33.** $3a - 3\sqrt{ab}$ **35.** $x - 6\sqrt{x} + 9$ **37.** $26 + 10\sqrt{5}$ **39.** $2x - 9\sqrt{x} + 10$ **41.** $15x - 22y\sqrt{x} + 8y^2$ **43.** $2 - y$ **45.** 19 **47.** $10x + 13\sqrt{xy} + 4y$ **49.** Less than **53.** **a.** No **b.** Yes **c.** No **d.** No **55.** 4 **57.** 7 **59.** 3 **61.** $x\sqrt{5}$ **63.** $\frac{a^2}{7}$ **65.** $\frac{\sqrt{3}}{3}$ **67.** $\sqrt{3}$ **69.** $\frac{3\sqrt{x}}{x}$ **71.** $\frac{\sqrt{3x}}{x}$ **73.** $\frac{\sqrt{2x}}{x}$ **75.** $\frac{2x\sqrt{y}}{3y}$ **77.** $\frac{1}{2}$ **79.** $\frac{\sqrt{2ab}}{2b}$ **81.** $\frac{y\sqrt{3}}{3}$ **83.** $-\frac{\sqrt{2} + 3}{7}$ **85.** $\frac{15 - 3\sqrt{5}}{20}$ **87.** $\frac{x\sqrt{y} + y\sqrt{x}}{x - y}$ **89.** $\sqrt{2y}$ **91.** $-\frac{\sqrt{10} + 5}{3}$ **93.** $\frac{x - 3\sqrt{x}}{x - 9}$ **95.** $-\frac{1}{2}$ **97.** 7 **99.** -5 **101.** $7 + 4\sqrt{3}$ **103.** $3 + \sqrt{6}$ **105.** $-\frac{20 + 7\sqrt{3}}{23}$ **107.** $\frac{6 + 5\sqrt{x} + x}{4 - x}$ **109.** $\frac{6 + 7\sqrt{y} + 2y}{4 - y}$ **111.** $\frac{x + 2\sqrt{xy} + y}{x - y}$ **113.** $\frac{5}{8}$ **115.** $\frac{5}{4}$ **117.** $-\frac{5}{2}$ **119.** $(62 + 17\sqrt{2})\ \text{m}^2$

SECTION 10.4

1. ii and iii **3.** Sometimes true **5.** Sometimes true **7.** 2; 2; $x - 2$; 25; 27; 2 **9.** 25 **11.** 144 **13.** 5 **15.** 16 **17.** 8 **19.** No solution **21.** 6 **23.** 24 **25.** $-\frac{2}{5}$ **27.** $\frac{4}{3}$ **29.** -1 **31.** 15 **33.** 5 **35.** 1 **37.** 1 **39.** 1 **41.** 2 **43.** No solution **45.** iii **47.** The number is 1. **49.** 90°; legs **53.** The length of the hypotenuse is 10.30 cm. **55.** The length of the other leg of the triangle is 9.75 ft. **57.** The length of the diagonal is 12.1 mi. **59.** **a.** Yes **b.** Yes **c.** Yes **d.** Yes **61.** The depth of the water is 12.5 ft. **63.** The pitcher's mound is less than halfway between home plate and second base. **65.** Yes, she will be able to call a friend on the dock. **67.** Yes, the ladder will be long enough to reach the gutters. **69.** The length of the pendulum is 7.30 ft. **71.** The bridge is 64 ft high. **73.** The height of the screen is 35.5 in. **75.** 15 **77.** 30 units **79.** $r^2(\pi - 2)$ **83.** **a.** $x\sqrt{2}$ inches **b.** $2x^2$ square inches **85.** $5\sqrt{2}$ units

CHAPTER 10 REVIEW EXERCISES

1. $-11\sqrt{3}$ [10.2.1] **2.** $-x\sqrt{3} - x\sqrt{5}$ [10.3.2] **3.** $x + 8$ [10.1.2] **4.** No solution [10.4.1] **5.** 1 [10.3.2] **6.** $7\sqrt{7}$ [10.2.1] **7.** $12a + 28\sqrt{ab} + 15b$ [10.3.1] **8.** $7x^2 + 42x + 63$ [10.1.2] **9.** 12 [10.1.1] **10.** 16 [10.4.1] **11.** $4x\sqrt{5}$ [10.2.1] **12.** $5ab - 7$ [10.3.1] **13.** 3 [10.4.1] **14.** $\sqrt{35}$ [10.1.1] **15.** $7x^2y\sqrt{15xy}$ [10.2.1] **16.** $\frac{x}{3y^2}$ [10.3.2] **17.** $9x - 6\sqrt{xy} + y$ [10.3.1] **18.** $a + 4$ [10.1.2] **19.** $20\sqrt{3}$ [10.1.1] **20.** $26\sqrt{3x}$ [10.2.1] **21.** $\frac{8\sqrt{x} + 24}{x - 9}$ [10.3.2] **22.** $3a\sqrt{2} + 2a\sqrt{3}$ [10.3.1] **23.** $9a^2\sqrt{2ab}$ [10.1.2] **24.** $-6\sqrt{30}$ [10.1.1] **25.** $-4a^2b^4\sqrt{5ab}$ [10.2.1] **26.** $7x^2y^4$ [10.3.2] **27.** $\frac{3}{5}$ [10.4.1] **28.** c^9 [10.1.2] **29.** $15\sqrt{2}$ [10.1.1] **30.** $18a\sqrt{5b} + 5a\sqrt{b}$ [10.2.1] **31.** $\frac{16\sqrt{a}}{a}$ [10.3.2] **32.** 8 [10.4.1] **33.** $a^5b^3c^2$ [10.3.1] **34.** $21\sqrt{70}$ [10.1.1] **35.** $2y^4\sqrt{6}$ [10.1.2] **36.** 5 [10.3.2] **37.** 5 [10.4.1] **38.** $8y + 10\sqrt{5y} - 15$ [10.3.1] **39.** 99.499 [10.1.1] **40.** 7 [10.3.1] **41.** $25x^4\sqrt{6x}$ [10.1.2] **42.** $-6x^3y^2\sqrt{2y}$ [10.2.1] **43.** $3a$ [10.3.2] **44.** $20\sqrt{10}$ [10.1.1] **45.** 20 [10.4.1] **46.** $\frac{7a}{3}$ [10.3.2] **47.** 10 [10.3.1] **48.** $6x^2y^2\sqrt{y}$ [10.1.2] **49.** $36x^8y^5\sqrt{3xy}$ [10.1.2] **50.** 3 [10.4.1] **51.** 20 [10.1.1] **52.** $-2\sqrt{2x}$ [10.2.1] **53.** 6 [10.4.1] **54.** 3 [10.3.1] **55.** The larger integer is 51. [10.4.2] **56.** The explorer's weight on the surface of Earth is 144 lb. [10.4.2] **57.** The length of the other leg is 14.25 cm. [10.4.2] **58.** The radius of the sharpest corner is 25 ft. [10.4.2] **59.** The depth is 100 ft. [10.4.2] **60.** The length of the guy wire is 26.25 ft. [10.4.2]

CHAPTER 10 TEST

1. $11x^4y$ [10.1.2, Example 4A] **2.** $-5\sqrt{2}$ [10.2.1, Example 1B] **3.** $6x^2\sqrt{xy}$ [10.3.1, Example 1] **4.** $3\sqrt{5}$ [10.1.1, Example 1] **5.** $6x^3y\sqrt{2x}$ [10.1.2, Example 4B] **6.** $6\sqrt{2y} + 3\sqrt{2x}$ [10.2.1, Example 2B] **7.** $y + 8\sqrt{y} + 15$ [10.3.1, Example 3] **8.** $\sqrt{2}$ [10.3.2, 2nd Focus On, page 481] **9.** 9 [10.3.2, Example 5A] **10.** 11 [10.4.1, Example 2A] **11.** 22.361 [10.1.1, Focus On, part B, page 467] **12.** $4a^2b^5\sqrt{2ab}$ [10.1.2, Example 4B] **13.** $a - \sqrt{ab}$ [10.3.1, Problem 2] **14.** $4x^2y^2\sqrt{5y}$ [10.3.1, Example 1] **15.** $7ab\sqrt{a}$ [10.3.2, Example 5A] **16.** 25 [10.4.1, Example 1] **17.** $8x^6y^2\sqrt{3xy}$ [10.1.2, Example 4B] **18.** $4ab\sqrt{2ab} - 5ab\sqrt{ab}$ [10.2.1, Example 2B]

19. $a - 4$ [10.3.1, Example 4] **20.** $-6 - 3\sqrt{5}$ [10.3.2, Example 5B] **21.** 5 [10.4.1, Example 2A] **22.** $3\sqrt{2} - x\sqrt{3}$ [10.3.1, Example 2] **23.** $-6\sqrt{a}$ [10.2.1, Example 1A] **24.** $6\sqrt{3}$ [10.1.1, Example 1] **25.** 7.937 [10.1.1, Focus On, part B, page 467] **26.** $6ab\sqrt{a}$ [10.3.2, Example 5A] **27.** No solution [10.4.1, Example 3] **28.** The smaller integer is 40. [10.4.2, Example 4] **29.** The length of the pendulum is 5.07 ft. [10.4.2, Problem 5] **30.** The ladder reaches 15.2 ft high on the building. [10.4.2, Problem 4]

CUMULATIVE REVIEW EXERCISES

1. $-\frac{1}{12}$ [1.4.2] **2.** $2x + 18$ [2.2.5] **3.** $\frac{1}{13}$ [3.3.3] **4.** $x \le -\frac{9}{2}$ [3.4.3] **5.** $-\frac{4}{3}$ [5.3.1] **6.** $y = \frac{1}{2}x - 2$ [5.4.1/5.4.2] **7.** -18 [5.5.1] **8.**

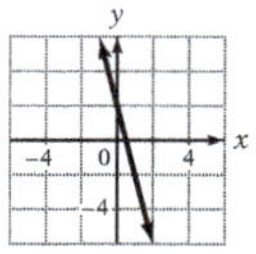

[5.5.2] **9.**

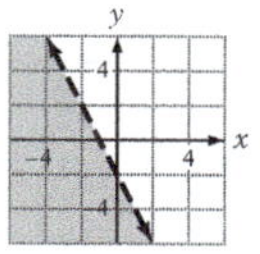

[5.6.1]

10. $(2, -1)$ [6.1.1] **11.** $(1, 1)$ [6.2.1] **12.** $(3, -2)$ [6.3.1] **13.** $\frac{6x^5}{y^3}$ [7.4.1] **14.** $-2b^2 + 1 - \frac{1}{3b^2}$ [7.5.1] **15.** $3x^2y^2(4x - 3y)$ [8.1.1] **16.** $(3b + 5)(3b - 4)$ [8.3.1/8.3.2] **17.** $2a(a - 3)(a - 5)$ [8.2.2] **18.** $\frac{1}{4(x + 1)}$ [9.1.2] **19.** $\frac{x - 5}{x - 3}$ [9.4.1] **20.** $\frac{x + 3}{x - 3}$ [9.3.2] **21.** $\frac{5}{3}$ [9.5.1] **22.** $38\sqrt{3a} - 35\sqrt{a}$ [10.2.1] **23.** 8 [10.3.2] **24.** 14 [10.4.1] **25.** The largest integer that satisfies the inequality is -33. [4.7.1] **26.** The cost is \$24.50. [4.3.1] **27.** 56 oz of pure water must be added. [4.5.2] **28.** The two numbers are 6 and 15. [4.1.2] **29.** It would take the small pipe 48 h to fill the tank. [9.8.1] **30.** The smaller integer is 24. [10.4.2]

Answers to Chapter 11 Selected Exercises

PREP TEST

1. 41 [2.1.1] **2.** $-\frac{1}{5}$ [3.3.1] **3.** $(x + 4)(x - 3)$ [8.2.1] **4.** $(2x - 3)^2$ [8.4.1] **5.** Yes [8.4.1] **6.** 3 [9.5.2] **7.**

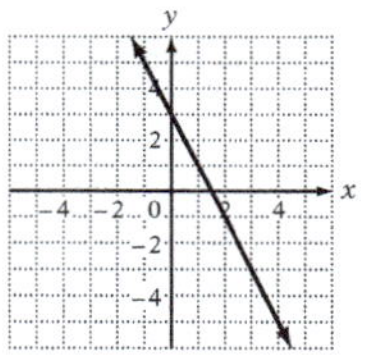

[5.2.2] **8.** $2\sqrt{7}$ [10.1.1] **9.** $|a|$ [10.1.2] **10.** The hiking trail is 3.6 mi long. [9.8.2]

SECTION 11.1

1. quadratic **3.** standard **5.** $a = 3, b = -4, c = 1$ **7.** $a = 2, b = 0, c = -5$ **9.** $a = 6, b = -3, c = 0$ **11.** $x^2 - 3x - 8 = 0$ **13.** $x^2 - 16 = 0$ **15.** $2x^2 + 12x + 13 = 0$ **21.** $x + 6$; $x - 6$; 0; 0; -6; 6 **23.** -3 and 5 **25.** 0 and 7 **27.** $-\frac{5}{2}$ and $\frac{1}{3}$ **29.** -5 and 3 **31.** 1 and 3 **33.** -1 and -2 **35.** 3 **37.** 0 and $\frac{3}{2}$ **39.** -2 and 5 **41.** $\frac{2}{3}$ and 1 **43.** -3 and $\frac{1}{3}$ **45.** $-\frac{3}{2}$ and $\frac{4}{3}$ **47.** -3 and $\frac{4}{5}$ **49.** $\frac{1}{3}$ **51.** -4 and 4 **53.** $-\frac{2}{3}$ and $\frac{2}{3}$ **55.** $\frac{2}{3}$ and 2 **57.** -6 and $\frac{3}{2}$ **59.** $\frac{2}{3}$ **61.** -3 and 6 **63.** 0 and 12 **65. a.** Two positive **b.** One positive and one negative **c.** One positive and one negative **67.** 9 **69.** -6 and 6 **71.** -1 and 1 **73.** $-\frac{7}{2}$ and $\frac{7}{2}$ **75.** $-\frac{2}{3}$ and $\frac{2}{3}$ **77.** $-\frac{3}{4}$ and $\frac{3}{4}$ **79.** $2\sqrt{3}$ and $-2\sqrt{3}$ **81.** $2\sqrt{6}$ and $-2\sqrt{6}$ **83.** -5 and 7 **85.** -7 and -3 **87.** -6 and 4 **89.** -9 and -1 **91.** $-\frac{9}{2}$ and $-\frac{3}{2}$ **93.** $-\frac{1}{3}$ and $\frac{7}{3}$ **95.** $4 + 2\sqrt{5}$ and $4 - 2\sqrt{5}$ **97.** $-1 + 3\sqrt{2}$ and $-1 - 3\sqrt{2}$ **99.** $\frac{1}{2} + \sqrt{6}$ and $\frac{1}{2} - \sqrt{6}$ **101.** One **103.** Two **105.** 11 and 35 **107.** $-\frac{1}{2}$ **109.** There are 7 teams in the league. **111.** The speed is 10 m/s.

SECTION 11.2

1. perfect-square trinomial **3.** perfect-square trinomial **5.** 18; 81 **7.** $x^2 + 12x + 36; (x + 6)^2$ **9.** $x^2 + 10x + 25; (x + 5)^2$ **11.** $x^2 - x + \frac{1}{4}; \left(x - \frac{1}{2}\right)^2$ **13.** -3 and 1 **15.** $-2 + \sqrt{3}$ and $-2 - \sqrt{3}$ **17.** 2 **19.** $1 + \sqrt{2}$ and $1 - \sqrt{2}$ **21.** $\frac{-3 + \sqrt{13}}{2}$ and $\frac{-3 - \sqrt{13}}{2}$ **23.** -8 and 1 **25.** $-3 + \sqrt{5}$ and $-3 - \sqrt{5}$ **27.** $4 + \sqrt{14}$ and $4 - \sqrt{14}$ **29.** 1 and 2 **31.** $\frac{3 + \sqrt{29}}{2}$ and $\frac{3 - \sqrt{29}}{2}$ **33.** $\frac{1 + \sqrt{5}}{2}$ and $\frac{1 - \sqrt{5}}{2}$ **35.** $-5 + 4\sqrt{2}$ and $-5 - 4\sqrt{2}$ **37.** $\frac{-3 + \sqrt{5}}{2}$ and $\frac{-3 - \sqrt{5}}{2}$ **39.** $\frac{1 + \sqrt{17}}{2}$ and $\frac{1 - \sqrt{17}}{2}$ **41.** $\frac{1}{2}$ and 1 **43.** -3 and $\frac{1}{2}$ **45.** $\frac{1 + \sqrt{2}}{2}$ and $\frac{1 - \sqrt{2}}{2}$ **47.** $\frac{2 + \sqrt{5}}{2}$ and $\frac{2 - \sqrt{5}}{2}$ **49.** $3 + \sqrt{2}$ and $3 - \sqrt{2}$ **51.** $1 + \sqrt{13}$ and $1 - \sqrt{13}$ **53.** $\frac{3 + 3\sqrt{3}}{2}$ and $\frac{3 - 3\sqrt{3}}{2}$ **55.** $4 + \sqrt{7}$ and $4 - \sqrt{7}$ **57.** Both negative **59.** 1.193 and -4.193 **61.** 2.766 and -1.266 **63.** 0.151 and -1.651 **65.** -3 or 3 **67.** $22 + 12\sqrt{2}$ and $22 - 12\sqrt{2}$ **69.** -3 **71.** $6 + \sqrt{58}$ and $6 - \sqrt{58}$ **73.** 26

SECTION 11.3

1. $\frac{1 + \sqrt{13}}{2}$ and $\frac{1 - \sqrt{13}}{2}$ **3.** True **5.** False **9.** $4x; 2; x^2 - 4x - 2; 1; -4; -2$ **11.** -7 and 1 **13.** -3 and 6 **15.** $1 + \sqrt{6}$ and $1 - \sqrt{6}$ **17.** $-3 + \sqrt{10}$ and $-3 - \sqrt{10}$ **19.** $\frac{-3 + \sqrt{29}}{2}$ and $\frac{-3 - \sqrt{29}}{2}$ **21.** $2 + \sqrt{13}$ and $2 - \sqrt{13}$ **23.** 0 and 1 **25.** $\frac{1 + \sqrt{2}}{2}$ and $\frac{1 - \sqrt{2}}{2}$ **27.** $-\frac{3}{2}$ and $\frac{3}{2}$ **29.** $\frac{3 + \sqrt{3}}{3}$ and $\frac{3 - \sqrt{3}}{3}$ **31.** $\frac{1 + \sqrt{10}}{3}$ and $\frac{1 - \sqrt{10}}{3}$ **33.** $\frac{4 + \sqrt{10}}{2}$ and $\frac{4 - \sqrt{10}}{2}$ **35.** $-\frac{2}{3}$ and 3 **37.** $\frac{5 + \sqrt{73}}{6}$ and $\frac{5 - \sqrt{73}}{6}$ **39.** $\frac{1 + \sqrt{37}}{6}$ and $\frac{1 - \sqrt{37}}{6}$ **41.** $\frac{1 + \sqrt{41}}{5}$ and $\frac{1 - \sqrt{41}}{5}$ **43.** -3 and $\frac{4}{5}$ **45.** $-3 + 2\sqrt{2}$ and $-3 - 2\sqrt{2}$ **47.** $\frac{3 + 2\sqrt{6}}{2}$ and $\frac{3 - 2\sqrt{6}}{2}$ **49.** $\frac{-1 + \sqrt{2}}{3}$ and $\frac{-1 - \sqrt{2}}{3}$ **51.** $-\frac{1}{2}$ and $\frac{2}{3}$ **53.** $\frac{-4 + \sqrt{5}}{2}$ and $\frac{-4 - \sqrt{5}}{2}$ **55.** $\frac{1 + 2\sqrt{3}}{2}$ and $\frac{1 - 2\sqrt{3}}{2}$ **57.** $\frac{-5 + \sqrt{2}}{3}$ and $\frac{-5 - \sqrt{2}}{3}$ **59.** $\frac{1 + \sqrt{19}}{3}$ and $\frac{1 - \sqrt{19}}{3}$ **61.** $\frac{5 + \sqrt{65}}{4}$ and $\frac{5 - \sqrt{65}}{4}$ **63.** $6 + \sqrt{11}$ and $6 - \sqrt{11}$ **65.** True **67.** 5.690 and -3.690 **69.** 7.690 and -1.690 **71.** 4.589 and -1.089 **73.** 0.118 and -2.118 **75.** 1.105 and -0.905 **77.** $2\sqrt{23}$ **79.** $3 + \sqrt{13}$ and $3 - \sqrt{13}$ **83.** Yes **85.** No. The solutions are $4 + \sqrt{30}$ and $4 - \sqrt{30}$.

SECTION 11.4

1. -1 **3.** 7 **5.** $6 - 5i$ **7.** $2 + 8i$ **9.** $\frac{1 + 6i}{1 + 6i}$ **11. a.** 5 **b.** -8 **13.** $9i$ **15.** $35i$ **17.** $i\sqrt{13}$ **19.** $5i\sqrt{3}$ **21.** $-12i\sqrt{6}$ **23.** $a - \sqrt{-b^2}$ **25.** $5 + 2i$ **27.** $-7 - 12i$ **29.** $5 + 12i\sqrt{3}$ **31.** $7 - 12i\sqrt{7}$ **33.** $3 + 5i$ **35.** $-1 + i\sqrt{6}$ **37.** $-\frac{2}{5} - \frac{12\sqrt{10}}{5}i$ **39.** $-2; 3; 4 - 5i$ **41.** $10i$ **43.** $-9i$ **45.** $7 + 9i$ **47.** $-8 + 7i$ **49.** $13 + 2i$ **51.** $2 - 10i$ **53.** $-8 - 2i$ **55.** 6 **57.** 0 **59.** False **61.** Negative **63.** $i^2; -1; -28$ **65.** -20 **67.** 18 **69.** $-20 + 10i$ **71.** $2 + 23i$ **73.** $-9 - 7i$ **75.** $-41 - i$ **77.** $-21 + 20i$ **79.** $-3 - 4i$ **81.** $29i$ **83.** 20 **85.** 17 **87.** False **89.** $i; i; 3i^2; -3$ **91.** $-\frac{9i}{2}$ **93.** $\frac{4i}{7}$ **95.** $1 - 6i$ **97.** $9 - 5i$ **99.** $-2 - 3i$ **101.** $-\frac{5}{3} - \frac{2}{3}i$ **103.** $\frac{5}{2} + \frac{2}{3}i$ **105.** $-\frac{4}{5} - \frac{8}{5}i$ **107.** $-3 + 3i$ **109.** $2 + 5i$ **111.** $2 + 5i$ **113.** $3 + i$ **115.** False **117.** $-3; -3; 1; -8; 1$ **119.** $5i$ and $-5i$ **121.** $6i$ and $-6i$ **123.** $2i\sqrt{10}$ and $-2i\sqrt{10}$ **125.** $i\sqrt{2}$ and $-i\sqrt{2}$ **127.** $3 + 9i$ and $3 - 9i$ **129.** $-4 + 3i\sqrt{2}$ and $-4 - 3i\sqrt{2}$ **131.** $5 + 2i\sqrt{2}$ and $5 - 2i\sqrt{2}$ **133.** $1 + 4i$ and $1 - 4i$ **135.** $1 + 2i$ and $1 - 2i$ **137.** $-2 + 2i$ and $-2 - 2i$ **139.** $1 + 3i$ and $1 - 3i$ **141.** $-1 + 2i\sqrt{3}$ and $-1 - 2i\sqrt{3}$ **143.** $-\frac{3}{2} + 2i$ and $-\frac{3}{2} - 2i$ **145.** Two imaginary solutions **147.** $5i$ **149.** $4 - 6i$ **151.** $-2 + 2i$ **153.** $14 - 2i$ **155.** $2i$ and $-2i$ **157.** $-5 + i\sqrt{3}$ and $-5 - i\sqrt{3}$ **159.** $3 + i\sqrt{3}$ and $3 - i\sqrt{3}$ **163. a.** 1 **b.** i **c.** -1 **d.** $-i$ **e.** -1 **f.** $-i$ **g.** 1 **165.** Linear; 2 **167.** Linear; 5

SECTION 11.5

1. Parabola **3.** The coefficient on x^2 is positive. **5.** Replace x by zero and solve for y. **7.** Up **9.** Down **11.** Up
13. a. -1; 0; 9 **b.** down **c.** 2; -4; 5 **15.** 4 **17.** -5 **19.** -42 **21.** 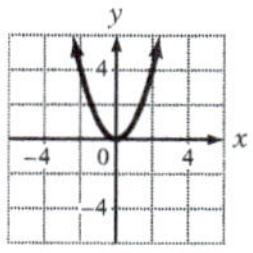 **23.** 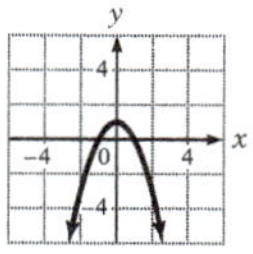

25. 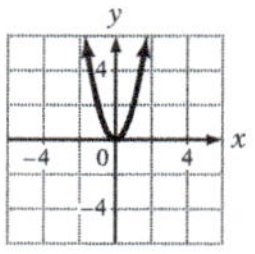 **27.** 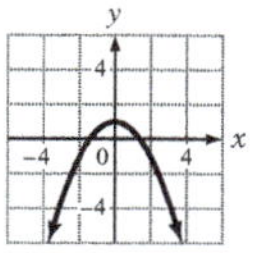 **29.** 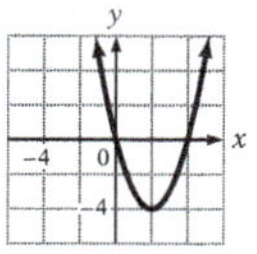 **31.** 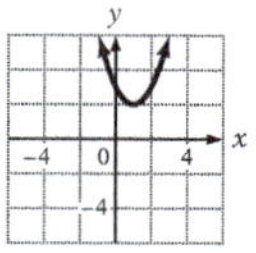 **33.** 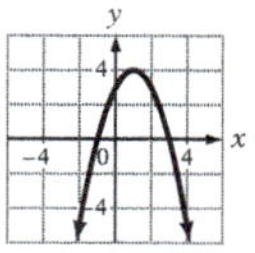 **35.** 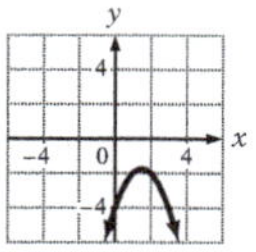

37. 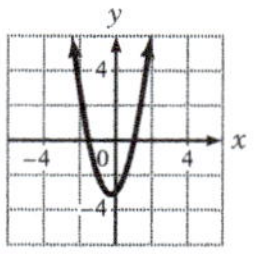 **39.** False **41.** The x-intercepts are (2, 0) and (3, 0). The y-intercept is (0, 6).

43. The x-intercepts are $(-3, 0)$ and (3, 0). The y-intercept is (0, 9). **45.** The x-intercepts are $(-1 + \sqrt{7}, 0)$ and $(-1 - \sqrt{7}, 0)$. The y-intercept is $(0, -6)$. **47.** The x-intercepts are $\left(\frac{3}{2}, 0\right)$ and $(-1, 0)$. The y-intercept is $(0, -3)$. **49.** 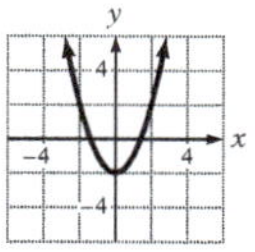

51. 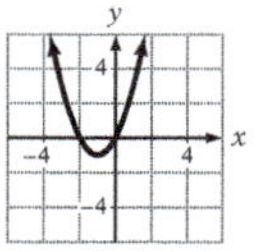 **53.** 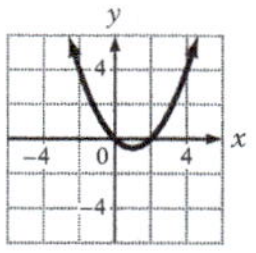 **55.** 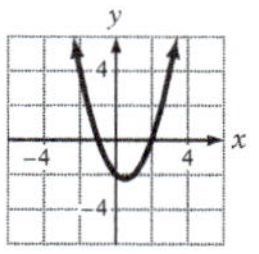 **57.** 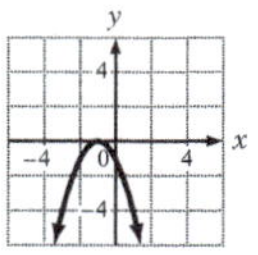

59. Linear function **61.** Neither **63.** Quadratic function **65. a.** $(-2, 0)$, (2, 0) **b.** $(0, -4)$ **c.** It is positive.
d. $(0, -4)$ **e.** The y-axis **f.** -3 **67.** Answers will vary. Sample answer: 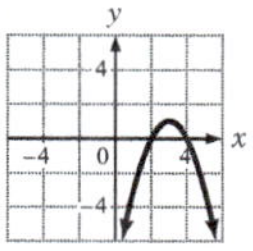 **69.** -2

SECTION 11.6

1. $2W + 3$ **3.** $r - 30$ **5. a.** $2w - 2$ **b.** $(2w - 2)w$ **c.** $(2w - 2)w$; 84 **7.** The height of the triangle is 10 m. The length of the base is 4 m. **9.** The length is 6 ft. The width is 4 ft. **11.** Working alone, the first drain would take 24 min. Working alone, the second drain would take 8 min. **13.** The length is 100 ft. The width is 50 ft. **15.** It would take the first engine 12 h and the second engine 6 h. **17.** The rate of the plane in calm air was 100 mph. **19.** The cyclist's rate during the first 150 mi was 50 mph. **21.** The arrow will be 32 ft above the ground at 1 s and at 2 s. **23.** The weed will cover an area of 10 m^2 after 25 days. **25.** There will be 50 million people aged 65 or older in the year 2017. **27.** The ball has been in the air for 1.58 s. **29.** The hang time is 5.5 s. **31.** False **33.** The lengths of the legs are 2 cm and 3 cm. **35.** The radius of the cone is 7.98 cm. **37.** The dimensions of the cardboard are 20 in. by 20 in. **39. a.** $12 = 2L + 2W$ **b.** $A = LW$ **c.** $L = 6 - W$; $A = (6 - W)W$; $A = -W^2 + 6W$ **d.** Length: 3 ft; width: 3ft

CHAPTER 11 REVIEW EXERCISES

1. 4 and -4 [11.1.1/11.1.2] **2.** $\frac{1 + \sqrt{13}}{2}$ and $\frac{1 - \sqrt{13}}{2}$ [11.2.1/11.3.1] **3.** $\frac{3 + \sqrt{29}}{2}$ and $\frac{3 - \sqrt{29}}{2}$ [11.2.1/11.3.1]
4. $\frac{5}{7}$ and $-\frac{5}{7}$ [11.1.1/11.1.2] **5.** 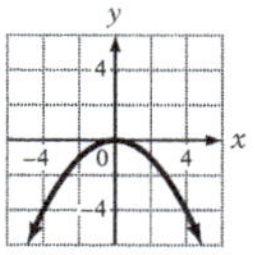 [11.5.1] **6.** 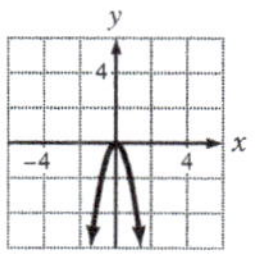 [11.5.1] **7.** $6i\sqrt{7}$ [11.4.1]

8. $-\frac{1}{2}$ and $-\frac{1}{3}$ [11.1.1] **9.** $\frac{3}{2}i$ and $-\frac{3}{2}i$ [11.4.5] **10.** $\frac{-10 + 2\sqrt{10}}{5}$ and $\frac{-10 - 2\sqrt{10}}{5}$ [11.2.1/11.3.1]

11. $2 - 4i$ [11.4.2] **12.** 6 and -5 [11.1.1] **13.** $\frac{4}{3}$ and $-\frac{7}{2}$ [11.1.1] **14.** $2\sqrt{10}$ and $-2\sqrt{10}$ [11.1.2]
15. $-3 + 5i$ [11.4.2] **16.** $1 + \sqrt{11}$ and $1 - \sqrt{11}$ [11.2.1/11.3.1] **17.** 3 and 9 [11.1.1]
18. 16 and -2 [11.1.2] **19.**

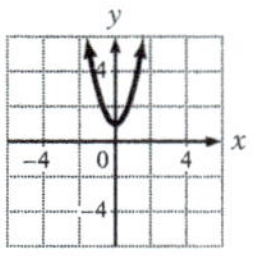

[11.5.1] **20.**

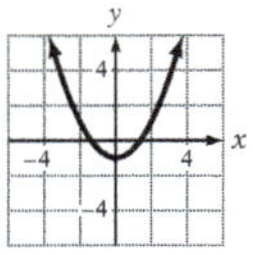

[11.5.1] **21.** 1 and -9 [11.1.2]

22. $\frac{-4 + \sqrt{23}}{2}$ and $\frac{-4 - \sqrt{23}}{2}$ [11.2.1/11.3.1] **23.** $-\frac{1}{6}$ and $-\frac{5}{4}$ [11.1.1] **24.** $2 + 2i$ and $2 - 2i$ [11.4.5]
25. $20 + 12i$ [11.4.3] **26.** $\frac{4}{7}$ [11.1.1] **27.** 2 and -1 [11.1.2] **28.** $-2 + 14i$ [11.4.3]
29. $\frac{-7 + \sqrt{61}}{2}$ and $\frac{-7 - \sqrt{61}}{2}$ [11.2.1/11.3.1] **30.** 2 and $\frac{5}{12}$ [11.1.1] **31.** $3 + \sqrt{5}$ and $3 - \sqrt{5}$ [11.1.2]
32. $-4 + \sqrt{19}$ and $-4 - \sqrt{19}$ [11.2.1/11.3.1] **33.**

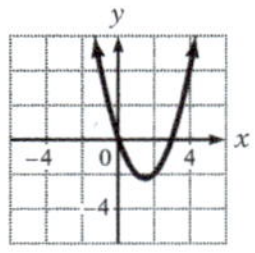

[11.5.1] **34.**

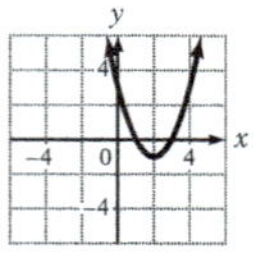

[11.5.1]

35. -7 and -10 [11.1.1] **36.** $\frac{1}{2} + \frac{1}{2}i$ [11.4.4] **37.** $-3 + i\sqrt{3}$ and $-3 - i\sqrt{3}$ [11.4.5]
38. $-3 + \sqrt{11}$ and $-3 - \sqrt{11}$ [11.2.1/11.3.1] **39.** 3 and $-\frac{1}{9}$ [11.1.1] **40.** $\frac{-5 + \sqrt{33}}{4}$ and $\frac{-5 - \sqrt{33}}{4}$ [11.2.1/11.3.1]
41.

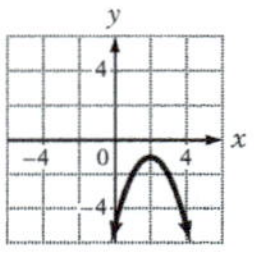

[11.5.1] **42.** $-3 - i$ [11.4.4] **43.** 1 and $\frac{5}{2}$ [11.1.1] **44.**

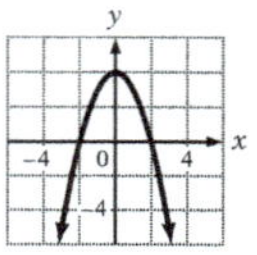

[11.5.1]

45. The rate of the air balloon in calm air is 25 mph. [11.6.1] **46.** The height is 10 m. The base is 4 m. [11.6.1]
47. The maximum speed is 151.6 km/h. [11.6.1] **48.** The rate of the boat in still water is 5 mph. [11.6.1] **49.** The object will be 12 ft above the ground at $\frac{1}{2}$ s and $1\frac{1}{2}$ s. [11.6.1] **50.** It would take the smaller drain 12 h and the larger drain 4 h. [11.6.1]

CHAPTER 11 TEST

1. $-4 + 2\sqrt{5}$ and $-4 - 2\sqrt{5}$ [11.1.2, Focus On, p. 507] **2.** $\frac{-4 + \sqrt{22}}{2}$ and $\frac{-4 - \sqrt{22}}{2}$ [11.2.1, Example 1/11.3.1, Example 1B] **3.** -4 and $\frac{5}{3}$ [11.1.1, Problem 1] **4.** $12i\sqrt{5}$ [11.4.1, Example 1D] **5.** $-2 + 2\sqrt{5}$ and $-2 - 2\sqrt{5}$ [11.2.1, Example 1/11.3.1, Example 1A] **6.**

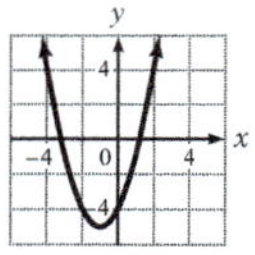

[11.5.1, Example 2B] **7.** $-2 + \sqrt{2}$ and $-2 - \sqrt{2}$ [11.2.1, Example 1/11.3.1, Example 1A] **8.** $-8 + 13i$ [11.4.2, Example 3B] **9.** $-\frac{1}{2}$ and 3 [11.1.1, Example 1]
10. $\frac{3 + \sqrt{7}}{2}$ and $\frac{3 - \sqrt{7}}{2}$ [11.2.1, Example 1/11.3.1, Example 1A] **11.** $24 + 30i$ [11.4.3, Example 4]
12. $5 + 3\sqrt{2}$ and $5 - 3\sqrt{2}$ [11.1.2, Example 3] **13.** $3 + \sqrt{14}$ and $3 - \sqrt{14}$ [11.2.1, Example 1/11.3.1, Example 1A]
14. $-18 + 13i$ [11.4.3, Example 5] **15.** $\frac{5 + \sqrt{33}}{2}$ and $\frac{5 - \sqrt{33}}{2}$ [11.2.1, Example 1/11.3.1, Example 1B]
16. $\frac{5}{2}$ and $\frac{1}{3}$ [11.1.1, Example 1] **17.** $\frac{-3 + \sqrt{37}}{2}$ and $\frac{-3 - \sqrt{37}}{2}$ [11.2.1, Example 1/11.3.1, Example 1A]
18. $-\frac{1}{2} - \frac{3}{2}i$ [11.4.4, Example 8] **19.** $4 + i$ and $4 - i$ [11.4.5, Example 11] **20.**

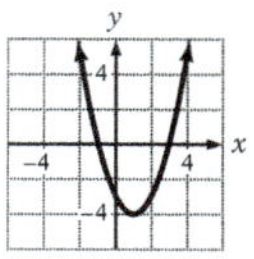

[11.5.1, Example 2A]

21. $3 + i$ [11.4.4, Example 9] **22.** The length is 8 ft. The width is 5 ft. [11.6.1, Problem 1] **23.** The rate of the boat in calm water is 11 mph. [11.6.1, in-text application problem, pp. 543–544] **24.** The middle odd integer is 5 or -5. [11.6.1, Problem 1] **25.** The rate for the last 8 mi was 4 mph. [11.6.1, Example 1]

CUMULATIVE REVIEW EXERCISES

1. $-28x + 27$ [2.2.5] **2.** $\frac{3}{2}$ [3.1.3] **3.** 3 [3.3.3] **4.** $x > \frac{1}{9}$ [3.4.3] **5.** (3, 0) and (0, -4) [5.2.3] **6.** $y = -\frac{4}{3}x - 2$ [5.4.1/5.4.2] **7.** Domain: $\{-2, -1, 0, 1, 2\}$; range: $\{-8, -1, 0, 1, 8\}$; yes [5.5.1] **8.** 37 [5.5.1] **9.**

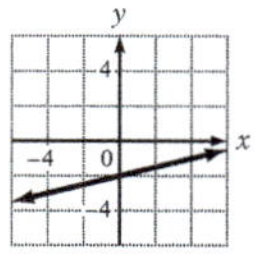

[5.2.2] **10.**

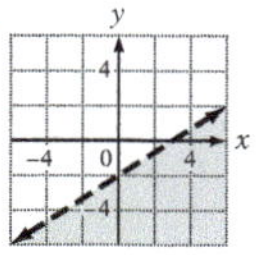

[5.6.1] **11.** (2, 1) [6.2.1] **12.** (2, -2) [6.3.1] **13.** $-\frac{4a}{3b^2}$ [7.4.1] **14.** $x + 2 - \frac{4}{x - 2}$ [7.5.2] **15.** $(x - 4)(4y - 3)$ [8.1.2] **16.** $x(3x - 4)(x + 2)$ [8.3.1/8.3.2] **17.** $\frac{9x^2(x - 2)(x - 4)}{(2x - 3)^2(x + 2)}$ [9.1.3] **18.** $\frac{x + 2}{2(x + 1)}$ [9.3.2] **19.** $\frac{x - 4}{2x + 5}$ [9.4.1] **20.** -3 and 6 [9.5.2] **21.** $a - 2$ [10.3.1] **22.** 5 [10.4.1] **23.** $\frac{1}{2}$ and 3 [11.1.1] **24.** $2 + 2\sqrt{3}$ and $2 - 2\sqrt{3}$ [11.1.2] **25.** $\frac{2 + \sqrt{19}}{3}$ and $\frac{2 - \sqrt{19}}{3}$ [11.2.1/11.3.1] **26.**

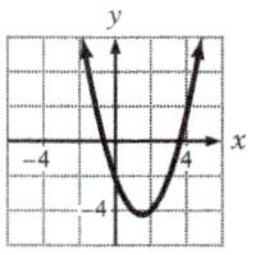

[11.5.1] **27.** The cost is \$72. [3.2.1] **28.** The cost is \$4.50 per pound. [4.5.1] **29.** To earn a dividend of \$752.50, 250 additieonal shares are required. [9.5.2] **30.** The rate of the plane in calm air is 200 mph. The rate of the wind is 40 mph. [6.4.1] **31.** The student must receive a score of 77 or better. [4.7.1] **32.** The length of the guy wire is 31.62 m. [10.4.2] **33.** The lngth is 78 ft. The width is 27 ft. [11.6.1]

Answers to Final Exam

1. -3 [1.1.2] **2.** -6 [1.2.2] **3.** 12.5% [1.3.4] **4.** -256 [1.4.1] **5.** -11 [1.4.2] **6.** $-\frac{15}{2}$ [2.1.1] **7.** $9x + 6y$ [2.2.2] **8.** $6z$ [2.2.3] **9.** $16x - 52$ [2.2.5] **10.** -50 [3.1.3] **11.** -3 [3.3.3] **12.** 15.2 [3.2.1] **13.** $x \le -3$ [3.4.3] **14.** $y \ge \frac{5}{2}$ [3.4.3] **15.** $\frac{2}{3}$ [5.3.1] **16.** $y = -\frac{2}{3}x - 2$ [5.4.1/5.4.2] **17.**

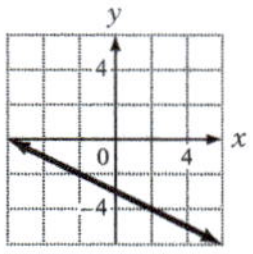

[5.3.2] **18.**

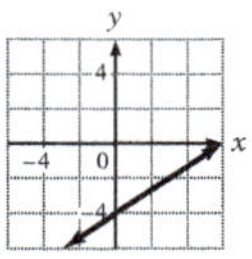

[5.5.2] **19.** $\{-1, 2, 5, 8, 11\}$ [5.5.1] **20.**

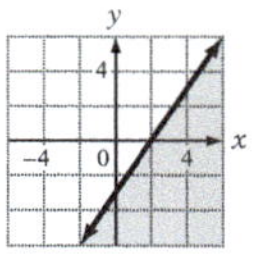

[5.6.1] **21.** (6, 17) [6.2.1] **22.** (2, -1) [6.3.1] **23.** $-3x^2 - 3x + 8$ [7.1.2] **24.** $81x^4y^{12}$ [7.2.2] **25.** $6x^3 + 7x^2 - 7x - 6$ [7.3.2] **26.** $-\frac{x^8y}{2}$ [7.4.1] **27.** $\frac{16y^7}{x^8}$ [7.4.1] **28.** $3x^2 - 4x - \frac{5}{x}$ [7.5.1] **29.** $5x - 12 + \frac{23}{x + 2}$ [7.5.2] **30.** $3.9 \cdot 10^{-8}$ [7.4.2] **31.** $2(4 - x)(a + 3)$ [8.1.2] **32.** $(x - 6)(x + 1)$ [8.2.1] **33.** $(2x - 3)(x + 1)$ [8.3.1/8.3.2] **34.** $(3x + 2)(2x - 3)$ [8.3.1/8.3.2] **35.** $4x(2x - 1)(x - 3)$ [8.3.1/8.3.2] **36.** $(5x + 4)(5x - 4)$ [8.4.1] **37.** $3y(5 + 2x)(5 - 2x)$ [8.5.1] **38.** $\frac{1}{2}$ and 3 [8.6.1/11.1.1] **39.** $\frac{2(x + 1)}{x - 1}$ [9.1.2] **40.** $\frac{-3x^2 + x - 25}{(2x - 5)(x + 3)}$ [9.3.2] **41.** $x + 1$ [9.4.1] **42.** 2 [9.5.1] **43.** $a = b$ [9.7.1] **44.** $7x^3$ [10.1.2] **45.** $38\sqrt{3a}$ [10.2.1] **46.** $\sqrt{15} + 2\sqrt{3}$ [10.3.2] **47.** 5 [10.4.1] **48.** $3 + \sqrt{7}$ and $3 - \sqrt{7}$ [11.1.2] **49.** $\frac{1 + \sqrt{5}}{4}$ and $\frac{1 - \sqrt{5}}{4}$ [11.2.1/11.3.1] **50.**

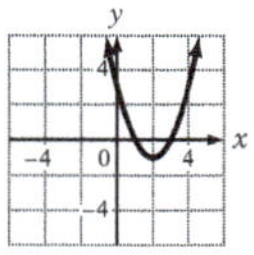

[11.5.1] **51.** $2x + 3(x - 2)$; $5x - 6$ [2.3.2]

52. The original value of the machine was \$3000. [3.2.1] **53.** The correct dosage is 20 mg. [9.6.1] **54.** The angles measure 60°, 50°, and 70°. [4.2.3] **55.** The markup rate is 65%. [4.3.1] **56.** An additional \$6000 must be invested at 11%. [4.4.1] **57.** The cost is \$3 per pound. [4.5.1] **58.** The percent concentration is 36%. [4.5.2] **59.** The plane traveled 215 km during the first hour. [4.6.1] **60.** The rate of the boat in calm water is 15 mph. The rate of the current is 5 mph. [6.4.1] **61.** The length is 10 m. The width is 5 m. [8.6.2] **62.** 16 oz of dye are required. [9.5.3] **63.** It would take them 0.6 h to prepare the dinner. [9.8.1] **64.** The length of the other leg is 11.5 cm. [10.4.2] **65.** The rate of the wind is 25 mph. [11.6.1]

Glossary

abscissa The first number in an ordered pair. It measures a horizontal distance and is also called the first coordinate. (Sec. 5.1)

absolute value of a number The distance of the number from zero on the number line. (Sec. 1.1)

acute angle An angle whose measure is between 0° and 90°. (Sec. 4.2)

addend In addition, a number being added. (Sec. 1.2)

addition The process of finding the total of two numbers. (Sec. 1.2)

addition method An algebraic method of finding an exact solution of a system of linear equations. (Sec. 6.3)

additive inverses Numbers that are the same distance from zero on the number line, but on opposite sides; also called opposites. (Sec. 1.1)

adjacent angles Two angles that share a common side. (Sec. 4.2)

alternate exterior angles Two nonadjacent angles that are on opposite sides of the transversal and outside the parallel lines. (Sec. 4.2)

alternate interior angles Two nonadjacent angles that are on opposite sides of the transversal and between the parallel lines. (Sec. 4.2)

analytic geometry Geometry in which a coordinate system is used to study the relationships between variables. (Sec. 5.1)

angle Figure formed when two rays start from the same point; it is measured in degrees. (Sec. 1.5)

area A measure of the amount of surface in a region. (Sec. 1.5)

arithmetic mean of values Average determined by calculating the sum of the values and then dividing that result by the number of values. (Sec. 1.2)

average of a set of numbers The sum of the numbers divided by the number of numbers in the set. (Sec. 1.2)

average rate of change The average rate of change of y with respect to x is the change in y divided by the change in x. (Sec. 5.1)

axes The two number lines that form a rectangular coordinate system; also called coordinate axes. (Sec. 5.1)

axis of symmetry of a parabola A line of symmetry that passes through the vertex of the parabola and is parallel to the y-axis for an equation of the form $y = ax^2 + bx + c$. (Sec. 11.5)

base In exponential notation, the factor that is multiplied the number of times shown by the exponent (Sec. 1.4)

base of a parallelogram One of the parallel sides of the parallelogram. (Sec. 1.5)

base of a triangle The side of a triangle that the triangle rests on. (Sec. 1.5)

binomial A polynomial of two terms. (Sec. 7.1)

binomial factor A factor that has two terms. (Sec. 8.1)

center of a circle The point from which all points on the circle are equidistant. (Sec. 1.5)

circle Plane figure in which all points are the same distance from its center. (Sec. 1.5)

circumference The distance around a circle. (Sec. 1.5)

clearing denominators Removing denominators from an equation that contains fractions by multiplying each side of the equation by the LCM of the denominators. (Sec. 3.3, 9.5)

coefficient The number part of a variable term. (Sec. 2.1)

combine like terms Use the Distributive Property to add the coefficients of like variable terms; add like terms of a variable expression. (Sec. 2.2)

common monomial factor A monomial factor that is a factor of each term in a polynomial. (Sec. 8.1)

complementary angles Two angles whose sum is 90°. (Sec. 1.5)

complete the square Add to a binomial the constant term that makes it a perfect-square trinomial. (Sec. 11.2)

complex conjugates Complex numbers of the form $a + bi$ and $a - bi$. (Sec. 11.4)

complex fraction A fraction whose numerator or denominator contains one or more fractions. (Sec. 9.4)

complex number A number of the form $a + bi$, where a and b are real numbers and $i = \sqrt{-1}$. (Sec. 11.4)

conjugates Binomial expressions that differ only in the sign of a term. The expressions $a + b$ and $a - b$ are conjugates. (Sec. 10.3)

consecutive even integers Even integers that follow one another in order. (Sec. 4.1)

consecutive integers Integers that follow one another in order. (Sec. 4.1)

consecutive odd integers Odd integers that follow one another in order. (Sec. 4.1)

constant of variation k in a variation equation. (Sec. 9.6)

constant term A term that includes no variable part; also called a constant. (Sec. 2.1)

coordinate axes The two number lines that form a rectangular coordinate system; also simply called axes. (Sec. 5.1)

coordinates of a point The numbers in an ordered pair that is associated with a point. (Sec. 5.1)

corresponding angles Two angles that are on the same side of the transversal and are both acute angles or are both obtuse angles. (Sec. 4.2)

cost The price that a business pays for a product. (Sec. 4.3)

decimal notation Notation in which a number consists of a whole-number part, a decimal point, and a decimal part. (Sec. 1.3)

degree A unit used to measure angles. (Sec. 1.5, 4.2)

degree of a polynomial in one variable In a polynomial, the largest exponent that appears on the variable. (Sec. 7.1)

dependent system of equations A system of equations that has an infinite number of solutions. (Sec. 6.1)

dependent variable In a function, the variable whose value depends on the value of another variable known as the independent variable. (Sec. 5.5)

descending order The terms of a polynomial in one variable arranged so that the exponents on the variable decrease from left to right. The polynomial $9x^5 - 2x^4 + 7x^3 + x^2 - 8x + 1$ is in descending order. (Sec. 7.1)

diameter of a circle A line segment that has endpoints on the circle and goes through the circle's center. (Sec. 1.5)

difference of two squares A polynomial of the form $a^2 - b^2$. (Sec. 8.4)

direct variation equation An equation of the form $y = kx$, where k is a constant called the constant of variation. (Sec. 9.6)

discount The amount by which a retailer reduces the regular price of a product for a promotional sale. (Sec. 4.3)

discount rate The percent of the regular price that the discount represents. (Sec. 4.3)

domain The set of first coordinates of the ordered pairs in a relation. (Sec. 5.5)

double root The two equal roots of a quadratic equation. (Sec. 11.1)

element of a set One of the objects in a set. (Sec. 1.1)

equation A statement of the equality of two mathematical expressions. (Sec. 3.1)

equilateral triangle A triangle in which all three sides are of equal length. (Sec. 1.5, 4.2)

evaluate a function Replace x in $f(x)$ with some value and then simplify the numerical expression that results. (Sec. 5.5)

evaluate a variable expression Replace each variable by its value and then simplify the resulting numerical expression. (Sec. 2.1)

even integer An integer that is divisible by 2. (Sec. 4.1)

exponent In exponential notation, the elevated number that indicates how many times the base occurs in the multiplication. (Sec. 1.4)

exponential form The expression 2^5 is in exponential form. Compare *factored form.* (Sec. 1.4)

exterior angle of a triangle An angle adjacent to an interior angle of the triangle. (Sec. 4.2)

factor In multiplication, a number being multiplied. (Sec. 1.2)

factor a polynomial To write the polynomial as a product of other polynomials. (Sec. 8.1)

factor a trinomial of the form $ax^2 + bx + c$ To express the trinomial as the product of two binomials. (Sec. 8.3)

factor by grouping Group and factor terms in a polynomial in such a way that a common binomial factor is found. (Sec. 8.1)

factor completely Write a polynomial as a product of factors that are nonfactorable over the integers. (Sec. 8.2)

factored form The expression $2 \cdot 2 \cdot 2 \cdot 2 \cdot 2$ is in factored form. Compare *exponential form.* (Sec. 1.4)

FOIL method A method of finding the product of two binomials in which the sum of the products of the first terms, of the outer terms, of the inner terms, and of the last terms is found. (Sec. 7.3)

formula A literal equation that states rules about measurements. (Sec. 9.7)

function A relation in which no two ordered pairs that have the same first coordinate have different second coordinates. (Sec. 5.5)

function notation Notation used for those equations that define functions. (Sec. 5.5)

graph of a function A graph of the ordered pairs (x, y) of the function. (Sec. 5.5)

graph of a relation The graph of the ordered pairs that belong to the relation. (Sec. 5.5)

graph of an equation in two variables A graph of the ordered-pair solutions of the equation. (Sec. 5.2)

graph of an integer A heavy dot directly above that number on the number line. (Sec. 1.1)

graph of an ordered pair The dot drawn at the coordinates of the point in the plane. (Sec. 5.1)

graph a point in the plane Place a dot at the location given by the ordered pair; plot a point in the plane. (Sec. 5.1)

graph of $x = a$ A vertical line passing through the point whose coordinates are $(a, 0)$. (Sec. 5.2)

graph of $y = b$ A horizontal line passing through the point whose coordinates are $(0, b)$. (Sec. 5.2)

greater than A number a is greater than another number b, written $a > b$, if a is to the right of b on the number line. (Sec. 1.1)

greater than or equal to The symbol $\geq$ means "is greater than or equal to." (Sec. 1.1)

greatest common factor of two monomials The product of the GCF of the coefficients of the monomials and the common variable factors. (Sec. 8.1)

greatest common factor of two numbers The greatest number that is a factor of the two numbers. (Sec. 8.1)

grouping symbols Parentheses ( ), brackets [ ], braces { }, the absolute value symbol, and the fraction bar. (Sec. 1.4)

half-plane The solution set of an inequality in two variables. (Sec. 5.6)

height of a parallelogram The distance between the base of the parallelogram and the opposite parallel side. It is perpendicular to the base. (Sec. 1.5)

height of a triangle A line segment perpendicular to the base of the triangle from the opposite vertex. (Sec. 1.5)

hypotenuse In a right triangle, the side opposite the 90° angle. (Sec. 10.4)

imaginary number A number of the form ai, where a is a real number and $i = \sqrt{-1}$. (Sec. 11.4)

imaginary part of a complex number For the complex number $a + bi$, b is the imaginary part. (Sec. 11.4)

imaginary unit The number whose square is -1, designated by the letter i. (Sec. 11.4)

inconsistent system of equations A system of equations that has no solution. (Sec. 6.1)

independent system of equations A system of equations that has one solution. (Sec. 6.1)

independent variable In a function, the variable that varies independently and whose value determines the value of the dependent variable. (Sec. 5.5)

inequality An expression that contains the symbol $>$, $<$, $\geq$ (is greater than or equal to), or $\leq$ (is less than or equal to). (Sec. 3.4)

integers The numbers ..., $-3, -2, -1, 0, 1, 2, 3$, ... (Sec. 1.1)

interior angle of a triangle An angle within the region enclosed by the triangle. (Sec. 4.2)

intersecting lines Lines that cross at a point in the plane. (Sec. 1.5)

inverse variation equation An equation of the form $y = k/x$, where k is a constant. (Sec. 9.6)

irrational number A number that cannot be written in the form a/b, where a and b are integers and b is not equal to zero. (Sec. 1.3, 10.1)

isosceles triangle A triangle that has two equal angles and two equal sides. (Sec. 1.5, 4.2)

least common denominator The smallest number that is a multiple of each denominator in question. (Sec. 9.2)

least common multiple of two numbers The smallest number that contains the prime factorization of each number. (Sec. 1.3, 9.2)

least common multiple of two polynomials The polynomial of least degree that contains the factors of each polynomial. (Sec. 9.2)

legs In a right triangle, the sides opposite the acute angles, or the two shortest sides of the triangle. (Sec. 10.4)

less than A number a is less than another number b, written $a < b$, if a is to the left of b on the number line. (Sec. 1.1)

less than or equal to The symbol $\leq$ means "is less than or equal to." (Sec. 1.1)

like terms Terms of a variable expression that have the same variable part. (Sec. 2.2)

line A line extends indefinitely in two directions in a plane; it has no width. (Sec. 1.5)

linear equation in two variables An equation of the form $y = mx + b$, where m is the coefficient of x and b is a constant; also called a linear function. (Sec. 5.2)

linear function An equation of the form $f(x) = mx + b$ or $y = mx + b$, where m is the coefficient of x and b is a constant; also called a linear equation in two variables. (Sec. 5.5)

line segment Part of a line; it has two endpoints. (Sec. 1.5)

literal equation An equation that contains more than one variable. (Sec. 9.7)

lowest common denominator The smallest number that is a multiple of each denominator in question. (Sec. 1.3)

markup The difference between selling price and cost. (Sec. 4.3)

markup rate The percent of a retailer's cost that the markup represents. (Sec. 4.3)

monomial A number, a variable, or a product of numbers and variables; a polynomial of one term. (Sec. 7.1)

multiplication The process of finding the product of two numbers. (Sec. 1.2)

multiplicative inverse of a number The reciprocal of the number. (Sec. 2.2)

natural numbers The numbers 1, 2, 3, ... (Sec. 1.1)

negative integers The numbers ..., $-4, -3, -2, -1$. (Sec. 1.1)

negative slope A property of a line that slants downward to the right. (Sec. 5.3)

nonfactorable over the integers A description of a polynomial that does not factor when only integers are used. (Sec. 8.2)

numerical coefficient The number part of a variable term. When the numerical coefficient is 1 or -1, the 1 is usually not written. (Sec. 2.1)

obtuse angle An angle whose measure is between 90° and 180°. (Sec. 4.2)

odd integer An integer that is not divisible by 2. (Sec. 4.1)

opposite numbers Two numbers that are the same distance from zero on the number line but lie on different sides of zero; also called additive inverses. (Sec. 1.1)

opposite of a polynomial The polynomial created when the sign of each term of the original polynomial is changed. (Sec. 7.1)

opposites Numbers that are the same distance from zero on the number line, but on opposite sides; also called additive inverses. (Sec. 1.1)

ordered pair A pair of numbers, such as (a, b), that can be used to identify a point in the plane determined by the axes of a rectangular coordinate system. (Sec. 5.1)

Order of Operations Agreement A set of rules that tells us in what order to perform the operations that occur in a numerical expression. (Sec. 1.4)

ordinate The second number in an ordered pair. It measures a vertical distance and is also called the second coordinate. (Sec. 5.1)

origin The point of intersection of the two coordinate axes that form a rectangular coordinate system. (Sec. 5.1)

parabola The name given to the graph of a quadratic equation in two variables. (Sec. 11.5)

parallel lines Lines that never meet; the distance between them is always the same. (Sec. 1.5, 4.2, 5.3)

parallelogram A four-sided plane figure with opposite sides parallel. (Sec. 1.5)

percent Parts of 100. (Sec. 1.3)

perfect cube The cube of an integer. (Sec. 8.4)

perfect square The square of an integer. (Sec. 10.1)

perfect-square trinomial A trinomial that is a product of a binomial and itself. (Sec. 8.4)

perimeter The distance around a plane geometric figure. (Sec. 1.5, 4.2)

perpendicular lines Intersecting lines that form right angles. (Sec. 1.5, 4.2, 5.3)

plane Flat surface determined by the intersection of two lines. (Sec. 1.5, 5.1)

plane figures Figures that lie entirely in a plane. (Sec. 1.5)

plot a point in the plane Place a dot at the location given by the ordered pair; graph a point in the plane. (Sec. 5.1)

point-slope formula If (x_1, y_1) is a point on a line with slope m, then $y - y_1 = m(x - x_1)$. (Sec. 5.4)

polynomial A variable expression in which the terms are monomials. (Sec. 7.1)

positive integers The integers 1, 2, 3, 4 ... . (Sec. 1.1)

positive slope A property of a line that slants upward to the right. (Sec. 5.3)

principal square root The positive square root of a number. (Sec. 10.1)

product In multiplication, the result of multiplying two numbers. (Sec. 1.2)

proportion An equation that states the equality of two ratios or rates. (Sec. 9.5)

Pythagorean Theorem If a and b are the lengths of the legs of a right triangle and c is the length of the hypotenuse, then $c^2 = a^2 + b^2$. (Sec. 10.4)

quadrant One of the four regions into which the two axes of a rectangular coordinate system divide the plane. (Sec. 5.1)

quadratic equation An equation of the form $ax^2 + bx + c$, $a \neq 0$; also called a second-degree equation. (Sec. 8.6, 11.1)

quadratic equation in two variables An equation of the form $y = ax^2 + bx + c$, where a is not equal to zero. (Sec. 11.5)

quadratic function A function of the form $f(x) = ax^2 + bx + c$, where a is not equal to zero. (Sec. 11.5)

radical equation An equation that contains a variable expression in a radicand. (Sec. 10.4)

radical sign The symbol $\sqrt{\ }$, which is used to indicate the positive, or principal, square root of a number. (Sec. 10.1)

radicand In a radical expression, the expression under the radical sign. (Sec. 10.1)

radius of a circle A line segment from the center of a circle to a point on the circle. (Sec. 1.5)

range The set of second coordinates of the ordered pairs in a relation. (Sec. 5.5)

rate The quotient of two quantities that have different units. (Sec. 9.5)

rate of work That part of a task that is completed in one unit of time. (Sec. 9.8)

ratio The quotient of two quantities that have the same unit. (Sec. 9.5)

rational expression A fraction in which the numerator and denominator are polynomials. (Sec. 9.1)

rational number A number that can be written in the form a/b, where a and b are integers and b is not equal to zero. (Sec. 1.3)

rationalize the denominator Remove a radical from the denominator of a fraction. (Sec. 10.3)

ray A figure that starts at a point and extends indefinitely in one direction. (Sec. 1.5)

real numbers The rational numbers and the irrational numbers. (Sec. 1.3)

real part of a complex number For the complex number $a + bi$, a is the real part. (Sec. 11.4)

reciprocal of a fraction The fraction that results when the numerator and denominator of a fraction are interchanged. (Sec. 1.3, 2.2)

reciprocal of a rational expression The rational expression that results when the numerator and denominator of a rational expression are interchanged. (Sec. 9.1)

rectangle A parallelogram that has four right angles. (Sec. 1.5)

rectangular coordinate system System formed by two number lines, one horizontal and one vertical, that intersect at the zero point of each line. (Sec. 5.1)

relation Any set of ordered pairs. (Sec. 5.5)

repeating decimal A decimal formed when dividing the numerator of a fraction by its denominator, in which a digit or a sequence of digits in the decimal repeats infinitely. (Sec. 1.3)

right angle An angle whose measure is 90°. (Sec. 1.5, 4.2)

right triangle A triangle that contains a 90° angle. (Sec. 10.4)

roster method Method of writing a set by enclosing a list of the elements in braces. (Sec. 1.1)

scatter diagram A graph of collected data as points in a coordinate system. (Sec. 5.1)

scientific notation Notation in which a number is expressed as the product of two factors, one a number between 1 and 10 and the other a power of 10. (Sec. 7.4)

second-degree equation An equation of the form $ax^2 + bx + c = 0$, $a \neq 0$; also called a quadratic equation. (Sec. 11.1)

selling price The price for which a business sells a product to a customer. (Sec. 4.3)

set A collection of objects. (Sec. 1.1)

similar objects Objects that have the same shape but not necessarily the same size. (Sec. 9.5)

similar triangles Triangles that have the same shape but not necessarily the same size. (Sec. 9.5)

simplest form of a fraction A fraction in which the numerator and denominator have no common factors other than 1. (Sec. 1.3, 9.1, 10.1)

slope The measure of the slant of a line. The symbol for slope is m. (Sec. 5.3)

slope-intercept form The slope-intercept form of an equation of a straight line is $y = mx + b$. (Sec. 5.3)

solids Objects in space. (Sec. 1.5)

solution of an equation A number that, when substituted for the variable, results in a true equation. (Sec. 3.1)

solution of a linear equation in two variables An ordered pair whose coordinates make the equation a true statement. (Sec. 5.2)

solution of a system of equations in two variables An ordered pair that is a solution of each equation of the system. (Sec. 6.1)

solution set of an inequality A set of numbers, each element of which, when substituted for the variable, results in a true inequality. (Sec. 3.4)

solve an equation Find solutions of the equation. (Sec. 3.1)

square A rectangle that has four equal sides. (Sec. 1.5)

square of a binomial An expression of the form $(a + b)^2$. (Sec. 7.3)

square root A square root of a positive number x is a number a for which $a^2 = x$. (Sec. 10.1)

standard form of a quadratic equation A quadratic equation written with the polynomial in descending order and equal to zero. $ax^2 + bx + c = 0$ is in standard form. (Sec. 8.6, 11.1)

straight angle An angle whose measure is 180°. (Sec. 1.5, 4.2)

substitution method An algebraic method of finding an exact solution of a system of equations. (Sec. 6.2)

subtraction The process of finding the difference between two numbers. (Sec. 1.2)

sum In addition, the total of two or more numbers. (Sec. 1.2)

sum and difference of two terms An expression of the form $(a + b)(a - b)$. (Sec. 7.3)

supplementary angles Two angles whose sum is 180°. (Sec. 1.5)

system of equations Equations that are considered together. (Sec. 6.1)

terminating decimal A decimal that is formed when dividing the numerator of its fractional counterpart by the denominator results in a remainder of zero. (Sec. 1.3)

terms of a variable expression The addends of the expression. (Sec. 2.1)

transversal A line intersecting two other lines at two different points. (Sec. 4.2)

triangle A three-sided closed figure. (Sec. 1.5, 4.2)

trinomial A polynomial of three terms. (Sec. 7.1)

undefined slope A property of a vertical line. (Sec. 5.3)

uniform motion The motion of a moving object whose speed and direction do not change. (Sec. 3.1, 4.6)

value of a function The result of evaluating a variable expression, represented by the symbol $f(x)$. (Sec. 5.5)

variable A letter of the alphabet used to stand for a number that is unknown or that can change. (Sec. 1.1, 2.1)

variable expression An expression that contains one or more variables. (Sec. 2.1)

variable part In a variable term, the variable or variables and their exponents. (Sec. 2.1)

variable term A term composed of a numerical coefficient and a variable part. (Sec. 2.1)

vertex of an angle Point at which the two rays that form the angle meet. (Sec. 1.5)

vertex of a parabola The lowest point on the graph of a parabola if the parabola opens up. The highest point on the graph of a parabola if the parabola opens down. (Sec. 11.5)

vertical angles Two angles that are on opposite sides of the intersection of two lines. (Sec. 4.2)

whole numbers The whole numbers are 0, 1, 2, 3 ... . (Sec. 1.1)

***x*-axis** The horizontal axis in an *xy*-coordinate system. (Sec. 5.1)

***x*-coordinate** The abscissa in an *xy*-coordinate system. (Sec. 5.1)

***x*-intercept** The point at which a graph crosses the *x*-axis. (Sec. 5.2)

***y*-axis** The vertical axis in an *xy*-coordinate system. (Sec. 5.1)

***y*-coordinate** The ordinate in an *xy*-coordinate system. (Sec. 5.1)

***y*-intercept** The point at which a graph crosses the *y*-axis. (Sec. 5.2)

zero slope A property of a horizontal line. (Sec. 5.3)

Digital Vision

Index

Index of Applications

CHAPTER 4

SOLVING EQUATIONS AND INEQUALITIES: APPLICATIONS

4.2 Integer, Coin, and Stamp Problems

OBJECTIVE 1 Consecutive integer problems

Recall that the integers are the numbers ..., −4, −3, −2, −1, 0, 1, 2, 3, 4, ...

An **even integer** is an integer that is divisible by 2. Examples of even integers are −8, 0, and 22.

An **odd integer** is an integer that is not divisible by 2. Examples of odd integers are −17, 1, and 39.

Consecutive integers are integers that follow one another in order. Examples of consecutive integers are shown at the right. (Assume the variable n represents an integer.)

11, 12, 13
−8, −7, −6
$n, n + 1, n + 2$

Examples of **consecutive even integers** are shown at the right. (Assume the variable n represents an even integer.)

24, 26, 28
−10, −8, −6
$n, n + 2, n + 4$

Examples of **consecutive odd integers** are shown at the right. (Assume the variable n represents an odd integer.)

19, 21, 23,
−1, 1, 3
$n, n + 2, n + 4$

Solve: The sum of three consecutive odd integers is 51. Find the integers.

STRATEGY for solving a consecutive integer problem

- Let a variable represent one of the integers. Express each of the other integers in terms of that variable. Remember that consecutive integers differ by 1. Consecutive even or consecutive odd integers differ by 2.

First odd integer: n
Second odd integer: $n + 2$
Third odd integer: $n + 4$

- Determine the relationship among the integers.

The sum of the three odd integers is 51.

$$n + (n + 2) + (n + 4) = 51$$
$$3n + 6 = 51$$
$$3n = 45$$
$$n = 15$$

$n + 2 = 15 + 2 = 17$
$n + 4 = 15 + 4 = 19$

• **Substitute the value of n into the variable expressions for the second and third integers.**

The three consecutive odd integers are 15, 17, and 19.

EXAMPLE 1 Find three consecutive even integers such that three times the second integer is six more than the sum of the first and third integers.

Strategy
- First even integer: n
  Second even integer: $n + 2$
  Third even integer: $n + 4$
- Three times the second integer equals six more than the sum of the first and third integers.

Solution

$$3(n + 2) = n + (n + 4) + 6$$
$$3n + 6 = 2n + 10$$
$$n + 6 = 10$$
$$n = 4$$

• The first even integer is 4.

$$n + 2 = 4 + 2 = 6$$
$$n + 4 = 4 + 4 = 8$$

• Substitute the value of n into the variable expressions for the second and third integers.

The three consecutive even integers are 4, 6, and 8.

PROBLEM 1 Find three consecutive integers whose sum is -12.

Solution See page S8.

OBJECTIVE 2

Coin and stamp problems

$3 \cdot 25 = 75$ cents

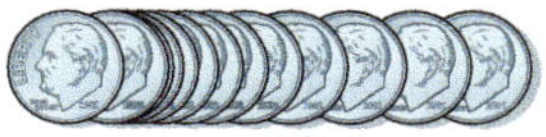

$d \cdot 10 = 10d$ cents

In solving problems that deal with coins or stamps of different values, it is necessary to represent the values of the coins or stamps in the same unit of money. Frequently, the unit of money is cents.

The value of 3 quarters in cents is $3 \cdot 25$ or 75 cents.
The value of 4 nickels in cents is $4 \cdot 5$ or 20 cents.
The value of d dimes in cents is $d \cdot 10$ or $10d$ cents.

Solve: A coin bank contains $1.20 in dimes and quarters. In all, there are nine coins in the bank. Find the number of quarters in the bank.

STRATEGY for solving a coin problem

- For each denomination of coin, write a numerical or variable expression for the number of coins, the value of the coin in cents, and the total value of the coins in cents. The results can be recorded in a table.

Take Note

Use the information given in the problem to fill in the number and value columns of the table. Fill in the total value column by multiplying the two expressions you wrote in each row.

The total number of coins is 9.

Number of quarters: x
Number of dimes: $9 - x$

Coin	Number of coins	$\cdot$	Value of coin in cents	$=$	Total value in cents
Quarter	x	$\cdot$	25	$=$	$25x$
Dime	$9 - x$	$\cdot$	10	$=$	$10(9 - x)$

▶ Determine the relationship between the total values of the different coins. Use the fact that the sum of the total values of the different denominations of coins is equal to the total value of all the coins.

The sum of the total values of the different denominations of coins is equal to the total value of all the coins (120 cents).

$$25x + 10(9 - x) = 120$$
$$25x + 90 - 10x = 120$$
$$15x + 90 = 120$$
$$15x = 30$$
$$x = 2$$

• **The total value of the quarters plus the total value of the dimes equals the amount of money in the bank.**

There are 2 quarters in the bank.

EXAMPLE 2 A collection of stamps consists of 3¢ stamps and 8¢ stamps. The number of 8¢ stamps is five more than three times the number of 3¢ stamps. The total value of all the stamps is \$1.75. Find the number of each type of stamp in the collection.

Strategy ▶ Number of 3¢ stamps: x
Number of 8¢ stamps: $3x + 5$

Stamp	Number	Value	Total value
3¢	x	3	$3x$
8¢	$3x + 5$	8	$8(3x + 5)$

▶ The sum of the total values of the different types of stamps equals the total value of all the stamps (175 cents).

Solution

$$3x + 8(3x + 5) = 175$$
$$3x + 24x + 40 = 175$$
$$27x + 40 = 175$$
$$27x = 135$$
$$x = 5$$

• **The total value of the 3¢ stamps plus the total value of the 8¢ stamps equals the total value of all the stamps.**

$$3x + 5 = 3(5) + 5$$
$$= 15 + 5 = 20$$

• **Find the number of 8¢ stamps.**

There are five 3¢ stamps and twenty 8¢ stamps in the collection.

PROBLEM 2 A coin bank contains nickels, dimes, and quarters. There are five times as many nickels as dimes and six more quarters than dimes. The total value of all the coins is \$6.30. Find the number of each kind of coin in the bank.

Solution See page S8.

4.2 Concept Review

Determine whether the statement is always true, sometimes true, or never true.

1. An even integer is a multiple of 2.

2. When $10d$ is used to represent the value of d dimes, 10 is written as the coefficient of d because it is the number of cents in one dime.

3. Given the consecutive odd integers -5 and -3, the next consecutive odd integer is -1.

4. If the first of three consecutive odd integers is n, then the second and third consecutive odd integers are represented as $n + 1$ and $n + 3$.

5. You have a total of 20 coins in nickels and dimes. If n represents the number of nickels you have, then $n - 20$ represents the number of dimes you have.

6. You have a total of seven coins in dimes and quarters. The total value of the coins is \$1.15. If d represents the number of dimes you have, then $10d + 25(7 - d) = 1.15$.

4.2 Exercises

1 Consecutive integer problems

1. Integers that follow one another in order are called __?__ integers.

2. Two consecutive integers differ by __?__. Two consecutive even integers differ by __?__. Two consecutive odd integers differ by __?__.

3. Explain how to represent three consecutive integers using only one variable.

4. Explain why both consecutive even integers and consecutive odd integers are represented algebraically as $n, n + 2, n + 4, \ldots$.

5. The sum of three consecutive integers is 54. Find the integers.

6. The sum of three consecutive integers is 75. Find the integers.

7. The sum of three consecutive even integers is 84. Find the integers.

8. The sum of three consecutive even integers is 48. Find the integers.

9. The sum of three consecutive odd integers is 57. Find the integers.

10. The sum of three consecutive odd integers is 81. Find the integers.

11. Find two consecutive even integers such that five times the first integer is equal to four times the second integer.

12. Find two consecutive even integers such that six times the first integer equals three times the second integer.

13. Nine times the first of two consecutive odd integers equals seven times the second. Find the integers.

14. Five times the first of two consecutive odd integers is three times the second. Find the integers.

15. Find three consecutive integers whose sum is negative twenty-four.

16. Find three consecutive even integers whose sum is negative twelve.

17. Three times the smallest of three consecutive even integers is two more than twice the largest. Find the integers.

18. Twice the smallest of three consecutive odd integers is five more than the largest. Find the integers.

19. Find three consecutive odd integers such that three times the middle integer is six more than the sum of the first and third integers.

20. Find three consecutive odd integers such that four times the middle integer is equal to two less than the sum of the first and third integers.

21. Which of the following could *not* be used to represent three consecutive even integers?

a. $n + 2, n + 4, n + 6$ **b.** $n, n + 4, n + 6$
c. $n + 1, n + 3, n + 5$ **d.** $n - 2, n, n + 2$

22. If n is an odd integer, which expression represents the third consecutive odd integer after n?

a. $n + 6$ **b.** $n + 4$ **c.** $n + 3$ **d.** $n + 2$

2 Coin and stamp problems

23. Explain how to represent the total value of x nickels using the equation

Number of coins · Value of the coin in cents = Total value in cents

24. Suppose a coin purse contains only dimes and quarters. In the context of this situation, explain the meaning of the statement "The sum of the total values of the different denominations of coins is equal to the total value of all the coins."

25. A bank contains 32 coins in nickels and dimes. Let n represent the number of nickels. Complete the following table.

Coin	Number of coins	$\cdot$	Value of the coin in cents	$=$	Total value of the coins in cents
Nickel	n	$\cdot$	___?___	$=$	___?___
Dime	___?___	$\cdot$	___?___	$=$	___?___

26. The total value of the 32 coins in Exercise 25 is \$2.70. The total value of the coins in cents is ___?___. Use this information and the information in the table in Exercise 25 to write an equation that can be solved to find the number of nickels in the bank: ___?___ + ___?___ = ___?___.

27. **a.** How many pennies are in d dollars?
b. How many nickels are in d dollars?
c. How many quarters are in d dollars?

28. **a.** How many cents are n 39¢ stamps worth?
b. How many dollars are n 39¢ stamps worth?

29. A bank contains 27 coins in dimes and quarters. The coins have a total value of \$4.95. Find the number of dimes and quarters in the bank.

30. A coin purse contains 18 coins in nickels and dimes. The coins have a total value of \$1.15. Find the number of nickels and dimes in the coin purse.

31. A business executive bought 40 stamps for \$14.40. The purchase included 39¢ stamps and 24¢ stamps. How many of each type of stamp were bought?

32. A postal clerk sold some 34¢ stamps and some 23¢ stamps. Altogether, 15 stamps were sold for a total cost of \$4.44. How many of each type of stamp were sold?

33. A drawer contains 29¢ stamps and 3¢ stamps. The number of 29¢ stamps is four less than three times the number of 3¢ stamps. The total value of all the stamps is \$1.54. How many 29¢ stamps are in the drawer?

34. The total value of the dimes and quarters in a bank is \$6.05. There are six more quarters than dimes. Find the number of each type of coin in the bank.

35. A child's piggy bank contains 44 coins in quarters and dimes. The coins have a total value of \$8.60. Find the number of quarters in the bank.

36. A coin bank contains nickels and dimes. The number of dimes is 10 less than twice the number of nickels. The total value of all the coins is \$2.75. Find the number of each type of coin in the bank.

37. A total of 26 bills are in a cash box. Some of the bills are one-dollar bills, and the rest are five-dollar bills. The total amount of cash in the box is \$50. Find the number of each type of bill in the cash box.

38. A bank teller cashed a check for \$200 using twenty-dollar bills and ten-dollar bills. In all, 12 bills were handed to the customer. Find the number of twenty-dollar bills and the number of ten-dollar bills.

39. A coin bank contains pennies, nickels, and dimes. There are six times as many nickels as pennies and four times as many dimes as pennies. The total amount of money in the bank is \$7.81. Find the number of pennies in the bank.

40. A coin bank contains pennies, nickels, and quarters. There are seven times as many nickels as pennies and three times as many quarters as pennies. The total amount of money in the bank is \$5.55. Find the number of pennies in the bank.

41. A collection of stamps consists of 22¢ stamps and 40¢ stamps. The number of 22¢ stamps is three more than four times the number of 40¢ stamps. The total value of the stamps is \$8.34. Find the number of 22¢ stamps in the collection.

42. A collection of stamps consists of 2¢ stamps, 8¢ stamps, and 14¢ stamps. The number of 2¢ stamps is five more than twice the number of 8¢ stamps. The number of 14¢ stamps is three times the number of 8¢ stamps. The total value of the stamps is \$2.26. Find the number of each type of stamp in the collection.

43. A collection of stamps consists of 3¢ stamps, 7¢ stamps, and 12¢ stamps. The number of 3¢ stamps is five less than the number of 7¢ stamps. The number of 12¢ stamps is one-half the number of 7¢ stamps. The total value of all the stamps is \$2.73. Find the number of each type of stamp in the collection.

44. A collection of stamps consists of 2¢ stamps, 5¢ stamps, and 7¢ stamps. There are nine more 2¢ stamps than 5¢ stamps and twice as many 7¢ stamps as 5¢ stamps. The total value of the stamps is \$1.44. Find the number of each type of stamp in the collection.

45. A collection of stamps consists of 6¢ stamps, 8¢ stamps, and 15¢ stamps. The number of 6¢ stamps is three times the number of 8¢ stamps. There are six more 15¢ stamps than there are 6¢ stamps. The total value of all the stamps is \$5.16. Find the number of each type of stamp.

46. A child's piggy bank contains nickels, dimes, and quarters. There are twice as many nickels as dimes and four more quarters than nickels. The total value of all the coins is $9.40. Find the number of each type of coin.

4.2 Applying Concepts

47. Find four consecutive even integers whose sum is −36.

48. Find four consecutive odd integers whose sum is −48.

49. A coin bank contains only dimes and quarters. The number of quarters in the bank is two less than twice the number of dimes. There are 34 coins in the bank. How much money is in the bank?

50. A postal clerk sold twenty stamps to a customer. The number of 39¢ stamps purchased was two more than twice the number of 24¢ stamps purchased. If the customer bought only 39¢ stamps and 24¢ stamps, how much money did the clerk collect from the customer?

51. Find three consecutive odd integers such that the sum of the first and third integers is twice the second integer.

52. Find four consecutive integers such that the sum of the first and fourth integers equals the sum of the second and third integers.

Adding fractions using Least Common Denominators (LCD)

I) Common denominators

Five gallons and two tablespoons make what? What about subtracting 3 volts from 7 amperes?

Adding or subtracting amounts of different units (where "unit" can be: 1 inch, 1 meter, 1 gallon, 1 tablespoon, 1 volt or 1 ampere) is not always meaningful; in actuality it can be even downright absurd. In the examples above, you can say that five gallons and two tablespoons are seven volume units (better yet: "some volume"); subtracting 3 volts from 7 amperes could be 4 "things" (but actually it's nothing meaningful at all!) As you can see, you can still add or subtract the numbers, but the units are lost, become vague.

However, sometimes you can find common ground. For example, let's find out what is 10 inches added to 3 meters. If we go for a vague unit, we can say that 10 in + 3 m = 13 "some units of length", but this is not helpful. Units of length can be as tiny as microns or as enormous as light-years. On the other hand, it is known that 1 inch equals precisely 2.54 centimeters; also, 1 meter equals precisely 100 centimeters. Using these facts we get:

$$10 \text{ in} + 3 \text{ m} = 25.4 \text{ cm} + 300 \text{ cm} = 325.4 \text{ cm}$$

At the expense of multiplying our original numbers with conversion factors (2.54 for inches and 100 for meters), we now have common unit, and addition is meaningful.

A fraction should always be viewed as an amount of equal parts of a division of 1. For example, 1/2 is one half, where "one" is the amount, and "half" is the part of 1; 2/3 is two thirds, where again "two" is the amount, and "third" the part of 1. It's quite important to realize the different roles the numerator and the denominator play. **The denominator will always have the role of the unit, while the numerator tells how much of that unit is provided**.

When we add or subtract fractions with the same denominator, the situation is certainly simple, as we are dealing with amounts of the same unit.

Example 1.

$$\frac{13}{5}+\frac{8}{5}=\frac{13+8}{5}=\frac{21}{5}$$
$$\frac{6}{35}-\frac{4}{35}=\frac{6-4}{35}=\frac{2}{35}$$

How about adding or subtracting two fractions, with different denominators? Different denominators will mean different units, and we need to use a trick, similar to the inches added to meters computation.

A **half** is precisely equal to 2 quarters; 3 sixths; 4 eighths; and so on. Thus, we can always change one half to a fraction having **a multiple of 2** in the denominator.

$$\frac{1}{2}=\frac{1\times 2}{2\times 2}=\frac{2}{4} \qquad \frac{1}{2}=\frac{1\times 3}{2\times 3}=\frac{3}{6}$$

$$\frac{1}{2}=\frac{1\times 4}{2\times 4}=\frac{4}{8} \qquad \frac{1}{2}=\frac{1\times 5}{2\times 5}=\frac{5}{10}$$

Similarly, a **third** equals precisely 2 sixths; 3 ninths; 4 twelfths; and so on. We get fractions whose denominator is a **multiple of 3**.

$$\frac{2}{3}=\frac{2\times 2}{3\times 2}=\frac{4}{6} \qquad \frac{2}{3}=\frac{2\times 3}{3\times 3}=\frac{6}{9}$$

$$\frac{2}{3}=\frac{2\times 4}{3\times 4}=\frac{8}{12} \qquad \frac{2}{3}=\frac{2\times 5}{3\times 5}=\frac{10}{15}$$

A quick glance tells us that **sixth** is common unit for both fractions.

Adding 1/2 and 2/3 could then be interpreted as 1+2=3 "some parts of 1" (vague description). Alternatively - and much better! - the sum can be written as three sixths (or 3/6) added to four sixths (4/6), by using that new, common unit.

$$\frac{1}{2}+\frac{2}{3}=\frac{3}{6}+\frac{4}{6}=\frac{3+4}{6}=\frac{7}{6}$$

<u>Observation:</u> An interesting fact, often overlooked (probably since it is expected!), is that *adding or subtracting fractions always results in a new fraction*.

Do we always have to write all the possible equivalent fractions of the given terms, to find out what the common denominator is? Actually, no – you might have noticed that **6** (from "sixth") **is exactly equal to 2 times 3** (2 from "half" and 3 from "third"). 6 must be at the same time a multiple of 2 and of 3, and the easiest way to obtain such a common multiple is by multiplying the two denominators.

Example 2.

$$\frac{3}{4}+\frac{5}{7}=\frac{3\times 7}{4\times 7}+\frac{5\times 4}{7\times 4}=\frac{21}{28}+\frac{20}{28}=\frac{21+20}{28}=\frac{41}{28}$$

$$\frac{3}{8}+\frac{5}{12}=\frac{3\times 12}{8\times 12}+\frac{5\times 8}{12\times 8}=\frac{36}{96}+\frac{40}{96}=\frac{36+40}{96}=\frac{76}{96}$$

In conclusion, when involving two (or, for that matter, more) fractions in additions and subtractions, a common unit is obtained by multiplying all the denominators together. Then, in order to add and/or subtract the fractions, replace the fractions with the equivalent fractions having the common denominator, at which point all fractions will have the same unit, and we only need to add or subtract the numerators, which now is meaningful.

II) Least common denominator (LCD)

In the second example above, an interesting, and undesired, phenomenon is happening: although ninety-sixths is a common unit for eighths and twelfths, another unit is common, and much smaller. Let's check:

$$\frac{3}{8}=\frac{3\times 2}{8\times 2}=\frac{6}{16} \qquad \frac{3}{8}=\frac{3\times 3}{8\times 3}=\frac{9}{24}$$

$$\frac{3}{8}=\frac{3\times 4}{8\times 4}=\frac{12}{32} \qquad \frac{3}{8}=\frac{3\times 5}{8\times 5}=\frac{15}{40}$$

$$\frac{5}{12}=\frac{5\times 2}{12\times 2}=\frac{10}{24} \qquad \frac{5}{12}=\frac{5\times 3}{12\times 3}=\frac{15}{36}$$

$$\frac{5}{12}=\frac{5\times 4}{12\times 4}=\frac{20}{48} \qquad \frac{5}{12}=\frac{5\times 5}{12\times 5}=\frac{25}{60}$$

Notice that twenty-fourths (denominator 24) is a common unit already!

$$\frac{3}{8}+\frac{5}{12}=\frac{3\times 3}{8\times 3}+\frac{5\times 2}{12\times 2}=\frac{9}{24}+\frac{10}{24}=\frac{9+10}{24}=\frac{19}{24}$$

We get the same result, although at first glance it is not obvious:

$$\frac{76}{96}=\frac{19\times 4}{24\times 4}=\frac{19}{24}$$

The result we got first was correct, however it was **not in lowest terms**. If we want to be more efficient when adding fractions – and you will see many problems and applications requiring you to have the answer in lowest terms! – simply getting a common denominator is not enough; we need to find the **smallest** common denominator, which we shall call **LCD=Least Common Denominator**. In the example above it seems that in order to do so we have to go back to listing all the denominators! But no, we can actually devise an algorithm. We will present two such algorithms.

A) The price of efficiency is investing some work first. We have to decompose the denominators into prime factors. Remember that prime numbers are those natural numbers that cannot be evenly divided by any natural number, different from 1 and themselves (exception is 1, which has a very good reason **not** to be prime). The list of prime numbers includes: 2, 3, 5, 7, 11, 13, 17, 19, 23, 29.

Take the number 180. Since the number is even, it can be divided by 2, so we can write 180=2*90. 90 is still even, so 90=2*45. 45 ends in 5, so we can divide it by 5, 45=5*9. 9 is divisible by 3, so 9=3*3. 3 is prime, so we stop. Putting all these together we get:

$$180 = 2 \times 2 \times 3 \times 3 \times 5$$

<u>Example 3.</u>

For some numbers you need to attempt (and fail) the division of the given number by quite a few of the primes above. For example, take 2717. Division by 2, 3, 5, 7 will not give an integer (check this!) However, we strike lucky with 11:

2717=11*247

247 will not be divisible by 2, 3, 5 or 7 (otherwise 2717 would have been!), so we continue by attempting to divide by 11 (not working), 13 (which works): 247=13*19. 19 is prime, so

$$2717 = 11 \times 13 \times 19.$$

Going back to our denominators, let us find the least common denominator for 3/8 and 5/12. Start by factoring the numbers into prime numbers:

$$8 = 2 \times 2 \times 2$$
$$12 = 2 \times 2 \times 3$$

Notice that both 8 and 12 already have a common part: 2x2; we just need to complete that existing common part to a full common multiple. What we do is simply multiply all the extra factors to the common part:

$$8 = \underbrace{2 \times 2} \times 2 \quad \rightarrow \textit{last 2 is extra}$$

$$12 = \overbrace{2 \times 2} \times 3 \quad \rightarrow 3 \textit{ is extra}$$

$$LCD = \overbrace{2 \times 2} \times 2 \times 3 = 24$$

<u>Example 4.</u>

$$2100 = 2 \times 2 \times 3 \times 5 \times 5 \times 7$$
$$378 = 2 \times 3 \times 3 \times 3 \times 7$$

Let's figure out the common part (which involves a bit of shuffling of factors), and then complete it to the full common multiple:

$$2100 = \underbrace{2 \times 3 \times 7} \times 2 \times 5 \times 5$$

$$378 = \overbrace{2 \times 3 \times 7} \times 3 \times 3$$

$$LCD = \overbrace{2 \times 3 \times 7} \times (2 \times 5 \times 5) \times (3 \times 3) = 18900$$

What to do for several denominators (three or more)? Start with two denominators, and find their least common denominator, which will now replace the original two denominators. Repeat the process with this new denominator and the third. Now all three denominators are replaced by a common one. And so on.

<u>Example 5.</u>

Perform the LCD algorithm to compute:

$$\frac{2}{9} + \frac{5}{12} - \frac{3}{8}$$

First we need to factor out the denominators:

$$9 = 3 \times 3 \qquad 12 = 2 \times 2 \times 3 \qquad 8 = 2 \times 2 \times 2$$

Start with 9 and 12:

$$9 = \underbrace{3}_{common} \times 3$$

$$12 = \overbrace{3}^{common} \times 2 \times 2$$

The common part is 3, and we complete it to a full common multiple:
$LCD = 3 \times (3) \times (2 \times 2) = 36$

We still need to involve 8, and we will do it now; instead of 9 and 12 we will use 36 though.

For 36 and 8 we have:

$$36 = \underbrace{2 \times 2}_{common} \times 3 \times 3$$

$$8 = \overbrace{2 \times 2}^{common} \times 2$$

Thus, the LCD will be obtained by completing the common part with the remaining factors:

$$LCD = 2 \times 2 \times (3 \times 3) \times (2) = 72$$

Interesting to note that we need to use two numbers at a time. In this example, if you check, there is **no common part for all three numbers,** however the LCD is not the product of the three numbers (which is 864; check it!)

Another interesting thing is that it doesn't matter which numbers we start with, we will always get the same outcome.

B) Alternatively, you can use the following algorithm (useful for smaller numbers).

Let us add/subtract:

$$\frac{3}{16} + \frac{7}{20} - \frac{9}{50}$$

The trick here is again to find the primes that will make up the LCD. In order to do so, we must check whether a chosen prime divides (or not) at least one of the numbers. If indeed it does, it will be selected, and **the denominators that can be divided by it will be divided**. Repeat the procedure until we get "all ones".

We will run the three denominators 16, 20 and 50 through these divisions with primes. If you remember the primes list (see above) we will start with the prime 2 and see if it divides exactly into at least one of the three denominators. Then choose 3, then 5, then 7, and so on.

- ✓ Since all three numbers are even, they all divide evenly by 2. **Select 2**, and perform the divisions:

$$16 \div 2 = 8 \qquad 20 \div 2 = 10 \qquad 50 \div 2 = 25$$

- ✓ Replace the original three denominators by the results of the division: 8, 10 and 25
- ✓ Since the first two are still even, both divide evenly again by 2. **Select 2**, and perform the divisions that are possible (we will not divide 25 by 2!)

$$8 \div 2 = 4 \qquad 10 \div 2 = 5 \qquad 25 \textit{ unchanged}$$

- ✓ Replace the three numbers with the results of the divisions: 4, 5 and 25
- ✓ 4 is still even. **Select 2** and perform the possible divisions (now both 5 and 25 will pass unchanged)

$$4 \div 2 = 2 \qquad 5 \textit{ unchanged} \qquad 25 \textit{ unchanged}$$

- ✓ Replace again the numbers: 2, 5, 25
- ✓ 2 is even. **Select 2** and perform the (final) division by 2

$$2 \div 2 = 1 \qquad 5 \textit{ unchanged} \qquad 25 \textit{ unchanged}$$

- ✓ Replace the numbers by: 1, 5, 25
- ✓ None of the numbers is even, we cannot divide by 2 anymore, since all numbers are odd. *Skip 2*
- ✓ Check division by 3 – none divide evenly by 3. *Skip 3*
- ✓ Check division by 5: both 5 and 25 work. **Select 5** and perform the divisions that are possible:

$$1 \textit{ unchanged} \qquad 5 \div 5 = 1 \qquad 25 \div 5 = 5$$

- ✓ Replace the numbers by: 1, 1 and 5
- ✓ We can divide once more by 5. **Select 5** and divide:

$$1 \textit{ unchanged} \qquad 1 \textit{ unchanged} \qquad 5 \div 5 = 1$$

We reached (1, 1, 1), and we stop. Now put together all the numbers that were selected:

$$2 \times 2 \times 2 \times 2 \times 5 \times 5 = 400$$

This is the LCD and it will allow us to perform the addition/subtraction:

$$\frac{3}{16}-\frac{7}{20}+\frac{9}{50}=\frac{3\times 25}{16\times 25}-\frac{7\times 20}{20\times 20}+\frac{9\times 8}{50\times 8}=\frac{75}{400}-\frac{140}{400}+\frac{72}{400}=\frac{75-140+72}{400}=\frac{7}{400}$$

Example 5.

Add the following fractions. (We will have a twist at the end!)

$$\frac{4}{35}+\frac{2}{63}+\frac{8}{45}$$

- ✓ The denominators are 35, 63, 45. We cannot divide either number by 2; *skip 2*. However, we can divide at least one by 3; **select 3**:

$$35 \textit{ unchanged} \qquad 63\div 3=21 \qquad 45\div 3=15$$

- ✓ The numbers are now 35, 21, 15. Divide again by 3; **select 3**:

$$35 \textit{ unchanged} \qquad 21\div 3=7 \qquad 15\div 3=5$$

- ✓ The numbers are now 35, 7, 5. We cannot divide again by 3; *skip now 3*; we can divide by 5; **select 5**:

$$35\div 5=7 \qquad 7 \textit{ unchanged} \qquad 5\div 5=1$$

- ✓ The numbers are now 7, 7, 1. None can be divided by 5 again; skip *now 5*. We can divide by 7; **select 7**:

$$7\div 7=1 \qquad 7\div 7=1 \qquad 1 \textit{ unchanged}$$

We reached "all ones", so we stop. The LCD is: $3\times 3\times 5\times 7=315$. We go back to our addition:

$$\frac{4}{35}+\frac{2}{63}+\frac{8}{45}=\frac{4\times 9}{35\times 9}+\frac{2\times 5}{63\times 5}+\frac{8\times 7}{45\times 7}=\frac{36}{315}+\frac{10}{315}+\frac{56}{315}=\frac{36+10+56}{315}=\frac{102}{315}$$

Even though we used the LCD, when adding or subtracting **three or more** fractions the result **might not be in the simplest form**. Above, 102 and 315 both are divisible by 3, so we must rewrite the fraction:

$$\frac{102}{315}=\frac{34\times 3}{105\times 3}=\frac{34}{105}$$

Read the problems carefully, to see if you need to check for lowest terms answer!

EXERCISES

Find the least common multiple of the given numbers using the LCD algorithms.

1. 14, 20
2. 12, 27
3. 36, 7, 24
4. 10, 30, 500
5. 68, 102, 32

Add or subtract by bringing all fractions to the least common denominator (LCD). If necessary (that is, for three or more fractions), bring the result to the lowest terms.

6. $\frac{1}{8}+\frac{1}{6}$
7. $\frac{7}{24}+\frac{11}{18}$
8. $\frac{7}{60}-\frac{6}{75}$
9. $4\frac{2}{5}+\frac{7}{15}$
10. $\frac{1}{12}+\frac{2}{15}-\frac{3}{20}$